College Trigonometry

SIXTH EDITION

Richard N. Aufmann

Vernon C. Barker

Richard D. Nation

Palomar College

HOUGHTON MIFFLIN COMPANY
Boston New York

Publisher: Richard Stratton
Senior Sponsoring Editor: Molly Taylor
Senior Marketing Manager: Jennifer Jones
Marketing Associate: Mary Legere
Associate Editor: Noel Kamm
Editorial Associate: Andrew Lipsett
Senior Project Editor: Carol Merrigan
Editorial Assistant: Anthony D'Aries
Art and Design Manager: Gary Crespo
Cover Design Manager: Anne S. Katzeff
Photo Editor: Jennifer Meyer Dare
Composition Buyer: Chuck Dutton

Cover photograph: © Dan Hill/cityofsound.com

PHOTO CREDITS

Chapter 1: *p. 1:* John Foxx/Stockbyte Silver/Getty Images. **Chapter 2:** *p. 117:* Andrew Brookes/CORBIS; *p. 128:* Bryan Mullennix/Iconica/Getty Images; *p. 132:* Courtesy of NASA and JPL; *p. 145:* Tony Craddock/Getty Images; *p. 159:* Reuters/New Media Inc./CORBIS. **Chapter 3:** *p. 254:* Courtesy of Richard Nation. **Chapter 4:** *p. 307:* McDuff/Everton/CORBIS. **Chapter 5:** *p. 334:* Dennis De Mars/Fractal Domains/www.fractaldomains.com; *p. 334:* Steve Allen/Alamy; *p. 351:* The Granger Collection. **Chapter 6:** *p. 371:* 1998 International Conference on Quality Control by Artificial Vision – QCAV '98, Kagawa Convention Center, Takamatsu, Kagawa, Japan, November 10–12, 1998, pp. 521–528; *p. 372:* Ian Morison/Jodrell Bank Conservatory; *p. 374:* Courtesy of Michael Levin, Opti-Gone International. Reprinted by permission; *p. 389:* Hugh Rooney/Eye Ubiquitous/CORBIS; *p. 397:* The Granger Collection. **Chapter 7:** *p. 451:* AP/Wide World Photos; *p. 458:* Bettmann/CORBIS; *p. 463:* Charles O'Rear/CORBIS; *p. 464:* David James/Getty Images; *p. 467:* Bettmann/CORBIS; *p. 474:* Jan Halaska/Index Stock Imagery/Jupiter Images; *p. 513:* Tom Brakefield/CORBIS; *p. 521:* Bettmann/CORBIS; *p. 525:* AP/Wide World.

Printed in the U.S.A.

Library of Congress Control Number: 2006933003

ISBN 10: 0-618-82507-X
ISBN 13: 978-0-618-82507-3

3456789-WC-10 09 08 07

Contents

Preface

College Trigonometry, *Sixth Edition* builds on the strong pedagogical features of the previous edition. Always with an eye toward supporting student success, we have increased our emphasis on conceptual understanding, quantitative reasoning, and applications.

Applications

We have retained our basic philosophy, which is to deliver a comprehensive and mathematically sound treatment of the topics considered essential for a college algebra course. To help students master these concepts, we have maintained a dynamic balance among theory, application, modeling, and drill. Carefully developed mathematics is complemented by abundant, relevant, and contemporary applications, many of which feature real data and tables, graphs, and charts.

Ever mindful of the motivating influence that contemporary and appropriate applications have on students, we have included many new application exercises from a wide range of disciplines, and in new formats. For example, a new *Quantitative Reasoning* feature is found at the end of each chapter. Students are urged to investigate concepts and apply those concepts to a variety of contexts while testing the reasonableness of their answers. Applications require students to use problem-solving strategies and newly learned skills to solve practical problems, demonstrating the value of algebra. Many application exercises are accompanied by a diagram that helps students visualize the mathematics of the application.

Technology

Technology is introduced very naturally to illustrate or enhance conceptual understanding of appropriate topics. We integrate technology into a discussion when it can be used to foster and promote better understanding of a concept. The optional graphing calculator exercises, optional *Integrating Technology* boxes, and optional *Exploring Concepts with Technology* features are designed to instill in students an appreciation for both the power and limitations of technology. Optional *Modeling* sections, which use real data, rely heavily on the use of a graphing calculator, and serve to motivate students, are incorporated throughout the text.

Aufmann Interactive Method (AIM)

By incorporating many interactive learning techniques, including the key features outlined below, ***College Trigonometry*** helps students to understand concepts, work independently, and obtain greater mathematical fluency.

- ***Try Exercise*** references follow all worked examples. This feature encourages students to test their understanding by working an exercise similar to the worked example. An icon and a page reference are given below the example, making it easy for the student to navigate to the suggested exercise and back. The complete solution to the *Try Exercise* can be found in the *Solutions to the Try Exercises* appendix. This interaction among the examples, the *Try Exercises*, the *Solutions to the Try Exercises*, and the exercise sets serves as a checkpoint for students as they read the text, do their homework, and study a section.

- **Annotated Examples** are provided throughout each section, and are titled so that students can see at a glance the type of problem being illustrated. The annotated steps assist the student in moving from step to step, and help explain the solution.
- **Question/Answer** In each section, we pose at least one question that encourages the reader to pause and think about the current discussion. To ensure that the student does not miss important information, the answer to the question is provided as a footnote on the same page.

CHANGES IN THE SIXTH EDITION

Overall changes

- **NEW! *Quantitative Reasoning*** After each set of *Chapter Review Exercises*, a new *Quantitative Reasoning* scenario uses concepts from the chapter to explore an application in more depth or extend a mathematical concept from the chapter.
- **NEW!** Most definitions are now immediately followed by an example, to enhance conceptual understanding.
- **NEW!** A ***Calculus Connection*** icon alerts students to a connection between the current topic and calculus. This feature identifies topics that will be revisited in a subsequent calculus course or other advanced course.
- **Revised! *Review Notes,*** which help students recognize the prerequisite skills needed to understand new concepts, are featured more prominently throughout the text, encouraging students to use them more frequently. These example-specific notes direct students to the appropriate page(s) for review, thus decreasing student frustration and creating more opportunities for comprehension.
- **Revised!** We have thoroughly reviewed each exercise set. In addition to updating and adding contemporary applications, we have focused our revisions on providing a smooth progression from routine exercises to exercises that are more challenging.
- **Revised!** New chapter openers demonstrate how the mathematics developed in each chapter is applied.
- **Moved! *Prepare for This Section*** exercises, formerly called *Prepare for the Next Section* exercises, have been moved from the end of the section to the beginning. An up-front review gives students a chance to test their understanding of prerequisite skills and concepts before proceeding to a new topic.
- **Revised! *Assessing Concepts*** exercises, found at the end of each chapter, have been enhanced with more question types, including fill-in-the blank, multiple choice, matching, and true/false.
- **Revised!** We have highlighted more of the important points within the body of the text, to enhance conceptual understanding.
- **Enhanced!** Technology program

Changes in each chapter

In addition to updating and adding new examples, applications, and exercises throughout, we have made a number of chapter-specific changes. Here are some of them:

Chapter 1: Functions and Graphs

- Added concepts involving the solution of literal equations
- Added an introductory discussion of asymptotes

Chapter 2: Trigonometric Functions

- Added exercises that involve finding a trigonometric function that can be used to model an application
- Rewrote the material on the linear and angular speed of a point moving on a circular path

Chapter 3: Trigonometric Identities and Equations

- Expanded the coverage concerning the verification of trigonometric identities
- Expanded the coverage concerning the use of power-reducing identities

Chapter 4: Applications of Trigonometry

- Added exercises that use the graph of a vector to find its components
- Added exercises that involve the equilibrium of forces

Chapter 5: Complex Numbers

- Included introductory material that involves the use of complex numbers and iteration to produce fractal images
- Increased the coverage concerning the Mandelbrot iteration procedure and the Mandelbrot set

Chapter 6: Topics in Analytic Geometry

- Included new figures to illustrate concepts involving the standard form of the equation of a parabola and its graph
- Included new exercises that involve matching the graph of a conic section with its equation
- Included a new example that illustrates a technique that can be used to write some polar equations in rectangular form
- Increased the parametric equation coverage concerning the use of time as a parameter and the simulation of motion

Chapter 7: Exponential and Logarithmic Functions

- Added exercises that use translations and/or reflections to graph exponential functions
- Added exercises that involve the evaluation of a logarithm, without using a calculator

- Added exercises that involve finding the domain of a logarithmic function
- Included a proof of the Product Property of logarithms. The proof of the Quotient Property and the Power Property of logarithms are given as exercises.
- Provided additional coverage and additional exercises involving the expanding and condensing of logarithmic expressions
- Added application exercises that can be solved by using exponential or logarithmic equations
- Included guidelines for selecting the type of mathematical function that models a given application

ACKNOWLEDGEMENTS

We would especially like to thank the users of the previous edition for their helpful suggestions on improving the text. Also, we sincerely appreciate the time, effort, and suggestions of the reviewers of this edition.

Jan Archibald – *Ventura College, CA*

Steve Armstrong – *LeTourneau University, TX*

Rick Bailey – *Midlands Technical College, SC*

Sandy Dieckman – *Northeast Community College, NE*

Robert Gardner – *East Tennessee State University, TN*

Jim Hendrickson – *Indiana University, IN*

Masoud F. Kazemi - *Lewis-Clark State College, ID*

Linda Kuroski – *Erie Community College, NY*

Betty Larson – *South Dakota State University, SD*

Helen Medley – *Kent State University, OH*

Scott Metcalf – *Eastern Kentucky University, KY*

Suellen Robinson – *North Shore Community College, MA*

Gary Wardall – *University of Wisconsin, Green Bay, WI*

College Trigonometry, *Sixth Edition* is designed to enhance conceptual understanding and quantitative reasoning through its motivating opening features, its *Interactive Method* (AIM), its features for student success, its exercises, its contemporary applications, and its use of technology.

Enhance Conceptual Understanding and Quantitative Reasoning: Using Motivating Features

Revised! Chapter Openers

New **Chapter Openers** demonstrate how the mathematics developed in each chapter is applied.

The icons SSG, , and Online Study Center at the bottom of the page let students know of additional resources.

4 Applications of Trigonometry

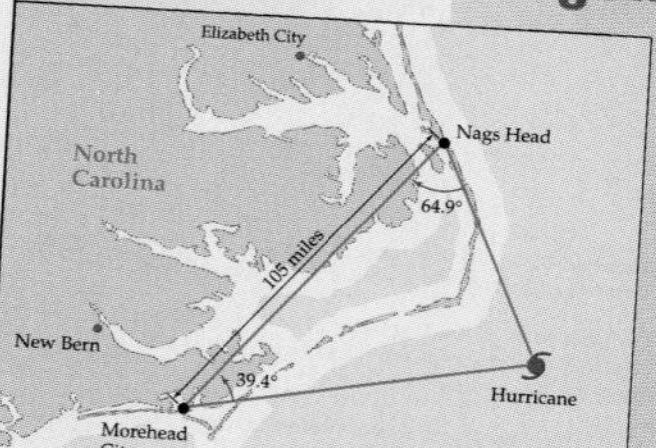

Trigonometry and Indirect Measurement

In Chapter 2 we used trigonometric functions to find the unknown length of a side of a given *right triangle*. In this chapter we develop theorems that can be used to find the length of a side or the measure of an angle of *any* triangle, even if it is not a right triangle. These theorems are often used in the areas of navigation, surveying, and building design. Meterologists use these theorems to estimate the distance from an approaching hurricane to cities in the projected path of the hurricane. For instance, in the diagram on the left, the distance from the hurricane to Nags Head can be determined using the Law of Sines, a theorem presented in this chapter.

See Exercises 30 and 31 on page 299 for additional applications that can be solved by using the Law of Sines.

Online Study Center
For online student resources, such as section quizzes, visit this website:
college.hmco.com/info/aufmannCAT

292

Each **Chapter Opener** ends with a reference to a particular exercise within the chapter that asks the student to solve a problem related to the chapter opener topic.

30. Naval Maneuvers The distance between an aircraft carrier and a Navy destroyer is 7620 feet. The angle of elevation from the destroyer to a helicopter is 77.2°, and the angle of elevation from the aircraft carrier to the helicopter is 59.0°. The helicopter is in the same vertical plane as the two ships, as shown in the following figure. Use this data to determine the distance x from the helicopter to the aircraft carrier.

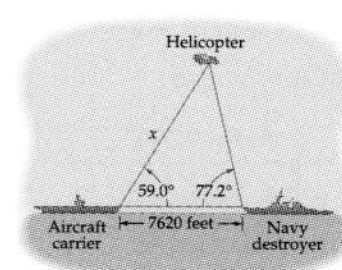

31. Choosing a Golf Strategy The following diagram shows two ways to play a golf hole. One is to hit the ball down the fairway on your first shot and then hit an approach shot to the green on your second shot. A second way is to hit directly toward the pin. Due to the water hazard, this is a more risky strategy. The distance AB is 165 yards, BC is 155 yards, and angle $A = 42.0°$. Find the distance AC from the tee directly to the pin. Assume that angle B is an obtuse angle.

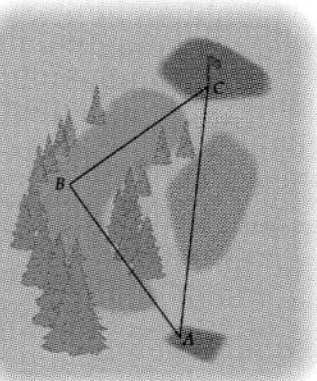

Prepare for This Section

Each section opens with review exercises, titled **Prepare for This Section,** which gives students a chance to test their understanding of prerequisite skills and concepts before proceeding to a new topic. An outline of the section's contents is also provided in the margin as a study aid.

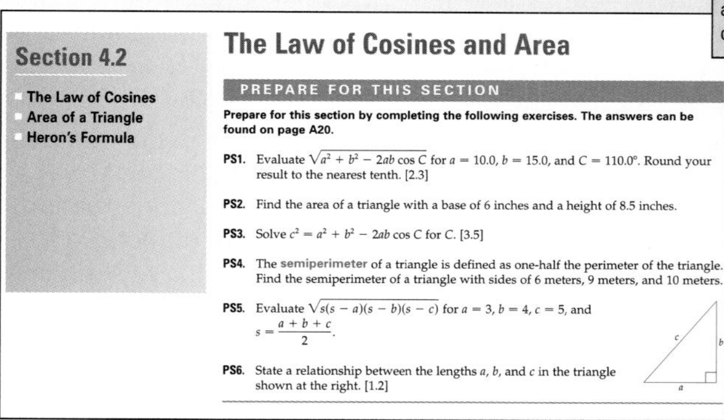

Section 4.2

- The Law of Cosines
- Area of a Triangle
- Heron's Formula

The Law of Cosines and Area

PREPARE FOR THIS SECTION

Prepare for this section by completing the following exercises. The answers can be found on page A20.

PS1. Evaluate $\sqrt{a^2 + b^2 - 2ab\cos C}$ for $a = 10.0$, $b = 15.0$, and $C = 110.0°$. Round your result to the nearest tenth. [2.3]

PS2. Find the area of a triangle with a base of 6 inches and a height of 8.5 inches.

PS3. Solve $c^2 = a^2 + b^2 - 2ab\cos C$ for C. [3.5]

PS4. The semiperimeter of a triangle is defined as one-half the perimeter of the triangle. Find the semiperimeter of a triangle with sides of 6 meters, 9 meters, and 10 meters.

PS5. Evaluate $\sqrt{s(s-a)(s-b)(s-c)}$ for $a = 3$, $b = 4$, $c = 5$, and $s = \dfrac{a+b+c}{2}$.

PS6. State a relationship between the lengths a, b, and c in the triangle shown at the right. [1.2]

NEW! A Calculus Connection icon Calculus Connection identifies topics that will be revisited in a subsequent calculus course.

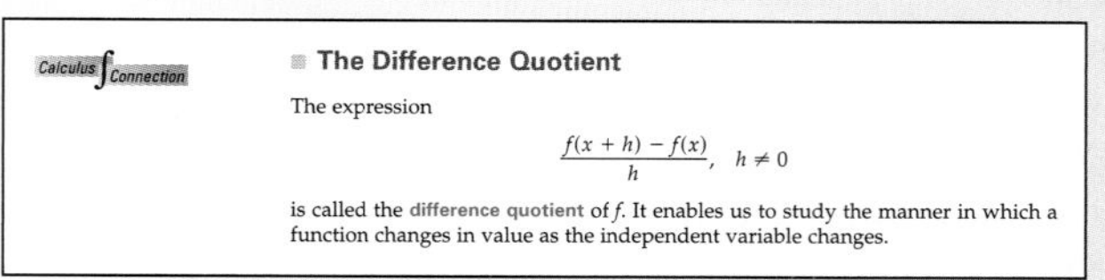

Calculus Connection

The Difference Quotient

The expression

$$\frac{f(x+h) - f(x)}{h}, \quad h \neq 0$$

is called the difference quotient of f. It enables us to study the manner in which a function changes in value as the independent variable changes.

Enhance Conceptual Understanding and Quantitative Reasoning: Using Contemporary Applications

NEW! Quantitative Reasoning

After each set of **Chapter Review Exercises,** a new **Quantitative Reasoning** scenario explores an application in more depth or to extend a mathematical concept from the chapter.

»»» Quantitative Reasoning: *Public Key Cryptography* »»»

As mentioned in the Chapter Opener, performing financial transactions over the Internet requires secure transmissions between two sites, the sender and the receiver. One method of creating secure transmissions is to use a *modular function*.

A modular function is one that gives, in integer form, the remainder when one number is divided by another. We write $a \equiv b \bmod m$ to mean that a is the remainder when b is divided by m. Here are some examples.

$4 \equiv 22 \bmod 6$ because $22 \div 6 = 3$ remainder 4.

$1 \equiv 37 \bmod 4$ because $37 \div 4 = 9$ remainder 1.

$0 \equiv 55 \bmod 11$ because $55 \div 11 = 5$ remainder 0.

$17 \equiv 17 \bmod 31$ because $17 \div 31 = 0$ remainder 17.

QR1. Find the value of each expression.
a. 15 mod 4 **b.** 37 mod 5 **c.** 52 mod 321

Public key cryptography uses a modular function to encrypt a message—say, a person's name or credit card number—so that only the receiver of the message can decrypt it. The message is decrypted by using the inverse of the modular function that was used to encrypt the message. Inverse functions were discussed in Section 1.6.

take note
To encrypt a message means to use a secret code to change the message so that it cannot be understood by an unauthorized user. To decrypt a message means to change a coded message back to its original form.

92 **Chapter 1** Functions and Graphs

35. $f(x) = -2x + 5$

36. $f(x) = -x + 3$

37. $f(x) = \frac{2x}{x-1}, \quad x \neq 1$

38. $f(x) = \frac{x}{x-2}, \quad x \neq 2$

39. $f(x) = \frac{x-1}{x+1}, \quad x \neq -1$

40. $f(x) = \frac{2x-1}{x+3}, \quad x \neq -3$

41. $f(x) = x^2 + 1, \quad x \geq 0$

42. $f(x) = x^2 - 4, \quad x \geq 0$

43. $f(x) = \sqrt{x-2}, \quad x \geq 2$

44. $f(x) = \sqrt{4-x}, \quad x \leq 4$

45. $f(x) = x^2 + 4x, \quad x \geq -2$

46. $f(x) = x^2 - 6x, \quad x \leq 3$

47. $f(x) = x^2 + 4x - 1, \quad x \leq -2$

48. $f(x) = x^2 - 6x + 1, \quad x \geq 3$

49. GEOMETRY The volume of a cube is given by $V(x) = x^3$, where x is the measure of the length of a side of the cube. Find $V^{-1}(x)$ and explain what it represents.

50. UNIT CONVERSIONS The function $f(x) = 12x$ converts feet, x, into inches, $f(x)$. Find $f^{-1}(x)$ and explain what it determines.

51. FAHRENHEIT TO CELSIUS The function

$$f(x) = \frac{5}{9}(x - 32)$$

is used to convert x degrees Fahrenheit to an equivalent Celsius temperature. Find f^{-1} and explain how it is used.

52. RETAIL SALES A clothing merchant uses the function

$$S(x) = \frac{3}{2}x + 18$$

to determine the retail selling price S, in dollars, of a winter coat for which she has paid a wholesale price of x dollars.

a. The merchant paid a wholesale price of \$96 for the winter coat. Use S to determine the retail selling price she will charge for this coat.

b. Find S^{-1} and use it to determine the merchant's wholesale price for a coat that retails at \$399.

53. FASHION The function $s(x) = 2x + 24$ can be used to convert a U.S. women's shoe size into an Italian women's shoe size. Determine the function $s^{-1}(x)$ that can be used to convert an Italian women's shoe size to its equivalent U.S. shoe size.

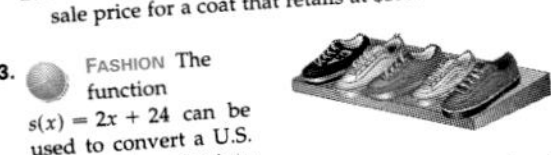

54. FASHION The function $K(x) = 1.3x - 4.7$ converts a men's shoe size in the United States to the equivalent shoe size in the United Kingdom. Determine the function $K^{-1}(x)$ that can be used to convert a United Kingdom men's shoe size to its equivalent U.S. shoe size.

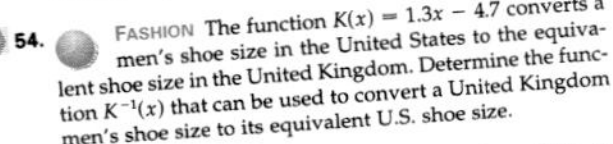

55. COMPENSATION The monthly earnings $E(s)$, in dollars, of a software sales executive is given by $E(s) = 0.05s + 2500$, where s is the value, in dollars, of the software sold by the executive during the month. Find $E^{-1}(s)$ and explain how the executive could use this function.

56. POSTAGE Does the first-class postage rate function given below have an inverse function? Explain your answer.

Weight (in ounces)	Cost
$0 < w \leq 1$	\$.39
$1 < w \leq 2$	\$.63
$2 < w \leq 3$	\$.87
$3 < w \leq 4$	\$1.11

57. THE BIRTHDAY PROBLEM A famous problem called the *birthday problem* goes like this: Suppose there is a randomly selected group of n people in a room. What is the probability that at least two of the people have a birthday on the same day of the year? It may surprise you that for a group of 23 people, the probability that at least two of the people share a birthday is about 50.7%. The following graph can be used to estimate shared birthday probabilities for $1 \leq n \leq 60$.

Updated! Applications

Carefully developed mathematics is complemented by abundant, relevant, and contemporary applications, many of which feature real data and tables, graphs, and charts. Note that applications using real data are identified by .

Applications demonstrate to students the value of algebra and cover topics from a wide variety of disciplines—including agriculture, business, chemistry, construction, earth sciences, economics, education, manufacturing, medicine, nutrition, real estate, and sociology. Besides providing motivation to study mathematics, applications assist students in developing good problem-solving skills.

»»» Projects »»»

MEDIAN–MEDIAN LINE Another linear model of data is called the median–median line. This line employs *summary points* calculated using the medians of subsets of the independent and dependent variables. The median of a data set is the middle number or the average of the two middle numbers for a data set arranged in numerical order. For instance, to find the median of {8, 12, 6, 7, 9}, first arrange the data in numerical order.

6, 7, 8, 9, 12

The median is 8, the number in the middle. To find the median of {15, 12, 20, 9, 13, 10}, arrange the numbers in numerical order.

9, 10, 12, 13, 15, 20

The median is 12.5, the average of the two middle numbers.

$$\text{Median} = \frac{12 + 13}{2} = 12.5$$

The median–median line is determined by dividing a data set into three equal groups. (If the set cannot be divided into three equal groups, the first and third groups should be equal. For instance, if there are 11 data points, divide the set into groups of 4, 3, and 4.) The slope of the median–median line is the slope of the line through the x-medians and y-medians of the first and third sets of points. The median–median line passes through the average of the x-and y-medians of all three sets.

x	y
2	3
3	5
4	4
5	7
6	8
7	9
8	12
9	12
10	14
11	15
12	14

3. Consider the data set {(1, 3), (2, 5), (3, 7), (4, 9), (5, 11), (6, 13), (7, 15), (8, 17)}.

a. Find the equation of the linear regression line for these data.

b. Find the equation of the median–median line for these data.

c. What conclusion might you draw from the answers to **a.** and **b.**?

Projects

Projects are designed to engage the student in mathematics. At the end of each section, students are asked to do one or more of the following types of projects:

- solve a more involved application problem
- investigate a concept in greater depth
- write a proof of a statement

With some projects, students are asked to chronicle the procedure used to solve it, and to suggest an extension to the project. These projects are ideal candidates for small group assignments.

Enhance Conceptual Understanding and Quantitative Reasoning: Using Technology

328 Chapter 4 Applications of Trigonometry

Exploring Concepts with Technology

Optimal Branching of Arteries

The physiologist Jean Louis Poiseuille (1799–1869) developed several laws concerning the flow of blood. One of his laws states that the resistance R of a blood vessel of length l and radius r is given by

$$R = k\frac{l}{r^4} \quad (1)$$

The number k is a variation constant that depends on the viscosity of the blood. **Figure 4.41** shows a large artery with radius r_1 and a smaller artery with radius r_2. The branching angle between the arteries is θ. Make use of Poiseuille's Law, Equation (1), to show that the resistance R of the blood along the path $P_1P_2P_3$ is

$$R = k\left(\frac{a - b\cot\theta}{(r_1)^4} + \frac{b\csc\theta}{(r_2)^4}\right) \quad (2)$$

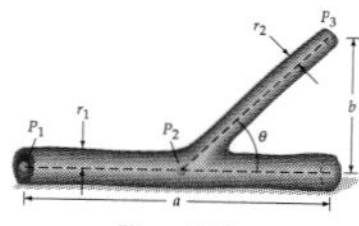

Figure 4.41

Use a graphing utility to graph R with $k = 0.0563$, $a = 8$ centimeters, $b = 4$ centimeters, $r_1 = 0.4$ centimeter, and $r_2 = \frac{3}{4}r_1 = 0.3$ centimeter. Then estimate (to the nearest degree) the angle θ that minimizes R. By using calculus, it can be demonstrated that R is minimized when

$$\cos\theta = \left(\frac{r_2}{r_1}\right)^4 \quad (3)$$

This equation is remarkable because it is much simpler than Equation (2) and because it does not involve the distance a or b. Solve Equation (3) for θ, with $r_2 = \frac{3}{4}r_1$. How does this value of θ compare with the value of θ you obtained by graphing?

Exploring Concepts with Technology

Optional **Exploring Concepts with Technology** problems extend ideas from the chapter, encouraging students to use calculators or computers to investigate solutions to computationally unpleasant problems. In this way, calculators and computers have expanded the limits of the types of problems that can be solved at this level. In addition, students are challenged to think about the pitfalls of computational solutions.

Integrating Technology

Using optional **Integrating Technology** boxes, technology is integrated into a discussion when it can be used to foster and promote a better conceptual understanding of a concept. Additionally, optional graphing calculator examples and exercises (identified by ) are presented throughout the text.

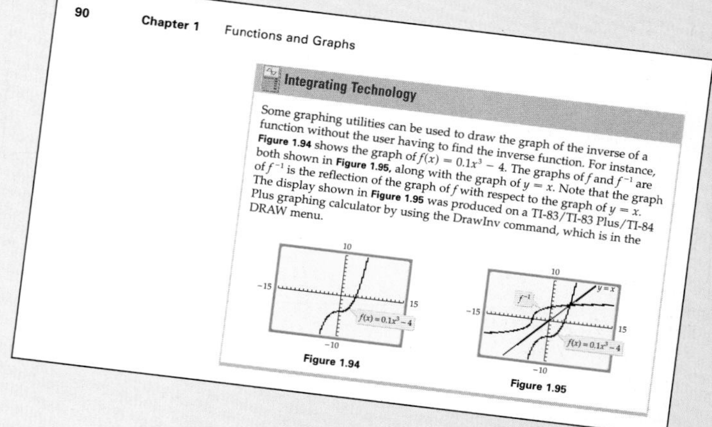

90 Chapter 1 Functions and Graphs

Integrating Technology

Some graphing utilities can be used to draw the graph of the inverse of a function without the user having to find the inverse function. For instance, **Figure 1.94** shows the graph of $f(x) = 0.1x^3 - 4$. The graphs of f and f^{-1} are both shown in **Figure 1.95**, along with the graph of $y = x$. Note that the graph of f^{-1} is the reflection of the graph of f with respect to the graph of $y = x$. The display shown in **Figure 1.95** was produced on a TI-83/TI-83 Plus/TI-84 Plus graphing calculator by using the DrawInv command, which is in the DRAW menu.

Figure 1.94

Figure 1.95

Linear Regression Models

The data in the table below show the population of selected states and the number of professional sports teams (Major League Baseball, National Football League, National Basketball Association, Women's National Basketball Association, National Hockey League) in those states. A scatter diagram of the data is shown in **Figure 1.96** on page 96.

Number of Professional Sports Teams for Selected States

State	Populations (in millions)	Number of Teams	State	Populations (in millions)	Number of Teams
Arizona	5.9	5	Minnesota	5.1	5
California	36.1	17	New Jersey	8.7	3
Colorado	4.7	4	New York	19.3	10
Florida	17.8	11	North Carolina	9.7	3
Illinois	12.8	5	Pennsylvania	12.4	7
Indiana	6.3	3	Texas	22.9	9
Michigan	10.1	5	Wisconsin	5.5	3

Modeling

Special modeling sections, which rely heavily on the use of a graphing calculator, are incorporated throughout the text. These optional sections introduce the idea of a mathematical model, using various real-world data sets that further motivate students and help them see the relevance of mathematical concepts.

Enhance Conceptual Understanding and Quantitative Reasoning: Using the Aufmann Interactive Method (AIM)

By incorporating many interactive learning techniques, including the key features outlined below, ***College Trigonometry*** uses the proven **Aufmann Interactive Method (AIM)** to help students understand concepts, work independently, and obtain greater mathematical fluency.

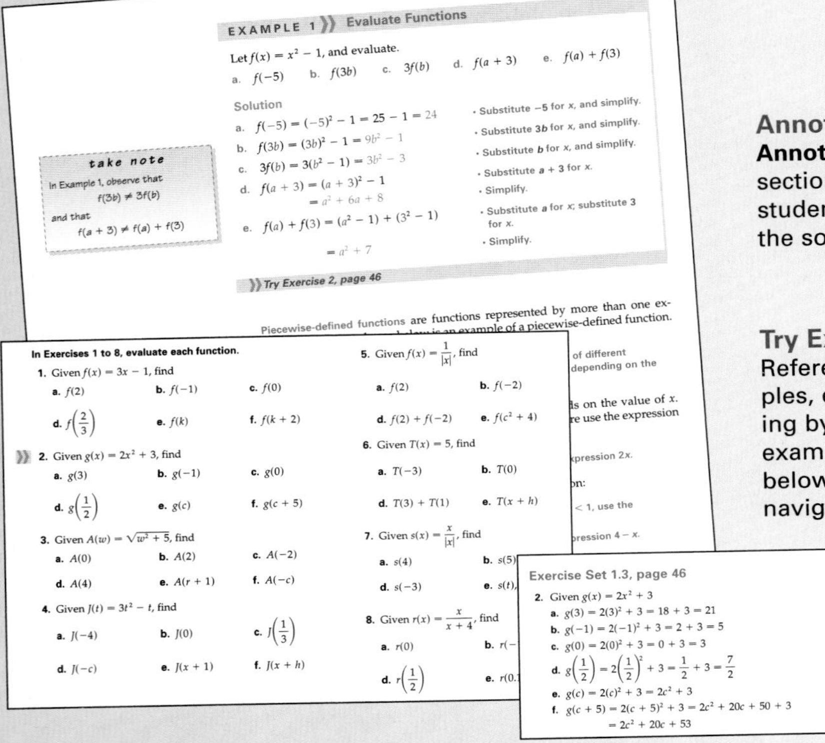

EXAMPLE 1 Evaluate Functions

Let $f(x) = x^2 - 1$, and evaluate.

a. $f(-5)$ b. $f(3b)$ c. $3f(b)$ d. $f(a + 3)$ e. $f(a) + f(3)$

Solution

a. $f(-5) = (-5)^2 - 1 = 25 - 1 = 24$ • Substitute −5 for x, and simplify.

b. $f(3b) = (3b)^2 - 1 = 9b^2 - 1$ • Substitute 3b for x, and simplify.

c. $3f(b) = 3(b^2 - 1) = 3b^2 - 3$ • Substitute b for x, and simplify.

d. $f(a + 3) = (a + 3)^2 - 1$ • Substitute a + 3 for x.
$= a^2 + 6a + 8$ • Simplify.

e. $f(a) + f(3) = (a^2 - 1) + (3^2 - 1)$ • Substitute a for x; substitute 3 for x.
$= a^2 + 7$ • Simplify.

take note
In Example 1, observe that $f(3b) \neq 3f(b)$ and that $f(a + 3) \neq f(a) + f(3)$

Try Exercise 2, page 46

Piecewise-defined functions are functions represented by more than one ex- ... example of a piecewise-defined function.

... of different ... depending on the ... s on the value of x. ... re use the expression ... xpression 2x. ... on: ... < 1, use the ... pression 4 − x.

In Exercises 1 to 8, evaluate each function.

1. Given $f(x) = 3x - 1$, find
 a. $f(2)$ b. $f(-1)$ c. $f(0)$ d. $f\left(\frac{2}{3}\right)$ e. $f(k)$ f. $f(k + 2)$
2. Given $g(x) = 2x^2 + 3$, find
 a. $g(3)$ b. $g(-1)$ c. $g(0)$ d. $g\left(\frac{1}{2}\right)$ e. $g(c)$ f. $g(c + 5)$
3. Given $A(w) = \sqrt{w^2 + 5}$, find
 a. $A(0)$ b. $A(2)$ c. $A(-2)$ d. $A(4)$ e. $A(r + 1)$ f. $A(-c)$
4. Given $J(t) = 3t^2 - t$, find
 a. $J(-4)$ b. $J(0)$ c. $J\left(\frac{1}{3}\right)$ d. $J(-c)$ e. $J(x + 1)$ f. $J(x + h)$
5. Given $f(x) = \frac{1}{|x|}$, find
 a. $f(2)$ b. $f(-2)$ d. $f(2) + f(-2)$ e. $f(c^2 + 4)$
6. Given $T(x) = 5$, find
 a. $T(-3)$ b. $T(0)$ d. $T(3) + T(1)$ e. $T(x + h)$
7. Given $s(x) = \frac{x}{|x|}$, find
 a. $s(4)$ b. $s(5)$ d. $s(-3)$ e. $s(t)$
8. Given $r(x) = \frac{x}{x + 4}$, find
 a. $r(0)$ b. $r(-$ d. $r\left(\frac{1}{2}\right)$ e. $r(0.$

Exercise Set 1.3, page 46

2. Given $g(x) = 2x^2 + 3$
 a. $g(3) = 2(3)^2 + 3 = 18 + 3 = 21$
 b. $g(-1) = 2(-1)^2 + 3 = 2 + 3 = 5$
 c. $g(0) = 2(0)^2 + 3 = 0 + 3 = 3$
 d. $g\left(\frac{1}{2}\right) = 2\left(\frac{1}{2}\right)^2 + 3 = \frac{1}{2} + 3 = \frac{7}{2}$
 e. $g(c) = 2(c)^2 + 3 = 2c^2 + 3$
 f. $g(c + 5) = 2(c + 5)^2 + 3 = 2c^2 + 20c + 50 + 3 = 2c^2 + 20c + 53$

Annotated Examples

Annotated Examples are provided throughout each section and are titled. The annotated steps assist the student in moving from step to step and help explain the solution.

Try Exercises

References to **Try Exercises** follow all worked examples, encouraging students to test their understanding by working an exercise similar to the worked example. An icon and a page reference are given below the example, making it easy for the student to navigate to the suggested exercise and back.

Solutions to Try Exercises

The complete solution to the **Try Exercises** can be found in the **Solutions to the Try Exercises** appendix. This interaction among the examples, the **Try Exercises,** the **Solutions to the Try Exercises,** and the exercise sets serves as a checkpoint for students as they read the text, do their homework, and study a section.

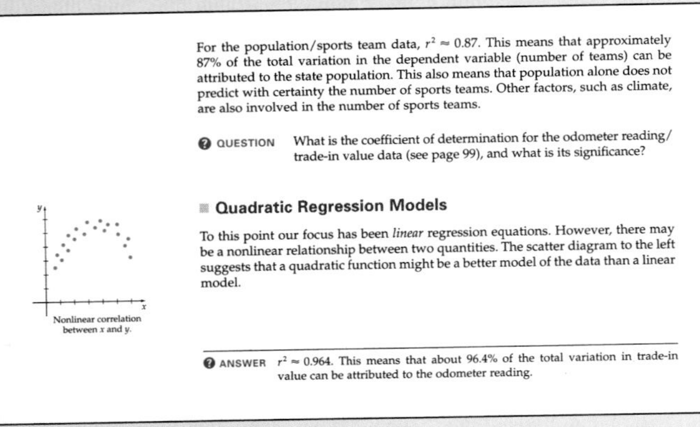

For the population/sports team data, $r^2 \approx 0.87$. This means that approximately 87% of the total variation in the dependent variable (number of teams) can be attributed to the state population. This also means that population alone does not predict with certainty the number of sports teams. Other factors, such as climate, are also involved in the number of sports teams.

QUESTION What is the coefficient of determination for the odometer reading/trade-in value data (see page 99), and what is its significance?

Quadratic Regression Models

To this point our focus has been *linear* regression equations. However, there may be a nonlinear relationship between two quantities. The scatter diagram to the left suggests that a quadratic function might be a better model of the data than a linear model.

Nonlinear correlation between x and y.

ANSWER $r^2 \approx 0.964$. This means that about 96.4% of the total variation in trade-in value can be attributed to the odometer reading.

Question/Answer

In each section, the authors pose at least one question that encourages students to pause and think about the concepts presented in the current discussion. To ensure that students do not miss important information, the answer to the question is provided as a footnote on the same page.

Enhance Conceptual Understanding and Quantitative Reasoning: Using Features for Student Success

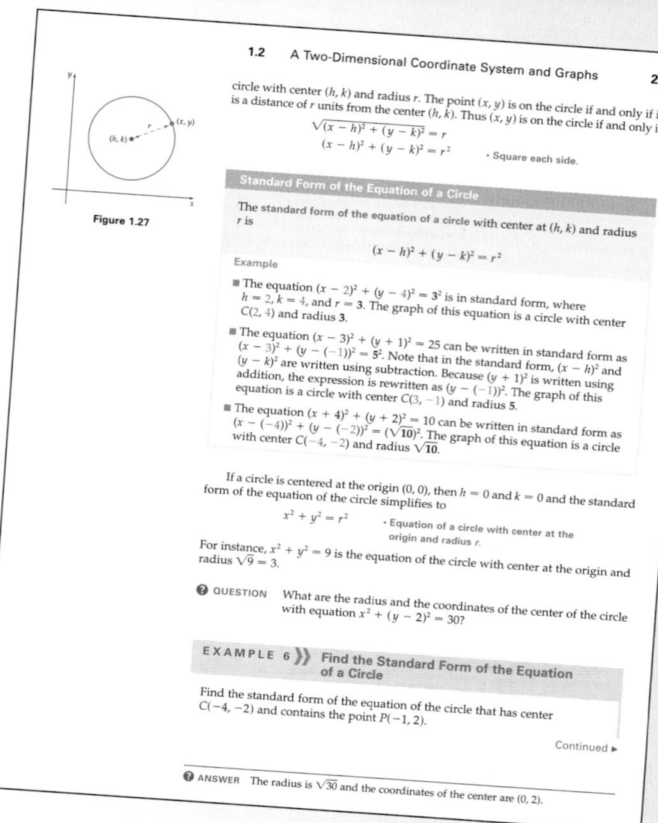

1.2 A Two-Dimensional Coordinate System and Graphs 25

circle with center (h, k) and radius r. The point (x, y) is on the circle if and only if it is a distance of r units from the center (h, k). Thus (x, y) is on the circle if and only if

$$\sqrt{(x-h)^2+(y-k)^2}=r$$
$$(x-h)^2+(y-k)^2=r^2 \quad \text{• Square each side.}$$

Figure 1.27

Standard Form of the Equation of a Circle

The standard form of the equation of a circle with center at (h, k) and radius r is

$$(x-h)^2+(y-k)^2=r^2$$

Example

- The equation $(x-2)^2+(y-4)^2=3^2$ is in standard form, where $h=2$, $k=4$, and $r=3$. The graph of this equation is a circle with center $C(2, 4)$ and radius 3.
- The equation $(x-3)^2+(y+1)^2=25$ can be written in standard form as $(x-3)^2+(y-(-1))^2=5^2$. Note that in the standard form, $(x-h)^2$ and $(y-k)^2$ are written using subtraction. Because $(y+1)^2$ is written using addition, the expression is rewritten as $(y-(-1))^2$. The graph of this equation is a circle with center $C(3, -1)$ and radius 5.
- The equation $(x+4)^2+(y+2)^2=10$ can be written in standard form as $(x-(-4))^2+(y-(-2))^2=(\sqrt{10})^2$. The graph of this equation is a circle with center $C(-4, -2)$ and radius $\sqrt{10}$.

If a circle is centered at the origin (0, 0), then $h=0$ and $k=0$ and the standard form of the equation of the circle simplifies to

$$x^2+y^2=r^2 \quad \text{• Equation of a circle with center at the origin and radius } r.$$

For instance, $x^2+y^2=9$ is the equation of the circle with center at the origin and radius $\sqrt{9}=3$.

QUESTION What are the radius and the coordinates of the center of the circle with equation $x^2+(y-2)^2=30$?

EXAMPLE 6 Find the Standard Form of the Equation of a Circle

Find the standard form of the equation of the circle that has center $C(-4, -2)$ and contains the point $P(-1, 2)$.

Continued ▸

ANSWER The radius is $\sqrt{30}$ and the coordinates of the center are (0, 2).

NEW! Immediate Examples of Definitions and Concepts

Immediate examples of many definitions enhance understanding.

Margin Notes

Take Notes alert students to a point requiring special attention or are used to amplify the concept under discussion. And **Math Matters** contain interesting sidelights about mathematics, its history, or its application.

take note

$f^{-1}(x)$ does not mean $\dfrac{1}{f(x)}$. For $f(x)=2x$, $f^{-1}(x)=\dfrac{1}{2}x$ but $\dfrac{1}{f(x)}=\dfrac{1}{2x}$.

Visualize the Solution

For appropriate examples, both algebraic and graphical solutions are provided to help the student visualize the mathematics of the example and to create a link between the algebraic and visual components of a solution.

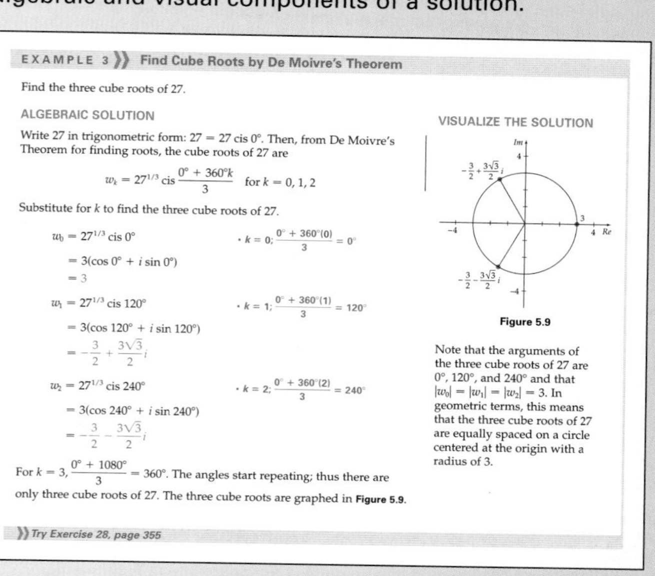

EXAMPLE 3 Find Cube Roots by De Moivre's Theorem

Find the three cube roots of 27.

ALGEBRAIC SOLUTION

Write 27 in trigonometric form: $27 = 27 \operatorname{cis} 0°$. Then, from De Moivre's Theorem for finding roots, the cube roots of 27 are

$$w_k = 27^{1/3} \operatorname{cis} \frac{0° + 360°k}{3} \quad \text{for } k = 0, 1, 2$$

Substitute for k to find the three cube roots of 27.

$$w_0 = 27^{1/3} \operatorname{cis} 0° \quad \text{• } k=0;\ \frac{0° + 360°(0)}{3} = 0°$$
$$= 3(\cos 0° + i \sin 0°)$$
$$= 3$$
$$w_1 = 27^{1/3} \operatorname{cis} 120° \quad \text{• } k=1;\ \frac{0° + 360°(1)}{3} = 120°$$
$$= 3(\cos 120° + i \sin 120°)$$
$$= -\frac{3}{2} + \frac{3\sqrt{3}}{2}i$$
$$w_2 = 27^{1/3} \operatorname{cis} 240° \quad \text{• } k=2;\ \frac{0° + 360°(2)}{3} = 240°$$
$$= 3(\cos 240° + i \sin 240°)$$
$$= -\frac{3}{2} - \frac{3\sqrt{3}}{2}i$$

For $k=3$, $\dfrac{0° + 1080°}{3} = 360°$. The angles start repeating; thus there are only three cube roots of 27. The three cube roots are graphed in Figure 5.9.

VISUALIZE THE SOLUTION

Figure 5.9

Note that the arguments of the three cube roots of 27 are 0°, 120°, and 240° and that $|w_0| = |w_1| = |w_2| = 3$. In geometric terms, this means that the three cube roots of 27 are equally spaced on a circle centered at the origin with a radius of 3.

Try Exercise 28, page 355

Updated! To Review Note

To Review notes help students recognize the prerequisite skills needed to understand new concepts. These notes direct students to the appropriate page(s) for review, thus decreasing student frustration and creating more opportunities for comprehension.

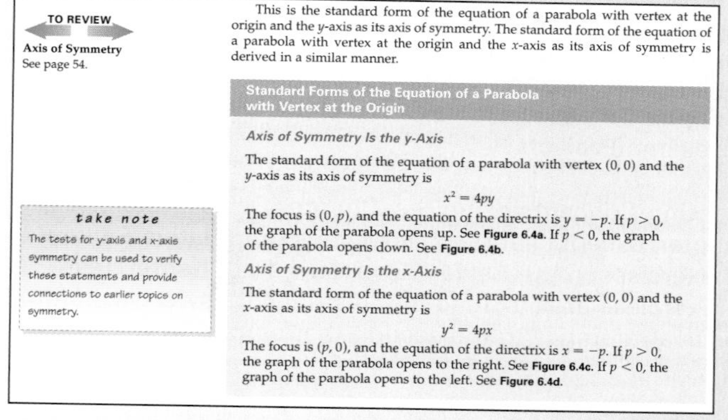

TO REVIEW
Axis of Symmetry
See page 54.

This is the standard form of the equation of a parabola with vertex at the origin and the y-axis as its axis of symmetry. The standard form of the equation of a parabola with vertex at the origin and the x-axis as its axis of symmetry is derived in a similar manner.

Standard Forms of the Equation of a Parabola with Vertex at the Origin

Axis of Symmetry Is the y-Axis

The standard form of the equation of a parabola with vertex (0, 0) and the y-axis as its axis of symmetry is

$$x^2 = 4py$$

The focus is $(0, p)$, and the equation of the directrix is $y = -p$. If $p > 0$, the graph of the parabola opens up. See Figure 6.4a. If $p < 0$, the graph of the parabola opens down. See Figure 6.4b.

Axis of Symmetry Is the x-Axis

The standard form of the equation of a parabola with vertex (0, 0) and the x-axis as its axis of symmetry is

$$y^2 = 4px$$

The focus is $(p, 0)$, and the equation of the directrix is $x = -p$. If $p > 0$, the graph of the parabola opens to the right. See Figure 6.4c. If $p < 0$, the graph of the parabola opens to the left. See Figure 6.4d.

take note

The tests for y-axis and x-axis symmetry can be used to verify these statements and provide connections to earlier topics on symmetry.

Enhance Conceptual Understanding and Quantitative Reasoning: Using Well-Developed Exercise Sets

Connecting Concepts

Each end-of-section exercise set features **Connecting Concepts** problems which include material from previous sections, are extensions of topics in the section, require data analysis, and offer challenge problems or problems of the form "prove or disprove."

Connecting Concepts

83. For $\mathbf{u} = \langle -1, 1\rangle$, $\mathbf{v} = \langle 2, 3\rangle$, and $\mathbf{w} = \langle 5, 5\rangle$, find the sum of the three vectors geometrically by using the triangle method of adding vectors.

84. For $\mathbf{u} = \langle 1, 2\rangle$, $\mathbf{v} = \langle 3, -2\rangle$, and $\mathbf{w} = \langle -1, 4\rangle$, find $\mathbf{u} + \mathbf{v} - \mathbf{w}$ geometrically by using the triangle method of adding vectors.

85. Find a vector that has initial point $(3, -1)$ and is equivalent to $\mathbf{v} = 2\mathbf{i} - 3\mathbf{j}$.

86. Find a vector that has initial point $(-2, 4)$ and is equivalent to $\mathbf{v} = \langle -1, 3\rangle$.

87. If $\mathbf{v} = 2\mathbf{i} - 5\mathbf{j}$ and $\mathbf{w} = 5\mathbf{i} + 2\mathbf{j}$ have the same initial point, is $\mathbf{v}$ perpendicular to $\mathbf{w}$? Why or why not?

88. If $\mathbf{v} = \langle 5, 6\rangle$ and $\mathbf{w} = \langle 6, 5\rangle$ have the same initial point, is $\mathbf{v}$ perpendicular to $\mathbf{w}$? Why or why not?

89. Let $\mathbf{v} = \langle -2, 7\rangle$. Find a vector perpendicular to $\mathbf{v}$.

90. Let $\mathbf{w} = 4\mathbf{i} + \mathbf{j}$. Find a vector perpendicular to $\mathbf{w}$.

93. Prove that $c(\mathbf{v} \cdot \mathbf{w}) = (c\mathbf{v}) \cdot \mathbf{w}$.

94. Show that the dot product of two nonzero vectors is positive if the angle between the vectors is an acute angle and negative if the angle between the two vectors is an obtuse angle.

95. COMPARISON OF WORK DONE Consider the following two situations. (1) A rope is being used to pull a box up a ramp inclined at an angle α. The rope exerts a force **F** on the box, and the rope makes an angle θ with the ramp. The box is pulled s feet. (2) A rope is being used to pull the same box along a level floor. The rope exerts the same force **F** on the box. The box is pulled the same s feet. In which case is more work done?

Revised! Assessing Concepts

Assessing Concepts exercises, found at the end of each chapter, have been enhanced with more question types—including fill-in-the-blank, multiple choice, matching, and true/false.

Chapter 4 Assessing Concepts

1. What is an oblique triangle?
2. In triangle ABC, $a = 4.5$, $b = 6.2$, and $C = 107°$. Which law, the Law of Sines or the Law of Cosines, can be used to find c?
3. Which of the following cases,

   ASA, AAS, SSA, SSS, or SAS

   is known as the ambiguous case of the Law of Sines?
4. In Heron's formula, what does the variable s represent?
5. Is the dot product of two vectors a vector or a scalar?
6. Let $\mathbf{v}$ and $\mathbf{w}$ be nonzero vectors. Is $\text{proj}_{\mathbf{w}}\mathbf{v}$ a vector or a scalar?
7. True or false: The vector $\left\langle \frac{12}{13}, -\frac{5}{13} \right\rangle$ is a unit vector.
8. True or false: $\mathbf{i} \cdot \mathbf{j} = 0$.
9. True or false: The Law of Sines can be used to solve any triangle, given two angles and any side.
10. True or false: If two nonzero vectors are orthogonal, then their dot product is 0.

Topics for Discussion

Each section ends with **Topics for Discussion** exercises to be used for group discussion or as writing exercises. These exercises are frequently conceptual and focus on key ideas in the section.

Topics for Discussion

1. Discuss the meaning of symmetry of a graph with respect to a line. How do you determine whether a graph has symmetry with respect to the x-axis? with respect to the y-axis?
2. Discuss the meaning of symmetry of a graph with respect to a point. How do you determine whether a graph has symmetry with respect to the origin?
3. What does it mean to reflect a graph across the x-axis or across the y-axis?
4. Explain how the graphs of y › $2x^3 - x^2$ and y › $2(-x)^3 - (-x)^2$ are related.
5. Given the graph of the function y › $f(x)$, explain how to obtain the graph of the function y › $f(x - 3) + 1$.
6. The graph of the *step function* y › $[\![x]\!]$ has steps that are 1 unit wide. Determine how wide the steps are for the graph of y › $\left[\!\left[\frac{1}{3}x\right]\!\right]$.

Extensive Exercises

The authors thoroughly reviewed each exercise set to update applications and to ensure a smooth progression from routine exercises to exercises that are more challenging. The exercises illustrate the many facets of topics discussed in the text. Each exercise set emphasizes skill building, skill maintenance, conceptual understanding, quantitative reasoning, and, as appropriate, applications. Each exercise set is directly proceeded by **Topics for Discussion** exercises and directly followed by **Connecting Concepts** exercises and **Projects.**

Each chapter ends with **Assessing Concepts** exercises, **Chapter Review Exercises,** and a **Chapter Test.** Each chapter, except chapter P, includes a **Cumulative Review Exercise Set.** Answers to all exercises in the **Review Exercises,** the **Chapter Test,** and the **Cumulative Review Exercises** are included in the student answer section. Along with the answer, there is a reference to the section that pertains to each exercise.

Additional Resources • Get More Value from Your Textbook!

college.hmco.com/info/aufmannCAT

Instructor Resources

Online Teaching Center *Online Teaching Center*
This website offers instructors prep and chapter tests, instructor's solutions, and more.

HM Testing™ (powered by Diploma®)
"Testing the way you want it" HM Testing offers all the tools needed to create, author, deliver, and customize multiple types of tests—including authoring and editing algorithmic questions.

Student Resources

Student Study Guide SSG
Contains complete solutions to all odd-numbered exercises and all of the solutions to the end-of-chapter material.

Online Study Center *Online Study Center*
This free website gives students access to ACE practice tests, and other study resources.

Instructional DVDs
Hosted by Dana Mosely, these text-specific DVDs cover all sections of the text and provide explanations of key concepts, examples, exercises, and applications in a lecture-based format. DVDs are now closed-captioned for the hearing-impaired.

Eduspace® (powered by Blackboard®)
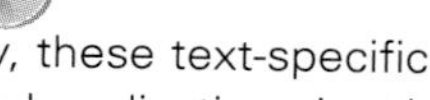
Eduspace is a web-based learning system that provides instructors with powerful course management tools and students with text-specific content to support all of their online teaching and learning needs. Eduspace makes it easy to deliver all or part of a course online. Resources such as algorithmic, automatically-graded homework exercises, tutorials, instructional video clips, an online multimedia eBook, live, online tutoring with SMARTHINKING™, and additional study materials all come ready-to-use. Instructors can choose to use the content as is, modify it, or even add their own.

- Visit *www.eduspace.com* for more information.

SMARTHINKING™ Live, Online Tutoring

SMARTHINKING provides an easy-to-use and effective online, text-specific tutoring service. A dynamic whiteboard and graphing calculator function enables students and e-structors to collaborate easily. SMARTHINKING offers three levels of service:

- Text-specific Tutoring provides real-time, one-on-one instruction with a specially qualified e-structor.
- Questions Any Time allows students to submit questions to the tutor outside the scheduled hours and receive a reply within 24 hours.
- Independent Study Resources connect students with around-the-clock access to additional educational services, including interactive websites, diagnostic tests, and Frequently Asked Questions posed to SMARTHINKING e-structors.

Online Course Content for Blackboard®, WebCT®, and eCollege®
Deliver program or text-specific Houghton Mifflin content online using your institution's local course management system. Houghton Mifflin offers homework, tutorials, videos, and other resources formatted for Blackboard®, WebCT®, eCollege®, and other course management systems. Add to an existing online course or create a new one by selecting from a wide range of powerful learning and instructional materials.

For more information, visit **college.hmco.com/info/aufmannCAT**
or contact your Houghton Mifflin sales representative.

1 Functions and Graphs

The Internet, the World Wide Web, and Modular Functions

In 1965, Lawrence Roberts connected two computers, one at MIT and one at UCLA, using a telephone line. This computer connection was the first demonstration of the feasibility of computer networks and led to the eventual establishment of the Internet, first known as ARPANET after the Advanced Research Projects Agency (ARPA), a group founded by the Department of Defense. Originally, ARPANET connected computers at four universities: the University of California, Los Angeles; Stanford University; the University of California, Santa Barbara; and the University of Utah. The first message was sent from UCLA to Stanford in 1969.

To expand the Internet, computer programs had to be written that allowed the interaction of many computers. This set of programs, called protocols, was created by Robert Kahn and Vint Cerf. The programs were called TCP/IP (Transmission Control Protocol/Internet Protocol). TCP/IP was universally adopted in 1983 and is still used today to control communication among computers on the Internet. A rough analog to TCP/IP programs is our telephone system. Each telephone has a unique telephone number. Similarly, each computer on the Internet has a unique IP (Internet Protocol) address. TCP programs are similar to telephone services such as call waiting and caller ID. They enable a computer to know when other computers are trying to communicate with it, and what kind of communication is incoming.

In 1991, British physicist Tim Berners-Lee developed a program that allowed physicists to exchange information in an efficient manner. His creation was called the World Wide Web. Two functions that can be performed using the Internet and the World Wide Web are credit card purchases and banking transactions. These transactions must be carried out in a secure mode using programs that encrypt credit card numbers and bank account numbers. One method of encryption makes use of a *modular function* and the *inverse* of that function. Modular functions are discussed in the Quantitative Reasoning feature on page 114. A basic example of the use of inverse functions to encrypt and decrypt messages is given in Exercise 59 on page 93.

Online Study Center
For online student resources, such as section quizzes, visit this website: college.hmco.com/info/aufmannCAT

Section 1.1

- The Real Numbers
- Absolute Value and Distance
- Linear and Quadratic Equations
- Inequalities
- Solving Inequalities by the Critical Value Method
- Absolute Value Inequalities

Math Matters

Archimedes (c. 287–212 B.C.) was the first to calculate π with any degree of precision. He was able to show that

$$3\frac{10}{71} < \pi < 3\frac{1}{7}$$

from which we get the approximation $3\frac{1}{7} \approx \pi$. The use of the symbol π for this quantity was introduced by Leonhard Euler (1707–1783) in 1739, approximately 2000 years after Archimedes.

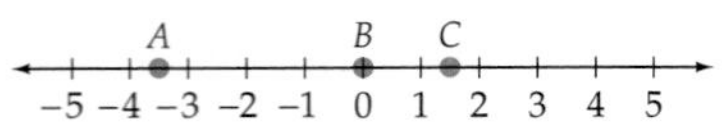

Figure 1.1

take note

The interval notation $[1, \infty)$ represents all real numbers greater than or equal to 1. The interval notation $(-\infty, 4)$ represents all real numbers less than 4.

Equations and Inequalities

The Real Numbers

The real numbers are used extensively in mathematics. The set of real numbers is quite comprehensive and contains several unique sets of numbers.

The **integers** are the set of numbers

$$\{\ldots, -4, -3, -2, -1, 0, 1, 2, 3, 4, \ldots\}$$

Recall that the brace symbols, { }, are used to identify a set. The positive integers are called **natural numbers.**

The **rational numbers** are the set of numbers of the form $\frac{a}{b}$, where a and b are integers and $b \neq 0$. Thus the rational numbers include $-\frac{3}{4}$ and $\frac{5}{2}$. Because each integer can be expressed in the form $\frac{a}{b}$ with denominator $b = 1$, the integers are included in the set of rational numbers. Every rational number can be written as either a terminating or a repeating decimal.

A number written in decimal form that does not repeat or terminate is called an **irrational number.** Some examples of irrational numbers are $0.141141114\ldots$, $\sqrt{2}$, and π. These numbers cannot be expressed as quotients of integers. The set of **real numbers** is the union of the sets of rational and irrational numbers.

A real number can be represented geometrically on a **coordinate axis** called a **real number line.** Each point on this line is associated with a real number called the **coordinate** of the point. Conversely, each real number can be associated with a point on a real number line. In **Figure 1.1,** the coordinate of A is $-\frac{7}{2}$, the coordinate of B is 0, and the coordinate of C is $\sqrt{2}$.

Given any two real numbers a and b, we say that a is **less than** b, denoted by $a < b$, if $a - b$ is a negative number. Similarly, we say that a is **greater than** b, denoted by $a > b$, if $a - b$ is a positive number. When a **equals** b, $a - b$ is zero. The symbols $<$ and $>$ are called **inequality symbols.** Two other inequality symbols, $\leq$ (less than or equal to) and $\geq$ (greater than or equal to), are also used.

The inequality symbols can be used to designate sets of real numbers. If $a < b$, the **interval notation** (a, b) is used to indicate the set of real numbers between a and b. This set of numbers also can be described using **set-builder notation:**

$$(a, b) = \{x \mid a < x < b\}$$

When reading a set written in set-builder notation, we read $\{x \mid$ as "the set of x such that." The expression that follows the vertical bar designates the elements in the set.

The set (a, b) is called an **open interval.** The graph of the open interval consists of all the points on the real number line between a and b, not including a and b. A **closed interval,** denoted by $[a, b]$, consists of all points between a and b, including a and b. We can also discuss **half-open intervals.** An example of each type of interval is shown in **Figure 1.2.**

$$(-2, 4) = \{x \mid -2 < x < 4\} \qquad \text{An open interval}$$
$$[1, 5] = \{x \mid 1 \le x \le 5\} \qquad \text{A closed interval}$$
$$[-4, 0) = \{x \mid -4 \le x < 0\} \qquad \text{A half-open interval}$$
$$(-5, -2] = \{x \mid -5 < x \le -2\} \qquad \text{A half-open interval}$$

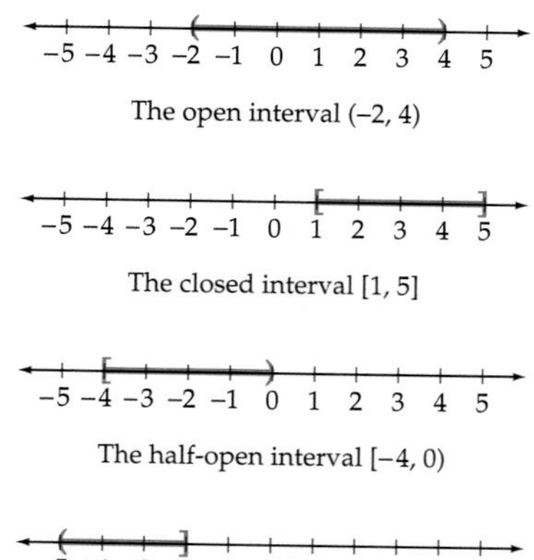
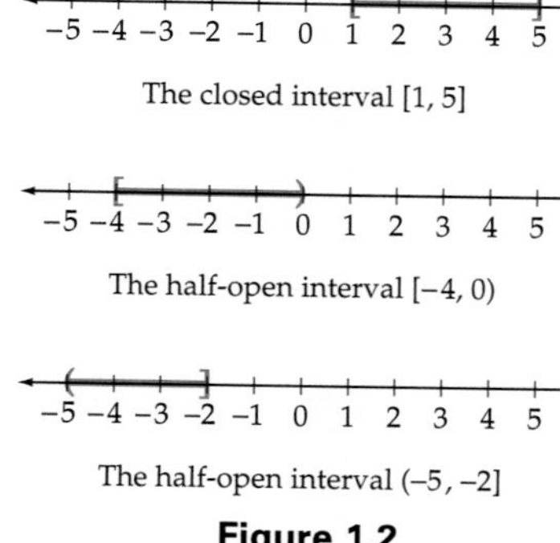

Figure 1.2

Absolute Value and Distance

The *absolute value* of a real number is a measure of the distance from zero to the point associated with the number on a real number line. Therefore, the absolute value of a real number is always positive or zero. We now give a more formal definition of absolute value.

Definition of Absolute Value

For a real number a, the **absolute value** of a, denoted by $|a|$, is

$$|a| = \begin{cases} a & \text{if } a \ge 0 \\ -a & \text{if } a < 0 \end{cases}$$

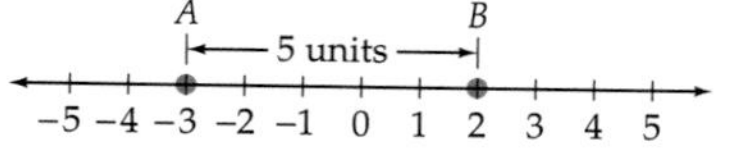

Figure 1.3

The distance d between the points A and B with coordinates -3 and 2, respectively, on a real number line is the absolute value of the difference between the coordinates. See **Figure 1.3**.

$$d = |2 - (-3)| = 5$$

Because the absolute value is used, we could also write

$$d = |(-3) - 2| = 5$$

In general, we define the *distance* between any two points A and B on a real number line as the absolute value of the difference between the coordinates of the points.

Definition of the Distance Between Two Points on a Real Number Line

Let a and b be the coordinates of the points A and B, respectively, on a real number line. Then the **distance** between A and B, denoted by $d(A, B)$, is

$$d(A, B) = |a - b|$$

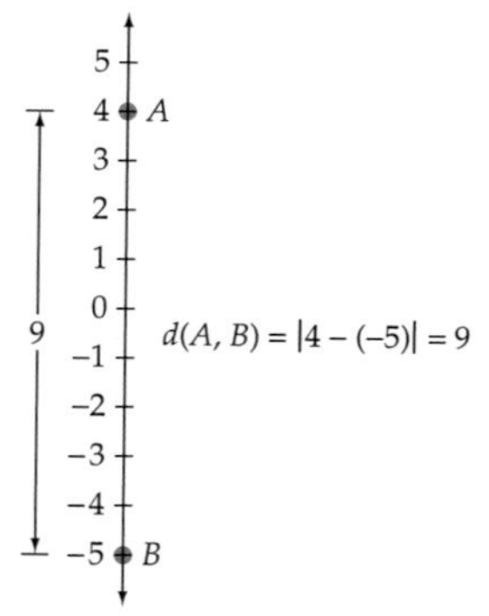

Figure 1.4

This formula applies to any real number line. It can be used to find the distance between two points on a vertical real number line, as shown in **Figure 1.4**.

Linear and Quadratic Equations

An **equation** is a statement about the equality of two expressions. Examples of equations follow.

$$7 = 2 + 5 \qquad x^2 = 4x + 5 \qquad 3x - 2 = 2(x + 1) + 3$$

The values of the variable that make an equation a true statement are the **roots** or **solutions** of the equation. To **solve** an equation means to find the solutions of the equation. The number 2 is said to **satisfy** the equation $2x + 1 = 5$ because substituting 2 for x produces $2(2) + 1 = 5$, which is a true statement.

Definition of a Linear Equation

A **linear equation** in the single variable x is an equation of the form $ax + b = 0$, where $a \neq 0$.

To solve a linear equation in one variable, isolate the variable on one side of the equals sign.

EXAMPLE 1 Solve a Linear Equation

Solve: $3x - 5 = 2$

Solution

$$3x - 5 = 2$$
$$3x - 5 + 5 = 2 + 5 \qquad \text{• Add 5 to each side of the equation.}$$
$$3x = 7$$
$$\frac{3x}{3} = \frac{7}{3} \qquad \text{• Divide each side of the equation by 3.}$$
$$x = \frac{7}{3}$$

The solution is $\frac{7}{3}$.

Try Exercise 6, page 12

An equation may contain more than one variable. For these equations, called **literal equations**, we may choose to solve for any one of the variables.

EXAMPLE 2 Solve a Literal Equation

Solve $\dfrac{by}{c - y} = ax$ for y.

Solution

$$\frac{by}{c-y} = ax$$

$$\frac{by}{c-y}(c-y) = ax(c-y)$$ • Multiply each side of the equation by $c - y$.

$$by = axc - axy$$ • Simplify.

$$by + axy = axc$$ • Add axy to each side of the equation.

$$y(b + ax) = axc$$ • Factor y from the left side of the equation.

$$y = \frac{axc}{b + ax}$$ • Divide each side of the equation by $b + ax$.

›› *Try Exercise 20, page 12*

Math Matters

The term *quadratic* is derived from the Latin word *quadrāre*, which means "to make square." Because the area of a square that measures x units on each side is x^2, we refer to equations that can be written in the form $ax^2 + bx + c = 0$ as equations that are quadratic in x.

Definition of a Quadratic Equation

An equation of the form $ax^2 + bx + c = 0, a \neq 0$, is a **quadratic equation in** x.

A quadratic equation can be solved by using the **quadratic formula.**

The Quadratic Formula

The solution of the quadratic equation $ax^2 + bx + c = 0, a \neq 0$, is given by

$$x = \frac{-b \pm \sqrt{b^2 - 4ac}}{2a}$$

QUESTION For $2x^2 - 3x - 1 = 0$, what are the values of a, b, and c?

Math Matters

There exists a general procedure to solve "by radicals" the general cubic

$$ax^3 + bx^2 + cx + d = 0$$

and the general quartic

$$ax^4 + bx^3 + cx^2 + dx + e = 0$$

However, it has been proved that there are no general procedures that can be used to solve "by radicals" general equations of degree 5 or larger.

EXAMPLE 3 ›› Solve a Quadratic Equation

Solve by using the quadratic formula: $2x^2 - 4x + 1 = 0$

Solution

We have $a = 2$, $b = -4$, and $c = 1$.

$$x = \frac{-(-4) \pm \sqrt{(-4)^2 - 4(2)(1)}}{2(2)} = \frac{4 \pm \sqrt{16 - 8}}{4}$$

$$= \frac{4 \pm \sqrt{8}}{4} = \frac{4 \pm 2\sqrt{2}}{4} = \frac{2 \pm \sqrt{2}}{2}$$

The solutions are $\frac{2 + \sqrt{2}}{2}$ and $\frac{2 - \sqrt{2}}{2}$.

›› *Try Exercise 30, page 12*

ANSWER $a = 2, b = -3, c = -1$

Although every quadratic equation can be solved using the quadratic formula, it is sometimes easier to factor and use the **zero product principle.**

Zero Product Principle

If a and b are algebraic expressions, then $ab = 0$ if and only if $a = 0$ or $b = 0$.

Example

To solve $2x^2 + x - 6 = 0$, first factor the polynomial.

$$2x^2 + x - 6 = 0$$
$$(2x - 3)(x + 2) = 0$$
$$2x - 3 = 0 \quad \text{or} \quad x + 2 = 0 \qquad \text{• Zero product principle}$$
$$x = \frac{3}{2} \qquad\qquad x = -2$$

The solutions are $\frac{3}{2}$ and -2.

EXAMPLE 4 Solve by Using the Zero Product Principle

Solve: $(2x - 1)(x - 3) = x^2 + x - 4$

Solution

$$(2x - 1)(x - 3) = x^2 + x - 4$$
$$2x^2 - 7x + 3 = x^2 + x - 4 \qquad \text{• Expand the binomial product.}$$
$$x^2 - 8x + 7 = 0 \qquad \text{• Write as } ax^2 + bx + c = 0.$$
$$(x - 7)(x - 1) = 0 \qquad \text{• Factor.}$$
$$x - 7 = 0 \quad \text{or} \quad x - 1 = 0 \qquad \text{• Apply the zero product principle.}$$
$$x = 7 \qquad\qquad x = 1$$

The solutions are 1 and 7.

Try Exercise 44, page 12

Inequalities

A statement that contains the symbol $<$, $>$, $\leq$, or $\geq$ is called an **inequality.** An inequality expresses the relative order of two mathematical expressions. The **solution set of an inequality** is the set of real numbers each of which, when substituted for the variable, results in a true inequality. The inequality $x > 4$ is true for any value of x greater than 4. For instance, 5, $\sqrt{21}$, and $\frac{17}{3}$ are all solutions of $x > 4$. The solution set of the inequality can be written in set-builder notation as $\{x \mid x > 4\}$ or in interval notation as $(4, \infty)$.

Equivalent inequalities have the same solution set. We solve an inequality by producing *simpler* but equivalent inequalities until the solutions are found. To produce these simpler but equivalent inequalities, we apply the following properties.

Properties of Inequalities

Let a, b, and c be real numbers.

1. ***Addition Property*** Adding the same real number to each side of an inequality preserves the direction of the inequality symbol.

   $a < b$ and $a + c < b + c$ are equivalent inequalities.

2. ***Multiplication Properties***
   a. Multiplying each side of an inequality by the same *positive* real number *preserves* the direction of the inequality symbol.

      If $c > 0$, then $a < b$ and $ac < bc$ are equivalent inequalities.

   b. Multiplying each side of an inequality by the same *negative* real number *changes* the direction of the inequality symbol.

      If $c < 0$, then $a < b$ and $ac > bc$ are equivalent inequalities.

Note the difference between Property 2a and Property 2b. Property 2a states that an equivalent inequality is produced when each side of a given inequality is multiplied by the same *positive* real number and that the direction of the inequality symbol is *not* changed. By contrast, Property 2b states that when each side of a given inequality is multiplied by a *negative* real number, we must *reverse* the direction of the inequality symbol to produce an equivalent inequality.

For instance, $-2b < 6$ and $b > -3$ are equivalent inequalities. (We multiplied each side of the first inequality by $-\frac{1}{2}$, and we changed the "less than" symbol to a "greater than" symbol.)

Because subtraction is defined in terms of addition, subtracting the same real number from each side of an inequality does not change the direction of the inequality symbol.

Because division is defined in terms of multiplication, dividing each side of an inequality by the same *positive* real number does *not* change the direction of the inequality symbol, and dividing each side of an inequality by a *negative* real number *changes* the direction of the inequality symbol.

take note

Solutions of inequalities are often stated using set-builder notation or interval notation. For instance, the real numbers that are solutions of the inequality in Example 5 can be written in set notation as $\{x \mid x > -2\}$ or in interval notation as $(-2, \infty)$.

EXAMPLE 5 ›› Solve an Inequality

Solve $2(x + 3) < 4x + 10$. Write the solution set in set-builder notation.

Solution

$$2(x + 3) < 4x + 10$$

$$2x + 6 < 4x + 10$$ • Use the distributive property.

$$-2x < 4$$ • Subtract $4x$ and 6 from each side of the inequality.

$$x > -2$$ • Divide each side by -2 and reverse the inequality symbol.

The solution set is $\{x \mid x > -2\}$.

›› *Try Exercise 58, page 12*

Solving Inequalities by the Critical Value Method

Any value of x that causes a polynomial in x to equal zero is called a **zero of the polynomial.** For example, -4 and 1 are both zeros of the polynomial $x^2 + 3x - 4$, because $(-4)^2 + 3(-4) - 4 = 0$ and $1^2 + 3 \cdot 1 - 4 = 0$.

A Sign Property of Polynomials

Nonzero polynomials in x have the property that for any value of x between two consecutive real zeros, either all values of the polynomial are positive or all values of the polynomial are negative.

In our work with inequalities that involve polynomials, the real zeros of the polynomial are also referred to as **critical values of the inequality,** because on a number line they separate the real numbers that make the inequality true from those that make it false. In Example 6 we use critical values and the sign property of polynomials to solve an inequality.

EXAMPLE 6 ›› Solve a Polynomial Inequality

Solve: $x^2 + 3x - 4 < 0$

Solution

Factoring the polynomial $x^2 + 3x - 4$ produces the equivalent inequality

$$(x + 4)(x - 1) < 0$$

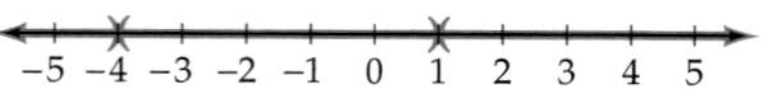

Figure 1.5

Thus the zeros of the polynomial $x^2 + 3x - 4$ are -4 and 1. They are the critical values of the inequality $x^2 + 3x - 4 < 0$. They separate the real number line into the three intervals shown in **Figure 1.5**.

To determine the intervals on which $x^2 + 3x - 4 < 0$, pick a number called a **test value** from each of the three intervals and then determine whether $x^2 + 3x - 4 < 0$ for each of these test values. For example, in the interval $(-\infty, -4)$, pick a test value of, say, -5. Then

$$x^2 + 3x - 4 = (-5)^2 + 3(-5) - 4 = 6$$

Because 6 is not less than 0, by the sign property of polynomials, no number in the interval $(-\infty, -4)$ makes $x^2 + 3x - 4 < 0$.

Now pick a test value from the interval $(-4, 1)$, say, 0. When $x = 0$,

$$x^2 + 3x - 4 = 0^2 + 3(0) - 4 = -4$$

Because -4 is less than 0, by the sign property of polynomials, all numbers in the interval $(-4, 1)$ make $x^2 + 3x - 4 < 0$.

If we pick a test value of 2 from the interval $(1, \infty)$, then

$$x^2 + 3x - 4 = (2)^2 + 3(2) - 4 = 6$$

Because 6 is not less than 0, by the sign property of polynomials, no number in the interval $(1, \infty)$ makes $x^2 + 3x - 4 < 0$.

The following table is a summary of our work.

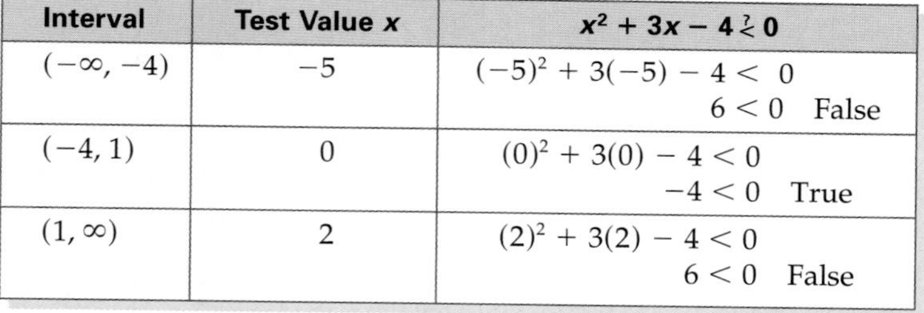

Interval	Test Value x	$x^2 + 3x - 4 \overset{?}{<} 0$
$(-\infty, -4)$	-5	$(-5)^2 + 3(-5) - 4 < 0$ $6 < 0$ False
$(-4, 1)$	0	$(0)^2 + 3(0) - 4 < 0$ $-4 < 0$ True
$(1, \infty)$	2	$(2)^2 + 3(2) - 4 < 0$ $6 < 0$ False

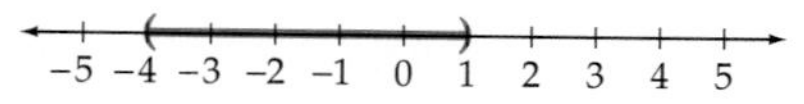

Figure 1.6

In interval notation, the solution set of $x^2 + 3x - 4 < 0$ is $(-4, 1)$. The solution set is graphed in **Figure 1.6**. Note that in this case the critical values -4 and 1 are not included in the solution set because they do not make $x^2 + 3x - 4$ less than 0.

Try Exercise 66, page 12

To avoid the arithmetic in Example 6, we often use a *sign diagram*. For example, note that the factor $(x + 4)$ is negative for all $x < -4$ and positive for all $x > -4$. The factor $(x - 1)$ is negative for all $x < 1$ and positive for all $x > 1$. These results are shown in **Figure 1.7**.

(x + 4) − O + + + + | + + + +
(x − 1) − | − − − − O + + + +
−5 −4 −3 −2 −1 0 1 2 3 4 5

Figure 1.7

To determine on which interval(s) the product $(x + 4)(x - 1)$ is negative, we examine the sign diagram to see where the factors have opposite signs. This occurs only on the interval $(-4, 1)$, where $(x + 4)$ is positive and $(x - 1)$ is negative. Thus the original inequality is true only on the interval $(-4, 1)$.

Following is a summary of the steps used to solve polynomial inequalities by the critical value method.

Solving a Polynomial Inequality by the Critical Value Method

1. Write the inequality so that one side of the inequality is a nonzero polynomial and the other side is 0.
2. Find the real zeros of the polynomial. They are the critical values of the original inequality.
3. Use test values to determine which of the intervals formed by the critical values are to be included in the solution set.
4. Any critical value that satisfies the original inequality is an element of the solution set.

Absolute Value Inequalities

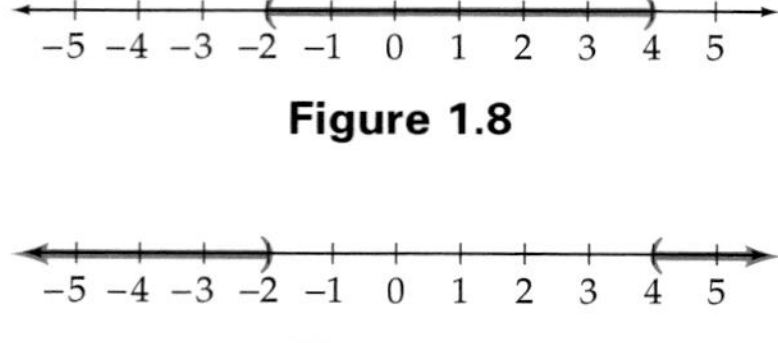

Figure 1.8

Figure 1.9

The solution set of the absolute value inequality $|x - 1| < 3$ is the set of all real numbers whose distance from 1 is *less than* 3. Therefore, the solution set consists of all numbers between -2 and 4. See **Figure 1.8**. In interval notation, the solution set is $(-2, 4)$.

The solution set of the absolute value inequality $|x - 1| > 3$ is the set of all real numbers whose distance from 1 is *greater than* 3. Therefore, the solution set consists of all real numbers less than -2 *or* greater than 4. See **Figure 1.9.** In interval notation, the solution set is $(-\infty, -2) \cup (4, \infty)$.

The following properties are used to solve absolute value inequalities.

Properties of Absolute Value Inequalities

For any variable expression E and any nonnegative real number k,

$$|E| \le k \quad \text{if and only if} \quad -k \le E \le k$$

$$|E| \ge k \quad \text{if and only if} \quad E \le -k \quad \text{or} \quad E \ge k$$

EXAMPLE 7 ›› Solve an Absolute Value Inequality

Solve: $|2 - 3x| < 7$

Solution

$|2 - 3x| < 7$ implies $-7 < 2 - 3x < 7$. Solve this compound inequality.

$$-7 < 2 - 3x < 7$$

$$-9 < -3x < 5$$ • Subtract 2 from each of the three parts of the inequality.

$$3 > x > -\frac{5}{3}$$ • Multiply each part of the inequality by $-\frac{1}{3}$ and reverse the inequality symbols.

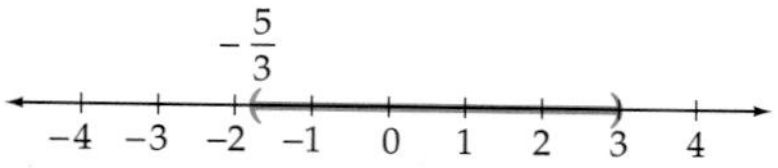

In interval notation, the solution set is given by $\left(-\frac{5}{3}, 3\right)$. See **Figure 1.10.**

Figure 1.10

Try Exercise 80, page 13

> **take note**
>
> Some inequalities have a solution set that consists of all real numbers. For example, $|x + 9| \geq 0$ is true for all values of x. Because an absolute value is always nonnegative, the equation is always true.

EXAMPLE 8 Solve an Absolute Value Inequality

Solve: $|4x - 3| \geq 5$

Solution

$|4x - 3| \geq 5$ implies $4x - 3 \leq -5$ or $4x - 3 \geq 5$. Solving each of these inequalities produces

$$4x - 3 \leq -5 \quad \text{or} \quad 4x - 3 \geq 5$$

$$4x \leq -2 \qquad\qquad 4x \geq 8$$

$$x \leq -\frac{1}{2} \qquad\qquad x \geq 2$$

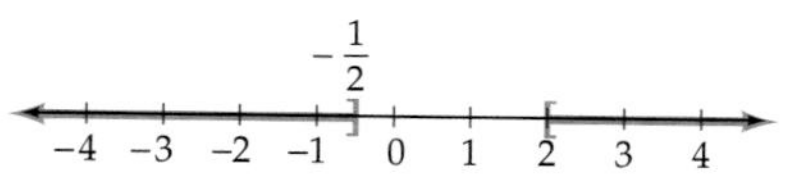

Figure 1.11

Therefore, the solution set is $\left(-\infty, -\frac{1}{2}\right] \cup [2, \infty)$. See **Figure 1.11.**

Try Exercise 78, page 12

Topics for Discussion

1. Discuss the similarities and differences among natural numbers, integers, rational numbers, and real numbers.
2. Discuss the differences among an equation, an inequality, and an expression.
3. Is it possible for an equation to have no solution? If not, explain why. If so, give an example of an equation with no solution.
4. Is the statement $|x| = -x$ ever true? Explain why or why not.
5. How do quadratic equations in one variable differ from linear equations in one variable? Explain how the method used to solve an equation depends on whether it is a linear or a quadratic equation.

Exercise Set 1.1

In Exercises 1 to 18, solve and check each equation.

1. $2x + 10 = 40$
2. $-3y + 20 = 2$
3. $5x + 2 = 2x - 10$
4. $4x - 11 = 7x + 20$
5. $2(x - 3) - 5 = 4(x - 5)$

6. $6(5s - 11) - 12(2s + 5) = 0$
7. $\frac{3}{4}x + \frac{1}{2} = \frac{2}{3}$
8. $\frac{x}{4} - 5 = \frac{1}{2}$
9. $\frac{2}{3}x - 5 = \frac{1}{2}x - 3$
10. $\frac{1}{2}x + 7 - \frac{1}{4}x = \frac{19}{2}$
11. $0.2x + 0.4 = 3.6$
12. $0.04x - 0.2 = 0.07$
13. $\frac{3}{5}(n + 5) - \frac{3}{4}(n - 11) = 0$
14. $-\frac{5}{7}(p + 11) + \frac{2}{5}(2p - 5) = 0$
15. $3(x + 5)(x - 1) = (3x + 4)(x - 2)$
16. $5(x + 4)(x - 4) = (x - 3)(5x + 4)$
17. $0.08x + 0.12(4000 - x) = 432$
18. $0.075y + 0.06(10{,}000 - y) = 727.50$

In Exercises 19 to 26, solve each equation for the indicated variable.

19. $x + 2y = 8; y$
20. $3x - 5y = 15; y$
21. $2x + 5y = 10; x$
22. $5x - 4y = 10; x$
23. $ay - by = c; y$
24. $ax + by = c; y$
25. $x = \frac{y}{1 - y}; y$
26. $x = \frac{2y - 3}{y - 1}; y$

In Exercises 27 to 40, solve by using the quadratic formula.

27. $x^2 - 2x - 15 = 0$
28. $x^2 - 5x - 24 = 0$
29. $x^2 + x - 1 = 0$
30. $x^2 + x - 2 = 0$
31. $2x^2 + 4x + 1 = 0$
32. $2x^2 + 4x - 1 = 0$
33. $3x^2 - 5x - 3 = 0$
34. $3x^2 - 5x - 4 = 0$
35. $\frac{1}{2}x^2 + \frac{3}{4}x - 1 = 0$
36. $\frac{2}{3}x^2 - 5x + \frac{1}{2} = 0$
37. $\sqrt{2}x^2 + 3x + \sqrt{2} = 0$
38. $2x^2 + \sqrt{5}x - 3 = 0$
39. $x^2 = 3x + 5$
40. $-x^2 = 7x - 1$

In Exercises 41 to 48, solve each quadratic equation by factoring and applying the zero product property.

41. $x^2 - 2x - 15 = 0$
42. $y^2 + 3y - 10 = 0$
43. $8y^2 + 189y - 72 = 0$
44. $12w^2 - 41w + 24 = 0$
45. $3x^2 - 7x = 0$
46. $5x^2 = -8x$
47. $(x - 5)^2 - 9 = 0$
48. $(3x + 4)^2 - 16 = 0$

In Exercises 49 to 58, use the properties of inequalities to solve each inequality. Write answers using interval notation.

49. $2x + 3 < 11$
50. $3x - 5 > 16$
51. $x + 4 > 3x + 16$
52. $5x + 6 < 2x + 1$
53. $-6x + 1 \geq 19$
54. $-5x + 2 \leq 37$
55. $-3(x + 2) \leq 5x + 7$
56. $-4(x - 5) \geq 2x + 15$
57. $-4(3x - 5) > 2(x - 4)$
58. $3(x + 7) \leq 5(2x - 8)$

In Exercises 59 to 66, use the critical value method to solve each inequality. Use interval notation to write each solution set.

59. $x^2 + 7x > 0$
60. $x^2 - 5x \leq 0$
61. $x^2 + 7x + 10 < 0$
62. $x^2 + 5x + 6 < 0$
63. $x^2 - 3x \geq 28$
64. $x^2 < -x + 30$
65. $6x^2 - 4 \leq 5x$
66. $12x^2 + 8x \geq 15$

In Exercises 67 to 84, use interval notation to express the solution set of each inequality.

67. $|x| < 4$
68. $|x| > 2$
69. $|x - 1| < 9$
70. $|x - 3| < 10$
71. $|x + 3| > 30$
72. $|x + 4| < 2$
73. $|2x - 1| > 4$
74. $|2x - 9| < 7$
75. $|x + 3| \geq 5$
76. $|x - 10| \geq 2$
77. $|3x - 10| \leq 14$
78. $|2x - 5| \geq 1$

79. $|4 - 5x| \geq 24$

80. $|3 - 2x| \leq 5$

81. $|x - 5| \geq 0$

82. $|x - 7| \geq 0$

83. $|x - 4| \leq 0$

84. $|2x + 7| \leq 0$

85. GEOMETRY The perimeter of a rectangle is 27 centimeters, and its area is 35 square centimeters. Find the length and width of the rectangle.

86. GEOMETRY The perimeter of a rectangle is 34 feet and its area is 60 square feet. Find the length and width of the rectangle.

87. RECTANGULAR ENCLOSURE A gardener wishes to use 600 feet of fencing to enclose a rectangular region and subdivide the region into two smaller rectangles. The total enclosed area is 15,000 square feet. Find the dimensions of the enclosed region.

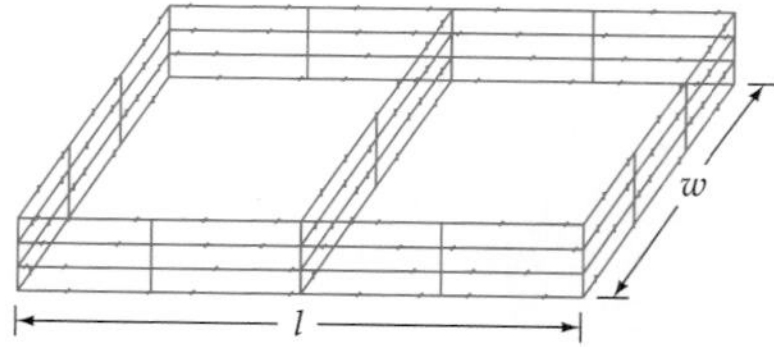

88. RECTANGULAR ENCLOSURE A farmer wishes to use 400 yards of fencing to enclose a rectangular region and subdivide the region into three smaller rectangles. If the total enclosed area is 400 square yards, find the dimensions of the enclosed region.

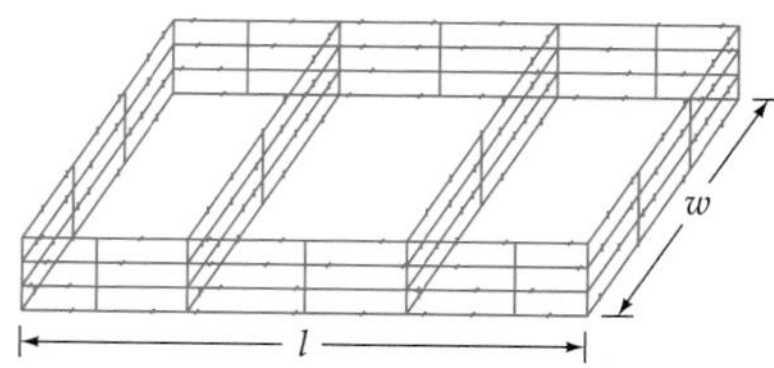

89. PERSONAL FINANCE A bank offers two checking account plans. The monthly fee and charge per check for each plan are shown in the table at the top of the next column. Under what conditions is it less expensive to use the LowCharge plan?

Account Plan	Monthly Fee	Charge per Check
LowCharge	\$5.00	\$.01
FeeSaver	\$1.00	\$.08

90. PERSONAL FINANCE You can rent a car for the day from company A for \$29.00 plus \$0.12 a mile. Company B charges \$22.00 plus \$0.21 a mile. Find the number of miles m (to the nearest mile) per day for which it is cheaper to rent from company A.

91. PERSONAL FINANCE A sales clerk has a choice between two payment plans. Plan A pays \$100.00 a week plus \$8.00 a sale. Plan B pays \$250.00 a week plus \$3.50 a sale. How many sales per week must be made for plan A to yield the greater paycheck?

92. PERSONAL FINANCE A video store offers two rental plans. The yearly membership fee and the daily charge per video are shown below. How many videos can be rented per year if the No-fee plan is to be the less expensive of the plans?

THE VIDEO STORE

Rental Plan	Yearly Fee	Daily Charge per Video
Low-rate	\$15.00	\$1.49
No-fee	None	\$1.99

93. AVERAGE TEMPERATURES The average daily minimum-to-maximum temperature range for the city of Palm Springs during the month of September is 68°F to 104°F. What is the corresponding temperature range measured on the Celsius temperature scale? (*Hint:* Let F be the average daily temperature. Then $68 \leq F \leq 104$. Now substitute $\frac{9}{5}C + 32$ for F and solve the resulting inequality for C.)

Connecting Concepts

94. A GOLDEN RECTANGLE The ancient Greeks defined a rectangle as a *golden rectangle* if its length l and its width w satisfied the equation

$$\frac{l}{w} = \frac{w}{l - w}$$

a. Solve this formula for w.

b. If the length of a golden rectangle is 101 feet, determine its width. Round to the nearest hundredth of a foot.

95. SUM OF NATURAL NUMBERS The sum S of the first n natural numbers $1, 2, 3, \ldots, n$ is given by the formula

$$S = \frac{n}{2}(n + 1)$$

How many consecutive natural numbers starting with 1 produce a sum of 253?

96. NUMBER OF DIAGONALS The number of diagonals D of a polygon with n sides is given by the formula

$$D = \frac{n}{2}(n - 3)$$

a. Determine the number of sides of a polygon with 464 diagonals.

b. Can a polygon have 12 diagonals? Explain.

97. REVENUE The monthly revenue R for a product is given by $R = 420x - 2x^2$, where x is the price in dollars of each unit produced. Find the interval in terms of x for which the monthly revenue is greater than zero.

98. ABSOLUTE VALUE INEQUALITIES Write an absolute value inequality to represent all real numbers within

a. 8 units of 3

b. k units of j (assume $k > 0$)

99. HEIGHT OF A PROJECTILE The equation

$$s = -16t^2 + v_0t + s_0$$

gives the height s, in feet above ground level, at the time t seconds after an object is thrown directly upward from a height s_0 feet above the ground with an initial velocity of v_0 feet per second. A ball is thrown directly upward from ground level with an initial velocity of 64 feet per second. Find the time interval during which the ball attains a height of than 48 feet.

100. HEIGHT OF A PROJECTILE A ball is thrown directly upward from a height of 32 feet above a stream with an initial velocity of 80 feet per second. Find the time interval during which the ball will be more than 96 feet above the stream. (*Hint:* See Exercise 99.)

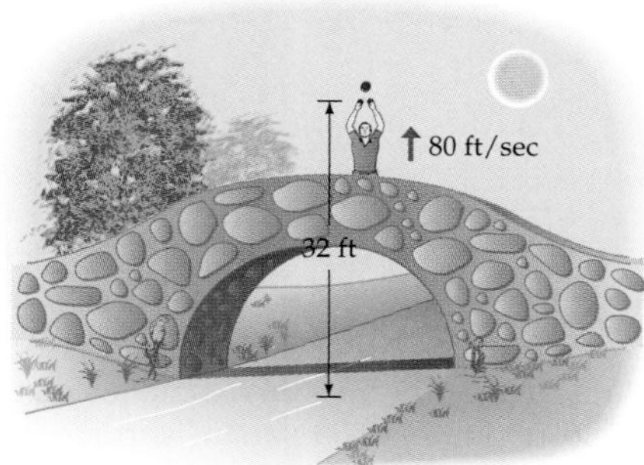

101. GEOMETRY The length of the side of a square has been measured accurately to within 0.01 foot. This measured length is 4.25 feet.

a. Write an absolute value inequality that describes the relationship between the actual length of each side of the square s and its measured length.

b. Solve the absolute value inequality you found in **a.** for s.

Projects

1. TEACHING MATHEMATICS Prepare a lesson that you could use to explain to someone how to solve linear and quadratic equations. Be sure to include an explanation of the differences between these two types of equations and the different methods that are used to solve them.

2. CUBIC EQUATIONS Write an essay on the development of the solution of the cubic equation. An excellent source of information is the chapter "Cardano and the Solution of the Cubic" in *Journey Through Genius* by William Dunham (New York: Wiley, 1990). Another excellent source is *A History of Mathematics: An Introduction* by Victor J. Katz (New York: Harper Collins, 1993).

Section 1.2

- Cartesian Coordinate Systems
- The Distance and Midpoint Formulas
- Graph of an Equation
- Intercepts
- Circles, Their Equations, and Their Graphs

A Two-Dimensional Coordinate System and Graphs

PREPARE FOR THIS SECTION

Prepare for this section by completing the following exercises. The answers can be found on page A1.

PS1. Evaluate $\frac{x_1 + x_2}{2}$ when $x_1 = 4$ and $x_2 = -7$.

PS2. Simplify $\sqrt{50}$.

PS3. Is $y = 3x - 2$ a true equation when $y = 5$ and $x = -1$? [1.1]

PS4. If $y = x^2 - 3x - 2$, find y when $x = -3$. [1.1]

PS5. Evaluate $|-x - y|$ when $x = 3$ and $y = -1$. [1.1]

PS6. Evaluate $\sqrt{b^2 - 4ac}$ when $a = -2$, $b = -3$, and $c = 2$.

Cartesian Coordinate Systems

Each point on a coordinate axis is associated with a number called its **coordinate.** Each point on a flat, two-dimensional surface, called a **coordinate plane** or xy-plane, is associated with an **ordered pair** of numbers called **coordinates** of the point. Ordered pairs are denoted by (a, b), where the real number a is the **x-coordinate** or **abscissa** and the real number b is the **y-coordinate** or **ordinate.**

The coordinates of a point are determined by the point's position relative to a horizontal coordinate axis called the **x-axis** and a vertical coordinate axis called the **y-axis.** The axes intersect at the point $(0, 0)$, called the **origin.** In **Figure 1.12,** the axes are labeled such that positive numbers appear to the right of the origin on the x-axis and above the origin on the y-axis. The four regions formed by the axes are called **quadrants** and are numbered counterclockwise. This two-dimensional coordinate system is referred to as a **Cartesian coordinate system** in honor of René Descartes.

> **take note**
>
> *Abscissa* comes from the same root word as *scissors*. An open pair of scissors looks like an x.

Math Matters

The concepts of *analytic geometry* developed over an extended period of time, culminating in 1637 with the publication of two works: *Discourse on the Method for Rightly Directing One's Reason and Searching for Truth in the Sciences* by René Descartes (1596–1650) and *Introduction to Plane and Solid Loci* by Pierre de Fermat. Each of these works was an attempt to integrate the study of geometry with the study of algebra. Of the two mathematicians, Descartes is usually given most of the credit for developing analytic geometry. In fact, Descartes became so famous in La Haye, the city in which he was born, that it was renamed La Haye-Descartes.

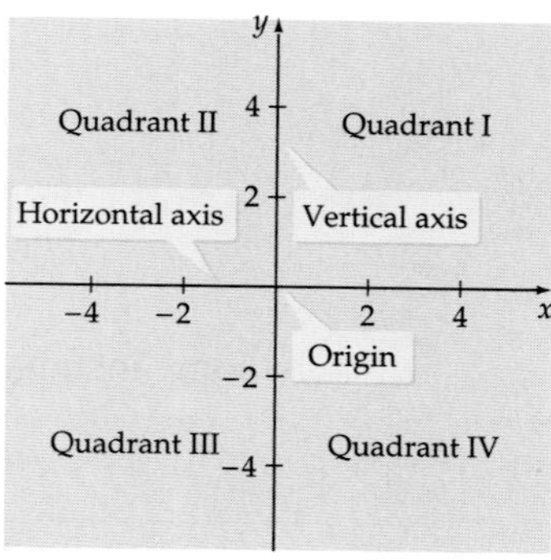

Figure 1.12

take note

The notation (a, b) was used earlier to denote an interval on a one-dimensional number line. In this section, (a, b) denotes an ordered pair in a two-dimensional plane. This should not cause confusion in future sections because as each mathematical topic is introduced, it will be clear whether a one-dimensional or a two-dimensional coordinate system is involved.

To **plot a point** $P(a, b)$ means to draw a dot at its location in the coordinate plane. In **Figure 1.13** we have plotted the points $(4, 3)$, $(-3, 1)$, $(-2, -3)$, $(3, -2)$, $(0, 1)$, $(1, 3)$, and $(3, 1)$. Note that $(1, 3)$ and $(3, 1)$ are not the same point. The order in which the coordinates of an ordered pair are listed is important.

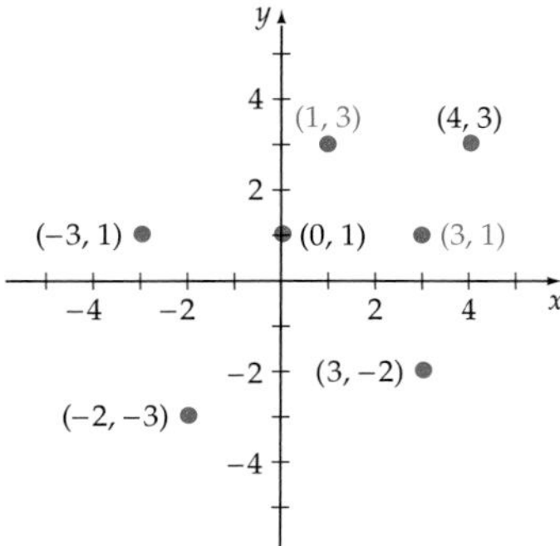

Figure 1.13

Data often are displayed in visual form as a set of points called a **scatter diagram** or **scatter plot**. For instance, the scatter diagram in **Figure 1.14** shows the current and projected revenues of Web-filtering software vendors. (Web-filtering software allows businesses to control which Internet sites are available to employees while at work.) The point whose coordinates are approximately (2005, 520) means that in the year 2005, approximately \$520 million in revenues were generated by companies that supplied this software. The line segments that connect the points in **Figure 1.14** help illustrate trends.

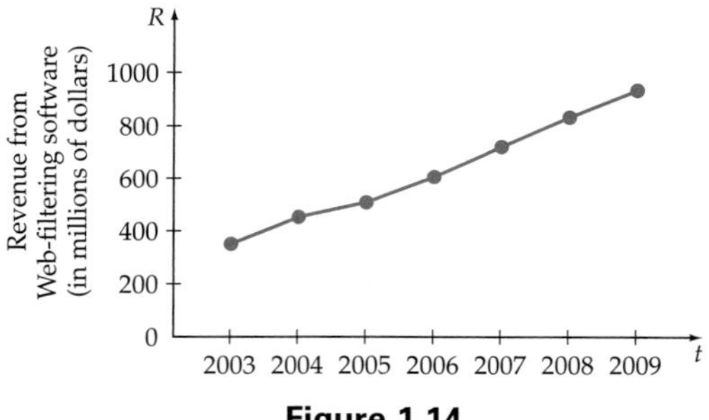

Figure 1.14

Source: IDC, 2005

QUESTION From **Figure 1.14,** will the revenues from Web-filtering software in 2009 be more or less than twice the revenues in 2003?

In some instances, it is important to know when two ordered pairs are equal.

ANSWER More. The revenue in 2003 was about \$350 million. The projected revenue in 2009 is about \$925 million, more than twice \$350 million.

Definition of the Equality of Ordered Pairs

The ordered pairs (a, b) and (c, d) are equal if and only if $a = c$ and $b = d$.

Example

If $(3, y) = (x, -2)$, then $x = 3$ and $y = -2$.

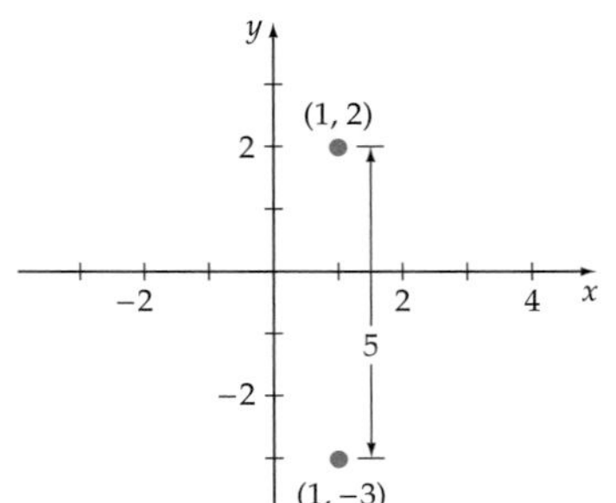

Figure 1.15

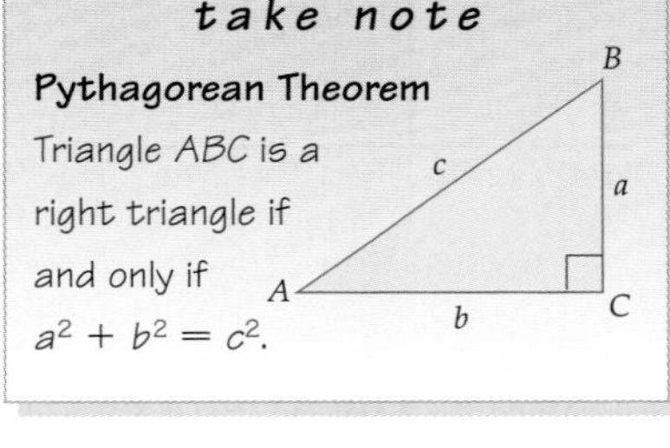

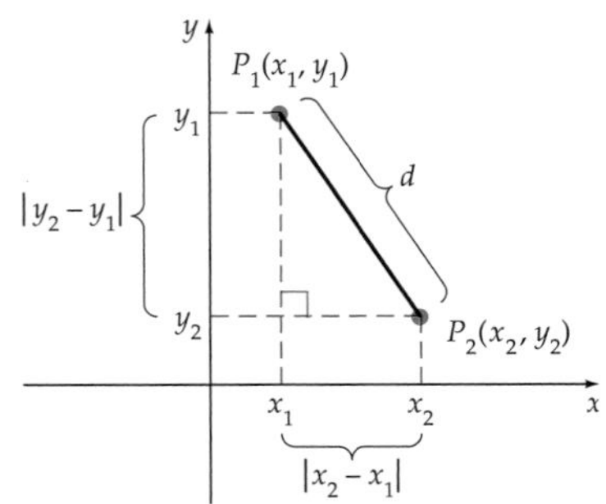

Figure 1.16

The Distance and Midpoint Formulas

The Cartesian coordinate system makes it possible to combine the concepts of algebra and geometry into a branch of mathematics called *analytic geometry.*

The distance between two points on a horizontal line is the absolute value of the difference between the x-coordinates of the two points. The distance between two points on a vertical line is the absolute value of the difference between the y-coordinates of the two points. For example, as shown in **Figure 1.15,** the distance d between the points with coordinates $(1, 2)$ and $(1, -3)$ is $d = |2 - (-3)| = 5$.

If two points are not on a horizontal or vertical line, then a *distance formula* for the distance between the two points can be developed as follows.

The distance between the points $P_1(x_1, y_1)$ and $P_2(x_2, y_2)$ in **Figure 1.16** is the length of the hypotenuse of a right triangle whose sides are horizontal and vertical line segments that measure $|x_2 - x_1|$ and $|y_2 - y_1|$, respectively. Applying the Pythagorean Theorem to this triangle produces

$$d^2 = |x_2 - x_1|^2 + |y_2 - y_1|^2$$

$$d = \sqrt{|x_2 - x_1|^2 + |y_2 - y_1|^2}$$

• **Take the square root of each side of the equation. Because d is nonnegative, the negative root is not listed.**

$$= \sqrt{(x_2 - x_1)^2 + (y_2 - y_1)^2}$$

• **Because $|x_2 - x_1|^2 = (x_2 - x_1)^2$ and $|y_2 - y_1|^2 = (y_2 - y_1)^2$**

Thus we have established the following theorem.

The Distance Formula

The distance $d(P_1, P_2)$ between the points $P_1(x_1, y_1)$ and $P_2(x_2, y_2)$ is

$$d(P_1, P_2) = \sqrt{(x_2 - x_1)^2 + (y_2 - y_1)^2}$$

Example

The distance between $P_1(-3, 4)$ and $P_2(7, 2)$ is given by

$$d(P_1, P_2) = \sqrt{(x_2 - x_1)^2 + (y_2 - y_1)^2}$$
$$= \sqrt{[7 - (-3)]^2 + (2 - 4)^2}$$
$$= \sqrt{10^2 + (-2)^2}$$
$$= \sqrt{104} = 2\sqrt{26} \approx 10.2$$

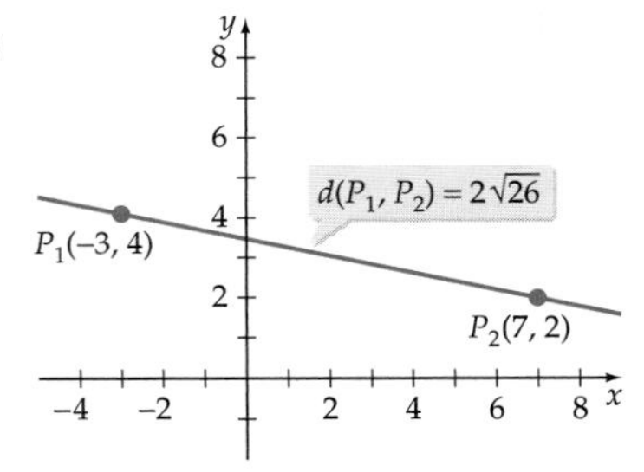

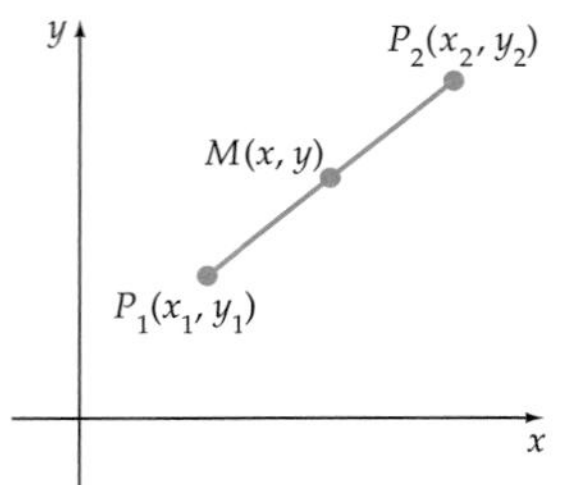

Figure 1.17

The **midpoint** M of a line segment is the point on the line segment that is equidistant from the endpoints $P_1(x_1, y_1)$ and $P_2(x_2, y_2)$ of the segment. See **Figure 1.17**.

The Midpoint Formula

The midpoint M of the line segment from $P_1(x_1, y_1)$ to $P_2(x_2, y_2)$ is given by

$$\left(\frac{x_1 + x_2}{2}, \frac{y_1 + y_2}{2}\right)$$

Example

The midpoint of the line segment between $P_1(-2, 6)$ and $P_2(3, 4)$ is given by

$$M = \left(\frac{x_1 + x_2}{2}, \frac{y_1 + y_2}{2}\right)$$
$$= \left(\frac{(-2) + 3}{2}, \frac{6 + 4}{2}\right)$$
$$= \left(\frac{1}{2}, 5\right)$$

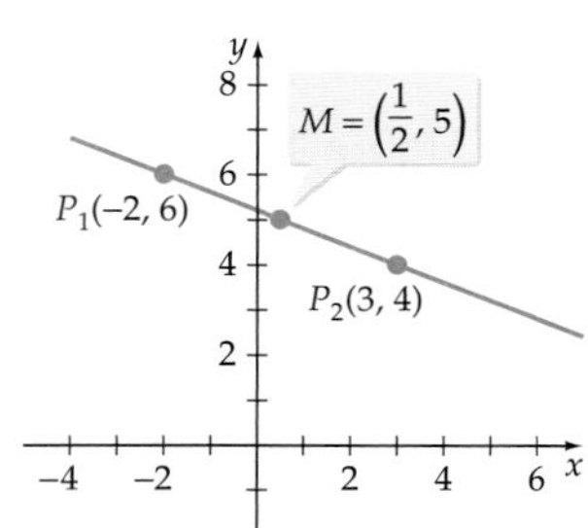

The midpoint formula states that the x-coordinate of the midpoint of a line segment is the *average* of the x-coordinates of the endpoints of the line segment and that the y-coordinate of the midpoint of a line segment is the *average* of the y-coordinates of the endpoints of the line segment.

EXAMPLE 1 » Find the Midpoint and Length of a Line Segment

Find the midpoint and the length of the line segment connecting the points whose coordinates are $P_1(-4, 3)$ and $P_2(4, -2)$.

Solution

$$\text{Midpoint} = \left(\frac{x_1 + x_2}{2}, \frac{y_1 + y_2}{2}\right)$$
$$= \left(\frac{-4 + 4}{2}, \frac{3 + (-2)}{2}\right)$$
$$= \left(0, \frac{1}{2}\right)$$

$$d(P_1, P_2) = \sqrt{(x_2 - x_1)^2 + (y_2 - y_1)^2}$$
$$= \sqrt{(4 - (-4))^2 + (-2 - 3)^2} = \sqrt{(8)^2 + (-5)^2}$$
$$= \sqrt{64 + 25} = \sqrt{89}$$

» *Try Exercise 6, page 28*

Graph of an Equation

The equations below are equations in two variables.

$$y = 3x^3 - 4x + 2 \qquad x^2 + y^2 = 25 \qquad y = \frac{x}{x+1}$$

Math Matters

Maria Agnesi (1718–1799) wrote *Foundations of Analysis for the Use of Italian Youth,* one of the most successful textbooks of the 18th century. The French Academy authorized a translation into French in 1749, noting that "there is no other book, in any language, which would enable a reader to penetrate as deeply, or as rapidly, into the fundamental concepts of analysis." A curve that she discusses in her text is given by the equation $y = \dfrac{a^3}{x^2 + a^2}$. Unfortunately, due to a translation error from Italian to English, the curve became known as the "witch of Agnesi."

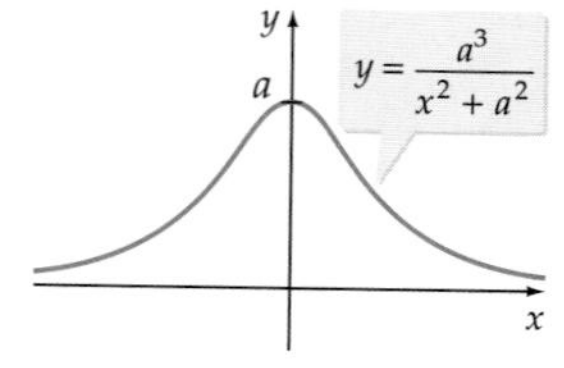

The solution of an equation in two variables is an ordered pair (x, y) whose coordinates satisfy the equation. For instance, the ordered pairs $(3, 4)$, $(4, -3)$, and $(0, 5)$ are some of the solutions of $x^2 + y^2 = 25$. Generally, there are an infinite number of solutions of an equation in two variables. These solutions can be displayed in a *graph.*

Definition of the Graph of an Equation

The **graph of an equation** in the two variables x and y is the set of all points (x, y) whose coordinates satisfy the equation.

Consider $y = 2x - 1$. Substituting various values of x into the equation and solving for y produces some of the ordered pairs that satisfy the equation. It is convenient to record the results in a table similar to the one shown below. The graph of the ordered pairs is shown in **Figure 1.18.**

x	$y = 2x - 1$	y	(x, y)
−2	2(−2) − 1	−5	(−2, −5)
−1	2(−1) − 1	−3	(−1, −3)
0	2(0) − 1	−1	(0, −1)
1	2(1) − 1	1	(1, 1)
2	2(2) − 1	3	(2, 3)

Choosing some noninteger values of x produces more ordered pairs to graph, such as $\left(-\frac{3}{2}, -4\right)$ and $\left(\frac{5}{2}, 4\right)$, as shown in **Figure 1.19.** Using still other values of x would result in more and more ordered pairs to graph. The result would be so many dots that the graph would appear as the straight line shown in **Figure 1.20,** which is the graph of $y = 2x - 1$.

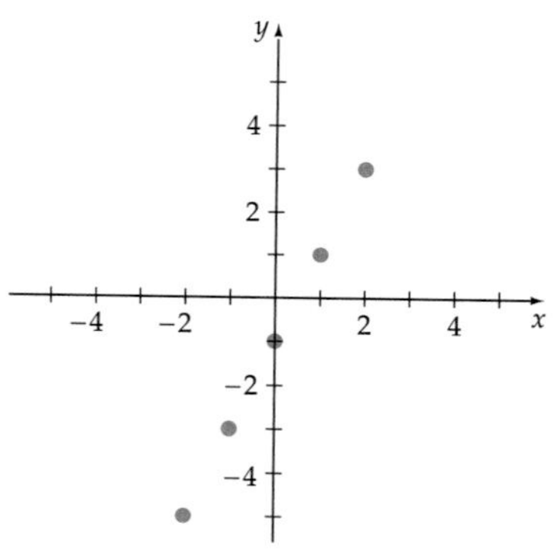

Figure 1.18

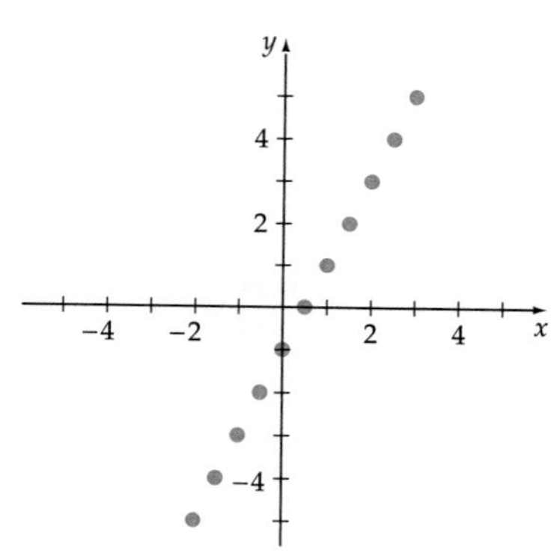

Figure 1.19

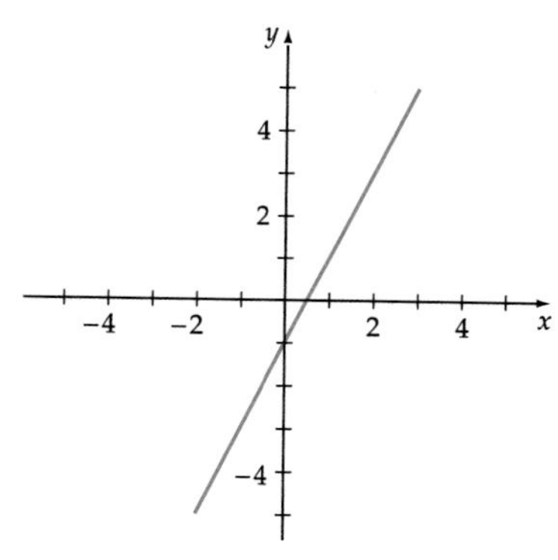

Figure 1.20

EXAMPLE 2 Draw a Graph by Plotting Points

Graph: $-x^2 + y = 1$

Solution

Solve the equation for y.

$$-x^2 + y = 1$$
$$y = x^2 + 1 \qquad \text{• Add } x^2 \text{ to each side.}$$

Select values of x and use the equation to calculate y. Choose enough values of x so that an accurate graph can be drawn. Plot the points and draw a smooth curve through them. See **Figure 1.21**.

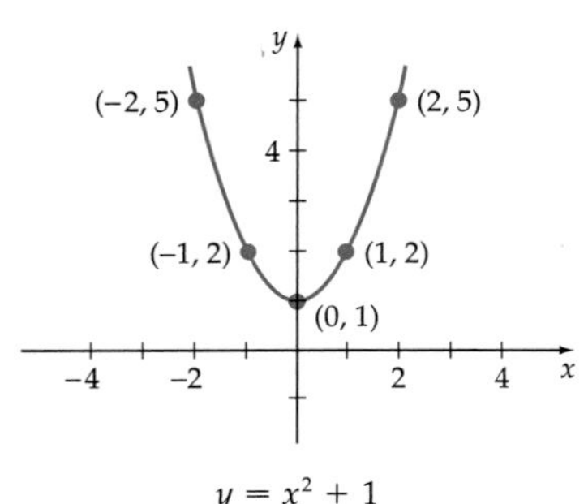

$y = x^2 + 1$

Figure 1.21

x	$y = x^2 + 1$	y	(x, y)
−2	$(-2)^2 + 1$	5	(−2, 5)
−1	$(-1)^2 + 1$	2	(−1, 2)
0	$(0)^2 + 1$	1	(0, 1)
1	$(1)^2 + 1$	2	(1, 2)
2	$(2)^2 + 1$	5	(2, 5)

Try Exercise 26, page 28

Integrating Technology

Some graphing calculators, such as the TI-83/TI-83 Plus/TI-84 Plus, have a TABLE feature that allows you to create a table similar to the one shown in Example 2. Enter the equation to be graphed, the first value for x, and the increment (the difference between successive values of x). For instance, entering $y_1 = x^2 + 1$, an initial value of x of -2, and an increment of 1 yields a display similar to the one in **Figure 1.22**. Changing the initial value to -6 and the increment to 2 gives the table in **Figure 1.23**.

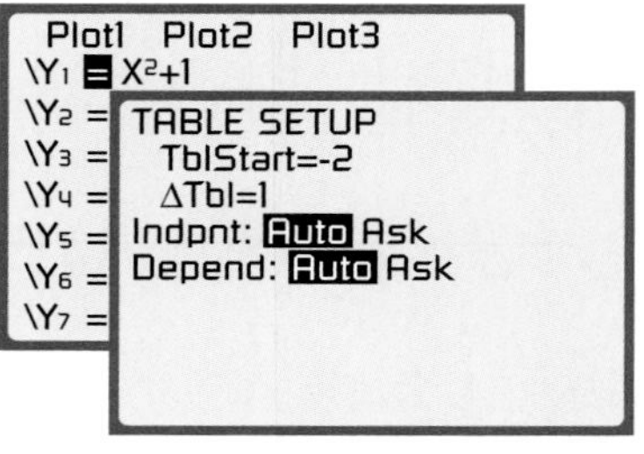

X	Y1
-2	5
-1	2
0	1
1	2
2	5
3	10
4	17

X=-2

Figure 1.22

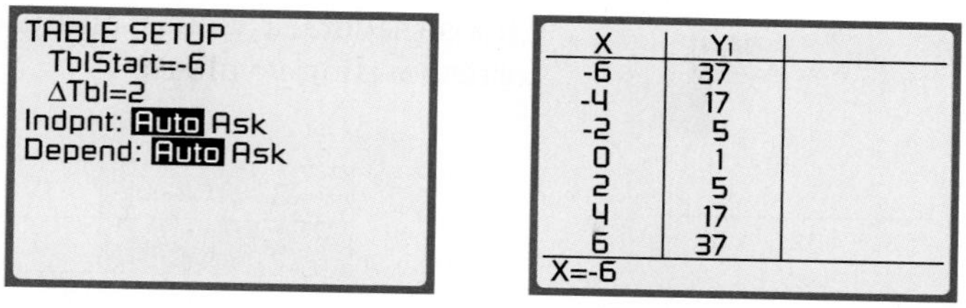

TABLE SETUP
TblStart=-6
ΔTbl=2
Indpnt: Auto Ask
Depend: Auto Ask
X Y1
-6 37
-4 17
-2 5
0 1
2 5
4 17
6 37
X=-6

Figure 1.23

With some calculators, you can scroll through the table by using the up- or down-arrow keys. In this way, you can determine many more ordered pairs of the graph.

EXAMPLE 3 ⟫ Graph by Plotting Points

Graph: $y = |x - 2|$

Solution

This equation is already solved for y, so start by choosing an x value and using the equation to determine the corresponding y value. For example, if $x = -3$, then $y = |(-3) - 2| = |-5| = 5$. Continuing in this manner produces the following table.

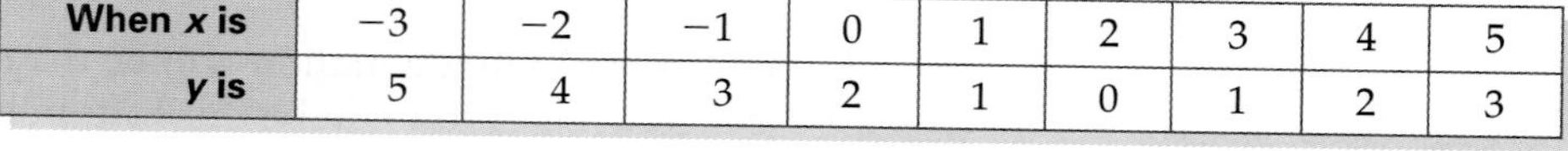

When x is	−3	−2	−1	0	1	2	3	4	5
y is	5	4	3	2	1	0	1	2	3

y
x
5
−2
5
(−3, 5)
(−2, 4)
(−1, 3)
(0, 2)
(1, 1)
(2, 0)
(3, 1)
(4, 2)
(5, 3)

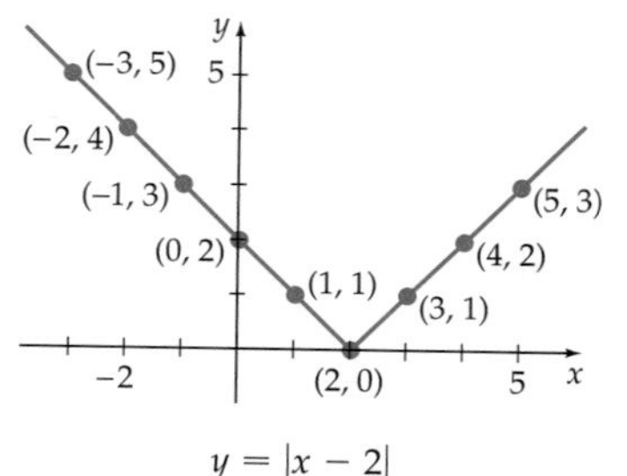

$y = |x - 2|$

Figure 1.24

Now plot the points listed in the table. Connecting the points forms a V shape, as shown in **Figure 1.24.**

⟫ *Try Exercise 30, page 28*

EXAMPLE 4 ⟫ Graph by Plotting Points

Graph: $y^2 = x$

Solution

Solve the equation for y.

$$y^2 = x$$
$$y = \pm\sqrt{x} \quad \text{• Take the square root of each side.}$$

Continued ▸

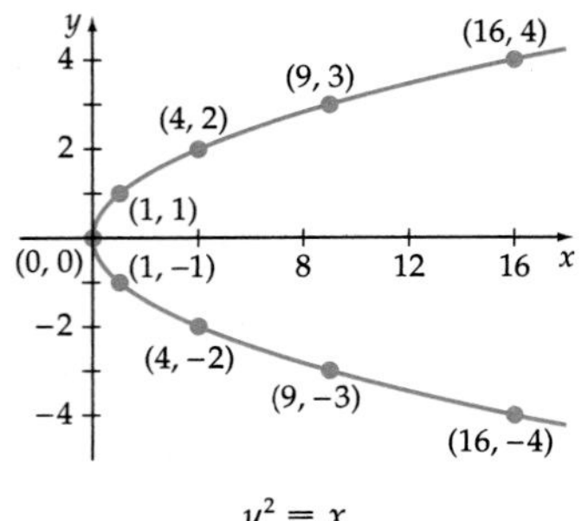

$y^2 = x$

Figure 1.25

Choose several x values, and use the equation to determine the corresponding y values.

When x is	0	1	4	9	16
y is	0	±1	±2	±3	±4

Plot the points as shown in **Figure 1.25.** The graph is a *parabola.*

Try Exercise 32, page 28

Integrating Technology

A graphing calculator or computer graphing software can be used to draw the graphs in Examples 3 and 4. These graphing utilities graph a curve in much the same way as you would, by selecting values of x and calculating the corresponding values of y. A curve is then drawn through the points.

If you use a graphing utility to graph $y = |x - 2|$, you will need to use the *absolute value* function that is built into the utility. The equation you enter will look similar to Y1=abs(X−2).

To graph the equation in Example 4, you will enter two equations. The equations you enter will be similar to

Y1=√(X)

Y2=−√(X)

The graph of the first equation will be the top half of the parabola; the graph of the second equation will be the bottom half.

Intercepts

Any point that has an x- or a y-coordinate of zero is called an **intercept** of the graph of an equation because it is at these points that the graph intersects the x- or the y-axis.

Definitions of x-Intercepts and y-Intercepts

If $(x_1, 0)$ satisfies an equation, then the point $(x_1, 0)$ is called an **x-intercept** of the graph of the equation.

If $(0, y_1)$ satisfies an equation, then the point $(0, y_1)$ is called a **y-intercept** of the graph of the equation.

To find the x-intercepts of the graph of an equation, let $y = 0$ and solve the equation for x. To find the y-intercepts of the graph of an equation, let $x = 0$ and solve the equation for y.

EXAMPLE 5 Find x- and y-Intercepts

Find the x- and y-intercepts of the graph of $y = x^2 - 2x - 3$.

ALGEBRAIC SOLUTION

To find the y-intercept, let $x = 0$ and solve for y.

$$y = 0^2 - 2(0) - 3 = -3$$

To find the x-intercepts, let $y = 0$ and solve for x.

$$0 = x^2 - 2x - 3$$
$$0 = (x - 3)(x + 1)$$
$$(x - 3) = 0 \quad \text{or} \quad (x + 1) = 0$$
$$x = 3 \quad \text{or} \quad x = -1$$

Because $y = -3$ when $x = 0$, $(0, -3)$ is a y-intercept. Because $x = 3$ or -1 when $y = 0$, $(3, 0)$ and $(-1, 0)$ are x-intercepts. **Figure 1.26** confirms that these three points are intercepts.

VISUALIZE THE SOLUTION

The graph of $y = x^2 - 2x - 3$ is shown below. Observe that the graph intersects the x-axis at $(-1, 0)$ and $(3, 0)$, the x-intercepts. The graph also intersects the y-axis at $(0, -3)$, the y-intercept.

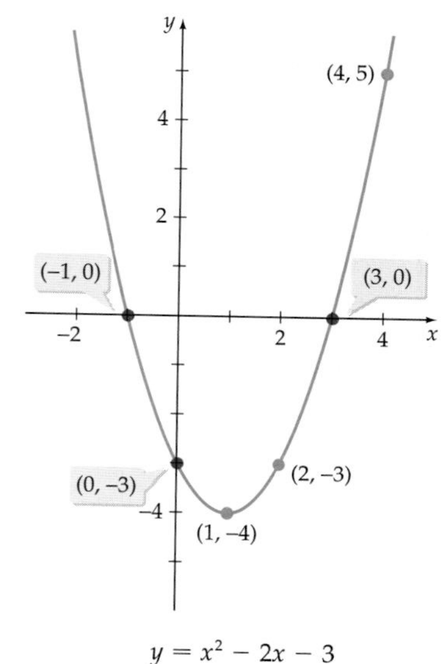

$y = x^2 - 2x - 3$

Figure 1.26

Try Exercise 40, page 28

Integrating Technology

In Example 5 it was possible to find the x-intercepts by solving a quadratic equation. In some instances, however, solving an equation to find the intercepts may be very difficult. In these cases, a graphing calculator can be used to estimate the x-intercepts.

The x-intercepts of the graph of $y = x^3 + x + 4$ can be estimated using the ZERO feature of a TI-83/TI-83 Plus/TI-84 Plus calculator. The keystrokes and some sample screens for this procedure are shown on page 24.

Continued ▶

Press [Y=]. Now enter X^3+X+4. Press [ZOOM] and select the standard viewing window. Press [ENTER].

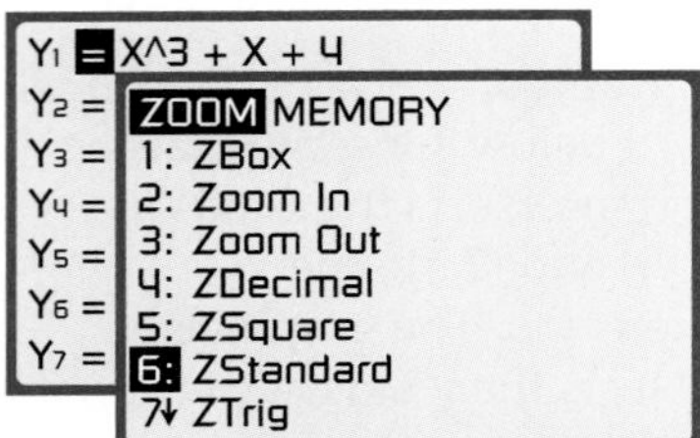

Press [2nd] CALC to access the CALCULATE menu. The y-coordinate of an x-intercept is zero. Therefore, select 2:zero. Press [ENTER].

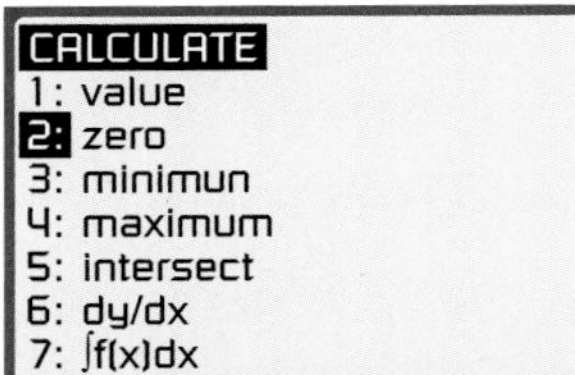

The "Left Bound?" shown on the bottom of the screen means to move the cursor until it is to the left of an x-intercept. Press [ENTER].

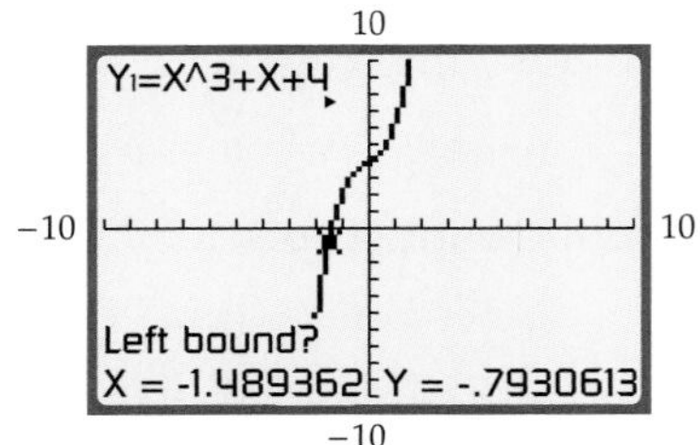

The "Right Bound?" shown on the bottom of the screen means to move the cursor until it is to the right of the desired x-intercept. Press [ENTER].

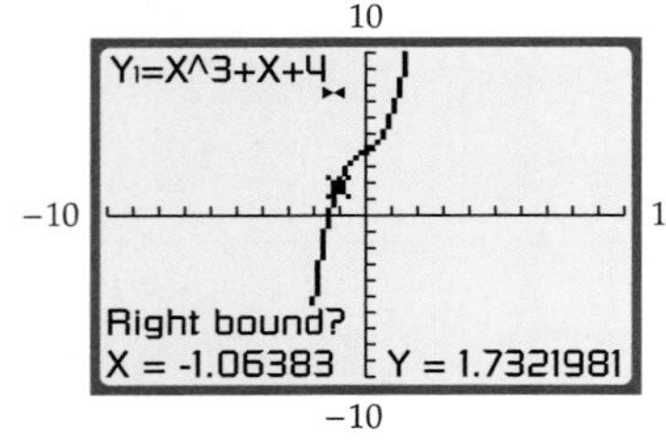

"Guess?" is shown on the bottom of the screen. Move the cursor until it is approximately on the x-intercept. Press [ENTER].

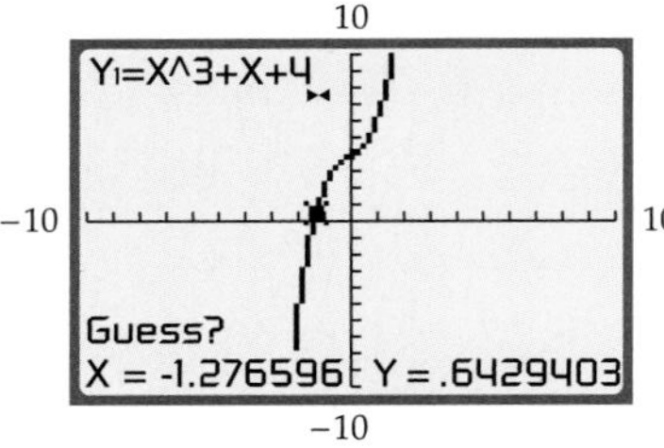

The "Zero" shown on the bottom of the screen means that the value of y is 0 when $x = -1.378797$. The x-intercept is about $(-1.378797, 0)$.

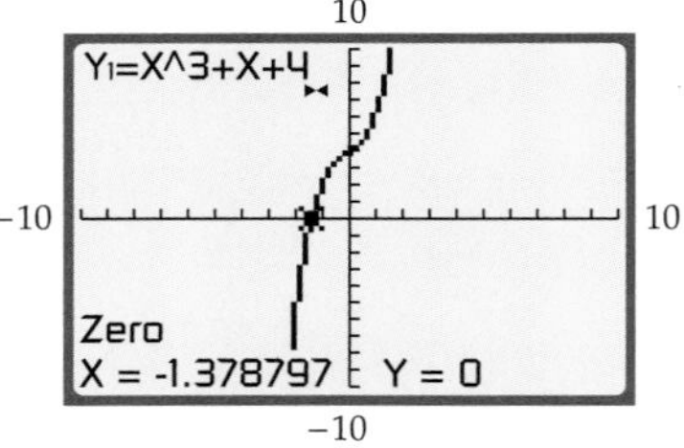

Circles, Their Equations, and Their Graphs

Frequently you will sketch graphs by plotting points. However, some graphs can be sketched merely by recognizing the form of the equation. A *circle* is an example of a curve whose graph you can sketch after you have inspected its equation.

Definition of a Circle

A **circle** is the set of points in a plane that are a fixed distance from a specified point. The fixed distance is the **radius** of the circle, and the specified point is the **center** of the circle.

The standard form of the equation of a circle is derived by using this definition. To derive the standard form, we use the distance formula. **Figure 1.27** is a

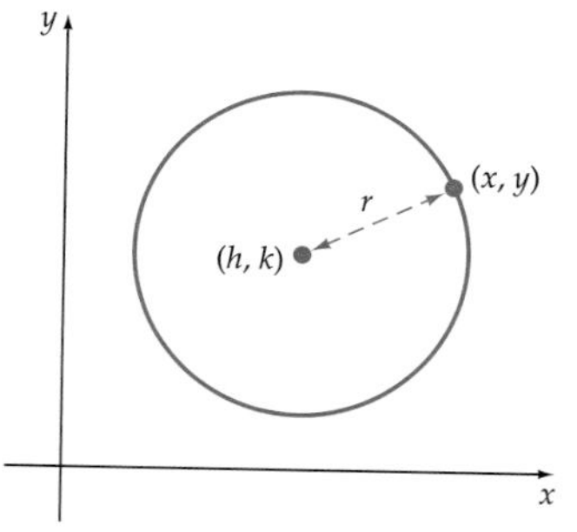

Figure 1.27

circle with center (h, k) and radius r. The point (x, y) is on the circle if and only if it is a distance of r units from the center (h, k). Thus (x, y) is on the circle if and only if

$$\sqrt{(x-h)^2 + (y-k)^2} = r$$

$$(x-h)^2 + (y-k)^2 = r^2 \quad \text{• Square each side.}$$

Standard Form of the Equation of a Circle

The **standard form of the equation of a circle** with center at (h, k) and radius r is

$$(x-h)^2 + (y-k)^2 = r^2$$

Example

- The equation $(x-2)^2 + (y-4)^2 = \mathbf{3}^2$ is in standard form, where $h = 2$, $k = 4$, and $r = \mathbf{3}$. The graph of this equation is a circle with center $C(2, 4)$ and radius **3**.
- The equation $(x-3)^2 + (y+1)^2 = 25$ can be written in standard form as $(x-3)^2 + (y-(-1))^2 = \mathbf{5}^2$. Note that in the standard form, $(x-h)^2$ and $(y-k)^2$ are written using subtraction. Because $(y+1)^2$ is written using addition, the expression is rewritten as $(y-(-1))^2$. The graph of this equation is a circle with center $C(3, -1)$ and radius **5**.
- The equation $(x+4)^2 + (y+2)^2 = 10$ can be written in standard form as $(x-(-4))^2 + (y-(-2))^2 = (\sqrt{\mathbf{10}})^2$. The graph of this equation is a circle with center $C(-4, -2)$ and radius $\sqrt{\mathbf{10}}$.

If a circle is centered at the origin $(0, 0)$, then $h = 0$ and $k = 0$ and the standard form of the equation of the circle simplifies to

$$x^2 + y^2 = r^2 \quad \text{• Equation of a circle with center at the origin and radius } r.$$

For instance, $x^2 + y^2 = 9$ is the equation of the circle with center at the origin and radius $\sqrt{9} = 3$.

QUESTION What are the radius and the coordinates of the center of the circle with equation $x^2 + (y-2)^2 = 30$?

EXAMPLE 6 » Find the Standard Form of the Equation of a Circle

Find the standard form of the equation of the circle that has center $C(-4, -2)$ and contains the point $P(-1, 2)$.

Continued ▶

ANSWER The radius is $\sqrt{30}$ and the coordinates of the center are $(0, 2)$.

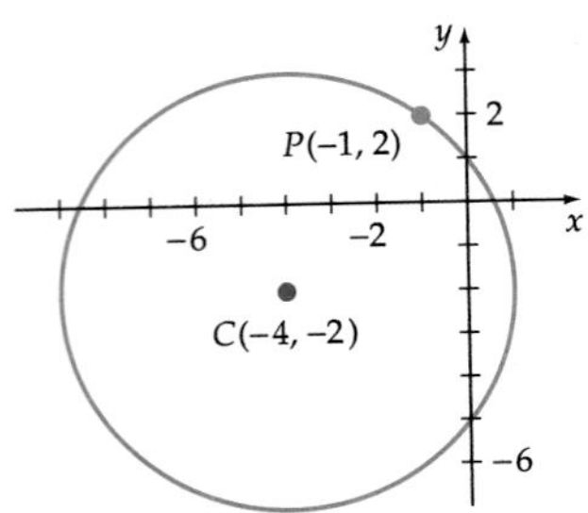

Figure 1.28

Solution

See the graph of the circle in **Figure 1.28**. Because the point P is on the circle, the radius r of the circle must equal the distance from C to P. Thus

$$r = \sqrt{(-1 - (-4))^2 + (2 - (-2))^2}$$
$$= \sqrt{9 + 16} = \sqrt{25} = 5$$

Using the standard form with $h = -4$, $k = -2$, and $r = 5$, we obtain

$$(x + 4)^2 + (y + 2)^2 = 5^2$$

Try Exercise 64, page 28

If we rewrite $(x + 4)^2 + (y + 2)^2 = 5^2$ by squaring and combining like terms, we produce

$$x^2 + 8x + 16 + y^2 + 4y + 4 = 25$$
$$x^2 + y^2 + 8x + 4y - 5 = 0$$

This form of the equation is known as the **general form of the equation of a circle.** By completing the square, it is always possible to write the general form $x^2 + y^2 + Ax + By + C = 0$ in the standard form

$$(x - h)^2 + (y - k)^2 = s$$

for some number s. If $s > 0$, the graph is a circle with radius $r = \sqrt{s}$. If $s = 0$, the graph is the point (h, k). If $s < 0$, the equation has no real solutions and there is no graph.

EXAMPLE 7 Find the Center and Radius of a Circle by Completing the Square

Find the center and radius of the circle given by

$$x^2 + y^2 - 6x + 4y - 3 = 0$$

Solution

First rearrange and group the terms as shown.

$$(x^2 - 6x) + (y^2 + 4y) = 3$$

Complete the square of $(x^2 - 6x)$ by adding 9, and complete the square of $(y^2 + 4y)$ by adding 4.

$$(x^2 - 6x + 9) + (y^2 + 4y + 4) = 3 + 9 + 4$$
$$(x - 3)^2 + (y + 2)^2 = 16$$
$$(x - 3)^2 + (y - (-2))^2 = 4^2$$

• **Add 9 and 4 to each side of the equation.**

This equation is the standard form of the equation of a circle and indicates that the graph of the original equation is a circle centered at $(3, -2)$ with radius 4. See **Figure 1.29**.

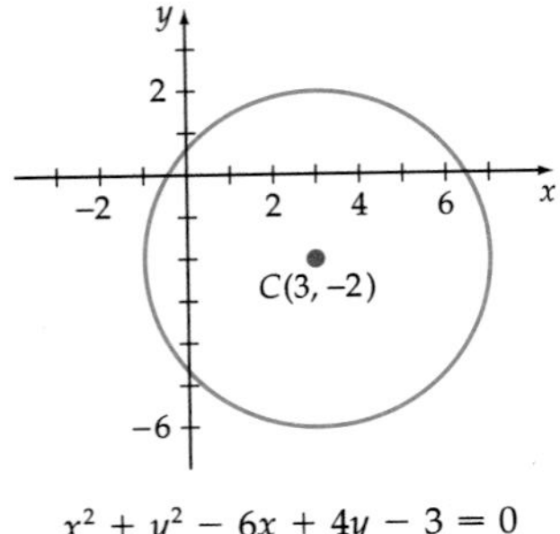

$x^2 + y^2 - 6x + 4y - 3 = 0$

Figure 1.29

Try Exercise 66, page 29

Topics for Discussion

1. The distance formula states that the distance d between the points $P_1(x_1, y_1)$ and $P_2(x_2, y_2)$ is $d = \sqrt{(x_2 - x_1)^2 + (y_2 - y_1)^2}$. Can the distance formula also be written as follows? Explain.

$$d = \sqrt{(x_1 - x_2)^2 + (y_1 - y_2)^2}$$

2. Does the equation $(x - 3)^2 + (y + 4)^2 = -6$ have a graph that is a circle? Explain.
3. Explain why the graph of $|x| + |y| = 1$ does not contain any points that have
   a. a y-coordinate that is greater than 1 or less than -1
   b. an x-coordinate that is greater than 1 or less than -1
4. Discuss the graph of $xy = 0$.
5. Explain how to determine the x- and y-intercepts of a graph defined by an equation (without using the graph).

Exercise Set 1.2

In Exercises 1 and 2, plot the points whose coordinates are given on a Cartesian coordinate system.

1. $(2, 4), (0, -3), (-2, 1), (-5, -3)$

2. $(-3, -5), (-4, 3), (0, 2), (-2, 0)$

3. PER CAPITA INCOME The following graph, based on data from the Bureau of Economic Analysis, shows the annual per capita income (total income earned divided by population) in the United States for selected years.

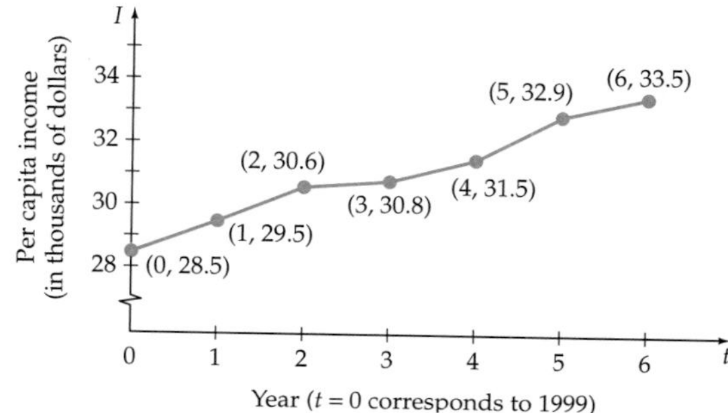

a. From the graph, what was the per capita income in 2003?

b. If the increase in the per capita income from 2005 to 2006 were the same as the increase from 2004 to 2005, what would be the per capita income in 2006?

c. If the percent increase in the per capita income from 2005 to 2006 were the same as the percent increase in the per capita income from 2004 to 2005, what would be the per capita income in 2006?

4. COMPUTER GAMES The graph below shows the results of market research conducted by a developer of computer games. It shows the projected numbers of sales N, in millions, of a game for selected selling prices p in dollars per game.

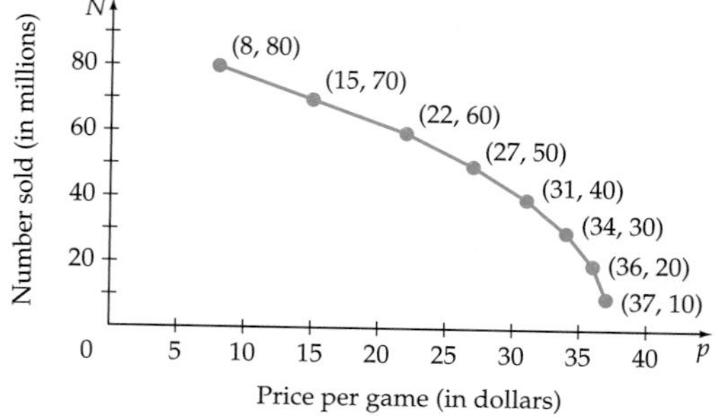

a. Explain the meaning of the ordered pair (22, 60) in the context of this problem.

b. Based on the graph, does the projected numbers of sales increase or decrease as the price of this game increases?

(Continued)

c. The product of the coordinates of the ordered pairs, $R = p \cdot N$, indicates the revenue R to the company generated by the sale of N games at p dollars per game. Create a scatter diagram of (p, R).

d. Based on the scatter diagram in **c.**, what happens to the revenue as the price of the game increases?

In Exercises 5 to 16, find the distance between the points whose coordinates are given.

5. $(6, 4), (-8, 11)$

6. $(-5, 8), (-10, 14)$

7. $(-4, -20), (-10, 15)$

8. $(40, 32), (36, 20)$

9. $(5, -8), (0, 0)$

10. $(0, 0), (5, 13)$

11. $(\sqrt{3}, \sqrt{8}), (\sqrt{12}, \sqrt{27})$

12. $(\sqrt{125}, \sqrt{20}), (6, 2\sqrt{5})$

13. $(a, b), (-a, -b)$

14. $(a - b, b), (a, a + b)$

15. $(x, 4x), (-2x, 3x)$, given that $x < 0$

16. $(x, 4x), (-2x, 3x)$, given that $x > 0$

17. Find all points on the x-axis that are 10 units from the point $(4, 6)$. (*Hint:* First write the distance formula with $(4, 6)$ as one of the points and $(x, 0)$ as the other point.)

18. Find all points on the y-axis that are 12 units from the point $(5, -3)$.

In Exercises 19 to 24, find the midpoint of the line segment having the given endpoints.

19. $(1, -1), (5, 5)$

20. $(-5, -2), (6, 10)$

21. $(6, -3), (6, 11)$

22. $(4, 7), (-10, 7)$

23. $(1.75, 2.25), (-3.5, 5.57)$

24. $(-8.2, 10.1), (-2.4, -5.7)$

In Exercises 25 to 38, graph each equation by plotting points that satisfy the equation.

25. $x - y = 4$

26. $2x + y = -1$

27. $y = 0.25x^2$

28. $3x^2 + 2y = -4$

29. $y = -2|x - 3|$

30. $y = |x + 3| - 2$

31. $y = x^2 - 3$

32. $y = x^2 + 1$

33. $y = \frac{1}{2}(x - 1)^2$

34. $y = 2(x + 2)^2$

35. $y = x^2 + 2x - 8$

36. $y = x^2 - 2x - 8$

37. $y = -x^2 + 2$

38. $y = -x^2 - 1$

In Exercises 39 to 48, find the x- and y-intercepts of the graph of each equation. Use the intercepts and additional points as needed to draw the graph of the equation.

39. $2x + 5y = 12$

40. $3x - 4y = 15$

41. $x = -y^2 + 5$

42. $x = y^2 - 6$

43. $x = |y| - 4$

44. $x = y^3 - 2$

45. $x^2 + y^2 = 4$

46. $x^2 = y^2$

47. $|x| + |y| = 4$

48. $|x - 4y| = 8$

In Exercises 49 to 56, determine the center and radius of the circle with the given equation.

49. $x^2 + y^2 = 36$

50. $x^2 + y^2 = 49$

51. $(x - 1)^2 + (y - 3)^2 = 49$

52. $(x - 2)^2 + (y - 4)^2 = 25$

53. $(x + 2)^2 + (y + 5)^2 = 25$

54. $(x + 3)^2 + (y + 5)^2 = 121$

55. $(x - 8)^2 + y^2 = \frac{1}{4}$

56. $x^2 + (y - 12)^2 = 1$

In Exercises 57 to 64, find an equation of a circle that satisfies the given conditions. Write your answer in standard form.

57. Center $(4, 1)$, radius $r = 2$

58. Center $(5, -3)$, radius $r = 4$

59. Center $\left(\frac{1}{2}, \frac{1}{4}\right)$, radius $r = \sqrt{5}$

60. Center $\left(0, \frac{2}{3}\right)$, radius $r = \sqrt{11}$

61. Center $(0, 0)$, passing through $(-3, 4)$

62. Center $(0, 0)$, passing through $(5, 12)$

63. Center $(1, 3)$, passing through $(4, -1)$

64. Center $(-2, 5)$, passing through $(1, 7)$

In Exercises 65 to 72, find the center and radius of the graph of the circle. The equations of the circles are written in general form.

65. $x^2 + y^2 - 6x + 5 = 0$

66. $x^2 + y^2 - 6x - 4y + 12 = 0$

67. $x^2 + y^2 - 14x + 8y + 56 = 0$

68. $x^2 + y^2 - 10x + 2y + 25 = 0$

69. $4x^2 + 4y^2 + 4x - 63 = 0$

70. $9x^2 + 9y^2 - 6y - 17 = 0$

71. $x^2 + y^2 - x + 3y - \frac{15}{4} = 0$

72. $x^2 + y^2 + 3x - 5y + \frac{25}{4} = 0$

73. Find an equation of a circle that has a diameter with endpoints $(2, 3)$ and $(-4, 11)$. Write your answer in standard form.

74. Find an equation of a circle that has a diameter with endpoints $(7, -2)$ and $(-3, 5)$. Write your answer in standard form.

75. Find an equation of a circle that has its center at $(7, 11)$ and is tangent to the x-axis. Write your answer in standard form.

76. Find an equation of a circle that has its center at $(-2, 3)$ and is tangent to the y-axis. Write your answer in standard form.

»»» Connecting Concepts

In Exercises 77 to 86, graph the set of all points whose *x*- and *y*-coordinates satisfy the given conditions.

77. $x = 1, y \geq 1$

78. $y = -3, x \geq -2$

79. $y \leq 3$

80. $x \geq 2$

81. $xy \geq 0$

82. $|y| \geq 1, \frac{x}{y} \leq 0$

83. $|x| = 2, |y| = 3$

84. $|x| = 4, |y| = 1$

85. $|x| \leq 2, y \geq 2$

86. $x \geq 1, |y| \leq 3$

In Exercises 87 to 90, find the other endpoint of the line segment that has the given endpoint and midpoint.

87. Endpoint $(5, 1)$, midpoint $(9, 3)$

88. Endpoint $(4, -6)$, midpoint $(-2, 11)$

89. Endpoint $(-3, -8)$, midpoint $(2, -7)$

90. Endpoint $(5, -4)$, midpoint $(0, 0)$

91. Find a formula for the set of all points (x, y) for which the distance from (x, y) to $(3, 4)$ is 5.

92. Find a formula for the set of all points (x, y) for which the distance from (x, y) to $(-5, 12)$ is 13.

93. Find a formula for the set of all points (x, y) for which the sum of the distances from (x, y) to $(4, 0)$ and from (x, y) to $(-4, 0)$ is 10.

94. Find a formula for the set of all points for which the absolute value of the difference of the distances from (x, y) to $(0, 4)$ and from (x, y) to $(0, -4)$ is 6.

95. Find an equation of a circle that is tangent to both axes, has its center in the second quadrant, and has a radius of 3.

96. Find an equation of a circle that is tangent to both axes, has its center in the third quadrant, and has a diameter of $\sqrt{5}$.

Projects

1. **Verify a Geometric Theorem** Use the midpoint formula and the distance formula to prove that the midpoint M of the hypotenuse of a right triangle is equidistant from each of the vertices of the triangle. (*Hint:* Label the vertices of the triangle as shown in the figure at the right.)

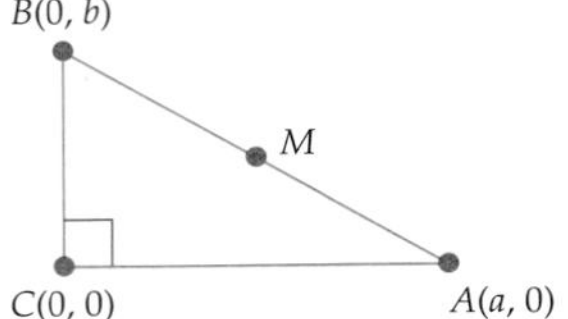

2. **Solve a Quadratic Equation Geometrically** In the 17th century, Descartes (and others) solved equations by using both algebra and geometry. This project outlines the method Descartes used to solve certain quadratic equations.

   a. Consider the equation $x^2 = 2ax + b^2$. Construct a right triangle ABC with $AC = a$ and $CB = b$. Now draw a circle with center at A and radius a. Let P be the point at which the circle intersects the hypotenuse of the right triangle and Q the point at which an extension of the hypotenuse intersects the circle. Your drawing should be similar to the one at the right.

   b. Show that QB is a solution of the equation $x^2 = 2ax + b^2$.

   c. Show that PB is a solution of the equation $x^2 = -2ax + b^2$.

   d. Construct a line parallel to line segment AC and passing through B. Let S and T be the points at which the line intersects the circle. Show that SB and TB are solutions of the equation $x^2 = 2ax - b^2$.

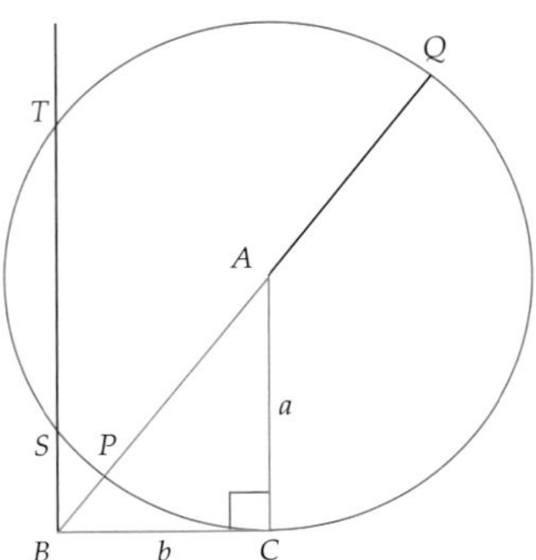

Section 1.3

- Relations
- Functions
- Function Notation
- Identifying Functions
- Graphs of Functions
- The Greatest Integer Function (Floor Function)
- Applications

Introduction to Functions

PREPARE FOR THIS SECTION

Prepare for this section by completing the following exercises. The answers can be found on page A2.

PS1. Let $y = 3x + 12$. Find x when $y = 0$. [1.1]

PS2. Let $y = x^2 - 4x + 3$. Find x when $y = 0$. [1.1]

PS3. Let (x, y) be ordered pairs for $y^2 = x$. Determine values of a and b for the ordered pairs $(9, a)$ and $(9, b)$. [1.2]

PS4. Find the length of the line segment connecting $P_1(-4, 1)$ and $P_2(3, -2)$. [1.2]

PS5. What is the greatest integer that is less than -3.2? [1.1]

PS6. Suppose $y = -2x + 6$. As the values of x increase, do the values of y increase or decrease? [1.2]

Relations

In many situations in science, business, and mathematics, a correspondence exists between two sets. The correspondence is often defined by a *table*, an *equation*, or a *graph*, each of which can be viewed from a mathematical perspective as a set of ordered pairs. In mathematics, any set of ordered pairs is called a **relation**.

Table 1.1

Score	Grade
[90, 100]	A
[80, 90)	B
[70, 80)	C
[60, 70)	D
[0, 60)	F

Table 1.1 defines a correspondence between a set of percent scores and a set of letter grades. For each score from 0 to 100, there corresponds only one letter grade. The score 94% corresponds to the letter grade of A. Using ordered-pair notation, we record this correspondence as (94, A).

The *equation* $d = 16t^2$ indicates that the distance d that a rock falls (neglecting air resistance) corresponds to the time t that it has been falling. For each nonnegative value t, the equation assigns only one value for the distance d. According to this equation, in 3 seconds a rock will fall 144 feet, which we record as (3, 144). Some of the other ordered pairs determined by $d = 16t^2$ are (0, 0), (1, 16), (2, 64), and (2.5, 100).

Equation:	$d = 16t^2$
If $t = 3$, then	$d = 16(3)^2 = 144$

The *graph* in **Figure 1.30** defines a correspondence between the length of a pendulum and the time it takes the pendulum to complete one oscillation. For each nonnegative pendulum length, the graph yields only one time. According to the graph, a pendulum length of 2 feet yields an oscillation time of 1.6 seconds, and a length of 4 feet yields an oscillation time of 2.2 seconds, where the time is measured to the nearest tenth of a second. These results can be recorded as the ordered pairs (2, 1.6) and (4, 2.2).

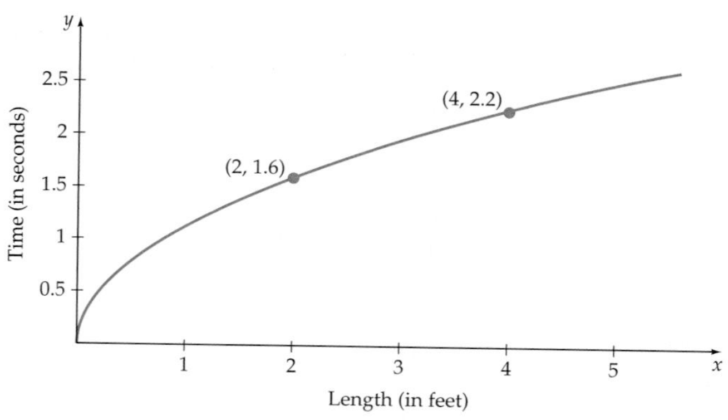

Figure 1.30

Functions

The preceding table, equation, and graph each determines a special type of relation called a *function*.

Math Matters

It is generally agreed among historians that Leonhard Euler (1707–1783) was the first person to use the word *function*. His definition of function occurs in his book *Introduction to Analysis of the Infinite*, published in 1748. Euler contributed to many areas of mathematics and was one of the most prolific expositors of mathematics.

Definition of a Function

A **function** is a set of ordered pairs in which no two ordered pairs have the same first coordinate and different second coordinates.

Although every function is a relation, not every relation is a function. For instance, consider (94, A) from the grading correspondence. The first coordinate, 94, is paired with a second coordinate of A. It would not make sense to have 94 paired with A, (94, A), and 94 paired with B, (94, B). The same first coordinate would be paired with two different second coordinates. This would mean that two students with the same score received different grades, one student an A and the other a B!

Functions may have ordered pairs with the same second coordinate. For instance, (94, A) and (95, A) are both ordered pairs that belong to the function defined by **Table 1.1**. A function may have different first coordinates and the same second coordinate.

The equation $d = 16t^2$ represents a function because for each value of t there is only one value of d. Not every equation, however, represents a function. For instance, $y^2 = 25 - x^2$ does not represent a function. The ordered pairs $(-3, 4)$ and $(-3, -4)$ are both solutions of the equation. However, these ordered pairs do not satisfy the definition of a function: there are two ordered pairs with the same first coordinate but *different* second coordinates.

QUESTION Does the set $\{(0, 0), (1, 0), (2, 0), (3, 0), (4, 0)\}$ define a function?

The **domain** of a function is the set of all the first coordinates of the ordered pairs. The **range** of a function is the set of all the second coordinates. In the function determined by the grading correspondence in **Table 1.1**, the domain is the interval $[0, 100]$. The range is {A, B, C, D, F}. In a function, each domain element is paired with one and only one range element.

If a function is defined by an equation, the variable that represents elements of the domain is the **independent variable**. The variable that represents elements of the range is the **dependent variable**. For the free-fall of a rock situation, we used the equation $d = 16t^2$. The elements of the domain represented the time the rock fell, and the elements of the range represented the distance the rock fell. Thus, in $d = 16t^2$, the independent variable is t and the dependent variable is d.

The specific letters used for the independent and dependent variables are not important. For example, $y = 16x^2$ represents the same function as $d = 16t^2$. Traditionally, x is used for the independent variable and y for the dependent variable. Anytime we use the phrase "y is a function of x" or a similar phrase with different letters, the variable that follows "function of" is the independent variable.

ANSWER Yes. There are no two ordered pairs with the same first coordinate and different second coordinates.

Function Notation

Functions can be named by using a letter or a combination of letters, such as f, g, A, log, or tan. If x is an element of the domain of f, then $f(x)$, which is read "f of x" or "the value of f at x," is the element in the range of f that corresponds to the domain element x. The notation "f" and the notation "$f(x)$" mean different things. "f" is the name of the function, whereas "$f(x)$" is the value of the function at x. Finding the value of $f(x)$ is referred to as *evaluating* f at x. To evaluate $f(x)$ at $x = a$, substitute a for x and simplify.

take note

In Example 1, observe that

$$f(3b) \neq 3f(b)$$

and that

$$f(a+3) \neq f(a) + f(3)$$

EXAMPLE 1 ❯❯ Evaluate Functions

Let $f(x) = x^2 - 1$, and evaluate.

a. $f(-5)$ b. $f(3b)$ c. $3f(b)$ d. $f(a+3)$ e. $f(a) + f(3)$

Solution

a. $f(-5) = (-5)^2 - 1 = 25 - 1 = 24$ • **Substitute −5 for x, and simplify.**

b. $f(3b) = (3b)^2 - 1 = 9b^2 - 1$ • **Substitute $3b$ for x, and simplify.**

c. $3f(b) = 3(b^2 - 1) = 3b^2 - 3$ • **Substitute b for x, and simplify.**

d. $f(a+3) = (a+3)^2 - 1$ • **Substitute $a + 3$ for x.**

$= a^2 + 6a + 8$ • **Simplify.**

e. $f(a) + f(3) = (a^2 - 1) + (3^2 - 1)$ • **Substitute a for x; substitute 3 for x.**

$= a^2 + 7$ • **Simplify.**

❯❯ *Try Exercise 2, page 46*

Piecewise-defined functions are functions represented by more than one expression. The function shown below is an example of a piecewise-defined function.

$$f(x) = \begin{cases} 2x, & x < -2 \\ x^2, & -2 \le x < 1 \\ 4 - x, & x \ge 1 \end{cases}$$

• **This function is made up of different *pieces*, $2x$, x^2, and $4 - x$, depending on the value of x.**

The expression that is used to evaluate this function depends on the value of x. For instance, to find $f(-3)$, we note that $-3 < -2$ and therefore use the expression $2x$ to evaluate the function.

$$f(-3) = 2(-3) = -6$$ • **When $x < -2$, use the expression $2x$.**

Here are some additional instances of evaluating this function:

$$f(-1) = (-1)^2 = 1$$ • **When x satisfies $-2 \le x < 1$, use the expression x^2.**

$$f(4) = 4 - 4 = 0$$ • **When $x \ge 1$, use the expression $4 - x$.**

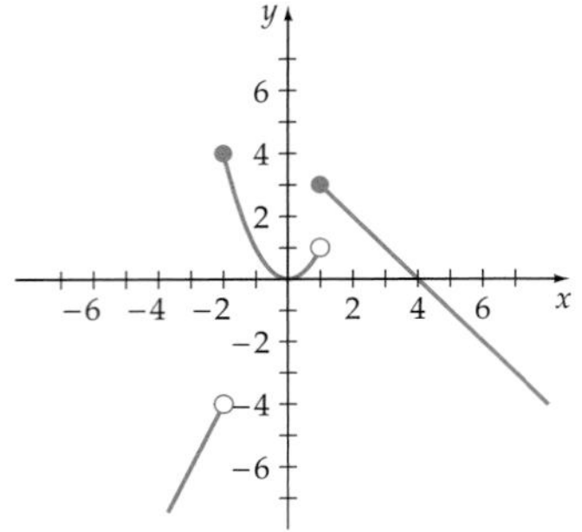

Figure 1.31

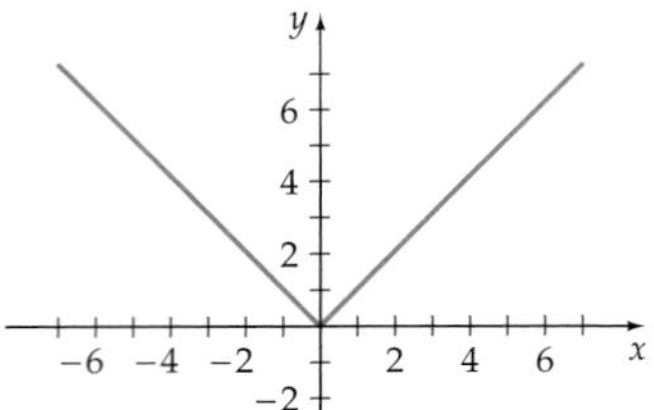

Figure 1.32

The graph of this function is shown in **Figure 1.31**. Note the use of the open and closed circles at the endpoints of the intervals. These circles are used to show the evaluation of the function at the endpoints of each interval. For instance, because -2 is in the interval $-2 \le x < 1$, the value of the function at -2 is 4 $[f(-2) = (-2)^2 = 4]$. Therefore a closed dot is placed at $(-2, 4)$. Similarly, when $x = 1$, because 1 is in the interval $x \ge 1$, the value of the function at 1 is 3 $[f(1) = 4 - 1 = 3]$.

QUESTION Evaluate the function f defined at the bottom of page 33 when $x = 0.5$.

The absolute value function is another example of a piecewise-defined function. Below is the definition of this function, which is sometimes abbreviated abs(x). Its graph **(Figure 1.32)** is shown at the left.

$$\text{abs}(x) = \begin{cases} -x, & x < 0 \\ x, & x \ge 0 \end{cases}$$

EXAMPLE 2 Evaluate a Piecewise-Defined Function

Figure 1.33 shows the distance from home plate to the base of the outfield fence of PETCO Park, the stadium for the San Diego Padres baseball team.

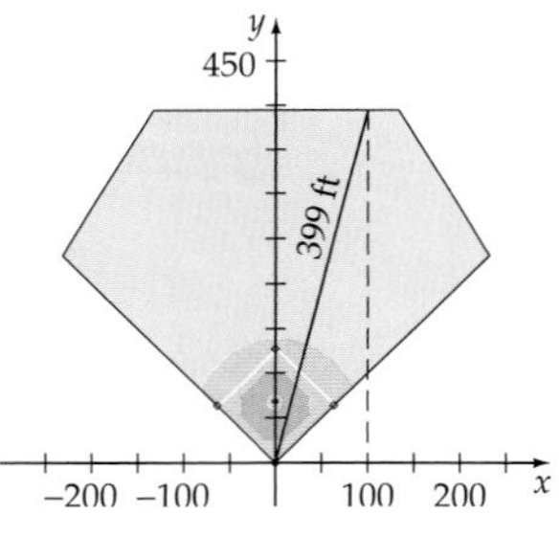

Figure 1.33

The distance $T(x)$, in feet, from home plate to the base of the outfield fences can be approximated by the piecewise function

$$T(x) = \begin{cases} \sqrt{4.298x^2 - 2332.11x + 412{,}253}, & 141 \le x < 228 \\ \sqrt{1.000x^2 - 6.3208x + 149{,}886}, & -104 \le x < 141 \\ \sqrt{2.356x^2 + 1169.88x + 257{,}818}, & -236 \le x < -104 \end{cases}$$

where x is the horizontal distance between the vertical axis and a place at the base of the outfield fence.

ANSWER 0.5 is in the interval $-2 \le x < 1$. Therefore, $f(0.5) = 0.5^2 = 0.25$.

For instance, when $x = 100$, the distance to the base of the fence is approximately 399 feet. Use this function to find, to the nearest foot, the distance from home plate to the outfield fence for the following values of x.

a. $x = -100$ feet b. $x = 200$ feet

Solution

a. Because -100 is in the interval $-104 \le x < 141$, evaluate $T(x) = \sqrt{1.000x^2 - 6.3208x + 149{,}886}$ when $x = -100$.

$$T(-100) = \sqrt{1.000(-100)^2 - 6.3208(-100) + 149{,}886} \approx 401$$

The distance to the base of the fence is approximately 401 feet.

b. Because 200 is in the interval $141 \le x < 228$, evaluate $T(x) = \sqrt{4.298x^2 - 2332.11x + 412{,}253}$ when $x = 200$.

$$T(200) = \sqrt{4.298(200^2) - 2332.11(200) + 412{,}253} \approx 343$$

The distance to the base of the fence is approximately 343 feet.

Try Exercise 10, page 47

Identifying Functions

Recall that although every function is a relation, not every relation is a function. In the next example we examine four relations to determine which are functions.

EXAMPLE 3 Identify Functions

State whether the relation defines y as a function of x.

a. $\{(2, 3), (4, 1), (4, 5)\}$ b. $3x + y = 1$ c. $-4x^2 + y^2 = 9$

d. The correspondence between the x values and the y values in **Figure 1.34** on page 36.

Solution

a. There are two ordered pairs, $(4, 1)$ and $(4, 5)$, with the same first coordinate and different second coordinates. This set does not define y as a function of x.

b. Solving $3x + y = 1$ for y yields $y = -3x + 1$. Because $-3x + 1$ is a unique real number for each x, this equation defines y as a function of x.

Continued ▶

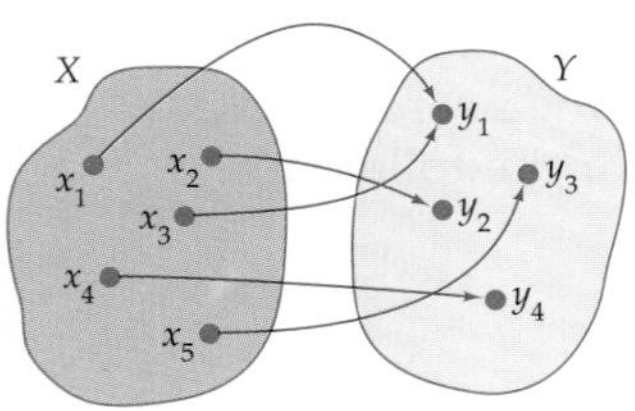

Figure 1.34

c. Solving $-4x^2 + y^2 = 9$ for y yields $y = \pm\sqrt{4x^2 + 9}$. The right side $\pm\sqrt{4x^2 + 9}$ produces two values of y for each value of x. For example, when $x = 0$, $y = 3$ or $y = -3$. Thus $-4x^2 + y^2 = 9$ does not define y as a function of x.

d. Each x is paired with one and only one y. The correspondence in **Figure 1.34** defines y as a function of x.

Try Exercise 14, page 47

take note

You may indicate the domain of a function using set notation or interval notation. For instance, the domain of $f(x) = \sqrt{x - 3}$ may be given in each of the following ways:

Set notation: $\{x \mid x \geq 3\}$

Interval notation: $[3, \infty)$

Sometimes the domain of a function is stated explicitly. For example, each of f, g, and h below is given by an equation followed by a statement that indicates the domain of the function.

$$f(x) = x^2, x > 0 \qquad g(t) = \frac{1}{t^2 + 4}, 0 \leq t \leq 5 \qquad h(x) = x^2, x = 1, 2, 3$$

Although f and h have the same equation, they are different functions because they have different domains. If the domain of a function is not explicitly stated, then its domain is determined by the following convention.

Domain of a Function

Unless otherwise stated, the domain of a function is the set of all real numbers for which the function makes sense and yields real numbers.

EXAMPLE 4 Determine the Domain of a Function

Determine the domain of each function.

a. $G(t) = \dfrac{1}{t - 4}$ b. $f(x) = \sqrt{x + 1}$

c. $A(s) = s^2$, where $A(s)$ is the area of a square whose sides are s units.

Solution

a. The number 4 is not an element of the domain because G is undefined when the denominator $t - 4$ equals 0. The domain of G is all real numbers except 4. In interval notation the domain is $(-\infty, 4) \cup (4, \infty)$.

b. The radical $\sqrt{x + 1}$ is a real number only when $x + 1 \geq 0$ or when $x \geq -1$. Thus, in set notation, the domain of f is $\{x \mid x \geq -1\}$.

c. Because s represents the length of a side of a square, s must be positive. In interval notation the domain of A is $(0, \infty)$.

Try Exercise 28, page 47

Graphs of Functions

If a is an element of the domain of a function f, then $(a, f(a))$ is an ordered pair that belongs to the function.

Definition of the Graph of a Function

The **graph of a function** is the graph of all the ordered pairs that belong to the function.

EXAMPLE 5 Graph a Function by Plotting Points

Graph each function. State the domain and the range of each function.

a. $f(x) = |x - 1|$ b. $n(x) = \begin{cases} 2, & \text{if } x \le 1 \\ x, & \text{if } x > 1 \end{cases}$

Solution

a. The domain of f is the set of all real numbers. Write the function as $y = |x - 1|$. Evaluate the function for several domain values. We have used $x = -3, -2, -1, 0, 1, 2, 3$, and 4.

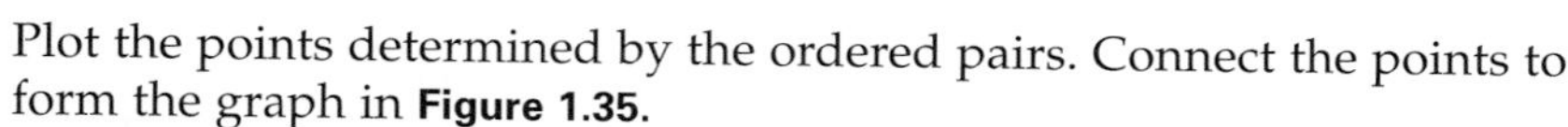

x	−3	−2	−1	0	1	2	3	4
$y = \|x - 1\|$	4	3	2	1	0	1	2	3

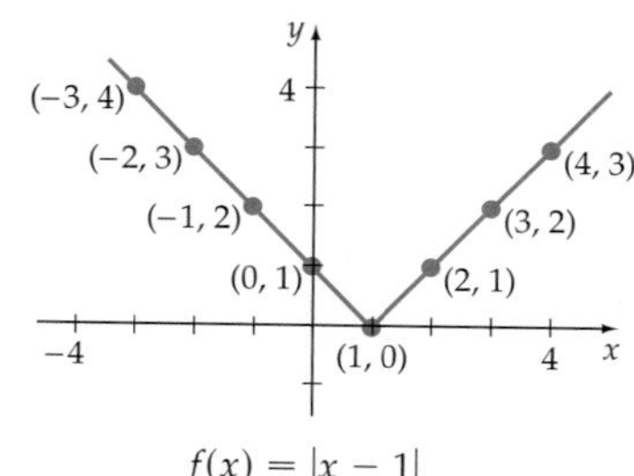

$f(x) = |x - 1|$

Figure 1.35

Plot the points determined by the ordered pairs. Connect the points to form the graph in **Figure 1.35.**

Because $|x - 1| \ge 0$, we can conclude that the graph of f extends from a height of 0 upward, so the range is $\{y \mid y \ge 0\}$.

b. The domain is the union of the inequalities $x \le 1$ and $x > 1$. Thus the domain of n is the set of all real numbers. For $x \le 1$, graph $n(x) = 2$. This results in the horizontal ray in **Figure 1.36.** The solid circle indicates that the point $(1, 2)$ *is* part of the graph. For $x > 1$, graph $n(x) = x$. This produces the second ray in **Figure 1.36.** The open circle indicates that the point $(1, 1)$ *is not* part of the graph.

Examination of the graph shows that it includes only points whose y values are greater than 1. Thus the range of n is $\{y \mid y > 1\}$.

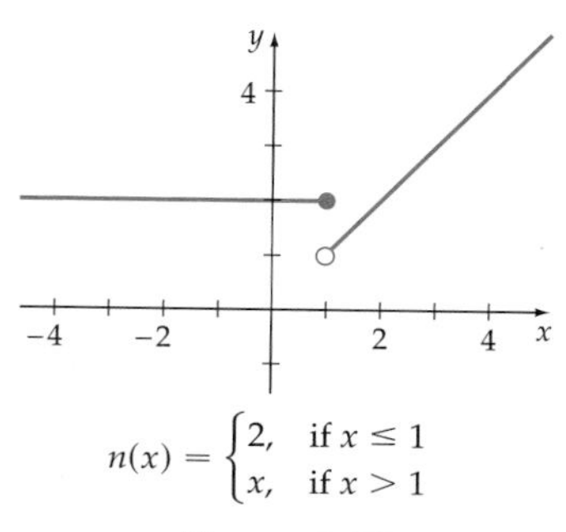
$n(x) = \begin{cases} 2, & \text{if } x \le 1 \\ x, & \text{if } x > 1 \end{cases}$

Figure 1.36

Try Exercise 40, page 47

Integrating Technology

A graphing utility can be used to draw the graph of a function. For instance, to graph $f(x) = x^2 - 1$, enter the equation Y1=X²–1. The graph is shown in **Figure 1.37**.

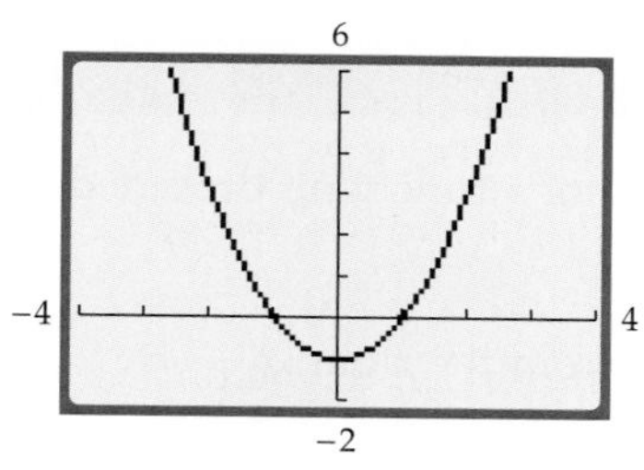

Figure 1.37

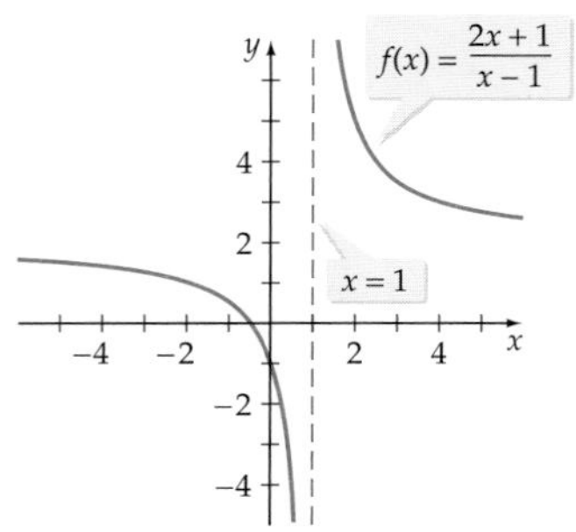

Figure 1.38

The graph of $f(x) = \dfrac{2x + 1}{x - 1}$ is shown in **Figure 1.38**. The domain of f is all real numbers except 1. Because $f(1)$ is undefined, there is no point on the graph when $x = 1$. However, there are points on the graph for values of x close to 1. The table below shows some values of the function when x is *less than* 1 but close to 1.

x	0.9	0.95	0.99	0.995	0.999	0.9999
$f(x)$	−28	−58	−298	−598	−2998	−29,998

It appears from the graph and the table that as x gets closer and closer to 1, $f(x)$ becomes smaller and smaller. That is, as x approaches 1 using values of x that are less than 1, $f(x)$ approaches $-\infty$. The notation

$$f(x) \to -\infty \text{ as } x \to 1^-$$

is used to describe this situation. This notation is read "$f(x)$ approaches negative infinity as x approaches 1 from the left." The negative superscript tells us to use values of x that are less than 1—that is, to the left of 1 on the x-axis.

take note

We can also write, "As $x \to 1^-$, $f(x) \to -\infty$." This is read, "As x approaches 1 from the left, $f(x)$ approaches negative infinity."

Next we focus on values of x that are close to 1 but *greater than* 1.

x	1.1	1.05	1.01	1.005	1.001	1.0001
$f(x)$	32	62	302	602	3002	30,002

This time, from **Figure 1.38** and the table, it appears that as x gets closer and closer to 1, $f(x)$ becomes larger and larger. That is, as x approaches 1 using values of x that are greater than 1, $f(x)$ approaches ∞. The notation

$$f(x) \to \infty \text{ as } x \to 1^+$$

is used to describe this situation. This notation is read "$f(x)$ approaches infinity as x approaches 1 from the right." The positive superscript tells us to use values of x that are greater than 1—that is, to the right of 1 on the x-axis.

The preceding analysis shows that the graph of $f(x) = \dfrac{2x + 1}{x - 1}$ approaches, but does not cross, the vertical line $x = 1$. The dashed line in **Figure 1.38** is called a *vertical asymptote* of the graph.

take note

Generally, we show the asymptotes of the graph of a function by using dashed lines.

Definition of Vertical Asymptote

The line $x = a$ is a **vertical asymptote** of the graph of a function f provided

$$f(x) \to -\infty \quad \text{or} \quad f(x) \to \infty$$

as x approaches a from either the left, $x \to a^-$, or from the right, $x \to a^+$.

As an example of this definition, let $r(x) = \dfrac{5x}{(x-2)^2}$. The domain of r is all real numbers except 2. The table below shows the behavior of $r(x)$ as $x \to 2^-$.

x	1.9	1.95	1.99	1.995
$r(x)$	950	3900	99,500	399,000

From the table, it appears that $r(x) \to \infty$ as $x \to 2^-$.

The following table shows the behavior of $r(x)$ as $x \to 2^+$.

x	2.1	2.05	2.01	2.005
$r(x)$	1050	4100	100,500	401,000

From the table, it appears that $r(x) \to \infty$ as $x \to 2^+$.

Because $r(x) \to \infty$ as $x \to 2^-$ and as $x \to 2^+$, the line $x = 2$ is a vertical asymptote. The graph of $r(x) = \dfrac{5x}{(x-2)^2}$, along with the vertical asymptote, is shown in **Figure 1.39**. In this case, the value of the function is approaching positive infinity on both sides of the vertical asymptote.

Figure 1.39

The definition of a function as a set of ordered pairs in which no two ordered pairs that have the same first coordinate have different second coordinates implies that any vertical line intersects the graph of a function at no more than one point. This is known as the *vertical line test.*

The Vertical Line Test for Functions

A graph is the graph of a function if and only if no vertical line intersects the graph at more than one point.

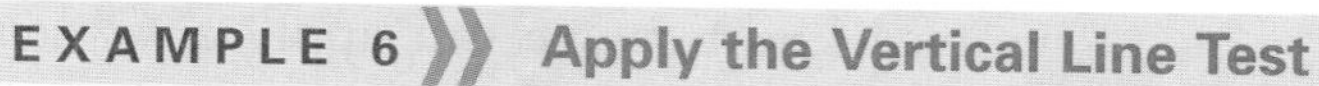

EXAMPLE 6 Apply the Vertical Line Test

State whether the graph is the graph of a function.

a.

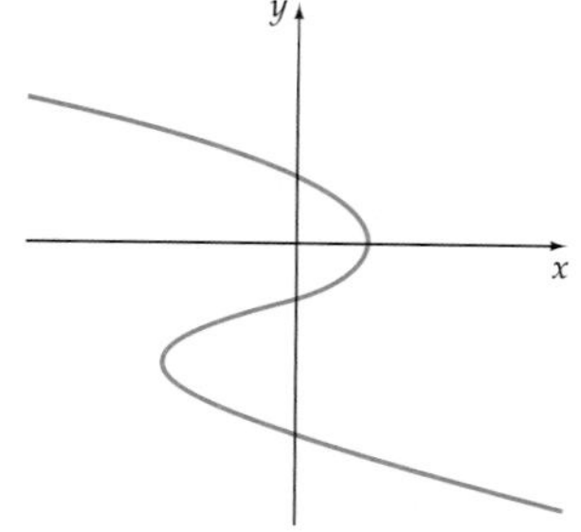

b.

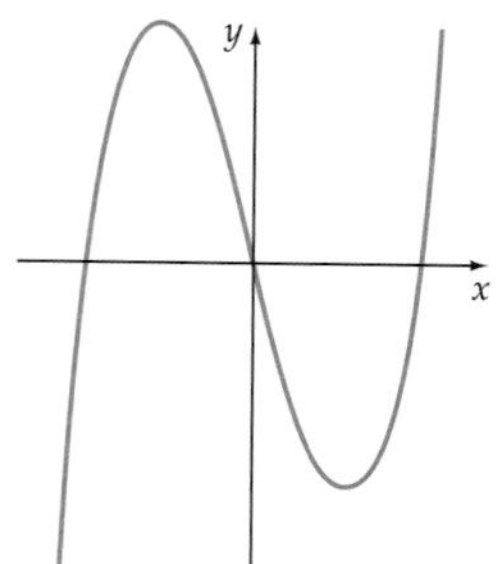

Continued ▶

Solution

a. This graph *is not* the graph of a function because some vertical lines intersect the graph in more than one point.

b. This graph *is* the graph of a function because every vertical line intersects the graph in at most one point.

Try Exercise 56, page 48

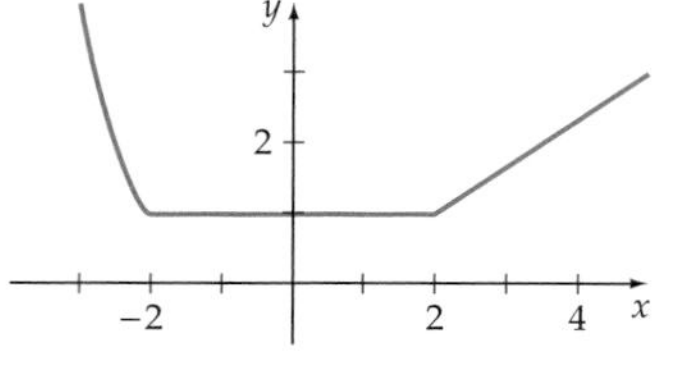

Figure 1.40

Consider the graph in **Figure 1.40**. As a point on the graph moves from left to right, this graph falls for values of $x \leq -2$, remains the same height from $x = -2$ to $x = 2$, and rises for $x \geq 2$. The function represented by the graph is said to be *decreasing* on the interval $(-\infty, -2]$, *constant* on the interval $[-2, 2]$, and *increasing* on the interval $[2, \infty)$.

Definition of Increasing, Decreasing, and Constant Functions

If a and b are elements of an interval I that is a subset of the domain of a function f, then

- f is **increasing** on I if $f(a) < f(b)$ whenever $a < b$.
- f is **decreasing** on I if $f(a) > f(b)$ whenever $a < b$.
- f is **constant** on I if $f(a) = f(b)$ for all a and b.

Recall that a function is a relation in which no two ordered pairs that have the same first coordinate have different second coordinates. This means that given any x, there is only one y that can be paired with that x. A **one-to-one function** satisfies the additional condition that given any y, there is only one x that can be paired with that given y. In a manner similar to applying the vertical line test, we can apply a *horizontal line test* to identify one-to-one functions.

Horizontal Line Test for a One-To-One Function

If every horizontal line intersects the graph of a function at most once, then the graph is the graph of a one-to-one function.

For example, some horizontal lines intersect the graph in **Figure 1.41** at more than one point. It is *not* the graph of a one-to-one function. Every horizontal line intersects the graph in **Figure 1.42** at most once. This is the graph of a one-to-one function.

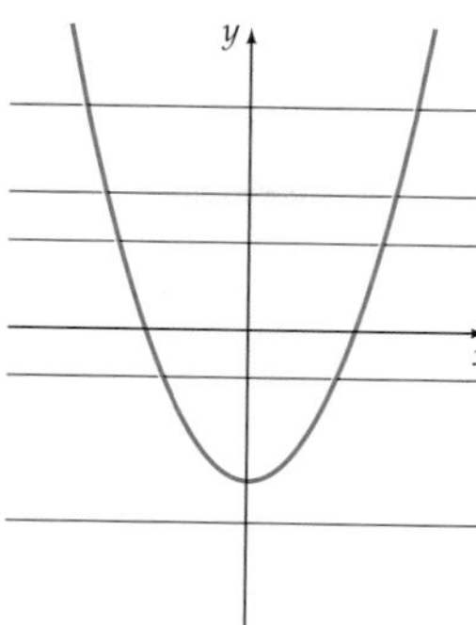

Figure 1.41
Some horizontal lines intersect this graph at more than one point. It is *not* the graph of a one-to-one function.

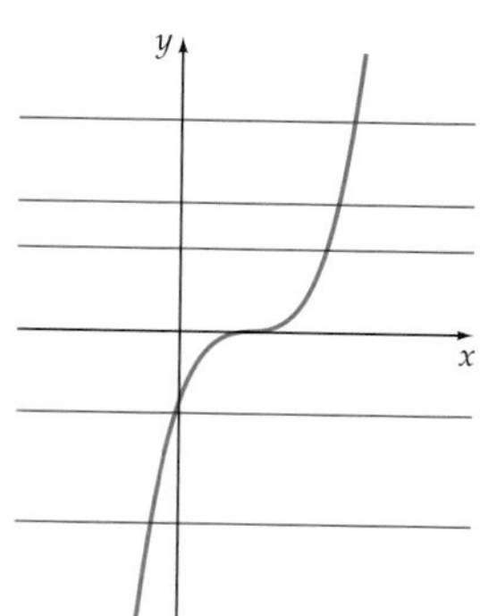

Figure 1.42
Every horizontal line intersects this graph at most once. It is the graph of a one-to-one function.

The Greatest Integer Function (Floor Function)

take note

The greatest integer function is an important function that is often used in advanced mathematics and computer science.

The graphs of some functions do not have any breaks or gaps. These functions, whose graphs can be drawn without lifting the pencil off the paper, are called *continuous functions*. The graphs of other functions do have breaks or *discontinuities*. One such function is the **greatest integer function** or **floor function**. This function is denoted by various symbols such as $[\![x]\!]$, $\lfloor x \rfloor$, and $\text{int}(x)$.

The value of the greatest integer function at x is the greatest integer that is less than or equal to x. For instance,

$$\lfloor -1.1 \rfloor = -2 \qquad [\![-3]\!] = -3 \qquad \text{int}\left(\frac{5}{2}\right) = 2 \qquad \lfloor 5 \rfloor = 5 \qquad [\![\pi]\!] = 3$$

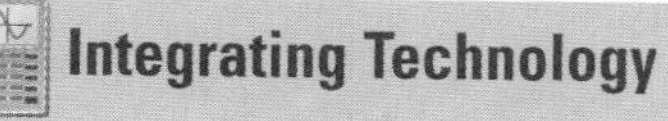

Many graphing calculators use the notation $\text{int}(x)$ for the greatest integer function. Here are screens from a TI-83 Plus/TI-84 Plus.

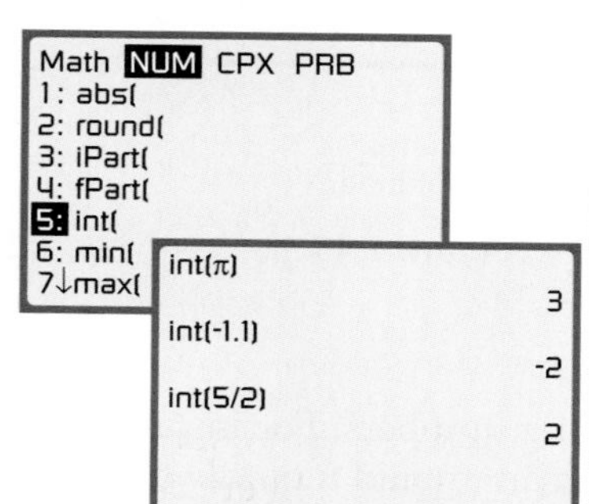

QUESTION Evaluate. **a.** $\text{int}\left(-\frac{3}{2}\right)$ **b.** $\lfloor 2 \rfloor$

To graph the floor function, first observe that the value of the floor function is constant between any two consecutive integers. For instance, between the integers 1 and 2, we have

$$\text{int}(1.1) = 1 \qquad \text{int}(1.35) = 1 \qquad \text{int}(1.872) = 1 \qquad \text{int}(1.999) = 1$$

Between -3 and -2, we have

$$\text{int}(-2.98) = -3 \qquad \text{int}(-2.4) = -3 \qquad \text{int}(-2.35) = -3 \qquad \text{int}(-2.01) = -3$$

Using this property of the floor function, we can create a table of values and then graph the floor function. See **Figure 1.43** on page 42.

ANSWER **a.** Because -2 is the greatest integer less than or equal to $-\frac{3}{2}$, $\text{int}\left(-\frac{3}{2}\right) = -2$.
b. Because 2 is the greatest integer less than or *equal* to 2, $\lfloor 2 \rfloor = 2$.

x	$y = \text{int}(x)$
$-5 \le x < -4$	-5
$-4 \le x < -3$	-4
$-3 \le x < -2$	-3
$-2 \le x < -1$	-2
$-1 \le x < 0$	-1
$0 \le x < 1$	0
$1 \le x < 2$	1
$2 \le x < 3$	2
$3 \le x < 4$	3
$4 \le x < 5$	4

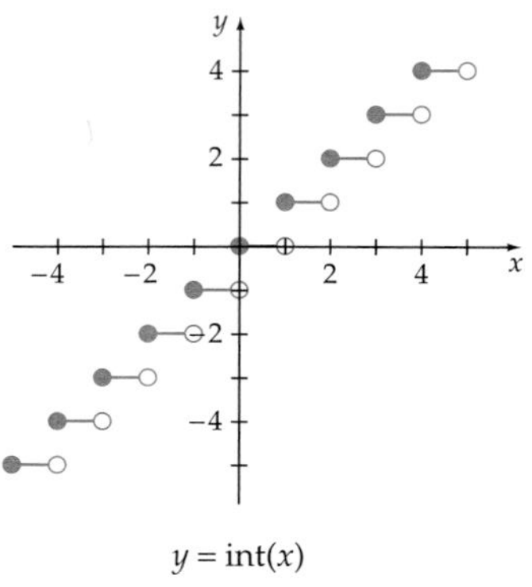

$y = \text{int}(x)$

Figure 1.43

The graph of the floor function has discontinuities (breaks) whenever x is an integer. The domain of the floor function is the set of real numbers; the range is the set of integers. Because the graph appears to be a series of steps, sometimes the floor function is referred to as a **step function.**

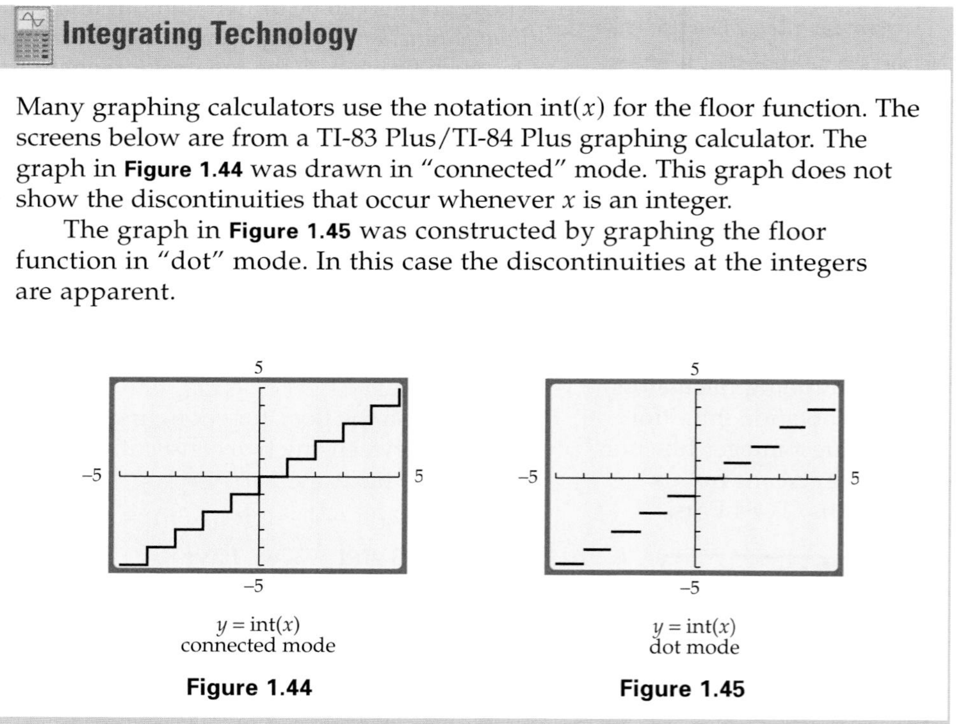

Integrating Technology

Many graphing calculators use the notation int(x) for the floor function. The screens below are from a TI-83 Plus/TI-84 Plus graphing calculator. The graph in **Figure 1.44** was drawn in "connected" mode. This graph does not show the discontinuities that occur whenever x is an integer.

The graph in **Figure 1.45** was constructed by graphing the floor function in "dot" mode. In this case the discontinuities at the integers are apparent.

$y = \text{int}(x)$
connected mode

Figure 1.44

$y = \text{int}(x)$
dot mode

Figure 1.45

One application of the floor function is rounding numbers. For instance, suppose a credit card company charges 1.5% interest on an unpaid monthly balance of \$237.84. Then the interest charge I is

$$I = 237.84 \cdot 0.015 = 3.5676$$

Thus the interest charge is \$3.57. Note that the result was rounded to the nearest cent, or hundredth. The computer program that determines the interest charge uses the floor function to calculate the rounding. To round a number N to the nearest kth decimal place, we use the following formula.

> **take note**
>
> Observe the effect of adding 0.5. In this case, it increases the units digit that will occupy the hundredths place.
>
> Now suppose that the monthly balance was \$237.57. Then $100(237.57 \cdot 0.015) + 0.5$ is $356.355 + 0.5 = 356.855$. In this case, the units digit is not changed and the final interest charge is \$3.56.

$$N \text{ to the nearest } k\text{th decimal} = \frac{\text{int}[10^k(N) + 0.5]}{10^k}$$

Here is the calculation for the interest owed. In this case, $N = 237.84 \cdot 0.015$ and $k = 2$ (round to the second decimal place).

$$\begin{aligned} I &= \frac{\text{int}[10^2(237.84 \cdot 0.015) + 0.5]}{10^2} \\ &= \frac{\text{int}[100(3.5676) + 0.5]}{100} \\ &= \frac{\text{int}[356.76 + 0.5]}{100} \\ &= \frac{\text{int}[357.26]}{100} \\ &= \frac{357}{100} = 3.57 \end{aligned}$$

Example 7 gives another application of the floor function.

EXAMPLE 7 Use the Greatest Integer Function to Model Expenses

The cost of parking in a garage is \$3 for the first hour or any part of the hour and \$2 for each additional hour or any part of an hour thereafter. If x is the time in hours that you park your car, then the cost is given by

$$C(x) = 3 - 2\,\text{int}(1 - x), \quad x > 0$$

a. Evaluate $C(2)$ and $C(2.5)$. b. Graph $y = C(x)$ for $0 < x \leq 5$.

Solution

a.
$$\begin{aligned} C(2) &= 3 - 2\,\text{int}(1 - 2) \\ &= 3 - 2\,\text{int}(-1) \\ &= 3 - 2(-1) \\ &= \$5 \end{aligned} \qquad \begin{aligned} C(2.5) &= 3 - 2\,\text{int}(1 - 2.5) \\ &= 3 - 2\,\text{int}(-1.5) \\ &= 3 - 2(-2) \\ &= \$7 \end{aligned}$$

b. To graph $C(x)$ for $0 < x \leq 5$, consider the value of $\text{int}(1 - x)$ for each of the intervals $0 < x \leq 1$, $1 < x \leq 2$, $2 < x \leq 3$, $3 < x \leq 4$, and $4 < x \leq 5$. For instance, when $0 < x \leq 1$, $0 \leq 1 - x < 1$. Thus $\text{int}(1 - x) = 0$ when $0 < x \leq 1$. Now consider $1 < x \leq 2$. When $1 < x \leq 2$, $-1 \leq 1 - x < 0$. Thus $\text{int}(1 - x) = -1$ when $1 < x \leq 2$. Applying the same reasoning to each of the other intervals gives the following table of values and the graph of C shown in **Figure 1.46**.

Continued ▶

x	$C(x) = 3 - 2\,\text{int}(1 - x)$
$0 < x \le 1$	3
$1 < x \le 2$	5
$2 < x \le 3$	7
$3 < x \le 4$	9
$4 < x \le 5$	11

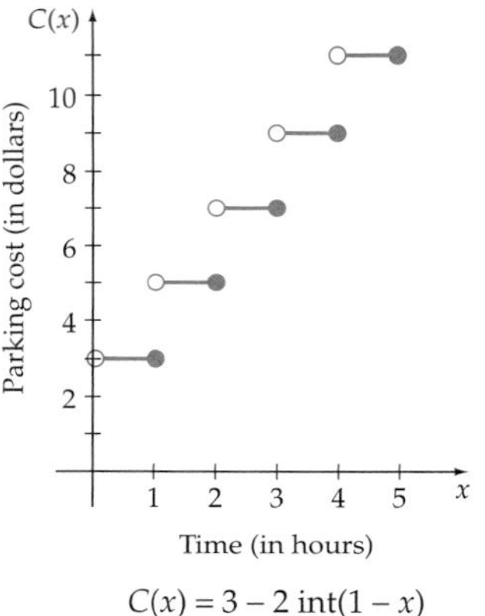

Figure 1.46

Because $C(1) = 3$, $C(2) = 5$, $C(3) = 7$, $C(4) = 9$, and $C(5) = 11$, we can use a solid circle at the right endpoint of each "step" and an open circle at each left endpoint.

❯❯ *Try Exercise 54, page 48*

Integrating Technology

The function graphed in Example 7 is an example of a function for which a graphing calculator may not produce a graph that is a good representation of the function. You may be required to *make adjustments* in the MODE, SET UP, or WINDOW of the graphing calculator so that it will produce a better representation of the function. A graph may also require some *fine tuning*, such as open or solid circles at particular points, to accurately represent the function.

Applications

EXAMPLE 8 ❯❯ Solve an Application

A car was purchased for \$16,500. Assuming the car depreciates at a constant rate of \$2200 per year (*straight-line depreciation*) for the first 7 years, write the value v of the car as a function of time, and calculate the value of the car 3 years after purchase.

Solution

Let t represent the number of years that have passed since the car was purchased. Then $2200t$ is the amount by which the value of the car has depreciated after t years. The value of the car at time t is given by

$$v(t) = 16{,}500 - 2200t, \quad 0 \le t \le 7$$

When $t = 3$, the value of the car is

$$v(3) = 16{,}500 - 2200(3) = 16{,}500 - 6600 = \$9900$$

» Try Exercise 72, page 49

Often in applied mathematics, formulas are used to determine the functional relationship that exists between two variables.

EXAMPLE 9 » Solve an Application

A lighthouse is 2 miles south of a port. A ship leaves port and sails east at a rate of 7 mph. Express the distance d between the ship and the lighthouse as a function of time, given that the ship has been sailing for t hours.

Solution

Draw a diagram and label it as shown in **Figure 1.47**. Note that because distance = (rate)(time) and the rate is 7, in t hours the ship has sailed a distance of $7t$.

Figure 1.47

$$[d(t)]^2 = (7t)^2 + 2^2$$ • **The Pythagorean Theorem**

$$[d(t)]^2 = 49t^2 + 4$$

$$d(t) = \sqrt{49t^2 + 4}$$ • **The ± sign is not used because the distance $d(t)$ must be nonnegative.**

» Try Exercise 78, page 50

EXAMPLE 10 » Solve an Application

An open box is to be made from a square piece of cardboard that measures 40 inches on each side. To construct the box, squares that measure x inches on each side are cut from each corner of the cardboard as shown in **Figure 1.48.**

a. Express the volume V of the box as a function of x.

b. Determine the domain of V.

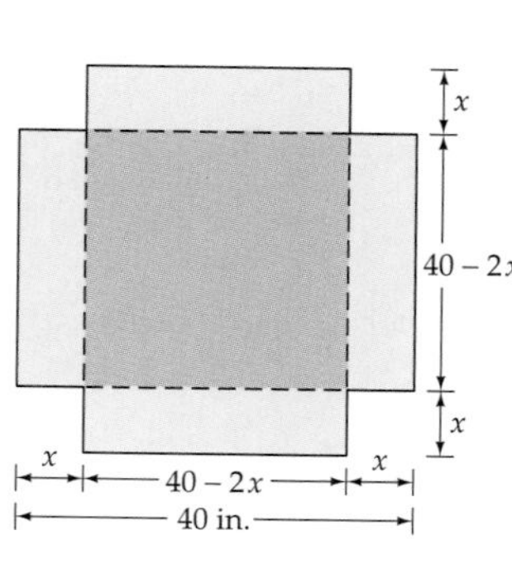

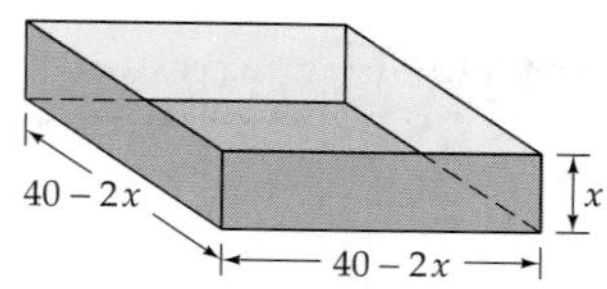

Figure 1.48

Solution

a. The length l of the box is $40 - 2x$. The width w is also $40 - 2x$. The height of the box is x. The volume V of a box is the product of its length, its width, and its height. Thus

$$V = (40 - 2x)^2 x$$

b. The squares that are cut from each corner require x to be larger than 0 inches but less than 20 inches. Thus the domain is $\{x \mid 0 < x < 20\}$.

» Try Exercise 74, page 49

Topics for Discussion

1. Discuss the definition of *function*. Give some examples of relationships that are functions and some that are not functions.
2. What is the difference between the domain and range of a function?
3. How many y-intercepts can a function have? How many x-intercepts can a function have?
4. Discuss how the vertical line test is used to determine whether or not a graph is the graph of a function. Explain why the vertical line test works.
5. What is the domain of $f(x) = \dfrac{\sqrt{1-x}}{x^2-9}$? Explain.
6. Is 2 in the range of $g(x) = \dfrac{6x-5}{3x+1}$? Explain the process you used to make your decision.
7. Suppose that f is a function and that $f(a) = f(b)$. Does this imply that $a = b$? Explain your answer.

Exercise Set 1.3

In Exercises 1 to 8, evaluate each function.

1. Given $f(x) = 3x - 1$, find
   a. $f(2)$ b. $f(-1)$ c. $f(0)$
   d. $f\left(\frac{2}{3}\right)$ e. $f(k)$ f. $f(k+2)$

2. Given $g(x) = 2x^2 + 3$, find
   a. $g(3)$ b. $g(-1)$ c. $g(0)$
   d. $g\left(\frac{1}{2}\right)$ e. $g(c)$ f. $g(c+5)$

3. Given $A(w) = \sqrt{w^2 + 5}$, find
   a. $A(0)$ b. $A(2)$ c. $A(-2)$
   d. $A(4)$ e. $A(r+1)$ f. $A(-c)$

4. Given $J(t) = 3t^2 - t$, find
   a. $J(-4)$ b. $J(0)$ c. $J\left(\frac{1}{3}\right)$
   d. $J(-c)$ e. $J(x+1)$ f. $J(x+h)$

5. Given $f(x) = \dfrac{1}{|x|}$, find
   a. $f(2)$ b. $f(-2)$ c. $f\left(\frac{-3}{5}\right)$
   d. $f(2) + f(-2)$ e. $f(c^2 + 4)$ f. $f(2+h)$

6. Given $T(x) = 5$, find
   a. $T(-3)$ b. $T(0)$ c. $T\left(\frac{2}{7}\right)$
   d. $T(3) + T(1)$ e. $T(x+h)$ f. $T(3k+5)$

7. Given $s(x) = \dfrac{x}{|x|}$, find
   a. $s(4)$ b. $s(5)$ c. $s(-2)$
   d. $s(-3)$ e. $s(t), t > 0$ f. $s(t), t < 0$

8. Given $r(x) = \dfrac{x}{x+4}$, find
   a. $r(0)$ b. $r(-1)$ c. $r(-3)$
   d. $r\left(\frac{1}{2}\right)$ e. $r(0.1)$ f. $r(10{,}000)$

In Exercises 9 and 10, evaluate each piecewise-defined function for the indicated values.

9. $P(x) = \begin{cases} 3x + 1, & \text{if } x < 2 \\ -x^2 + 11, & \text{if } x \geq 2 \end{cases}$

a. $P(-4)$ **b.** $P(\sqrt{5})$

c. $P(c), \quad c < 2$ **d.** $P(k + 1), \quad k \geq 1$

10. $Q(t) = \begin{cases} 4, & \text{if } 0 \leq t \leq 5 \\ -t + 9, & \text{if } 5 < t \leq 8 \\ \sqrt{t - 7}, & \text{if } 8 < t \leq 11 \end{cases}$

a. $Q(0)$ **b.** $Q(a), \quad 6 < a < 7$

c. $Q(n), \quad 1 < n < 2$ **d.** $Q(m^2 + 7), \quad 1 < m \leq 2$

In Exercises 11 to 20, state whether the equation defines *y* as a function of *x*.

11. $2x + 3y = 7$

12. $5x + y = 8$

13. $-x + y^2 = 2$

14. $x^2 - 2y = 2$

15. $y = 4 \pm \sqrt{x}$

16. $x^2 + y^2 = 9$

17. $y = \sqrt[3]{x}$

18. $y = |x| + 5$

19. $y^2 = x^2$

20. $y^3 = x^3$

In Exercises 21 to 26, state whether the set of ordered pairs (*x*, *y*) defines *y* as a function of *x*.

21. $\{(2, 3), (5, 1), (-4, 3), (7, 11)\}$

22. $\{(5, 10), (3, -2), (4, 7), (5, 8)\}$

23. $\{(4, 4), (6, 1), (5, -3)\}$

24. $\{(2, 2), (3, 3), (7, 7)\}$

25. $\{(1, 0), (2, 0), (3, 0)\}$

26. $\left\{\left(-\frac{1}{3}, \frac{1}{4}\right), \left(-\frac{1}{4}, \frac{1}{3}\right), \left(\frac{1}{4}, \frac{2}{3}\right)\right\}$

In Exercises 27 to 38, determine the domain of the function represented by the given equation.

27. $f(x) = 3x - 4$

28. $f(x) = -2x + 1$

29. $f(x) = x^2 + 2$

30. $f(x) = 3x^2 + 1$

31. $f(x) = \dfrac{4}{x + 2}$

32. $f(x) = \dfrac{6}{x - 5}$

33. $f(x) = \sqrt{7 + x}$

34. $f(x) = \sqrt{4 - x}$

35. $f(x) = \sqrt{4 - x^2}$

36. $f(x) = \sqrt{12 - x^2}$

37. $f(x) = \dfrac{1}{\sqrt{x + 4}}$

38. $f(x) = \dfrac{1}{\sqrt{5 - x}}$

In Exercises 39 to 46, graph each function. Insert solid circles or open circles where necessary to indicate the true nature of the function.

39. $f(x) = \begin{cases} |x|, & \text{if } x \leq 1 \\ 2, & \text{if } x > 1 \end{cases}$

40. $g(x) = \begin{cases} -4, & \text{if } x \leq 0 \\ x^2 - 4, & \text{if } 0 < x \leq 1 \\ -x, & \text{if } x > 1 \end{cases}$

41. $J(x) = \begin{cases} 4, & \text{if } x \leq -1 \\ x^2, & \text{if } -1 < x < 1 \\ -x + 5, & \text{if } x \geq 1 \end{cases}$

42. $K(x) = \begin{cases} 1, & \text{if } x \leq -2 \\ x^2 - 4, & \text{if } -2 < x < 2 \\ \frac{1}{2}x, & \text{if } x \geq 2 \end{cases}$

43. $L(x) = \left[\!\left[\frac{1}{3}x\right]\!\right]$ for $-6 \leq x \leq 6$

44. $M(x) = [\![x]\!] + 2$ for $0 \leq x \leq 4$

45. $N(x) = \text{int}(-x)$ for $-3 \leq x \leq 3$

46. $P(x) = \text{int}(x) + x$ for $0 \leq x \leq 4$

In Exercises 47 to 52, use the floor function to write and then evaluate an expression that can be used to round the given number to the given place value.

47. $N = 2.3458$; hundredths

48. $N = 34.567$; tenths

49. $N = 34.05622$; thousandths

50. $N = 109.83$; whole number

51. $N = 0.08951$; ten-thousandths

52. $N = 2.98245$; thousandths

53. **FIRST-CLASS MAIL** In 2006, the cost to mail a first-class letter was given by

$$C(w) = 0.39 - 0.34\,\text{int}(1 - w), \quad w > 0$$

where C is in dollars and w is the weight of the letter in ounces.

a. What was the cost (in 2006) to mail a letter that weighed 2.8 ounces?

b. Graph C for $0 < w \le 5$.

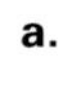
54. **INCOME TAX** The amount of federal income tax $T(x)$ a person owed in 2006 is given by

$$T(x) = \begin{cases} 0.10x, & 0 \le x < 7550 \\ 0.15(x - 7550) + 755, & 7550 \le x < 30{,}650 \\ 0.25(x - 30{,}650) + 4220, & 30{,}650 \le x < 74{,}200 \\ 0.28(x - 74{,}200) + 15{,}107.50, & 74{,}200 \le x < 154{,}800 \\ 0.33(x - 154{,}800) + 37{,}675.50, & 154{,}800 \le x < 336{,}550 \\ 0.35(x - 336{,}550) + 97{,}653, & x \ge 336{,}550 \end{cases}$$

where x is the adjusted gross income tax of the taxpayer.

a. What is the domain of this function?

b. Find the income tax owed by a taxpayer whose adjusted gross income was \$31,250.

c. Find the income tax owed by a taxpayer whose adjusted gross income was \$78,900.

55. Use the vertical line test to determine which of the following graphs are graphs of functions.

a.

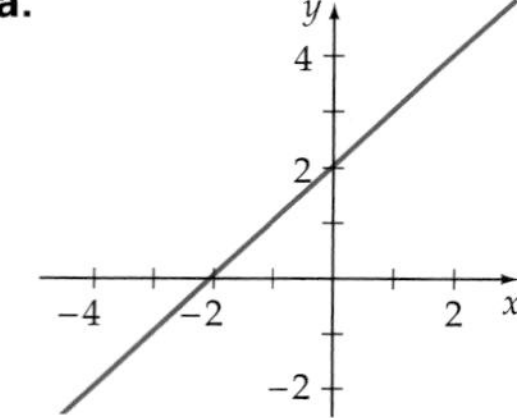

b.

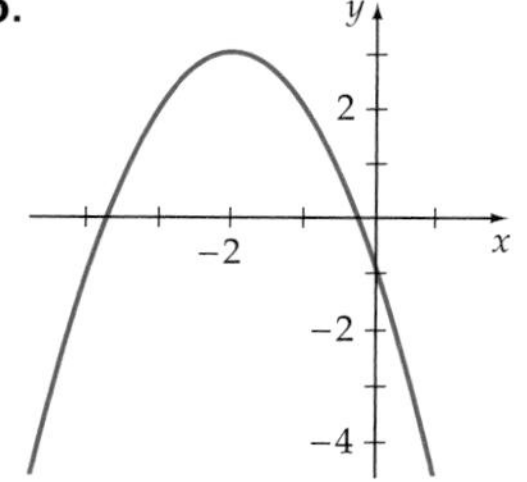

c.

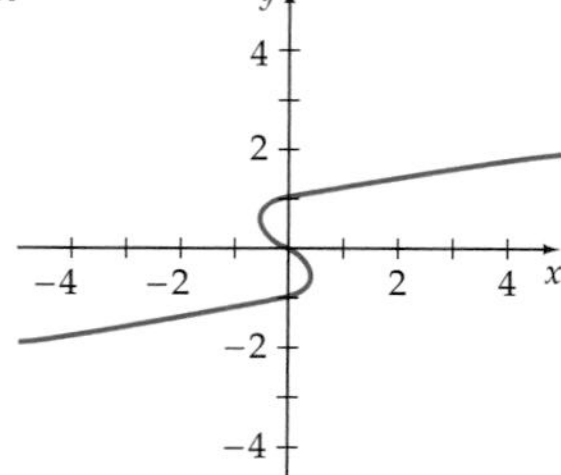

d.

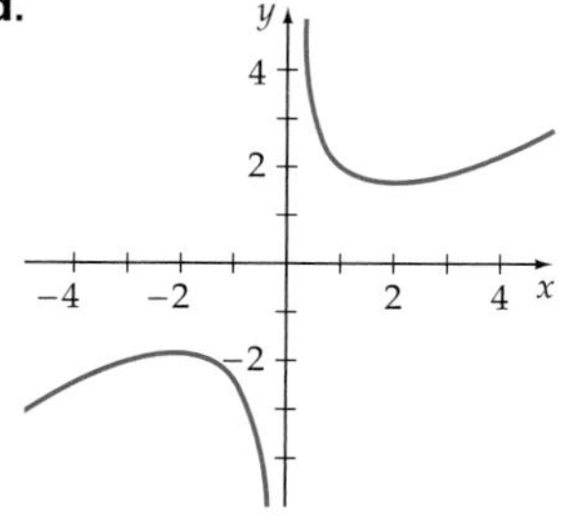

56. Use the vertical line test to determine which of the following graphs are graphs of functions.

a.

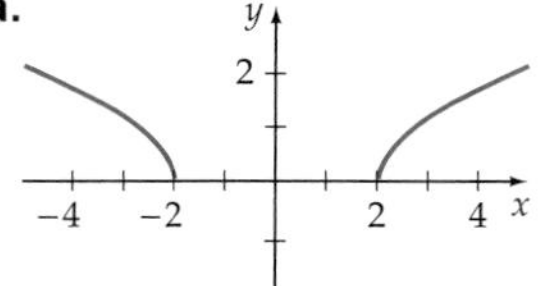

b.

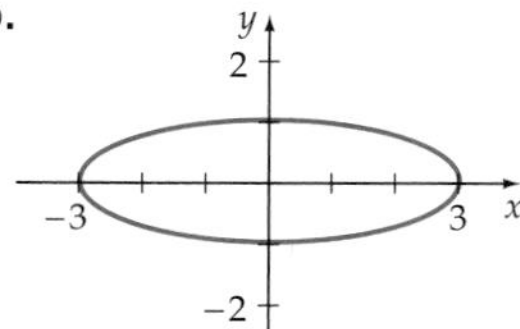

c.

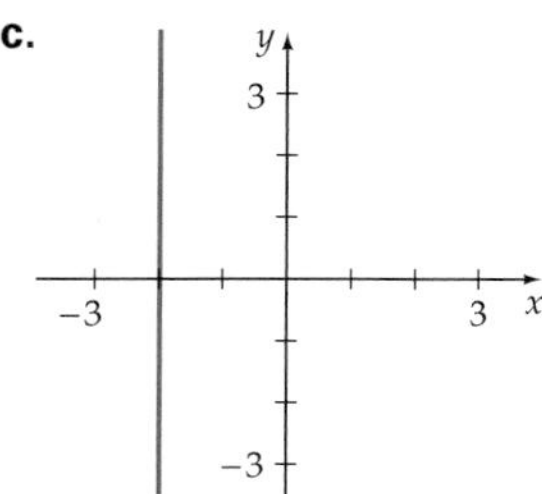

d.

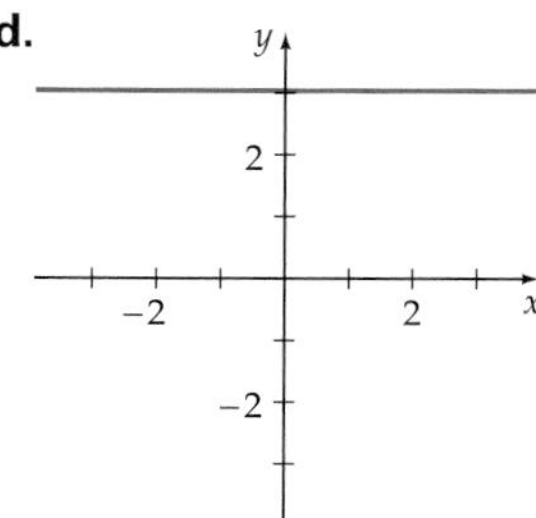

In Exercises 57 to 66, use the indicated graph to identify the intervals over which the function is increasing, constant, or decreasing.

57.

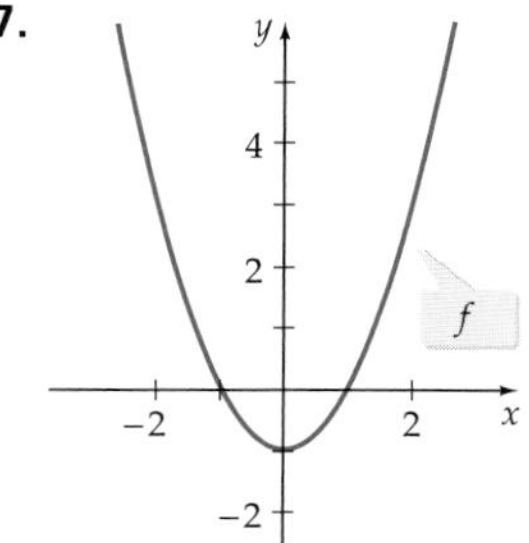

58.

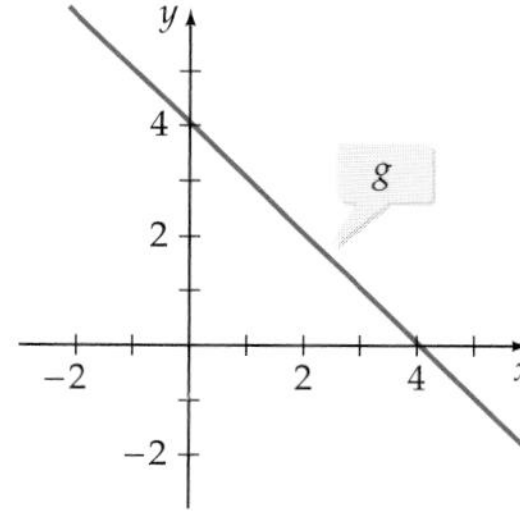

59.

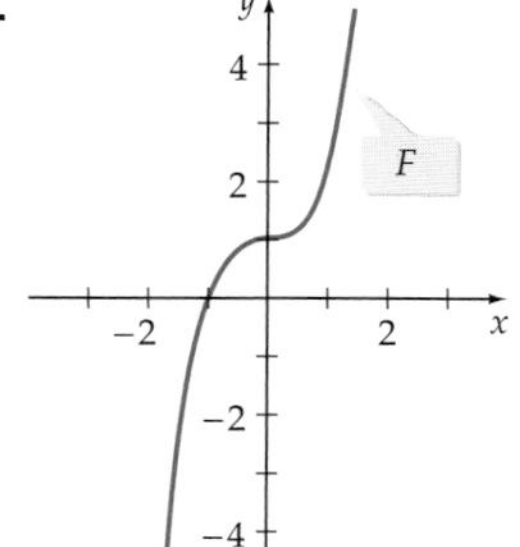

60.

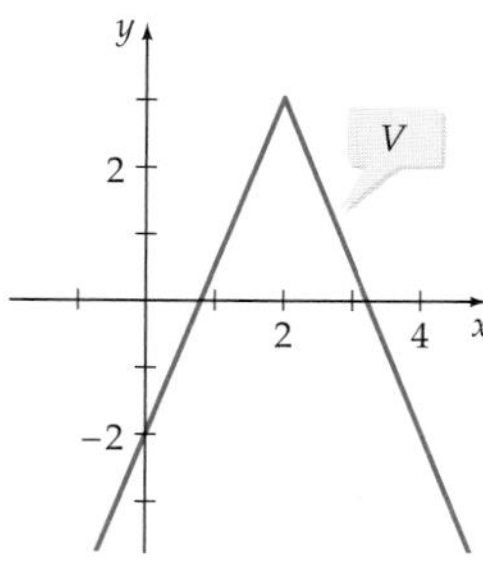

61.

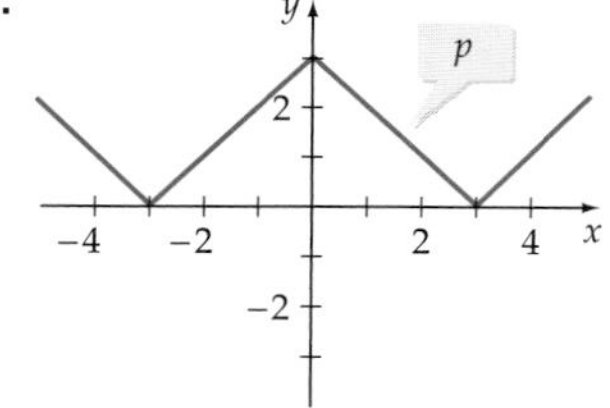

62.

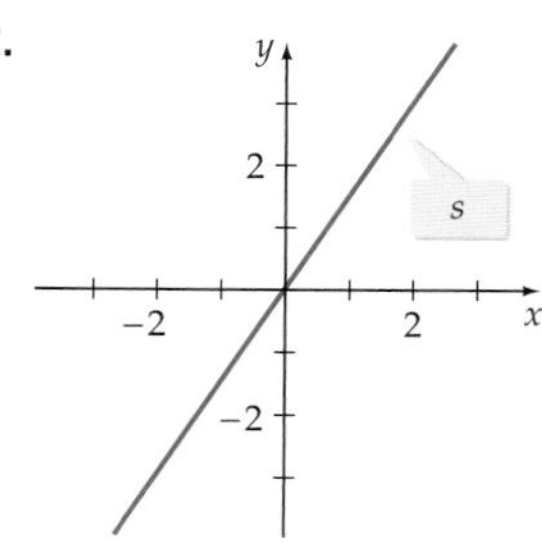

63.

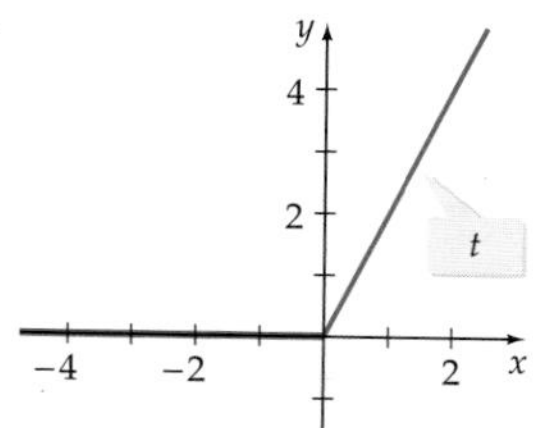

64.

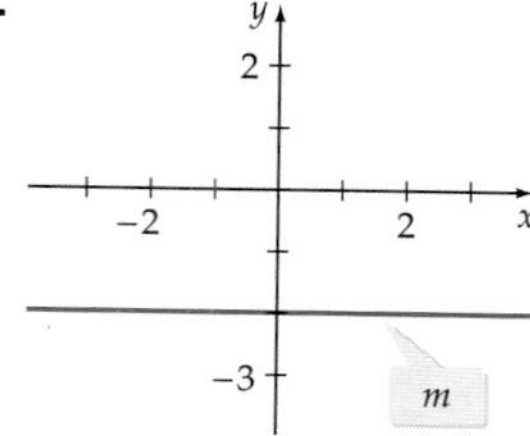

65.

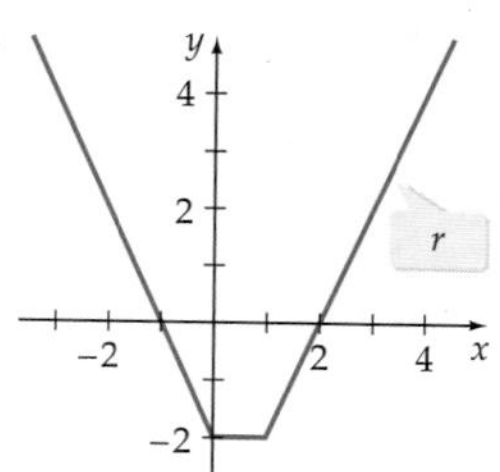

66.

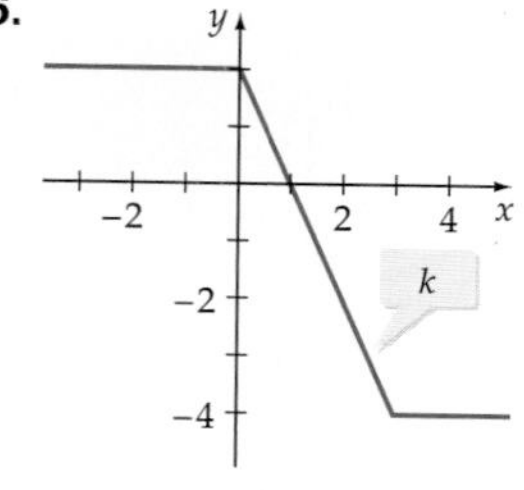

67. Use the horizontal line test to determine which of the following functions are one-to-one.

f as shown in Exercise 57
g as shown in Exercise 58
F as shown in Exercise 59
V as shown in Exercise 60
p as shown in Exercise 61

68. Use the horizontal line test to determine which of the following functions are one-to-one.

s as shown in Exercise 62
t as shown in Exercise 63
m as shown in Exercise 64
r as shown in Exercise 65
k as shown in Exercise 66

69. **Geometry** A rectangle has a length of l feet and a perimeter of 50 feet.

a. Write the width w of the rectangle as a function of its length.

b. Write the area A of the rectangle as a function of its length.

70. **Length of a Shadow** A child 4 feet tall is standing near a street lamp that is 12 feet high. The light from the lamp casts a shadow as shown in the following diagram.

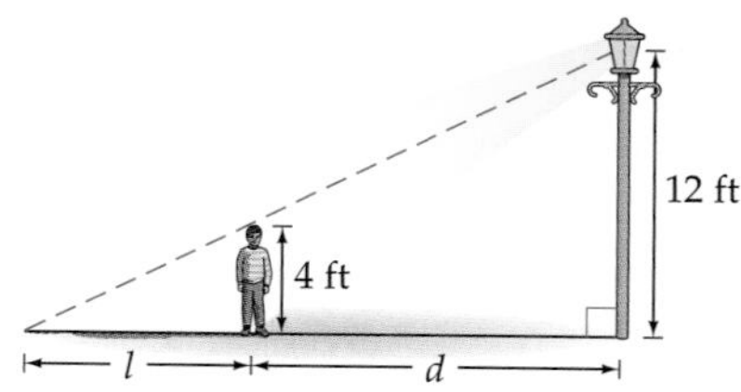

a. Find the length l of the shadow as a function of the distance d of the child from the lamppost. (*Suggestion*: Use the fact that the ratios of corresponding sides of similar triangles are equal.)

b. What is the domain of the function?

c. What is the length of the shadow when the child is 8 feet from the base of the lamppost?

71. **Depreciation** A bus was purchased for \$80,000. Assuming the bus depreciates at a rate of \$6500 per year (*straight-line depreciation*) for the first 10 years, write the value v of the bus as a function of the time t (measured in years) for $0 \le t \le 10$.

72. **Depreciation** A boat was purchased for \$44,000. Assuming the boat depreciates at a rate of \$4200 per year (*straight-line depreciation*) for the first 8 years, write the value v of the boat as a function of the time t (measured in years) for $0 \le t \le 8$.

73. **Cost, Revenue, and Profit** A manufacturer produces a product at a cost of \$22.80 per unit. The manufacturer has a fixed cost of \$400.00 per day. Each unit retails for \$37.00. Let x represent the number of units produced in a 5-day period.

a. Write the total cost C as a function of x.

b. Write the revenue R as a function of x.

c. Write the profit P as a function of x. (*Hint:* The profit function is given by $P(x) = R(x) - C(x)$.)

74. **Volume of a Box** An open box is to be made from a square piece of cardboard having dimensions 30 inches by 30 inches by cutting out squares of area x^2 from each corner, as shown in the figure.

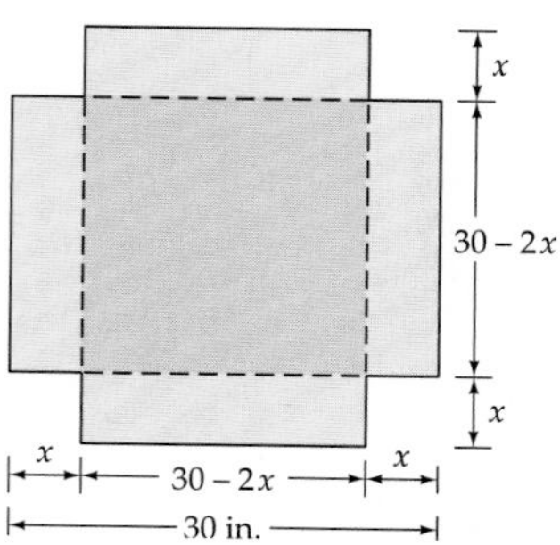

a. Express the volume V of the box as a function of x.

b. State the domain of V.

75. Height of an Inscribed Cylinder A cone has an altitude of 15 centimeters and a radius of 3 centimeters. A right circular cylinder of radius r and height h is inscribed in the cone as shown in the figure. Use similar triangles to write h as a function of r.

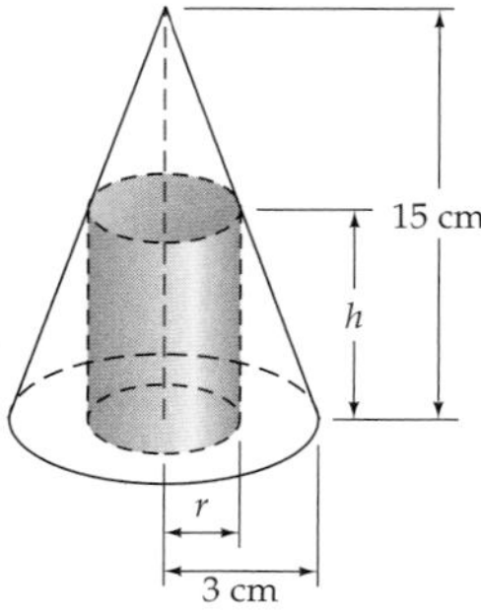

76. Volume of Water Water is flowing into a conical drinking cup that has an altitude of 4 inches and a radius of 2 inches, as shown in the figure.

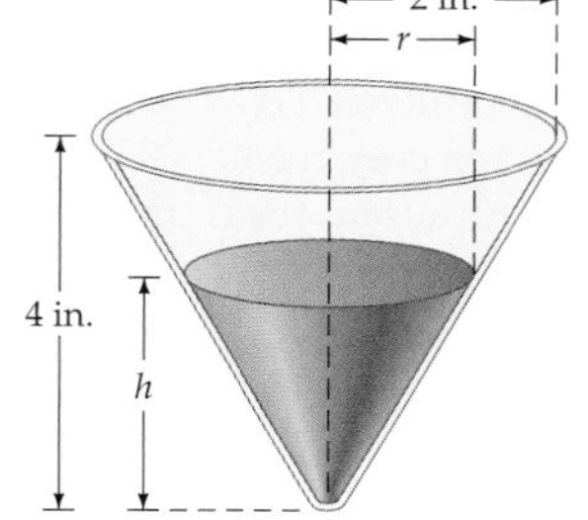

a. Write the radius r of the surface of the water as a function of its depth h.

b. Write the volume V of the water as a function of its depth h.

77. Distance from a Balloon For the first minute of flight, a hot air balloon rises vertically at a rate of 3 meters per second. If t is the time in seconds that the balloon has been airborne, write the distance d between the balloon and a point on the ground 50 meters from the point of lift-off as a function of t.

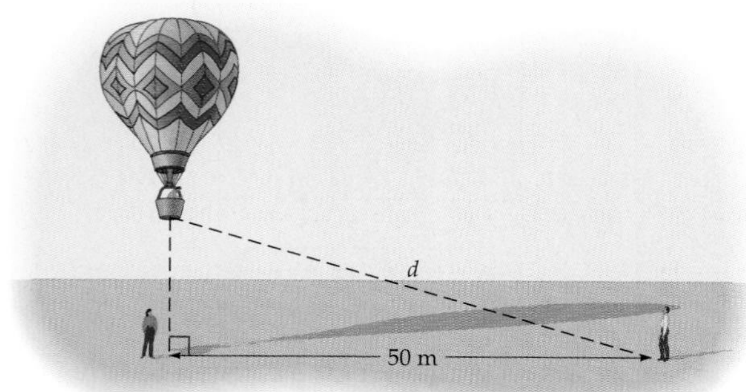

78. Time for a Swimmer An athlete swims from point A to point B at a rate of 2 mph and runs from point B to point C at a rate of 8 mph. Use the dimensions in the figure to write the time t required to reach point C as a function of x.

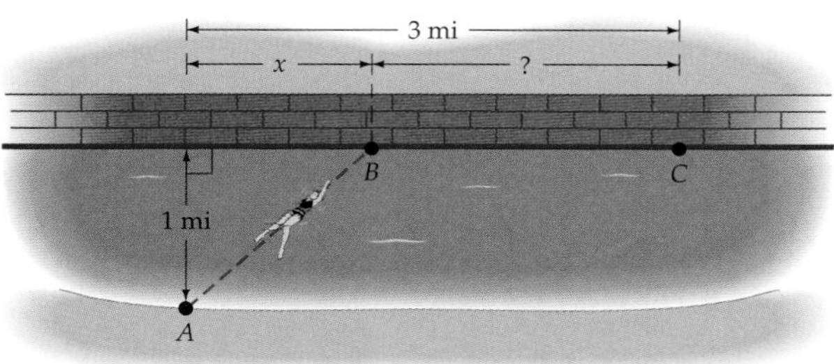

79. Distance Between Ships At 12:00 noon Ship A is 45 miles due south of Ship B and is sailing north at a rate of 8 mph. Ship B is sailing east at a rate of 6 mph. Write the distance d between the ships as a function of the time t, where $t = 0$ represents 12:00 noon.

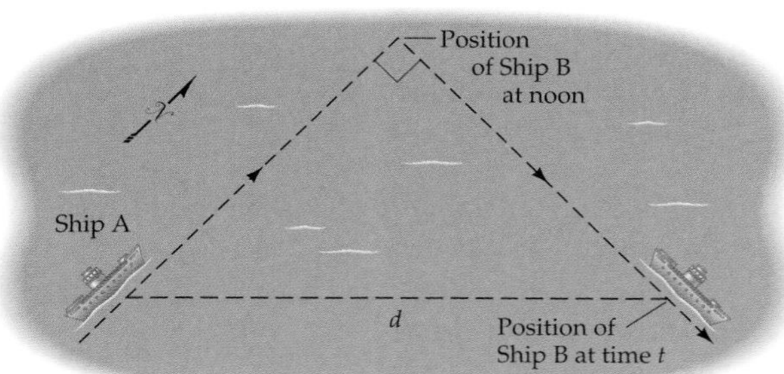

80. Area A rectangle is bounded by the x- and y-axes and the graph of $y = -\frac{1}{2}x + 4$.

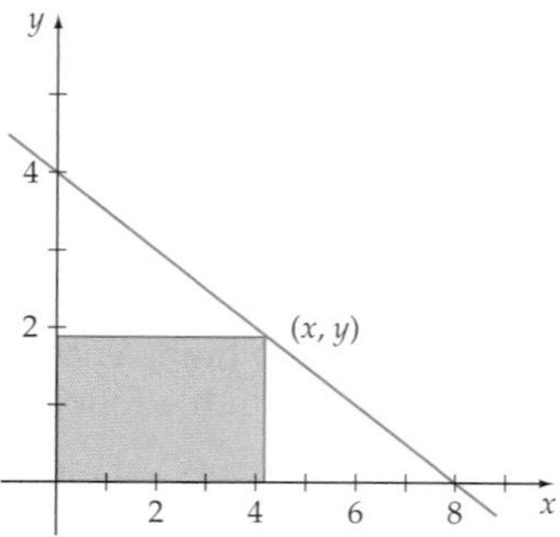

a. Find the area of the rectangle as a function of x.

b. Complete the following table.

x	Area
1	
2	
4	
6	
7	

c. What is the domain of this function?

81. **AREA** A piece of wire 20 centimeters long is cut at a point x centimeters from the left end. The left-hand piece is formed into the shape of a circle and the right-hand piece is formed into a square.

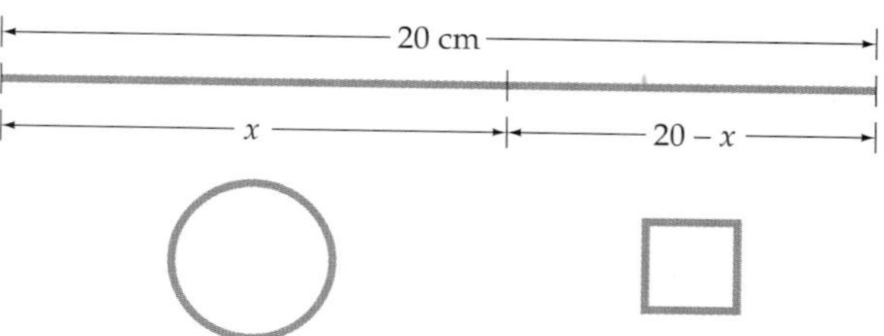

a. Find the area enclosed by the two figures as a function of x.

b. Complete the following table. Round the area to the nearest hundredth of a square centimeter.

x	Total Area Enclosed
0	
4	
8	
12	
16	
20	

c. What is the domain of this function?

82. **AREA** A triangle is bounded by the x- and y-axes and must pass through $P(2, 2)$, as shown below.

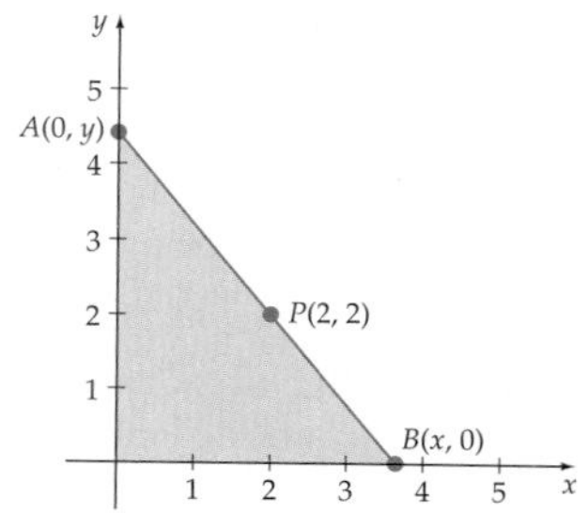

a. Find the area of the triangle as a function of x. (*Suggestion:* Let C be the point $(0, 2)$ and D be the point $(2, 0)$. Use the fact that ACP and PDB are similar triangles.)

b. What is the domain of the function you found in **a.**?

83. **LENGTH** Two guy wires are attached to utility poles that are 40 feet apart, as shown in the following diagram.

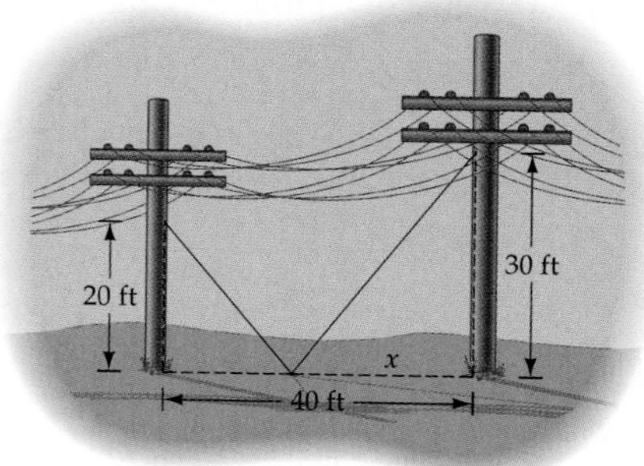

a. Find the total length of the two guy wires as a function of x.

b. Complete the following table. Round the length to the nearest hundredth of a foot.

x	Total Length of Wires
0	
10	
20	
30	
40	

c. What is the domain of this function?

84. **SALES VS. PRICE** A business finds that the number of feet f of pipe it can sell per week is a function of the price p in cents per foot as given by

$$f(p) = \frac{320{,}000}{p + 25}, \quad 40 \le p \le 90$$

Complete the following table by evaluating f (to the nearest hundred feet) for the indicated values of p.

p	40	50	60	75	90
f(p)					

85. **MODEL YIELD** The yield Y of apples per tree is related to the amount x of a particular type of fertilizer applied (in pounds per year) by the function

$$Y(x) = 400[1 - 5(x - 1)^{-2}], \quad 5 \le x \le 20$$

Complete the following table by evaluating Y (to the nearest apple) for the indicated amounts of fertilizer.

x	5	10	12.5	15	20
$Y(x)$					

86. **Model Cost** A manufacturer finds that the cost C in dollars of producing x items of a product is given by

$$C(x) = \left(225 + 1.4\sqrt{x}\right)^2, \quad 100 \le x \le 1000$$

Complete the following table by evaluating C (to the nearest dollar) for the indicated numbers of items.

x	100	200	500	750	1000
$C(x)$					

87. If $f(x) = x^2 - x - 5$ and $f(c) = 1$, find c.

88. If $g(x) = -2x^2 + 4x - 1$ and $g(c) = -4$, find c.

89. Determine whether 1 is in the range of $f(x) = \dfrac{x-1}{x+1}$.

90. Determine whether 0 is in the range of $g(x) = \dfrac{1}{x-3}$.

In Exercises 91 to 96, use a graphing utility.

91. Graph $f(x) = \dfrac{[\![x]\!]}{|x|}$ for $-4.7 \le x \le 4.7$ and $x \ne 0$.

92. Graph $f(x) = \dfrac{[\![2x]\!]}{|x|}$ for $-4 \le x \le 4$ and $x \ne 0$.

93. Graph: $f(x) = x^2 - 2|x| - 3$

94. Graph: $f(x) = x^2 - |2x - 3|$

95. Graph: $f(x) = |x^2 - 1| - |x - 2|$

96. Graph: $f(x) = |x^2 - 2x| - 3$

Connecting Concepts

The notation $f(x)|_a^b$ is used to denote the difference $f(b) - f(a)$. That is,

$$f(x)|_a^b = f(b) - f(a)$$

In Exercises 97 to 100, evaluate $f(x)|_a^b$ for the given function f and the indicated values of a and b.

97. $f(x) = x^2 - x; f(x)|_2^3$

98. $f(x) = -3x + 2; f(x)|_4^7$

99. $f(x) = 2x^3 - 3x^2 - x; f(x)|_0^2$

100. $f(x) = \sqrt{8 - x}; f(x)|_0^8$

In Exercises 101 to 104, each function has two or more independent variables.

101. Given $f(x, y) = 3x + 5y - 2$, find

a. $f(1, 7)$ **b.** $f(0, 3)$ **c.** $f(-2, 4)$

d. $f(4, 4)$ **e.** $f(k, 2k)$ **f.** $f(k+2, k-3)$

102. Given $g(x, y) = 2x^2 - |y| + 3$, find

a. $g(3, -4)$ **b.** $g(-1, 2)$

c. $g(0, -5)$ **d.** $g\left(\dfrac{1}{2}, -\dfrac{1}{4}\right)$

e. $g(c, 3c), c > 0$ **f.** $g(c + 5, c - 2), c < 0$

103. **Area of a Triangle** The area of a triangle with sides a, b, and c is given by the function

$$A(a, b, c) = \sqrt{s(s-a)(s-b)(s-c)}$$

where s is the semiperimeter

$$s = \frac{a+b+c}{2}$$

Find $A(5, 8, 11)$.

104. **Cost of a Painter** The cost in dollars to hire a house painter is given by the function

$$C(h, g) = 15h + 14g$$

where h is the number of hours it takes to paint the house and g is the number of gallons of paint required to paint the house. Find $C(18, 11)$.

A *fixed point* of a function is a number *a* such that $f(a) = a$. In Exercises 105 and 106, find all fixed points for the given function.

105. $f(x) = x^2 + 3x - 3$

106. $g(x) = \dfrac{x}{x + 5}$

In Exercises 107 and 108, sketch the graph of the piecewise-defined function.

107. $s(x) = \begin{cases} 1, & \text{if } x \text{ is an integer} \\ 2, & \text{if } x \text{ is not an integer} \end{cases}$

108. $v(x) = \begin{cases} 2x - 2, & \text{if } x \neq 3 \\ 1, & \text{if } x = 3 \end{cases}$

»»» Projects

1. DAY OF THE WEEK A formula known as Zeller's Congruence makes use of the greatest integer function $[\![x]\!]$ to determine the day of the week on which a given day fell or will fall. To use Zeller's Congruence, we first compute the integer z given by

$$z = \left[\!\!\left[\frac{13m - 1}{5}\right]\!\!\right] + \left[\!\!\left[\frac{y}{4}\right]\!\!\right] + \left[\!\!\left[\frac{c}{4}\right]\!\!\right] + d + y - 2c$$

The variables c, y, d, and m are defined as follows:

c = the century
y = the year of the century
d = the day of the month
m = the month, using 1 for March, 2 for April, . . . , 10 for December. January and February are assigned the values 11 and 12 of the previous year.

For example, for the date September 12, 2001, we use $c = 20$, $y = 1$, $d = 12$, and $m = 7$. The remainder of z divided by 7 gives the day of the week. A remainder of 0 represents a Sunday, a remainder of 1 a Monday, . . . , and a remainder of 6 a Saturday.

a. Verify that December 7, 1941 was a Sunday.

b. Verify that January 1, 2010 will fall on a Friday.

c. Determine on what day of the week Independence Day (July 4, 1776) fell.

d. Determine on what day of the week you were born.

Section 1.4

- Symmetry
- Even and Odd Functions
- Translations of Graphs
- Reflections of Graphs
- Compressing and Stretching of Graphs

Properties of Graphs

PREPARE FOR THIS SECTION

Prepare for this section by completing the following exercises. The answers can be found on page A3.

PS1. Graph $f(x) = x^2$ and $g(x) = x^2 + 1$ on the same coordinate grid. How is the graph of g related to the graph of f? [1.2]

PS2. For $f(x) = \dfrac{3x^4}{x^2 + 1}$, show that $f(-3) = f(3)$. [1.3]

PS3. For $f(x) = 2x^3 - 5x$, show that $f(-2) = -f(2)$. [1.3]

PS4. Let $f(x) = x^2$ and $g(x) = x + 3$. Find $f(a) - g(a)$ for $a = -2, -1, 0, 1, 2$. [1.3]

PS5. What is the midpoint of the line segment between $P(-a, b)$ and $Q(a, b)$? [1.2]

PS6. What is the midpoint of the line segment between $P(-a, -b)$ and $Q(a, b)$? [1.2]

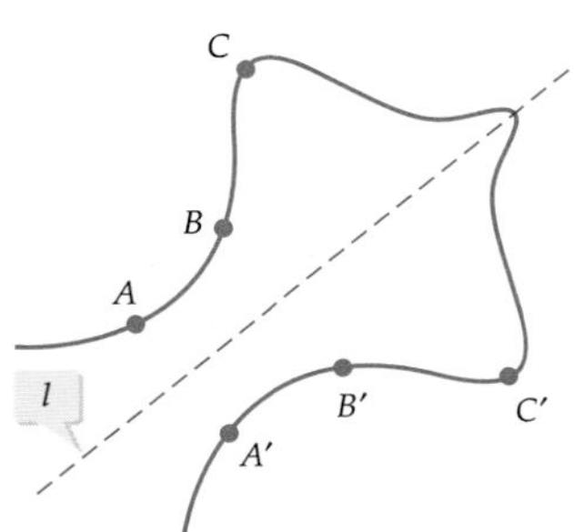

Figure 1.49

Symmetry

The graph in **Figure 1.49** is symmetric with respect to the line l. Note that the graph has the property that if the paper is folded along the dotted line l, the point A' will coincide with the point A, the point B' will coincide with the point B, and the point C' will coincide with the point C. One part of the graph is a *mirror image* of the rest of the graph across the line l.

A graph is **symmetric with respect to the y-axis** if, whenever the point given by (x, y) is on the graph, then $(-x, y)$ is also on the graph. The graph in **Figure 1.50** is symmetric with respect to the y-axis. A graph is **symmetric with respect to the x-axis** if, whenever the point given by (x, y) is on the graph, then $(x, -y)$ is also on the graph. The graph in **Figure 1.51** is symmetric with respect to the x-axis.

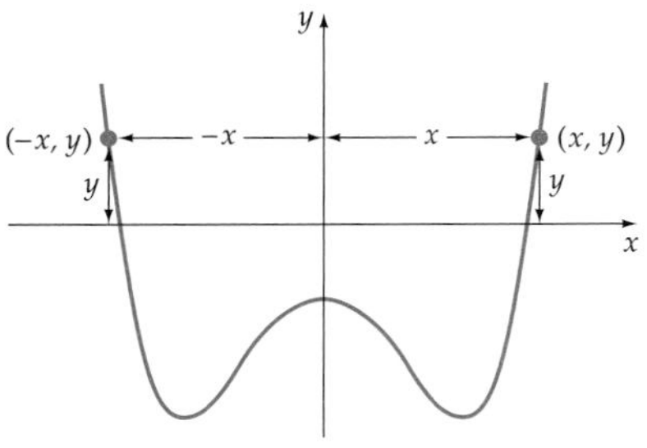

Figure 1.50
Symmetry with respect to the y-axis

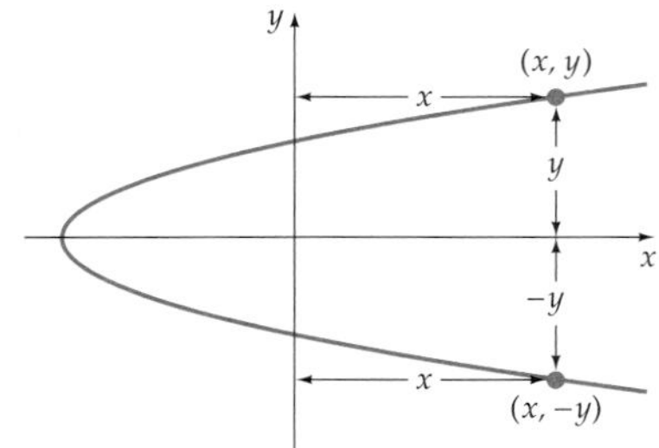

Figure 1.51
Symmetry with respect to the x-axis

Tests for Symmetry with Respect to a Coordinate Axis

The graph of an equation is symmetric with respect to

- the y-axis if the replacement of x with $-x$ leaves the equation unaltered.
- the x-axis if the replacement of y with $-y$ leaves the equation unaltered.

QUESTION Which of the graphs below, I, II, or III, is **a.** symmetric with respect to the x-axis? **b.** symmetric with respect to the y-axis?

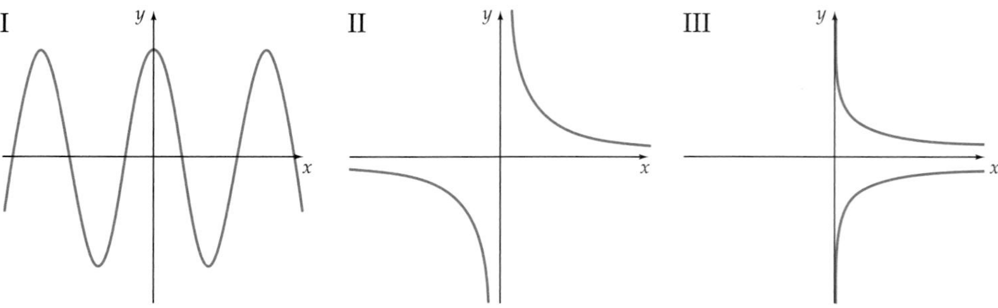

ANSWER **a.** III is symmetric with respect to the x-axis.
b. I is symmetric with respect to the y-axis.

EXAMPLE 1 》 Determine Symmetries of a Graph

Determine whether the graph of the given equation has symmetry with respect to either the x- or the y-axis.

a. $y = x^2 + 2$ b. $x = |y| - 2$

Solution

a. The equation $y = x^2 + 2$ *is unaltered* by the replacement of x with $-x$. That is, the simplification of $y = (-x)^2 + 2$ yields the original equation $y = x^2 + 2$. Thus the graph of $y = x^2 + 2$ is symmetric with respect to the y-axis. However, the equation $y = x^2 + 2$ *is altered* by the replacement of y with $-y$. That is, the simplification of $-y = x^2 + 2$, which is $y = -x^2 - 2$, *does not* yield the original equation $y = x^2 + 2$. The graph of $y = x^2 + 2$ is not symmetric with respect to the x-axis. See **1.52**.

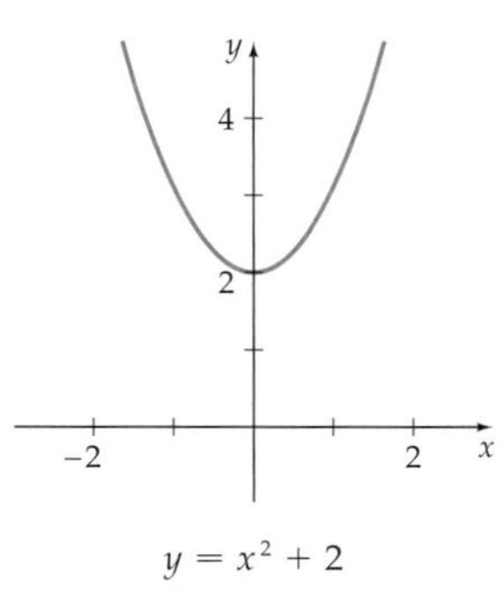

$y = x^2 + 2$

Figure 1.52

b. The equation $x = |y| - 2$ *is altered* by the replacement of x with $-x$. That is, the simplification of $-x = |y| - 2$, which is $x = -|y| + 2$, *does not* yield the original equation $x = |y| - 2$. This implies that the graph of $x = |y| - 2$ is not symmetric with respect to the y-axis. However, the equation $x = |y| - 2$ *is unaltered* by the replacement of y with $-y$. That is, the simplification of $x = |-y| - 2$ yields the original equation $x = |y| - 2$. The graph of $x = |y| - 2$ is symmetric with respect to the x-axis. See **Figure 1.53**.

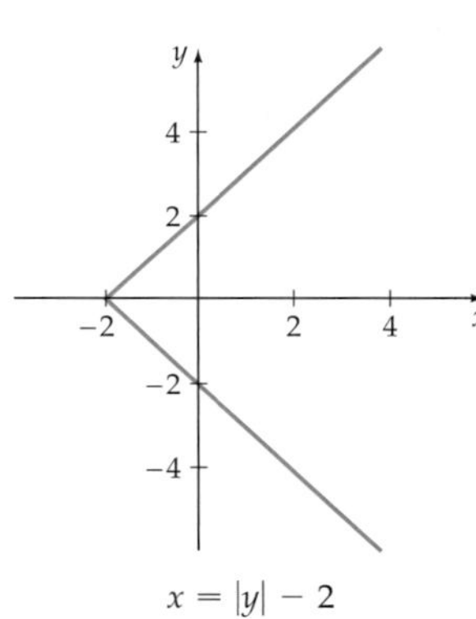

$x = |y| - 2$

Figure 1.53

》 *Try Exercise 14, page 64*

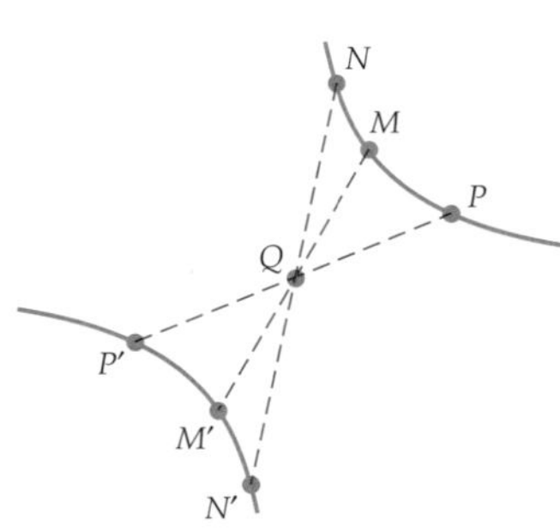

Figure 1.54

Definition of Symmetry with Respect to a Point

A graph is **symmetric with respect to a point** Q if for each point P on the graph there is a point P' on the graph such that Q is the midpoint of the line segment PP'.

The graph in **Figure 1.54** is symmetric with respect to the point Q. For any point P on the graph, there exists a point P' on the graph such that Q is the midpoint of $P'P$.

When we discuss symmetry with respect to a point, we frequently use the origin. A graph is symmetric with respect to the origin if, whenever the point given by (x, y) is on the graph, then $(-x, -y)$ is also on the graph. The graph in **Figure 1.55** is symmetric with respect to the origin.

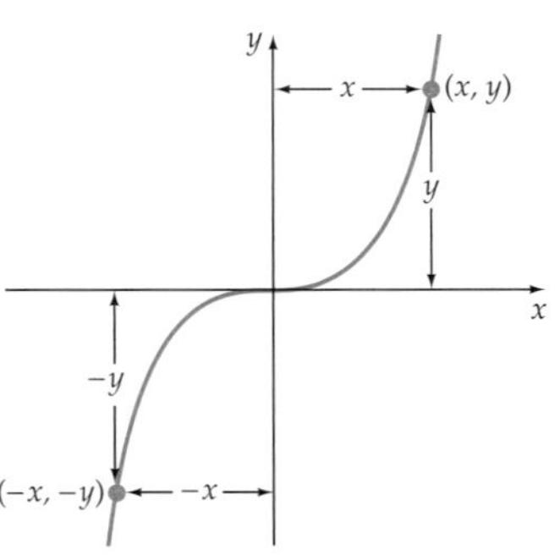

Figure 1.55

Symmetry with respect to the origin

Test for Symmetry with Respect to the Origin

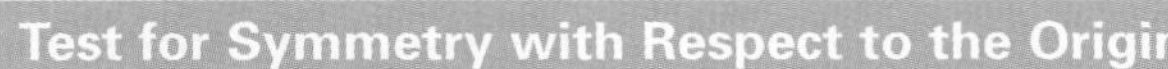

The graph of an equation is symmetric with respect to the origin if the replacement of x with $-x$ and of y with $-y$ leaves the equation unaltered.

EXAMPLE 2 Determine Symmetry with Respect to the Origin

Determine whether the graph of each equation has symmetry with respect to the origin.

a. $xy = 4$ **b.** $y = x^3 + 1$

Solution

a. The equation $xy = 4$ is unaltered by the replacement of x with $-x$ and y with $-y$. That is, the simplification of $(-x)(-y) = 4$ yields the original equation $xy = 4$. Thus the graph of $xy = 4$ is symmetric with respect to the origin. See **Figure 1.56.**

b. The equation $y = x^3 + 1$ *is altered* by the replacement of x with $-x$ and y with $-y$. That is, the simplification of $-y = (-x)^3 + 1$, which is $y = x^3 - 1$, *does not* yield the original equation $y = x^3 + 1$. Thus the graph of $y = x^3 + 1$ is not symmetric with respect to the origin. See **Figure 1.57.**

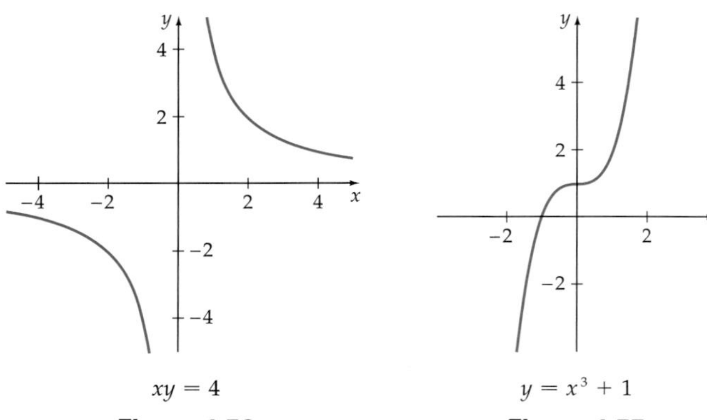

$xy = 4$

Figure 1.56

$y = x^3 + 1$

Figure 1.57

Try Exercise 24, page 64

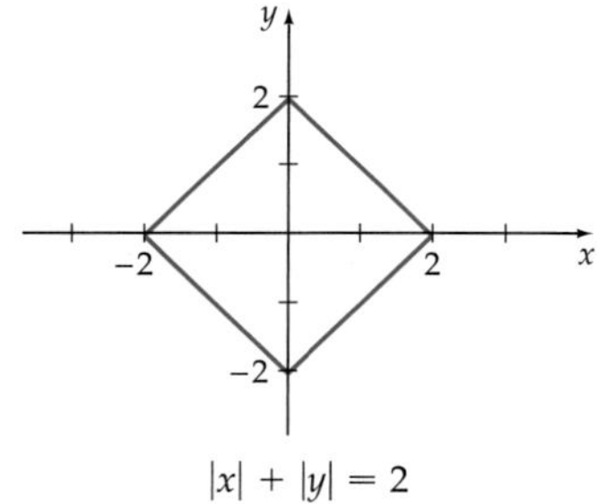

$|x| + |y| = 2$

Figure 1.58

Some graphs have more than one symmetry. For example, the graph of $|x| + |y| = 2$ has symmetry with respect to the x-axis, the y-axis, and the origin. **Figure 1.58** is the graph of $|x| + |y| = 2$.

Even and Odd Functions

Some functions are classified as either *even* or *odd*.

Definition of Even and Odd Functions

The function f is an **even function** if

$$f(-x) = f(x) \quad \text{for all } x \text{ in the domain of } f$$

The function f is an **odd function** if

$$f(-x) = -f(x) \quad \text{for all } x \text{ in the domain of } f$$

EXAMPLE 3 Identify Even or Odd Functions

Determine whether each function is even, odd, or neither.

a. $f(x) = x^3$ b. $F(x) = |x|$ c. $h(x) = x^4 + 2x$

Solution

Replace x with $-x$ and simplify.

a. $f(-x) = (-x)^3 = -x^3 = -(x^3) = -f(x)$
Because $f(-x) = -f(x)$, this function is an odd function.

b. $F(-x) = |-x| = |x| = F(x)$
Because $F(-x) = F(x)$, this function is an even function.

c. $h(-x) = (-x)^4 + 2(-x) = x^4 - 2x$
This function is neither an even nor an odd function because

$$h(-x) = x^4 - 2x,$$

which is not equal to either $h(x)$ or $-h(x)$.

Try Exercise 44, page 64

The following properties are results of the tests for symmetry:

- The graph of an even function is symmetric with respect to the y-axis.
- The graph of an odd function is symmetric with respect to the origin.

The graph of f in **Figure 1.59** is symmetric with respect to the y-axis. It is the graph of an even function. The graph of g in **Figure 1.60** is symmetric with respect to the origin. It is the graph of an odd function. The graph of h in **Figure 1.61** is not symmetric with respect to the y-axis and is not symmetric with respect to the origin. It is neither an even nor an odd function.

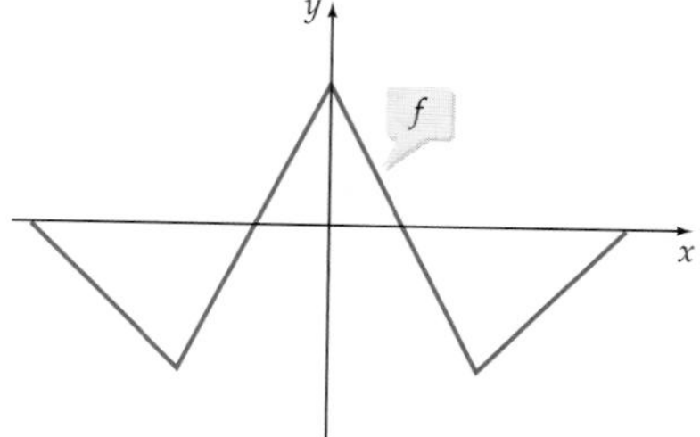

Figure 1.59
The graph of an even function is symmetric with respect to the y-axis.

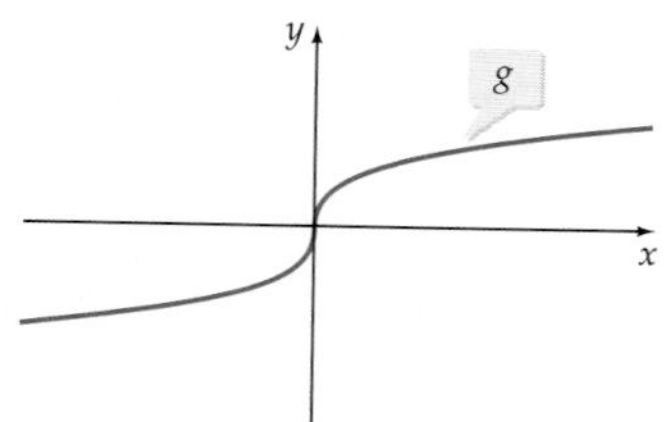

Figure 1.60
The graph of an odd function is symmetric with respect to the origin.

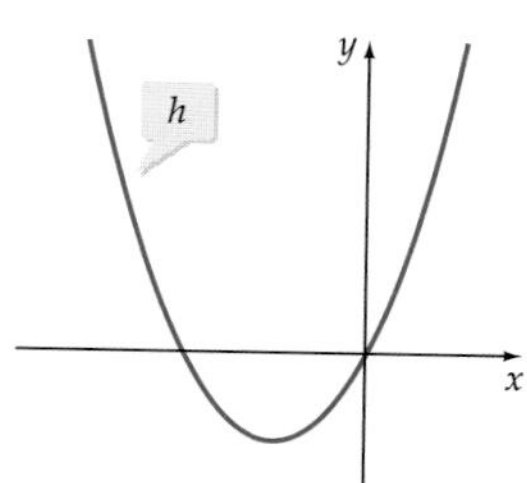

Figure 1.61
The graph of a function that is neither even nor odd is not symmetric with respect to the y-axis or the origin.

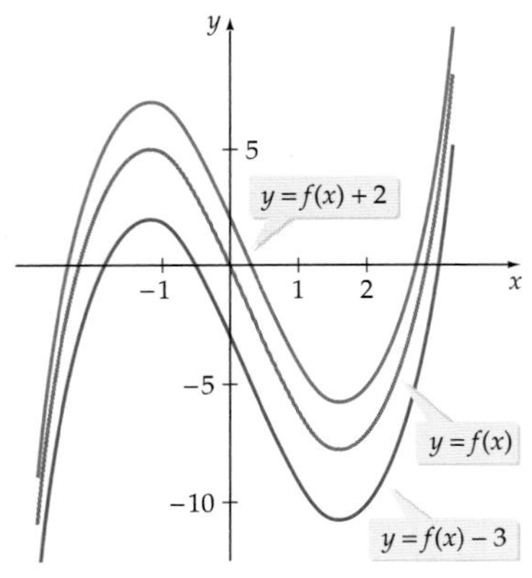

Figure 1.62

Translations of Graphs

The shape of a graph may be exactly the same as the shape of another graph; only their positions in the *xy*-plane may differ. For example, the graph of $y = f(x) + 2$ is the graph of $y = f(x)$ with each point moved up vertically 2 units. The graph of $y = f(x) - 3$ is the graph of $y = f(x)$ with each point moved down vertically 3 units. See **Figure 1.62**.

The graphs of $y = f(x) + 2$ and $y = f(x) - 3$ in **Figure 1.62** are called *vertical translations* of the graph of $y = f(x)$.

Vertical Translations

If f is a function and c is a positive constant, then the graph of

- $y = f(x) + c$ is the graph of $y = f(x)$ shifted up *vertically* c units.
- $y = f(x) - c$ is the graph of $y = f(x)$ shifted down *vertically* c units.

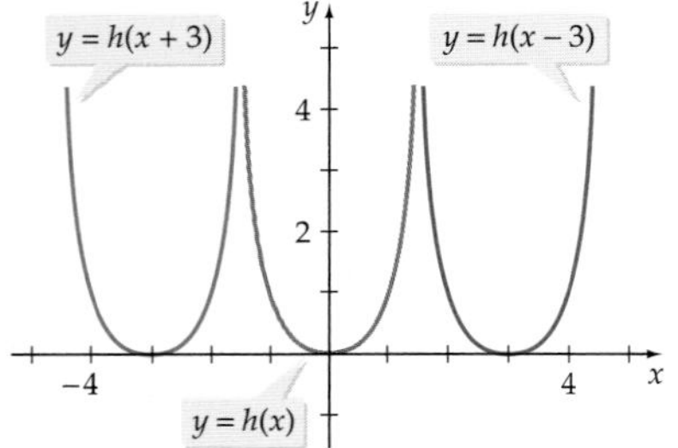

Figure 1.63

In **Figure 1.63**, the graph of $y = h(x + 3)$ is the graph of $y = h(x)$ with each point shifted to the left horizontally 3 units. Similarly, the graph of $y = h(x - 3)$ is the graph of $y = h(x)$ with each point shifted to the right horizontally 3 units.

The graphs of $y = h(x + 3)$ and $y = h(x - 3)$ in **Figure 1.63** are called *horizontal translations* of the graph of $y = h(x)$.

Horizontal Translations

If f is a function and c is a positive constant, then the graph of

- $y = f(x + c)$ is the graph of $y = f(x)$ shifted left *horizontally* c units.
- $y = f(x - c)$ is the graph of $y = f(x)$ shifted right *horizontally* c units.

Integrating Technology

A graphing calculator can be used to draw the graphs of a *family* of functions. For instance, $f(x) = x^2 + c$ constitutes a family of functions with **parameter** c. The only feature of the graph that changes is the value of c.

A graphing calculator can be used to produce the graphs of a family of curves for specific values of the parameter. The LIST feature of the calculator can be used. For instance, to graph $f(x) = x^2 + c$ for $c = -2, 0$, and 1, we will create a list and use that list to produce the family of curves. The keystrokes for a TI-83/TI-83 Plus/TI-84 Plus calculator are given below.

2nd { -2 , 0 , 1 2nd } STO 2nd L1

Now use the Y= key to enter

Y= X x^2 + 2nd L1 ZOOM 6

Sample screens for the keystrokes and graphs are shown here. You can use similar keystrokes for Exercises 75–82 of this section.

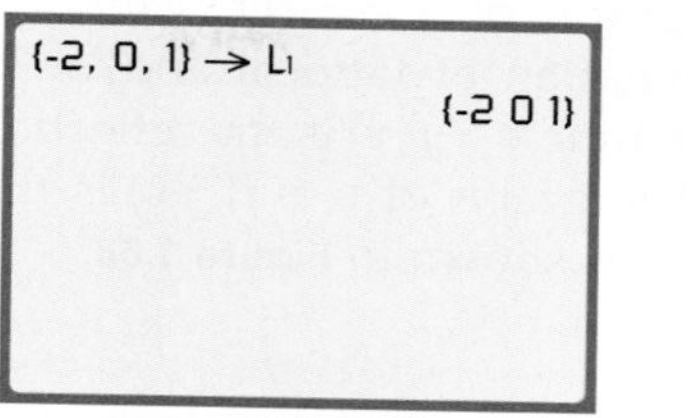

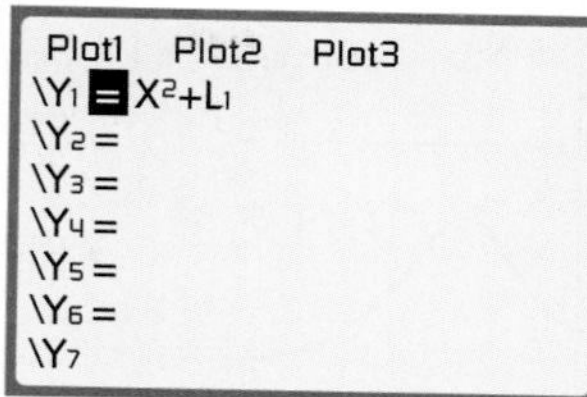

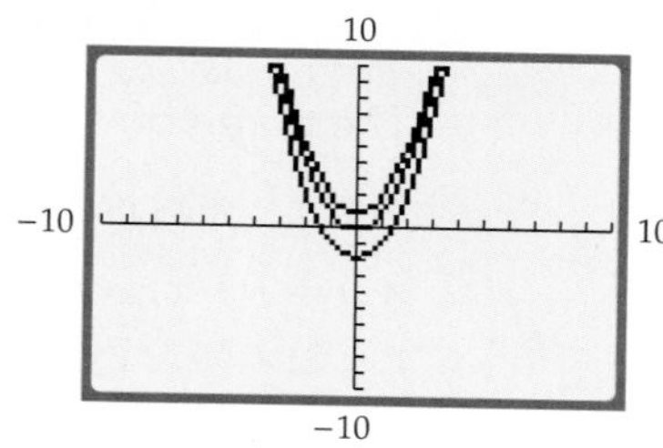

EXAMPLE 4 Graph by Using Translations

Use vertical and horizontal translations of the graph of $f(x) = x^3$, shown in **Figure 1.64,** to graph

a. $g(x) = x^3 - 2$ b. $h(x) = (x + 1)^3$

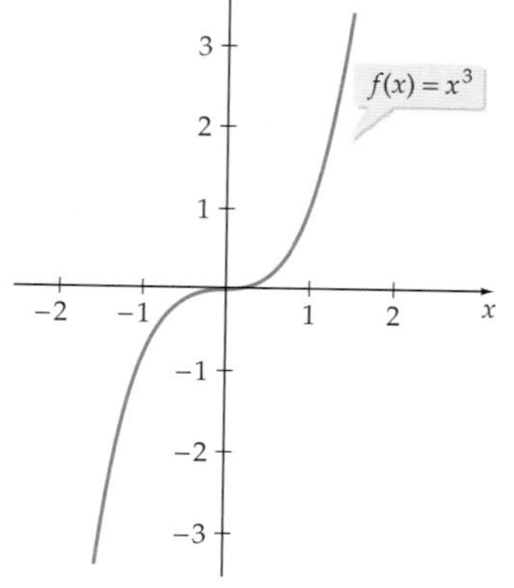

Figure 1.64

Solution

a. The graph of $g(x) = x^3 - 2$ is the graph of $f(x) = x^3$ shifted down vertically 2 units. See **Figure 1.65.**

b. The graph of $h(x) = (x + 1)^3$ is the graph of $f(x) = x^3$ shifted to the left horizontally 1 unit. See **Figure 1.66.**

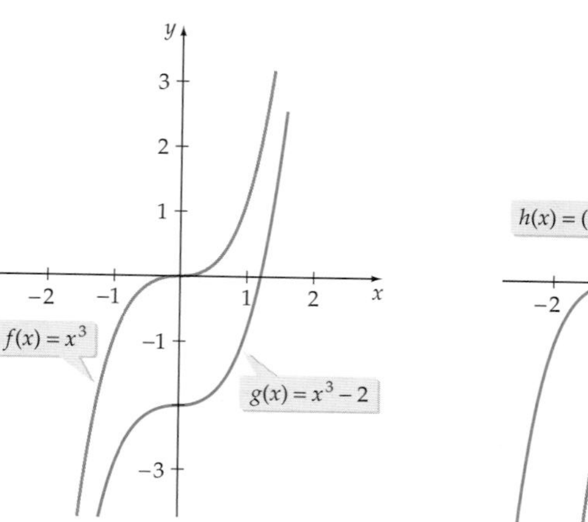

Figure 1.65

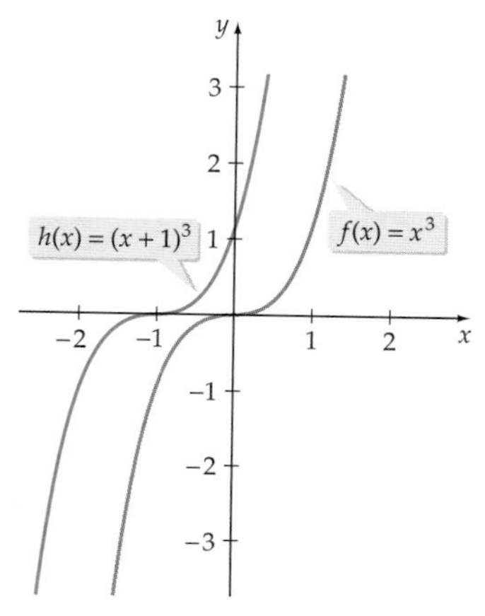

Figure 1.66

Try Exercise 58, page 65

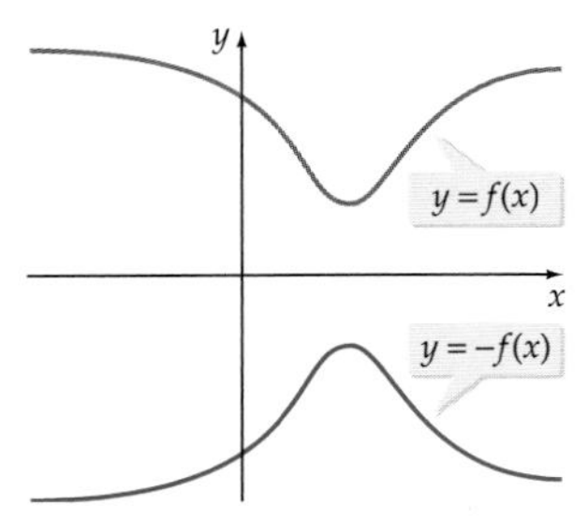

Figure 1.67

Reflections of Graphs

The graph of $y = -f(x)$ cannot be obtained from the graph of $y = f(x)$ by a combination of vertical and/or horizontal shifts. **Figure 1.67** illustrates that the graph of $y = -f(x)$ is the reflection of the graph of $y = f(x)$ across the x-axis.

The graph of $y = f(-x)$ is the reflection of the graph of $y = f(x)$ across the y-axis, as shown in **Figure 1.68**.

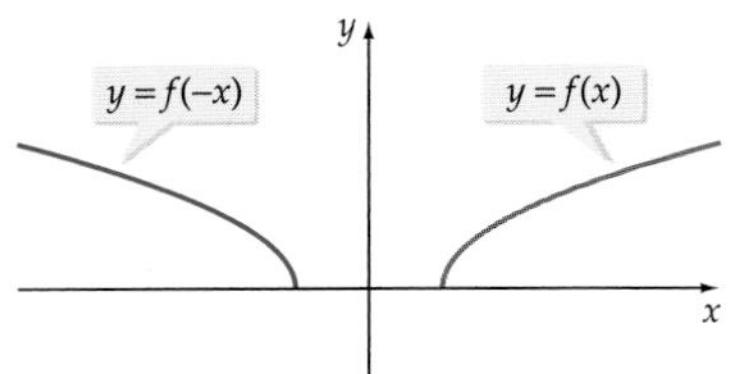

Figure 1.68

Reflections

The graph of

- $y = -f(x)$ is the graph of $y = f(x)$ reflected across the x-axis.
- $y = f(-x)$ is the graph of $y = f(x)$ reflected across the y-axis.

EXAMPLE 5 » Graph by Using Reflections

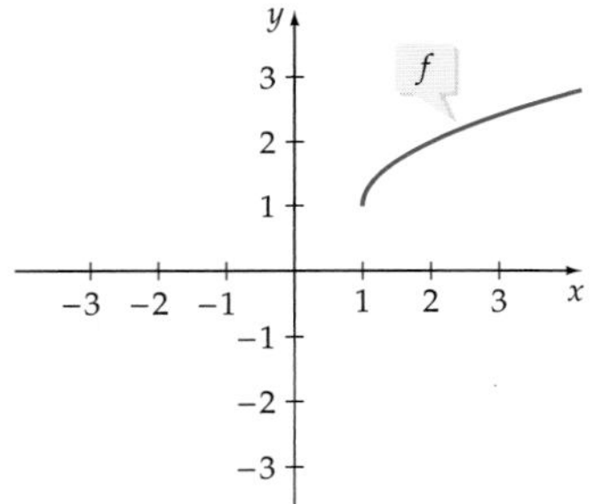

Figure 1.69

Use reflections of the graph of $f(x) = \sqrt{x-1} + 1$, shown in **Figure 1.69**, to graph

a. $g(x) = -\left(\sqrt{x-1} + 1\right)$ b. $h(x) = \sqrt{-x-1} + 1$

Solution

a. Because $g(x) = -f(x)$, the graph of g is the graph of f reflected across the x-axis. See **Figure 1.70**.

b. Because $h(x) = f(-x)$, the graph of h is the graph of f reflected across the y-axis. See **Figure 1.71**.

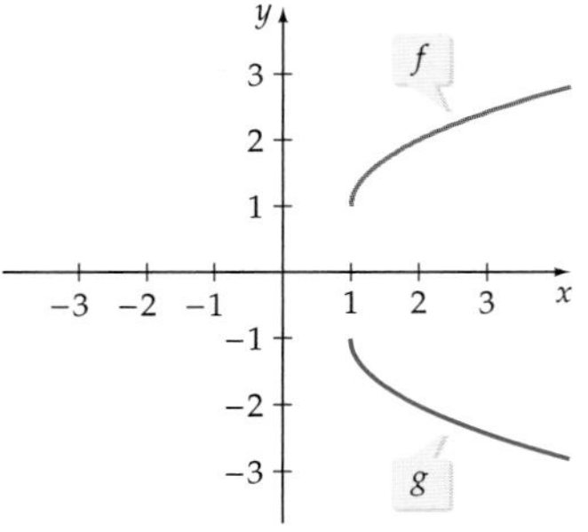

Figure 1.70

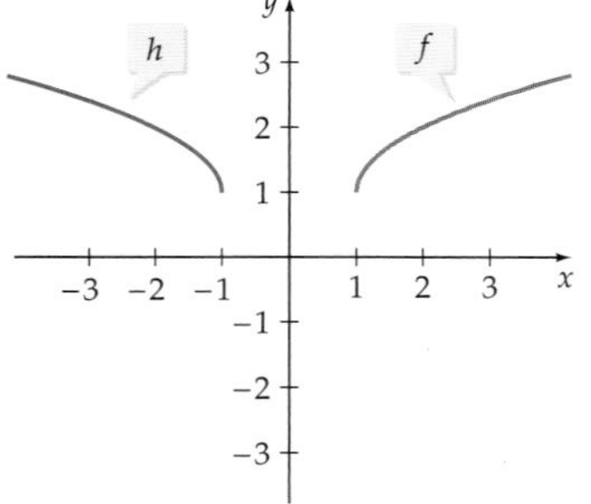

Figure 1.71

» *Try Exercise 68, page 66*

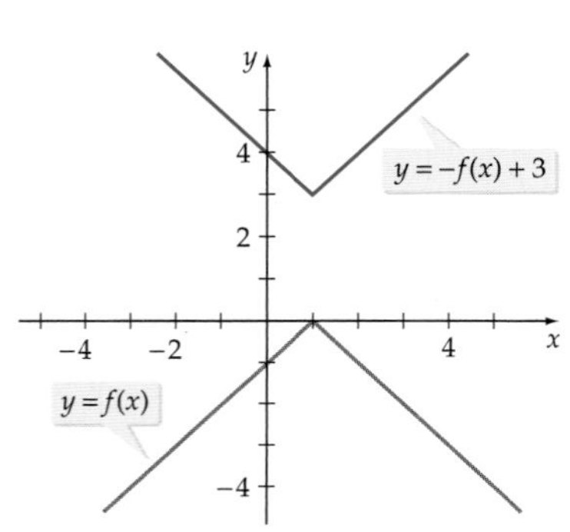

Figure 1.72

Some graphs of functions can be constructed by using a combination of translations and reflections. For instance, the graph of $y = -f(x) + 3$ in **Figure 1.72** was obtained by reflecting the graph of $y = f(x)$ in **Figure 1.72** across the x-axis and then shifting that graph up vertically 3 units.

Compressing and Stretching of Graphs

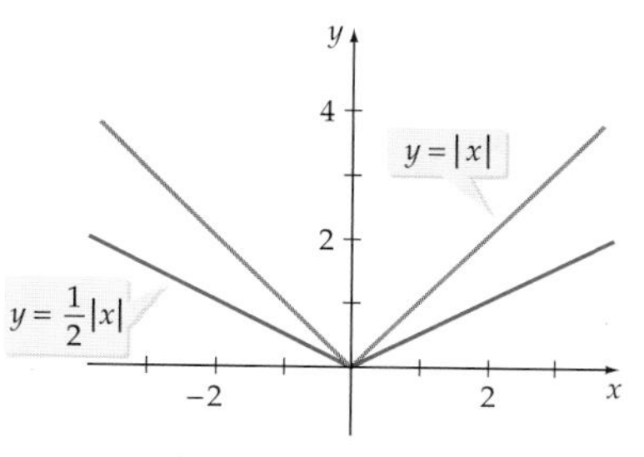

Figure 1.73

The graph of the equation $y = c \cdot f(x)$ for $c \neq 1$ vertically compresses or stretches the graph of $y = f(x)$. To determine the points on the graph of $y = c \cdot f(x)$, multiply the y-coordinate of each point on the graph of $y = f(x)$ by c. For example, **Figure 1.73** shows that the graph of $y = \frac{1}{2}|x|$ can be obtained by plotting points that have a y-coordinate that is one-half of the y-coordinate of those points that make up the graph of $y = |x|$.

If $0 < c < 1$, then the graph of $y = c \cdot f(x)$ is obtained by *compressing* the graph of $y = f(x)$. **Figure 1.73** illustrates the vertical compressing of the graph of $y = |x|$ toward the x-axis to form the graph of $y = \frac{1}{2}|x|$.

If $c > 1$, then the graph of $y = c \cdot f(x)$ is obtained by *stretching* the graph of $y = f(x)$. For example, if $f(x) = |x|$, then we obtain the graph of

$$y = 2f(x) = 2|x|$$

by stretching the graph of f away from the x-axis. See **Figure 1.74.**

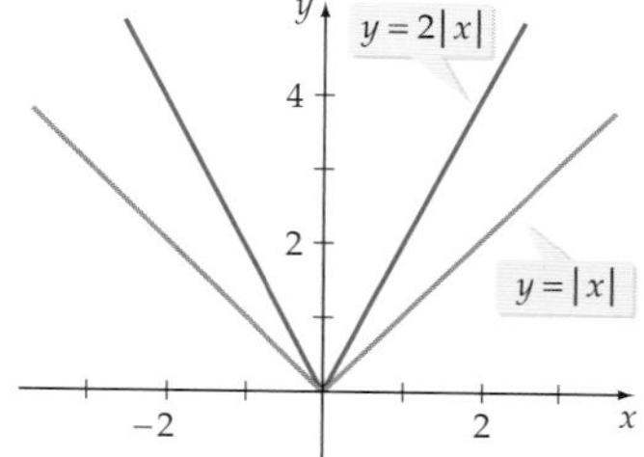

Figure 1.74

Vertical Stretching and Compressing of Graphs

If f is a function and c is a positive constant, then

- if $c > 1$, the graph of $y = c \cdot f(x)$ is the graph of $y = f(x)$ *stretched* vertically away from the x-axis by a factor of c.
- if $0 < c < 1$, the graph of $y = c \cdot f(x)$ is the graph of $y = f(x)$ *compressed* vertically toward the x-axis by a factor of c.

EXAMPLE 6 ⟫ Graph by Using Vertical Compressing and Shifting

Graph: $H(x) = \frac{1}{4}|x| - 3$

Solution

The graph of $y = |x|$ has a V shape that has its lowest point at $(0, 0)$ and passes through $(4, 4)$ and $(-4, 4)$. The graph of $y = \frac{1}{4}|x|$ is a compression of the graph of $y = |x|$. The y-coordinates of the ordered pairs $(0, 0)$, $(4, 1)$, and $(-4, 1)$ are obtained by multiplying the y-coordinates of the ordered pairs $(0, 0)$, $(4, 4)$, and $(-4, 4)$ by $\frac{1}{4}$. To find the points on the graph of H, we still need to subtract 3 from each y-coordinate. Thus the graph of H is a V shape that has its lowest point at $(0, -3)$ and passes through $(4, -2)$ and $(-4, -2)$. See **Figure 1.75**.

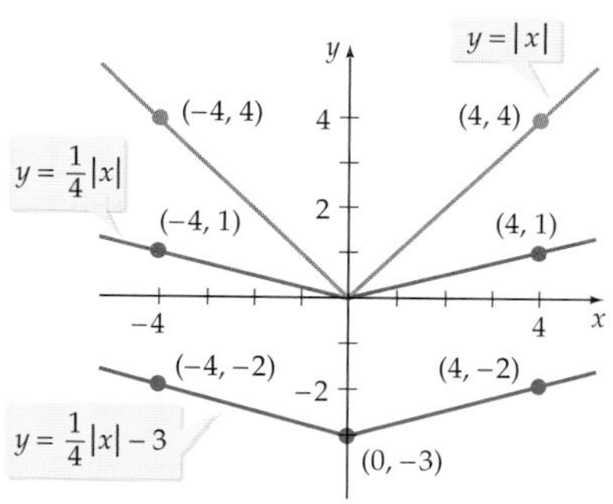

Figure 1.75

⟫ *Try Exercise 70, page 66*

EXAMPLE 7 » Use a Composite Function to Solve an Application

A graphic artist has drawn a 3-inch by 2-inch rectangle on a computer screen. The artist has been scaling the size of the rectangle for t seconds in such a way that the upper right corner of the original rectangle is moving to the right at the rate of 0.5 inch per second and downward at the rate of 0.2 inch per second. See **Figure 1.82.**

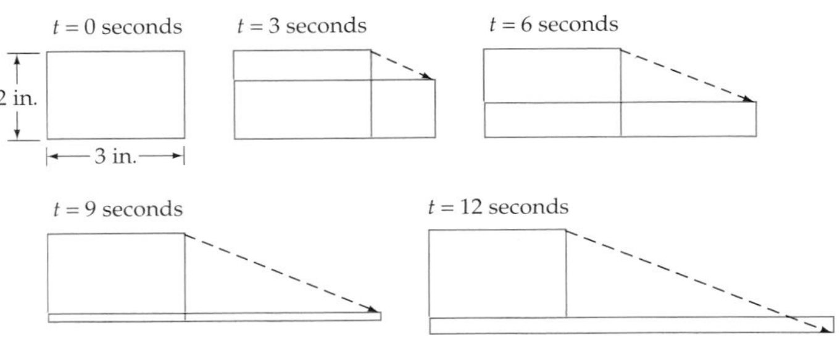

Figure 1.82

a. Write the length l and the width w of the scaled rectangles as functions of t.

b. Write the area A of the scaled rectangle as a function of t.

c. Find the intervals on which A is an increasing function for $0 \leq t \leq 14$. Also find the intervals on which A is a decreasing function.

d. Find the value of t (where $0 \leq t \leq 14$) that maximizes $A(t)$.

Solution

a. Because *distance* = *rate* · *time,* we see that the change in l is given by $0.5t$. Therefore, the length at any time t is $l = 3 + 0.5t$. For $0 \leq t \leq 10$, the width is given by $w = 2 - 0.2t$. For $10 < t \leq 14$, the width is $w = -2 + 0.2t$. In either case the width can be determined by finding $w = |2 - 0.2t|$. (The absolute value symbol is needed to keep the width positive for $10 < t \leq 14$.)

b. $A = lw = (3 + 0.5t)|2 - 0.2t|$

c. Use a graphing utility to determine that A is increasing on $[0, 2]$ and on $[10, 14]$ and that A is decreasing on $[2, 10]$. See **Figure 1.83.**

d. The highest point on the graph of A occurs when $t = 14$ seconds. See **Figure 1.83.**

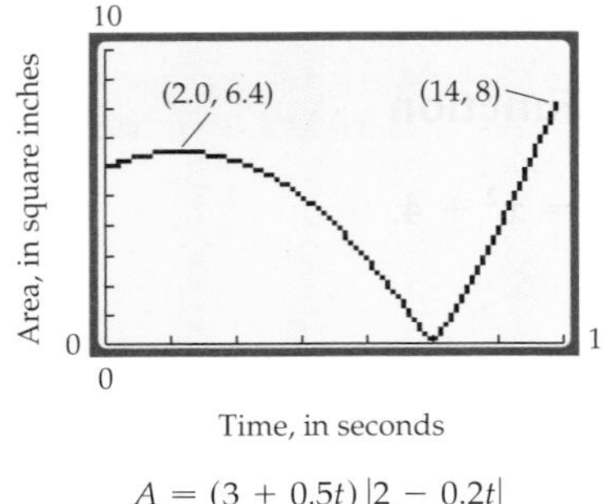

$A = (3 + 0.5t)|2 - 0.2t|$

Figure 1.83

» *Try Exercise 66, page 78*

You may be inclined to think that if the area of a rectangle is decreasing, then its perimeter is also decreasing, but this is not always the case. For example, the area of the scaled rectangle in Example 7 was shown to decrease on [2, 10] even though its perimeter is always increasing. See Exercise 68 in Exercise Set 1.5.

Topics for Discussion

1. The domain of $f + g$ consists of all real numbers formed by the *union* of the domain of f and the domain of g. Do you agree?
2. Given $f(x) = 3x - 2$ and $g(x) = \frac{1}{3}x + \frac{2}{3}$, determine $f \circ g$ and $g \circ f$. Do your results show that composition of functions is a commutative operation?
3. A tutor states that the difference quotient of $f(x) = x^2$ and the difference quotient of $g(x) = x^2 + 4$ are the same. Do you agree?
4. A classmate states that the difference quotient of any linear function $f(x) = mx + b$ is always m. Do you agree?
5. When we use a difference quotient to determine an average velocity, we generally replace the variable h with the variable Δt. What does Δt represent?

Exercise Set 1.5

In Exercises 1 to 12, use the given functions f and g to find $f + g$, $f - g$, fg, and $\frac{f}{g}$. State the domain of each.

1. $f(x) = x^2 - 2x - 15, \quad g(x) = x + 3$
2. $f(x) = x^2 - 25, \quad g(x) = x - 5$
3. $f(x) = 2x + 8, \quad g(x) = x + 4$
4. $f(x) = 5x - 15, \quad g(x) = x - 3$
5. $f(x) = x^3 - 2x^2 + 7x, \quad g(x) = x$
6. $f(x) = x^2 - 5x - 8, \quad g(x) = -x$
7. $f(x) = 4x - 7, \quad g(x) = 2x^2 + 3x - 5$
8. $f(x) = 6x + 10, \quad g(x) = 3x^2 + x - 10$
9. $f(x) = \sqrt{x - 3}, \quad g(x) = x$
10. $f(x) = \sqrt{x - 4}, \quad g(x) = -x$
11. $f(x) = \sqrt{4 - x^2}, \quad g(x) = 2 + x$
12. $f(x) = \sqrt{x^2 - 9}, \quad g(x) = x - 3$

In Exercises 13 to 28, evaluate the indicated function, where $f(x) = x^2 - 3x + 2$ and $g(x) = 2x - 4$.

13. $(f + g)(5)$
14. $(f + g)(-7)$
15. $(f + g)\left(\frac{1}{2}\right)$
16. $(f + g)\left(\frac{2}{3}\right)$
17. $(f - g)(-3)$
18. $(f - g)(24)$
19. $(f - g)(-1)$
20. $(f - g)(0)$
21. $(fg)(7)$
22. $(fg)(-3)$
23. $(fg)\left(\frac{2}{5}\right)$
24. $(fg)(-100)$
25. $\left(\frac{f}{g}\right)(-4)$
26. $\left(\frac{f}{g}\right)(11)$
27. $\left(\frac{f}{g}\right)\left(\frac{1}{2}\right)$
28. $\left(\frac{f}{g}\right)\left(\frac{1}{4}\right)$

In Exercises 29 to 36, find the difference quotient of the given function.

29. $f(x) = 2x + 4$
30. $f(x) = 4x - 5$
31. $f(x) = x^2 - 6$
32. $f(x) = x^2 + 11$

33. $f(x) = 2x^2 + 4x - 3$

34. $f(x) = 2x^2 - 5x + 7$

35. $f(x) = -4x^2 + 6$

36. $f(x) = -5x^2 - 4x$

In Exercises 37 to 48, find $(g \circ f)(x)$ and $(f \circ g)(x)$ for the given functions f and g.

37. $f(x) = 3x + 5, \quad g(x) = 2x - 7$

38. $f(x) = 2x - 7, \quad g(x) = 3x + 2$

39. $f(x) = x^2 + 4x - 1, \quad g(x) = x + 2$

40. $f(x) = x^2 - 11x, \quad g(x) = 2x + 3$

41. $f(x) = x^3 + 2x, \quad g(x) = -5x$

42. $f(x) = -x^3 - 7, \quad g(x) = x + 1$

43. $f(x) = \dfrac{2}{x+1}, \quad g(x) = 3x - 5$

44. $f(x) = \sqrt{x+4}, \quad g(x) = \dfrac{1}{x}$

45. $f(x) = \dfrac{1}{x^2}, \quad g(x) = \sqrt{x-1}$

46. $f(x) = \dfrac{6}{x-2}, \quad g(x) = \dfrac{3}{5x}$

47. $f(x) = \dfrac{3}{|5-x|}, \quad g(x) = -\dfrac{2}{x}$

48. $f(x) = |2x + 1|, \quad g(x) = 3x^2 - 1$

In Exercises 49 to 64, evaluate each composite function, where $f(x) = 2x + 3$, $g(x) = x^2 - 5x$, and $h(x) = 4 - 3x^2$.

49. $(g \circ f)(4)$

50. $(f \circ g)(4)$

51. $(f \circ g)(-3)$

52. $(g \circ f)(-1)$

53. $(g \circ h)(0)$

54. $(h \circ g)(0)$

55. $(f \circ f)(8)$

56. $(f \circ f)(-8)$

57. $(h \circ g)\left(\dfrac{2}{5}\right)$

58. $(g \circ h)\left(-\dfrac{1}{3}\right)$

59. $(g \circ f)(\sqrt{3})$

60. $(f \circ g)(\sqrt{2})$

61. $(g \circ f)(2c)$

62. $(f \circ g)(3k)$

63. $(g \circ h)(k + 1)$

64. $(h \circ g)(k - 1)$

65. Water Tank A water tank has the shape of a right circular cone with height 16 feet and radius 8 feet. Water is running into the tank so that the radius r (in feet) of the surface of the water is given by $r = 1.5t$, where t is the time (in minutes) that the water has been running.

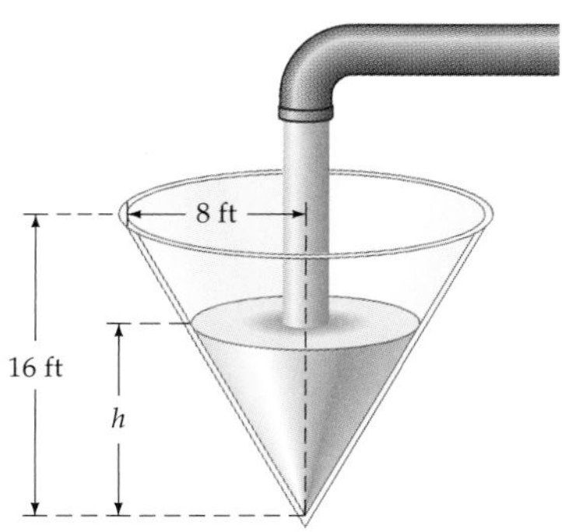

a. The area A of the surface of the water is $A = \pi r^2$. Find $A(t)$ and use it to determine the area of the surface of the water when $t = 2$ minutes.

b. The volume V of the water is given by $V = \dfrac{1}{3}\pi r^2 h$. Find $V(t)$ and use it to determine the volume of the water when $t = 3$ minutes. (*Hint:* The height of the water in the cone is always twice the radius of the surface of the water.)

66. Scaling a Rectangle Rework Example 7 of this section with the scaling as follows. The upper right corner of the original rectangle is pulled to the *left* at 0.5 inch per second and downward at 0.2 inch per second.

67. Towing a Boat A boat is towed by a rope that runs through a pulley that is 4 feet above the point where the rope is tied to the boat. The length (in feet) of the rope from the boat to the pulley is given by $s = 48 - t$, where t is the time in seconds that the boat has been in tow. The horizontal distance from the pulley to the boat is d.

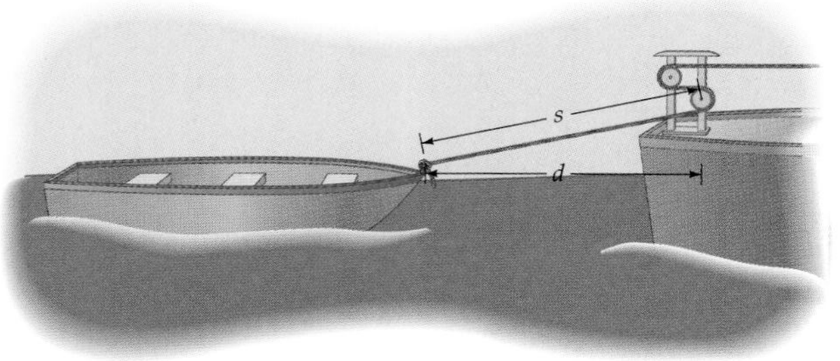

a. Find $d(t)$. **b.** Evaluate $s(35)$ and $d(35)$.

68. **PERIMETER OF A SCALED RECTANGLE** Show by a graph that the perimeter

$$P = 2(3 + 0.5t) + 2|2 - 0.2t|$$

of the scaled rectangle in Example 7 of this section is an increasing function over $0 \le t \le 14$.

69. **CONVERSION FUNCTIONS** The function $F(x) = \dfrac{x}{12}$ converts x inches to feet. The function $Y(x) = \dfrac{x}{3}$ converts x feet to yards. Explain the meaning of $(Y \circ F)(x)$.

70. **CONVERSION FUNCTIONS** The function $F(x) = 3x$ converts x yards to feet. The function $I(x) = 12x$ converts x feet to inches. Explain the meaning of $(I \circ F)(x)$.

71. **CONCENTRATION OF A MEDICATION** The concentration $C(t)$ (in milligrams per liter) of a medication in a patient's blood is given by the data in the following table.

Concentration of Medication in Patient's Blood

t hours	$C(t)$ mg/l
0	0
0.25	47.3
0.50	78.1
0.75	94.9
1.00	99.8
1.25	95.7
1.50	84.4
1.75	68.4
2.00	50.1
2.25	31.6
2.50	15.6
2.75	4.3

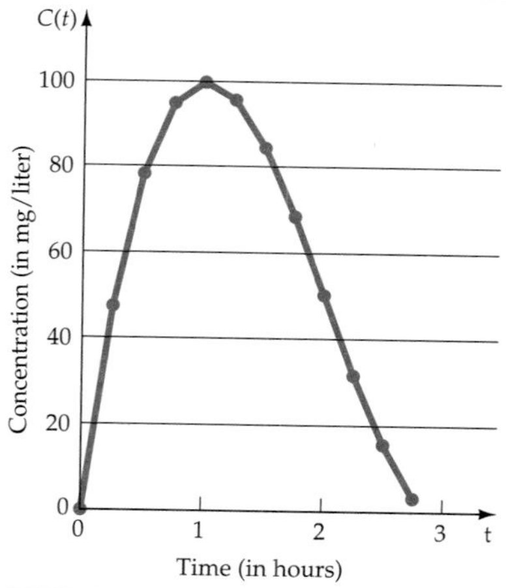

The **average rate of change** of the concentration over the time interval from $t = a$ to $t = a + \Delta t$ is

$$\frac{C(a + \Delta t) - C(a)}{\Delta t}$$

Use the data in the table to evaluate the average rate of change for each of the following time intervals.

a. $[0, 1]$ (*Hint:* In this case, $a = 0$ and $\Delta t = 1$.) Compare this result to the slope of the line through $(0, C(0))$ and $(1, C(1))$.

b. $[0, 0.5]$ c. $[1, 2]$ d. $[1, 1.5]$ e. $[1, 1.25]$

f. The data in the table can be modeled by the function $Con(t) = 25t^3 - 150t^2 + 225t$. Use $Con(t)$ to verify that the average rate of change over $[1, 1 + \Delta t]$ is $-75(\Delta t) + 25(\Delta t)^2$. What does the average rate of change over $[1, 1 + \Delta t]$ seem to approach as Δt approaches 0?

72. **BALL ROLLING ON A RAMP** The distance traveled by a ball rolling down a ramp is given by $s(t) = 6t^2$, where t is the time in seconds after the ball is released, and $s(t)$ is measured in feet. The ball travels 6 feet in 1 second and 24 feet in 2 seconds. Use the difference quotient for average velocity given on page 71 to evaluate the average velocity for each of the following time intervals.

a. $[2, 3]$ (*Hint:* In this case, $a = 2$ and $\Delta t = 1$.) Compare this result to the slope of the line through $(2, s(2))$ and $(3, s(3))$.

b. $[2, 2.5]$ c. $[2, 2.1]$ d. $[2, 2.01]$ e. $[2, 2.001]$

f. Verify that the average velocity over $[2, 2 + \Delta t]$ is $24 + 6(\Delta t)$. What does the average velocity seem to approach as Δt approaches 0?

Connecting Concepts

In Exercises 73 to 76, show that $(f \circ g)(x) = (g \circ f)(x)$.

73. $f(x) = 2x + 3;\ g(x) = 5x + 12$

74. $f(x) = 4x - 2;\ g(x) = 7x - 4$

75. $f(x) = \dfrac{6x}{x - 1};\ g(x) = \dfrac{5x}{x - 2}$

76. $f(x) = \dfrac{5x}{x + 3};\ g(x) = -\dfrac{2x}{x - 4}$

In Exercises 77 to 82, show that

$$(g \circ f)(x) = x \quad \text{and} \quad (f \circ g)(x) = x$$

77. $f(x) = 2x + 3, \quad g(x) = \dfrac{x-3}{2}$

78. $f(x) = 4x - 5, \quad g(x) = \dfrac{x+5}{4}$

79. $f(x) = \dfrac{4}{x+1}, \quad g(x) = \dfrac{4-x}{x}$

80. $f(x) = \dfrac{2}{1-x}, \quad g(x) = \dfrac{x-2}{x}$

81. $f(x) = x^3 - 1, \quad g(x) = \sqrt[3]{x+1}$

82. $f(x) = -x^3 + 2, \quad g(x) = \sqrt[3]{2-x}$

Projects

1. A GRAPHING UTILITY PROJECT For any two different real numbers x and y, the larger of the two numbers is given by

$$\text{Maximum}(x, y) = \frac{x+y}{2} + \frac{|x-y|}{2} \quad (1)$$

a. Verify Equation (1) for $x = 5$ and $y = 9$.

b. Verify Equation (1) for $x = 201$ and $y = 80$.

For any two different function values $f(x)$ and $g(x)$, the larger of the two is given by

$$\text{Maximum}(f(x), g(x)) = \frac{f(x)+g(x)}{2} + \frac{|f(x)-g(x)|}{2} \quad (2)$$

To illustrate how we might make use of Equation (2), consider the functions $y_1 = x^2$ and $y_2 = \sqrt{x}$ on the interval from Xmin $= -1$ to Xmax $= 6$. The graphs of y_1 and y_2 are shown below.

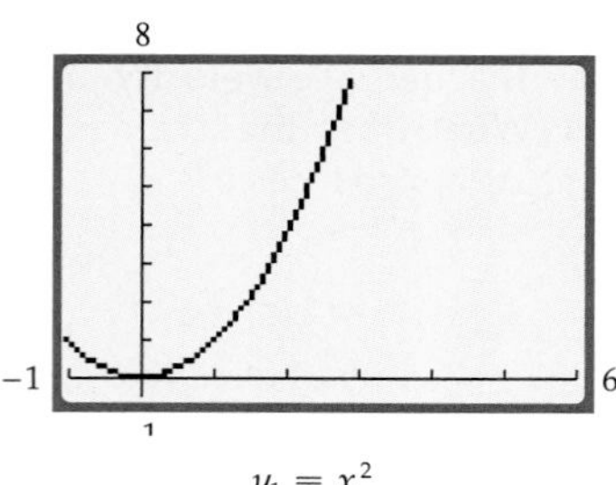

$y_1 = x^2$

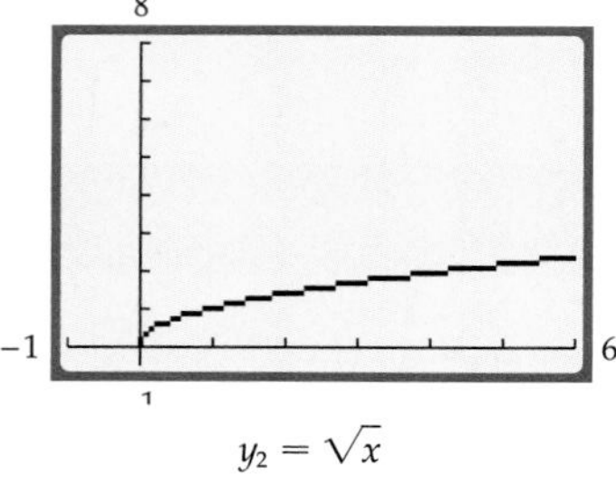

$y_2 = \sqrt{x}$

Now consider the function

$$y_3 = (y_1 + y_2)/2 + (\text{abs}(y_1 - y_2))/2$$

where "abs" represents the absolute value function. The graph of y_3 is shown below.

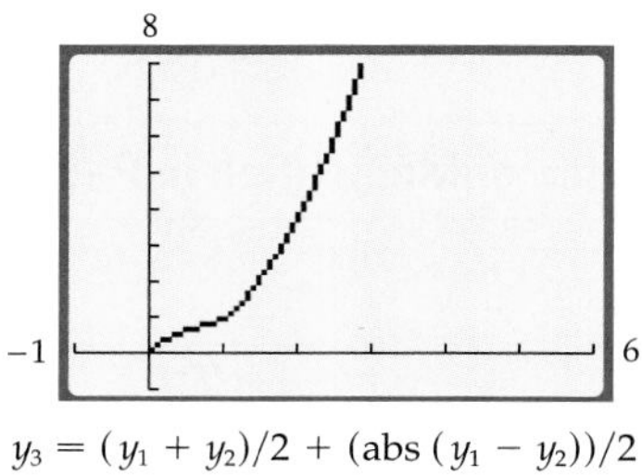

$y_3 = (y_1 + y_2)/2 + (\text{abs}\,(y_1 - y_2))/2$

c. Write a sentence or two that explains why the graph of y_3 is as shown.

d. What is the domain of y_1? of y_2? of y_3? Write a sentence that explains how to determine the domain of y_3, given the domain of y_1 and the domain of y_2.

e. Determine a formula for the function

$$\text{Minimum}(f(x), g(x))$$

2. THE NEVER-NEGATIVE FUNCTION The author J. D. Murray describes a function f_+ that is defined in the following manner.[1]

$$f_+ = \begin{cases} f, & \text{if } f \ge 0 \\ 0, & \text{if } f < 0 \end{cases}$$

We will refer to this function as a **never-negative** function. Never-negative functions can be graphed by using Equation (2) in Project 1. For example, if we let $g(x) = 0$, then Equation (2) simplifies to

$$\text{Maximum}(f(x), 0) = \frac{f(x)}{2} + \frac{|f(x)|}{2} \quad (3)$$

[1] *Mathematical Biology* (New York: Springer-Verlag, 1989), p. 101.

The graph of $y = \text{Maximum}(f(x), 0)$ is the graph of $y = f(x)$ provided that $f(x) \geq 0$, and it is the graph of $y = 0$ provided that $f(x) < 0$.

An Application The mosquito population per acre of a large resort is controlled by spraying on a monthly basis. A biologist has determined that the mosquito population can be approximated by the never-negative function M_+ with

$$M(t) = -35{,}400(t - \text{int}(t))^2 + 35{,}400(t - \text{int}(t)) - 4000$$

Here t represents the month, and $t = 0$ corresponds to June 1, 2004.

a. Use a graphing utility to graph M for $0 \leq t \leq 3$.

b. Use a graphing utility to graph M_+ for $0 \leq t \leq 3$.

c. Write a sentence or two that explains how the graph of M_+ differs from the graph of M.

d. What is the maximum mosquito population per acre for $0 \leq t \leq 3$? When does this maximum mosquito population occur?

e. Explain when would be the best time to visit the resort, provided that you wished to minimize your exposure to mosquitos.

Section 1.6

- Introduction to Inverse Functions
- Graphs of Inverse Functions
- Composition of a Function and Its Inverse
- Find an Inverse Function

Inverse Functions

PREPARE FOR THIS SECTION

Prepare for this section by completing the following exercises. The answers can be found on page A6.

PS1. Solve $2x + 5y = 15$ for y. [1.1]

PS2. Solve $x = \dfrac{y + 1}{y}$ for y. [1.1]

PS3. Given $f(x) = \dfrac{2x^2}{x - 1}$, find $f(-1)$. [1.3]

PS4. If f is a function and $f(3) = 7$, write an ordered pair of the function. [1.3]

PS5. Let $f(x) = 3x - 7$. What is the domain of f? [1.3]

PS6. Let $f(x) = \sqrt{x + 2}$. What is the domain of f? [1.3]

Introduction to Inverse Functions

Consider the "doubling function" $f(x) = 2x$ that doubles every input. Some of the ordered pairs of this function are

$$\left\{(-4, -8), (-1.5, -3), (1, 2), \left(\frac{5}{3}, \frac{10}{3}\right), (7, 14)\right\}$$

Now consider the "halving function" $g(x) = \dfrac{1}{2}x$ that takes one-half of every input. Some of the ordered pairs of this function are

$$\left\{(-8, -4), (-3, -1.5), (2, 1), \left(\frac{10}{3}, \frac{5}{3}\right), (14, 7)\right\}$$

> **take note**
>
> In this section our primary interest concerns finding the inverse of a function; however, we can also find the inverse of a relation. Recall that a relation r is any set of ordered pairs. The inverse of r is the set of ordered pairs formed by reversing the order of the coordinates of the ordered pairs in r.

Observe that the ordered pairs of g are the ordered pairs of f with the order of the coordinates reversed. The following two examples illustrate this concept.

$f(5) = 2(5) = 10$ Ordered pair: (5, 10)

$g(10) = \frac{1}{2}(10) = 5$ Ordered pair: (10, 5)

$f(a) = 2(a) = 2a$ Ordered pair: (a, $2a$)

$g(2a) = \frac{1}{2}(2a) = a$ Ordered pair: ($2a$, a)

The function g is said to be the *inverse function* of f.

Definition of an Inverse Function

If the ordered pairs of a function g are the ordered pairs of a function f with the order of the coordinates reversed, then g is the **inverse function** of f.

Consider a function f and its inverse function g. Because the ordered pairs of g are the ordered pairs of f with the order of the coordinates reversed, the domain of the inverse function g is the range of f, and the range of g is the domain of f.

Not all functions have an inverse that is a function. Consider, for instance, the "square function" $S(x) = x^2$. Some of the ordered pairs of S are

$$\{(-3, 9), (-1, 1), (0, 0), (1, 1), (3, 9), (5, 25)\}$$

If we reverse the coordinates of the ordered pairs, we have

$$\{(9, -3), (1, -1), (0, 0), (1, 1), (9, 3), (25, 5)\}$$

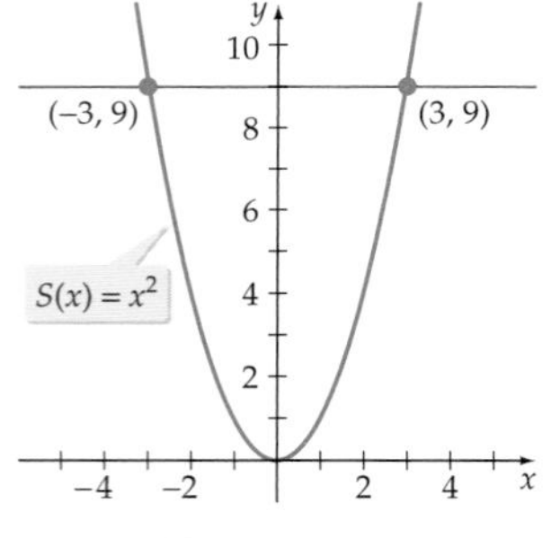

Figure 1.84

This set of ordered pairs is not a function because there are ordered pairs, for instance $(9, -3)$ and $(9, 3)$, with the same first coordinate and different second coordinates. In this case, S has an inverse *relation* but not an inverse *function*.

A graph of S is shown in **Figure 1.84**. Note that $x = -3$ and $x = 3$ produce the same value of y. Thus the graph of S fails the horizontal line test, and therefore S is not a one-to-one function. This observation is used in the following theorem.

Condition for an Inverse Function

A function f has an inverse function if and only if f is a one-to-one function.

Recall that increasing functions and decreasing functions are one-to-one functions. Thus we can state the following theorem.

Alternative Condition for an Inverse Function

If f is an increasing or a decreasing function, then f has an inverse function.

QUESTION Which of the functions graphed below has an inverse function?

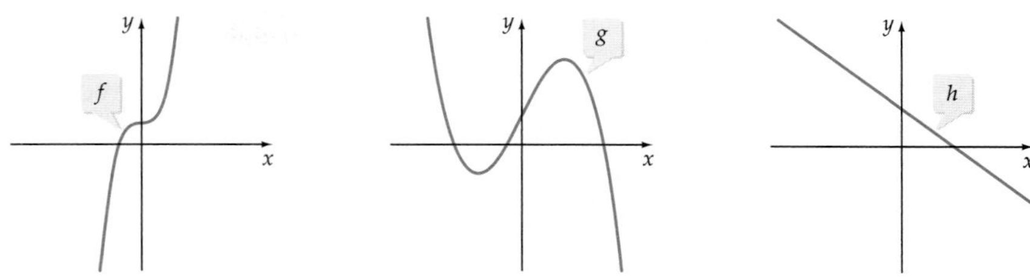

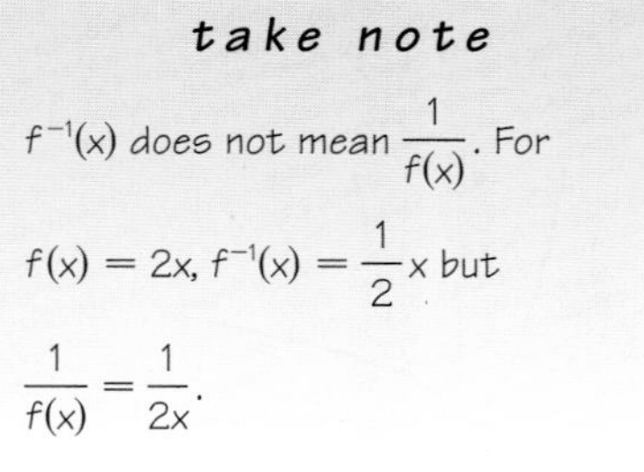

take note

$f^{-1}(x)$ does not mean $\frac{1}{f(x)}$. For $f(x) = 2x$, $f^{-1}(x) = \frac{1}{2}x$ but $\frac{1}{f(x)} = \frac{1}{2x}$.

If a function g is the inverse of a function f, we usually denote the inverse function by f^{-1} rather than g. For the doubling and halving functions f and g discussed on page 81, we write

$$f(x) = 2x \qquad f^{-1}(x) = \frac{1}{2}x$$

Graphs of Inverse Functions

Because the coordinates of the ordered pairs of the inverse of a function f are the ordered pairs of f with the order of the coordinates reversed, we can use them to create a graph of f^{-1}.

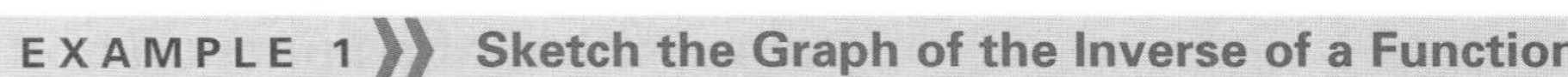

EXAMPLE 1 Sketch the Graph of the Inverse of a Function

Sketch the graph of f^{-1} given that f is the function shown in **Figure 1.85.**

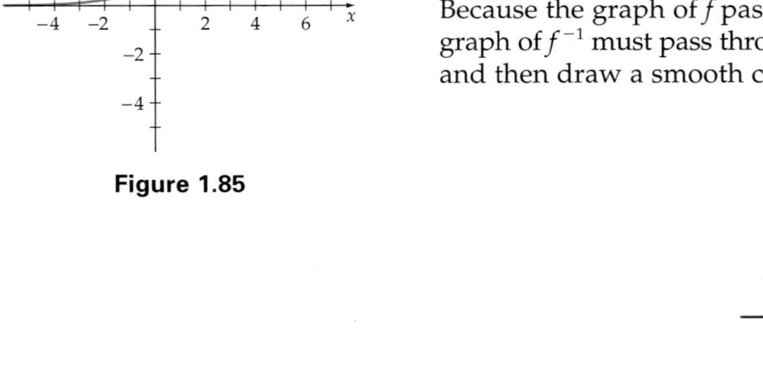

Figure 1.85

Solution

Because the graph of f passes through $(-1, 0.5)$, $(0, 1)$, $(1, 2)$, and $(2, 4)$, the graph of f^{-1} must pass through $(0.5, -1)$, $(1, 0)$, $(2, 1)$, and $(4, 2)$. Plot the points and then draw a smooth curve through the points, as shown in **Figure 1.86.**

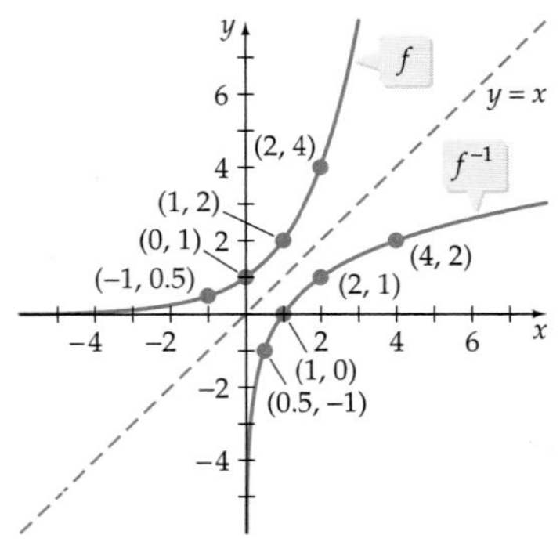

Figure 1.86

Try Exercise 10, page 91

ANSWER The graph of f is the graph of an increasing function. Therefore, f is a one-to-one function and has an inverse function. The graph of h is the graph of a decreasing function. Therefore, h is a one-to-one function and has an inverse function. The graph of g is not the graph of a one-to-one function. g does not have an inverse function.

The graph from the solution to Example 1 is shown again in **Figure 1.87**. Note that the graph of f^{-1} is symmetric to the graph of f with respect to the graph of $y = x$. If the graph were folded along the dashed line, the graph of f would lie on top of the graph of f^{-1}. This is a characteristic of all graphs of functions and their inverses. In **Figure 1.88,** although S does not have an inverse that is a function, the graph of the inverse relation S^{-1} is symmetric to S with respect to the graph of $y = x$.

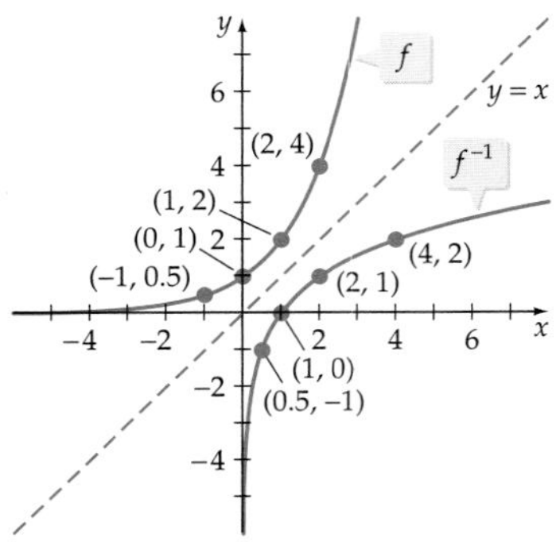

Figure 1.87

Figure 1.88

Composition of a Function and Its Inverse

Observe the effect of forming the composition of $f(x) = 2x$ and $g(x) = \frac{1}{2}x$.

$$f(x) = 2x \qquad\qquad g(x) = \frac{1}{2}x$$

$$f[g(x)] = 2\left[\frac{1}{2}x\right] \quad \text{• Replace } x \text{ by } g(x). \qquad g[f(x)] = \frac{1}{2}[2x] \quad \text{• Replace } x \text{ by } f(x).$$

$$f[g(x)] = x \qquad\qquad g[f(x)] = x$$

This property of the composition of inverse functions always holds true. When taking the composition of inverse functions, the inverse function reverses the effect of the original function. For the two functions above, f doubles a number, and g halves a number. If you double a number and then take one-half of the result, you are back to the original number.

take note

If we think of a function as a machine, then the Composition of Inverse Functions Property can be represented as shown below. Take any input x for f. Use the output of f as the input for f^{-1}. The result is the original input, x.

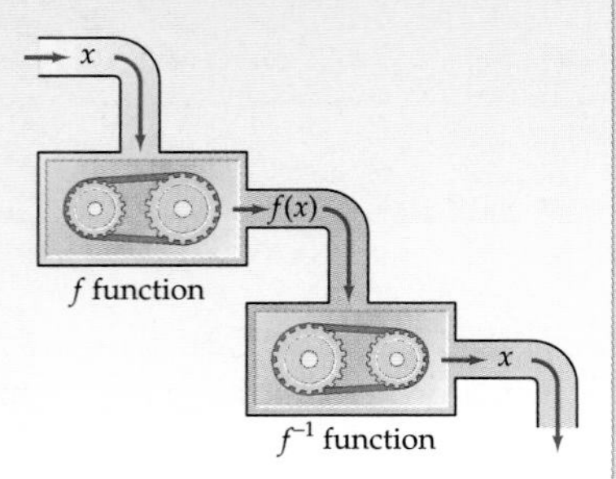

Composition of Inverse Functions Property

If f is a one-to-one function, then f^{-1} is the inverse function of f if and only if

$$(f \circ f^{-1})(x) = f[f^{-1}(x)] = x \quad \text{for all } x \text{ in the domain of } f^{-1}$$

and

$$(f^{-1} \circ f)(x) = f^{-1}[f(x)] = x \quad \text{for all } x \text{ in the domain of } f.$$

EXAMPLE 2 » Use the Composition of Inverse Functions Property

Use composition of functions to show that $f^{-1}(x) = 3x - 6$ is the inverse function of $f(x) = \frac{1}{3}x + 2$.

Solution

We must show that $f[f^{-1}(x)] = x$ and $f^{-1}[f(x)] = x$.

$$f(x) = \frac{1}{3}x + 2 \qquad f^{-1}(x) = 3x - 6$$

$$f[f^{-1}(x)] = \frac{1}{3}[3x - 6] + 2 \qquad f^{-1}[f(x)] = 3\left[\frac{1}{3}x + 2\right] - 6$$

$$f[f^{-1}(x)] = x \qquad f^{-1}[f(x)] = x$$

» *Try Exercise 20, page 91*

Integrating Technology

In the standard viewing window of a calculator, the distance between two tick marks on the x-axis is not equal to the distance between two tick marks on the y-axis. As a result, the graph of $y = x$ does not appear to bisect the first and third quadrants. See **Figure 1.89**. This anomaly is important if a graphing calculator is being used to check whether two functions are inverses of one another. Because the graph of $y = x$ does not appear to bisect the first and third quadrants, the graphs of f and f^{-1} will not appear to be symmetric about the graph of $y = x$. The graphs of $f(x) = \frac{1}{3}x + 2$ and $f^{-1}(x) = 3x - 6$ from Example 2 are shown in **Figure 1.90**. Notice that the graphs do not appear to be quite symmetric about the graph of $y = x$.

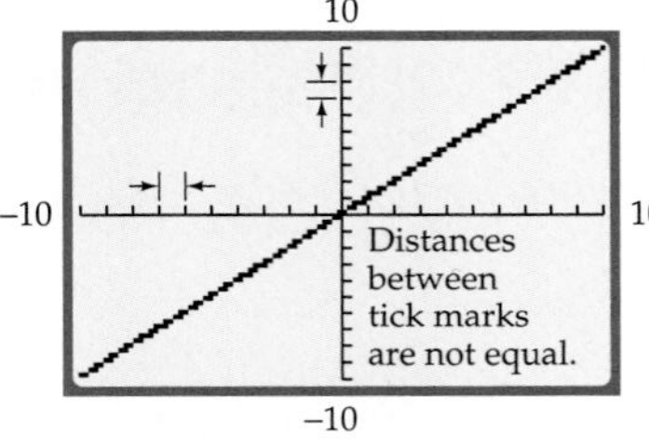

$y = x$ in the standard viewing window

Figure 1.89

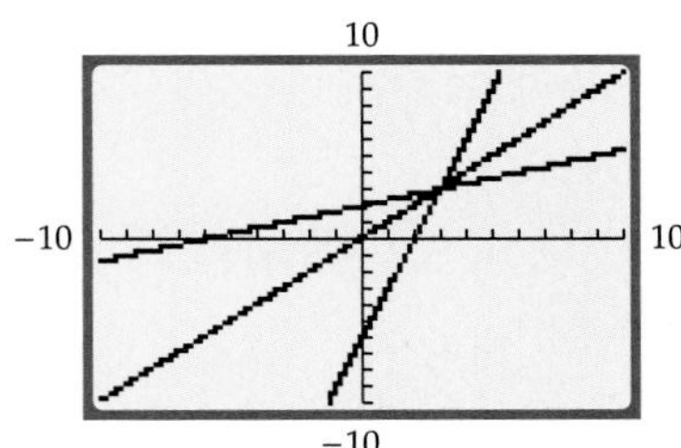

f, f^{-1}, and $y = x$ in the standard viewing window

Figure 1.90

Continued ►

To get a better view of a function and its inverse, it is necessary to use the SQUARE viewing window, as in **Figure 1.91**. In this window, the distance between two tick marks on the x-axis is equal to the distance between two tick marks on the y-axis.

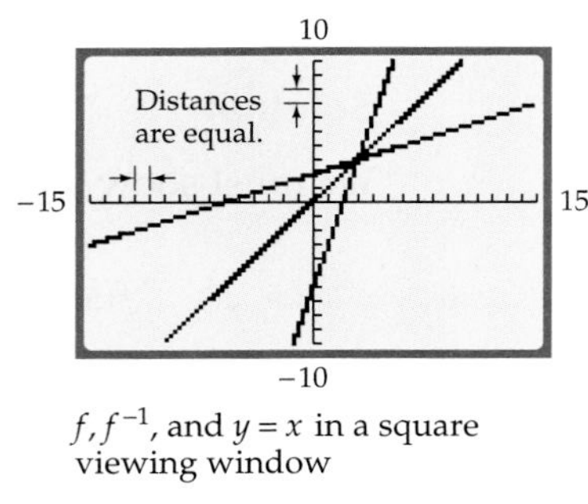

f, f^{-1}, and $y = x$ in a square viewing window

Figure 1.91

Find an Inverse Function

If a one-to-one function f is defined by an equation, then we can use the following method to find the equation for f^{-1}.

> **take note**
>
> If the ordered pairs of f are given by (x, y), then the ordered pairs of f^{-1} are given by (y, x). That is, x and y are interchanged. This is the reason for Step 2 at the right.

Steps for Finding the Inverse of a Function

To find the equation of the inverse f^{-1} of the one-to-one function f:

1. Substitute y for $f(x)$.
2. Interchange x and y.
3. Solve, if possible, for y in terms of x.
4. Substitute $f^{-1}(x)$ for y.

EXAMPLE 3 ⟫ Find the Inverse of a Function

Find the inverse of $f(x) = 3x + 8$.

Solution

$$f(x) = 3x + 8$$

$$y = 3x + 8 \qquad \text{• Replace } f(x) \text{ by } y.$$

$$x = 3y + 8 \qquad \text{• Interchange } x \text{ and } y.$$

$$x - 8 = 3y \qquad \text{• Solve for } y.$$

$$\frac{x-8}{3} = y$$

$$\frac{1}{3}x - \frac{8}{3} = f^{-1}(x) \qquad \text{• Replace } y \text{ by } f^{-1}.$$

The inverse function is given by $f^{-1}(x) = \frac{1}{3}x - \frac{8}{3}$.

Try Exercise 32, page 91

In the next example we find the inverse of a rational function.

EXAMPLE 4 Find the Inverse of a Function

Find the inverse of $f(x) = \frac{2x+1}{x}$, $x \neq 0$.

Solution

$$f(x) = \frac{2x+1}{x}$$

$$y = \frac{2x+1}{x} \qquad \text{• Replace } f(x) \text{ by } y.$$

$$x = \frac{2y+1}{y} \qquad \text{• Interchange } x \text{ and } y.$$

$$xy = 2y + 1 \qquad \text{• Solve for } y.$$

$$xy - 2y = 1$$

$$y(x-2) = 1 \qquad \text{• Factor the left side.}$$

$$y = \frac{1}{x-2}$$

$$f^{-1}(x) = \frac{1}{x-2}, x \neq 2 \qquad \text{• Replace } y \text{ by } f^{-1}.$$

Try Exercise 38, page 92

QUESTION If f is a one-to-one function and $f(4) = 5$, what is $f^{-1}(5)$?

The graph of $f(x) = x^2 + 4x + 3$ is shown in **Figure 1.92a**. The function f is not a one-to-one function and therefore does not have an inverse function. However,

---

ANSWER Because (4, 5) is an ordered pair of f, (5, 4) must be an ordered pair of f^{-1}. Therefore, $f^{-1}(5) = 4$.

the function given by $G(x) = x^2 + 4x + 3$, shown in **Figure 1.92b**, for which the domain is restricted to $\{x|x \geq -2\}$, is a one-to-one function and has an inverse function G^{-1}. This is shown in Example 5.

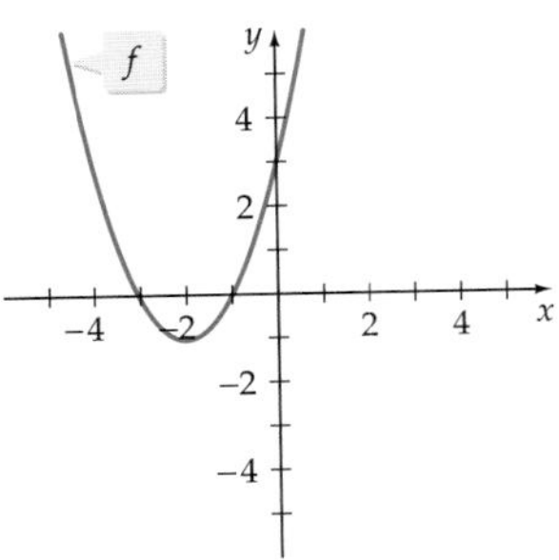

Figure 1.92a

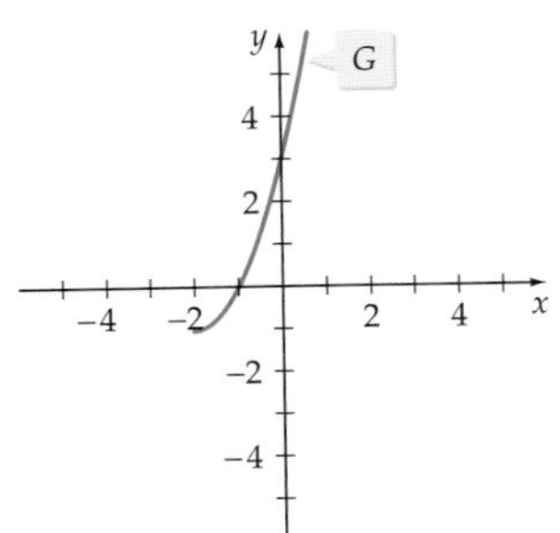

Figure 1.92b

EXAMPLE 5 » Find the Inverse of a Function with a Restricted Domain

Find the inverse of $G(x) = x^2 + 4x + 3$, where the domain of G is $\{x|x \geq -2\}$.

take note

Recall that the range of a function f is the domain of f^{-1}, and the domain of f is the range of f^{-1}.

Solution

$$G(x) = x^2 + 4x + 3$$

$$y = x^2 + 4x + 3$$ • Replace $G(x)$ by y.

$$x = y^2 + 4y + 3$$ • Interchange x and y.

$$x = (y^2 + 4y + 4) - 4 + 3$$ • Solve for y by completing the square of $y^2 + 4y$.

$$x = (y + 2)^2 - 1$$ • Factor.

$$x + 1 = (y + 2)^2$$ • Add 1 to each side of the equation.

$$\sqrt{x + 1} = \sqrt{(y + 2)^2}$$ • Take the square root of each side of the equation.

$$\pm\sqrt{x + 1} = y + 2$$ • Recall that if $a^2 = b$, then $a = \pm\sqrt{b}$.

$$\pm\sqrt{x + 1} - 2 = y$$

Because the domain of G is $\{x|x \geq -2\}$, the range of G^{-1} is $\{y|y \geq -2\}$. This means that we must choose the positive value of $\pm\sqrt{x + 1}$. Thus $G^{-1}(x) = \sqrt{x + 1} - 2$. See **Figure 1.93**.

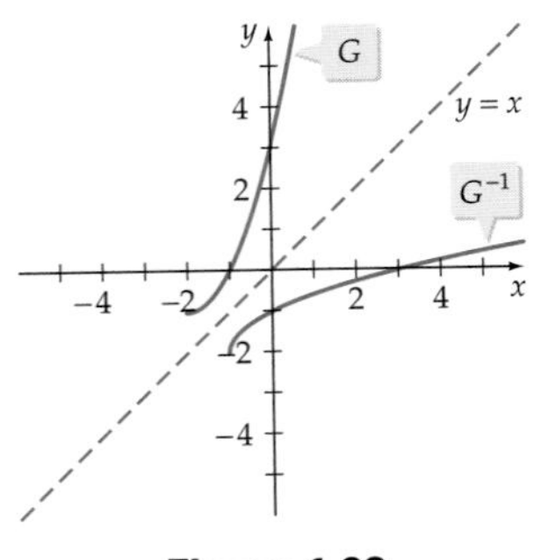

Figure 1.93

» *Try Exercise 44, page 92*

In Example 6b, we use an inverse function to determine the wholesale price of a gold bracelet for which we know the retail price.

EXAMPLE 6 Solve an Application

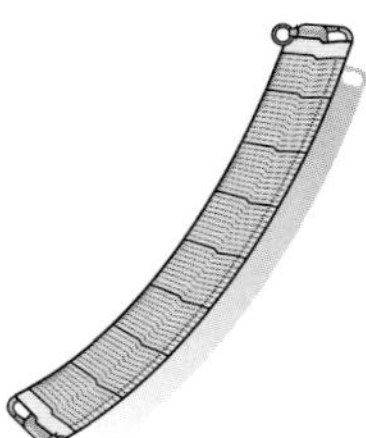

A merchant uses the function

$$S(x) = \frac{4}{3}x + 100$$

to determine the retail selling price S, in dollars, of a gold bracelet for which she has paid a wholesale price of x dollars.

a. The merchant paid a wholesale price of \$672 for a gold bracelet. Use S to determine the retail selling price of this bracelet.

b. Find S^{-1} and use it to determine the merchant's wholesale price for a gold bracelet that retails at \$1596.

Solution

a. $S(672) = \frac{4}{3}(672) + 100 = 896 + 100 = 996$

The merchant charges \$996 for a bracelet that has a wholesale price of \$672.

b. To find S^{-1}, begin by substituting y for $S(x)$.

$$S(x) = \frac{4}{3}x + 100$$

$$y = \frac{4}{3}x + 100 \quad \text{• Replace } S(x) \text{ with } y.$$

$$x = \frac{4}{3}y + 100 \quad \text{• Interchange } x \text{ and } y.$$

$$x - 100 = \frac{4}{3}y \quad \text{• Solve for } y.$$

$$\frac{3}{4}(x - 100) = y$$

$$\frac{3}{4}x - 75 = y$$

Using inverse notation, the above equation can be written as

$$S^{-1}(x) = \frac{3}{4}x - 75$$

Substitute 1596 for x to determine the wholesale price.

$$S^{-1}(1596) = \frac{3}{4}(1596) - 75$$
$$= 1197 - 75$$
$$= 1122$$

A gold bracelet that the merchant retails at \$1596 has a wholesale price of \$1122.

Try Exercise 54, page 92

Integrating Technology

Some graphing utilities can be used to draw the graph of the inverse of a function without the user having to find the inverse function. For instance, **Figure 1.94** shows the graph of $f(x) = 0.1x^3 - 4$. The graphs of f and f^{-1} are both shown in **Figure 1.95,** along with the graph of $y = x$. Note that the graph of f^{-1} is the reflection of the graph of f with respect to the graph of $y = x$. The display shown in **Figure 1.95** was produced on a TI-83/TI-83 Plus/TI-84 Plus graphing calculator by using the DrawInv command, which is in the DRAW menu.

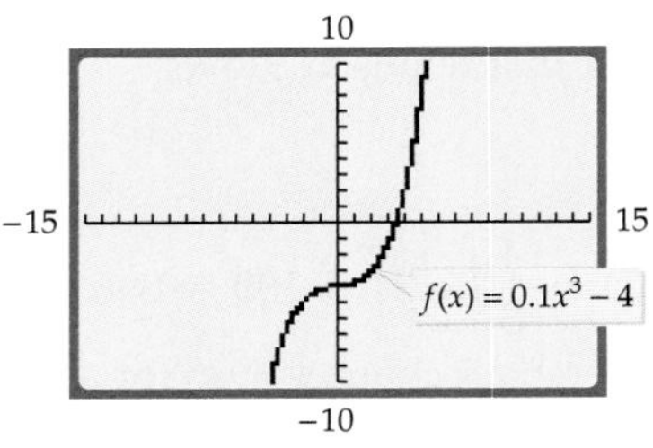

Figure 1.94

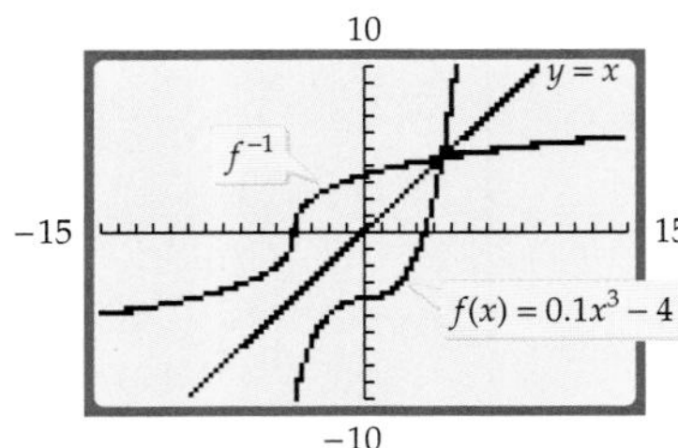

Figure 1.95

Topics for Discussion

1. If $f(x) = 3x + 1$, what are the values of $f^{-1}(2)$ and $[f(2)]^{-1}$?
2. How are the domain and range of a one-to-one function f related to the domain and range of the inverse function of f?
3. How is the graph of the inverse of a function f related to the graph of f?
4. The function $f(x) = -x$ is its own inverse. Find two other functions that are their own inverses.
5. What are the steps in finding the inverse of a one-to-one function?

Exercise Set 1.6

In Exercises 1 to 4, assume that the given function has an inverse function.

1. Given $f(3) = 7$, find $f^{-1}(7)$.

2. Given $g(-3) = 5$, find $g^{-1}(5)$.

3. Given $h^{-1}(-3) = -4$, find $h(-4)$.

4. Given $f^{-1}(7) = 0$, find $f(0)$.

5. If 3 is in the domain of f^{-1}, find $f[f^{-1}(3)]$.

6. If f is a one-to-one function and $f(0) = 5$, $f(1) = 2$, and $f(2) = 7$, find:

a. $f^{-1}(5)$ **b.** $f^{-1}(2)$

7. The domain of the inverse function f^{-1} is the ________ of f.

8. The range of the inverse function f^{-1} is the ________ of f.

In Exercises 9 to 16, draw the graph of the inverse relation. Is the inverse relation a function?

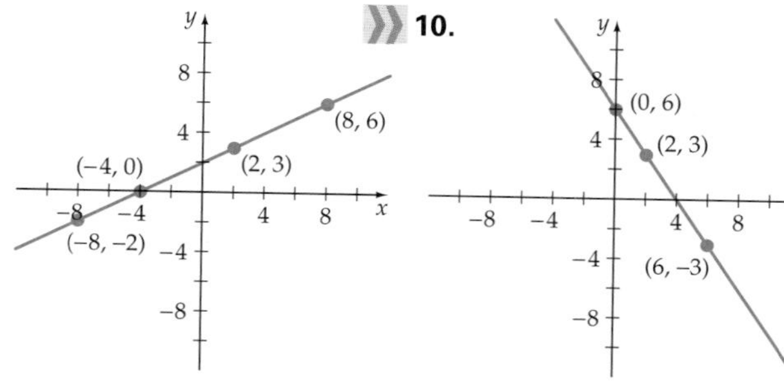

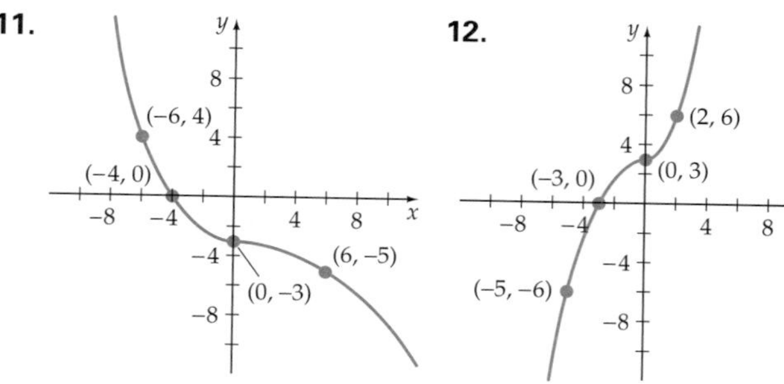

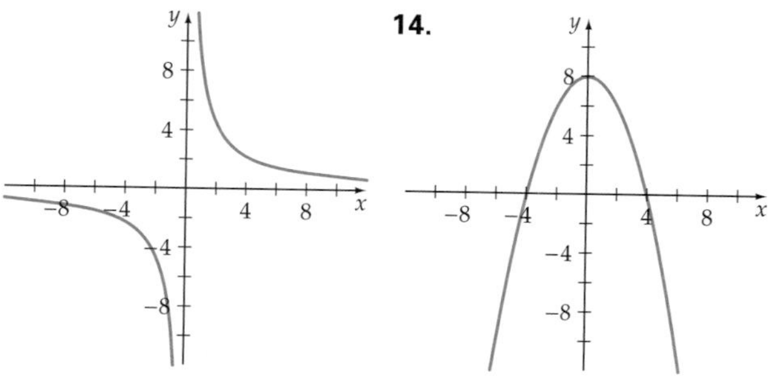

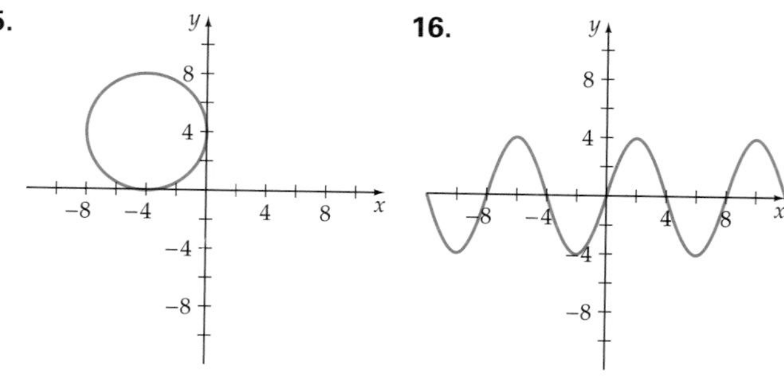

In Exercises 17 to 26, use composition of functions to determine whether *f* and *g* are inverses of one another.

17. $f(x) = 4x;\ g(x) = \dfrac{x}{4}$

18. $f(x) = 3x;\ g(x) = \dfrac{1}{3x}$

19. $f(x) = 4x - 1;\ g(x) = \dfrac{1}{4}x + \dfrac{1}{4}$

20. $f(x) = \dfrac{1}{2}x - \dfrac{3}{2};\ g(x) = 2x + 3$

21. $f(x) = -\dfrac{1}{2}x - \dfrac{1}{2};\ g(x) = -2x + 1$

22. $f(x) = 3x + 2;\ g(x) = \dfrac{1}{3}x - \dfrac{2}{3}$

23. $f(x) = \dfrac{5}{x - 3};\ g(x) = \dfrac{5}{x} + 3$

24. $f(x) = \dfrac{2x}{x - 1};\ g(x) = \dfrac{x}{x - 2}$

25. $f(x) = x^3 + 2;\ g(x) = \sqrt[3]{x - 2}$

26. $f(x) = (x + 5)^3;\ g(x) = \sqrt[3]{x} - 5$

In Exercises 27 to 30, find the inverse of the function. If the function does not have an inverse function, write "no inverse function."

27. $\{(-3, 1), (-2, 2), (1, 5), (4, -7)\}$

28. $\{(-5, 4), (-2, 3), (0, 1), (3, 2), (7, 11)\}$

29. $\{(0, 1), (1, 2), (2, 4), (3, 8), (4, 16)\}$

30. $\{(1, 0), (10, 1), (100, 2), (1000, 3), (10{,}000, 4)\}$

In Exercises 31 to 48, find $f^{-1}(x)$. State any restrictions on the domain of $f^{-1}(x)$.

31. $f(x) = 2x + 4$

32. $f(x) = 4x - 8$

33. $f(x) = 3x - 7$

34. $f(x) = -3x - 8$

35. $f(x) = -2x + 5$

36. $f(x) = -x + 3$

37. $f(x) = \frac{2x}{x - 1}, \quad x \neq 1$

38. $f(x) = \frac{x}{x - 2}, \quad x \neq 2$

39. $f(x) = \frac{x - 1}{x + 1}, \quad x \neq -1$

40. $f(x) = \frac{2x - 1}{x + 3}, \quad x \neq -3$

41. $f(x) = x^2 + 1, \quad x \geq 0$

42. $f(x) = x^2 - 4, \quad x \geq 0$

43. $f(x) = \sqrt{x - 2}, \quad x \geq 2$

44. $f(x) = \sqrt{4 - x}, \quad x \leq 4$

45. $f(x) = x^2 + 4x, \quad x \geq -2$

46. $f(x) = x^2 - 6x, \quad x \leq 3$

47. $f(x) = x^2 + 4x - 1, \quad x \leq -2$

48. $f(x) = x^2 - 6x + 1, \quad x \geq 3$

49. **Geometry** The volume of a cube is given by $V(x) = x^3$, where x is the measure of the length of a side of the cube. Find $V^{-1}(x)$ and explain what it represents.

50. **Unit Conversions** The function $f(x) = 12x$ converts feet, x, into inches, $f(x)$. Find $f^{-1}(x)$ and explain what it determines.

51. **Fahrenheit to Celsius** The function

$$f(x) = \frac{5}{9}(x - 32)$$

is used to convert x degrees Fahrenheit to an equivalent Celsius temperature. Find f^{-1} and explain how it is used.

52. **Retail Sales** A clothing merchant uses the function

$$S(x) = \frac{3}{2}x + 18$$

to determine the retail selling price S, in dollars, of a winter coat for which she has paid a wholesale price of x dollars.

a. The merchant paid a wholesale price of \$96 for the winter coat. Use S to determine the retail selling price she will charge for this coat.

b. Find S^{-1} and use it to determine the merchant's wholesale price for a coat that retails at \$399.

53. **Fashion** The function $s(x) = 2x + 24$ can be used to convert a U.S. women's shoe size into an Italian women's shoe size. Determine the function $s^{-1}(x)$ that can be used to convert an Italian women's shoe size to its equivalent U.S. shoe size.

54. **Fashion** The function $K(x) = 1.3x - 4.7$ converts a men's shoe size in the United States to the equivalent shoe size in the United Kingdom. Determine the function $K^{-1}(x)$ that can be used to convert a United Kingdom men's shoe size to its equivalent U.S. shoe size.

55. **Compensation** The monthly earnings $E(s)$, in dollars, of a software sales executive is given by $E(s) = 0.05s + 2500$, where s is the value, in dollars, of the software sold by the executive during the month. Find $E^{-1}(s)$ and explain how the executive could use this function.

56. **Postage** Does the first-class postage rate function given below have an inverse function? Explain your answer.

Weight (in ounces)	Cost
$0 < w \leq 1$	\$.39
$1 < w \leq 2$	\$.63
$2 < w \leq 3$	\$.87
$3 < w \leq 4$	\$1.11

57. **The Birthday Problem** A famous problem called the *birthday problem* goes like this: Suppose there is a randomly selected group of n people in a room. What is the probability that at least two of the people have a birthday on the same day of the year? It may surprise you that for a group of 23 people, the probability that at least two of the people share a birthday is about 50.7%. The following graph can be used to estimate shared birthday probabilities for $1 \leq n \leq 60$.

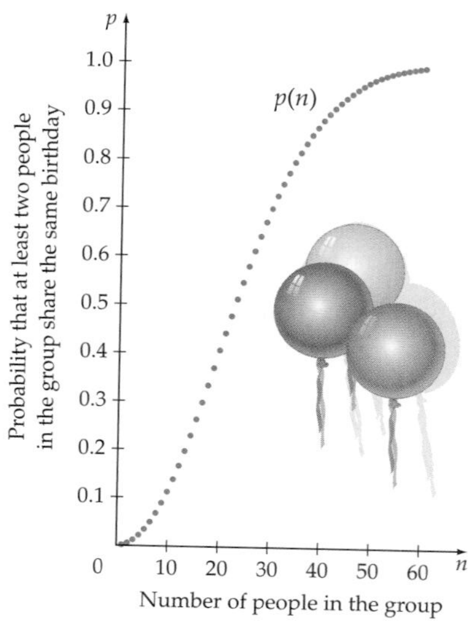

a. Use the graph of p to estimate $p(10)$ and $p(30)$.

b. Consider the function p with $1 \leq n \leq 60$, as shown in the graph. Explain how you can tell that p has an inverse that is a function.

c. Write a sentence that explains the meaning of $p^{-1}(0.223)$ in the context of this application.

58. Medication Level The function L shown in the following graph models the level of pseudoephedrine hydrochloride, in milligrams, in the bloodstream of a patient t hours after 30 milligrams of the medication have been administered.

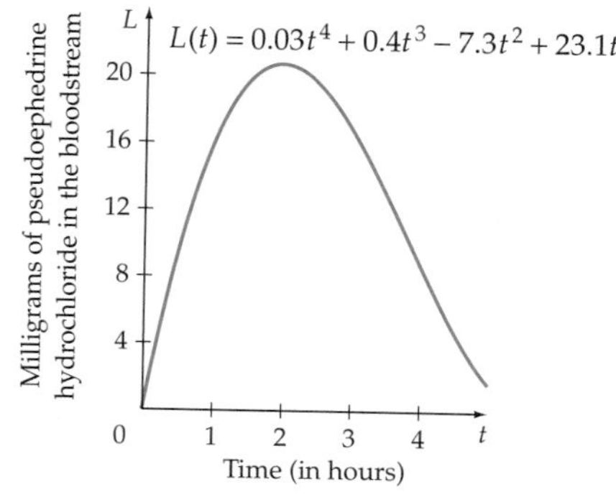

a. Use the graph of L to estimate two different values of t for which the pseudoephedrine hydrochloride levels are the same.

b. Does L have an inverse that is a function? Explain.

59. Cryptology Cryptology is the study of making and breaking secret codes. Secret codes are often used to send messages over the Internet. By devising a code that is difficult to break, the sender hopes to prevent the messages from being read by an unauthorized person.

In practice, very complicated one-to-one functions and their inverses are used to encode and decode messages. The following procedure uses the simple function $f(x) = 2x - 1$ to illustrate the basic concepts that are involved.

Assign to each letter of the alphabet, and a blank space, a two-digit numerical value, as shown below.

A	10	H	17	O	24	V	31
B	11	I	18	P	25	W	32
C	12	J	19	Q	26	X	33
D	13	K	20	R	27	Y	34
E	14	L	21	S	28	Z	35
F	15	M	22	T	29		36
G	16	N	23	U	30		

Note: **A blank space is represented by the numerical value 36.**

Using these numerical values, the message MEET YOU AT NOON would be represented by

22 14 14 29 36 34 24 30 36 10 29 36 23 24 24 23

Let $f(x) = 2x - 1$ define a coding function. The above message can be encoded by finding $f(22)$, $f(14)$, $f(14)$, $f(29)$, $f(36)$, $f(34)$, $f(24)$, $\ldots$, $f(23)$, which yields

43 27 27 57 71 67 47 59 71 19 57 71 45 47 47 45

The inverse of f, which is

$$f^{-1}(x) = \frac{x+1}{2}$$

is used by the receiver of the message to decode the message. For instance,

$$f^{-1}(43) = \frac{43+1}{2} = 22$$

which represents M, and

$$f^{-1}(27) = \frac{27+1}{2} = 14$$

which represents E.

a. Use the above coding procedure to encode the message DO YOUR HOMEWORK.

b. Use $f^{-1}(x)$ as defined above to decode the message 49 33 47 45 27 71 33 47 43 27.

c. Explain why it is important to use a one-to-one function to encode a message.

60. CRYPTOGRAPHY A friend is using the letter-number correspondence in Exercise 59 and the coding function $g(x) = 2x + 3$. Your friend sends you the coded message

59 31 39 73 31 75 61 37 31 75 29 23 71

Use $g^{-1}(x)$ to decode this message.

In Exercises 61 to 66, answer the question without finding the equation of the linear function.

61. Suppose that f is a linear function, $f(2) = 7$, and $f(5) = 12$. If $f(4) = c$, then is c less than 7, between 7 and 12, or greater than 12? Explain your answer.

62. Suppose that f is a linear function, $f(1) = 13$, and $f(4) = 9$. If $f(3) = c$, then is c less than 9, between 9 and 13, or greater than 13? Explain your answer.

63. Suppose that f is a linear function, $f(2) = 3$, and $f(5) = 9$. Between which two numbers is $f^{-1}(6)$?

64. Suppose that f is a linear function, $f(5) = -1$, and $f(9) = -3$. Between which two numbers is $f^{-1}(-2)$?

65. Suppose that g is a linear function, $g^{-1}(3) = 4$, and $g^{-1}(7) = 8$. Between which two numbers is $g(5)$?

66. Suppose that g is a linear function, $g^{-1}(-2) = 5$, and $g^{-1}(0) = -3$. Between which two numbers is $g(0)$?

Connecting Concepts

67. Consider the linear function $f(x) = mx + b$, $m \neq 0$. The graph of f has a slope of m and a y-intercept of $(0, b)$. What are the slope and y-intercept of the graph of f^{-1}?

68. Find the inverse of $f(x) = ax^2 + bx + c$, $a \neq 0$, $x \geq -\dfrac{b}{2a}$.

69. Use a graph of $f(x) = -x + 3$ to explain why f is its own inverse.

70. Use a graph of $f(x) = \sqrt{16 - x^2}$, with $0 \leq x \leq 4$, to explain why f is its own inverse.

Only one-to-one functions have inverses that are functions. In Exercises 71 to 74, determine if the given function is a one-to-one function.

71. $p(t) = \sqrt{9 - t}$

72. $v(t) = \sqrt{16 + t}$

73. $F(x) = |x| + x$

74. $T(x) = |x^2 - 6|$, $x \geq 0$

Projects

1. INTERSECTION POINTS FOR THE GRAPHS OF f AND f^{-1}. For each of the following, graph f and its inverse.

i. $f(x) = 2x - 4$

ii. $f(x) = -x + 2$

iii. $f(x) = x^3 + 1$

iv. $f(x) = x - 3$

v. $f(x) = -3x + 2$

vi. $f(x) = \dfrac{1}{x}$

a. Do the graphs of a function and its inverse always intersect?

b. If the graphs of a function and its inverse intersect at one point, what is true about the coordinates of the point of intersection?

c. Can the graphs of a function and its inverse intersect at more than one point?

Section 1.7

- Linear Regression Models
- Correlation Coefficient
- Quadratic Regression Models

Modeling Data Using Regression

PREPARE FOR THIS SECTION

Prepare for this section by completing the following exercises. The answers can be found on page A6.

PS1. Find the slope and y-intercept of the graph of $y = -\frac{3}{4}x + 4$.

PS2. Find the slope and y-intercept of the graph of $3x - 4y = 12$.

PS3. Find an equation of the line that has slope -0.45 and y-intercept $(0, 2.3)$.

PS4. If $f(x) = 3x^2 + 4x - 1$, find $f(2)$. [1.3]

PS5. Find the distance between $P_1(3, 5)$ and $P_2(3, f(3))$, where $f(x) = 2x + 5$. [1.2/1.3]

PS6. You are given $P_1(2, -1)$ and $P_2(4, 14)$. If $f(x) = x^2 - 3$, find $|f(x_1) - y_1| + |f(x_2) - y_2|$. [1.3]

Linear Regression Models

The data in the table below show the population of selected states and the number of professional sports teams (Major League Baseball, National Football League, National Basketball Association, Women's National Basketball Association, National Hockey League) in those states. A scatter diagram of the data is shown in **Figure 1.96** on page 96.

Number of Professional Sports Teams for Selected States

State	Populations (in millions)	Number of Teams	State	Populations (in millions)	Number of Teams
Arizona	5.9	5	Minnesota	5.1	5
California	36.1	17	New Jersey	8.7	3
Colorado	4.7	4	New York	19.3	10
Florida	17.8	11	North Carolina	9.7	3
Illinois	12.8	5	Pennsylvania	12.4	7
Indiana	6.3	3	Texas	22.9	9
Michigan	10.1	5	Wisconsin	5.5	3

Although there is no one line that passes through every point, we could find an approximate linear model for these data. For instance, the line shown in **Figure 1.97** in blue approximates the data better than the line shown in red. However, as **Figure 1.98** shows, there are many other lines we could draw that seem to approximate the data.

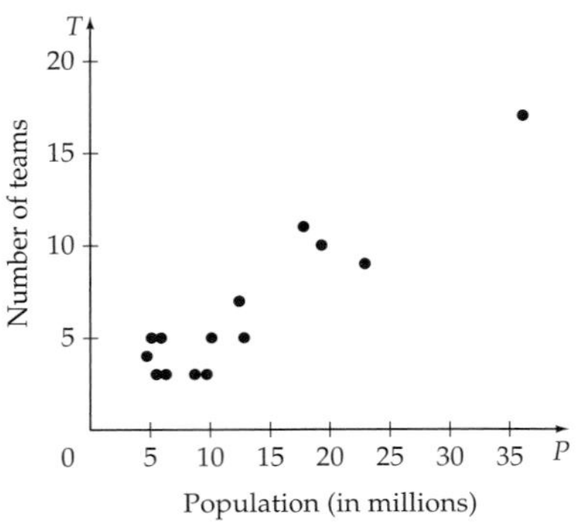

Figure 1.96

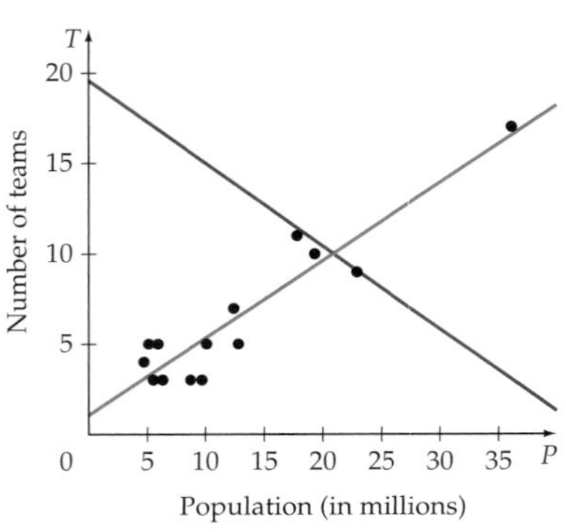

Figure 1.97

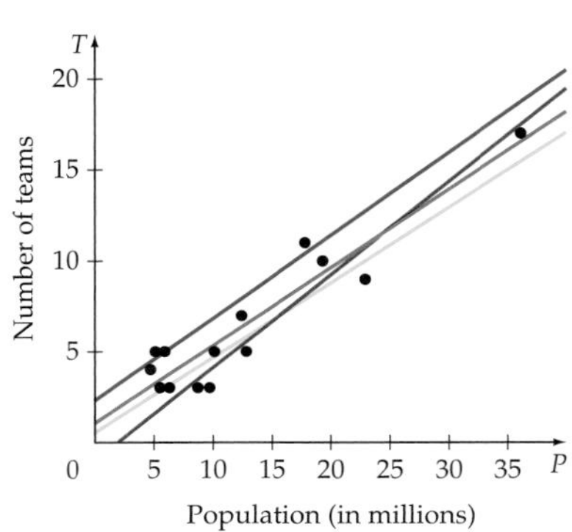

Figure 1.98

To find the line that "best" approximates the data, **regression analysis** is used. This type of analysis produces the linear function whose graph is called the **line of best fit** or the **least-squares regression line.**[2]

Definition of the Least-Squares Regression Line

The **least-squares regression line** is the line that minimizes the sum of the squares of the vertical deviations of all data points from the line.

To help understand this definition, consider the data set

$$S = \{(1, 2), (2, 3), (3, 3), (4, 4), (5, 7)\}$$

as shown in **Figure 1.99.** As we will show later, the least-squares line for this data set is $y = 1.1x + 0.5$, also shown in **Figure 1.99.** If we evaluate this linear function at the x-coordinates of the data set S, we obtain the set of ordered pairs $T = \{(1, 1.6), (2, 2.7), (3, 3.8), (4, 4.9), (5, 6)\}$. The vertical deviations are the differences between the y-coordinates in S and the y-coordinates in T. From the definition, we must calculate the sum of the squares of these deviations.

$$(2 - 1.6)^2 + (3 - 2.7)^2 + (3 - 3.8)^2 + (4 - 4.9)^2 + (7 - 6)^2 = 2.7$$

Because $y = 1.1x + 0.5$ is the least squares regression line, for no other line is the sum of the squares of the deviations less than 2.7. For instance, if we consider the equation $y = 1.25x + 0.75$, which is the equation of the line through the two

[2]The least-squares regression line is also called the *least-squares line* and the *regression line.*

points $P_1(1, 2)$ and $P_2(5, 7)$ of the data set, the sum of the squared deviations is larger than 2.7. See **Figure 1.100.**

$$(2 - 2)^2 + (3 - 3.25)^2 + (3 - 4.5)^2 + (4 - 5.75)^2 + (7 - 7)^2 = 5.375$$

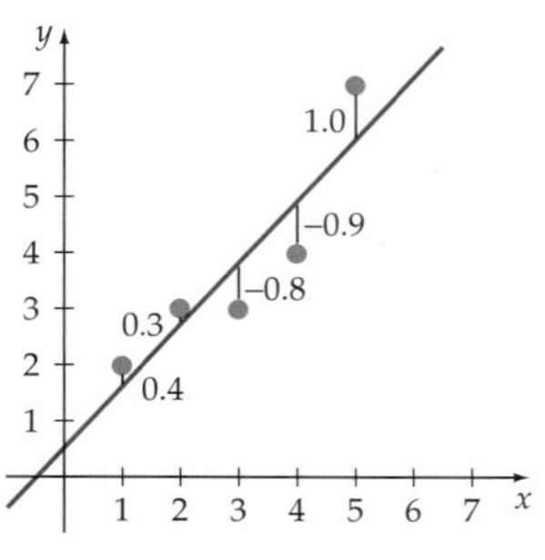

Figure 1.99

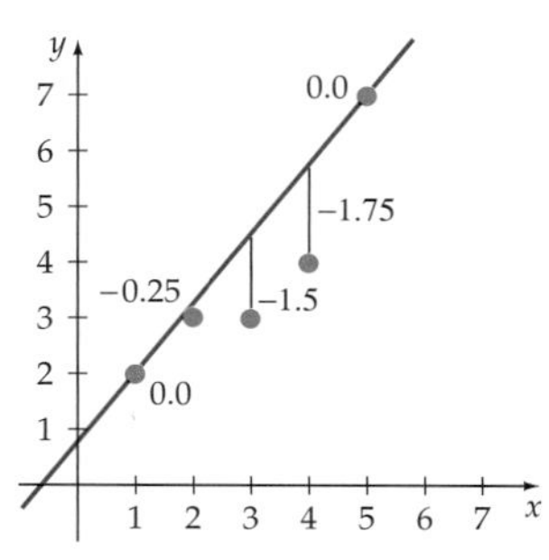

Figure 1.100

Integrating Technology

The equations used to calculate a regression line are somewhat cumbersome. Fortunately, these equations are preprogrammed into most graphing calculators. We will now illustrate the technique for a TI-83/TI-83 Plus/TI-84 Plus calculator using data set S given on page 96.

Press STAT. Select EDIT.
Press ENTER.

```
EDIT CALC TESTS
1:Edit...
2:SortA(
3:SortD(
4:ClrList
5:SetUpEditor
```

If necessary, delete any data in L1 and L2.
Enter the given data.

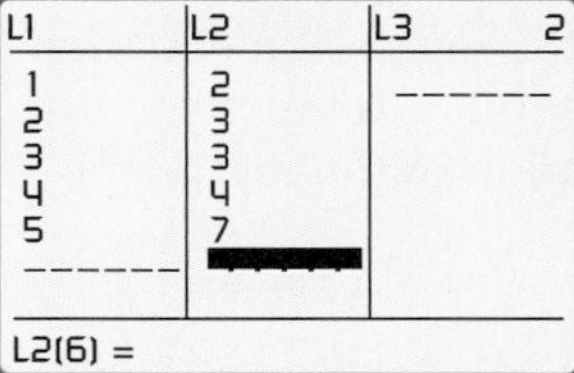

Press STAT. Select CALC.
Select 4, LinReg(ax+b).
Press ENTER.

```
EDIT CALC TESTS
1:1-Var Stats
2:1-Var Stats
3:Med-Med
4:LinReg(ax+b)
5:QuadReg
6:CubicReg
7↓QuartReg
```

Press VARS.

```
LinReg(ax+b)
```

Select Y-VARS.
Press ENTER.

```
VARS Y-VARS
1:Function...
2:Parametric...
3:Polar...
4:On/Off...
```

Select 1. Press ENTER.

```
FUNCTION
1:Y1
2:Y2
3:Y3
4:Y4
5:Y5
6:Y6
7↓Y7
```

Press ENTER.

```
LinReg(ax+b) Y1
```

View the results.

```
LinReg
 y=ax+b
 a=1.1
 b=.5
 r²=.8175675676
 r=.9041944302
```

Continued ►

From the last screen, the equation of the regression line is $y = 1.1x + 0.5$. Your last screen may not look exactly like ours. The information provided on our screen requires that DiagnosticOn be enabled. This is accomplished using the following keystrokes:

[2ND] CATALOG (Scroll to DiagnosticOn) [ENTER] [ENTER]

With DiagnosticOn enabled, in addition to the values for the regression equation, two other values, r^2 and r, are given. We will discuss these values later in this section.

If you used the keystrokes we have shown here, the regression line will be stored in Y1. This is helpful if you wish to graph the regression line. However, if it is not necessary to graph the regression line, then instead of pressing [VARS] at step 4, just press [ENTER]. The result will be the last screen showing the results of the regression calculations.

EXAMPLE 1 ›› Find a Regression Equation

Find the regression equation for the data on the population of a state and the number of professional sports teams in that state. How many sports teams are predicted for Indiana, whose population is approximately 6.3 million? Round to the nearest whole number.

Solution

Using your calculator, enter the data from the table on page 95. Then have the calculator produce the values for the regression equation. Your results should be similar to those shown in **Figure 1.101.** The equation of the regression line is

```
LinReg
 y=ax+b
 a=.4285944776
 b=1.000728509
 r²=.8725655709
 r=.9341121833
```

Figure 1.101

$$y = 0.4285944776x + 1.000728509$$

To find the number of sports teams the regression equation predicts for Indiana, evaluate the regression equation for $x = 6.3$.

$$\begin{aligned} y &= 0.4285944776x + 1.000728509 \\ &= 0.4285944776(6.3) + 1.000728509 \\ &\approx 3.7008737 \end{aligned}$$

The equation predicts that Indiana should have four sports teams.

›› Try Exercise 18, page 104

Integrating Technology

If you followed the steps we gave on page 97 and stored the regression equation from Example 1 in Y1, then you can evaluate the regression equation using the following keystrokes:

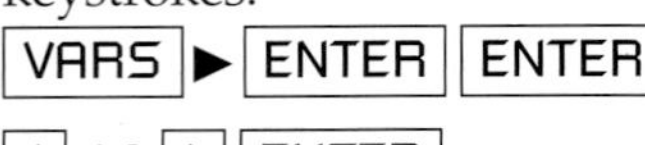
[VARS] ▶ [ENTER] [ENTER] [(] 6.3 [)] [ENTER]

Correlation Coefficient

Figure 1.102

The scatter diagram of a state's population and the corresponding number of professional sports teams is shown in **Figure 1.102**, along with the graph of the regression line. Note that the slope of the regression line is positive. This indicates that as a state's population increases, the number of teams increases. Note also that for these data the value of r on the regression calculation screen was positive, $r \approx 0.9341$.

Now consider the data in the table below, which shows the trade-in value of a 2003 Corvette for various odometer readings. The graph in **Figure 1.103** shows the scatter diagram and the regression line.

> **take note**
>
> The data for the Corvette were created assuming that the condition of the car was excellent. The only variable that changed was the odometer reading.

Trade-in Value of 2003 Corvette Coupe, January 2006

Odometer Reading, in thousands of miles	Trade-in Value in dollars
25	29,100
30	28,000
40	27,175
45	26,450
55	24,225

Source: Kelley Blue Book website, January 2006.

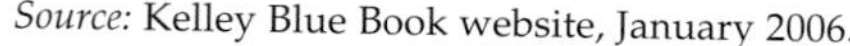

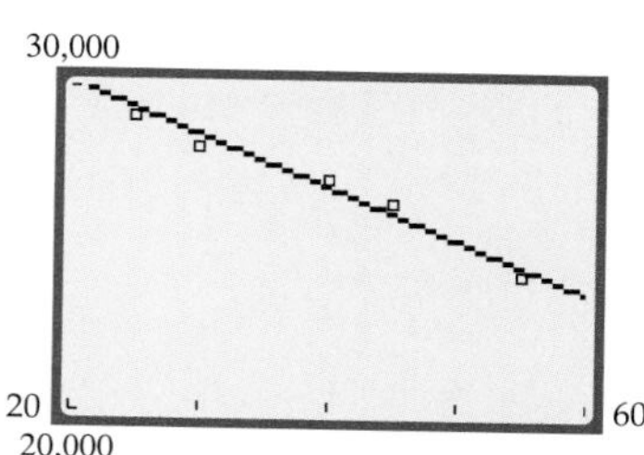

Figure 1.103

```
LinReg
 y=ax+b
 a=-150.745614
 b=32869.07895
 r²=.9635182628
 r=-.9815896611
```

Figure 1.104

In this case the slope of the regression line is negative. This means that as the odometer reading increases, the trade-in value of the car decreases. Note also that the value of r is negative, $r \approx -0.9816$. See **Figure 1.104**.

> **Linear Correlation Coefficient**
>
> The **linear correlation coefficient** r is a measure of how close the points of a data set can be modeled by a straight line. If $r = -1$, then the points of the data set can be modeled *exactly* by a straight line with negative slope. If $r = 1$, then the data set can be modeled *exactly* by a straight line with positive slope. For all data sets, $-1 \le r \le 1$.

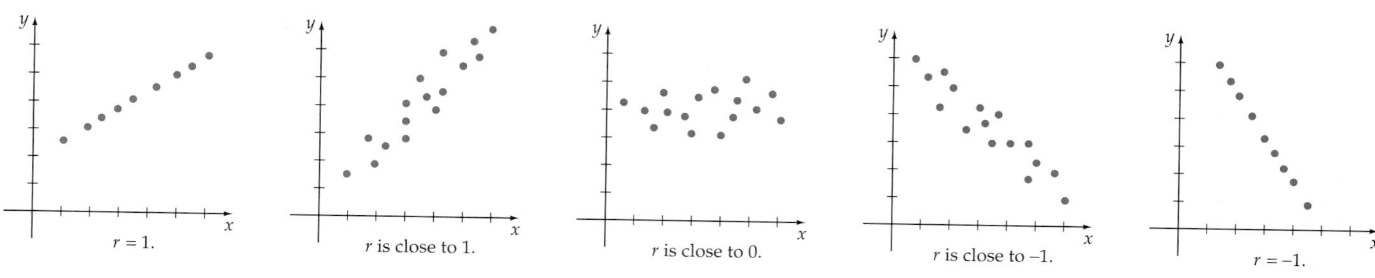

If $r \neq 1$ or $r \neq -1$, then the data set *cannot* be modeled exactly by a straight line. The further the value of r is from 1 or -1 (the closer the value of r to zero), the more the ordered pairs of the data set deviate from a straight line.

The graphs below show the points of the data sets and the graphs of the regression lines for the state population/sports teams data and the odometer reading/trade-in data. Note the values of r and the closeness of the data points to the regression lines.

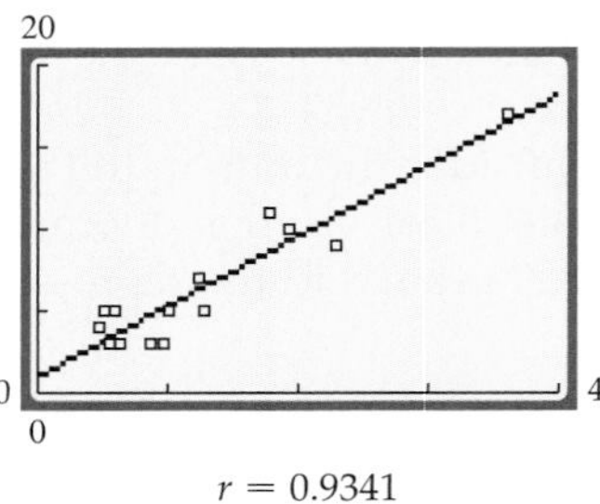

$r = 0.9341$

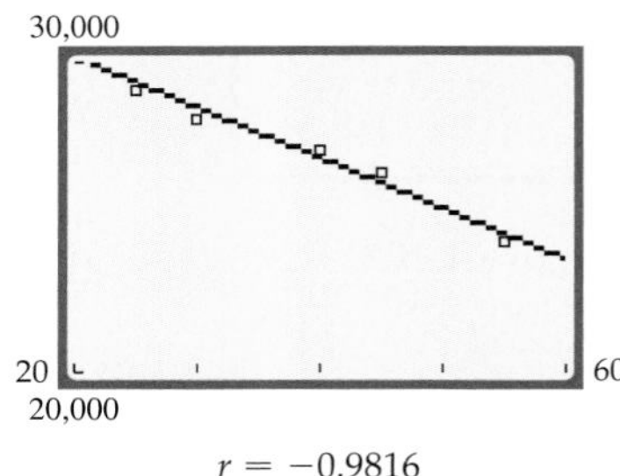

$r = -0.9816$

A researcher calculates a regression line to determine the relationship between two variables. The researcher wants to know whether a change in one variable produces a predictable change in the second variable. The value of r^2 tells the researcher the extent of that relationship.

Coefficient of Determination

The **coefficient of determination** r^2 measures the proportion of the variation in the dependent variable that is explained by the regression line.

For the population/sports team data, $r^2 \approx 0.87$. This means that approximately 87% of the total variation in the dependent variable (number of teams) can be attributed to the state population. This also means that population alone does not predict with certainty the number of sports teams. Other factors, such as climate, are also involved in the number of sports teams.

QUESTION What is the coefficient of determination for the odometer reading/trade-in value data (see page 99), and what is its significance?

Quadratic Regression Models

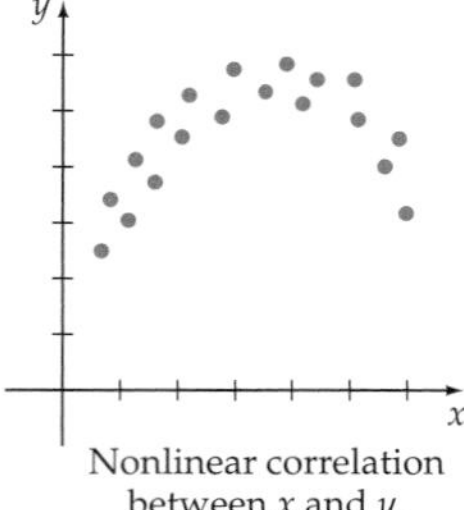

Nonlinear correlation between x and y.

To this point our focus has been *linear* regression equations. However, there may be a nonlinear relationship between two quantities. The scatter diagram to the left suggests that a quadratic function might be a better model of the data than a linear model.

ANSWER $r^2 \approx 0.964$. This means that about 96.4% of the total variation in trade-in value can be attributed to the odometer reading.

EXAMPLE 2 » Find a Quadratic Regression Model

The data in the table below were collected on five successive Saturdays. They show the average number of cars entering a shopping center parking lot. The value of t is the number of minutes after 9:00 A.M. The value of N is the number of cars that entered the parking lot in the 10 minutes prior to the value of t. Find a regression model for this data.

Average Number of Cars Entering a Shopping Center Parking Lot

t	N
20	70
40	135
60	178
80	210
100	260
120	280
140	301
160	298
180	284
200	286
220	260
240	195

Solution

1. **Construct a scatter diagram for these data.** Enter the data into your calculator as explained on page 97.

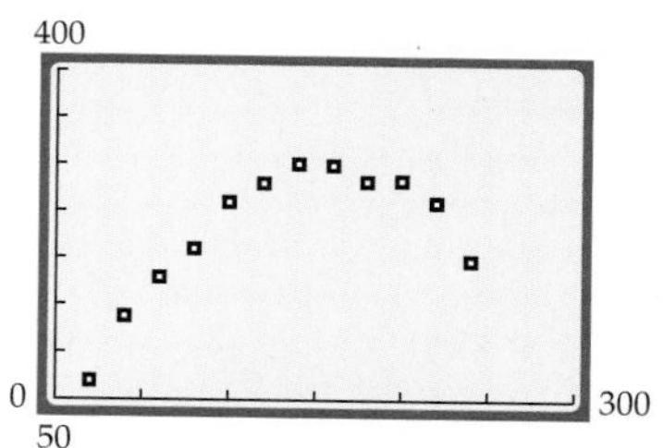

From the scatter diagram, it appears that there is a nonlinear relationship between the variables.

2. **Find the regression equation.** Try a quadratic regression model. For a TI-83/TI-83 Plus/T1-84 Plus calculator, press STAT ► 5 ENTER.

```
QuadReg
 y=ax²+bx+c
 a=-.0124881369
 b=3.904433067
 c=-7.25
 R²=.9840995401
```

3. **Examine the coefficient of determination.** The coefficient of determination is approximately 0.984. Because this number is fairly close to 1, the regression equation $y = -0.0124881369x^2 + 3.904433067x - 7.25$ provides a good model of the data.

take note

For nonlinear regression calculations, the value of r is not shown on a TI-83/T1-83 Plus/T1-84 Plus graphing calculator. In such cases, the coefficient of determination is used to determine how well the data fit the model.

Continued ►

The following calculator screen shows the scatter diagram and the parabola that is the graph of the regression equation from page 101.

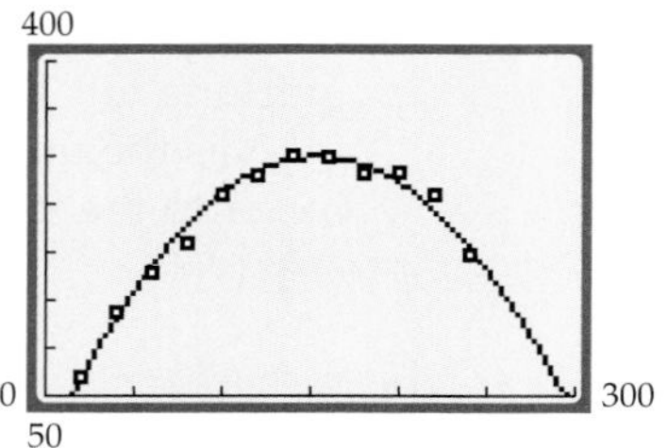

›› Try Exercise 32, page 107

```
LinReg
 y=ax+b
 a=.6575174825
 b=144.2727273
 r²=.4193509866
 r=.6475731515
```

In Example 2, we could have calculated the *linear* regression model for the data. The results are shown at the left. Note that the coefficient of determination for this calculation is approximately 0.419. Because this number is less than the coefficient of determination for the quadratic model, we choose a quadratic model of the data rather than a linear model.

A final note: The regression line equation does not *prove* that the changes in the dependent variable are *caused* by the independent variable. For instance, suppose various cities throughout the United States were randomly selected and the numbers of gas stations (independent variable) and restaurants (dependent variable) were recorded in a table. If we calculated the regression equation for these data, we would find that r would be close to 1. However, this does not mean that gas stations *cause* restaurants to be built. The primary cause is that there are fewer gas stations and restaurants in cities with small populations and greater numbers of gas stations and restaurants in cities with large populations.

Topics for Discussion

1. What is the purpose of calculating the equation of a regression line?
2. Discuss the implications of the following correlation coefficients: $r = -1$, $r = 0$, and $r = 1$.
3. Discuss the coefficient of determination and what its value says about a data set.
4. What are the implications of $r^2 = 1$ for a quadratic regression equation?

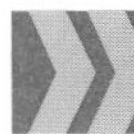

Exercise Set 1.7

Use a graphing calculator for this Exercise Set.

In Exercises 1 to 4, determine whether the scatter diagram suggests a linear relationship between *x* and *y*, a nonlinear relationship between *x* and *y*, or no relationship between *x* and *y*.

1.

2.

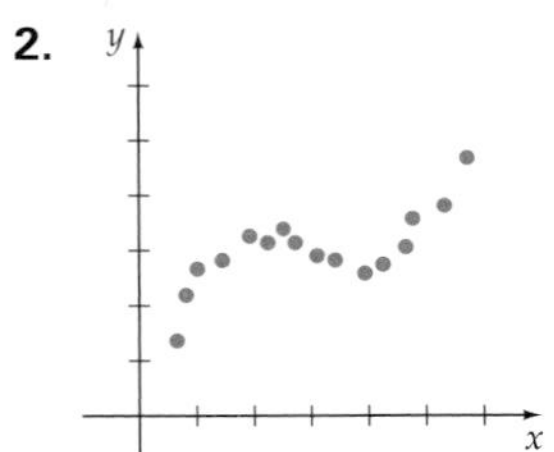

3.

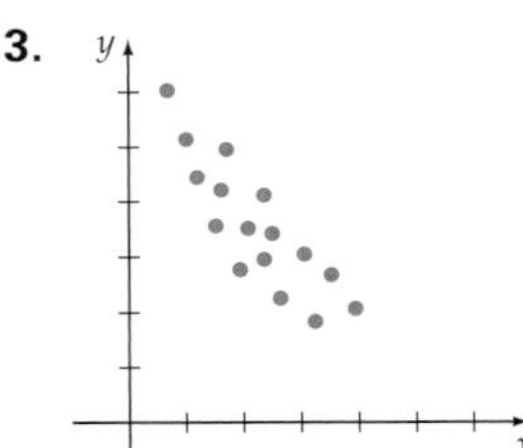

4.

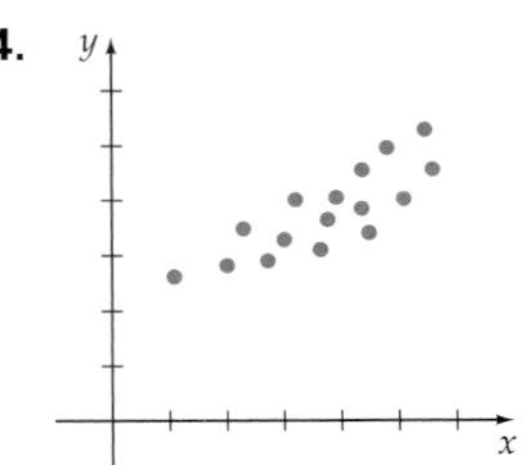

In Exercises 5 and 6, determine for which scatter diagram, A or B, the coefficient of determination is closer to 1.

5.

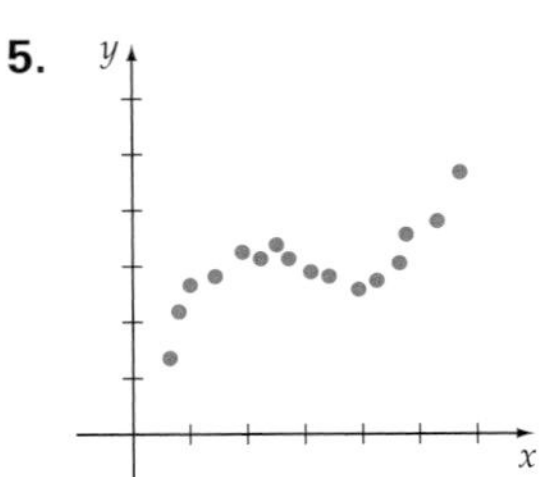

Figure A

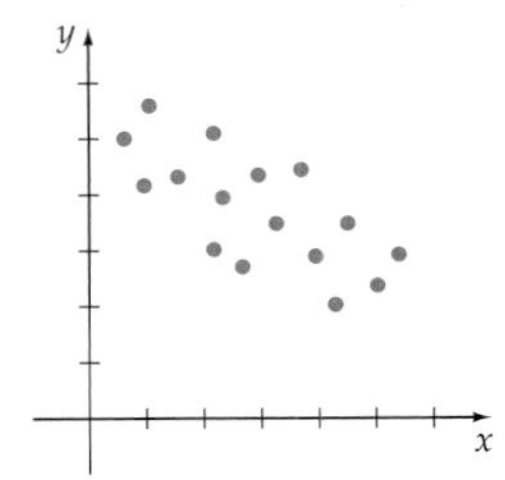

Figure B

6.

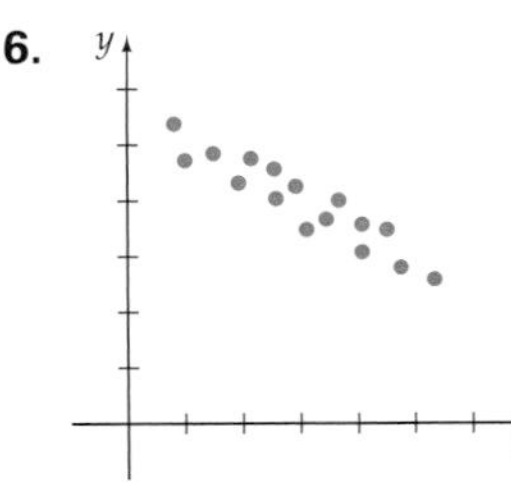

Figure A

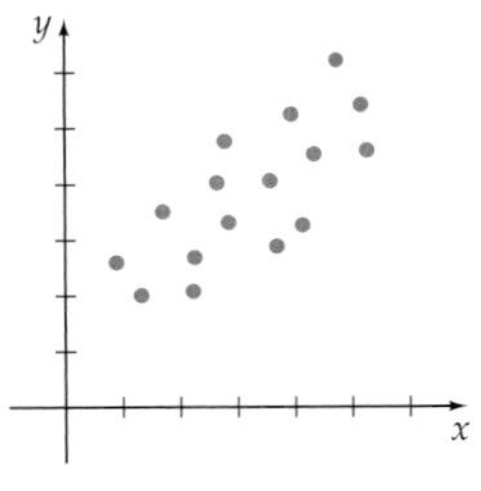

Figure B

In Exercises 7 to 12, find the linear regression equation for the given set.

7. $\{(2, 6), (3, 6), (4, 8), (6, 11), (8, 18)\}$

8. $\{(2, -3), (3, -4), (4, -9), (5, -10), (7, -12)\}$

9. $\{(-3, 11.8), (-1, 9.5), (0, 8.6), (2, 8.7), (5, 5.4)\}$

10. $\{(-7, -11.7), (-5, -9.8), (-3, -8.1), (1, -5.9), (2, -5.7)\}$

11. $\{(1.3, -4.1), (2.6, -0.9), (5.4, 1.2), (6.2, 7.6), (7.5, 10.5)\}$

12. $\{(-1.5, 8.1), (-0.5, 6.2), (3.0, -2.3), (5.4, -7.1), (6.1, -9.6)\}$

In Exercises 13 to 16, find a quadratic model for the given data.

13. $\{(1, -1), (2, 1), (4, 8), (5, 14), (6, 25)\}$

14. $\{(-2, -5), (-1, 0), (0, 1), (1, 4), (2, 4)\}$

15. $\{(1.5, -2.2), (2.2, -4.8), (3.4, -11.2), (5.1, -20.6), (6.3, -28.7)\}$

16. $\{(-2, -1), (-1, -3.1), (0, -2.9), (1, 0.8), (2, 6.8), (3, 15.9)\}$

17. ARCHEOLOGY The data below show the length of the humerus and the wingspan, in centimeters, of several pterosaurs, which are extinct flying reptiles of the order Pterosauria. (*Source:* Southwest Educational Development Laboratory.)

Pterosaur Data

Humerus (cm)	Wingspan (cm)	Humerus (cm)	Wingspan (cm)
24	600	20	500
32	750	27	570
22	430	15	300
17	370	15	310
13	270	9	240
4.4	68	4.4	55
3.2	53	2.9	50
1.5	24		

a. Compute the linear regression equation for these data.

b. On the basis of this model, what is the projected wingspan of the pterosaur *Quetzalcoatlus northropi,* which is thought to have been the largest of the prehistoric birds, if its humerus is 54 centimeters? Round to the nearest centimeter.

18. SPORTS The data in the table below show the distance, in feet, a ball travels for various swing speeds, in miles per hour, of a bat.

Bat speed (mph)	Distance (ft)
40	200
45	213
50	242
60	275
70	297
75	326
80	335

a. Find the linear regression equation for these data.

b. Using the regression model, what is the expected distance a ball will travel when the bat speed is 58 miles per hour? Round to the nearest foot.

19. SPORTS The table below shows the number of strokes per minute that a rower makes and the speed of the boat in meters per second.

Strokes per minute	Speed (m/s)
30	4.1
31	4.2
33	4.4
34	4.5
36	4.7
39	5.1

a. Find the linear regression equation for these data.

b. Using the regression model, what is the expected speed of the boat when the rowing rate is 32 strokes per minute? Round to the nearest tenth of a meter per second.

20. BIOLOGY A medical researcher wanted to determine the effect of pH (a measure of alkalinity or acidity, with pure water having a pH of 7) on the growth of a bacteria culture. The table below gives the measurements of different cultures, in thousands of bacteria, after 8 hours.

pH	Number of Bacteria (in thousands)
4	116
5	120
6	131
7	136
8	141
9	151
10	148
11	163

a. Find the linear regression equation for these data.

b. Using the regression model, what is the expected number of bacteria when the pH is 7.5? Round to the nearest thousand bacteria.

21. HEALTH The body mass index (BMI) of a person is a measure of the person's ideal body weight. The table below shows the BMI for different weights for a person 5 feet 6 inches tall. (*Source:* San Diego *Union-Tribune*, May 31, 2000)

BMI Data for Person 5′ 6″ Tall

Weight (lb)	BMI	Weight (lb)	BMI
110	17	160	25
120	19	170	27
125	20	180	29
135	21	190	30
140	22	200	32
145	23	205	33
150	24	215	34

a. Compute the linear regression equation for these data.

b. On the basis of the model, what is the estimated BMI for a person 5 feet 6 inches tall whose weight is 158 pounds?

22. HEALTH The BMI (see Exercise 21) of a person depends on height as well as weight. The table below shows the changes in BMI for a 150-pound person as height (in inches) changes. (*Source:* San Diego *Union-Tribune*, May 31, 2000)

BMI Data for 150-Pound Person

Height (in.)	BMI	Height (in.)	BMI
60	29	71	21
62	27	72	20
64	25	73	19
66	24	74	19
67	23	75	18
68	23	76	18
70	21		

a. Compute the linear regression equation for these data.

b. On the basis of the model, what is the estimated BMI for a 150-pound person who is 5 feet 8 inches tall?

23. INDUSTRIAL ENGINEERING Permanent-magnet direct-current motors are used in a variety of industrial applications. For these motors to be effective, there must be a strong linear relationship between the current (in amps, A) supplied to the motor and the resulting torque (in newton-centimeters, N-cm) produced by the motor. A randomly selected motor is chosen from a production line and tested, with the following results.

Direct-Current Motor Data at 12 Volts

Current, in A	Torque, in N-cm	Current, in A	Torque, in N-cm
7.3	9.4	8.5	8.6
11.9	2.8	7.9	4.3
5.6	5.6	14.5	9.5
14.2	4.9	12.7	8.3
7.9	7.0	10.6	4.7

Based on the data in this table, is the chosen motor effective? Explain.

24.  HEALTH SCIENCES The average remaining lifetime for men in the United States is given in the following table. (*Source:* National Institutes of Health.)

Average Remaining Lifetime for Men

Age	Years	Age	Years
0	73.6	65	15.9
15	59.4	75	9.9
35	40.8		

Based on the data in this table, is there a strong correlation between a man's age and the average remaining lifetime for that man? Explain.

25. HEALTH SCIENCES The average remaining lifetime for women in the United States is given in the following table. (*Source:* National Institutes of Health.)

Average Remaining Lifetime for Women

Age	Years	Age	Years
0	79.4	65	19.2
15	65.1	75	12.1
35	45.7		

a. Based on the data in this table, is there a strong correlation between a woman's age and the average remaining lifetime for that woman?

b. Compute the linear regression equation for these data.

c. On the basis of the model, what is the estimated remaining lifetime of a woman of age 25?

26. BIOLOGY The table below gives the body lengths, in centimeters, and the highest observed flying speeds, in meters per second, of various animals.

Species	Length (cm)	Flying speed (m/s)
Horsefly	1.3	6.6
Hummingbird	8.1	11.2
Dragonfly	8.5	10.0
Willow warbler	11	12.0
Common pintail	56	22.8

Based on these data, what is the flying speed of a Whimbrel whose length is 41 centimeters? Round to the nearest meter per second. (*Source:* Based on data from Leiva, *Algebra 2: Explorations and Applications,* p. 76, McDougall Littell, Boston; copyright 1997.)

27. HEALTH The table below shows the number of calories burned in 1 hour when running at various speeds.

Running Speed (mph)	Calories Burned
10	1126
10.9	1267
5	563
5.2	633
6	704
6.7	774
7	809
8	950
8.6	985
9	1056
7.5	880

a. Are the data positively or negatively correlated?

b. How many calories does this model predict a person who runs at 9.5 mph for 1 hour will burn? Round to the nearest calorie.

28. TRAFFIC SAFETY A traffic safety institute measured the braking distance, in feet, of a car traveling at certain speeds in miles per hour. The data from one of those tests are given in the following table.

Speed (mph)	Breaking Distance (ft)
20	23.9
30	33.7
40	40.0
50	41.7
60	46.8
70	48.9
80	49.0

a. Find the quadratic regression equation for these data.

b. Using the regression model, what is the expected braking distance when a car is traveling at 65 mph? Round to the nearest tenth of a foot.

29. BIOLOGY The survival of certain larvae after hatching depends on the temperature (in degrees Celsius) of the surrounding environment. The following table shows the number of larvae that survive at various temperatures. Find a quadratic model for these data.

Larvae Surviving for Various Temperatures

Temp. (°C)	Number Surviving	Temp. (°C)	Number Surviving
20	40	26	68
21	47	27	67
22	52	28	64
23	61	29	62
24	64	30	61
25	64		

30. METEOROLOGY The temperature at various times on a summer day at a resort in southern California is given in the following table. The variable t is the number of minutes after 6:00 A.M., and the variable T is the temperature in degrees Fahrenheit.

Temperatures at a Resort

t (min)	T (°F)	t (min)	T (°F)
20	59	240	86
40	65	280	88
80	71	320	86
120	78	360	85
160	81	400	80
200	83		

a. Find a quadratic model for these data.

b. Use the model to predict the temperature at 1:00 P.M. Round to the nearest tenth of a degree.

31. AUTOMOTIVE ENGINEERING The fuel efficiency, in miles per gallon, for a certain midsize car at various speeds, in miles per hour, is given in the table below.

Fuel Efficiency of a Midsize Car

mph	mpg	mph	mpg
25	29	55	31
30	32	60	28
35	33	65	24
40	35	70	19
45	34	75	17
50	33		

a. Find a quadratic model for these data.

b. Use the model to predict the fuel efficiency of this car when it is traveling at a speed of 50 mph.

32. BIOLOGY The data in the table at the right show the oxygen consumption, in milliliters per minute, of a bird flying level at various speeds in kilometers per hour.

a. Find a quadratic model for these data.

b. Use the model to predict the oxygen consumption of a bird flying at 40 kilometers per hour. Round to the nearest milliliter per minute.

Oxygen Consumption

Speed (km/h)	Consumption (ml/min)
20	32
25	27
28	22
35	21
42	26
50	34

Connecting Concepts

33. PHYSICS Galileo (1564–1642) studied acceleration due to gravity by allowing balls of various weights to roll down an incline. This allowed him to time the descent of a ball more accurately than by just dropping the ball. The data in the table show some possible results of such an experiment using balls of different masses. Time t is measured in seconds; distance s is measured in centimeters.

Distance Traveled for Balls of Various Weights

5-Pound Ball		10-Pound Ball		15-Pound Ball	
t	s	t	s	t	s
2	2	3	5	3	5
4	10	6	22	5	15
6	22	9	49	7	30
8	39	12	87	9	49
10	61	15	137	11	75
12	86	18	197	13	103
14	120			15	137
16	156				

a. Find a quadratic model for each of the balls.

b. On the basis of a similar experiment, Galileo concluded that if air resistance is excluded, all falling objects fall with the same acceleration. Explain how one could make such a conclusion from the regression equations.

34. ASTRONOMY In 1929, Edwin Hubble published a paper that revolutionized astronomy ("A Relationship Between Distance and Radial Velocity Among Extra-Galactic Nebulae," *Proceedings of the National Academy of Science*, 168). His paper dealt with the distance an extragalactic nebula was from the Milky Way galaxy and the nebula's velocity with respect to the Milky Way. The data are given in the table below. Distance is measured in megaparsecs (1 megaparsec equals 1.918×10^{19} miles), and velocity (called the *recession velocity*) is measured in kilometers per second. A negative velocity means the nebula is moving toward the Milky Way; a positive velocity means the nebula is moving away from the Milky Way.

Recession Velocities

Distance	Velocity
0.032	170
0.034	290
0.214	−130
0.263	−70
0.275	−185
0.275	−220
0.45	200
0.5	290
0.5	270
0.63	200
0.8	300
0.9	−30

Distance	Velocity
0.9	650
0.9	150
0.9	500
1.0	920
1.1	450
1.1	500
1.4	500
1.7	960
2.0	500
2.0	850
2.0	800
2.0	1090

a. Find the linear regression model for these data.

b. On the basis of this model, what is the recession velocity of a nebula that is 1.5 megaparsecs from the Milky Way?

Projects

Median–Median Line Another linear model of data is called the **median–median line**. This line employs *summary points* calculated using the medians of subsets of the independent and dependent variables. The **median** of a data set is the middle number or the average of the two middle numbers for a data set arranged in numerical order. For instance, to find the median of {8, 12, 6, 7, 9}, first arrange the data in numerical order.

$$6, 7, 8, 9, 12$$

The median is 8, the number in the middle. To find the median of {15, 12, 20, 9, 13, 10}, arrange the numbers in numerical order.

$$9, 10, 12, 13, 15, 20$$

The median is 12.5, the average of the two middle numbers.

$$\text{Median} = \frac{12 + 13}{2} = 12.5$$

The median–median line is determined by dividing a data set into three equal groups. (If the set cannot be divided into three equal groups, the first and third groups should be equal. For instance, if there are 11 data points, divide the set into groups of 4, 3, and 4.) The slope of the median–median line is the slope of the line through the x-medians and y-medians of the first and third sets of points. The median–median line passes through the average of the x-and y-medians of all three sets.

A graphing calculator can be used to find the equation of the median–median line. This line, along with the linear regression line, is shown in the next column for the data in the accompanying table.

1. Find the equation of the median–median line for the data in Exercise 17 on page 103.

2. Find the equation of the median–median line for the data in Exercise 18 on page 104.

x	y
2	3
3	5
4	4
5	7
6	8
7	9
8	12
9	12
10	14
11	15
12	14

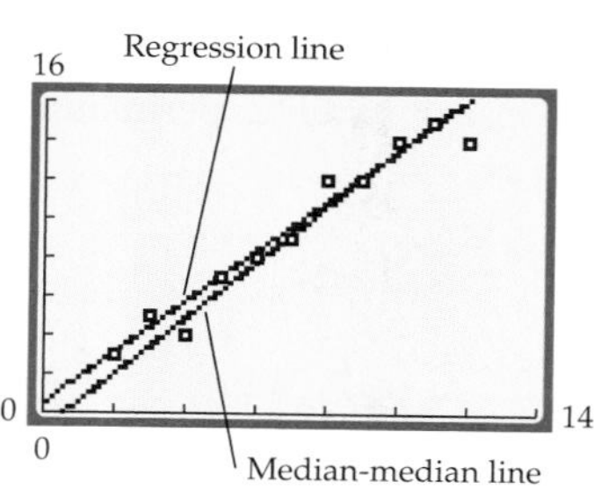

3. Consider the data set {(1, 3), (2, 5), (3, 7), (4, 9), (5, 11), (6, 13), (7, 15), (8, 17)}.
   a. Find the equation of the linear regression line for these data.
   b. Find the equation of the median–median line for these data.
   c. What conclusion might you draw from the answers to **a.** and **b.**?

4. For this exercise, use the data in the table above.
   a. Calculate the equation of the median–median line and the linear regression line.
   b. Change the entry (12, 14) to (12, 1) and then recalculate the equations of the median–median line and the linear regression line.
   c. Explain why there is more change in the linear regression line than in the median–median line.

Exploring Concepts with Technology

Graphing Piecewise Functions with a Graphing Calculator

A graphing calculator can be used to graph piecewise functions by including as part of the function the interval on which each piece of the function is defined. The method is based on the fact that a graphing calculator "evaluates" inequalities. For purposes of this Exploration, we will use keystrokes for a TI-83/TI-83 Plus/TI-84 Plus calculator.

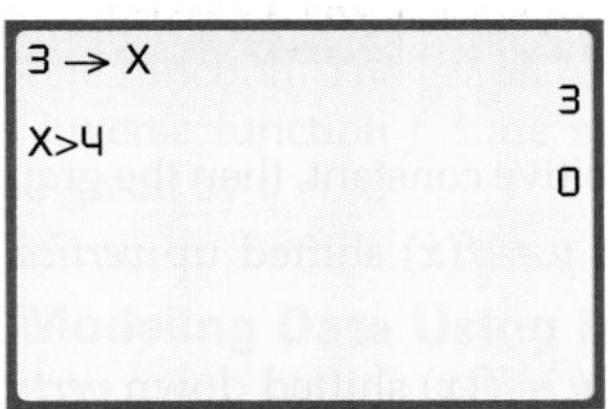

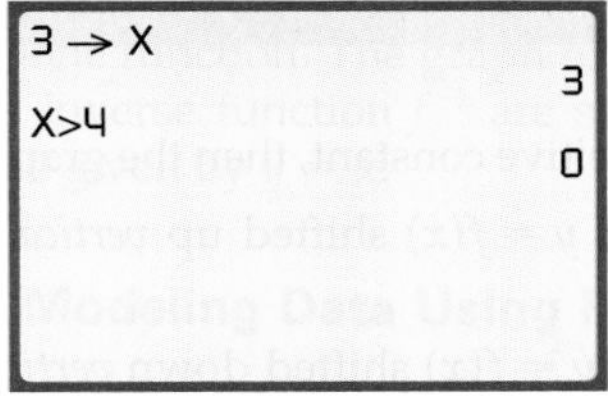

For instance, store 3 in X by pressing 3 STO▶ X,T,Θ,*n* ENTER. Now enter the inequality $x > 4$ by pressing X,T,Θ,*n* 2ND TEST 3 4 ENTER. Your screen should look like the one at the left. Note that the value of the inequality on the screen is 0. This occurs because the calculator replaced X by 3 and then determined whether the inequality $3 > 4$ was true or false. The calculator expresses the fact that the inequality is false by placing a zero on the screen. If we repeat the sequence of steps above, except that we store 5 in X instead of 3, the calculator will determine that the inequality is true and place a 1 on the screen.

This property of calculators is used to graph piecewise functions. Graphs of these functions work best when Dot mode rather than Connected mode is used. To switch to Dot mode, select MODE, use the arrow keys to highlight DOT, and then press ENTER.

Next we will graph the piecewise function defined by

$$f(x) = \begin{cases} x, & x \le -2 \\ x^2, & x > -2 \end{cases}$$

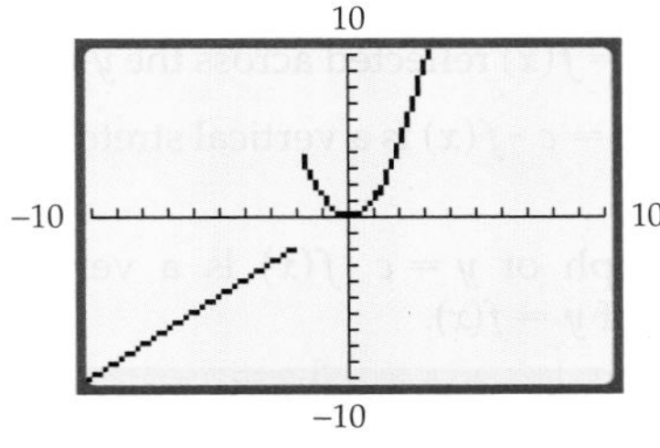

Enter the function[3] as $Y_1 = X*(X \le -2) + X^2*(X > -2)$ and graph it in the standard viewing window. Note that you are multiplying each piece of the function by its domain. The graph will appear as shown at the left.

To understand how the graph is drawn, we will consider two values of x, -8 and 2, and evaluate Y_1 for each of these values.

$Y_1 = X*(X \le -2) + X^2*(X > -2)$

$= -8(-8 \le -2) + (-8)^2(-8 > -2)$

$= -8(1) + 64(0) = -8$ • **When $x = -8$, the value assigned to $-8 \le -2$ is 1; the value assigned to $-8 > -2$ is 0.**

$Y_1 = X*(X \le -2) + X^2*(X > -2)$

$= 2(2 \le -2) + 2^2(2 > -2)$

$= 2(0) + 4(1) = 4$ • **When $x = 2$, the value assigned to $2 \le -2$ is 0; the value assigned to $2 > -2$ is 1.**

In a similar manner, for any value of x, $x \le -2$, the value assigned to $(X \le -2)$ is 1 and the value assigned to $(X > -2)$ is 0. Thus $Y_1 = X*1 + X^2*0 = X$ on that interval. This means that only the $f(x) = x$ piece of the function is graphed. When $x > -2$, the value assigned to $(X \le -2)$ is 0 and the value assigned to $(X > -2)$ is 1. Thus $Y_1 = X*0 + X^2*1 = X^2$ on that interval. This means that only the $f(x) = x^2$ piece of the function is graphed on that interval.

1. Graph: $f(x) = \begin{cases} x^2, & x < 2 \\ -x, & x \ge 2 \end{cases}$

2. Graph: $f(x) = \begin{cases} x^2 - x, & x < 2 \\ -x + 4, & x \ge 2 \end{cases}$

3. Graph: $f(x) = \begin{cases} -x^2 + 1, & x < 0 \\ x^2 - 1, & x \ge 0 \end{cases}$

4. Graph: $f(x) = \begin{cases} x^3 - 4x, & x < 1 \\ x^2 - x + 2, & x \ge 1 \end{cases}$

[3]Note that pressing 2ND TEST will display the inequality menu.

Chapter 1 Test

1. Solve: $4x - 2(2 - x) = 5 - 3(2x + 1)$

2. Solve: $6 - 3x \geq 3 - 4(2 - 2x)$

3. Solve: $2x^2 - 3x = 2$

4. Solve: $3x^2 - x = 2$

5. Solve: $|4 - 5x| > 6$

6. Find the distance between the points $(-2, 5)$ and $(4, -2)$.

7. Find the midpoint and the length of the line segment with endpoints $(-2, 3)$ and $(4, -1)$.

8. Determine the x- and y-intercepts of, and then graph, the equation $x = 2y^2 - 4$.

9. Graph the equation $y = |x + 2| + 1$.

10. Find the center and radius of the circle that has the general form $x^2 - 4x + y^2 + 2y - 4 = 0$.

11. Given $f(x) = -\sqrt{25 - x^2}$, evaluate $f(-3)$.

12. Determine the domain of the function defined by

$$f(x) = -\sqrt{x^2 - 16}.$$

13. Graph $f(x) = -2|x - 2| + 1$. Identify the intervals over which the function is increasing, constant, or decreasing.

14. Use the graph of $f(x) = |x|$ to graph $y = -f(x + 2) - 1$.

15. Which of the following define odd functions?

a. $f(x) = x^4 - x^2$ **b.** $f(x) = x^3 - x$

c. $f(x) = x - 1$

16. Let $f(x) = x^2 - 1$ and $g(x) = x - 2$. Find $(f + g)$ and $\left(\frac{f}{g}\right)$.

17. Find the difference quotient of the function

$$f(x) = x^2 + 1$$

18. Evaluate $(f \circ g)$, where

$$f(x) = x^2 - 2x + 1 \quad \text{and} \quad g(x) = \sqrt{x - 2}$$

19. Find the inverse of the function given by the equation

$$f(x) = \frac{x}{x + 1}$$

20. CALORIE CONTENT The table below shows the percent of water and the number of calories in various canned soups to which 100 grams of water are added.

Percent Water in Soups

% Water	Calories	% Water	Calories
93.2	28	89.6	56
92.3	26	90.5	36
91.9	39	91.9	32
89.5	56	91.7	32

a. Find the equation of the linear regression line for these data.

b. Using the linear model from **a.**, find the expected number of calories in a soup that is 89% water.

2 Trigonometric Functions

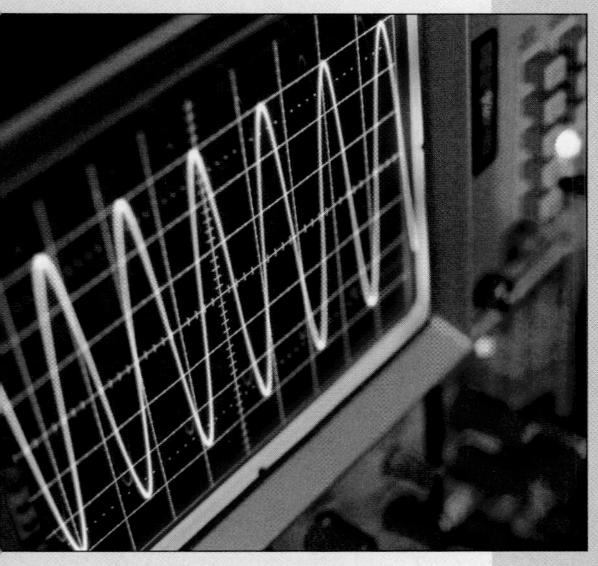

Applications of Trigonometric Functions

In this chapter we introduce an important group of functions called the *trigonometric functions*. These functions are often used in applications involving relationships among the sides and angles of triangles. Exercise 59 on page 143 illustrates this aspect of trigonometry.

In the seventeenth century, a unit circle approach was used to create *trigonometric functions of real numbers* or *circular functions*. These functions enable us to solve a wider variety of application problems. For instance, in Exercise 63 on page 178, a trigonometric function of a real number is used to model a sound wave and find the frequency of the sound wave.

An application that involves finding the time it takes for the music produced by two sound tracks to repeat is given in the Quantitative Reasoning exercises on page 214.

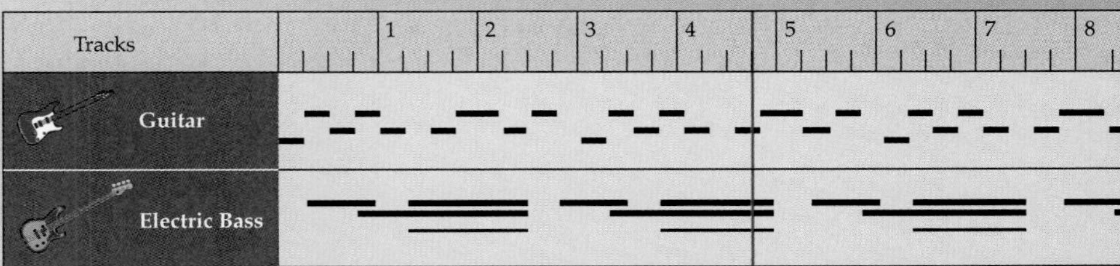

Online Study Center For online student resources, such as section quizzes, visit this website: college.hmco.com/info/aufmannCAT

Section 2.1

- Degree Measure
- Radian Measure
- Arcs and Arc Length
- Linear and Angular Speed

Angles and Arcs

A point P on a line separates the line into two parts, each of which is called a **half-line.** The union of point P and the half-line formed by P that includes point A is called a **ray,** and it is represented as $\overrightarrow{PA}$. The point P is the **endpoint** of ray $\overrightarrow{PA}$. **Figure 2.1** shows the ray $\overrightarrow{PA}$ and a second ray $\overrightarrow{QR}$.

In geometry, an *angle* is defined simply as the union of two rays that have a common endpoint. In trigonometry and many advanced mathematics courses, it is beneficial to define an angle in terms of a rotation.

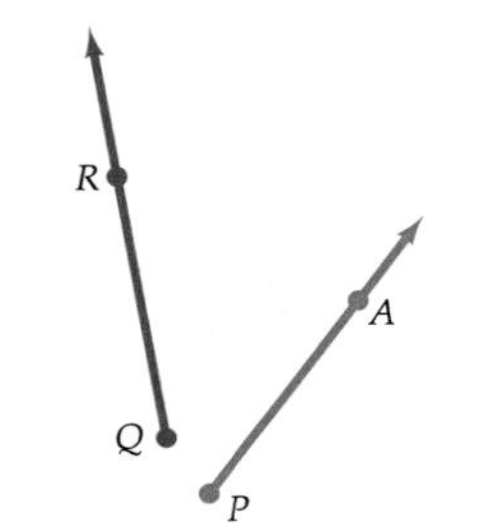

Figure 2.1

Definition of an Angle

An **angle** is formed by rotating a given ray about its endpoint to some terminal position. The original ray is the **initial side** of the angle, and the second ray is the **terminal side** of the angle. The common endpoint is the **vertex** of the angle.

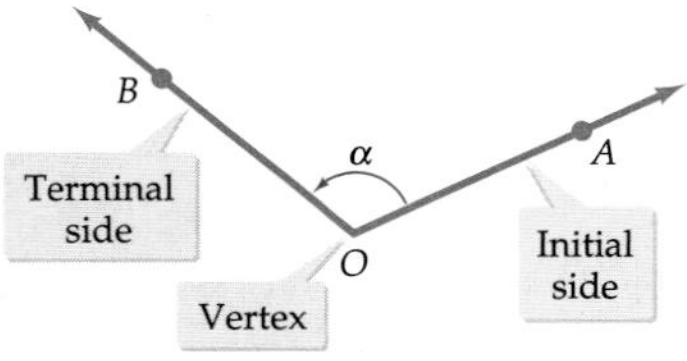

Figure 2.2

There are several methods used to name an angle. One way is to employ Greek letters. For example, the angle shown in **Figure 2.2** can be designated as α or as $\angle\alpha$. It also can be named $\angle O$, $\angle AOB$, or $\angle BOA$. If you name an angle by using three points, such as $\angle AOB$, it is traditional to list the vertex point between the other two points.

Angles formed by a counterclockwise rotation are considered **positive angles,** and angles formed by a clockwise rotation are considered **negative angles.** See **Figure 2.3.**

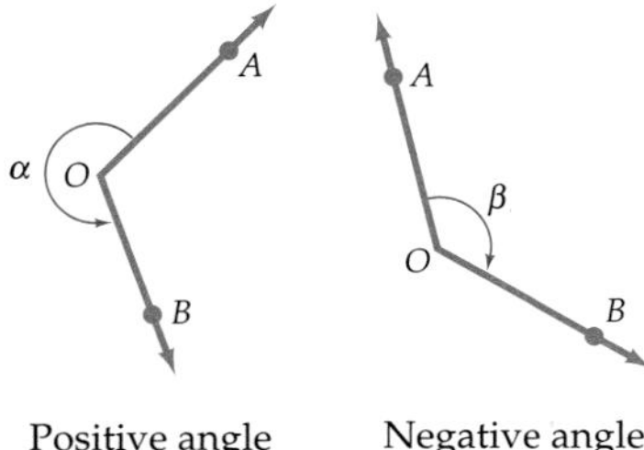

Figure 2.3

Degree Measure

The **measure** of an angle is determined by the amount of rotation of the initial side. An angle formed by rotating the initial side counterclockwise exactly once until it coincides with itself (one complete revolution) is defined to have a measure of 360 degrees, which is abbreviated as 360°.

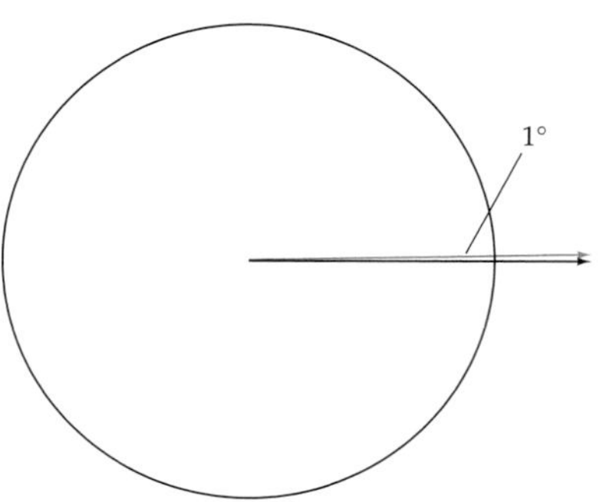

Figure 2.4
$1° = \frac{1}{360}$ of a revolution

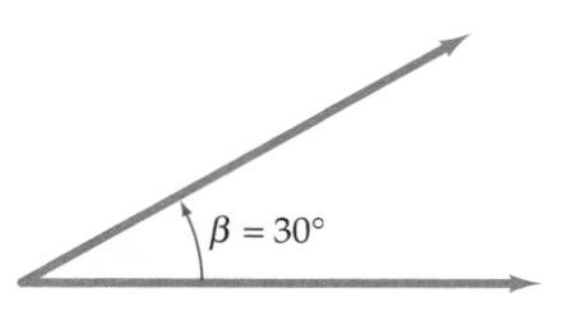

Figure 2.5

Definition of Degree

One **degree** is the measure of an angle formed by rotating a ray $\frac{1}{360}$ of a complete revolution. The symbol for degree is °.

The angle shown in **Figure 2.4** has a measure of 1°. The angle β shown in **Figure 2.5** has a measure of 30°. We will use the notation $\beta = 30°$ to denote that the measure of angle β is 30°. The protractor shown in **Figure 2.6** can be used to measure an angle in degrees or to draw an angle with a given degree measure.

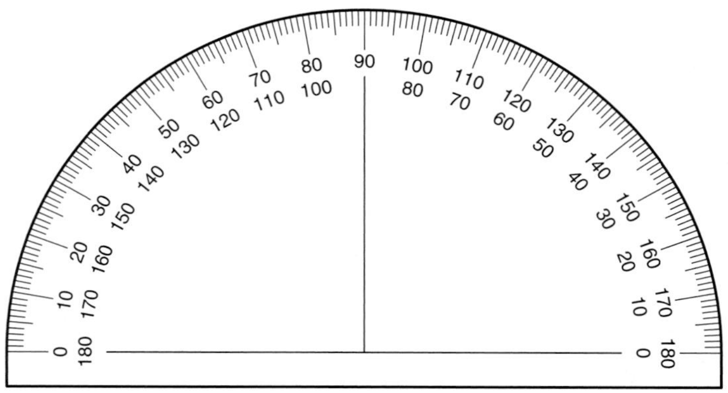

Figure 2.6
Protractor for measuring angles in degrees

Angles are often classified according to their measure.

- 180° angles are **straight angles.** See **Figure 2.7a.**
- 90° angles are **right angles.** See **Figure 2.7b.**
- Angles that have a measure greater than 0° but less than 90° are **acute angles.** See **Figure 2.7c.**
- Angles that have a measure greater than 90° but less than 180° are **obtuse angles.** See **Figure 2.7d.**

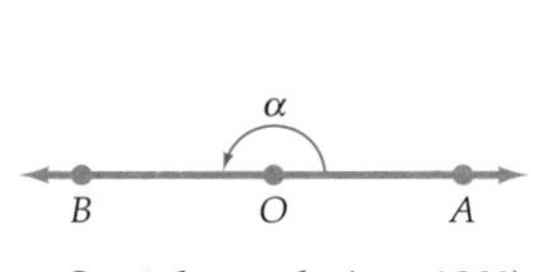

a. Straight angle ($\alpha = 180°$)

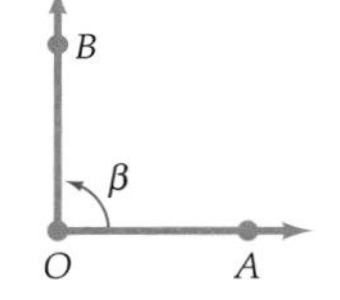

b. Right angle ($\beta = 90°$)

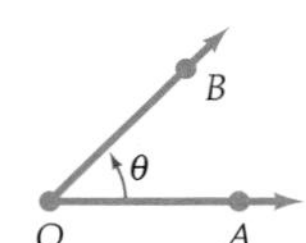

c. Acute angle ($0° < \theta < 90°$)

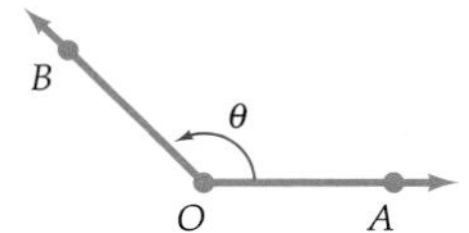

d. Obtuse angle ($90° < \theta < 180°$)

Figure 2.7

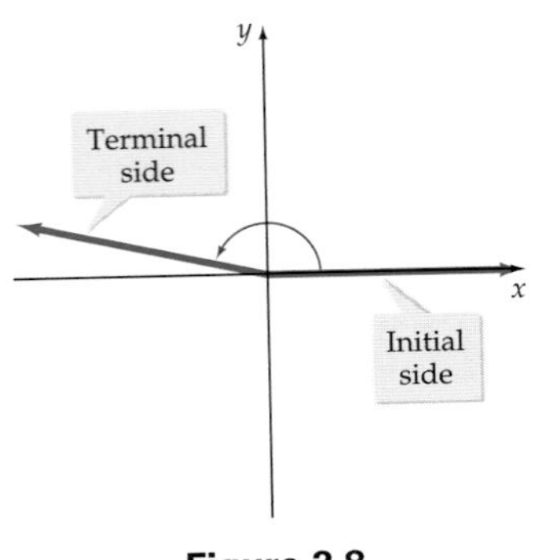

Figure 2.8
An angle in standard position

An angle superimposed in a Cartesian coordinate system is in **standard position** if its vertex is at the origin and its initial side is on the positive x-axis. See **Figure 2.8.**

Two positive angles are **complementary angles (Figure 2.9a)** if the sum of the measures of the angles is 90°. Each angle is the *complement* of the other angle. Two positive angles are **supplementary angles (Figure 2.9b)** if the sum of the measures of the angles is 180°. Each angle is the *supplement* of the other angle.

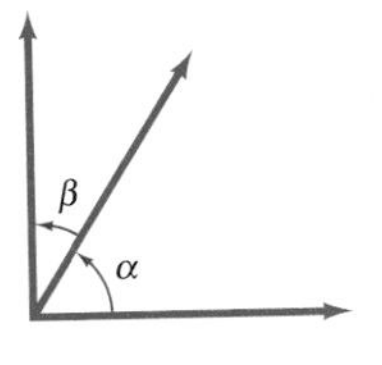

a. Complementary angles
$\alpha + \beta = 90°$

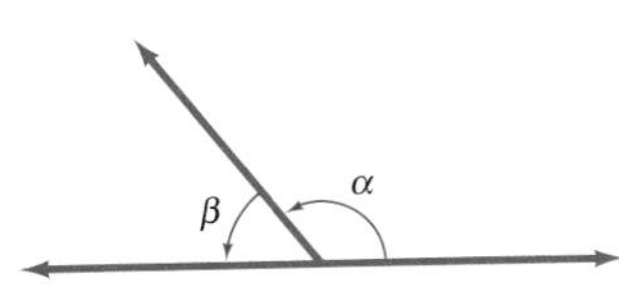

b. Supplementary angles
$\alpha + \beta = 180°$

Figure 2.9

EXAMPLE 1 ›› Find the Measure of the Complement and the Supplement of an Angle

For each angle, find the measure (if possible) of its complement and of its supplement.

a. $\theta = 40°$ **b.** $\theta = 125°$

Solution

a. **Figure 2.10** shows $\angle\theta = 40°$ in standard position. The measure of its complement is $90° - 40° = 50°$. The measure of its supplement is $180° - 40° = 140°$.

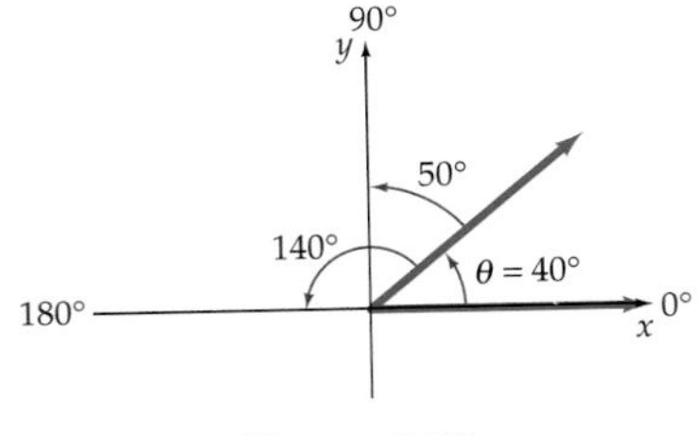

Figure 2.10

b. **Figure 2.11** shows $\angle\theta = 125°$ in standard position. Angle θ does not have a complement because there is no positive number x such that

$$x° + 125° = 90°$$

The measure of its supplement is $180° - 125° = 55°$.

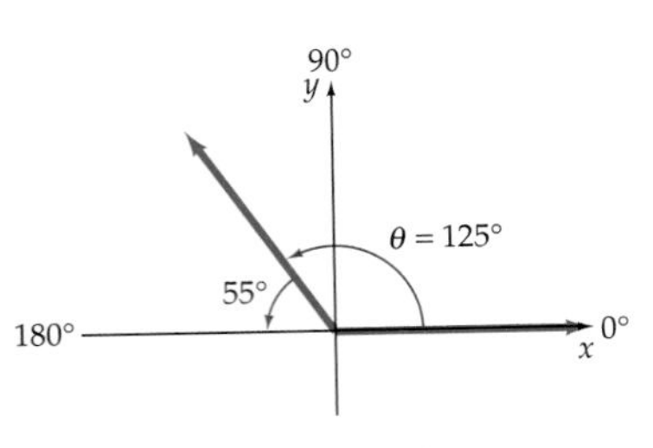

Figure 2.11

›› *Try Exercise 2, page 130*

QUESTION Are the two acute angles of any right triangle complementary angles? Explain.

Some angles have a measure greater than 360°. See **Figure 2.12a** and **Figure 2.12b.** The angle shown in **Figure 2.12c** has a measure less than −360°,

---

ANSWER Yes. The sum of the measures of the angles of any triangle is 180°. The right angle has a measure of 90°. Thus the sum of the measures of the two acute angles must be $180° - 90° = 90°$.

because it is formed by a clockwise rotation of more than one revolution of the initial side.

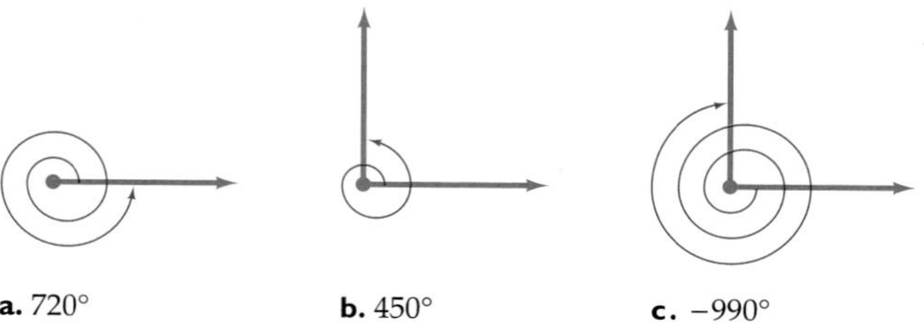

Figure 2.12

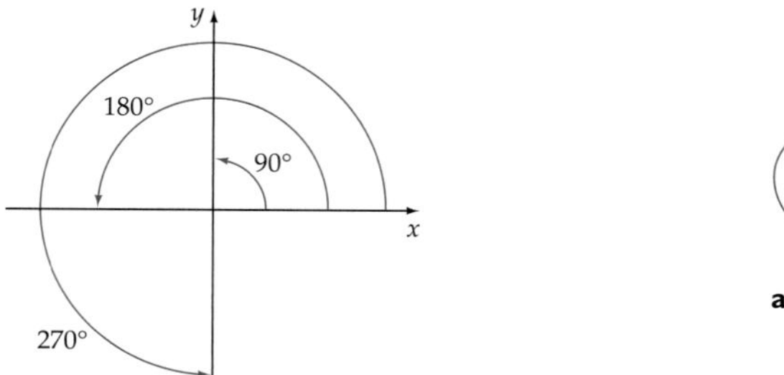

Figure 2.13

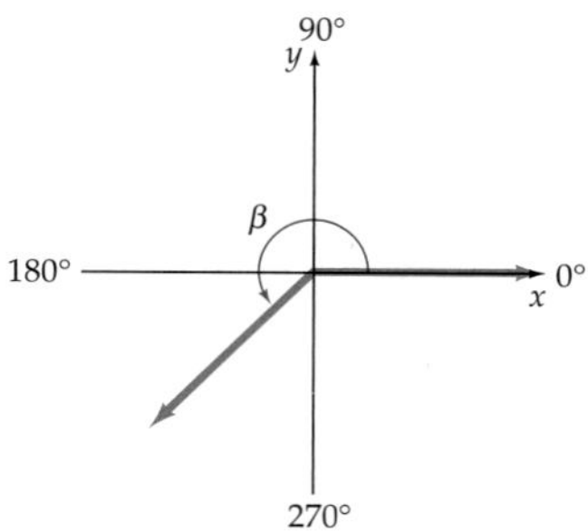

Figure 2.14

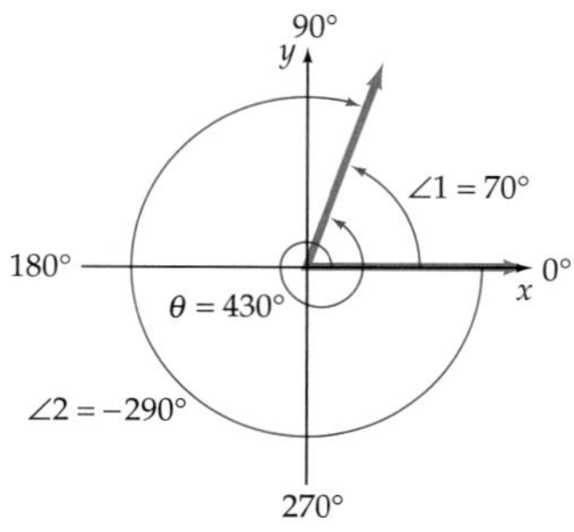

Figure 2.15

If the terminal side of an angle in standard position lies on a coordinate axis, then the angle is classified as a **quadrantal angle.** For example, the 90° angle, the 180° angle, and the 270° angle shown in **Figure 2.13** are all quadrantal angles.

If the terminal side of an angle in standard position does not lie on a coordinate axis, then the angle is classified according to the quadrant that contains the terminal side. For example, $\angle\beta$ in **Figure 2.14** is a Quadrant III angle.

Angles in standard position that have the same terminal sides are **coterminal angles.** Every angle has an unlimited number of coterminal angles. **Figure 2.15** shows $\angle\theta$ and two of its coterminal angles, labeled $\angle 1$ and $\angle 2$.

Measures of Coterminal Angles

Given $\angle\theta$ in standard position with measure $x°$, then the measures of the angles that are coterminal with $\angle\theta$ are given by

$$x° + k \cdot 360°$$

where k is an integer.

This theorem states that the measures of any two coterminal angles differ by an integer multiple of 360°. For instance, in **Figure 2.15,** $\theta = 430°$,

$$\angle 1 = 430° + (-1) \cdot 360° = 70°, \quad \text{and}$$
$$\angle 2 = 430° + (-2) \cdot 360° = -290°$$

If we add positive multiples of 360° to 430°, we find that the angles with measures 790°, 1150°, 1510°, ... are also coterminal with $\angle\theta$.

EXAMPLE 2 » Classify by Quadrant and Find a Coterminal Angle

Assume the following angles are in standard position. Determine the measure of the positive angle with measure less than 360° that is coterminal with the given angle and then classify the angle by quadrant.

a. $\alpha = 550°$ b. $\beta = -225°$ c. $\gamma = 1105°$

Continued ►

a.

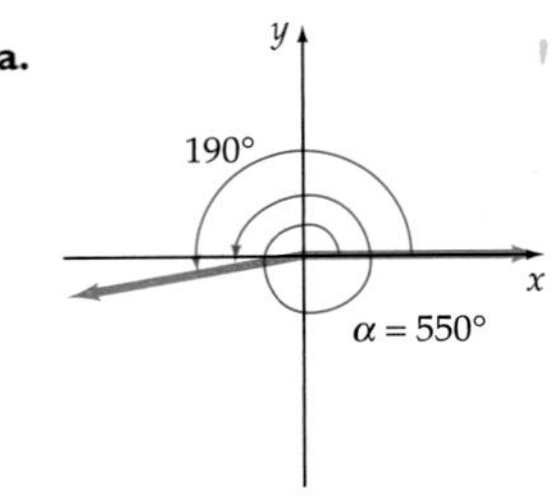

b.

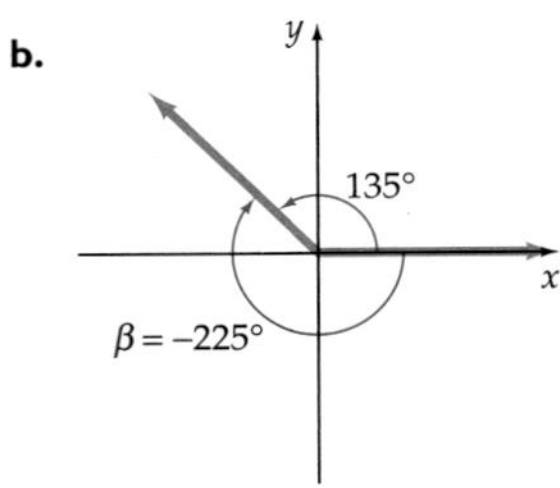

c.

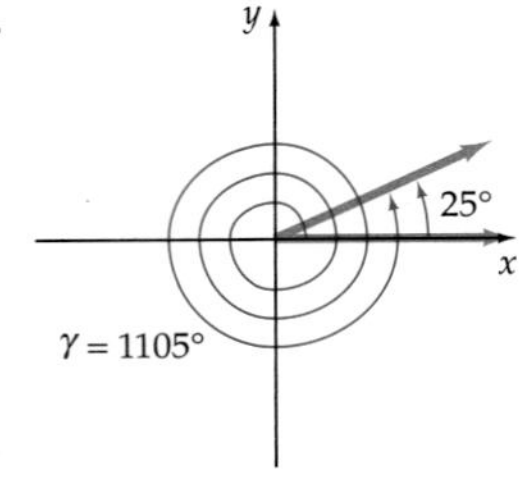

Figure 2.16

Solution

a. Because $550° = 190° + 360°$, $\angle\alpha$ is coterminal with an angle that has a measure of 190°. $\angle\alpha$ is a Quadrant III angle. See **Figure 2.16a.**

b. Because $-225° = 135° + (-1) \cdot 360°$, $\angle\beta$ is coterminal with an angle that has a measure of 135°. $\angle\beta$ is a Quadrant II angle. See **Figure 2.16b.**

c. $1105° \div 360° = 3\frac{5}{72}$. Thus $\angle\gamma$ is an angle formed by three complete counterclockwise rotations, plus $\frac{5}{72}$ of a rotation. To convert $\frac{5}{72}$ of a rotation to degrees, multiply $\frac{5}{72}$ times 360°.

$$\frac{5}{72} \cdot 360° = 25°$$

Thus $1105° = 25° + 3 \cdot 360°$. Hence $\angle\gamma$ is coterminal with an angle that has a measure of 25°. $\angle\gamma$ is a Quadrant I angle. See **Figure 2.16c.**

» ***Try Exercise 14, page 130***

There are two popular methods for representing a fractional part of a degree. One is the **decimal degree method.** For example, the measure 29.76° is a decimal degree. It means

29° plus 76 hundredths of 1°

A second method of measurement is known as the **DMS** (**D**egree, **M**inute, **S**econd) **method.** In the DMS method, a degree is subdivided into 60 equal parts, each of which is called a *minute,* denoted by ′. Thus $1° = 60'$. Furthermore, a minute is subdivided into 60 equal parts, each of which is called a *second,* denoted by ″. Thus $1' = 60''$ and $1° = 3600''$. The fractions

$$\frac{1°}{60'} = 1, \quad \frac{1'}{60''} = 1, \quad \text{and} \quad \frac{1°}{3600''} = 1$$

are another way of expressing the relationships among degrees, minutes, and seconds. Each of the fractions is known as a **unit fraction** or a **conversion factor.** Because all conversion factors are equal to 1, you can multiply a numerical value by a conversion factor and not change the numerical value, even though you change the units used to express the numerical value. The following illustrates the process of multiplying by conversion factors to write 126°12′27″ as a decimal degree.

$$\begin{aligned} 126°12'27'' &= 126° + 12' + 27'' \\ &= 126° + 12'\left(\frac{1°}{60'}\right) + 27''\left(\frac{1°}{3600''}\right) \\ &= 126° + 0.2° + 0.0075° = 126.2075° \end{aligned}$$

Integrating Technology

Many graphing calculators can be used to convert a decimal degree measure to its equivalent DMS measure, and vice versa. For instance, **Figure 2.17** shows that 31.57° is equivalent to 31°34′12″. On a TI-83/TI-83 Plus/TI-84 Plus graphing calculator, the degree symbol, °, and the DMS function are in the ANGLE menu.

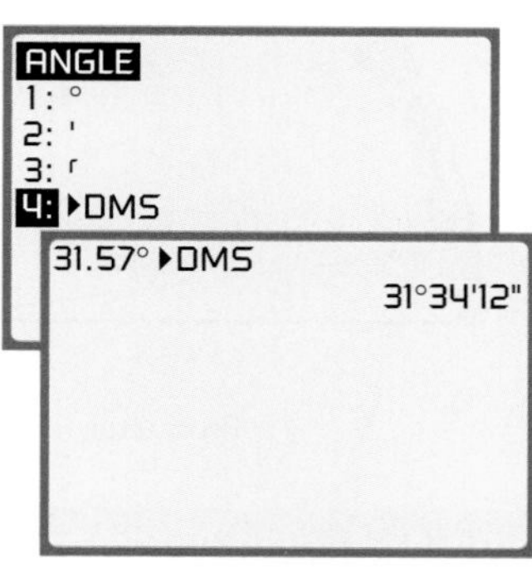

Figure 2.17

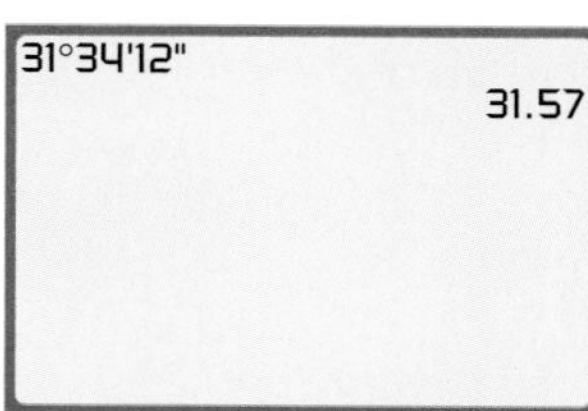

Figure 2.18

To convert a DMS measure to its equivalent decimal degree measure, enter the DMS measure and press ENTER. The calculator screen in **Figure 2.18** shows that 31°34′12″ is equivalent to 31.57°. A TI-83/TI-83 Plus/TI-84 Plus calculator needs to be in degree mode to produce the results displayed in **Figures 2.17** and **2.18.** On a TI-83/TI-83 Plus/TI-84 Plus calculator, the degree symbol, °, and the minute symbol, ′, are both in the ANGLE menu; however, the second symbol, ″, is entered by pressing ALPHA +.

Radian Measure

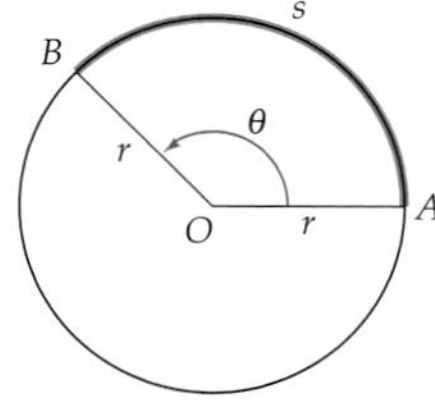

Figure 2.19

Another commonly used angle measurement is the *radian.* To define a radian, first consider a circle of radius r and two radii OA and OB. The angle θ formed by the two radii is a **central angle.** The portion of the circle between A and B is an **arc** of the circle and is written $\overset{\frown}{AB}$. We say that $\overset{\frown}{AB}$ *subtends* the angle θ. The length of $\overset{\frown}{AB}$ is s (see **Figure 2.19**).

Definition of a Radian

One **radian** is the measure of the central angle subtended by an arc of length r on a circle of radius r. See **Figure 2.20.**

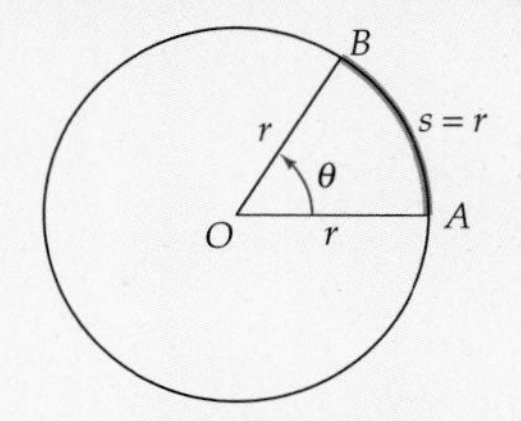

Figure 2.20
Central angle θ has a measure of 1 radian.

Figure 2.21 shows a protractor that can be used to measure angles in radians or to construct angles given in radian measure.

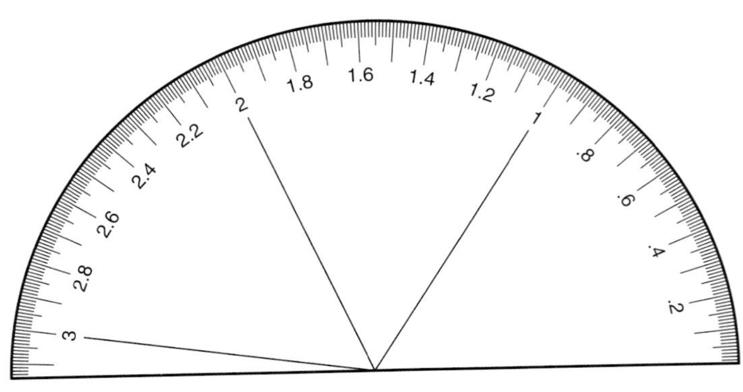

Figure 2.21
Protractor for measuring angles in radians

Definition of Radian Measure

Given an arc of length s on a circle of radius r, the measure of the central angle subtended by the arc is $\theta = \frac{s}{r}$ radians.

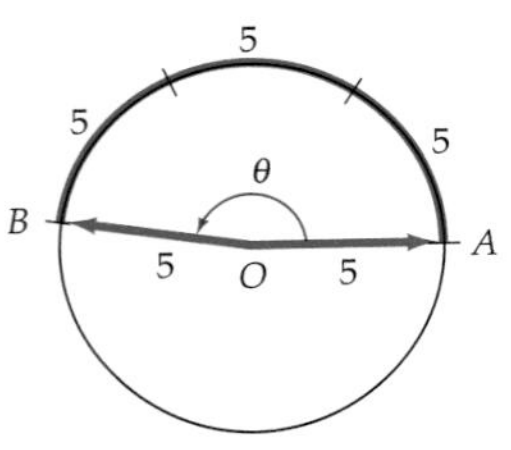

Figure 2.22
Central angle θ has a measure of 3 radians.

As an example, consider that an arc of length 15 centimeters on a circle with a radius of 5 centimeters subtends an angle of 3 radians, as shown in **Figure 2.22**. The same result can be found by dividing 15 centimeters by 5 centimeters.

To find the measure in radians of any central angle θ, divide the length s of the arc that subtends θ by the length of the radius of the circle. Using the formula for radian measure, we find that an arc of length 12 centimeters on a circle of radius 8 centimeters subtends a central angle θ whose measure is

$$\theta = \frac{s}{r} \text{ radians} = \frac{12 \text{ centimeters}}{8 \text{ centimeters}} \text{ radians} = \frac{3}{2} \text{ radians}$$

Note that the centimeter units are *not* part of the final result. The radian measure of a central angle formed by an arc of length 12 miles on a circle of radius 8 miles would be the same, $\frac{3}{2}$ radians. If an angle has a measure of t radians, where t is a real number, then the measure of the angle is often stated as t instead of t radians. For instance, if an angle θ has a measure of 2 radians, we can simply write $\theta = 2$ instead of $\theta = 2$ radians. There will be no confusion concerning whether an angle measure is in degrees or radians, because the degree symbol is *always* used for angle measurements that are in degrees.

Recall that the circumference of a circle is given by the equation $C = 2\pi r$. The radian measure of the central angle θ subtended by the circumference is $\theta = \frac{2\pi r}{r} = 2\pi$. In degree measure, the central angle θ subtended by the circumference is 360°. Thus we have the relationship $360° = 2\pi$ radians. Dividing each side of the equation by 2 gives $180° = \pi$ radians. From this last equation, we can establish the following conversion factors.

Integrating Technology

A calculator shows that

1 radian ≈ 57.29577951°

and

1° ≈ 0.017453293 radian

Radian-Degree Conversion

- To convert from radians to degrees, multiply by $\left(\dfrac{180°}{\pi \text{ radians}}\right)$.
- To convert from degrees to radians, multiply by $\left(\dfrac{\pi \text{ radians}}{180°}\right)$.

EXAMPLE 3 Convert from Degrees to Radians

Convert each angle in degrees to radians.

a. 60° b. 315° c. −150°

Solution

Multiply each degree measure by $\left(\dfrac{\pi \text{ radians}}{180°}\right)$ and simplify. In each case, the degree units in the numerator cancel with the degree units in the denominator.

a. $60° = 60\not{°}\left(\dfrac{\pi \text{ radians}}{180\not{°}}\right) = \dfrac{60\pi}{180} \text{ radians} = \dfrac{\pi}{3} \text{ radians}$

b. $315° = 315\not{°}\left(\dfrac{\pi \text{ radians}}{180\not{°}}\right) = \dfrac{315\pi}{180} \text{ radians} = \dfrac{7\pi}{4} \text{ radians}$

c. $-150° = -150\not{°}\left(\dfrac{\pi \text{ radians}}{180\not{°}}\right) = -\left(\dfrac{150\pi}{180}\right) \text{radians} = -\dfrac{5\pi}{6} \text{ radians}$

Try Exercise 32, page 130

EXAMPLE 4 Convert from Radians to Degrees

Convert each angle in radians to degrees.

a. $\dfrac{3\pi}{4}$ radians b. 1 radian c. $-\dfrac{5\pi}{2}$ radians

Solution

Multiply each radian measure by $\left(\dfrac{180°}{\pi \text{ radians}}\right)$ and simplify. In each case, the radian units in the numerator cancel with the radian units in the denominator.

a. $\dfrac{3\pi}{4} \text{ radians} = \left(\dfrac{3\cancel{\pi} \cancel{\text{radians}}}{4}\right)\left(\dfrac{180°}{\cancel{\pi} \cancel{\text{radians}}}\right) = \dfrac{3 \cdot 180°}{4} = 135°$

b. $1 \text{ radian} = (1 \cancel{\text{radian}})\left(\dfrac{180°}{\pi \cancel{\text{radians}}}\right) = \dfrac{180°}{\pi} \approx 57.3°$

c. $-\dfrac{5\pi}{2} \text{ radians} = \left(-\dfrac{5\cancel{\pi} \cancel{\text{radians}}}{2}\right)\left(\dfrac{180°}{\cancel{\pi} \cancel{\text{radians}}}\right) = -\dfrac{5 \cdot 180°}{2} = -450°$

Try Exercise 44, page 130

Section 2.2

- The Six Trigonometric Functions
- Trigonometric Functions of Special Angles
- Applications Involving Right Triangles

Right Triangle Trigonometry

PREPARE FOR THIS SECTION

Prepare for this section by completing the following exercises. The answers can be found on page A9.

PS1. Rationalize the denominator of $\frac{1}{\sqrt{3}}$.

PS2. Rationalize the denominator of $\frac{2}{\sqrt{2}}$.

PS3. Simplify: $a \div \left(\frac{a}{2}\right)$

PS4. Simplify: $\left(\frac{a}{2}\right) \div \left(\frac{\sqrt{3}}{2}a\right)$

PS5. Solve $\frac{\sqrt{2}}{2} = \frac{x}{5}$ for x. Round your answer to the nearest hundredth. [1.1]

PS6. Solve $\frac{\sqrt{3}}{3} = \frac{x}{18}$ for x. Round your answer to the nearest hundredth. [1.1]

The Six Trigonometric Functions

The study of trigonometry, which means "triangle measurement," began more than 2000 years ago, partially as a means of solving surveying problems. Early trigonometry used the length of a line segment between two points of a circle as the value of a *trigonometric function.* In the sixteenth century, right triangles were used to define a trigonometric function. We will use a modification of this approach.

When working with right triangles, it is convenient to refer to the side *opposite* an angle or the side *adjacent* to (next to) an angle. **Figure 2.29** shows the sides opposite and adjacent to the angle θ. **Figure 2.30** shows the sides opposite and adjacent to the angle β. In both cases, the hypotenuse remains the same.

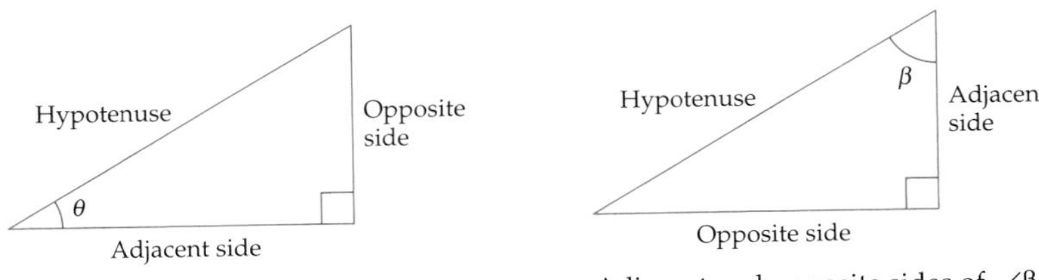

Adjacent and opposite sides of $\angle\theta$

Figure 2.29

Adjacent and opposite sides of $\angle\beta$

Figure 2.30

Six ratios can be formed by using two lengths of the three sides of a right triangle. Each ratio defines a value of a trigonometric function of a given acute angle θ. The functions are **sine** (sin), **cosine** (cos), **tangent** (tan), **cosecant** (csc), **secant** (sec), and **cotangent** (cot).

Definitions of Trigonometric Functions of an Acute Angle

Let θ be an acute angle of a right triangle. See **Figure 2.29.** The values of the six trigonometric functions of θ are

$$\sin\theta = \frac{\text{length of opposite side}}{\text{length of hypotenuse}} \qquad \cos\theta = \frac{\text{length of adjacent side}}{\text{length of hypotenuse}}$$

$$\tan\theta = \frac{\text{length of opposite side}}{\text{length of adjacent side}} \qquad \cot\theta = \frac{\text{length of adjacent side}}{\text{length of opposite side}}$$

$$\sec\theta = \frac{\text{length of hypotenuse}}{\text{length of adjacent side}} \qquad \csc\theta = \frac{\text{length of hypotenuse}}{\text{length of opposite side}}$$

We will write opp, adj, and hyp as abbreviations for *the length of the* opposite side, adjacent side, and hypotenuse, respectively.

EXAMPLE 1 Evaluate Trigonometric Functions

Find the values of the six trigonometric functions of θ for the triangle given in **Figure 2.31.**

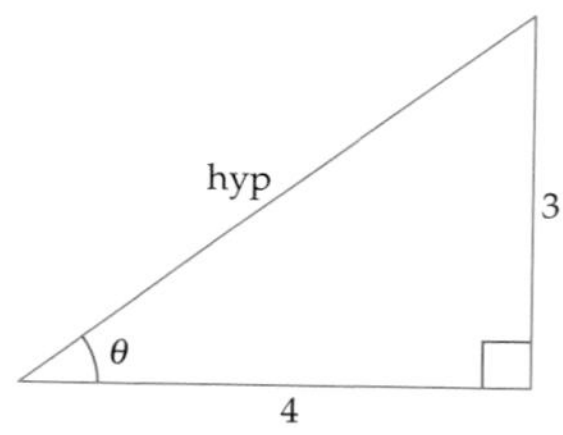

Figure 2.31

Solution

Use the Pythagorean Theorem to find the length of the hypotenuse.

$$\text{hyp} = \sqrt{3^2 + 4^2} = \sqrt{25} = 5$$

From the definitions of the trigonometric functions,

$$\sin\theta = \frac{\text{opp}}{\text{hyp}} = \frac{3}{5} \qquad \cos\theta = \frac{\text{adj}}{\text{hyp}} = \frac{4}{5} \qquad \tan\theta = \frac{\text{opp}}{\text{adj}} = \frac{3}{4}$$

$$\cot\theta = \frac{\text{adj}}{\text{opp}} = \frac{4}{3} \qquad \sec\theta = \frac{\text{hyp}}{\text{adj}} = \frac{5}{4} \qquad \csc\theta = \frac{\text{hyp}}{\text{opp}} = \frac{5}{3}$$

Try Exercise 6, page 142

Given the value of one trigonometric function of the acute angle θ, it is possible to find the value of any of the remaining trigonometric functions of θ.

EXAMPLE 2 Find the Value of a Trigonometric Function

Given that θ is an acute angle and $\cos\theta = \dfrac{5}{8}$, find $\tan\theta$.

Continued ▶

Solution

$$\cos\theta = \frac{5}{8} = \frac{\text{adj}}{\text{hyp}}$$

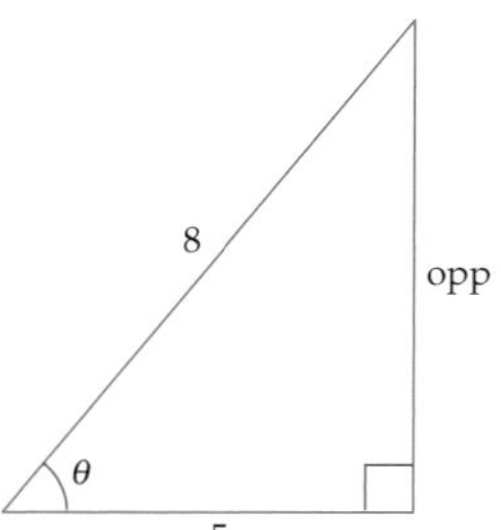

Figure 2.32

Sketch a right triangle with one leg of length 5 units and a hypotenuse of length 8 units. Label as θ the acute angle that has the leg of length 5 units as its adjacent side (see **Figure 2.32**). Use the Pythagorean Theorem to find the length of the opposite side.

$$\begin{aligned}(\text{opp})^2 + 5^2 &= 8^2\\ (\text{opp})^2 + 25 &= 64\\ (\text{opp})^2 &= 39\\ \text{opp} &= \sqrt{39}\end{aligned}$$

Therefore, $\tan\theta = \dfrac{\text{opp}}{\text{adj}} = \dfrac{\sqrt{39}}{5}$.

Try Exercise 18, page 142

Trigonometric Functions of Special Angles

In Example 1, the lengths of the legs of the triangle were given, and you were asked to find the values of the six trigonometric functions of the angle θ. Often we will want to find the value of a trigonometric function when we are given *the measure of an angle* rather than the measure of the sides of a triangle. For most angles, advanced mathematical methods are required to evaluate a trigonometric function. For some *special angles,* however, the value of a trigonometric function can be found by geometric methods. These special acute angles are 30°, 45°, and 60°.

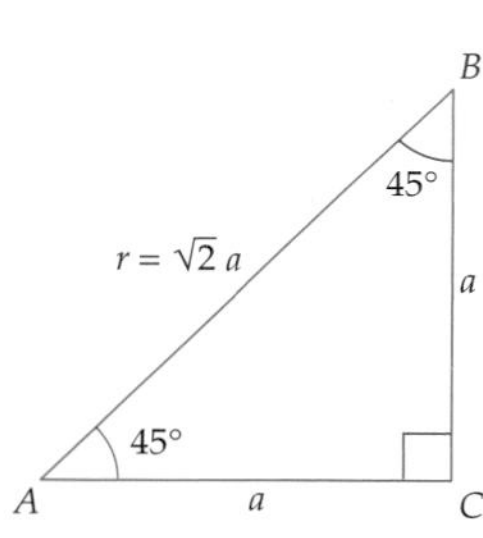

Figure 2.33

First, we will find the values of the six trigonometric functions of 45°. (This discussion is based on angles measured in degrees. Radian measure could have been used without changing the results.) **Figure 2.33** shows a right triangle with angles 45°, 45°, and 90°. Because $\angle A = \angle B$, the lengths of the sides opposite these angles are equal. Let the length of each equal side be denoted by a. From the Pythagorean Theorem,

$$\begin{aligned}r^2 &= a^2 + a^2 = 2a^2\\ r &= \sqrt{2a^2} = \sqrt{2}\,a\end{aligned}$$

The values of the six trigonometric functions of 45° are

$$\sin 45^\circ = \frac{a}{\sqrt{2}a} = \frac{1}{\sqrt{2}} = \frac{\sqrt{2}}{2} \qquad \cos 45^\circ = \frac{a}{\sqrt{2}a} = \frac{1}{\sqrt{2}} = \frac{\sqrt{2}}{2}$$

$$\tan 45^\circ = \frac{a}{a} = 1 \qquad \cot 45^\circ = \frac{a}{a} = 1$$

$$\sec 45^\circ = \frac{\sqrt{2}a}{a} = \sqrt{2} \qquad \csc 45^\circ = \frac{\sqrt{2}a}{a} = \sqrt{2}$$

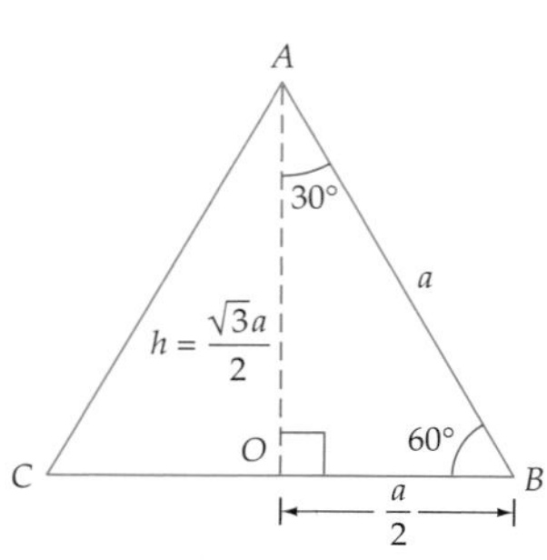

Figure 2.34

The values of the trigonometric functions of the special angles 30° and 60° can be found by drawing an equilateral triangle and bisecting one of the angles, as **Figure 2.34** shows. The angle bisector also bisects one of the sides. Thus the length of the side opposite the 30° angle is one-half the length of the hypotenuse of triangle OAB.

Let a denote the length of the hypotenuse. Then the length of the side opposite the 30° angle is $\frac{a}{2}$. The length of the side adjacent to the 30° angle, h, is found by using the Pythagorean Theorem.

$$a^2 = \left(\frac{a}{2}\right)^2 + h^2$$

$$a^2 = \frac{a^2}{4} + h^2$$

$$\frac{3a^2}{4} = h^2 \qquad \text{• Subtract } \frac{a^2}{4} \text{ from each side.}$$

$$h = \frac{\sqrt{3}a}{2} \qquad \text{• Solve for } h.$$

The values of the six trigonometric functions of 30° are

$$\sin 30^\circ = \frac{a/2}{a} = \frac{1}{2} \qquad \cos 30^\circ = \frac{\sqrt{3}a/2}{a} = \frac{\sqrt{3}}{2}$$

$$\tan 30^\circ = \frac{a/2}{\sqrt{3}a/2} = \frac{1}{\sqrt{3}} = \frac{\sqrt{3}}{3} \qquad \cot 30^\circ = \frac{\sqrt{3}a/2}{a/2} = \sqrt{3}$$

$$\sec 30^\circ = \frac{a}{\sqrt{3}a/2} = \frac{2}{\sqrt{3}} = \frac{2\sqrt{3}}{3} \qquad \csc 30^\circ = \frac{a}{a/2} = 2$$

The values of the trigonometric functions of 60° can be found by again using **Figure 2.34.** The length of the side opposite the 60° angle is $\frac{\sqrt{3}a}{2}$, and the length of the side adjacent to the 60° angle is $\frac{a}{2}$. The values of the trigonometric functions of 60° are

$$\sin 60^\circ = \frac{\sqrt{3}a/2}{a} = \frac{\sqrt{3}}{2} \qquad \cos 60^\circ = \frac{a/2}{a} = \frac{1}{2}$$

$$\tan 60^\circ = \frac{\sqrt{3}a/2}{a/2} = \sqrt{3} \qquad \cot 60^\circ = \frac{a/2}{\sqrt{3}a/2} = \frac{1}{\sqrt{3}} = \frac{\sqrt{3}}{3}$$

$$\sec 60^\circ = \frac{a}{a/2} = 2 \qquad \csc 60^\circ = \frac{a}{\sqrt{3}a/2} = \frac{2}{\sqrt{3}} = \frac{2\sqrt{3}}{3}$$

Table 2.2 summarizes the values of the trigonometric functions of the special angles 30° ($\pi/6$), 45° ($\pi/4$), and 60° ($\pi/3$).

take note

Memorizing the values given in Table 2.2 will prove to be extremely useful in the remaining trigonometry sections.

Table 2.2 Trigonometric Functions of Special Angles

θ	sin θ	cos θ	tan θ	csc θ	sec θ	cot θ
$30°; \frac{\pi}{6}$	$\frac{1}{2}$	$\frac{\sqrt{3}}{2}$	$\frac{\sqrt{3}}{3}$	2	$\frac{2\sqrt{3}}{3}$	$\sqrt{3}$
$45°; \frac{\pi}{4}$	$\frac{\sqrt{2}}{2}$	$\frac{\sqrt{2}}{2}$	1	$\sqrt{2}$	$\sqrt{2}$	1
$60°; \frac{\pi}{3}$	$\frac{\sqrt{3}}{2}$	$\frac{1}{2}$	$\sqrt{3}$	$\frac{2\sqrt{3}}{3}$	2	$\frac{\sqrt{3}}{3}$

QUESTION What is the measure, in degrees, of the acute angle θ for which $\sin\theta = \cos\theta$, $\tan\theta = \cot\theta$, and $\sec\theta = \csc\theta$?

EXAMPLE 3 Evaluate a Trigonometric Expression

Find the *exact* value of $\sin^2 45^\circ + \cos^2 60^\circ$.

Note: $\sin^2\theta = (\sin\theta)(\sin\theta) = (\sin\theta)^2$ and $\cos^2\theta = (\cos\theta)(\cos\theta) = (\cos\theta)^2$.

Solution

Substitute the values of sin 45° and cos 60° and simplify.

$$\sin^2 45^\circ + \cos^2 60^\circ = \left(\frac{\sqrt{2}}{2}\right)^2 + \left(\frac{1}{2}\right)^2 = \frac{2}{4} + \frac{1}{4} = \frac{3}{4}$$

Try Exercise 34, page 142

take note

The patterns in the following chart can be used to memorize the sine and cosine of 30°, 45°, and 60°.

$\sin 30^\circ = \frac{\sqrt{1}}{2}$	$\cos 30^\circ = \frac{\sqrt{3}}{2}$
$\sin 45^\circ = \frac{\sqrt{2}}{2}$	$\cos 45^\circ = \frac{\sqrt{2}}{2}$
$\sin 60^\circ = \frac{\sqrt{3}}{2}$	$\cos 60^\circ = \frac{\sqrt{1}}{2}$

From the definition of the sine and cosecant functions,

$$(\sin\theta)(\csc\theta) = \frac{\text{opp}}{\text{hyp}} \cdot \frac{\text{hyp}}{\text{opp}} = 1 \quad \text{or} \quad (\sin\theta)(\csc\theta) = 1$$

By rewriting the last equation, we find

$$\sin\theta = \frac{1}{\csc\theta} \quad \text{and} \quad \csc\theta = \frac{1}{\sin\theta}, \text{ provided } \sin\theta \neq 0$$

The sine and cosecant functions are called **reciprocal functions**. The cosine and secant are also reciprocal functions, as are the tangent and cotangent functions. **Table 2.3** shows each trigonometric function and its reciprocal. These relationships hold for all values of θ for which both of the functions are defined.

Table 2.3 Trigonometric Functions and Their Reciprocals

$\sin\theta = \frac{1}{\csc\theta}$	$\cos\theta = \frac{1}{\sec\theta}$	$\tan\theta = \frac{1}{\cot\theta}$
$\csc\theta = \frac{1}{\sin\theta}$	$\sec\theta = \frac{1}{\cos\theta}$	$\cot\theta = \frac{1}{\tan\theta}$

Integrating Technology

The values of the trigonometric functions of the special angles 30°, 45°, and 60° shown in **Table 2.2** are exact values. If an angle is not one of these special angles, then a graphing calculator often is used to approximate the value of a trigonometric function. For instance, to find sin 52.4° on a TI-83/TI-83 Plus/TI-84 Plus calculator, first check that the calculator is in degree mode. Then use the sine function key SIN to key in sin(52.4) and press ENTER. See **Figure 2.35**.

ANSWER 45°

To find sec 1.25, first check that the calculator is in radian mode. A TI-83/TI-83 Plus/TI-84 Plus calculator does not have a secant function key, but because the secant function is the reciprocal of the cosine function, we can evaluate sec 1.25 by evaluating 1/(cos 1.25). See **Figure 2.36.**

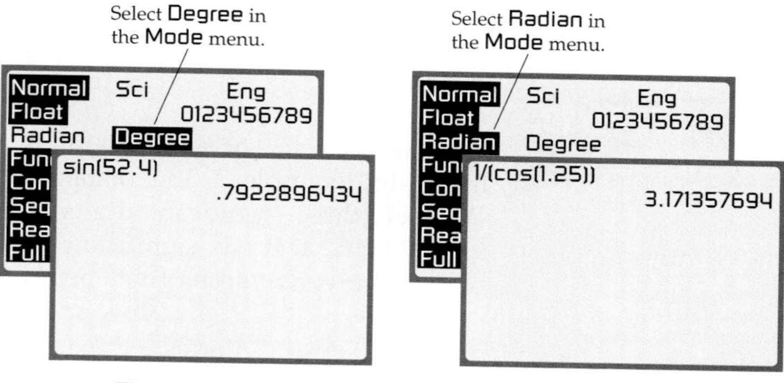

Figure 2.35 **Figure 2.36**

When you evaluate a trigonometric function with a calculator, be sure the calculator is in the correct mode. Many errors are made because the correct mode has not been selected.

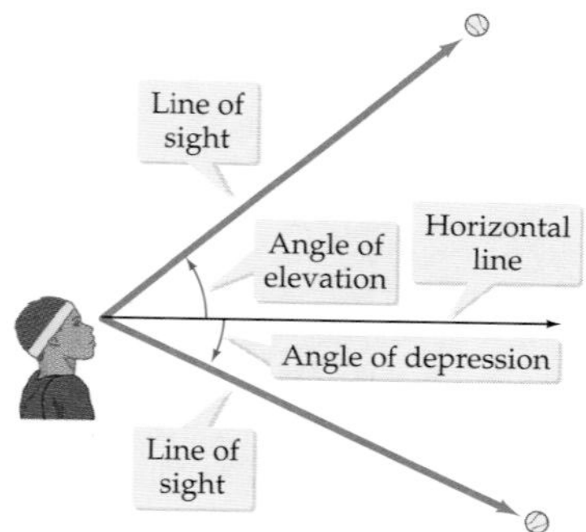

Figure 2.37

Applications Involving Right Triangles

Some applications concern an observer looking at an object. In these applications, angles of elevation or angles of depression are formed by a line of sight and a horizontal line. If the object being observed is above the observer, the acute angle formed by the line of sight and the horizontal line is an **angle of elevation.** If the object being observed is below the observer, the acute angle formed by the line of sight and the horizontal line is an **angle of depression.** See **Figure 2.37.**

EXAMPLE 4 ⟫ Solve an Angle-of-Elevation Application

From a point 115 feet from the base of a redwood tree, the angle of elevation to the top of the tree is 64.3°. Find the height of the tree to the nearest foot.

Solution

From **Figure 2.38,** the length of the adjacent side of the angle is known (115 feet). Because we need to determine the height of the tree (length of the opposite side), we use the tangent function. Let h represent the length of the opposite side.

$$\tan 64.3^\circ = \frac{\text{opp}}{\text{adj}} = \frac{h}{115}$$

$$h = 115 \tan 64.3^\circ \approx 238.952$$

• **Use a calculator to evaluate tan 64.3°.**

The height of the tree is approximately 239 feet.

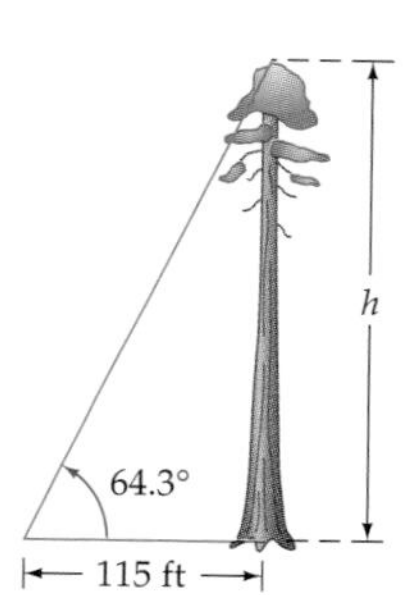

Figure 2.38

⟫ ***Try Exercise 56, page 143***

> **take note**
>
> The significant digits of an approximate number are
> - every nonzero digit.
> - the digit 0, provided it is between two nonzero digits or it is to the right of a nonzero digit in a number which includes a decimal point.
>
> For example, the approximate number:
>
> 502 has 3 significant digits.
> 3700 has 2 significant digits.
> 47.0 has 3 significant digits.
> 0.0023 has 2 significant digits.
> 0.00840 has 3 significant digits.

Because the cotangent function involves the sides adjacent to and opposite an angle, we could have solved Example 4 by using the cotangent function. The solution would have been

$$\cot 64.3° = \frac{\text{adj}}{\text{opp}} = \frac{115}{h}$$

$$h = \frac{115}{\cot 64.3°} \approx 238.952 \text{ feet}$$

The accuracy of a calculator is sometimes beyond the limits of measurement. In Example 4 the distance from the base of the tree was given as 115 feet (three significant digits), whereas the height of the tree was shown to be 238.952 feet (six significant digits). When using approximate numbers, we will use the conventions given below for calculating with trigonometric functions.

A Rounding Convention: Significant Digits for Trigonometric Calculations

Angle Measure to the Nearest	*Significant Digits of the Lengths*
Degree	Two
Tenth of a degree	Three
Hundredth of a degree	Four

EXAMPLE 5 ❯❯ Solve an Angle-of-Depression Application

DME (Distance Measuring Equipment) is standard avionic equipment on a commercial airplane. This equipment measures the distance from a plane to a radar station. If the distance from a plane to a radar station is 160 miles and the angle of depression is 33°, find the number of ground miles from a point directly below the plane to the radar station.

Solution

From **Figure 2.39**, the length of the hypotenuse is known (160 miles). The length of the side opposite the angle of 57° is unknown. The sine function involves the hypotenuse and the opposite side, x, of the 57°angle.

$$\sin 57° = \frac{x}{160}$$

$$x = 160 \sin 57° \approx 134.1873$$

Rounded to two significant digits, the plane is 130 ground miles from the radar station.

❯❯ *Try Exercise 58, page 143*

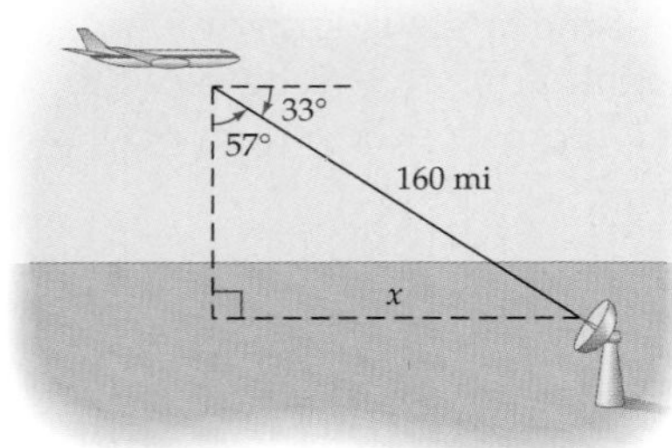

Figure 2.39

EXAMPLE 6 Solve an Angle-of-Elevation Application

An observer notes that the angle of elevation from point A to the top of a space shuttle is 27.2°. From a point 17.5 meters further from the space shuttle, the angle of elevation is 23.9°. Find the height of the space shuttle.

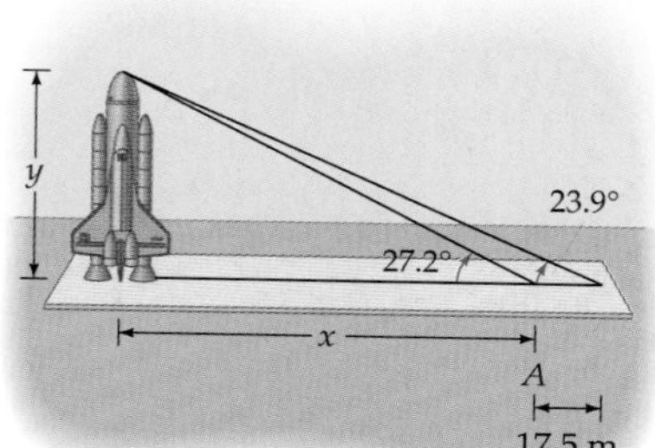

Figure 2.40

take note

The intermediate calculations in Example 6 were not rounded off. This ensures better accuracy for the final result. Using the rounding convention stated on page 140, we round off only the last result.

Solution

From **Figure 2.40**, let x denote the distance from point A to the base of the space shuttle, and let y denote the height of the space shuttle. Then

$$(1)\quad \tan 27.2^\circ = \frac{y}{x} \qquad \text{and} \qquad (2)\quad \tan 23.9^\circ = \frac{y}{x + 17.5}$$

Solving Equation (1) for x, $x = \dfrac{y}{\tan 27.2^\circ} = y \cot 27.2^\circ$, and substituting into Equation (2), we have

$$\tan 23.9^\circ = \frac{y}{y \cot 27.2^\circ + 17.5}$$

$$y = (\tan 23.9^\circ)(y \cot 27.2^\circ + 17.5) \qquad \text{• Solve for } y.$$

$$y - y \tan 23.9^\circ \cot 27.2^\circ = (\tan 23.9^\circ)(17.5)$$

$$y(1 - \tan 23.9^\circ \cot 27.2^\circ) = (\tan 23.9^\circ)(17.5)$$

$$y = \frac{(\tan 23.9^\circ)(17.5)}{1 - \tan 23.9^\circ \cot 27.2^\circ} \approx 56.2993$$

To three significant digits, the height of the space shuttle is 56.3 meters.

Try Exercise 66, page 145

Topics for Discussion

1. If θ is an acute angle of a right triangle for which $\cos \theta = \frac{3}{8}$, then it must be the case that $\sin \theta = \frac{5}{8}$. Do you agree? Explain.
2. A tutor claims that $\tan 30^\circ = \cot 60^\circ$. Do you agree?
3. Does $\sin 2\theta = 2 \sin \theta$? Explain.
4. How many significant digits are in each of the following measurements?

   a. 0.0042 inches b. 5.03 inches c. 62.00 inches

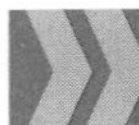

Exercise Set 2.2

In Exercises 1 to 12, find the values of the six trigonometric functions of θ for the right triangle with the given sides.

1.

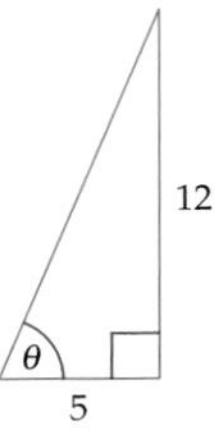

2.

3.

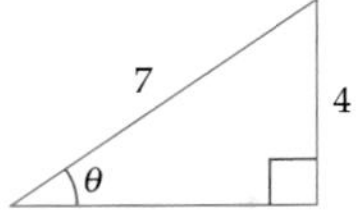

4.

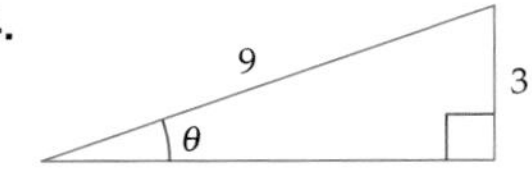

5.

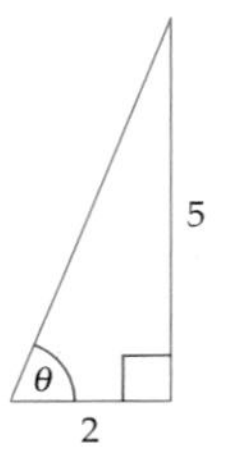

6.

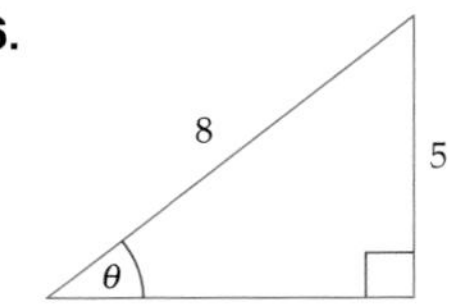

7.

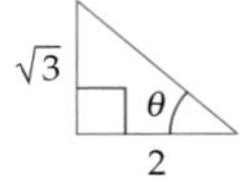

8.

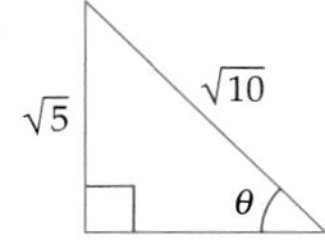

9.

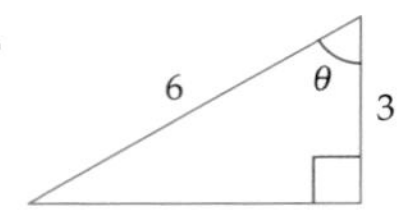

10.

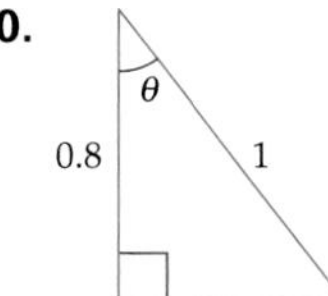

11.

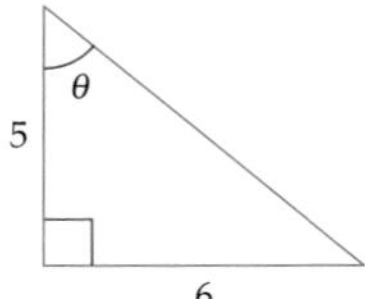

12.

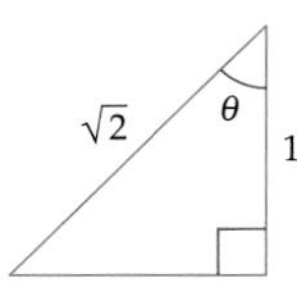

In Exercises 13 to 15, let θ be an acute angle of a right triangle for which $\sin\theta = \dfrac{3}{5}$. Find

13. $\tan\theta$ **14.** $\sec\theta$ **15.** $\cos\theta$

In Exercises 16 to 18, let θ be an acute angle of a right triangle for which $\tan\theta = \dfrac{4}{3}$. Find

16. $\sin\theta$ **17.** $\cot\theta$ **18.** $\sec\theta$

In Exercises 19 to 21, let β be an acute angle of a right triangle for which $\sec\beta = \dfrac{13}{12}$. Find

19. $\cos\beta$ **20.** $\cot\beta$ **21.** $\csc\beta$

In Exercises 22 to 24, let θ be an acute angle of a right triangle for which $\cos\theta = \dfrac{2}{3}$. Find

22. $\sin\theta$ **23.** $\sec\theta$ **24.** $\tan\theta$

In Exercises 25 to 38, find the *exact* value of each expression.

25. $\sin 45^\circ + \cos 45^\circ$

26. $\csc 45^\circ - \sec 45^\circ$

27. $\sin 30^\circ \cos 60^\circ - \tan 45^\circ$

28. $\csc 60^\circ \sec 30^\circ + \cot 45^\circ$

29. $\sin 30^\circ \cos 60^\circ + \tan 45^\circ$

30. $\sec 30^\circ \cos 30^\circ - \tan 60^\circ \cot 60^\circ$

31. $\sin\dfrac{\pi}{3} + \cos\dfrac{\pi}{6}$

32. $\csc\dfrac{\pi}{6} - \sec\dfrac{\pi}{3}$

33. $\sin\dfrac{\pi}{4} + \tan\dfrac{\pi}{6}$

34. $\sin\dfrac{\pi}{3}\cos\dfrac{\pi}{4} - \tan\dfrac{\pi}{4}$

35. $\sec\dfrac{\pi}{3}\cos\dfrac{\pi}{3} - \tan\dfrac{\pi}{6}$

36. $\cos\dfrac{\pi}{4}\tan\dfrac{\pi}{6} + 2\tan\dfrac{\pi}{3}$

37. $2\csc\dfrac{\pi}{4} - \sec\dfrac{\pi}{3}\cos\dfrac{\pi}{6}$

38. $3\tan\dfrac{\pi}{4} + \sec\dfrac{\pi}{6}\sin\dfrac{\pi}{3}$

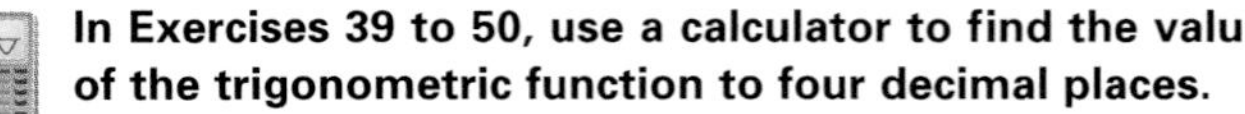

In Exercises 39 to 50, use a calculator to find the value of the trigonometric function to four decimal places.

39. $\tan 32^\circ$ **40.** $\sec 88^\circ$ **41.** $\cos 63^\circ 20'$

42. $\cot 55^\circ 50'$ **43.** $\cos 34.7^\circ$ **44.** $\tan 81.3^\circ$

45. sec 5.9°

46. $\sin \frac{\pi}{5}$

47. $\tan \frac{\pi}{7}$

48. $\sec \frac{3\pi}{8}$

49. csc 1.2

50. sin 0.45

51. Vertical Height from Slant Height A 12-foot ladder is resting against a wall and makes an angle of 52° with the ground. Find the height to which the ladder will reach on the wall.

52. Distance Across a Marsh Find the distance AB across the marsh shown in the accompanying figure.

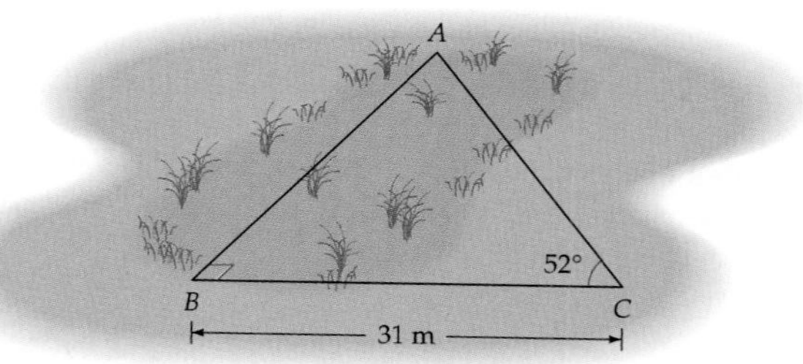

53. Width of a Ramp A skateboarder wishes to build a jump ramp that is inclined at a 19° angle and that has a maximum height of 32.0 inches. Find the horizontal width x of the ramp.

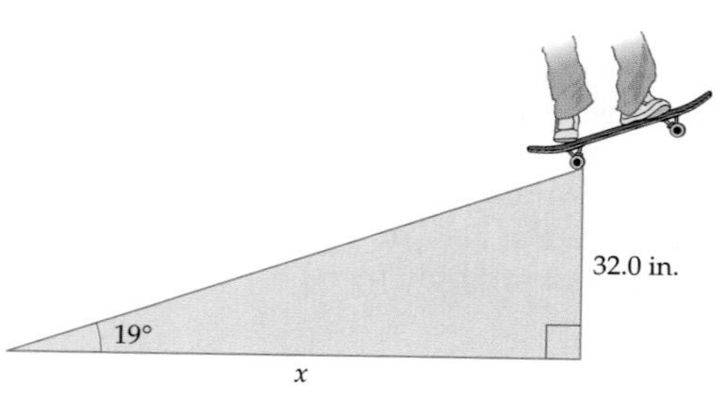

54. Time of Closest Approach At 3:00 P.M., a boat is 12.5 miles due west of a radar station and traveling at 11 mph in a direction that is 57.3° south of an east-west line. At what time will the boat be closest to the radar station?

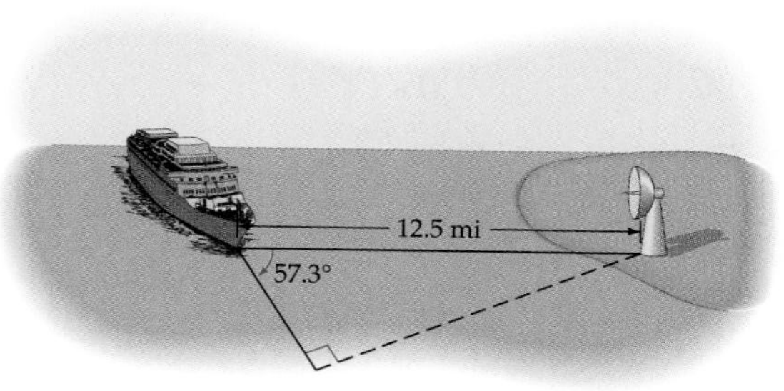

55. Placement of a Light For best illumination of a piece of art, a lighting specialist for an art gallery recommends that a ceiling-mounted light be 6 feet from the piece of art and that the angle of depression of the light be 38°. How far from a wall should the light be placed so that the recommendations of the specialist are met? Notice that the art extends outward 4 inches from the wall.

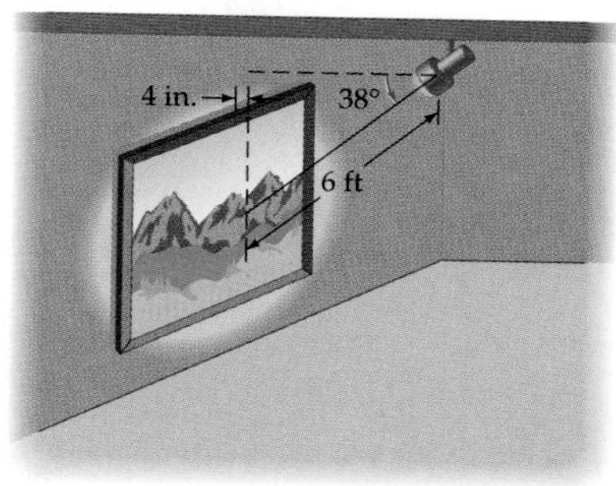

56. Height of the Eiffel Tower The angle of elevation from a point 116 meters from the base of the Eiffel Tower to the top of the tower is 68.9°. Find the approximate height of the tower.

57. Distance of a Descent An airplane traveling at 240 mph is descending at an angle of depression of 6°. How many miles will the plane descend in 4 minutes?

58. Time of a Descent A submarine traveling at 9.0 mph is descending at an angle of depression of 5°. How many minutes, to the nearest tenth, does it take the submarine to reach a depth of 80 feet?

59. Height of a Building A surveyor determines that the angle of elevation from a transit to the top of a building is 27.8°. The transit is positioned 5.5 feet above ground level and 131 feet from the building. Find the height of the building to the nearest tenth of a foot.

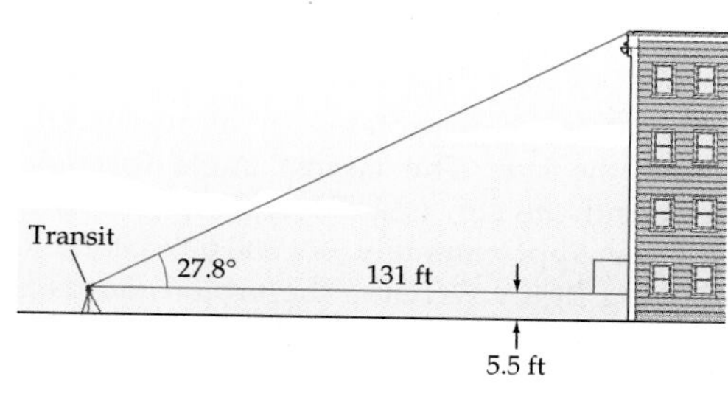

60. **Width of a Lake** The angle of depression to one side of a lake, measured from a balloon 2500 feet above the lake as shown in the accompanying figure, is 43°. The angle of depression to the opposite side of the lake is 27°. Find the width of the lake.

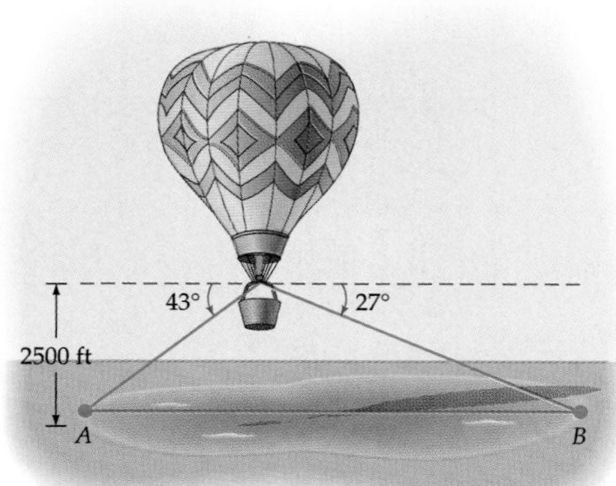

61. **Astronomy** The moon Europa rotates in a nearly circular orbit around Jupiter. The orbital radius of Europa is approximately 670,900 kilometers. During a revolution of Europa around Jupiter, an astronomer found that the maximum value of the angle θ formed by Europa, Earth, and Jupiter was 0.056°. Find the distance d between Earth and Jupiter at the time the astronomer found the maximum value of θ. Round to the nearest million kilometers.

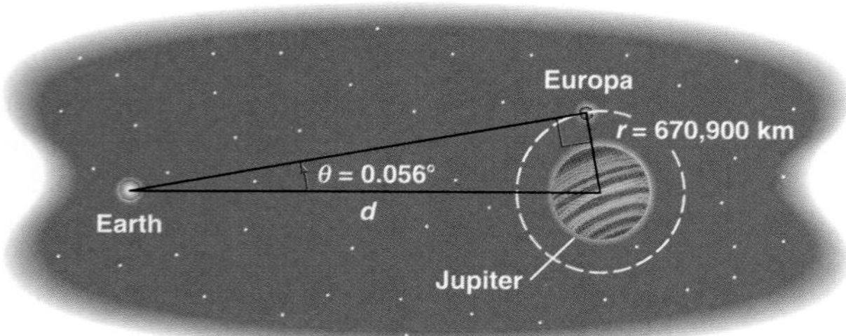

Not drawn to scale.

62. **Astronomy** Venus rotates in a nearly circular orbit around the sun. The largest angle formed by Venus, Earth, and the sun is 46.5°. The distance from Earth to the sun is approximately 149,000,000 kilometers. See the following figure. What is the orbital radius r of Venus? Round to the nearest million kilometers.

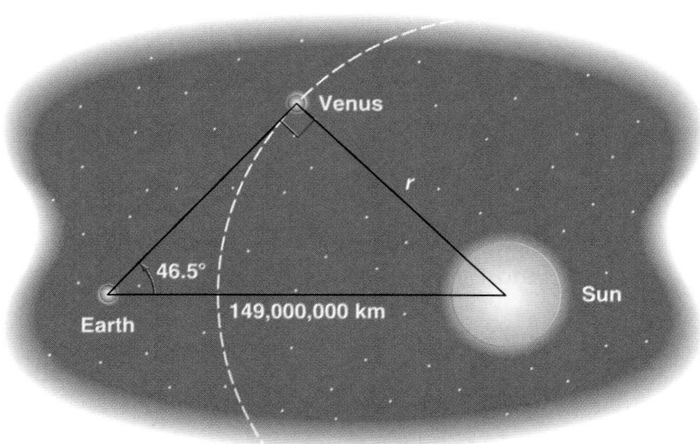

63. **Area of an Isosceles Triangle** Consider the following *isosceles* triangle. The length of each of the two equal sides of the triangle is a, and each of the base angles has a measure of θ. Verify that the area of the triangle is $A = a^2 \sin\theta\cos\theta$.

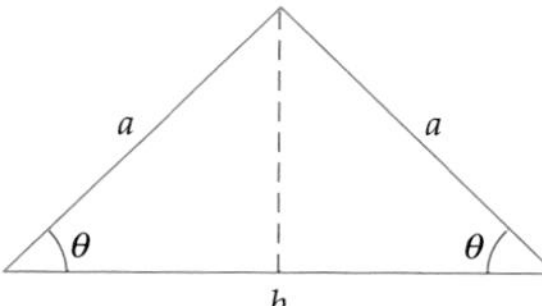

64. **Area of a Hexagon** Find the area of the hexagon. (*Hint:* The area consists of six isosceles triangles. Use the formula from Exercise 63 to compute the area of one of the triangles and multiply by 6.)

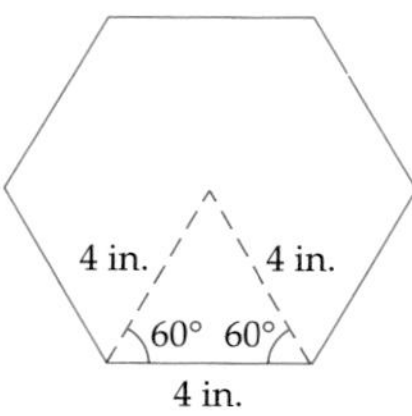

65. **Height of a Pyramid** The angle of elevation to the top of the Egyptian pyramid of Cheops is 36.4°, measured from a point 350 feet from the base of the pyramid. The angle of elevation from the base of a face of the pyramid is 51.9°. Find the height of the Cheops pyramid.

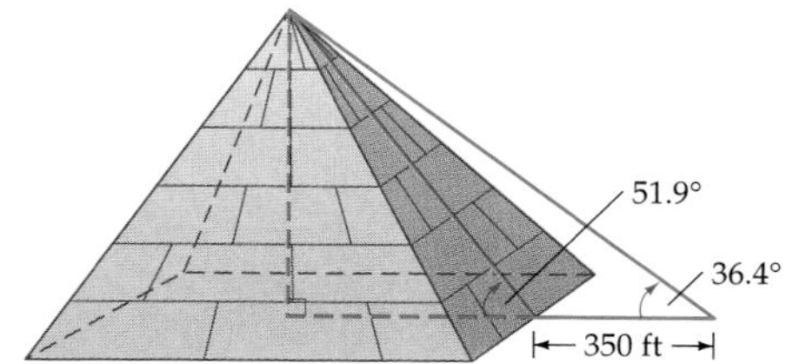

66. **Height of a Building** Two buildings are 240 feet apart. The angle of elevation from the top of the shorter building to the top of the other building is 22°. If the shorter building is 80 feet high, how high is the taller building?

67. **Height of the Washington Monument** From a point A on a line from the base of the Washington Monument, the angle of elevation to the top of the monument is 42.0°. From a point 100 feet away from A and on the same line, the angle to the top is 37.8°. Find the height, to the nearest foot, of the Washington Monument.

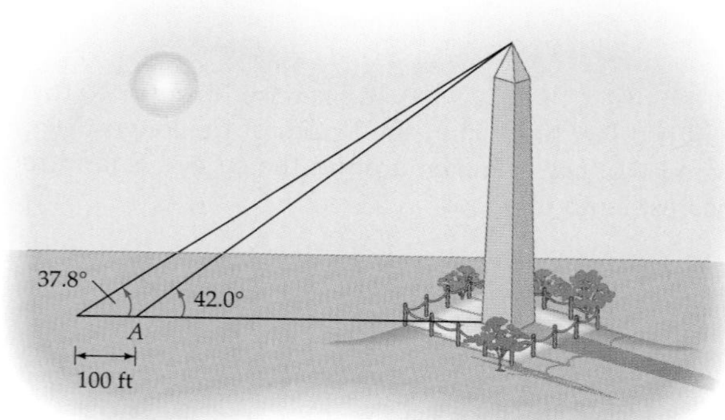

68. **Height of a Tower** The angle of elevation from a point A to the top of a tower is 32.1°. From point B, which is on the same line but 55.5 feet closer to the tower, the angle of elevation is 36.7°. Find the height of the tower.

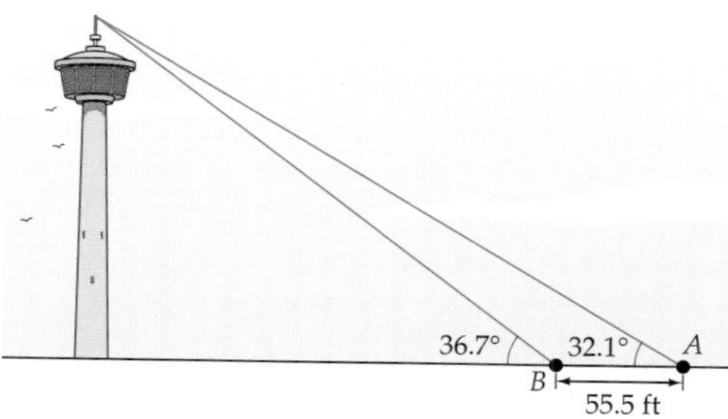

69. **The Petronas Towers** The Petronas Towers in Kuala Lumpur, Malaysia, are the world's tallest twin towers. Each tower is 1483 feet in height. The towers are connected by a skybridge at the forty-first floor. Note the information given in the accompanying figure.

  a. Determine the height of the skybridge.

  b. Determine the length of the skybridge.

$AB = 412$ feet
$\angle CAB = 53.6°$
$\overline{AB}$ is at ground level
$\angle CAD = 15.5°$

70. **An Eiffel Tower Replica** Use the information in the accompanying figure to estimate the height of the Eiffel Tower replica that stands in front of the Paris Las Vegas Hotel in Las Vegas, Nevada.

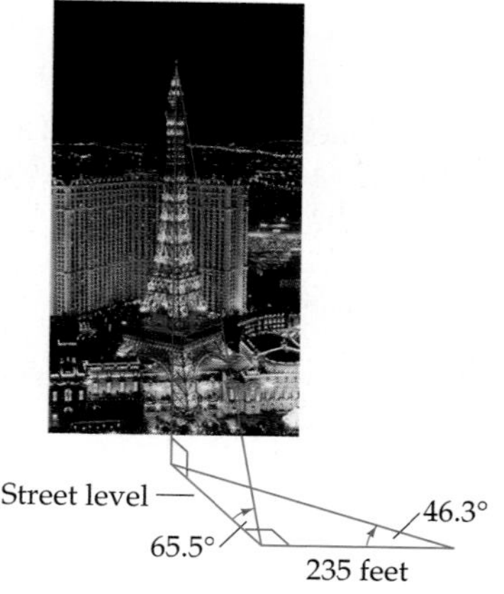

Connecting Concepts

71. A circle is inscribed in a regular hexagon with each side 6.0 meters long. Find the radius of the circle.

72. Show that the area A of the triangle given in the figure is $A = \frac{1}{2}ab \sin \theta$.

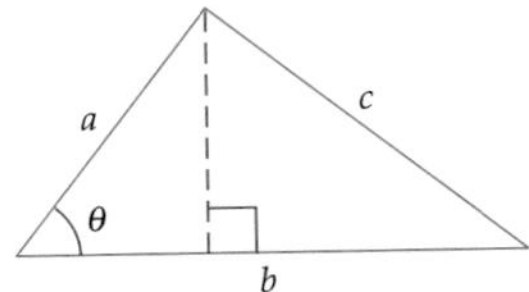

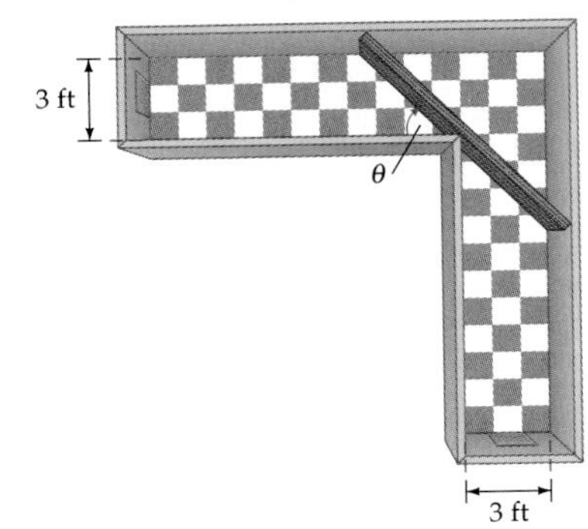

73. **Find a Maximum Length** Find the length of the longest piece of wood that can be slid around the corner of the hallway in the figure in the column at right. Round to the nearest tenth of a foot.

74. **Find a Maximum Length** In Exercise 73, suppose that the hall is 8 feet high. Find the length of the longest piece of wood that can be taken around the corner. Round to the nearest tenth of a foot.

Projects

1. a. **Perimeter of a Regular n-gon** Show that the perimeter P of a regular n-sided polygon (n-gon) inscribed in a circle of radius 1 is $P = 2n \sin \dfrac{180°}{n}$.

b. Let P_n denote the perimeter of a regular n-gon inscribed in a circle of radius 1. Use the result from **a.** to complete the following table.

n	10	50	100	1000	10,000
P_n					

Write a few sentences explaining why P_n approaches 2π as n increases without bound.

2. a. **Area of a Regular n-gon** Show that the area A of a regular n-gon inscribed in a circle of radius 1 is $A = \dfrac{n}{2} \sin \dfrac{360°}{n}$.

b. Let A_n denote the area of a regular n-gon inscribed in a circle of radius 1. Use the result from **a.** to complete the following table.

n	10	50	100	1000	10,000
A_n					

Write a few sentences explaining why A_n approaches π as n increases without bound.

Section 2.3

- Trigonometric Functions of Any Angle
- Trigonometric Functions of Quadrantal Angles
- Signs of Trigonometric Functions
- The Reference Angle

Trigonometric Functions of Any Angle

PREPARE FOR THIS SECTION

Prepare for this section by completing the following exercises. The answers can be found on page A9.

PS1. Find the reciprocal of $-\frac{3}{4}$.

PS2. Find the reciprocal of $\frac{2\sqrt{5}}{5}$.

PS3. Evaluate: $|120 - 180|$ [1.1]

PS4. Simplify: $2\pi - \frac{9\pi}{5}$

PS5. Simplify: $\frac{3}{2}\pi - \frac{\pi}{2}$

PS6. Simplify: $\sqrt{(-3)^2 + (-5)^2}$ [1.2]

Trigonometric Functions of Any Angle

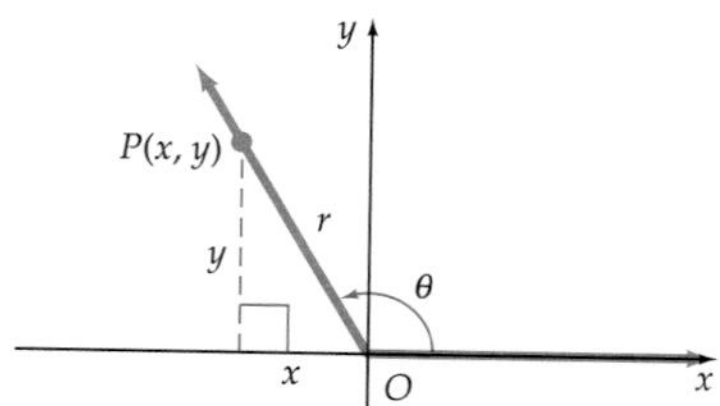

Figure 2.41

The applications of trigonometry would be quite limited if all angles had to be acute angles. Fortunately, this is not the case. In this section we extend the definition of a trigonometric function to include any angle.

Consider angle θ in **Figure 2.41** in standard position and a point $P(x, y)$ on the terminal side of the angle. We define the trigonometric functions of any angle according to the following definitions.

Definitions of the Trigonometric Functions of Any Angle

Let $P(x, y)$ be any point, except the origin, on the terminal side of an angle θ in standard position. Let $r = d(O, P)$, the distance from the origin to P. The six trigonometric functions of θ are

$$\sin\theta = \frac{y}{r} \qquad \cos\theta = \frac{x}{r} \qquad \tan\theta = \frac{y}{x},\quad x \neq 0$$

$$\csc\theta = \frac{r}{y},\quad y \neq 0 \qquad \sec\theta = \frac{r}{x},\quad x \neq 0 \qquad \cot\theta = \frac{x}{y},\quad y \neq 0$$

where $r = \sqrt{x^2 + y^2}$.

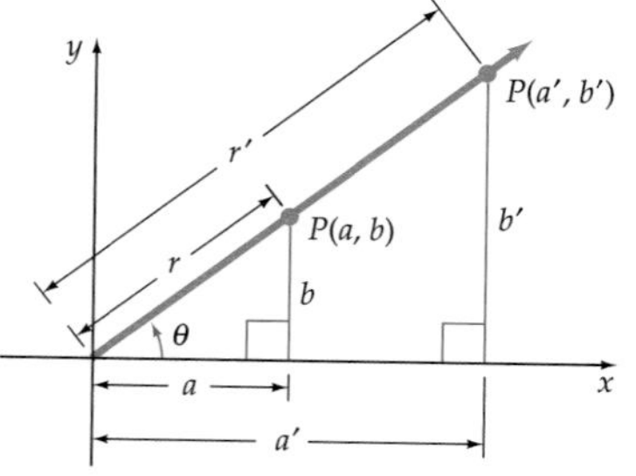

Figure 2.42

The value of a trigonometric function is independent of the point chosen on the terminal side of the angle. Consider any two points on the terminal side of an angle θ in standard position, as shown in **Figure 2.42**. The right triangles formed are similar triangles, so the ratios of the corresponding sides are equal. Thus, for example, $\frac{b}{a} = \frac{b'}{a'}$. Because $\tan\theta = \frac{b}{a} = \frac{b'}{a'}$, we have $\tan\theta = \frac{b'}{a'}$. Therefore, the value of the tangent function is independent of the point chosen on the terminal side of the angle. By a similar argument, we can show that the value of any trigonometric function is independent of the point chosen on the terminal side of the angle.

Any point in a rectangular coordinate system (except the origin) can determine an angle in standard position. For example, $P(-4, 3)$ in **Figure 2.43** is a point in the second quadrant and determines an angle θ in standard position with $r = \sqrt{(-4)^2 + 3^2} = 5$.

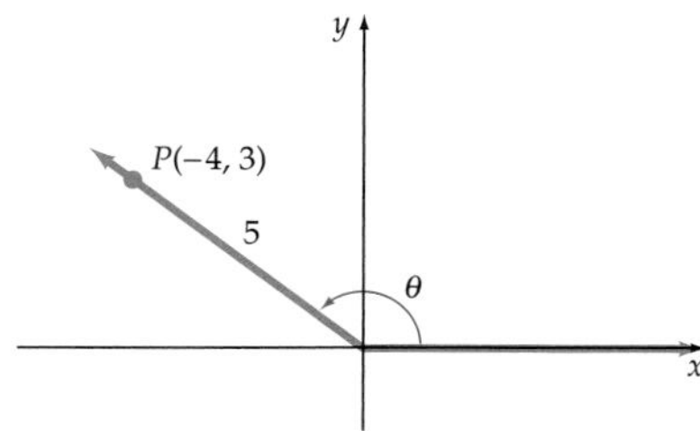

Figure 2.43

The values of the trigonometric functions of θ as shown in **Figure 2.43** are

$$\sin\theta = \frac{3}{5} \qquad \cos\theta = \frac{-4}{5} = -\frac{4}{5} \qquad \tan\theta = \frac{3}{-4} = -\frac{3}{4}$$

$$\csc\theta = \frac{5}{3} \qquad \sec\theta = \frac{5}{-4} = -\frac{5}{4} \qquad \cot\theta = \frac{-4}{3} = -\frac{4}{3}$$

EXAMPLE 1 ›› Evaluate Trigonometric Functions

Find the exact value of each of the six trigonometric functions of an angle θ in standard position whose terminal side contains the point $P(-3, -2)$.

Solution

The angle is sketched in **Figure 2.44.** Find r by using the equation $r = \sqrt{x^2 + y^2}$, where $x = -3$ and $y = -2$.

$$r = \sqrt{(-3)^2 + (-2)^2} = \sqrt{9 + 4} = \sqrt{13}$$

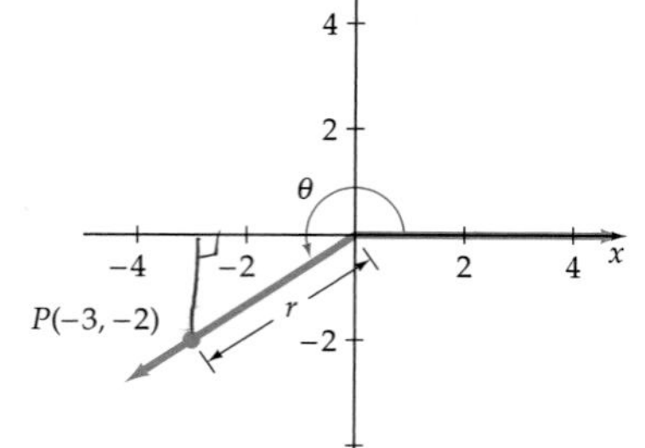

Figure 2.44

Now use the definitions of the trigonometric functions.

$$\sin\theta = \frac{-2}{\sqrt{13}} = -\frac{2\sqrt{13}}{13} \qquad \cos\theta = \frac{-3}{\sqrt{13}} = -\frac{3\sqrt{13}}{13} \qquad \tan\theta = \frac{-2}{-3} = \frac{2}{3}$$

$$\csc\theta = \frac{\sqrt{13}}{-2} = -\frac{\sqrt{13}}{2} \qquad \sec\theta = \frac{\sqrt{13}}{-3} = -\frac{\sqrt{13}}{3} \qquad \cot\theta = \frac{-3}{-2} = \frac{3}{2}$$

›› *Try Exercise 6, page 153*

Trigonometric Functions of Quadrantal Angles

Recall that a quadrantal angle is an angle whose terminal side coincides with the x- or y-axis. The value of a trigonometric function of a quadrantal angle can be found by choosing any point on the terminal side of the angle and then applying the definition of that trigonometric function.

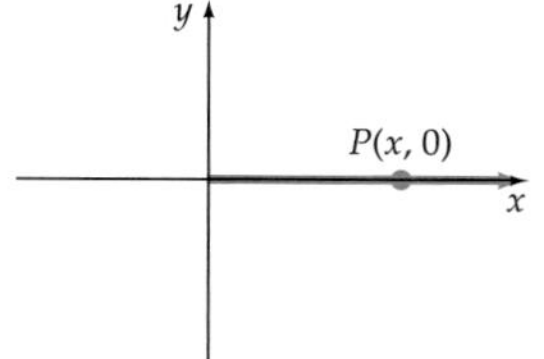

Figure 2.45

The terminal side of 0° coincides with the positive x-axis. Let $P(x, 0)$, $x > 0$, be a point on the x-axis, as shown in **Figure 2.45**. Then $y = 0$ and $r = x$. The values of the six trigonometric functions of 0° are

$$\sin 0° = \frac{0}{r} = 0 \qquad \cos 0° = \frac{x}{r} = \frac{x}{x} = 1 \qquad \tan 0° = \frac{0}{x} = 0$$

$$\csc 0° \text{ is undefined.} \qquad \sec 0° = \frac{r}{x} = \frac{x}{x} = 1 \qquad \cot 0° \text{ is undefined.}$$

QUESTION Why are csc 0° and cot 0° undefined?

In like manner, the values of the trigonometric functions of the other quadrantal angles can be found. The results are shown in **Table 2.4**.

Table 2.4 Values of Trigonometric Functions of Quadrantal Angles

θ	sin θ	cos θ	tan θ	csc θ	sec θ	cot θ
0°	0	1	0	undefined	1	undefined
90°	1	0	undefined	1	undefined	0
180°	0	−1	0	undefined	−1	undefined
270°	−1	0	undefined	−1	undefined	0

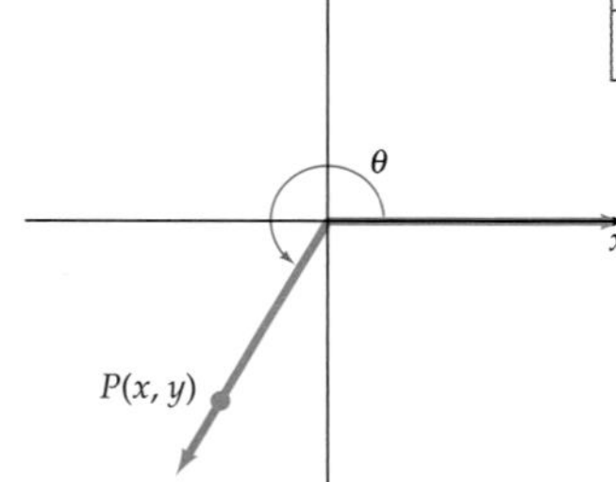

Figure 2.46

Signs of Trigonometric Functions

The sign of a trigonometric function depends on the quadrant in which the terminal side of the angle lies. For example, if θ is an angle whose terminal side lies in Quadrant III and $P(x, y)$ is on the terminal side of θ, then both x and y are negative, and therefore $\frac{y}{x}$ and $\frac{x}{y}$ are positive. See **Figure 2.46**. Because $\tan\theta = \frac{y}{x}$ and $\cot\theta = \frac{x}{y}$, the values of the tangent and cotangent functions are positive for any Quadrant III angle. The values of the other four trigonometric functions of any Quadrant III angle are all negative.

Table 2.5 lists the signs of the six trigonometric functions in each quadrant. **Figure 2.47** is a graphical display of the contents of **Table 2.5**.

Sine and cosecant positive | All functions positive
Tangent and cotangent positive | Cosine and secant positive

Figure 2.47

Table 2.5 Signs of the Trigonometric Functions

Sign of	Terminal Side of θ in Quadrant I	II	III	IV
sin θ and csc θ	positive	positive	negative	negative
cos θ and sec θ	positive	negative	negative	positive
tan θ and cot θ	positive	negative	positive	negative

ANSWER $P(x, 0)$ is a point on the terminal side of a 0° angle in standard position. Thus $\csc 0° = \frac{r}{0}$, which is undefined. Similarly, $\cot 0° = \frac{x}{0}$, which is undefined.

In the next example we are asked to evaluate two trigonometric functions of the angle θ. A key step is to use our knowledge about trigonometric functions and their signs to determine that θ is a Quadrant IV angle.

EXAMPLE 2 Evaluate Trigonometric Functions

Given $\tan\theta = -\dfrac{7}{5}$ and $\sin\theta < 0$, find $\cos\theta$ and $\csc\theta$.

Solution

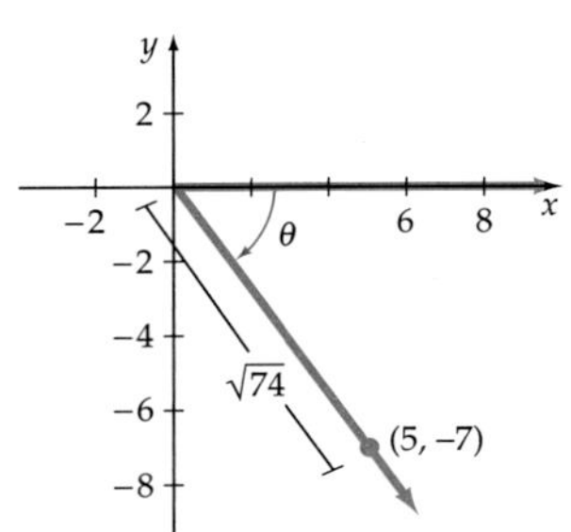

Figure 2.48

The terminal side of angle θ must lie in Quadrant IV; that is the only quadrant in which $\sin\theta$ and $\tan\theta$ are both negative. Because

$$\tan\theta = -\frac{7}{5} = \frac{y}{x} \qquad (1)$$

and the terminal side of θ is in Quadrant IV, we know that y must be negative and x must be positive. Thus Equation (1) is true for $y = -7$ and $x = 5$. Now $r = \sqrt{5^2 + (-7)^2} = \sqrt{74}$. See **Figure 2.48.** Hence

$$\cos\theta = \frac{x}{r} = \frac{5}{\sqrt{74}} = \frac{5\sqrt{74}}{74}$$

and

$$\csc\theta = \frac{r}{y} = \frac{\sqrt{74}}{-7} = -\frac{\sqrt{74}}{7}$$

> **take note**
> In this section r is a distance and hence nonnegative.

Try Exercise 30, page 153

The Reference Angle

We will often find it convenient to evaluate trigonometric functions by making use of the concept of a *reference angle.*

> **take note**
> The reference angle is a very important concept that will be used time and time again in the remaining trigonometry sections.

Definition of a Reference Angle

Given $\angle\theta$ in standard position, its **reference angle** θ' is the acute angle formed by the terminal side of $\angle\theta$ and the x-axis.

Figure 2.49 shows $\angle\theta$ and its reference angle θ' for four cases. In every case the reference angle θ' is formed by the terminal side of $\angle\theta$ and the x-axis (never the y-axis). The process of determining the measure of $\angle\theta'$ varies according to which quadrant contains the terminal side of $\angle\theta$.

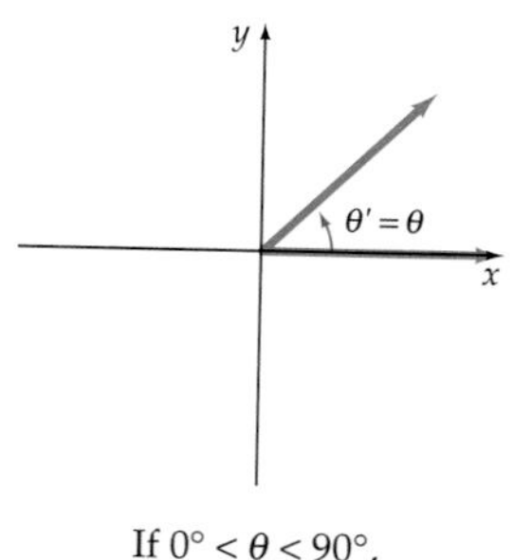

If 0° < θ < 90°, then θ′ = θ.

If 90° < θ < 180°, then θ′ = 180° − θ.

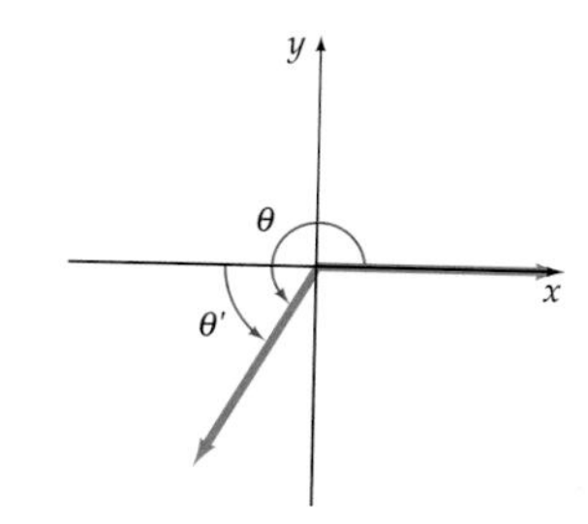

If 180° < θ < 270°, then θ′ = θ − 180°.

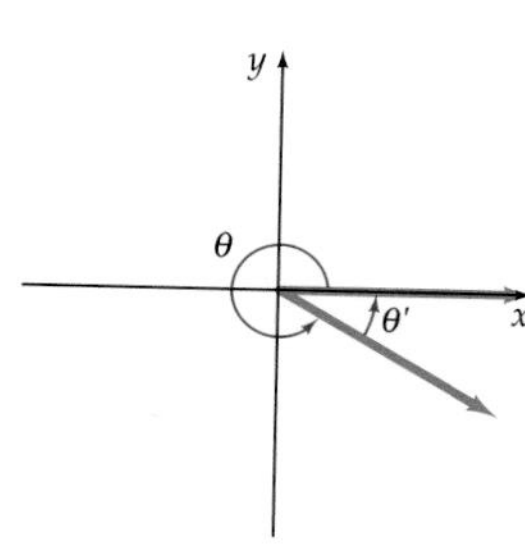

If 270° < θ < 360°, then θ′ = 360° − θ.

Figure 2.49

EXAMPLE 3 Find the Measure of a Reference Angle

Find the measure of the reference angle θ' for each angle.

a. $\theta = 120°$ b. $\theta = 345°$ c. $\theta = \dfrac{7\pi}{4}$ d. $\theta = \dfrac{13\pi}{6}$

Solution

For any angle in standard position, the measure of its reference angle is the measure of the acute angle formed by its terminal side and the x-axis.

a.

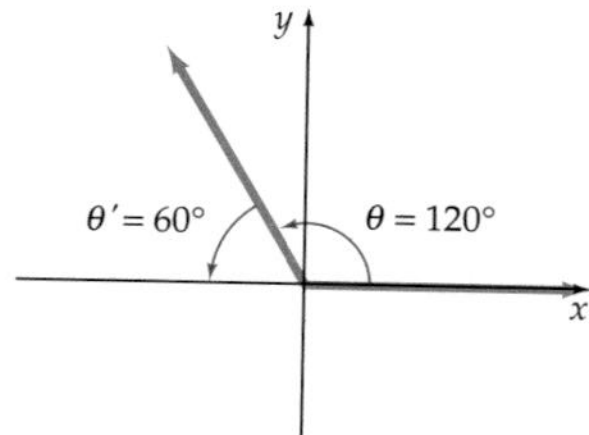

$$\theta' = 180° - 120° = 60°$$

b.

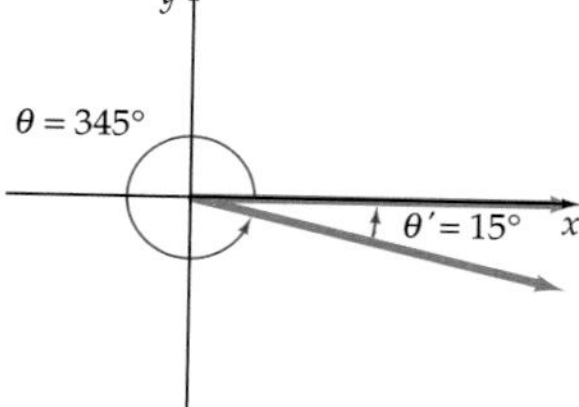

$$\theta' = 360° - 345° = 15°$$

c.

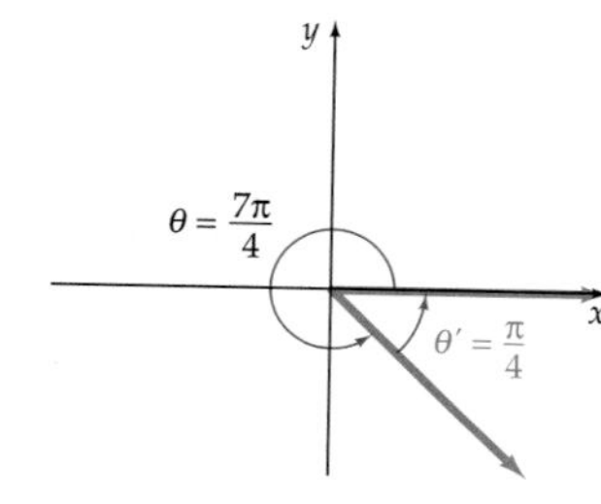

$$\theta' = 2\pi - \frac{7\pi}{4} = \frac{\pi}{4}$$

d.

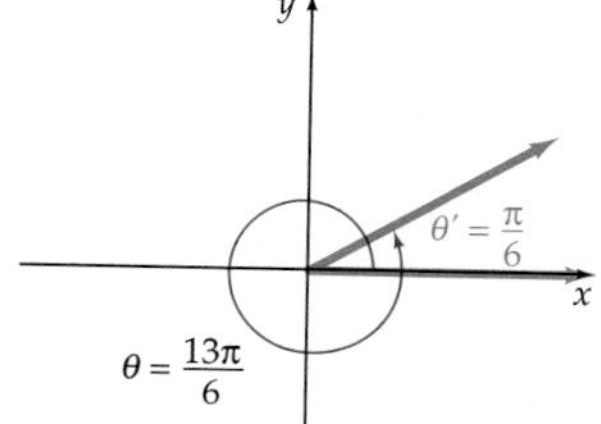

$$\theta' = \frac{13\pi}{6} - 2\pi = \frac{\pi}{6}$$

Try Exercise 38, page 153

Integrating Technology

A TI-83/TI-83 Plus/TI-84 Plus graphing calculator program is available to compute the measure of the reference angle for a given angle. This program, REFANG, can be found on our website at college.hmco.com/info/aufmannCAT

The following theorem states an important relationship that exists between $\sin\theta$ and $\sin\theta'$, where θ' is the reference angle for angle θ.

take note

The Reference Angle Theorem is also valid if the sine function is replaced by any other trigonometric function.

Reference Angle Theorem

To evaluate $\sin\theta$, determine $\sin\theta'$. Then use either $\sin\theta'$ or its opposite as the answer, depending on which has the correct sign.

In the following example, we illustrate how to evaluate a trigonometric function of θ by first evaluating the trigonometric function of θ'.

EXAMPLE 4 » Use the Reference Angle Theorem to Evaluate Trigonometric Functions

Determine the exact value of each function.

a. $\sin 210°$ b. $\cos 405°$ c. $\tan\frac{5\pi}{3}$

Solution

a. We know that $\sin 210°$ is negative (the sign chart is given in **Table 2.5** on page 149). The reference angle for $\theta = 210°$ is $\theta' = 30°$. By the Reference Angle Theorem, we know that $\sin 210°$ equals either

$$\sin 30° = \frac{1}{2} \qquad \text{or} \qquad -\sin 30° = -\frac{1}{2}$$

Thus $\sin 210° = -\frac{1}{2}$.

b. Because $\theta = 405°$ is a Quadrant I angle, we know that $\cos 405° > 0$. The reference angle for $\theta = 405°$ is $\theta' = 45°$. By the reference angle theorem, $\cos 405°$ equals either

$$\cos 45° = \frac{\sqrt{2}}{2} \qquad \text{or} \qquad -\cos 45° = -\frac{\sqrt{2}}{2}$$

Thus $\cos 405° = \frac{\sqrt{2}}{2}$.

c. Because $\theta = \frac{5\pi}{3}$ is a Quadrant IV angle, $\tan\frac{5\pi}{3} < 0$. The reference angle for $\theta = \frac{5\pi}{3}$ is $\theta' = \frac{\pi}{3}$. Hence $\tan\frac{5\pi}{3}$ equals either

$$\tan\frac{\pi}{3} = \sqrt{3} \qquad \text{or} \qquad -\tan\frac{\pi}{3} = -\sqrt{3}$$

Thus $\tan\frac{5\pi}{3} = -\sqrt{3}$.

» *Try Exercise 50, page 154*

Topics for Discussion

1. Is every reference angle an acute angle? Explain.
2. If θ' is the reference angle for the angle θ, then $\sin\theta = \sin\theta'$. Do you agree? Explain.
3. If $\sin\theta < 0$ and $\cos\theta > 0$, then the terminal side of the angle θ lies in which quadrant?
4. Explain how to find the measure of the reference angle θ' for the angle $\theta = \frac{19}{5}\pi$.

Exercise Set 2.3

In Exercises 1 to 8, find the value of each of the six trigonometric functions for the angle whose terminal side passes through the given point.

1. $P(2, 3)$
2. $P(3, 7)$
3. $P(-2, 3)$
4. $P(-3, 5)$
5. $P(-8, -5)$
6. $P(-6, -9)$
7. $P(-5, 0)$
8. $P(0, 2)$

In Exercises 9 to 20, evaluate the trignometric function of the quadrantal angle, or state that the function is undefined.

9. $\sin 180°$
10. $\cos 270°$
11. $\tan 180°$
12. $\sec 90°$
13. $\csc 90°$
14. $\cot 90°$
15 $\cos\frac{\pi}{2}$
16. $\sin\frac{3\pi}{2}$
17. $\tan\frac{\pi}{2}$
18 $\cot\pi$
19. $\sin\frac{\pi}{2}$
20. $\cos\pi$

In Exercises 21 to 26, let θ be an angle in standard position. State the quadrant in which the terminal side of θ lies.

21. $\sin\theta > 0, \quad \cos\theta > 0$
22. $\tan\theta < 0, \quad \sin\theta < 0$
23. $\cos\theta > 0, \quad \tan\theta < 0$
24. $\sin\theta < 0, \quad \cos\theta > 0$
25. $\sin\theta < 0, \quad \cos\theta < 0$
26. $\tan\theta < 0, \quad \cos\theta < 0$

In Exercises 27 to 36, find the exact value of each expression.

27. $\sin\theta = -\frac{1}{2}$, $180° < \theta < 270°$; find $\tan\theta$.
28. $\cot\theta = -1$, $90° < \theta < 180°$; find $\cos\theta$.
29. $\csc\theta = \sqrt{2}$, $\frac{\pi}{2} < \theta < \pi$; find $\cot\theta$.
30. $\sec\theta = \frac{2\sqrt{3}}{3}$, $\frac{3\pi}{2} < \theta < 2\pi$; find $\sin\theta$.
31. $\sin\theta = -\frac{1}{2}$ and $\cos\theta > 0$; find $\tan\theta$.
32. $\tan\theta = 1$ and $\sin\theta < 0$; find $\cos\theta$.
33. $\cos\theta = \frac{1}{2}$ and $\tan\theta = \sqrt{3}$; find $\csc\theta$.
34. $\tan\theta = 1$ and $\sin\theta = \frac{\sqrt{2}}{2}$; find $\sec\theta$.
35. $\cos\theta = -\frac{1}{2}$ and $\sin\theta = \frac{\sqrt{3}}{2}$; find $\cot\theta$.
36. $\sec\theta = \frac{2\sqrt{3}}{3}$ and $\sin\theta = -\frac{1}{2}$; find $\cot\theta$.

In Exercises 37 to 48, find the measure of the reference angle θ' for the given angle θ.

37. $\theta = 160°$
38. $\theta = 255°$
39. $\theta = 351°$
40. $\theta = 48°$
41. $\theta = \frac{11}{5}\pi$
42. $\theta = -6$
43. $\theta = \frac{8}{3}$
44. $\theta = \frac{18}{7}\pi$
45. $\theta = 1406°$
46. $\theta = 840°$
47. $\theta = -475°$
48. $\theta = -650°$

In Exercises 49 to 60, use the Reference Angle Theorem to find the exact value of each trigonometric function.

49. $\sin 225°$ **50.** $\cos 300°$ **51.** $\tan 405°$

52. $\sec 150°$ **53.** $\csc\left(\frac{4}{3}\pi\right)$ **54.** $\cot\left(\frac{7}{6}\pi\right)$

55. $\cos\frac{17\pi}{4}$ **56.** $\tan\left(-\frac{\pi}{3}\right)$ **57.** $\sec 765°$

58. $\csc(-510°)$ **59.** $\cot 540°$ **60.** $\cos 570°$

In Exercises 61 to 72, use a calculator to approximate the given trigonometric function to six significant digits.

61. $\sin 127°$ **62.** $\sin(-257°)$ **63.** $\cos(-116°)$

64. $\cot 398°$ **65.** $\sec 578°$ **66.** $\sec 740°$

67. $\sin\left(-\frac{\pi}{5}\right)$ **68.** $\cos\frac{3\pi}{7}$ **69.** $\csc\frac{9\pi}{5}$

70. $\tan(-4.12)$ **71.** $\sec(-4.45)$ **72.** $\csc 0.34$

In Exercises 73 to 80, find (without using a calculator) the exact value of each expression.

73. $\sin 210° - \cos 330° \tan 330°$

74. $\tan 225° + \sin 240° \cos 60°$

75. $\sin^2 30° + \cos^2 30°$

76. $\cos\pi \sin\frac{7\pi}{4} - \tan\frac{11\pi}{6}$

77. $\sin\left(\frac{3\pi}{2}\right)\tan\left(\frac{\pi}{4}\right) - \cos\left(\frac{\pi}{3}\right)$

78. $\cos\left(\frac{7\pi}{4}\right)\tan\left(\frac{4\pi}{3}\right) + \cos\left(\frac{7\pi}{6}\right)$

79. $\sin^2\left(\frac{5\pi}{4}\right) + \cos^2\left(\frac{5\pi}{4}\right)$

80. $\tan^2\left(\frac{7\pi}{4}\right) - \sec^2\left(\frac{7\pi}{4}\right)$

Connecting Concepts

In Exercises 81 to 86, find two values of θ, $0° \le \theta < 360°$, that satisfy the given trigonometric equation.

81. $\sin\theta = \frac{1}{2}$ **82.** $\tan\theta = -\sqrt{3}$

83. $\cos\theta = -\frac{\sqrt{3}}{2}$ **84.** $\tan\theta = 1$

85. $\csc\theta = -\sqrt{2}$ **86.** $\cot\theta = -1$

In Exercises 87 to 92, find two values of θ, $0 \le \theta < 2\pi$, that satisfy the given trigonometric equation.

87. $\tan\theta = -1$ **88.** $\cos\theta = \frac{1}{2}$

89. $\tan\theta = -\frac{\sqrt{3}}{3}$ **90.** $\sec\theta = -\frac{2\sqrt{3}}{3}$

91. $\sin\theta = \frac{\sqrt{3}}{2}$ **92.** $\cos\theta = -\frac{1}{2}$

If $P(x, y)$ is a point on the terminal side of an acute angle θ in standard position and $r = \sqrt{x^2 + y^2}$, then $\sin\theta = \frac{y}{r}$ and $\cos\theta = \frac{x}{r}$. Using these definitions, we find that

$$\cos^2\theta + \sin^2\theta = \left(\frac{x}{r}\right)^2 + \left(\frac{y}{r}\right)^2 = \frac{x^2}{r^2} + \frac{y^2}{r^2} = \frac{x^2 + y^2}{r^2}.$$

$$= \frac{r^2}{r^2} = 1$$

Hence $\cos^2\theta + \sin^2\theta = 1$ for all acute angles θ. This important identity is actually true for all angles θ. We will show this later. In the meantime, use the definitions of the trigonometric functions to prove that the equations are identities for the acute angle θ in Exercises 93 to 96.

93. $1 + \tan^2\theta = \sec^2\theta$ **94.** $\cot^2\theta + 1 = \csc^2\theta$

95. $\cos(90° - \theta) = \sin\theta$ **96.** $\sin(90° - \theta) = \cos\theta$

Projects

1. **FIND SUMS OR PRODUCTS** Determine the following sums or products. Do not use a calculator. (*Hint:* The Reference Angle Theorem may be helpful.) Explain to a classmate how you know you are correct.

   a. $\cos 0° + \cos 1° + \cos 2° + \cdots + \cos 178° + \cos 179° + \cos 180°$

   b. $\sin 0° + \sin 1° + \sin 2° + \cdots + \sin 358° + \sin 359° + \sin 360°$

   c. $\cot 1° + \cot 2° + \cot 3° + \cdots + \cot 177° + \cot 178° + \cot 179°$

   d. $(\cos 1°)(\cos 2°)(\cos 3°) \cdots (\cos 177°)(\cos 178°)(\cos 179°)$

   e. $\cos^2 1° + \cos^2 2° + \cos^2 3° + \cdots + \cos^2 357° + \cos^2 358° + \cos^2 359°$

Section 2.4

- The Wrapping Function
- Trigonometric Functions of Real Numbers
- Properties of Trigonometric Functions of Real Numbers
- Trigonometric Identities

Trigonometric Functions of Real Numbers

PREPARE FOR THIS SECTION

Prepare for this section by completing the following exercises. The answers can be found on page A10.

PS1. Determine whether the point (0, 1) is a point on the circle defined by $x^2 + y^2 = 1$. [1.1/1.2]

PS2. Determine whether the point $\left(\frac{1}{2}, \frac{\sqrt{3}}{2}\right)$ is a point on the circle defined by $x^2 + y^2 = 1$. [1.1/1.2]

PS3. Determine whether the point $\left(\frac{\sqrt{2}}{2}, \frac{\sqrt{3}}{2}\right)$ is a point on the circle defined by $x^2 + y^2 = 1$. [1.1/1.2]

PS4. Determine the circumference of a circle with a radius of 1. [2.1]

PS5. Determine whether $f(x) = x^2 - 3$ is an even function, an odd function, or neither. [1.4]

PS6. Determine whether $f(x) = x^3 - x^2$ is an even function, an odd function, or neither. [1.4]

The Wrapping Function

In the seventeenth century, applications of trigonometry were extended to problems in physics and engineering. These kinds of problems required trigonometric functions whose domains were sets of real numbers rather than sets of angles. This extension of the definitions of trigonometric functions to include real numbers was accomplished by using a correspondence between an angle and the length of an arc on a *unit circle.*

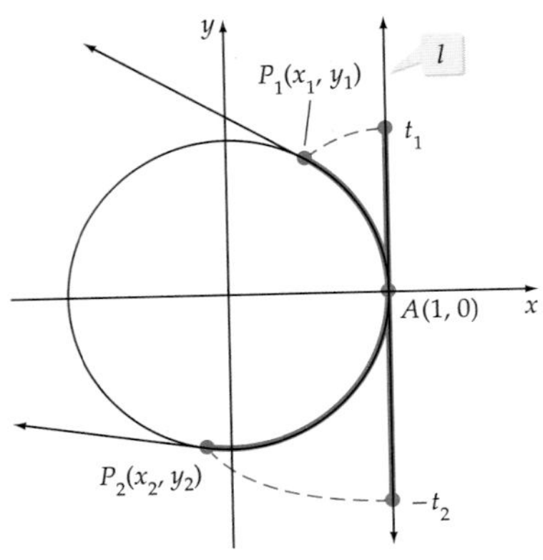

Figure 2.50

Consider a circle given by the equation $x^2 + y^2 = 1$, called a **unit circle,** and a vertical coordinate line l tangent to the unit circle at $(1, 0)$. We define a function W that pairs a real number t on the coordinate line with a point $P(x, y)$ on the unit circle. This function W is called the *wrapping function* because it is analogous to wrapping a line around a circle.

As shown in **Figure 2.50,** the positive part of the coordinate line is wrapped around the unit circle in a counterclockwise direction. The negative part of the coordinate line is wrapped around the circle in a clockwise direction. The wrapping function is defined by the equation $W(t) = P(x, y)$, where t is a real number and $P(x, y)$ is the point on the unit circle that corresponds to t.

Through the wrapping function, each real number t defines an arc $\widehat{AP}$ that subtends a central angle with a measure of θ radians. The length of the arc $\widehat{AP}$ is t (see **Figure 2.51**). From the equation $s = r\theta$ for the arc length of a circle, we have (with $t = s$) $t = r\theta$. For a unit circle, $r = 1$, and the equation becomes $t = \theta$. Thus, on a unit circle, *the measure of a central angle and the length of its arc can be represented by the same real number t.*

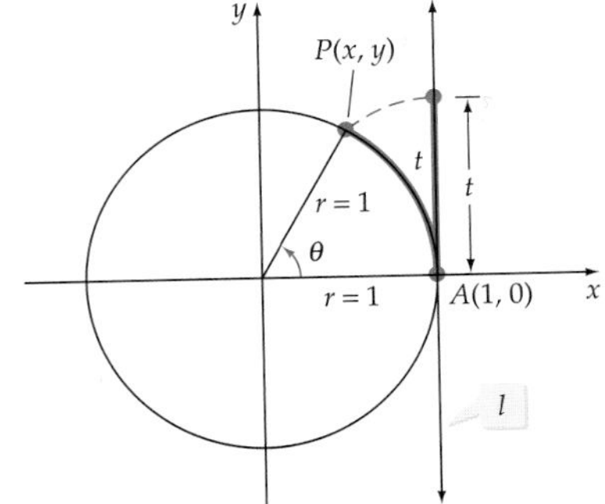

Figure 2.51

EXAMPLE 1 Evaluate the Wrapping Function

Evaluate $W\left(\frac{\pi}{3}\right)$.

Solution

The point $\frac{\pi}{3}$ on line l is shown in **Figure 2.52**. From the wrapping function, $W\left(\frac{\pi}{3}\right)$ is the point P on the unit circle for which arc $\widehat{AP}$ subtends an angle θ, the measure of which is $\frac{\pi}{3}$ radians. The coordinates of P can be determined from the definitions of $\cos\theta$ and $\sin\theta$ given in Section 2.3 and from **Table 2.2,** page 137.

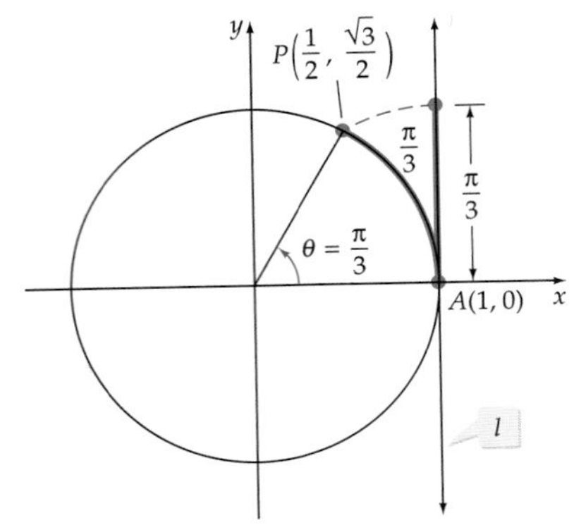

Figure 2.52

$$\cos\theta = \frac{x}{r} \qquad \sin\theta = \frac{y}{r}$$

$$\cos\frac{\pi}{3} = \frac{x}{1} = x \qquad \sin\frac{\pi}{3} = \frac{y}{1} = y \qquad \bullet\ \theta = \frac{\pi}{3};\ r = 1$$

$$\frac{1}{2} = x \qquad \frac{\sqrt{3}}{2} = y \qquad \bullet\ \cos\frac{\pi}{3} = \frac{1}{2};\ \sin\frac{\pi}{3} = \frac{\sqrt{3}}{2}$$

From these equations, $x = \frac{1}{2}$ and $y = \frac{\sqrt{3}}{2}$. Therefore, $W\left(\frac{\pi}{3}\right) = \left(\frac{1}{2}, \frac{\sqrt{3}}{2}\right)$.

Try Exercise 10, page 166

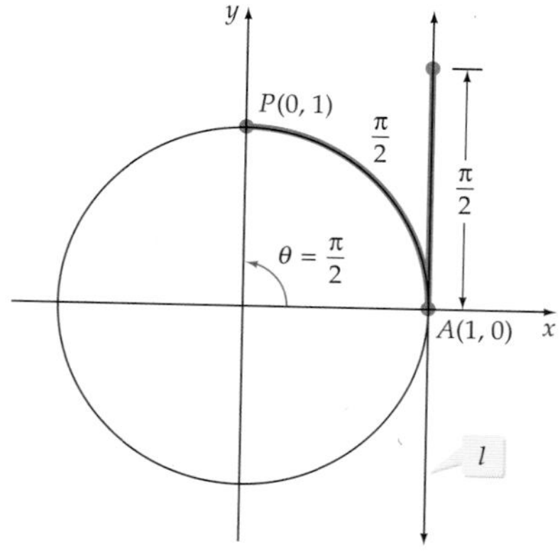

Figure 2.53

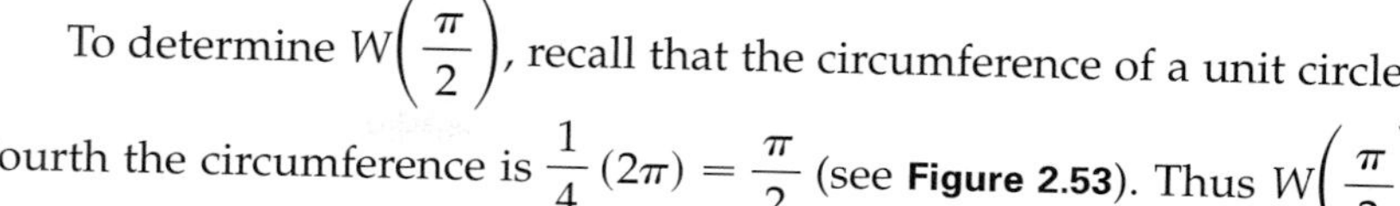

To determine $W\left(\frac{\pi}{2}\right)$, recall that the circumference of a unit circle is 2π. One-fourth the circumference is $\frac{1}{4}(2\pi) = \frac{\pi}{2}$ (see **Figure 2.53**). Thus $W\left(\frac{\pi}{2}\right) = P(0, 1)$.

Note from the last two examples that for the given real number t, $\cos t = x$ and $\sin t = y$. That is, for a real number t and $W(t) = P(x, y)$, the value of the cosine of t is the x-coordinate of P, and the value of the sine of t is the y-coordinate of P. See **Figure 2.54.**

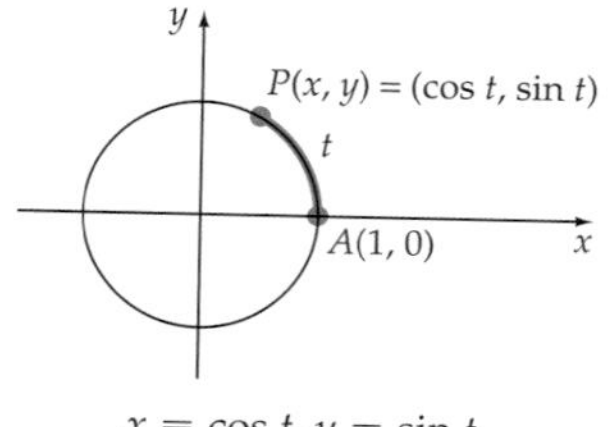

$x = \cos t$, $y = \sin t$

Figure 2.54

QUESTION What is the point defined by $W\left(\frac{\pi}{4}\right)$?

Trigonometric Functions of Real Numbers

The following definition makes use of the wrapping function $W(t)$ to define trigonometric functions of real numbers.

Definitions of the Trigonometric Functions of Real Numbers

Let W be the wrapping function, t be a real number, and $W(t) = P(x, y)$. Then

$$\sin t = y \qquad \cos t = x \qquad \tan t = \frac{y}{x},\quad x \neq 0$$

$$\csc t = \frac{1}{y},\quad y \neq 0 \qquad \sec t = \frac{1}{x},\quad x \neq 0 \qquad \cot t = \frac{x}{y},\quad y \neq 0$$

Trigonometric functions of real numbers are frequently called *circular functions* to distinguish them from trigonometric functions of angles.

The *trigonometric functions of real numbers* (or circular functions) look remarkably like the trigonometric functions defined in the last section. The difference between the two is that of domain: In one case, the domains are sets of *real numbers;* in the other case, the domains are sets of *angle measurements.* However, there are similarities between the two functions.

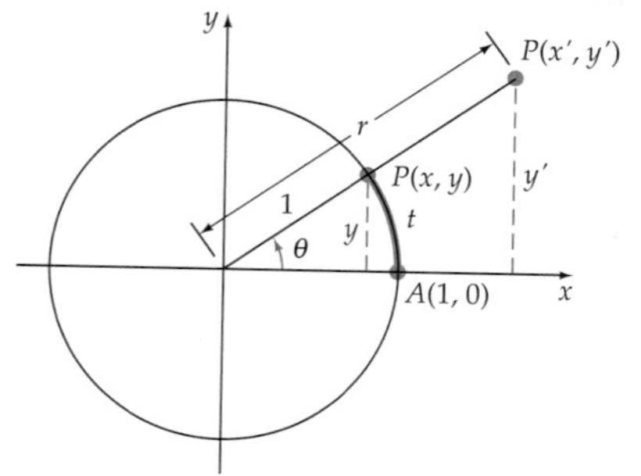

Figure 2.55

Consider an angle θ (measured in radians) in standard position, as shown in **Figure 2.55.** Let $P(x, y)$ and $P'(x', y')$ be two points on the terminal side of θ, where $x^2 + y^2 = 1$ and $(x')^2 + (y')^2 = r^2$. Let t be the length of the arc from $A(1, 0)$ to $P(x, y)$. Then

$$\sin \theta = \frac{y'}{r} = \frac{y}{1} = \sin t$$

Thus the value of the sine function of θ, measured in radians, is equal to the value of the sine of the real number t. Similar arguments can be given to show cor-

ANSWER $\left(\frac{\sqrt{2}}{2}, \frac{\sqrt{2}}{2}\right)$

responding results for the other five trigonometric functions. With this in mind, we can assert that the value of a trigonometric function at the real number t is its value at an angle of t radians.

EXAMPLE 2 Evaluate Trigonometric Functions of Real Numbers

Find the exact value of each function.

a. $\cos\dfrac{\pi}{4}$ b. $\sin\left(-\dfrac{7\pi}{6}\right)$ c. $\tan\left(-\dfrac{5\pi}{4}\right)$ d. $\sec\dfrac{5\pi}{3}$

Solution

The value of a trigonometric function at the real number t is its value at an angle of t radians. Using **Table 2.2** on page 137, we have

a. $\cos\dfrac{\pi}{4} = \dfrac{\sqrt{2}}{2}$

b. $\sin\left(-\dfrac{7\pi}{6}\right) = \sin\dfrac{\pi}{6} = \dfrac{1}{2}$

• Reference angle for $-\dfrac{7\pi}{6}$ is $\dfrac{\pi}{6}$ and $\sin t > 0$ in Quadrant II.

c. $\tan\left(-\dfrac{5\pi}{4}\right) = -\tan\dfrac{\pi}{4} = -1$

• Reference angle for $-\dfrac{5\pi}{4}$ is $\dfrac{\pi}{4}$ and $\tan t < 0$ in Quadrant II.

d. $\sec\dfrac{5\pi}{3} = \sec\dfrac{\pi}{3} = 2$

• Reference angle for $\dfrac{5\pi}{3}$ is $\dfrac{\pi}{3}$ and $\sec t > 0$ in Quadrant IV.

Try Exercise 16, page 166

In Example 3 we evaluate a trigonometric function of a real number to solve an application.

EXAMPLE 3 Determine a Height as a Function of Time

The Millennium Wheel, in London, is the world's largest Ferris wheel. It has a diameter of 450 feet. When the Millennium Wheel is in uniform motion, it completes one revolution every 30 minutes. The height h, in feet above the Thames River, of a person riding on the Millennium Wheel can be estimated by

$$h(t) = 255 - 225\cos\left(\frac{\pi}{15}t\right)$$

where t is the time in minutes since the person started the ride.

The Millennium Wheel, on the banks of the Thames River, London.

a. How high is the person at the start of the ride ($t = 0$)?

b. How high is the person after 18.0 minutes? Round to the nearest foot.

Solution

a. $$\begin{aligned} h(0) &= 255 - 225\cos\left(\frac{\pi}{15}\cdot 0\right) \\ &= 255 - 225 \\ &= 30 \end{aligned}$$

At the start of the ride, the person is 30 feet above the Thames.

b. $$\begin{aligned} h(18.0) &= 255 - 225\cos\left(\frac{\pi}{15}\cdot 18.0\right) \\ &\approx 255 - (-182) \\ &= 437 \end{aligned}$$

After 18.0 minutes, the person is about 437 feet above the Thames.

Try Exercise 82, page 167

TO REVIEW

Domain and Range
See page 32.

Properties of Trigonometric Functions of Real Numbers

The domain and range of the trigonometric functions of real numbers can be found from the definitions of these functions. If t is any real number and $P(x, y)$ is the point corresponding to $W(t)$, then by definition $\cos t = x$ and $\sin t = y$. Thus the domain of the sine and cosine functions is the set of real numbers.

Because the radius of the unit circle is 1, we have

$$-1 \le x \le 1 \quad \text{and} \quad -1 \le y \le 1$$

Therefore, with $x = \cos t$ and $y = \sin t$, we have

$$-1 \le \cos t \le 1 \quad \text{and} \quad -1 \le \sin t \le 1$$

The range of the cosine and sine functions is $[-1, 1]$.

Using the definitions of tangent and secant,

$$\tan t = \frac{y}{x} \quad \text{and} \quad \sec t = \frac{1}{x}$$

The domain of the tangent function is all real numbers t except those for which the x-coordinate of $W(t)$ is zero. The x-coordinate is zero when $t = \pm\frac{\pi}{2}$, $t = \pm\frac{3\pi}{2}$, $t = \pm\frac{5\pi}{2}$, and in general when $t = \frac{(2n + 1)\pi}{2}$, where n is an integer. Thus the domain of the tangent function is the set of all real numbers t except $t = \frac{(2n + 1)\pi}{2}$, where n is an integer. The range of the tangent function is all real numbers.

Similar methods can be used to find the domain and range of the cotangent, secant, and cosecant functions. The results are summarized in **Table 2.6** on the next page.

Table 2.6 Domain and Range of the Trigonometric Functions of Real Numbers (*n* is an integer)

Function	Domain	Range
$y = \sin t$	$\{t \mid -\infty < t < \infty\}$	$\{y \mid -1 \le y \le 1\}$
$y = \cos t$	$\{t \mid -\infty < t < \infty\}$	$\{y \mid -1 \le y \le 1\}$
$y = \tan t$	$\left\{t \mid -\infty < t < \infty, t \ne \frac{(2n+1)\pi}{2}\right\}$	$\{y \mid -\infty < y < \infty\}$
$y = \csc t$	$\{t \mid -\infty < t < \infty, t \ne n\pi\}$	$\{y \mid y \ge 1, y \le -1\}$
$y = \sec t$	$\left\{t \mid -\infty < t < \infty, t \ne \frac{(2n+1)\pi}{2}\right\}$	$\{y \mid y \ge 1, y \le -1\}$
$y = \cot t$	$\{t \mid -\infty < t < \infty, t \ne n\pi\}$	$\{y \mid -\infty < y < \infty\}$

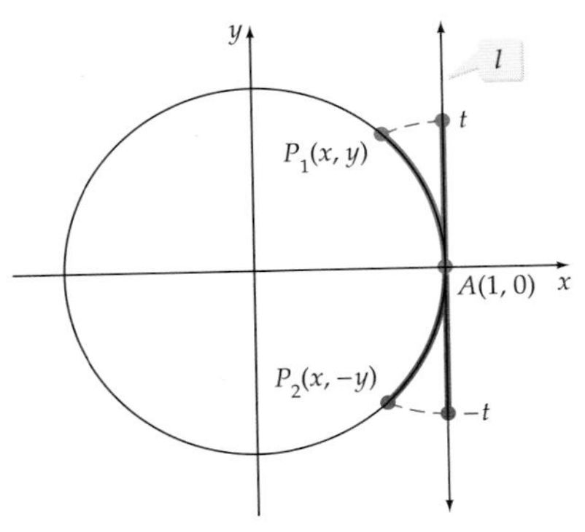

Figure 2.56

Consider the points t and $-t$ on the coordinate line l tangent to the unit circle at the point $(1, 0)$. The points $W(t)$ and $W(-t)$ are symmetric with respect to the x-axis. Therefore, if $P_1(x, y)$ are the coordinates of $W(t)$, then $P_2(x, -y)$ are the coordinates of $W(-t)$. See **Figure 2.56.**

From the definitions of the trigonometric functions, we have

$$\sin t = y \quad \text{and} \quad \sin(-t) = -y \qquad \text{and} \qquad \cos t = x \quad \text{and} \quad \cos(-t) = x$$

TO REVIEW

Odd and Even Functions
See page 56.

Substituting $\sin t$ for y and $\cos t$ for x yields

$$\sin(-t) = -\sin t \qquad \text{and} \qquad \cos(-t) = \cos t$$

Thus the sine is an odd function and the cosine is an even function. Because $\csc t = \dfrac{1}{\sin t}$ and $\sec t = \dfrac{1}{\cos t}$, it follows that

$$\csc(-t) = -\csc t \qquad \text{and} \qquad \sec(-t) = \sec t$$

These equations show that the cosecant is an odd function and the secant is an even function.

From the definition of the tangent function, we have $\tan t = \dfrac{y}{x}$ and $\tan(-t) = -\dfrac{y}{x}$. Substituting $\tan t$ for $\dfrac{y}{x}$ yields $\tan(-t) = -\tan t$. Because $\cot t = \dfrac{1}{\tan t}$, it follows that $\cot(-t) = -\cot t$. Thus the tangent and cotangent functions are odd functions.

Even and Odd Trigonometric Functions

The odd trigonometric functions are $y = \sin t$, $y = \csc t$, $y = \tan t$, and $y = \cot t$. The even trigonometric functions are $y = \cos t$ and $y = \sec t$. Thus for all t in their domain,

$$\sin(-t) = -\sin t \qquad \cos(-t) = \cos t \qquad \tan(-t) = -\tan t$$
$$\csc(-t) = -\csc t \qquad \sec(-t) = \sec t \qquad \cot(-t) = -\cot t$$

EXAMPLE 4 Determine Whether a Function Is Even, Odd, or Neither

Is $f(x) = x - \tan x$ an even function, an odd function, or neither?

Solution

Find $f(-x)$ and compare it to $f(x)$.

$$\begin{aligned} f(-x) &= (-x) - \tan(-x) = -x + \tan x & \bullet\ \tan(-x) = -\tan x \\ &= -(x - \tan x) \\ &= -f(x) \end{aligned}$$

The function $f(x) = x - \tan x$ is an odd function.

Try Exercise 44, page 166

We encounter many recurring patterns in everyday life. For instance, the time of day repeats every 24 hours. If $f(t)$ represents the present time of day, then 24 hours later the time of day will be exactly the same. Using mathematical notation, we can express this concept as

$$f(t + 24) = f(t)$$

The function f is said to be *periodic* in that it repeats itself over and over. The *period* of f is 24 hours, the time it takes to complete one full cycle.

Definition of a Periodic Function

A function f is **periodic** if there exists a positive constant p such that

$$f(t + p) = f(t)$$

for all t in the domain of f. The smallest such positive number p for which f is periodic is called the **period** of f.

The unit circle can be used to show that the sine and cosine functions are periodic functions. First note that the circumference of the unit circle is 2π. Thus, if we start at any point $P(x, y) = P(\cos t, \sin t)$ on the unit circle and travel a distance of 2π units around the circumference, we will be back at point P. Hence $(\cos(t + 2\pi), \sin(t + 2\pi)) = (\cos t, \sin t)$. Equating the first componenets gives us $\cos(t + 2\pi) = \cos t$ and equating the second components yields $\sin(t + 2\pi) = \sin t$.

The following equations illustrate that the secant and cosecant functions are also periodic functions.

$$\sec(t + 2\pi) = \frac{1}{\cos(t + 2\pi)} = \frac{1}{\cos t} = \sec t$$

$$\csc(t + 2\pi) = \frac{1}{\sin(t + 2\pi)} = \frac{1}{\sin t} = \csc t$$

Period of the Sine, Cosine, Secant, and Cosecant Functions

The sine, cosine, secant, and cosecant functions are periodic functions with a period of 2π.

$$\sin(t + 2\pi) = \sin t \qquad \cos(t + 2\pi) = \cos t$$
$$\sec(t + 2\pi) = \sec t \qquad \csc(t + 2\pi) = \csc t$$

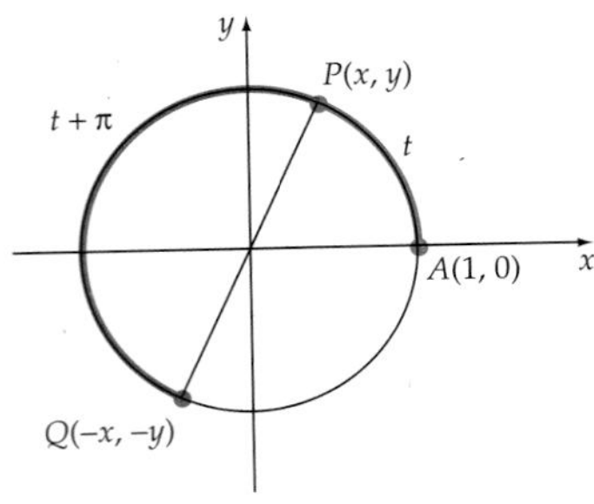

Figure 2.57

Although it is true that $\tan(t + 2\pi) = \tan t$, the period of the tangent function is not 2π. Recall that the period of a function is the *smallest* value of p for which $f(t + p) = f(t)$. Examine **Figure 2.57,** which shows that if you start at any point $P(x, y)$ on the unit circle and travel a distance of π units around the circumference, you will arrive at the point $(-x, -y)$. By definition,

$$\tan t = \frac{y}{x} \qquad \text{and} \qquad \tan(t + \pi) = \frac{-y}{-x} = \frac{y}{x} = \tan t$$

Thus we know that $\tan(t + \pi) = \tan t$ for all t. A similar argument can be used to show that $\cot(t + \pi) = \cot t$ for all t.

Period of the Tangent and Cotangent Functions

The tangent and cotangent functions are periodic functions with a period of π.

$$\tan(t + \pi) = \tan t \qquad \text{and} \qquad \cot(t + \pi) = \cot t$$

The following theorem illustrates the repetitive nature of each of the trigonometric functions of a real number.

Periodic Properties of the Trigonometric Functions of a Real Number

For any real number t and integer k,

$$\sin(t + 2k\pi) = \sin t \qquad \cos(t + 2k\pi) = \cos t$$
$$\sec(t + 2k\pi) = \sec t \qquad \csc(t + 2k\pi) = \csc t$$
$$\tan(t + k\pi) = \tan t \qquad \cot(t + k\pi) = \cot t$$

Trigonometric Identities

Recall that any equation that is true for every number in the domain of the equation is an identity. The statement

$$\csc t = \frac{1}{\sin t}, \quad \sin t \neq 0$$

is an identity because the two expressions produce the same result for all values of t for which both functions are defined.

The **ratio identities** are obtained by writing the tangent and cotangent functions in terms of the sine and cosine functions.

$$\tan t = \frac{y}{x} = \frac{\sin t}{\cos t} \quad \text{and} \quad \cot t = \frac{x}{y} = \frac{\cos t}{\sin t}$$ • $x = \cos t$ and $y = \sin t$

The **Pythagorean identities** are based on the equation of a unit circle, $x^2 + y^2 = 1$, and on the definitions of the sine and cosine functions.

$$x^2 + y^2 = 1$$

$$\cos^2 t + \sin^2 t = 1$$ • Replace x by $\cos t$ and y by $\sin t$.

Dividing each term of $\cos^2 t + \sin^2 t = 1$ by $\cos^2 t$, we have

$$\frac{\cos^2 t}{\cos^2 t} + \frac{\sin^2 t}{\cos^2 t} = \frac{1}{\cos^2 t}$$ • $\cos t \neq 0$

$$1 + \tan^2 t = \sec^2 t$$ • $\frac{\sin t}{\cos t} = \tan t$

Dividing each term of $\cos^2 t + \sin^2 t = 1$ by $\sin^2 t$, we have

$$\frac{\cos^2 t}{\sin^2 t} + \frac{\sin^2 t}{\sin^2 t} = \frac{1}{\sin^2 t}$$ • $\sin t \neq 0$

$$\cot^2 t + 1 = \csc^2 t$$ • $\frac{\cos t}{\sin t} = \cot t$

Here is a summary of the Fundamental Trigonometric Identities:

Fundamental Trigonometric Identities

The reciprocal identities are

$$\sin t = \frac{1}{\csc t} \qquad \cos t = \frac{1}{\sec t} \qquad \tan t = \frac{1}{\cot t}$$

The ratio identities are

$$\tan t = \frac{\sin t}{\cos t} \qquad \cot t = \frac{\cos t}{\sin t}$$

The Pythagorean identities are

$$\cos^2 t + \sin^2 t = 1 \qquad 1 + \tan^2 t = \sec^2 t \qquad 1 + \cot^2 t = \csc^2 t$$

EXAMPLE 5 Use the Unit Circle to Verify an Identity

Use the unit circle and the definitions of the trigonometric functions to show that $\sin(t + \pi) = -\sin t$.

Solution

Sketch the unit circle and let P be the point on the unit circle such that $W(t) = P(x, y)$, as shown in **Figure 2.58**. Draw a diameter from P and label the endpoint Q. For any line through the origin, if $P(x, y)$ is a point on the line, then $Q(-x, -y)$ is also a point on the line. Because line segment PQ is a diameter, the length of the arc from P to Q is π. Thus the length of the arc from A through P to Q is $t + \pi$. Therefore, $W(t + \pi) = Q(-x, -y)$. From the definition of $\sin t$, we have

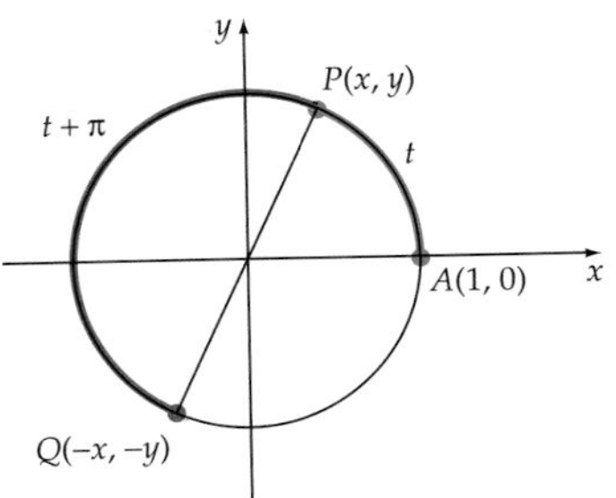

Figure 2.58

$$\sin t = y \qquad \text{and} \qquad \sin(t + \pi) = -y$$

Thus $\sin(t + \pi) = -\sin t$.

Try Exercise 56, page 167

Using identities and basic algebra concepts, we can rewrite trigonometric expressions in different forms.

EXAMPLE 6 Simplify a Trigonometric Expression

Write the expression $\dfrac{1}{\sin^2 t} + \dfrac{1}{\cos^2 t}$ as a single term.

Solution

Express each fraction in terms of a common denominator. The common denominator is $\sin^2 t \cos^2 t$.

$$\frac{1}{\sin^2 t} + \frac{1}{\cos^2 t} = \frac{1}{\sin^2 t} \cdot \frac{\cos^2 t}{\cos^2 t} + \frac{1}{\cos^2 t} \cdot \frac{\sin^2 t}{\sin^2 t}$$

$$= \frac{\cos^2 t + \sin^2 t}{\sin^2 t \cos^2 t} = \frac{1}{\sin^2 t \cos^2 t} \qquad \bullet\ \cos^2 t + \sin^2 t = 1$$

Try Exercise 72, page 167

take note

Because

$$\frac{1}{\sin^2 t \cos^2 t} = \frac{1}{\sin^2 t} \cdot \frac{1}{\cos^2 t} = (\csc^2 t)(\sec^2 t)$$

we could have written the answer to Example 6 in terms of the cosecant and secant functions.

EXAMPLE 7 ›› Write a Trigonometric Expression in Terms of a Given Function

For $\frac{\pi}{2} < t < \pi$, write $\tan t$ in terms of $\sin t$.

Solution

Write $\tan t = \frac{\sin t}{\cos t}$. Now solve $\cos^2 t + \sin^2 t = 1$ for $\cos t$.

$$\cos^2 t + \sin^2 t = 1$$
$$\cos^2 t = 1 - \sin^2 t$$
$$\cos t = \pm\sqrt{1 - \sin^2 t}$$

Because $\frac{\pi}{2} < t < \pi$, $\cos t$ is negative. Therefore, $\cos t = -\sqrt{1 - \sin^2 t}$.

Thus

$$\tan t = \frac{\sin t}{\cos t} = -\frac{\sin t}{\sqrt{1 - \sin^2 t}} \qquad \cdot \frac{\pi}{2} < t < \pi$$

›› *Try Exercise 78, page 167*

Topics for Discussion

1. Is $W(t)$ a number? Explain.
2. Explain why the equation $\cos^2 t + \sin^2 t = 1$ is called a Pythagorean identity.
3. Is $f(x) = \cos^3 x$ an even function or an odd function? Explain how you made your decision.
4. Explain how to make use of a unit circle to show that $\sin(-t) = -\sin t$.

Exercise Set 2.4

In Exercises 1 to 12, evaluate $W(t)$ for each given t.

1. $t = \frac{\pi}{6}$ **2.** $t = \frac{\pi}{4}$ **3.** $t = \frac{7\pi}{6}$

4. $t = \frac{4\pi}{3}$ **5.** $t = \frac{5\pi}{3}$ **6.** $t = -\frac{\pi}{6}$

7. $t = \frac{11\pi}{6}$ **8.** $t = 0$ **9.** $t = \pi$

10. $t = -\frac{7\pi}{4}$ **11.** $t = -\frac{2\pi}{3}$ **12.** $t = -\pi$

In Exercises 13 to 22, find the exact value of each function.

13. $\tan\left(\frac{11\pi}{6}\right)$ **14.** $\cot\left(\frac{2\pi}{3}\right)$

15. $\cos\left(-\frac{2\pi}{3}\right)$ **16.** $\sec\left(-\frac{5\pi}{6}\right)$

17. $\csc\left(-\frac{\pi}{3}\right)$ **18.** $\tan(12\pi)$

19. $\sin\left(\frac{3\pi}{2}\right)$ **20.** $\cos\left(\frac{7\pi}{3}\right)$

21. $\sec\left(-\frac{7\pi}{6}\right)$ **22.** $\sin\left(-\frac{5\pi}{3}\right)$

In Exercises 23 to 32, use a calculator to find an approximate value of each function. Round your answers to the nearest ten-thousandth.

23. $\sin 1.22$ **24.** $\cos 4.22$

25. $\csc(-1.05)$ **26.** $\sin(-0.55)$

27. $\tan\left(\frac{11\pi}{12}\right)$ **28.** $\cos\left(\frac{2\pi}{5}\right)$

29. $\cos\left(-\frac{\pi}{5}\right)$ **30.** $\csc 8.2$

31. $\sec 1.55$ **32.** $\cot 2.11$

In Exercises 33 to 40, use the unit circle to estimate the following values to the nearest tenth.

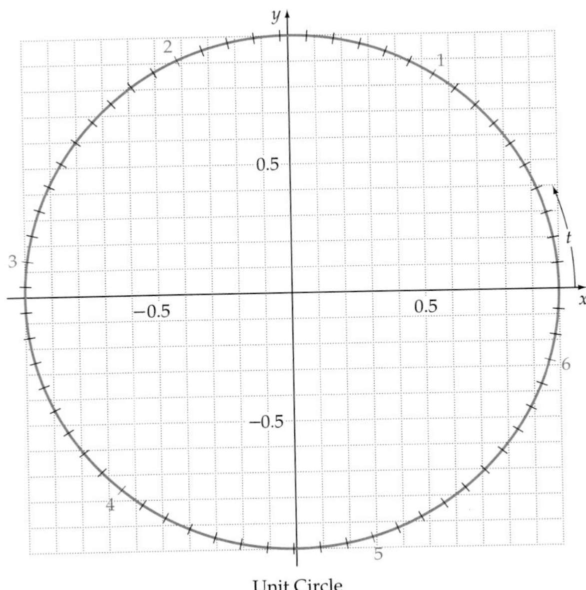

Unit Circle

33. a. $\sin 2$ **b.** $\cos 2$

34. a. $\sin 3$ **b.** $\cos 3$

35. a. $\sin 5.4$ **b.** $\cos 5.4$

36. a. $\sin 4.1$ **b.** $\cos 4.1$

37. All real numbers t between 0 and 2π for which $\sin t = 0.4$

38. All real numbers t between 0 and 2π for which $\cos t = 0.8$

39. All real numbers t between 0 and 2π for which $\sin t = -0.3$

40. All real numbers t between 0 and 2π for which $\cos t = -0.7$

In Exercises 41 to 48, determine whether the function is even, odd, or neither.

41. $f(x) = -4 \sin x$ **42.** $f(x) = -2 \cos x$

43. $G(x) = \sin x + \cos x$ **44.** $F(x) = \tan x + \sin x$

45. $S(x) = \dfrac{\sin x}{x}, x \neq 0$

46. $C(x) = \dfrac{\cos x}{x}, x \neq 0$

47. $v(x) = 2 \sin x \cos x$

48. $w(x) = x \tan x$

In Exercises 49 to 54, state the period of each function.

49. $f(t) = \sin t$

50. $f(t) = \cos t$

51. $f(t) = \tan t$

52. $f(t) = \cot t$

53. $f(t) = \sec t$

54. $f(t) = \csc t$

In Exercises 55 to 60, use the unit circle to verify each identity.

55. $\cos(-t) = \cos t$

56. $\tan(t - \pi) = \tan t$

57. $\cos(t + \pi) = -\cos t$

58. $\sin(-t) = -\sin t$

59. $\sin(t - \pi) = -\sin t$

60. $\sec(-t) = \sec t$

In Exercises 61 to 76, use trigonometric identities to write each expression in terms of a single trigonometric function or a constant. Answers may vary.

61. $\tan t \cos t$

62. $\cot t \sin t$

63. $\dfrac{\csc t}{\cot t}$

64. $\dfrac{\sec t}{\tan t}$

65. $1 - \sec^2 t$

66. $1 - \csc^2 t$

67. $\tan t - \dfrac{\sec^2 t}{\tan t}$

68. $\dfrac{\csc^2 t}{\cot t} - \cot t$

69. $\dfrac{1 - \cos^2 t}{\tan^2 t}$

70. $\dfrac{1 - \sin^2 t}{\cot^2 t}$

71. $\dfrac{1}{1 - \cos t} + \dfrac{1}{1 + \cos t}$

72. $\dfrac{1}{1 - \sin t} + \dfrac{1}{1 + \sin t}$

73. $\dfrac{\tan t + \cot t}{\tan t}$

74. $\dfrac{\csc t - \sin t}{\csc t}$

75. $\sin^2 t(1 + \cot^2 t)$

76. $\cos^2 t(1 + \tan^2 t)$

77. Write $\sin t$ in terms of $\cos t$, $0 < t < \dfrac{\pi}{2}$.

78. Write $\tan t$ in terms of $\sec t$, $\dfrac{3\pi}{2} < t < 2\pi$.

79. Write $\csc t$ in terms of $\cot t$, $\dfrac{\pi}{2} < t < \pi$.

80. Write $\sec t$ in terms of $\tan t$, $\pi < t < \dfrac{3\pi}{2}$.

81. **Path of a Satellite** A satellite is launched into space from Cape Canaveral. The directed distance, in miles, that the satellite is north or south of the equator is

$$d(t) = 1970 \cos\left(\frac{\pi}{64} t\right)$$

where t is the number of minutes since liftoff. A negative d value indicates that the satellite is south of the equator.

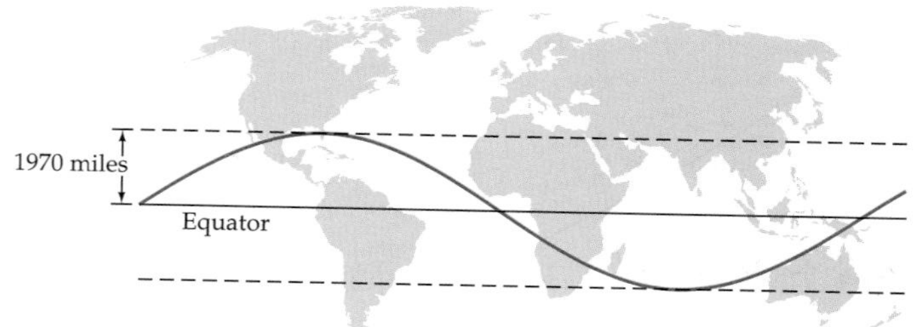

What distance, to the nearest 10 miles, is the satellite north of the equator 24 minutes after liftoff?

82. **Average High Temperature** The average high temperature T, in degrees Fahrenheit, for Fairbanks, Alaska, is given by

$$T(t) = -41 \cos\left(\frac{\pi}{6} t\right) + 36$$

where t is the number of months after January 5. Use the formula to estimate (to the nearest tenth of a degree Fahrenheit) the average high temperature in Fairbanks for March 5 and July 20.

In Exercises 83 to 94, perform the indicated operation and simplify.

83. $\cos t - \dfrac{1}{\cos t}$

84. $\tan t + \dfrac{1}{\tan t}$

85. $\cot t + \dfrac{1}{\cot t}$

86. $\sin t - \dfrac{1}{\sin t}$

87. $(1 - \sin t)^2$

88. $(1 - \cos t)^2$

89. $(\sin t - \cos t)^2$

90. $(\sin t + \cos t)^2$

91. $(1 - \sin t)(1 + \sin t)$

92. $(1 - \cos t)(1 + \cos t)$

93. $\dfrac{\sin t}{1 + \cos t} + \dfrac{1 + \cos t}{\sin t}$

94. $\dfrac{1 - \sin t}{\cos t} - \dfrac{1}{\tan t + \sec t}$

In Exercises 95 to 100, factor the expression.

95. $\cos^2 t - \sin^2 t$

96. $\sec^2 t - \csc^2 t$

97. $\tan^2 t - \tan t - 6$

98. $\cos^2 t + 3\cos t - 4$

99. $2\sin^2 t - \sin t - 1$

100. $4\cos^2 t + 4\cos t + 1$

Connecting Concepts

In Exercises 101 to 104, use trigonometric identities to find the value of the function.

101. Given $\csc t = \sqrt{2}, 0 < t < \frac{\pi}{2}$, find $\cos t$.

102. Given $\cos t = \frac{1}{2}, \frac{3\pi}{2} < t < 2\pi$, find $\sin t$.

103. Given $\sin t = \frac{1}{2}, \frac{\pi}{2} < t < \pi$, find $\tan t$.

104. Given $\cot t = \frac{\sqrt{3}}{3}, \pi < t < \frac{3\pi}{2}$, find $\cos t$.

In Exercises 105 to 108, simplify the first expression to the second expression.

105. $\dfrac{\sin^2 t + \cos^2 t}{\sin^2 t}$; $\csc^2 t$

106. $\dfrac{\sin^2 t + \cos^2 t}{\cos^2 t}$; $\sec^2 t$

107. $(\cos t - 1)(\cos t + 1)$; $-\sin^2 t$

108. $(\sec t - 1)(\sec t + 1)$; $\tan^2 t$

Projects

1.  VISUAL INSIGHT

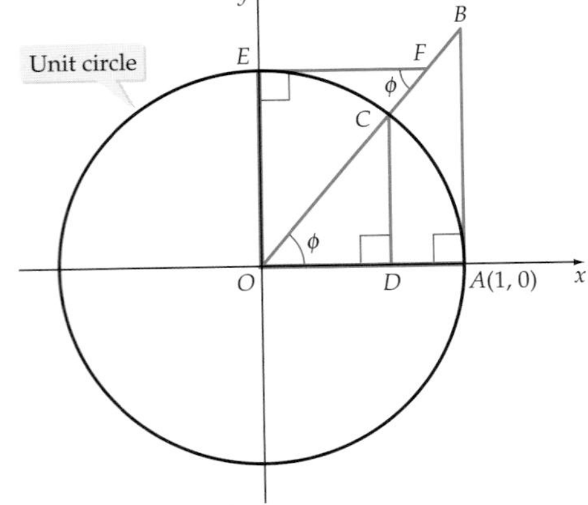

$OD = \cos\phi \quad DC = \sin\phi$

Make use of the above circle and similar triangles to explain why

a. AB is equal to $\tan\phi$

b. EF is equal to $\cot\phi$

c. OB is equal to $\sec\phi$

d. OF is equal to $\csc\phi$

2. Consider a square as shown. Start at the point $(1, 0)$ and travel counterclockwise around the square for a distance t $(t \geq 0)$.

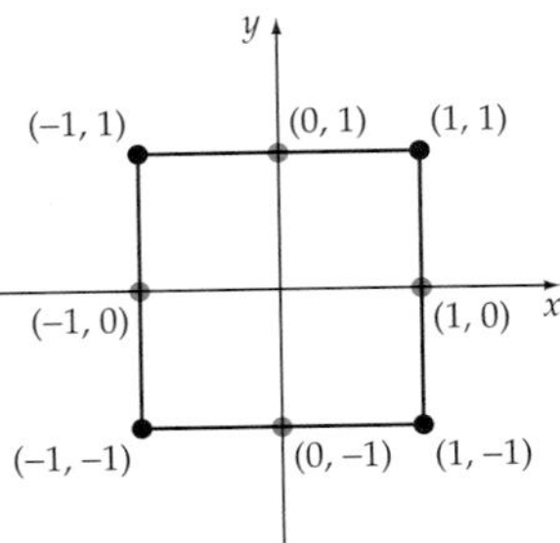

Let $WSQ(t) = P(x, y)$ be the point on the square determined by traveling counterclockwise a distance of t units from $(1, 0)$. For instance,

$$WSQ(0.5) = (1, 0.5)$$
$$WSQ(1.75) = (0.25, 1)$$

Find $WSQ(4.2)$ and $WSQ(6.4)$. We define the square sine of t, denoted by ssin t, to be the y-value of point P. The square cosine of t, denoted by scos t, is defined to be the x-value of point P. For example,

$$\text{ssin } 0.4 = 0.4 \qquad \text{scos } 0.4 = 1$$
$$\text{scos } 1.2 = 0.8 \qquad \text{scos } 5.3 = -0.7$$

The square tangent of t, denoted by stan t, is defined as

$$\text{stan } t = \frac{\text{ssin } t}{\text{scos } t}, \qquad \text{scos } t \neq 0$$

Find each of the following.

a. ssin 3.2 **b.** scos 4.4 **c.** stan 5.5

d. ssin 11.2 **e.** scos -5.2 **f.** stan -6.5

Section 2.5

- The Graph of the Sine Function
- The Graph of the Cosine Function

Graphs of the Sine and Cosine Functions

PREPARE FOR THIS SECTION

Prepare for this section by completing the following exercises. The answers can be found on page A10.

PS1.  Estimate, to the nearest tenth, $\sin \frac{3\pi}{4}$. [2.4]

PS2. Estimate, to the nearest tenth, $\cos \frac{5\pi}{4}$. [2.4]

PS3. Explain how to use the graph of $y = f(x)$ to produce the graph of $y = -f(x)$. [1.4]

PS4. Explain how to use the graph of $y = f(x)$ to produce the graph of $y = f(2x)$. [1.4]

PS5. Simplify: $\frac{2\pi}{1/3}$

PS6. Simplify: $\frac{2\pi}{2/5}$

The Graph of the Sine Function

The trigonometric functions can be graphed on a rectangular coordinate system by plotting the points whose coordinates belong to the function. We begin with the graph of the sine function.

Table 2.7 lists some ordered pairs (x, y) of the graph of $y = \sin x$ for $0 \leq x \leq 2\pi$.

Table 2.7 Ordered Pairs of the Graph of $y = \sin x$

x	0	$\frac{\pi}{6}$	$\frac{\pi}{3}$	$\frac{\pi}{2}$	$\frac{2\pi}{3}$	$\frac{5\pi}{6}$	π	$\frac{7\pi}{6}$	$\frac{4\pi}{3}$	$\frac{3\pi}{2}$	$\frac{5\pi}{3}$	$\frac{11\pi}{6}$	2π
$y = \sin x$	0	$\frac{1}{2}$	$\frac{\sqrt{3}}{2}$	1	$\frac{\sqrt{3}}{2}$	$\frac{1}{2}$	0	$-\frac{1}{2}$	$-\frac{\sqrt{3}}{2}$	-1	$-\frac{\sqrt{3}}{2}$	$-\frac{1}{2}$	0

In **Figure 2.59,** the points from the table are plotted and a smooth curve is drawn through the points. We could use decimal approximations for π on the x-axis, but it is more convenient to simply label the tick marks on the x-axis in terms of π. *Note:* The y-value $\frac{\sqrt{3}}{2} \approx 0.87$.

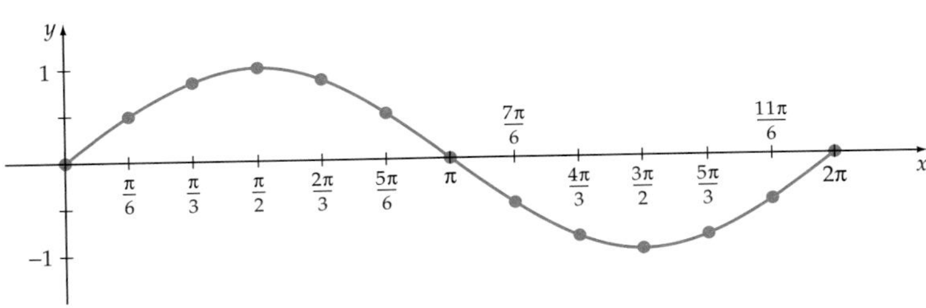

$y = \sin x, 0 \leq x \leq 2\pi$

Figure 2.59

Because the domain of the sine function is the real numbers and the period is 2π, the graph of $y = \sin x$ is drawn by repeating the portion shown in **Figure 2.59.** Any part of the graph that corresponds to one period (2π) is one **cycle** of the graph of $y = \sin x$ (see **Figure 2.60**).

The Graph of $y = \sin x$

Figure 2.60

Basic Properties

- **Domain:** All real numbers
- **Range:** $\{y \mid -1 \leq y \leq 1\}$
- **Period:** 2π
- **Symmetry:** With respect to the origin
- **x-intercepts:** At multiples of π

The maximum value M reached by $\sin x$ is 1, and the minimum value m is -1. The **amplitude** of the graph of $y = \sin x$ is given by

$$\text{Amplitude} = \frac{1}{2}(M - m)$$

? QUESTION What is the amplitude of $y = \sin x$?

Recall that the graph of $y = a \cdot f(x)$ is obtained by *stretching* ($|a| > 1$) or *shrinking* ($0 < |a| < 1$) the graph of $y = f(x)$. **Figure 2.61** shows the graph of $y = 3 \sin x$

? ANSWER Amplitude $= \frac{1}{2}(M - m) = \frac{1}{2}[1 - (-1)] = \frac{1}{2}(2) = 1$

that was drawn by stretching the graph of $y = \sin x$. The amplitude of $y = 3 \sin x$ is 3 because

$$\text{Amplitude} = \frac{1}{2}(M - m) = \frac{1}{2}[3 - (-3)] = 3$$

Note that for $y = \sin x$ and $y = 3 \sin x$, the amplitude of the graph is the coefficient of $\sin x$. This suggests the following theorem.

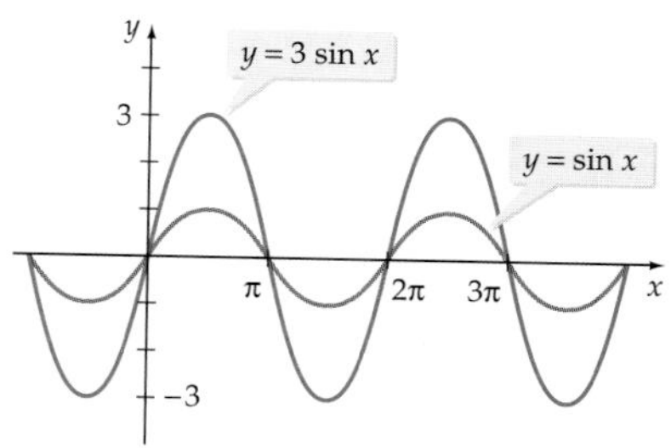

Figure 2.61

take note

The amplitude is defined to be half the difference between the maximum height and the minimum height. It may not be equal to the maximum height. For example, the graph of $y = 4 + \sin x$ has a maximum height of 5 and an amplitude of 1.

Amplitude of $y = a \sin x$

The amplitude of $y = a \sin x$ is $|a|$.

EXAMPLE 1 Graph $y = a \sin x$

Graph: $y = -2 \sin x$

Solution

The amplitude of $y = -2 \sin x$ is 2. The graph of $y = -f(x)$ is a *reflection* across the x-axis of $y = f(x)$. Thus the graph of $y = -2 \sin x$ is a reflection across the x-axis of $y = 2 \sin x$. See the following figure.

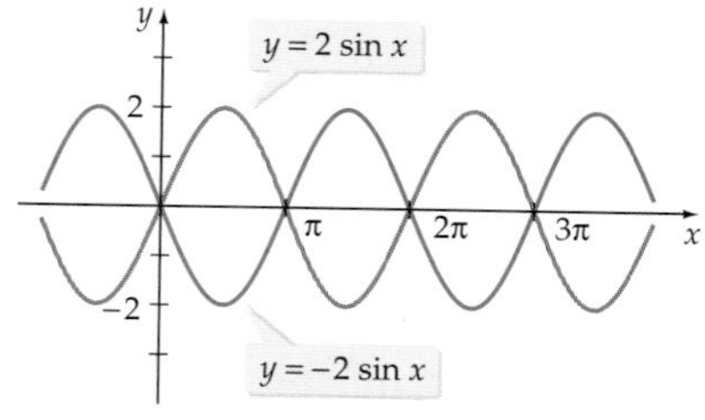

Try Exercise 22, page 177

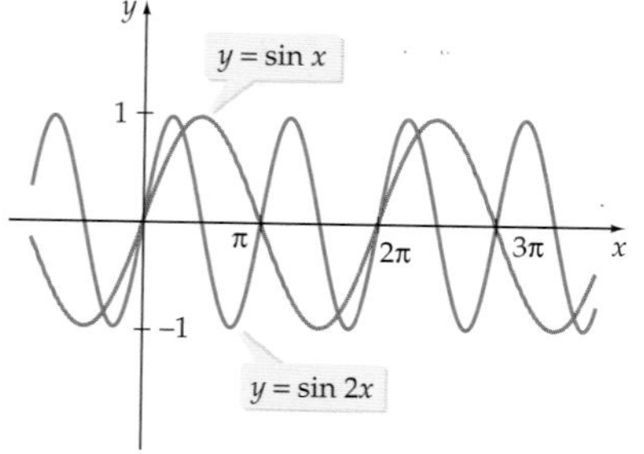

Figure 2.62

Figure 2.63

The graphs of $y = \sin x$ and $y = \sin 2x$ are shown in **Figure 2.62.** Because one cycle of the graph of $y = \sin 2x$ is completed in an interval of length π, the period of $y = \sin 2x$ is π.

The graphs of $y = \sin x$ and $y = \sin \frac{x}{2}$ are shown in **Figure 2.63.** Because one cycle of the graph of $y = \sin \frac{x}{2}$ is completed in an interval of length 4π, the period of $y = \sin \frac{x}{2}$ is 4π.

Generalizing the last two examples, one cycle of $y = \sin bx$, $b > 0$, is completed as bx varies from 0 to 2π. Therefore,

$$0 \le bx \le 2\pi$$
$$0 \le x \le \frac{2\pi}{b}$$

The length of the interval, $\frac{2\pi}{b}$, is the period of $y = \sin bx$. Now we consider the case when the coefficient of x is negative. If $b > 0$, then using the fact that the sine function is an odd function, we have $y = \sin(-bx) = -\sin bx$, and thus the period is still $\frac{2\pi}{b}$. This gives the following theorem.

Period of $y = \sin bx$

The period of $y = \sin bx$ is $\frac{2\pi}{|b|}$.

Table 2.8 gives the amplitude and period of several sine functions.

Table 2.8

Function	$y = a \sin bx$	$y = 3\sin(-2x)$	$y = -\sin \frac{x}{3}$	$y = -2\sin \frac{3x}{4}$
Amplitude	$\|a\|$	$\|3\| = 3$	$\|-1\| = 1$	$\|-2\| = 2$
Period	$\frac{2\pi}{\|b\|}$	$\frac{2\pi}{2} = \pi$	$\frac{2\pi}{1/3} = 6\pi$	$\frac{2\pi}{3/4} = \frac{8\pi}{3}$

EXAMPLE 2 ❯❯ Graph $y = \sin bx$

Graph: $y = \sin \pi x$

Solution

$$\text{Amplitude} = 1 \qquad \text{Period} = \frac{2\pi}{|b|} = \frac{2\pi}{\pi} = 2 \qquad \bullet\ b = \pi$$

The graph is sketched in **Figure 2.64.**

❯❯ *Try Exercise 32, page 177*

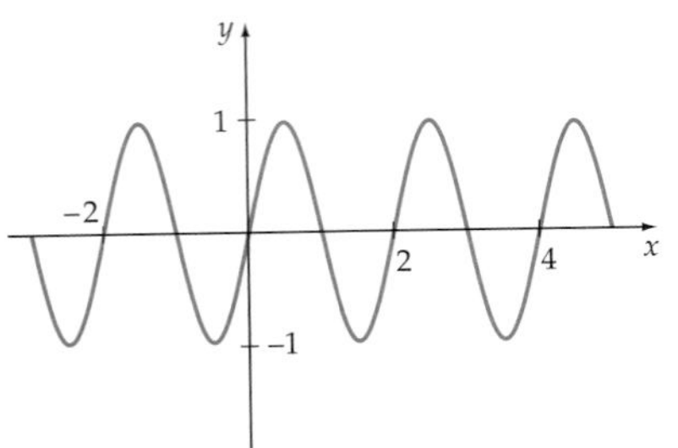

$y = \sin \pi x$

Figure 2.64

Figure 2.65 shows the graph of $y = a \sin bx$ for both a and b positive. Note from the graph the following properties of the function $y = a \sin bx$.

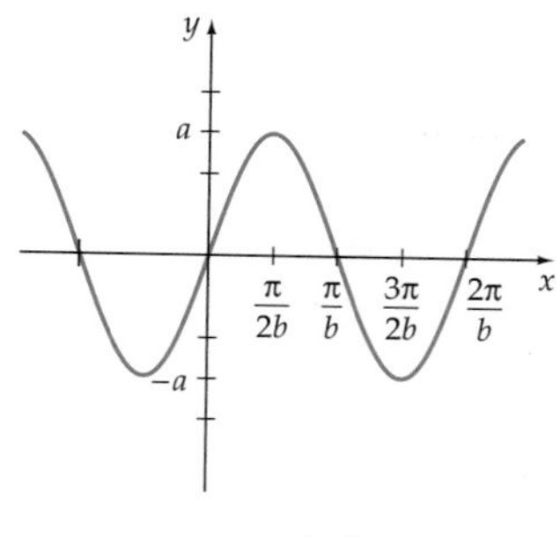

$y = a \sin bx$

Figure 2.65

- The amplitude is a.
- The period is $\frac{2\pi}{b}$.
- For $0 \le x \le \frac{2\pi}{b}$, the zeros are 0, $\frac{\pi}{b}$, and $\frac{2\pi}{b}$.
- The maximum value is a when $x = \frac{\pi}{2b}$.
- The minimum value is $-a$ when $x = \frac{3\pi}{2b}$.
- If $a < 0$, the graph is reflected across the x-axis.

EXAMPLE 3 ⟫ Graph $y = a \sin bx$

Graph: $y = -\frac{1}{2} \sin \frac{x}{3}$

Solution

$$\text{Amplitude} = \left| -\frac{1}{2} \right| = \frac{1}{2} \qquad \text{Period} = \frac{2\pi}{|1/3|} = 6\pi \qquad \bullet\ b = \frac{1}{3}$$

The zeros in the interval $0 \le x \le 6\pi$ are 0, $\frac{\pi}{1/3} = 3\pi$, and $\frac{2\pi}{1/3} = 6\pi$, so the graph has x-intercepts at $(0, 0)$, $(3\pi, 0)$, and $(6\pi, 0)$. Because $-\frac{1}{2} < 0$, the graph is the graph of $y = \frac{1}{2} \sin \frac{x}{3}$ reflected across the x-axis, as shown in **Figure 2.66**.

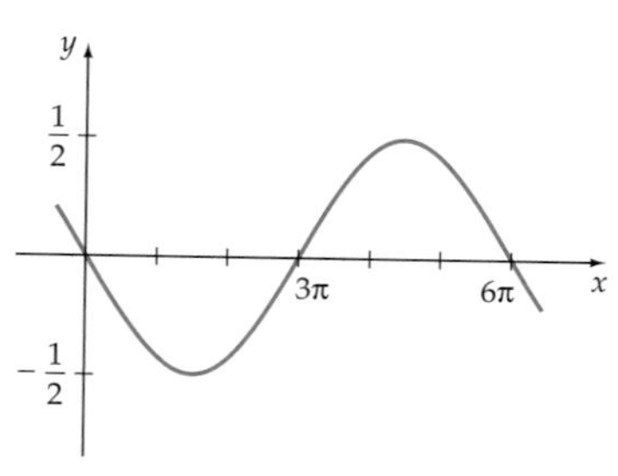

$y = -\frac{1}{2} \sin \frac{x}{3}$

Figure 2.66

⟫ *Try Exercise 40, page 178*

The Graph of the Cosine Function

Table 2.9 lists some ordered pairs (x, y) of the graph of $y = \cos x$ for $0 \le x \le 2\pi$.

Table 2.9 Ordered Pairs of the Graph of $y = \cos x$

x	0	$\frac{\pi}{6}$	$\frac{\pi}{3}$	$\frac{\pi}{2}$	$\frac{2\pi}{3}$	$\frac{5\pi}{6}$	π	$\frac{7\pi}{6}$	$\frac{4\pi}{3}$	$\frac{3\pi}{2}$	$\frac{5\pi}{3}$	$\frac{11\pi}{6}$	2π
$y = \cos x$	1	$\frac{\sqrt{3}}{2}$	$\frac{1}{2}$	0	$-\frac{1}{2}$	$-\frac{\sqrt{3}}{2}$	-1	$-\frac{\sqrt{3}}{2}$	$-\frac{1}{2}$	0	$\frac{1}{2}$	$\frac{\sqrt{3}}{2}$	0

In **Figure 2.67** on page 174, the points from the table are plotted and a smooth curve is drawn through the points.

$y = \cos x, 0 \le x \le 2\pi$

Figure 2.67

Because the domain of $y = \cos x$ is the real numbers and the period is 2π, the graph of $y = \cos x$ is drawn by repeating the portion shown in **Figure 2.67**. Any part of the graph corresponding to one period (2π) is one cycle of $y = \cos x$ (see **Figure 2.68**).

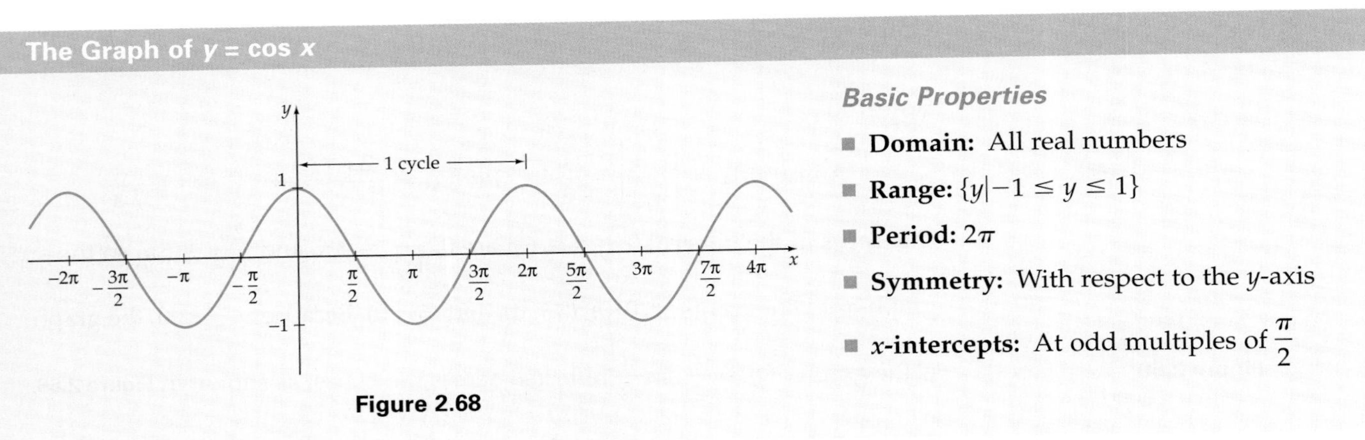

The Graph of $y = \cos x$

Basic Properties

- **Domain:** All real numbers
- **Range:** $\{y|-1 \le y \le 1\}$
- **Period:** 2π
- **Symmetry:** With respect to the y-axis
- **x-intercepts:** At odd multiples of $\dfrac{\pi}{2}$

Figure 2.68

The following two theorems concerning cosine functions can be developed using methods that are analogous to those we used to determine the amplitude and period of a sine function.

Amplitude of $y = a \cos x$

The amplitude of $y = a \cos x$ is $|a|$.

Period of $y = \cos bx$

The period of $y = \cos bx$ is $\dfrac{2\pi}{|b|}$.

Table 2.10 gives the amplitude and period of some cosine functions.

Table 2.10

Function	$y = a \cos bx$	$y = 2 \cos 3x$	$y = -3 \cos \dfrac{2x}{3}$
Amplitude	$\lvert a \rvert$	$\lvert 2 \rvert = 2$	$\lvert -3 \rvert = 3$
Period	$\dfrac{2\pi}{\lvert b \rvert}$	$\dfrac{2\pi}{3}$	$\dfrac{2\pi}{2/3} = 3\pi$

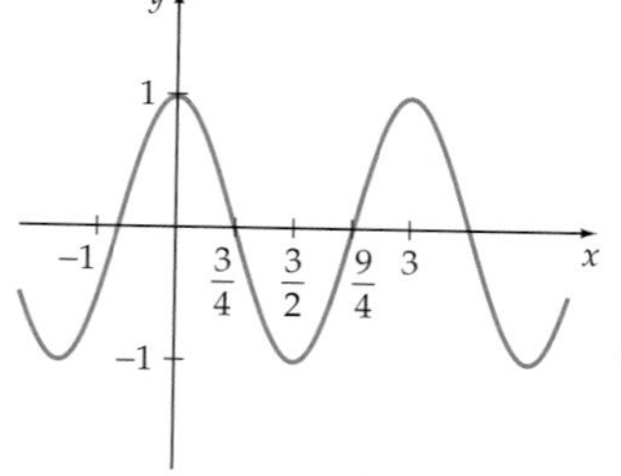

$y = \cos \dfrac{2\pi}{3} x$

Figure 2.69

EXAMPLE 4 Graph $y = \cos bx$

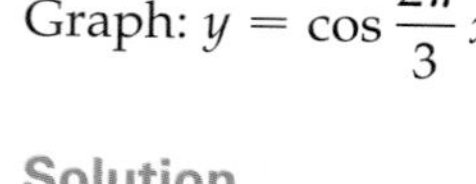

Graph: $y = \cos \dfrac{2\pi}{3} x$

Solution

$$\text{Amplitude} = 1 \qquad \text{Period} = \frac{2\pi}{|b|} = \frac{2\pi}{2\pi/3} = 3 \qquad \bullet\ b = \frac{2\pi}{3}$$

The graph is shown in **Figure 2.69**.

Try Exercise 34, page 177

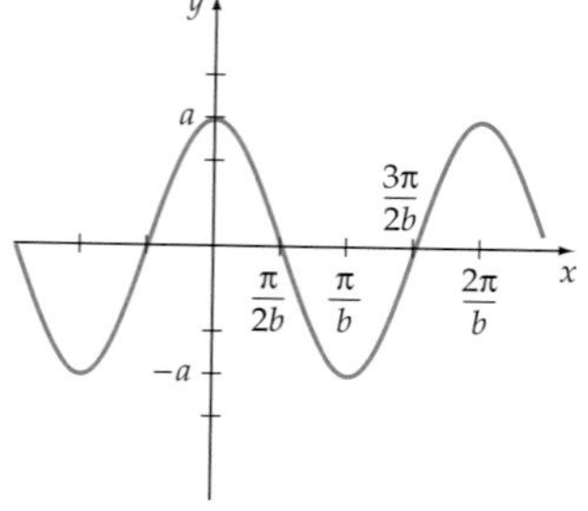

$y = a \cos bx$

Figure 2.70

Figure 2.70 shows the graph of $y = a \cos bx$ for both a and b positive. Note from the graph the following properties of the function $y = a \cos bx$.

- The amplitude is a.
- The period is $\dfrac{2\pi}{b}$.
- For $0 \le x \le \dfrac{2\pi}{b}$, the zeros are $\dfrac{\pi}{2b}$ and $\dfrac{3\pi}{2b}$.
- The maximum value is a when $x = 0$.
- The minimum value is $-a$ when $x = \dfrac{\pi}{b}$.
- If $a < 0$, then the graph is reflected across the x-axis.

EXAMPLE 5 Graph a Cosine Function

Graph: $y = -2 \cos \dfrac{\pi x}{4}$

Solution

$$\text{Amplitude} = |-2| = 2 \qquad \text{Period} = \frac{2\pi}{|\pi/4|} = 8 \qquad \bullet\ b = \frac{\pi}{4}$$

Continued ▶

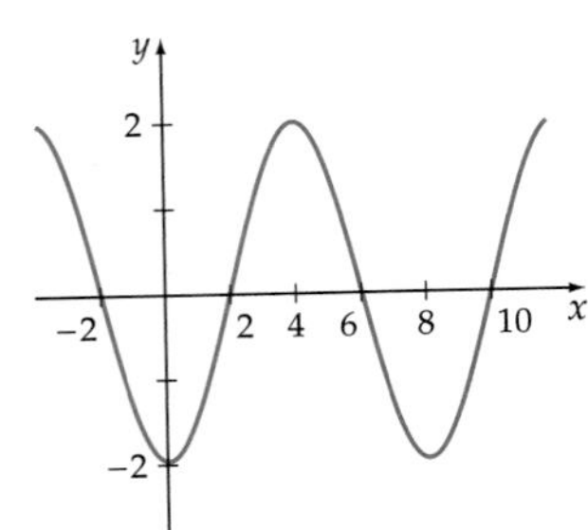

$y = -2\cos\frac{\pi x}{4}$

Figure 2.71

The zeros in the interval $0 \le x \le 8$ are $\frac{\pi}{2\pi/4} = 2$ and $\frac{3\pi}{2\pi/4} = 6$, so the graph has x-intercepts at (2, 0) and (6, 0). Because $-2 < 0$, the graph is the graph of $y = 2\cos\frac{\pi x}{4}$ reflected across the x-axis, as shown in **Figure 2.71**.

Try Exercise 48, page 178

EXAMPLE 6 Graph the Absolute Value of the Cosine Function

Graph $y = |\cos x|$, where $0 \le x \le 2\pi$.

Solution

Because $|\cos x| \ge 0$, the graph of $y = |\cos x|$ is drawn by reflecting the negative portion of the graph of $y = \cos x$ across the x-axis. The graph is the one shown in purple and light blue in **Figure 2.72**.

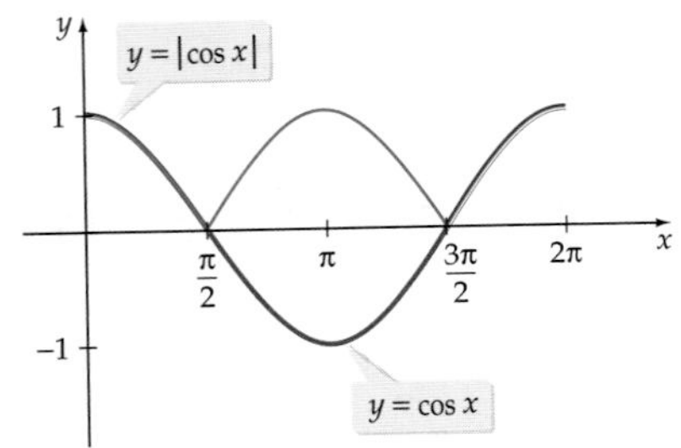

Figure 2.72

Try Exercise 54, page 178

In Example 7 we determine an equation for a given graph.

EXAMPLE 7 Find an Equation of a Graph

The graph at the right shows one cycle of the graph of a sine or cosine function. Find an equation for the graph.

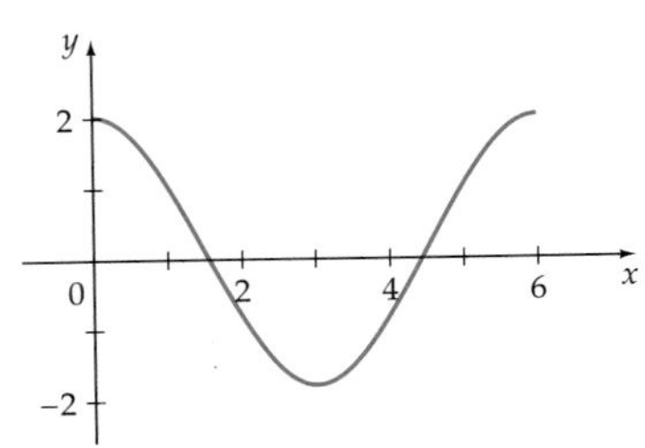

Solution

Because the graph obtains its maximum value at $x = 0$, start with an equation of the form $y = a\cos bx$. The graph completes one cycle in 6 units. Thus the period is 6. Use the equation $6 = \frac{2\pi}{|b|}$ to determine b.

$$6 = \frac{2\pi}{|b|}$$

$$6|b| = 2\pi \qquad \text{• Multiply each side by } |b|.$$

$$|b| = \frac{\pi}{3} \qquad \text{• Divide each side by 6.}$$

$$b = \pm\frac{\pi}{3}$$

We can use either $\frac{\pi}{3}$ or $-\frac{\pi}{3}$ as the b value. The graph has a maximum height of 2 and a minimum height of -2. Thus the amplitude is $a = 2$.

Substituting 2 for a and $\frac{\pi}{3}$ for b in $y = a \cos bx$ produces $y = 2 \cos \frac{\pi}{3}x$.

Try Exercise 60, page 178

Topics for Discussion

1. Is the graph of $f(x) = |\sin x|$ the same as the graph of $y = \sin|x|$? Explain.
2. Explain how the graph of $y = \cos 2x$ differs from the graph of $y = \cos x$.
3. Does the graph of $y = \sin(-2x)$ have the same period as the graph of $y = \sin 2x$? Explain.
4. The function $h(x) = a \sin bt$ has an amplitude of 3 and a period of 4. What are the possible values of a? What are the possible values of b?

Exercise Set 2.5

In Exercises 1 to 18, state the amplitude and period of the function defined by each equation.

1. $y = 2 \sin x$
2. $y = -\frac{1}{2} \sin x$
3. $y = \sin 2x$
4. $y = \sin \frac{2x}{3}$
5. $y = \frac{1}{2} \sin 2\pi x$
6. $y = 2 \sin \frac{\pi x}{3}$
7. $y = -2 \sin \frac{x}{2}$
8. $y = -\frac{1}{2} \sin \frac{x}{2}$
9. $y = \frac{1}{2} \cos x$
10. $y = -3 \cos x$
11. $y = \cos \frac{x}{4}$
12. $y = \cos 3x$
13. $y = 2 \cos \frac{\pi x}{3}$
14. $y = \frac{1}{2} \cos 2\pi x$
15. $y = -3 \cos \frac{2x}{3}$
16. $y = \frac{3}{4} \cos 4x$
17. $y = 4.7 \sin 0.8\pi t$
18. $y = 2.3 \cos 0.005\pi t$

In Exercises 19 to 56, graph one full period of the function defined by each equation.

19. $y = \frac{1}{2} \sin x$
20. $y = \frac{3}{2} \cos x$
21. $y = 3 \cos x$
22. $y = -\frac{3}{2} \sin x$
23. $y = -\frac{7}{2} \cos x$
24. $y = 3 \sin x$
25. $y = -4 \sin x$
26. $y = -5 \cos x$
27. $y = \cos 3x$
28. $y = \sin 4x$
29. $y = \sin \frac{3x}{2}$
30. $y = \cos \pi x$
31. $y = \cos \frac{\pi}{2} x$
32. $y = \sin \frac{3\pi}{4} x$
33. $y = \sin 2\pi x$
34. $y = \cos 3\pi x$
35. $y = 4 \cos \frac{x}{2}$
36. $y = 2 \cos \frac{3x}{4}$
37. $y = -2 \cos \frac{x}{3}$
38. $y = -\frac{4}{3} \cos 3x$

39. $y = 2 \sin \pi x$

40. $y = \frac{1}{2} \sin \frac{\pi x}{3}$

41. $y = \frac{3}{2} \cos \frac{\pi x}{2}$

42. $y = \cos \frac{\pi x}{3}$

43. $y = 4 \sin \frac{2\pi x}{3}$

44. $y = 3 \cos \frac{3\pi x}{2}$

45. $y = 2 \cos 2x$

46. $y = \frac{1}{2} \sin 2.5x$

47. $y = -2 \sin 1.5x$

48. $y = -\frac{3}{4} \cos 5x$

49. $y = \left|2 \sin \frac{x}{2}\right|$

50. $y = \left|\frac{1}{2} \sin 3x\right|$

51. $y = |-2 \cos 3x|$

52. $y = \left|-\frac{1}{2} \cos \frac{x}{2}\right|$

53. $y = -\left|2 \sin \frac{x}{3}\right|$

54. $y = -\left|3 \sin \frac{2x}{3}\right|$

55. $y = -|3 \cos \pi x|$

56. $y = -\left|2 \cos \frac{\pi x}{2}\right|$

In Exercises 57 to 62, one cycle of the graph of a sine or cosine function is shown. Find an equation of each graph.

57.

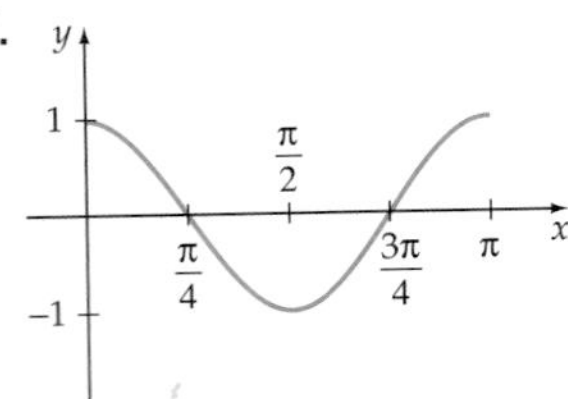

58.

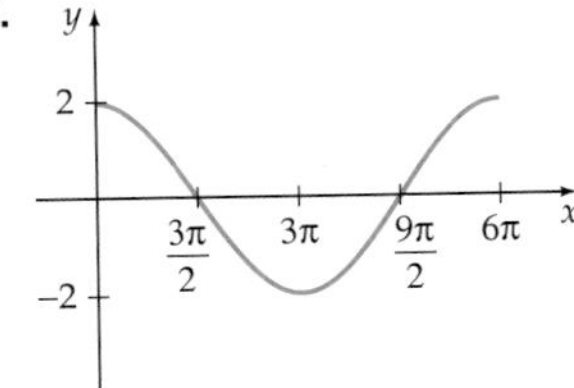

59.

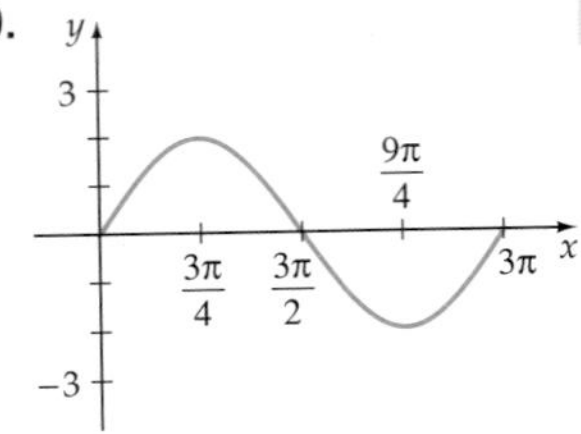

60.

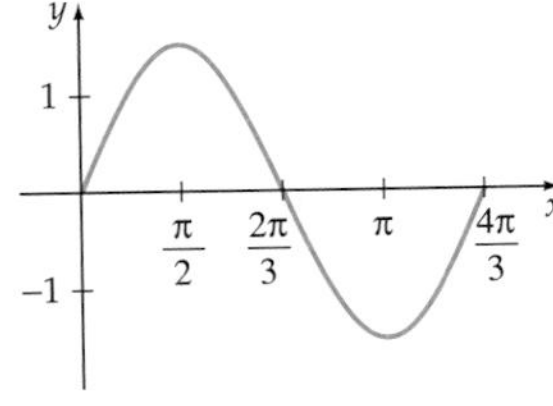

61.

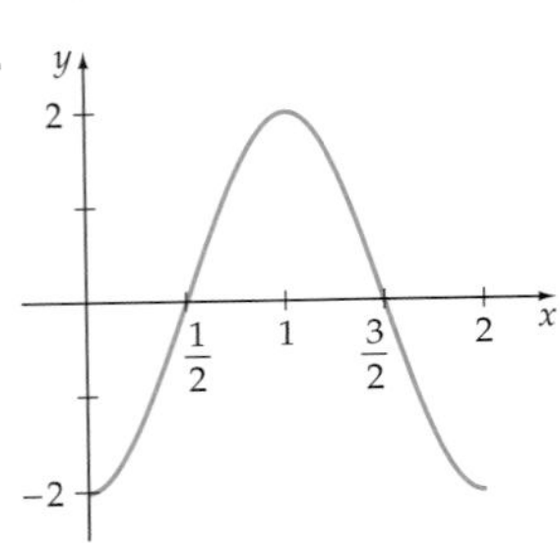

62.

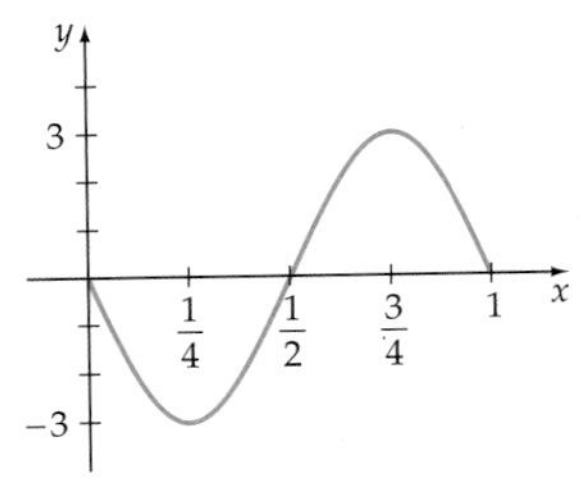

63. **Model a Sound Wave** The following oscilloscope screen displays the signal generated by the sound of a tuning fork during a time span of 8 milliseconds.

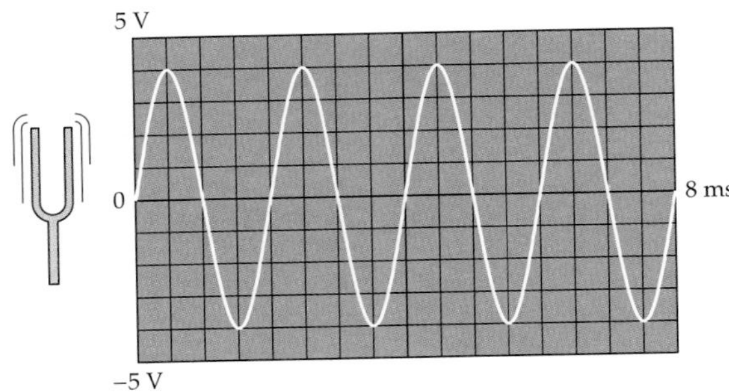

a. Write an equation of the form $V = a \sin bt$ that models the signal, where V is in volts and t is in milliseconds.

b. The **frequency of a sound wave** is defined as the reciprocal of the period of the sound wave. What is the frequency of the sound wave that produced the signal shown above? State your answer in cycles per millisecond.

64. **Model a Sound Wave** The following oscilloscope screen displays the signal generated by the sound of a tuning fork during a time span of 5 milliseconds.

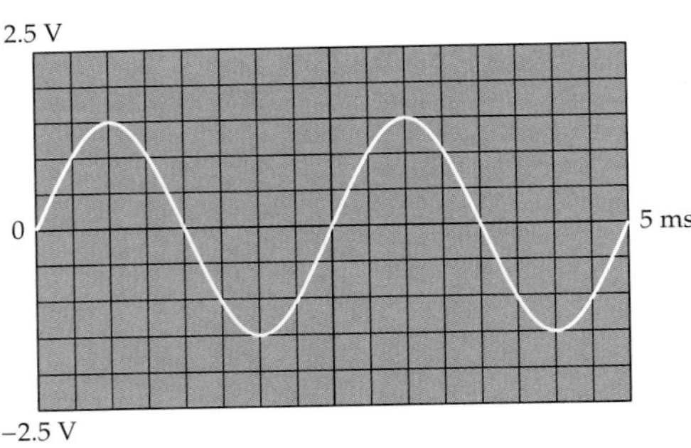

a. Write an equation of the form $V = a \sin bt$ that models the signal, where V is in volts and t is in milliseconds.

b. What is the frequency of the sound wave that produced the signal shown above? (*Hint:* See Exercise **63b**). State your answer in cycles per millisecond.

65. Sketch the graph of $y = 2 \sin \frac{2x}{3}$, $-3\pi \le x \le 6\pi$.

66. Sketch the graph of $y = -3 \cos \frac{3x}{4}$, $-2\pi \le x \le 4\pi$.

67. Sketch the graphs of

$$y_1 = 2 \cos \frac{x}{2} \quad \text{and} \quad y_2 = 2 \cos x$$

on the same set of axes for $-2\pi \le x \le 4\pi$.

68. Sketch the graphs of

$$y_1 = \sin 3\pi x \quad \text{and} \quad y_2 = \sin \frac{\pi x}{3}$$

on the same set of axes for $-2 \le x \le 4$.

In Exercises 69 to 72, use a graphing utility to graph each function.

69. $y = \frac{1}{2}x \sin x$

70. $y = \frac{1}{2}x + \sin x$

71. $y = -x \cos x$

72. $y = -x + \cos x$

73. Graph $y = e^{\sin x}$, where $e \approx 2.71828$. What is the maximum value of $e^{\sin x}$? What is the minimum value of $e^{\sin x}$? Is the function defined by $y = e^{\sin x}$ a periodic function? If so, what is the period?

74. Graph $y = e^{\cos x}$, where $e \approx 2.71828$. What is the maximum value of $e^{\cos x}$? What is the minimum value of $e^{\cos x}$? Is the function defined by $y = e^{\cos x}$ a periodic function? If so, what is the period?

Connecting Concepts

In Exercises 75 to 77, write an equation for a sine function using the given information.

75. Amplitude = 2; period = 3π

76. Amplitude = 4; period = 2

77. Amplitude = 2.5; period = 3.2

In Exercises 78 to 80, write an equation for a cosine function using the given information.

78. Amplitude = 3; period = $\frac{\pi}{2}$

79. Amplitude = 3; period = 2.5

80. Amplitude = 4.2; period = 1

Projects

1. CEPHEID VARIABLE STARS AND THE PERIOD-LUMINOSITY RELATIONSHIP A *variable star* is a star whose brightness increases and decreases in a periodic fashion. Variable stars known as *Cepheid variables* have very regular periods. The *light curve* below is for the star Delta Cephei. It shows how Delta Cephei's brightness varies every 5.4 days.

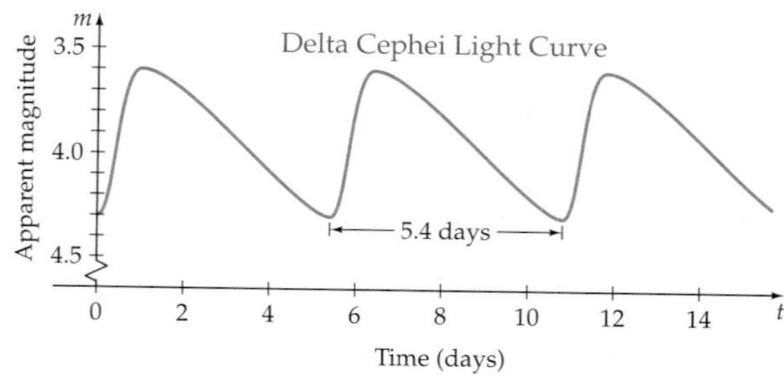

a. During the first decade of the 1900s, the astronomer Henrietta Leavitt (1868–1921) discovered the *period-luminosity relationship* for Cepheid variable stars. Write a few sentences that describe this relationship.

b. What is the major application of Leavitt's discovery? (*Note*: The Internet is a good source of information on the period-luminosity relationship.)

Section 2.6

- The Graph of the Tangent Function
- The Graph of the Cotangent Function
- The Graph of the Cosecant Function
- The Graph of the Secant Function

Graphs of the Other Trigonometric Functions

PREPARE FOR THIS SECTION

Prepare for this section by completing the following exercises. The answers can be found on page A11.

PS1. Estimate, to the nearest tenth, $\tan \frac{\pi}{3}$. [2.2]

PS2. Estimate, to the nearest tenth, $\cot \frac{\pi}{3}$. [2.2]

PS3. Explain how to use the graph of $y = f(x)$ to produce the graph of $y = 2f(x)$. [1.4]

PS4. Explain how to use the graph of $y = f(x)$ to produce the graph of $y = f(x - 2) + 3$. [1.4]

PS5. Simplify: $\dfrac{\pi}{1/2}$

PS6. Simplify: $\dfrac{\pi}{\left|-\frac{3}{4}\right|}$

The Graph of the Tangent Function

Table 2.11 lists some ordered pairs (x, y) of the graph of $y = \tan x$ for $0 \le x < \frac{\pi}{2}$.

Table 2.11 Ordered Pairs of the Graph of $y = \tan x$, $0 \le x < \frac{\pi}{2}$

x	0	$\frac{\pi}{12}$	$\frac{\pi}{6}$	$\frac{\pi}{4}$	$\frac{\pi}{3}$	$\frac{5\pi}{12}$	$\frac{\pi}{2}$
$y = \tan x$	0	≈ 0.27	$\frac{\sqrt{3}}{3} \approx 0.58$	1	$\sqrt{3} \approx 1.73$	≈ 3.7	undefined

In **Figure 2.73**, the points from the table are plotted and a smooth curve is drawn through the points. Notice that as x increases on $\left[0, \frac{\pi}{2}\right)$, $y = \tan x$ increases on $[0, \infty)$. The y values increase slowly at first and then more rapidly as $x \to \frac{\pi}{2}$ from the left. The line given by $x = \frac{\pi}{2}$ is a vertical asymptote of the graph.

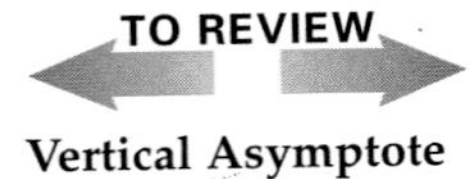

Vertical Asymptote
See page 38.

Because the tangent function is an odd function, its graph is symmetric with respect to the origin. We have used this property to produce the graph of $y = \tan x$ for $-\frac{\pi}{2} < x < \frac{\pi}{2}$ shown in **Figure 2.74.**

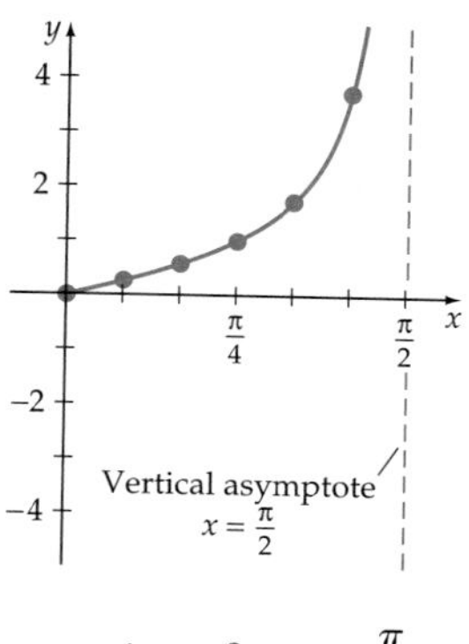

$y = \tan x, 0 \le x \le \frac{\pi}{2}$

Figure 2.73

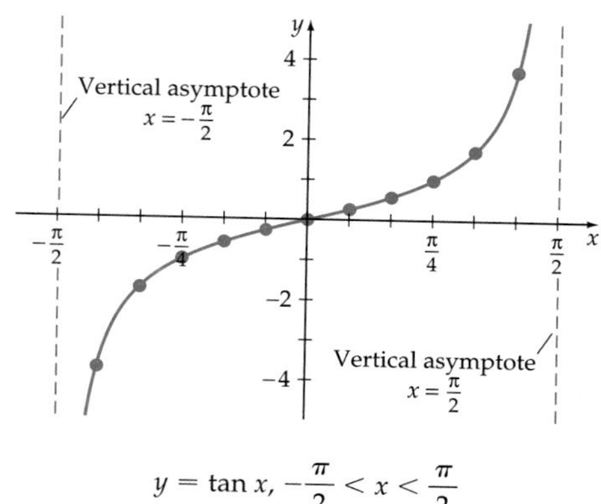

$y = \tan x, -\frac{\pi}{2} < x < \frac{\pi}{2}$

Figure 2.74

Recall from Section 2.4 that the period of $y = \tan x$ is π. Thus a complete graph of $y = \tan x$ can be produced by replicating the graph in **Figure 2.74** to the right and left, as shown in **Figure 2.75**.

The Graph of $y = \tan x$

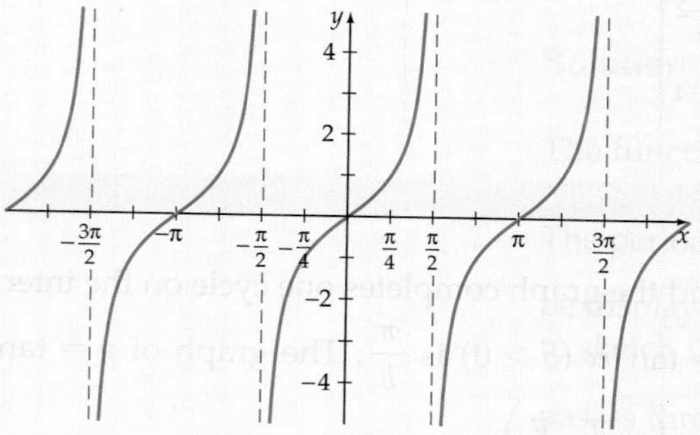

Figure 2.75

Basic Properties

- **Domain:** All real numbers except odd multiples of $\frac{\pi}{2}$
- **Range:** All real numbers
- **Period:** π
- **Symmetry:** With respect to the origin
- **x-intercepts:** At multiples of π
- **Vertical asymptotes:** At odd multiples of $\frac{\pi}{2}$
- **Key points:** $\left(\frac{\pi}{4} + k\pi, 1\right)$ and $\left(-\frac{\pi}{4} + k\pi, -1\right)$ when k is an integer.

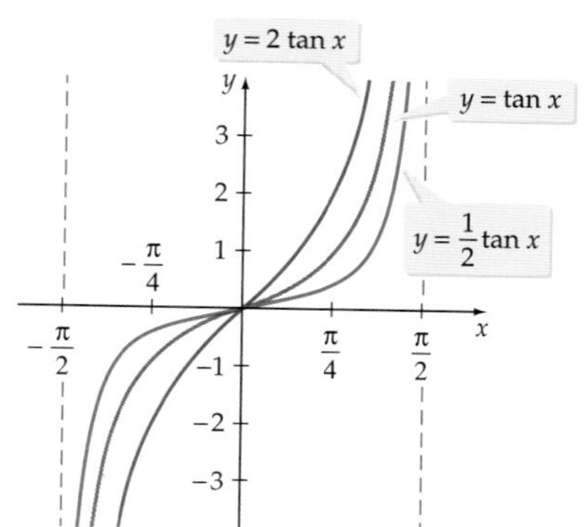

Figure 2.76

Because the tangent function is unbounded, it has no amplitude. The graph of $y = a \tan x$ can be drawn by stretching ($|a| > 1$) or shrinking ($|a| < 1$) the graph of $y = \tan x$. **Figure 2.76** shows the graph of three tangent functions. Because the point $\left(\frac{\pi}{4}, 1\right)$ is on the graph of $y = \tan x$, we can see that the point $\left(\frac{\pi}{4}, a\right)$ must be on the graph of $y = a \tan x$.

The Graph of the Cotangent Function

A graph of the cotangent function $y = \cot x$ is shown in **Figure 2.78**. Notice that its graph is similar to the graph of $y = \tan x$ in that it has a period of π and is symmetric with respect to the origin.

The Graph of $y = \cot x$

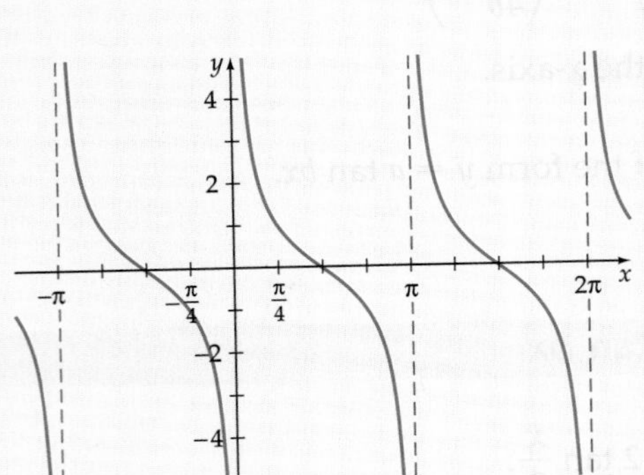

Figure 2.78

Basic Properties

- **Domain:** All real numbers except multiples of π
- **Range:** All real numbers
- **Period:** π
- **Symmetry:** With respect to the origin
- ***x*-intercepts:** At odd multiples of $\frac{\pi}{2}$
- **Vertical asymptotes:** At multiples of π
- **Key points:** $\left(\frac{\pi}{4} + k\pi, 1\right)$ and $\left(-\frac{\pi}{4} + k\pi, -1\right)$ where k is an integer

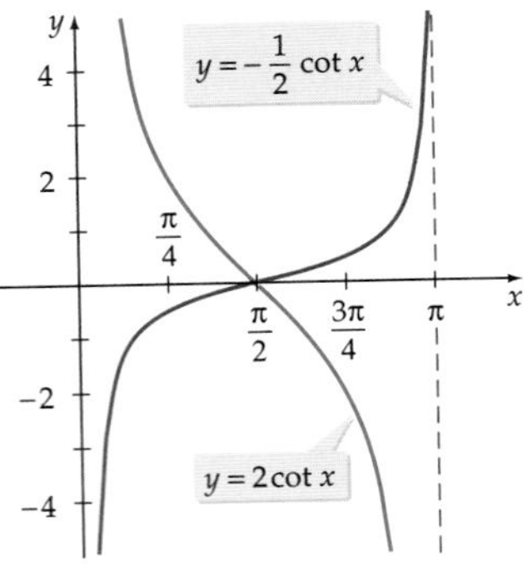

The graph of $y = a \cot x$ is drawn by stretching ($|a| > 1$) or shrinking ($|a| < 1$) the graph of $y = \cot x$. The graph is reflected across the x-axis when $a < 0$. **Figure 2.79** shows the graphs of two cotangent functions.

Figure 2.79

The period of $y = \cot x$ is π, and the period of $y = \cot bx$ is $\frac{\pi}{|b|}$. One cycle of the graph of $y = \cot bx$ is completed on the interval $\left(0, \frac{\pi}{b}\right)$.

Period of $y = \cot bx$

The period of $y = \cot bx$ is $\frac{\pi}{|b|}$.

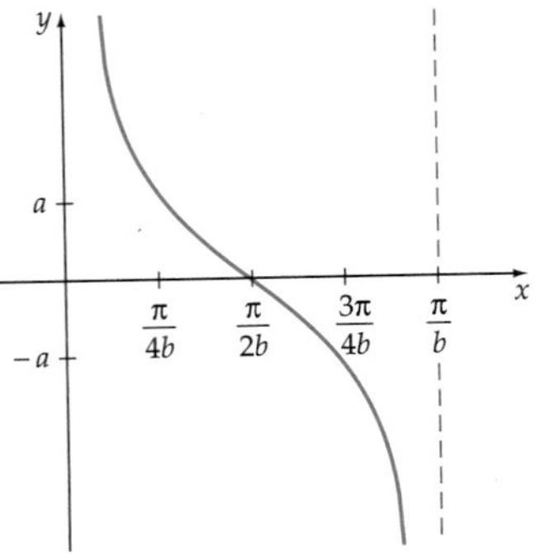

Figure 2.80 shows one cycle of the graph of $y = a \cot bx$ for both a and b positive. Note from the graph the following properties of the function $y = a \cot bx$.

- The period is $\frac{\pi}{b}$.
- $x = \frac{\pi}{2b}$ is a zero.
- The graph passes through $\left(\frac{\pi}{4b}, a\right)$ and $\left(\frac{3\pi}{4b}, -a\right)$.
- If $a < 0$, the graph is reflected across the x-axis.

$y = a \cot bx, 0 < x < \frac{\pi}{b}$

Figure 2.80

In Example 3 we graph a function of the form $y = a \cot bx$.

EXAMPLE 3 ❯❯ Graph $y = a \cot bx$

Graph one period of the function $y = 2 \cot \frac{x}{3}$.

Solution

The function $y = 2 \cot \frac{x}{3}$ is of the form $y = a \cot bx$, with $a = 2$ and $b = \frac{1}{3}$. The period is given by $\frac{\pi}{|b|} = \frac{\pi}{1/3} = 3\pi$. Thus one period of the graph will be displayed on any interval of length 3π. In the graph below, we have chosen to sketch the function over the interval $0 < x < 3\pi$. The graph passes through $\left(\frac{\pi}{4b}, a\right) = \left(\frac{\pi}{4(1/3)}, 2\right) = \left(\frac{3\pi}{4}, 2\right)$ and $\left(\frac{3\pi}{4b}, -a\right) = \left(\frac{3\pi}{4(1/3)}, -2\right) = \left(\frac{9\pi}{4}, -2\right)$. The graph has an x-intercept at $x = \frac{\pi}{2b} = \frac{\pi}{2(1/3)} = \frac{3}{2}\pi$.

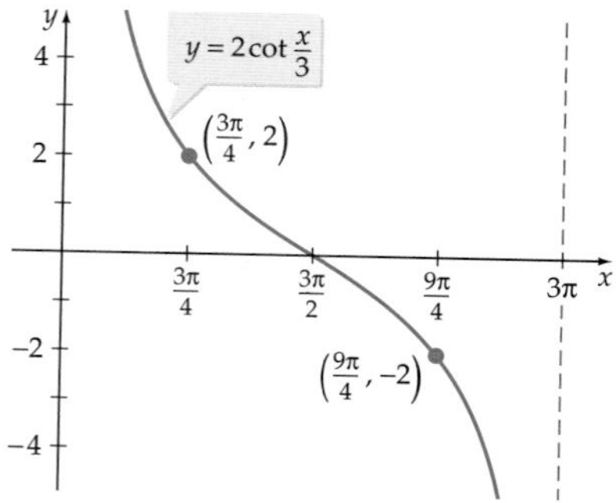

❯❯ *Try Exercise 34, page 190*

The Graph of the Cosecant Function

Because $\csc x = \frac{1}{\sin x}$, the value of $\csc x$ is the reciprocal of the value of $\sin x$. Therefore, $\csc x$ is undefined when $\sin x = 0$ or when $x = k\pi$, where k is an integer. The graph of $y = \csc x$ has vertical asymptotes at $k\pi$. Because $y = \csc x$ has period 2π, the graph will be repeated along the x-axis every 2π units. A graph of $y = \csc x$ is shown in **Figure 2.81** on page 186.

The graph of $y = \sin x$ is also shown in **Figure 2.81.** Note the relationships among the x-intercepts of $y = \sin x$ and the asymptotes of $y = \csc x$.

The Graph of $y = \csc x$

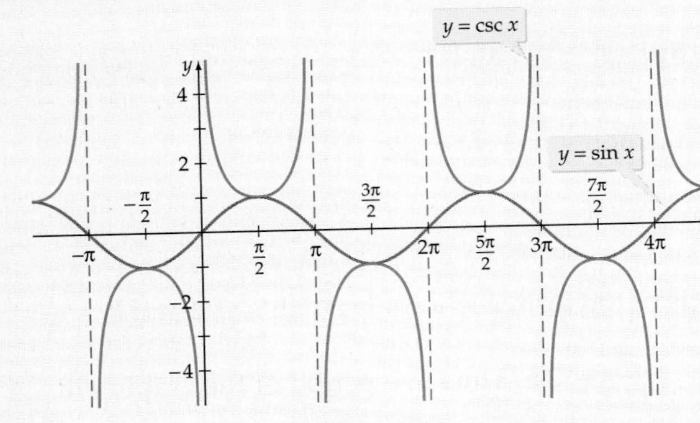

Figure 2.81

Basic Properties

- **Domain:** All real numbers except multiples of π
- **Range:** $\{y \mid y \geq 1, y \leq -1\}$
- **Period:** 2π
- **Symmetry:** With respect to the origin
- **x-intercepts:** None
- **Vertical asymptotes:** At multiples of π
- **Reciprocal relationship:** If (x, y), $y \neq 0$, is a point on the graph of $y = \sin x$, then $(x, 1/y)$ is a corresponding point on the graph of $y = \csc x$.

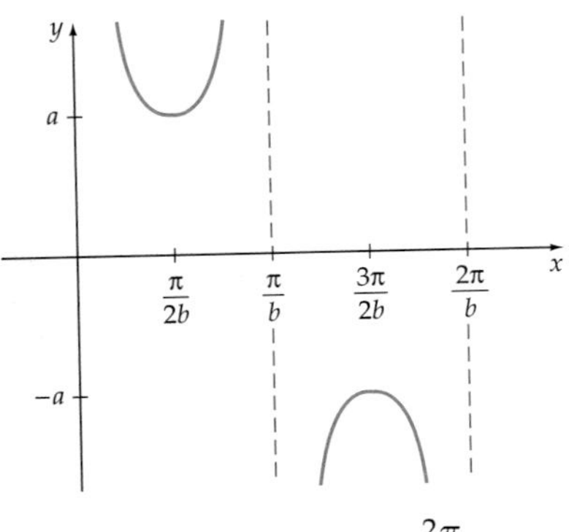

$y = a \csc bx, 0 < x < \dfrac{2\pi}{b}$

Figure 2.82

Figure 2.82 shows one cycle of the graph of $y = a \csc bx$ for both a and b positive. Note from the graph the following properties of the function $y = a \csc bx$.

- The period is $\dfrac{2\pi}{b}$.
- The vertical asymptotes of $y = a \csc bx$ are located at the zeros of $y = a \sin bx$.
- The graph passes through $\left(\dfrac{\pi}{2b}, a\right)$ and $\left(\dfrac{3\pi}{2b}, -a\right)$.
- If $a < 0$, then the graph is reflected across the x-axis.

One procedure for graphing $y = a \csc bx$ is to begin by graphing $y = a \sin bx$. For instance, in **Figure 2.83,** we have used the graph of $y = 2 \sin 4x$ to produce the graph of $y = 2 \csc 4x$. Observe that

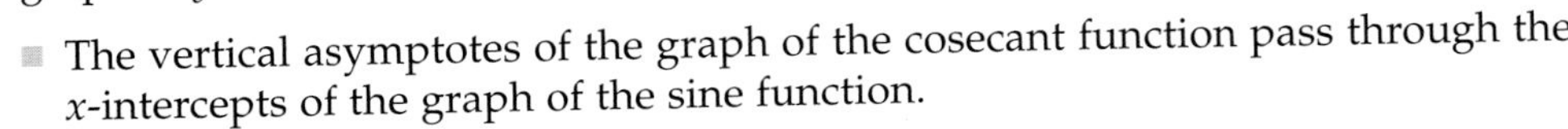

- The vertical asymptotes of the graph of the cosecant function pass through the x-intercepts of the graph of the sine function.

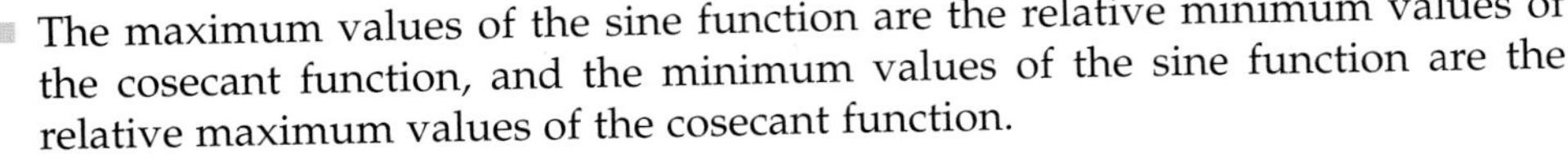

- The maximum values of the sine function are the relative minimum values of the cosecant function, and the minimum values of the sine function are the relative maximum values of the cosecant function.

> **take note**
>
> The **maximum value of a function** is its largest range value. The **minimum value of a function** is its smallest range value.
>
> The value $f(a)$ is called a **relative maximum of a function** f, provided $f(a)$ is greater than or equal to the values of f near $x = a$. The value of $f(b)$ is called a **relative minimum of a function** f, provided $f(b)$ is less than or equal to the values of f near $x = b$.

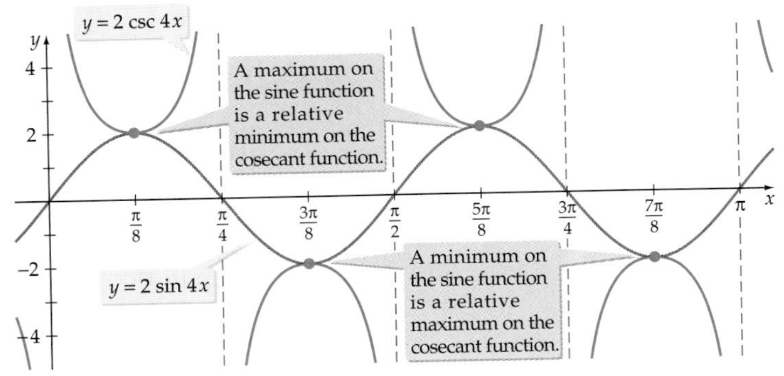

Figure 2.83

EXAMPLE 4 Graph $y = a \csc bx$

Graph one complete period of $y = 2 \csc \frac{\pi x}{2}$.

Solution

Graph one period of $y = 2 \sin \frac{\pi x}{2}$ and draw vertical asymptotes through the x-intercepts. Use the asymptotes as guides to draw the cosecant function. The maximum values of $y = 2 \sin \frac{\pi x}{2}$ are the relative minimum values of $y = 2 \csc \frac{\pi x}{2}$, and the minimum values of $y = 2 \sin \frac{\pi x}{2}$ are the relative maximum values of $y = 2 \csc \frac{\pi x}{2}$.

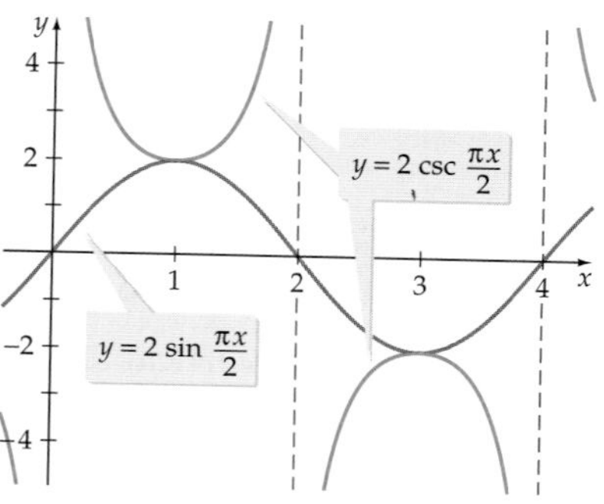

Try Exercise 40, page 190

The Graph of the Secant Function

Because $\sec x = \frac{1}{\cos x}$, the value of $\sec x$ is the reciprocal of the value of $\cos x$. Therefore, $\sec x$ is undefined when $\cos x = 0$ or when $x = \frac{\pi}{2} + k\pi$, k an integer. The graph of $y = \sec x$ has vertical asymptotes at $\frac{\pi}{2} + k\pi$. Because $y = \sec x$ has period 2π, the graph will be replicated along the x-axis every 2π units. A graph of $y = \sec x$ is shown in **Figure 2.84** on page 188.

The graph of $y = \cos x$ is also shown in **Figure 2.84**. Note the relationships among the x-intercepts of $y = \cos x$ and the asymptotes of $y = \sec x$.

The Graph of $y = \sec x$

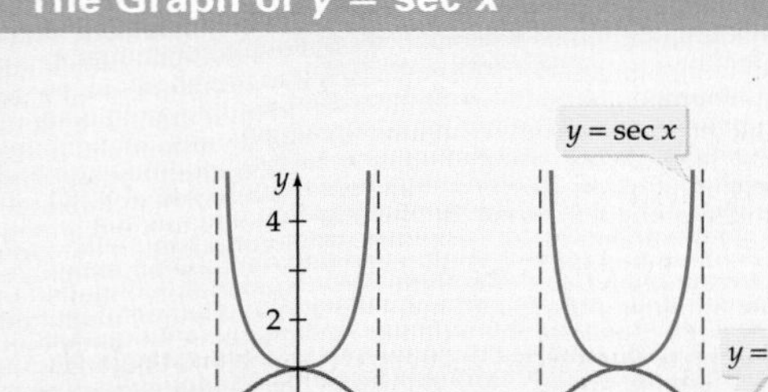

Figure 2.84

Basic Properties

- **Domain:** All real numbers except odd multiples of $\frac{\pi}{2}$
- **Range:** $\{y \mid y \geq 1, y \leq -1\}$
- **Period:** 2π
- **Symmetry:** With respect to the y-axis
- **x-intercepts:** None
- **Vertical asymptotes:** At odd multiples of $\frac{\pi}{2}$
- **Reciprocal relationship:** If (x, y), $y \neq 0$, is a point on the graph of $y = \cos x$, then $(x, 1/y)$ is a corresponding point on the graph of $y = \sec x$.

The procedure for graphing $y = a \sec bx$ is analogous to the procedure used to graph cosecant functions. First graph $y = a \cos bx$ to determine its x-intercepts and its maximum and minimum points.

- The vertical asymptotes of the graph of the secant function pass through the x-intercepts of the graph of the cosine function.
- The maximum values of the cosine function are the relative minimum values of the secant function, and the minimum values of the cosine function are the relative maximum values of the secant function.

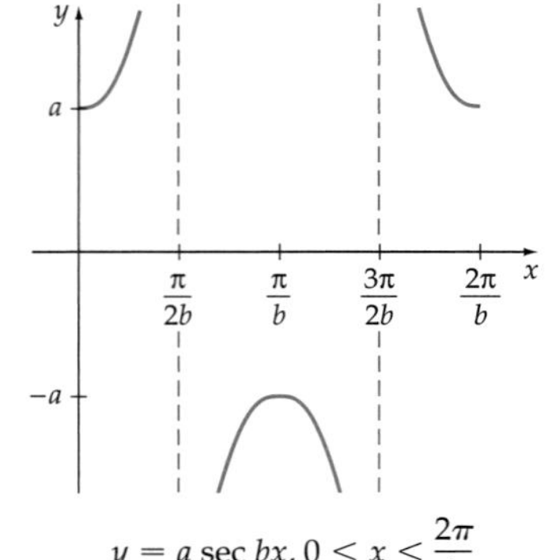

$y = a \sec bx, 0 < x < \frac{2\pi}{b}$

Figure 2.85

Figure 2.85 shows one cycle of the graph of $y = a \sec bx$ for both a and b positive. Note from the graph the following properties of the function $y = a \sec bx$.

- The period is $\frac{2\pi}{b}$.
- The vertical asymptotes of $y = a \sec bx$ are located at the x-intercepts of $y = a \cos bx$.
- The graph passes through $(0, a)$, $\left(\frac{\pi}{b}, -a\right)$, and $\left(\frac{2\pi}{b}, a\right)$.
- If $a < 0$, then the graph is reflected across the x-axis.

EXAMPLE 5 ⟫ Graph $y = a \sec bx$

Graph: $y = -3 \sec \frac{x}{2}$

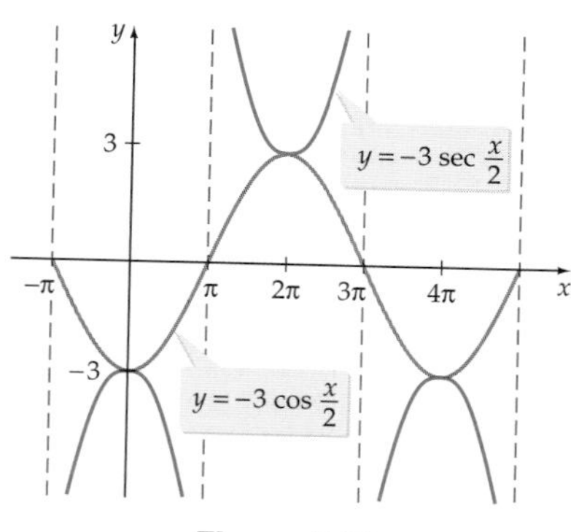

Figure 2.86

Solution

First sketch the graph of $y = -3 \cos \frac{x}{2}$ and draw vertical asymptotes through the x-intercepts. Now sketch the graph of $y = -3 \sec \frac{x}{2}$, using the asymptotes as guides for the graph, as shown in **Figure 2.86.**

Try Exercise 44, page 190

Topics for Discussion

1. Explain how you could use the graph of $y = \sin x$ to produce the graph of $y = \csc x$.
2. Explain how to determine the period of $y = \tan bx$.
3. What is the amplitude of the function $y = \cot \frac{x}{2}$? Explain.
4. Explain how to determine the location of the vertical asymptotes for the graph of $y = 2 \tan bx$.

Exercise Set 2.6

1. For what values of x is $y = \tan x$ undefined?

2. For what values of x is $y = \cot x$ undefined?

3. For what values of x is $y = \sec x$ undefined?

4. For what values of x is $y = \csc x$ undefined?

In Exercises 5 to 22, state the period of each function.

5. $y = \sec x$ **6.** $y = \cot x$ **7.** $y = \tan x$

8. $y = \csc x$ **9.** $y = 2 \tan \frac{x}{2}$ **10.** $y = \frac{1}{2} \cot 2x$

11. $y = \csc 3x$ **12.** $y = \csc \frac{x}{2}$

13. $y = -\tan 3x$ **14.** $y = -3 \cot \frac{2x}{3}$

15. $y = -3 \sec \frac{x}{4}$ **16.** $y = -\frac{1}{2} \csc 2x$

17. $y = \cot \pi x$ **18.** $y = \cot \frac{\pi x}{3}$

19. $y = 0.5 \tan \left(\frac{\pi}{5} t\right)$ **20.** $y = 1.6 \cot \left(\frac{\pi}{2} t\right)$

21. $y = 2.4 \csc \left(\frac{\pi}{4.25} t\right)$ **22.** $y = 4.5 \sec \left(\frac{\pi}{2.5} t\right)$

In Exercises 23 to 42, sketch one full period of the graph of each function.

23. $y = 3 \tan x$ **24.** $y = \frac{1}{3} \tan x$

25. $y = \frac{3}{2} \cot x$ **26.** $y = 4 \cot x$

27. $y = 2 \sec x$ **28.** $y = \frac{3}{4} \sec x$

29. $y = \frac{1}{2} \csc x$ **30.** $y = 2 \csc x$

31. $y = 2 \tan \dfrac{x}{2}$

32. $y = -3 \tan 3x$

33. $y = -3 \cot \dfrac{x}{2}$

34. $y = \dfrac{1}{2} \cot 2x$

35. $y = -2 \csc \dfrac{x}{3}$

36. $y = \dfrac{3}{2} \csc 3x$

37. $y = \dfrac{1}{2} \sec 2x$

38. $y = -3 \sec \dfrac{2x}{3}$

39. $y = -2 \sec \pi x$

40. $y = 3 \csc \dfrac{\pi x}{2}$

41. $y = 3 \tan 2\pi x$

42. $y = -\dfrac{1}{2} \cot \dfrac{\pi x}{2}$

43. Graph $y = 2 \csc 3x$ from -2π to 2π.

44. Graph $y = \sec \dfrac{x}{2}$ from -4π to 4π.

45. Graph $y = 3 \sec \pi x$ from -2 to 4.

46. Graph $y = \csc \dfrac{\pi x}{2}$ from -4 to 4.

47. Graph $y = 2 \cot 2x$ from $-\pi$ to π.

48. Graph $y = \dfrac{1}{2} \tan \dfrac{x}{2}$ from -4π to 4π.

49. Graph $y = 3 \tan \pi x$ from -2 to 2.

50. Graph $y = \cot \dfrac{\pi x}{2}$ from -4 to 4.

In Exercises 51 to 56, each blue graph displays one cycle of the graph of a trigonometric function. Find an equation of each blue graph.

51.

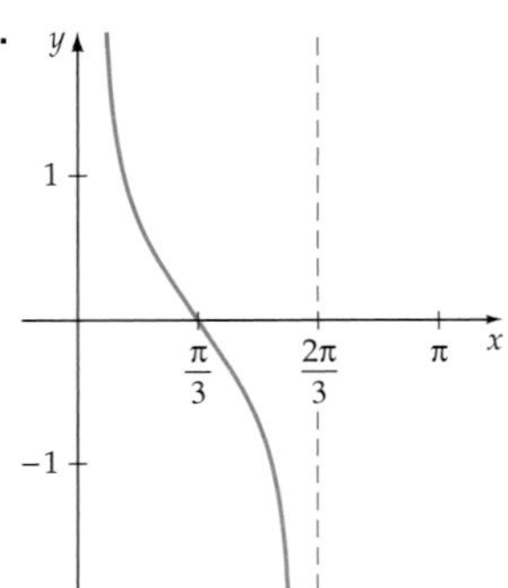

52.

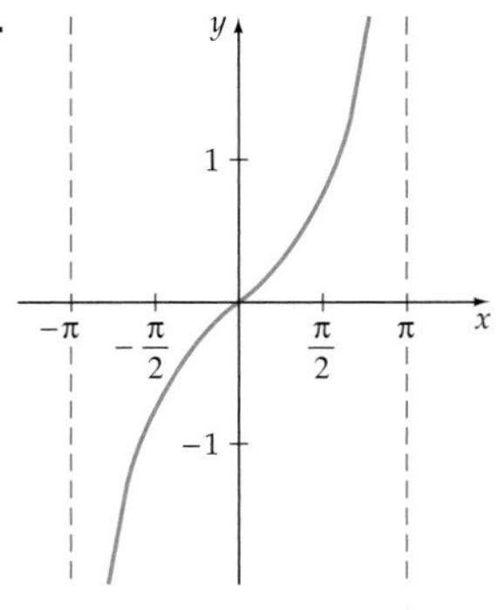

53.

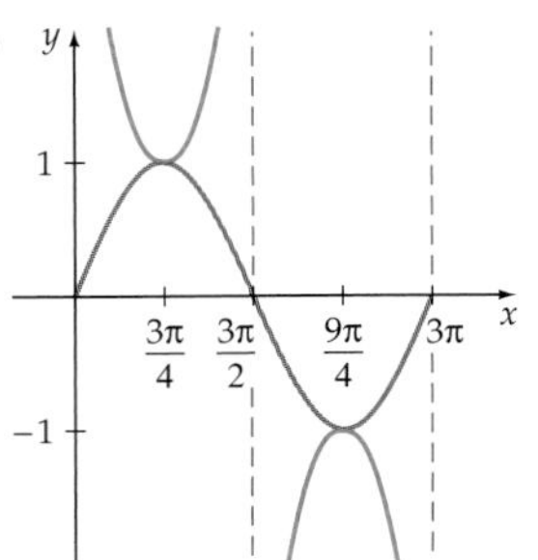

54.

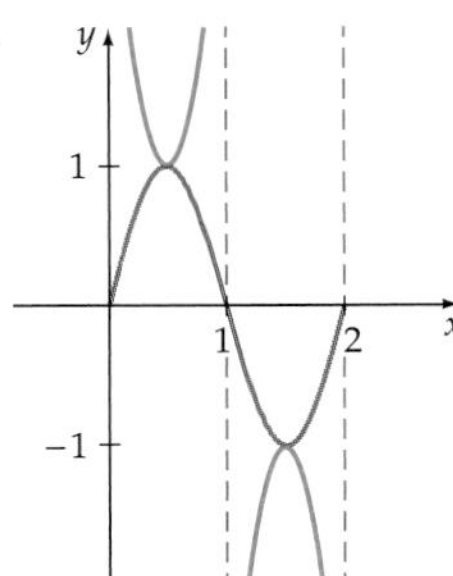

55.

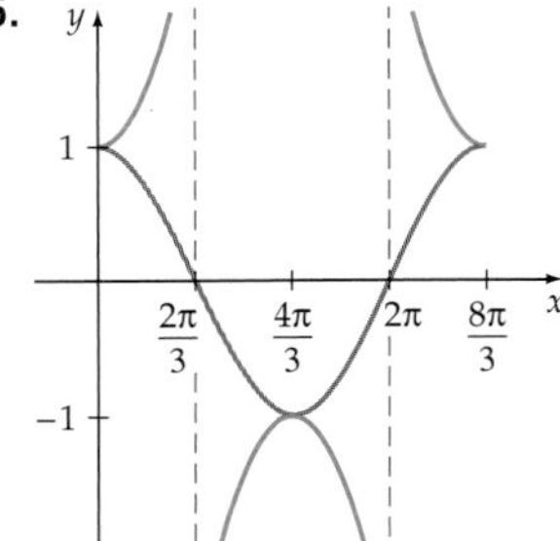

56.

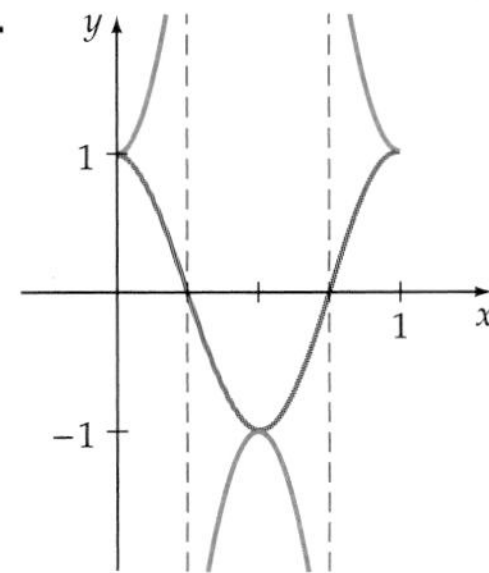

In Exercises 57 to 60, graph each equation.

57. $y = \tan |x|$

58. $y = \sec |x|$

59. $y = |\csc x|$

60. $y = |\cot x|$

61. ROCKET LAUNCH An observer is 1.4 miles from the launch pad of a rocket. The rocket is launched upward as shown in the accompanying figure.

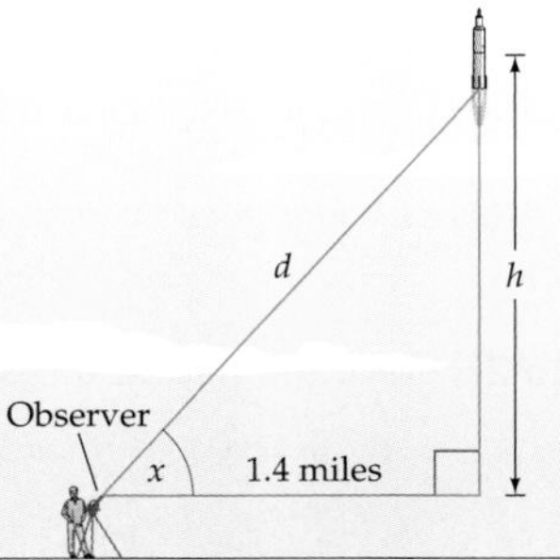

a. Write the height h, in miles, of the rocket as a function of the angle of elevation x.

b. Write the distance d, in miles, from the observer to the rocket as a function of angle x.

c. Graph both functions in the same viewing window with Xmin = 0, Xmax = $\pi/2$, Ymin = 0, and Ymax = 20.

d. Describe the relationship between the graphs in **c**.

62. AVIATION A helicopter maintains an elevation of 3.5 miles and is flying toward a control tower. The angle of elevation from the tower to the helicopter is x radians.

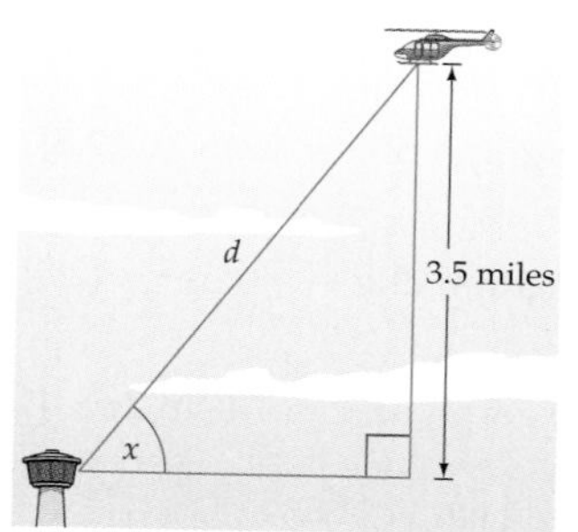

a. Write the distance d, in miles, between the tower and the helicopter as a function of x.

b. Find d when $x = 1$ and $x = 1.2$. Round to the nearest hundredth of a mile.

Connecting Concepts

In Exercises 63 to 70, write an equation of the form $y = \tan bx$, $y = \cot bx$, $y = \sec bx$, or $y = \csc bx$ that satisfies the given conditions.

63. Tangent, period $\frac{\pi}{3}$

64. Cotangent, period $\frac{\pi}{2}$

65. Secant, period $\frac{3\pi}{4}$

66. Cosecant, period $\frac{5\pi}{2}$

67. Cotangent, period 2

68. Tangent, period 0.5

69. Cosecant, period 1.5

70. Secant, period 3

Projects

1. A TECHNOLOGY QUESTION A student's calculator shows the display at the right. Note that the domain values 4.7123 and 4.7124 are close together, but the range values are over 100,000 units apart. Explain how this is possible.

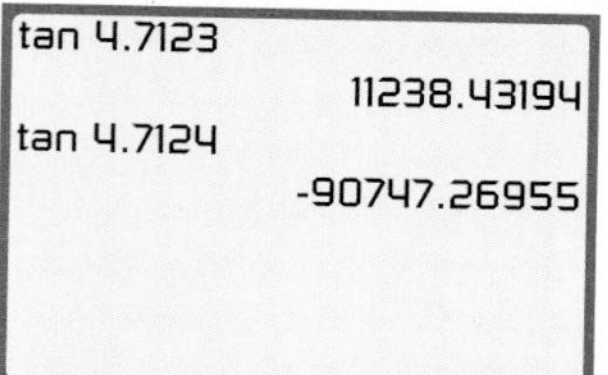

2. SOLUTIONS OF A TRIGONOMETRIC EQUATION How many solutions does $\tan \frac{1}{x} = 0$ have on the interval $-1 \leq x \leq 1$? Explain.

Section 2.7

- Translations of Trigonometric Functions
- Addition of Ordinates
- Damping Factor

Graphing Techniques

PREPARE FOR THIS SECTION

Prepare for this section by completing the following exercises. The answers can be found on page A12.

PS1. Find the amplitude and period of the graph of $y = 2 \sin 2x$. [2.5]

PS2. Find the amplitude and period of the graph of $y = \frac{2}{3} \cos \frac{x}{3}$. [2.5]

PS3. Find the amplitude and period of the graph of $y = -4 \sin 2\pi x$. [2.5]

PS4. What is the maximum value of $f(x) = 2 \sin x$? [2.5]

PS5. What is the minimum value of $f(x) = 3 \cos 2x$? [2.5]

PS6. Is the graph of $f(x) = \cos x$ symmetric with respect to the origin or with respect to the y-axis? [2.5]

Translations of Trigonometric Functions

TO REVIEW

Translations of Graphs
See page 58.

Recall that the graph of $y = f(x) \pm c$ is a *vertical translation* of the graph of $y = f(x)$. For $c > 0$, the graph of $y = f(x) - c$ is the graph of $y = f(x)$ shifted c units down; the graph of $y = f(x) + c$ is the graph of $y = f(x)$ shifted c units up. The graph in **Figure 2.87** is a graph of the equation $y = 2 \sin \pi x - 3$, which is a vertical translation of $y = 2 \sin \pi x$ down 3 units. Note that subtracting 3 from $y = 2 \sin \pi x$ changes neither its amplitude nor its period.

Also, the graph of $y = f(x \pm c)$ is a *horizontal translation* of the graph of $y = f(x)$. For $c > 0$, the graph of $y = f(x - c)$ is the graph of $y = f(x)$ shifted c units to the right; the graph of $y = f(x + c)$ is the graph of $y = f(x)$ shifted c units to the left. The graph in **Figure 2.88** is a graph of the equation $y = 2 \sin\left(x - \frac{\pi}{4}\right)$, which is the graph of $y = 2 \sin x$ translated $\frac{\pi}{4}$ units to the right. Note that neither the period nor the amplitude is affected. The horizontal shift of the graph of a trigonometric function is called its **phase shift**.

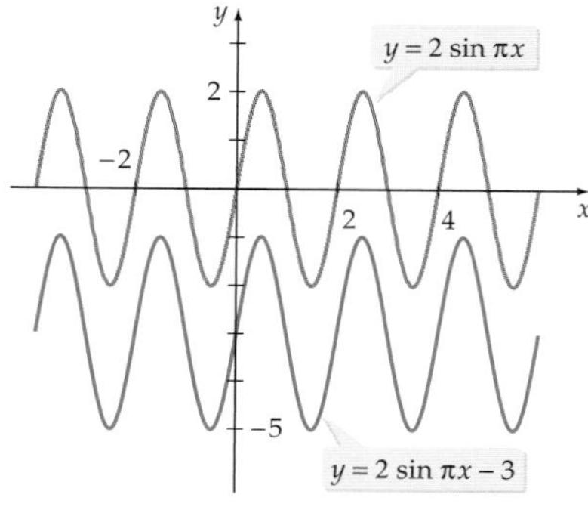

Figure 2.87

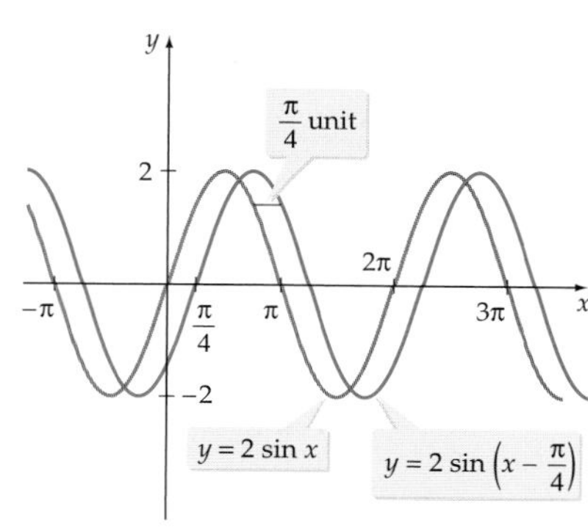

Figure 2.88

Because one cycle of $y = a \sin x$ is completed for $0 \le x \le 2\pi$, one cycle of the graph of $y = a \sin(bx + c)$, where $b > 0$, is completed for $0 \le bx + c \le 2\pi$. Solving this inequality for x, we have

$$0 \le bx + c \le 2\pi$$
$$-c \le bx \le -c + 2\pi$$
$$-\frac{c}{b} \le x \le -\frac{c}{b} + \frac{2\pi}{b}$$

The number $-\frac{c}{b}$ is the phase shift for $y = a \sin(bx + c)$. The graph of the equation $y = a \sin(bx + c)$ is the graph of $y = a \sin bx$ shifted $-\frac{c}{b}$ units horizontally. Similar arguments apply to the remaining trigonometric functions.

The Graphs of $y = a \sin(bx + c)$ and $y = a \cos(bx + c)$

The graphs of $y = a \sin(bx + c)$ and $y = a \cos(bx + c)$, have

Amplitude: $|a|$ Period: $\frac{2\pi}{|b|}$ Phase shift: $-\frac{c}{b}$

To graph $y = a \sin(bx + c)$, shift the graph of $y = a \sin bx$ horizontally $-\frac{c}{b}$ units.

To graph $y = a \cos(bx + c)$, shift the graph of $y = a \cos bx$ horizontally $-\frac{c}{b}$ units.

QUESTION What is the phase shift of the graph of $y = 3 \sin\left(\frac{1}{2}x - \frac{\pi}{6}\right)$?

EXAMPLE 1 Graph by Using a Translation

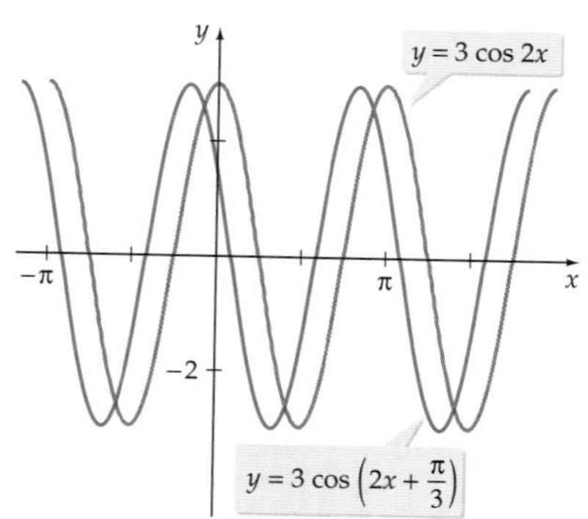

Figure 2.89

Graph: $y = 3 \cos\left(2x + \frac{\pi}{3}\right)$

Solution

The phase shift is $-\frac{c}{b} = -\frac{\pi/3}{2} = -\frac{\pi}{6}$. The graph of the equation $y = 3 \cos\left(2x + \frac{\pi}{3}\right)$ is the graph of $y = 3 \cos 2x$ shifted $\frac{\pi}{6}$ units to the left, as shown in **Figure 2.89**.

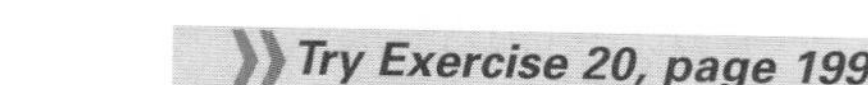

Try Exercise 20, page 199

ANSWER $\frac{\pi}{3}$

The Graphs of $y = a\tan(bx + c)$ and $y = a\cot(bx + c)$

The graphs of $y = a\tan(bx + c)$ and $y = a\cot(bx + c)$ have

$$\text{Period: } \frac{\pi}{|b|} \qquad \text{Phase shift: } -\frac{c}{b}$$

To graph $y = a\tan(bx + c)$, shift the graph of $y = a\tan bx$ horizontally $-\dfrac{c}{b}$ units.

To graph $y = a\cot(bx + c)$, shift the graph of $y = a\cot bx$ horizontally $-\dfrac{c}{b}$ units.

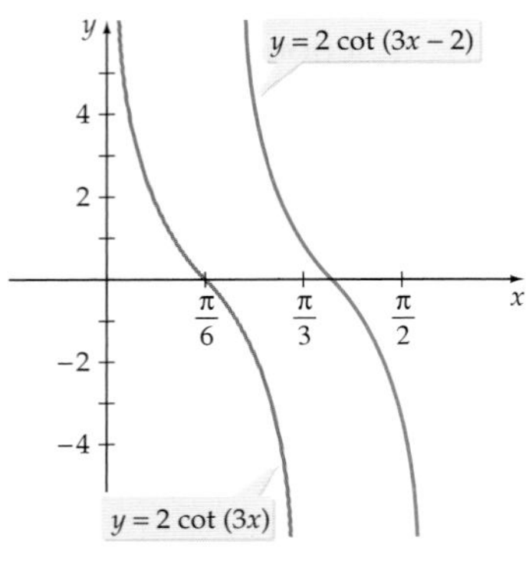

Figure 2.90

EXAMPLE 2 Graph by Using a Translation

Graph one period of the function $y = 2\cot(3x - 2)$.

Solution

The phase shift is

$$-\frac{c}{b} = -\frac{-2}{3} = \frac{2}{3} \qquad \bullet\ 3x - 2 = 3x + (-2)$$

The graph of $y = 2\cot(3x - 2)$ is the graph of $y = 2\cot(3x)$ shifted $\dfrac{2}{3}$ unit to the right, as shown in **Figure 2.90**.

Try Exercise 22, page 199

In Example 3 we use both a horizontal translation and a vertical translation to graph a function of the form $y = a\sin(bx + c) + d$.

EXAMPLE 3 Graph $y = a\sin(bx + c) + d$

Graph: $y = \dfrac{1}{2}\sin\left(x - \dfrac{\pi}{4}\right) - 2$

Solution

The phase shift is $-\dfrac{c}{b} = -\dfrac{-\pi/4}{1} = \dfrac{\pi}{4}$. The vertical shift is 2 units down.

The graph of $y = \dfrac{1}{2}\sin\left(x - \dfrac{\pi}{4}\right) - 2$ is the graph of $y = \dfrac{1}{2}\sin x$ shifted $\dfrac{\pi}{4}$ units to the right and 2 units down, as shown in **Figure 2.91**.

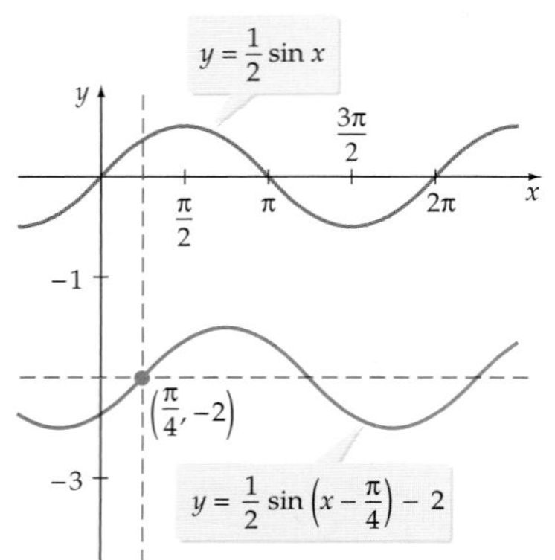

Figure 2.91

Try Exercise 40, page 199

In Example 4 we use both a horizontal translation and a vertical translation to graph a function of the form $y = a\cos(bx + c) + d$.

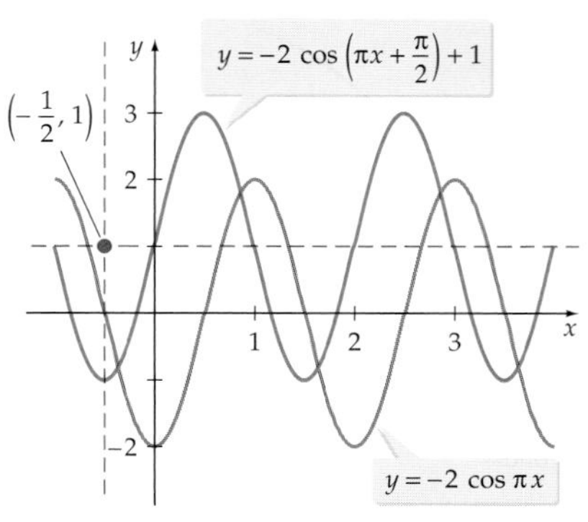

Figure 2.92

EXAMPLE 4 Graph $y = a\cos(bx + c) + d$

Graph: $y = -2\cos\left(\pi x + \dfrac{\pi}{2}\right) + 1$

Solution

The phase shift is $-\dfrac{c}{b} = -\dfrac{\pi/2}{\pi} = -\dfrac{1}{2}$. The vertical shift is 1 unit up. The graph of $y = -2\cos\left(\pi x + \dfrac{\pi}{2}\right) + 1$ is the graph of $y = -2\cos \pi x$ shifted $\dfrac{1}{2}$ unit to the left and 1 unit up, as shown in **Figure 2.92.**

Try Exercise 42, page 199

The following application involves a function of the form $y = \cos(bx + c) + d$.

EXAMPLE 5 A Mathematical Model of a Patient's Blood Pressure

The function $bp(t) = 32\cos\left(\dfrac{10\pi}{3}t - \dfrac{\pi}{3}\right) + 112, 0 \le t \le 20$, gives the blood pressure, in millimeters of mercury (mm Hg), of a patient during a 20-second interval.

a. Find the phase shift and the period of *bp*.

b. Graph one period of *bp*.

c. What are the patient's maximum (*systolic*) and minimum (*diastolic*) blood pressure readings during the given time interval?

d. What is the patient's pulse rate in beats per minute?

Solution

a. Phase shift $= -\dfrac{c}{b} = -\dfrac{\left(-\dfrac{\pi}{3}\right)}{\left(\dfrac{10\pi}{3}\right)} = 0.1$

Period $= \dfrac{2\pi}{|b|} = \dfrac{2\pi}{\left(\dfrac{10\pi}{3}\right)} = 0.6$ second

b. The graph of *bp* is the graph of $y_1 = 32\cos\left(\dfrac{10\pi}{3}t\right)$ shifted one tenth of a unit to the right, shown by $y_2 = 32\cos\left(\dfrac{10\pi}{3}t - \dfrac{\pi}{3}\right)$ in the graph at the top of the next page, and 112 units upward.

Continued ▶

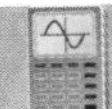

Exploring Concepts with Technology

Sinusoidal Families

Some graphing calculators have a feature that allows you to graph a family of functions easily. For instance, entering Y1={2,4,6}sin(X) in the Y= menu and pressing the GRAPH key on a TI-83/TI-83 Plus/TI-84 Plus calculator produces a graph of the three functions $y = 2 \sin x$, $y = 4 \sin x$, and $y = 6 \sin x$, all displayed in the same window.

1. Use a graphing calculator to graph Y1={2,4,6}sin(X). Write a sentence that indicates the similarities and differences among the three graphs.
2. Use a graphing calculator to graph Y1=sin({π,2π,4π}X). Write a sentence that indicates the similarities and differences among the three graphs.
3. Use a graphing calculator to graph Y1=sin(X+{π/4,π/6,π/12}). Write a sentence that indicates the similarities and differences among the three graphs.
4. A student has used a graphing calculator to graph Y1=sin(X+{π,3π,5π}) and expects to see three graphs. However, the student sees only one graph displayed in the graph window. Has the calculator displayed all three graphs? Explain.

Chapter 2 Summary

2.1 Angles and Arcs

- An angle is in standard position when its initial side is along the positive x-axis and its vertex is at the origin of the coordinate axes.
- Angle α is an acute angle when $0° < \alpha < 90°$; it is an obtuse angle when $90° < \alpha < 180°$.
- α and β are complementary angles when $\alpha + \beta = 90°$; they are supplementary angles when $\alpha + \beta = 180°$.
- One radian is the measure of a central angle subtended by an arc of length r on a circle of radius r.
- The length of the arc s that subtends the central angle θ (in radians) on a circle of radius r is given by $s = r\theta$.
- A point moves on a circular path with radius r at a constant rate of θ radians per unit of time t. Its linear speed is $v = \dfrac{s}{t}$, and its angular speed is $\omega = \dfrac{\theta}{t}$.

2.2 Right Triangle Trigonometry

- Let θ be an acute angle of a right triangle. The six trigonometric functions of θ are given by

$$\sin\theta = \frac{\text{opp}}{\text{hyp}} \qquad \cos\theta = \frac{\text{adj}}{\text{hyp}} \qquad \tan\theta = \frac{\text{opp}}{\text{adj}}$$

$$\csc\theta = \frac{\text{hyp}}{\text{opp}} \qquad \sec\theta = \frac{\text{hyp}}{\text{adj}} \qquad \cot\theta = \frac{\text{adj}}{\text{opp}}$$

2.3 Trigonometric Functions of Any Angle

- Let $P(x, y)$ be a point, except the origin, on the terminal side of an angle θ in standard position. The six trigonometric functions of θ are

$$\sin\theta = \frac{y}{r} \qquad \csc\theta = \frac{r}{y}, \quad y \neq 0$$

$$\cos\theta = \frac{x}{r} \qquad \sec\theta = \frac{r}{x}, \quad x \neq 0$$

$$\tan\theta = \frac{y}{x}, \quad x \neq 0 \qquad \cot\theta = \frac{x}{y}, \quad y \neq 0$$

- Given $\angle\theta$ in standard position, its reference angle θ' is the smallest positive angle formed by the terminal side of $\angle\theta$ and the x-axis.

2.4 Trigonometric Functions of Real Numbers

- The wrapping function pairs a real number with a point on the unit circle.

- Let W be the wrapping function, t be a real number, and $W(t) = P(x, y)$. Then the trigonometric functions of the real number t are defined as follows:

$$\sin t = y \qquad \csc t = \frac{1}{y}, \quad y \neq 0$$

$$\cos t = x \qquad \sec t = \frac{1}{x}, \quad x \neq 0$$

$$\tan t = \frac{y}{x}, \quad x \neq 0 \qquad \cot t = \frac{x}{y}, \quad y \neq 0$$

- $\sin t$, $\csc t$, $\tan t$, and $\cot t$ are odd functions.
- $\cos t$ and $\sec t$ are even functions.
- $\sin t$, $\cos t$, $\sec t$, and $\csc t$ have period 2π.
- $\tan t$ and $\cot t$ have period π.
- The domain and range of each trigonometric function is given in Table 2.6 on page 160.
- An identity is an equation that is true for every number in the domain of the equation. The Fundamental Trigonometric Identities are given on page 163.

2.5 Graphs of the Sine and Cosine Functions

- The graphs of $y = a \sin bx$ and $y = a \cos bx$ both have an amplitude of $|a|$ and a period of $\frac{2\pi}{|b|}$. The graph of each function for $a > 0$ and $b > 0$ is shown below.

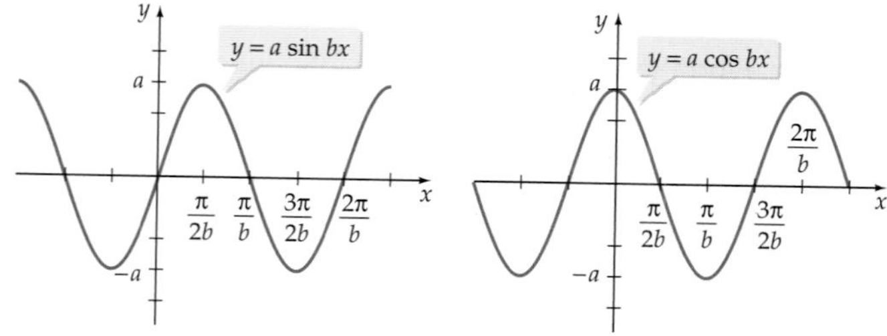

2.6 Graphs of the Other Trigonometric Functions

- The graphs of $y = a \tan bx$ and $y = a \cot bx$ both have a period of $\frac{\pi}{|b|}$. The graph of one period of each function for $a > 0$ and $b > 0$ is shown below.

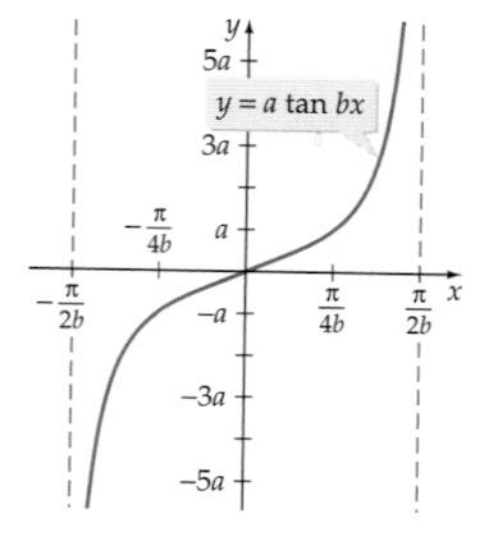

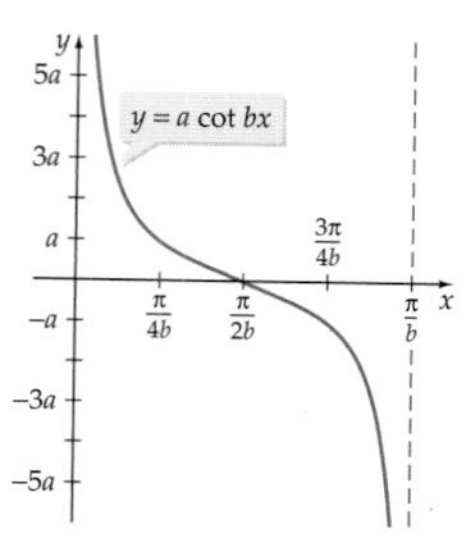

- The graphs of $y = a \csc bx$ and $y = a \sec bx$ both have a period of $\frac{2\pi}{|b|}$. The graphs of one period of each function, with $a > 0$ and $b > 0$, are shown below.

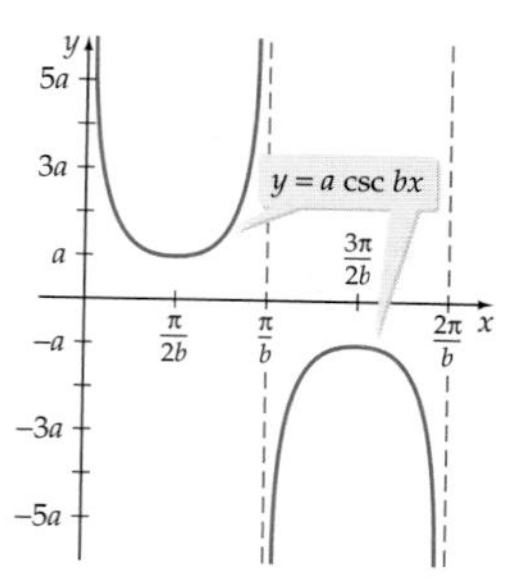

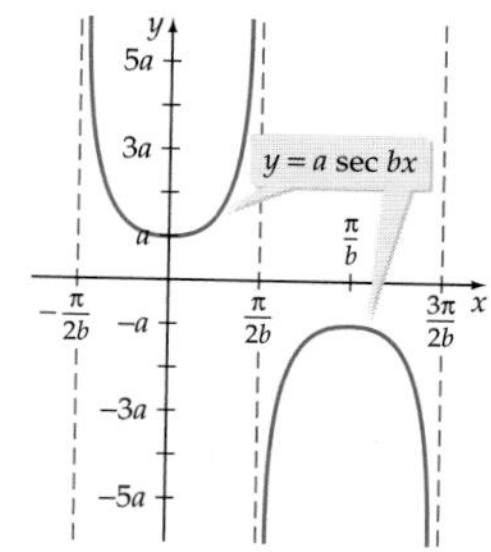

2.7 Graphing Techniques

- If $y = f(bx + c)$, where f is a trigonometric function, then the graph of $y = f(bx + c)$ can be produced by shifting the graph of $y = f(bx)$ horizontally $-\frac{c}{b}$ units. The value $-\frac{c}{b}$ is called the phase shift of $y = f(bx + c)$.
- Addition of ordinates is a method of graphing the sum of two functions by geometrically adding the values of their y-coordinates.
- The factor $g(x)$ in $f(x) = g(x) \cos x$ is called a damping factor. The graph of f lies on or between the graphs of the equations $y = g(x)$ and $y = -g(x)$.

2.8 Harmonic Motion—An Application of the Sine and Cosine Functions

- The equations of simple harmonic motion are

$$y = a \cos 2\pi ft \qquad \text{and} \qquad y = a \sin 2\pi ft$$

where $|a|$ is the amplitude, f is the frequency, y is the displacement, and t is the time.
- Functions of the form $f(t) = ae^{-kt} \cos \omega t$ are used to model some forms of damped harmonic motion.

Chapter 2 Assessing Concepts

1. True or False: In the formula $s = r\theta$, the measure of angle θ must be in radians.

2. True or False: $\sec^2 \theta + \tan^2 \theta = 1$ is an identity.

3. True or False: The measure of one radian differs depending on the radius of the circle used.

4. True or False: The graph of $y = \sin x$ is symmetric with respect to the origin.

5. What is the measure of the reference angle for the angle $\theta = \dfrac{3\pi}{4}$?

6. What is the point defined by $W\left(\dfrac{\pi}{2}\right)$?

7. What is the period of the graph of $y = \dfrac{1}{2}\cos\dfrac{3\pi}{4}x$?

8. Explain how to use the graph of $y_1 = \sin\dfrac{x}{2}$ to produce the graph of $y_2 = \sin\left(\dfrac{x}{2} + \dfrac{\pi}{4}\right)$.

9. What is the domain of the function $y = \cot x$?

10. Consider the graph of $y = \sec x$ on the interval $[0, 2\pi]$. What are the equations of the vertical asymptotes of the graph?

Chapter 2 Review Exercises

1. Find the complement and supplement of the angle θ whose measure is 65°.

2. Find the measure of the reference angle θ' for the angle θ whose measure is 980°.

3. Convert 2 radians to the nearest hundredth of a degree.

4. Convert 315° to radian measure.

5. Find the length (to the nearest hundredth of a meter) of the arc on a circle of radius 3 meters that subtends an angle of 75°.

6. Find the radian measure of the angle subtended by an arc of length 12 centimeters on a circle whose radius is 40 centimeters.

7. A car with a 16-inch-radius wheel is moving with a speed of 50 mph. Find the angular speed (to the nearest radian per second) of the wheel.

In Exercises 8 to 11, let θ be an acute angle of a right triangle and $\csc \theta = \dfrac{3}{2}$. Evaluate each function.

8. $\cos \theta$ **9.** $\cot \theta$ **10.** $\sin \theta$ **11.** $\sec \theta$

12. Find the values of the six trigonometric functions of an angle in standard position with the point $P(1, -3)$ on the terminal side of the angle.

13. Find the exact value of

a. $\sec 150°$ **b.** $\tan\left(-\dfrac{3\pi}{4}\right)$

c. $\cot(-225°)$ **d.** $\cos\left(\dfrac{2\pi}{3}\right)$

14. Find the value of each of the following to the nearest ten-thousandth.

a. $\cos 123°$ **b.** $\cot 4.22$

c. $\sec 612°$ **d.** $\tan\dfrac{2\pi}{5}$

15. Given $\cos \phi = -\dfrac{\sqrt{3}}{2}$, $180° < \phi < 270°$, find the exact value of

a. $\sin \phi$ **b.** $\tan \phi$

16. Given $\tan \phi = -\dfrac{\sqrt{3}}{3}$, $90° < \phi < 180°$, find the exact value of

a. $\sec \phi$ **b.** $\csc \phi$

17. Given $\sin \phi = -\dfrac{\sqrt{2}}{2}$, $270° < \phi < 360°$, find the exact value of

a. $\cos \phi$ **b.** $\cot \phi$

18. Let W be the wrapping function. Evaluate

a. $W(\pi)$ **b.** $W\left(-\frac{\pi}{3}\right)$ **c.** $W\left(\frac{5\pi}{4}\right)$ **d.** $W(28\pi)$

19. Is the function defined by $f(x) = \sin(x)\tan(x)$ even, odd, or neither?

In Exercises 20 and 21, use the unit circle to show that each equation is an identity.

20. $\cos(\pi + t) = -\cos t$

21. $\tan(-t) = -\tan t$

In Exercises 22 to 27, use trigonometric identities to write each expression in terms of a single trigonometric function or as a constant.

22. $1 + \dfrac{\sin^2 \phi}{\cos^2 \phi}$

23. $\dfrac{\tan \phi + 1}{\cot \phi + 1}$

24. $\dfrac{\cos^2 \phi + \sin^2 \phi}{\csc \phi}$

25. $\sin^2 \phi(\tan^2 \phi + 1)$

26. $1 + \dfrac{1}{\tan^2 \phi}$

27. $\dfrac{\cos^2 \phi}{1 - \sin^2 \phi} - 1$

In Exercises 28 to 33, state the amplitude (if it exists), period, and phase shift of the graph of each function.

28. $y = 3\cos(2x - \pi)$

29. $y = 2\tan 3x$

30. $y = -2\sin\left(3x + \frac{\pi}{3}\right)$

31. $y = \cos\left(2x - \frac{2\pi}{3}\right) + 2$

32. $y = -4\sec\left(4x - \frac{3\pi}{2}\right)$

33. $y = 2\csc\left(x - \frac{\pi}{4}\right) - 3$

In Exercises 34 to 51, graph each function.

34. $y = 2\cos \pi x$

35. $y = -\sin \dfrac{2x}{3}$

36. $y = 2\sin \dfrac{3x}{2}$

37. $y = \cos\left(x - \frac{\pi}{2}\right)$

38. $y = \frac{1}{2}\sin\left(2x + \frac{\pi}{4}\right)$

39. $y = 3\cos 3(x - \pi)$

40. $y = -\tan \dfrac{x}{2}$

41. $y = 2\cot 2x$

42. $y = \tan\left(x - \frac{\pi}{2}\right)$

43. $y = -\cot\left(2x + \frac{\pi}{4}\right)$

44. $y = -2\csc\left(2x - \frac{\pi}{3}\right)$

45. $y = 3\sec\left(x + \frac{\pi}{4}\right)$

46. $y = 3\sin 2x - 3$

47. $y = 2\cos 3x + 3$

48. $y = -\cos\left(3x + \frac{\pi}{2}\right) + 2$

49. $y = 3\sin\left(4x - \frac{2\pi}{3}\right) - 3$

50. $y = 2 - \sin 2x$

51. $y = \sin x - \sqrt{3}\cos x$

52. A car climbs a hill that has a constant angle of 4.5° for a distance of 1.14 miles. What is the car's increase in altitude?

53. Height of a Tree A tree casts a shadow of 8.55 feet when the angle of elevation of the sun is 55.3°. Find the height of the tree.

54. Linear Speeds on a Carousel A carousel has two circular rings of horses. The inner ring has a radius of 14.5 feet and the outer ring has a radius of 21.0 feet. The carousel makes one complete revolution every 24 seconds. How much greater, in feet per second, is the linear speed of a horse in the outer ring than the linear speed of a horse in the inner ring? Round to the nearest tenth of a foot per second.

55. Height of a Building Find the height of a building if the angle of elevation to the top of the building changes from 18° to 37° as an observer moves a distance of 80 feet toward the building.

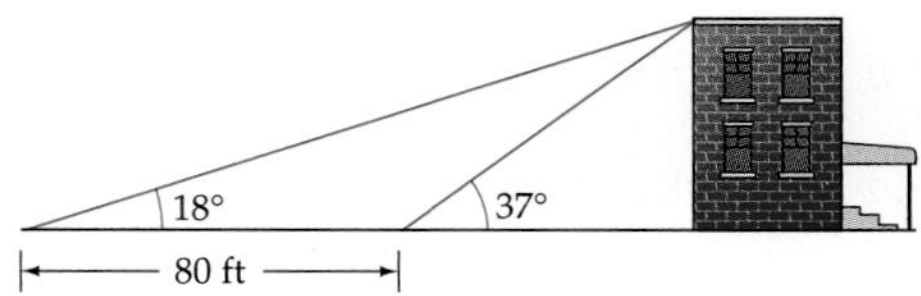

56. Find the amplitude, period, and frequency of the simple harmonic motion given by $y = 2.5\sin 50t$.

57. A mass of 5 kilograms is in equilibrium suspended from a spring. The mass is pulled down 0.5 foot and released. Find the period, frequency, and amplitude of the motion, assuming the mass oscillates in simple harmonic motion. Write an equation of motion. Assume $k = 20$.

58. Graph the damped harmonic motion that is modeled by

$$f(t) = 3e^{-0.75t}\cos \pi t$$

where t is in seconds. Use the graph to determine how long (to the nearest tenth of a second) it will be until the absolute value of the displacement of the mass is always less than 0.01.

Quantitative Reasoning: *Find the Periods of Trigonometric Functions and Combined Musical Sound Tracks*

QR1. Let f be a function with period j and let g be a function with period k. The function $f + g$ is a periodic function provided there exist natural numbers m and n such that $\frac{j}{k} = \frac{m}{n}$. Furthermore, if m and n have no common prime factors, then the period of $f + g$ is $n \cdot j = m \cdot k$. For each of the following, determine the period of $f + g$.

a. $f(x) = \sin x,\ g(x) = \cos \frac{2}{3}x$

b. $f(x) = \cos 3\pi x,\ g(x) = \sin \frac{1}{2}\pi x$

c. $f(x) = \tan 2x,\ g(x) = \sin 3x$

d. $f(x) = \cot \frac{2}{3}x,\ g(x) = \cos \frac{3}{4}x$

e. $f(x) = \sec \frac{4\pi}{5}x,\ g(x) = \cot \frac{2\pi}{3}x$

f. $f(x) = \csc \frac{5}{2}x,\ g(x) = \tan \frac{1}{4}x$

QR2. The following figure shows a guitar sound track with a period of 3 seconds and an electric bass track with a period of 2.5 seconds. Use the procedure in Exercise QR1 to find the period of the music produced when the two sound tracks are played together.

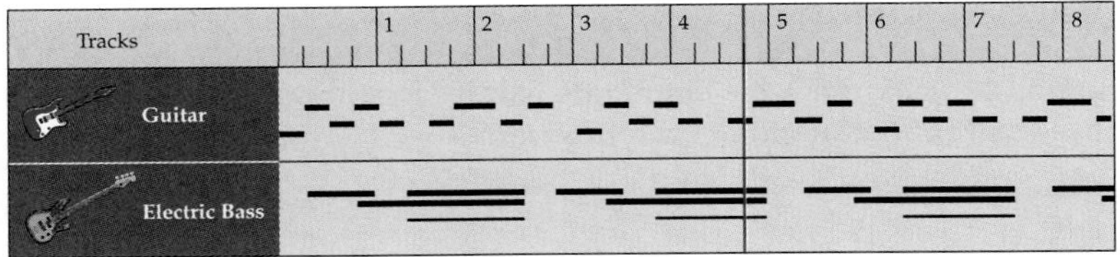

QR3. A tambourine sound track has a period of 1.25 seconds and a drum sound track has a period of 2.25 seconds. Use the procedure in Exercise QR1 to find the period of the music produced when the two sound tracks are played together.

QR4. An alto sax sound track has a period of 6 seconds, a piano sound track has a period of 4.5 seconds, and an electric guitar sound track has a period of 27 seconds. Find the period of the music produced when the three sound tracks are played together.

Chapter 2 Test

1. Convert 150° to exact radian measure.

2. Find the supplement of the angle whose radian measure is $\frac{11}{12}\pi$. Express your answer in terms of π.

3. Find the length (to the nearest tenth of a centimeter) of an arc that subtends a central angle of 75° in a circle of radius 10 centimeters.

4. **Angular Speed** A wheel is rotating at 6 revolutions per second. Find the angular speed in radians per second.

5. **Linear Speed** A wheel with a diameter of 16 centimeters is rotating at 10 radians per second. Find the linear speed (in centimeters per second) of a point on the edge of the wheel.

6. If θ is an acute angle of a right triangle and $\tan \theta = \frac{3}{7}$, find $\sec \theta$.

7. Use a calculator to find the value of csc 67° to the nearest ten-thousandth.

8. Find the exact value of $\tan \frac{\pi}{6} \cos \frac{\pi}{3} - \sin \frac{\pi}{2}$.

9. Find the exact coordinates of $W\left(\frac{11\pi}{6}\right)$.

10. Express $\frac{\sec^2 t - 1}{\sec^2 t}$ in terms of a single trigonometric function.

11. State the period of $y = -4 \tan 3x$.

12. State the amplitude, period, and phase shift for the function $y = -3 \cos\left(2x + \frac{\pi}{2}\right)$.

13. State the period and phase shift for the function $y = 2 \cot\left(\frac{\pi}{3}x + \frac{\pi}{6}\right)$.

14. Graph one full period of $y = 3 \cos \frac{1}{2}x$.

15. Graph one full period of $y = -2 \sec \frac{1}{2}x$.

16. Write a sentence that explains how to obtain the graph of $y = 2 \sin\left(2x - \frac{\pi}{2}\right) - 1$ from the graph of $y = 2 \sin 2x$.

17. Graph one full period of $y = 2 - \sin \frac{x}{2}$.

18. Graph one full period of $y = \sin x - \cos 2x$.

19. **Height of a Tree** The angle of elevation from point A to the top of a tree is 42.2°. From point B, which is 5.24 meters from A and on a line through the base of the tree and A, the angle of elevation to the top of the tree is 37.4°. Find the height of the tree.

20. Write the equation for simple harmonic motion given that the amplitude is 13 feet, the period is 5 seconds, and the displacement is zero when $t = 0$.

Cumulative Review Exercises

1. Find the distance between the points $P(-3, 2)$ and $Q(4, 1)$.

2. The hypotenuse of a right triangle has a length of 1 and one of its legs has a length of $\frac{1}{2}$. Find the length of the other leg.

3. Find the x- and the y-intercept(s) of the graph of $f(x) = x^2 - 9$.

4. Determine whether $f(x) = \frac{x}{x^2 + 1}$ is an even function or an odd function.

5. Find the inverse of $f(x) = \frac{x}{2x - 3}$.

6. Use interval notation to state the domain of $f(x) = \frac{2}{x - 4}$.

7. Solve: $x^2 + x - 6 = 0$

8. Explain how to use the graph of $y = f(x)$ to produce the graph of $y = f(x - 3)$.

9. Explain how to use the graph of $y = f(x)$ to produce the graph of $y = f(-x)$.

10. Convert 300° to radians.

11. Convert $\frac{5\pi}{4}$ to degrees.

12. Evaluate $f(x) = \sin\left(x + \frac{\pi}{6}\right)$ for $x = \frac{\pi}{3}$.

13. Evaluate $f(x) = \sin x + \sin \frac{\pi}{6}$ for $x = \frac{\pi}{3}$.

14. Find the exact value of $\cos^2 45° + \sin^2 60°$.

15. Determine the sign of $\tan \theta$ given that $\frac{\pi}{2} < \theta < \pi$.

16. What is the measure of the reference angle for the angle $\theta = 210°$?

17. What is the measure of the reference angle for the angle $\theta = \frac{2\pi}{3}$?

18. Use interval notation to state the domain of $f(x) = \sin x$, where x is a real number.

19. Use interval notation to state the range of $f(x) = \cos x$, where x is a real number.

20. If θ is an acute angle of a right triangle and $\tan \theta = \frac{3}{4}$, find $\sin \theta$.

3 Trigonometric Identities and Equations

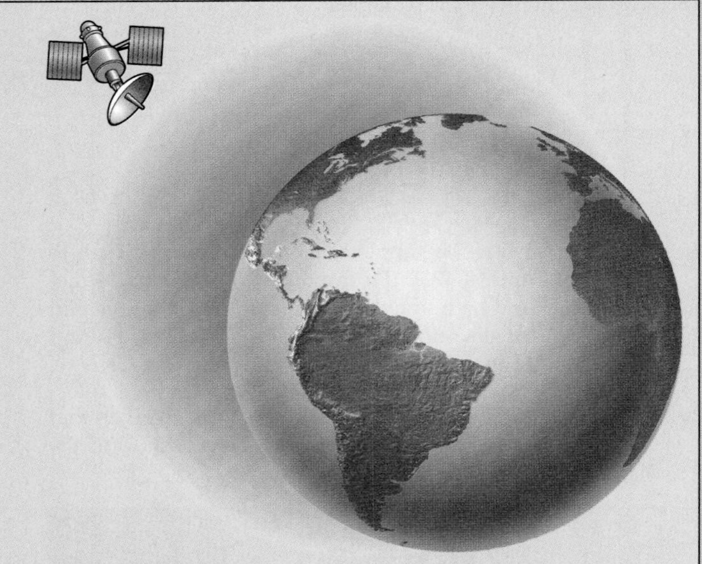

Trigonometric Equations and Applications

This chapter explores trigonometric identities, inverse trigonometric functions, and trigonometric equations. Understanding these concepts will provide you with a foundation for more advanced mathematical studies. You will also gain additional problem-solving skills and learn procedures that can be used to solve a wide variety of real-life applications. For example, in Exercise 83 on page 266, you will determine the minimum altitude a communications satellite must attain to provide service over a desired region. In Quantitative Reasoning Exercise 2a on page 289, you will find the launch angle that minimizes the velocity at which a basketball must be launched to make a shot from a given distance.

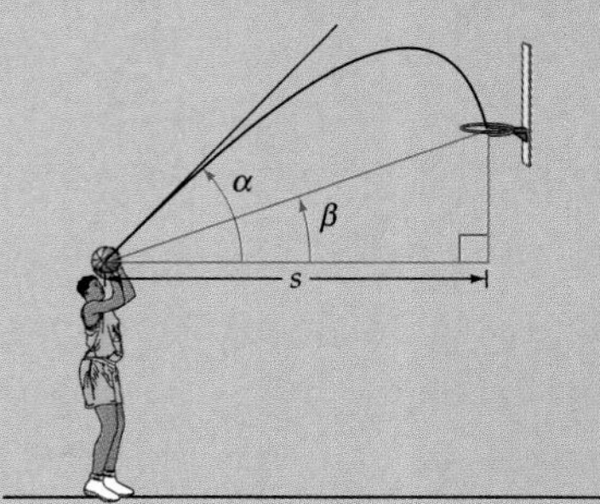

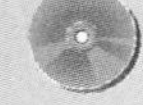

Online Study Center
For online student resources, such as section quizzes, visit this website: college.hmco.com/info/aufmannCAT

Section 3.1

- Fundamental Trigonometric Identities
- Verification of Trigonometric Identities

Verification of Trigonometric Identities

Fundamental Trigonometric Identities

The domain of an equation consists of all values of the variable for which every term is defined. For example, the domain of

$$\frac{\sin x \cos x}{\sin x} = \cos x \tag{1}$$

includes all real numbers x except $x = k\pi$, where k is an integer, because $\sin x = 0$ for $x = k\pi$, and division by 0 is undefined. An **identity** is an equation that is true for all of its domain values. **Table 3.1** lists identities that were introduced earlier.

Table 3.1 Fundamental Trigonometric Identities

Reciprocal identities	$\sin x = \dfrac{1}{\csc x}$	$\cos x = \dfrac{1}{\sec x}$	$\tan x = \dfrac{1}{\cot x}$
Ratio identities	$\tan x = \dfrac{\sin x}{\cos x}$	$\cot x = \dfrac{\cos x}{\sin x}$	
Pythagorean identities	$\sin^2 x + \cos^2 x = 1$	$\tan^2 x + 1 = \sec^2 x$	$1 + \cot^2 x = \csc^2 x$
Odd–even identities	$\sin(-x) = -\sin x$ $\cos(-x) = \cos x$	$\tan(-x) = -\tan x$ $\cot(-x) = -\cot x$	$\sec(-x) = \sec x$ $\csc(-x) = -\csc x$

Verification of Trigonometric Identities

To verify an identity, we show that one side of the identity can be rewritten in an equivalent form that is identical to the other side. There is no one method that can be used to verify every identity; however, the following guidelines should prove useful.

Guidelines for Verifying Trigonometric Identities

- If one side of the identity is more complex than the other, then it is generally best to try first to simplify the more complex side until it becomes identical to the other side.
- Perform indicated operations such as adding fractions or squaring a binomial. Also be aware of any factorization that may help you to achieve your goal of producing the expression on the other side.
- Make use of previously established identities that enable you to rewrite one side of the identity in an equivalent form.
- Rewrite one side of the identity so that it involves only sines and/or cosines.

Continued ▶

- Rewrite one side of the identity in terms of a single trigonometric function.
- Multiplying both the numerator and the denominator of a fraction by the same factor (such as the conjugate of the denominator or the conjugate of the numerator) may get you closer to your goal.
- Keep your goal in mind. Does it involve products, quotients, sums, radicals, or powers? Knowing exactly what your goal is may provide the insight you need to verify the identity.

In Example 1 we verify an identity by rewriting the more complicated side so that it involves only sines and cosines.

EXAMPLE 1 » Change to Sines and Cosines to Verify an Identity

Verify the identity $\sin x \cot x \sec x = 1$.

Solution

The left side of the identity is more complicated than the right side. We will try to verify the identity by rewriting the left side so that it involves only sines and cosines.

$$\sin x \cot x \sec x = \sin x \cdot \frac{\cos x}{\sin x} \cdot \frac{1}{\cos x}$$

- **Apply the fundamental identities $\cot x = \frac{\cos x}{\sin x}$ and $\sec x = \frac{1}{\cos x}$.**

$$= \frac{\cancel{\sin x}\,\cancel{\cos x}}{\cancel{\sin x}\,\cancel{\cos x}}$$

- **Multiply the fractions to produce a single fraction.**

$$= 1$$

- **Simplify by dividing out the common factors.**

We have rewritten the left side of the equation so that it is identical to the right side. Thus we have verified that the equation is an identity.

» *Try Exercise 2, page 222*

QUESTION Is $\cos(-x) = \cos x$ an identity?

EXAMPLE 2 » Use a Pythagorean Identity to Verify an Identity

Verify the identity $1 - 2\sin^2 x = 2\cos^2 x - 1$.

ANSWER Yes, $\cos(-x) = \cos x$ is one of the odd–even identities shown in **Table 3.1**.

take note

Each of the Pythagorean identities can be written in several different forms. For instance,

$$\sin^2 x + \cos^2 x = 1$$

also can be written as

$$\sin^2 x = 1 - \cos^2 x$$

and as

$$\cos^2 x = 1 - \sin^2 x$$

Solution

Rewrite the right side of the equation.

$$\begin{aligned} 2\cos^2 x - 1 &= 2(1 - \sin^2 x) - 1 && \bullet\ \cos^2 x = 1 - \sin^2 x \\ &= 2 - 2\sin^2 x - 1 \\ &= 1 - 2\sin^2 x && \bullet\ \text{Simplify.} \end{aligned}$$

Try Exercise 12, page 223

Figure 3.1 shows the graph of $f(x) = 1 - 2\sin^2 x$ and the graph of $g(x) = 2\cos^2 x - 1$ on the same coordinate axes. The fact that the graphs appear to be identical on the interval $[-2\pi, 2\pi]$ supports the verification in Example 2.

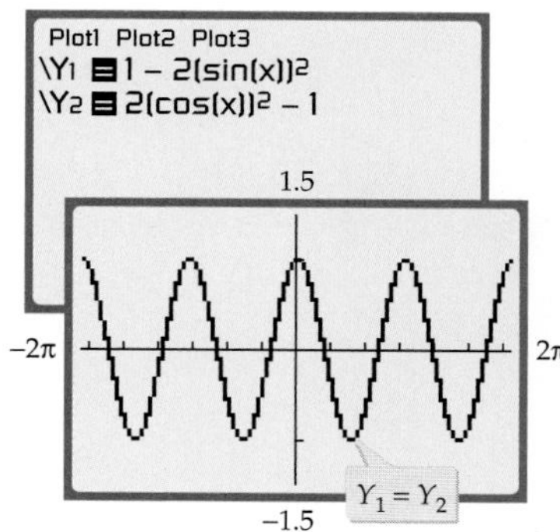

Figure 3.1

EXAMPLE 3 Factor to Verify an Identity

Verify the identity $\csc^2 x - \cos^2 x \csc^2 x = 1$.

Solution

Simplify the left side of the equation.

$$\begin{aligned} \csc^2 x - \cos^2 x \csc^2 x &= \csc^2 x(1 - \cos^2 x) && \bullet\ \text{Factor out } \csc^2 x. \\ &= \csc^2 x \sin^2 x && \bullet\ 1 - \cos^2 x = \sin^2 x \\ &= \frac{1}{\sin^2 x} \cdot \sin^2 x = 1 && \bullet\ \csc^2 x = \frac{1}{\sin^2 x} \end{aligned}$$

Try Exercise 24, page 223

In the next example we make use of the guideline that states that it may be helpful to multiply both the numerator and the denominator of a fraction by the same factor.

EXAMPLE 4 Multiply by a Conjugate to Verify an Identity

Verify the identity $\dfrac{\sin x}{1 + \cos x} = \dfrac{1 - \cos x}{\sin x}$.

Continued ▶

take note

The sum $a + b$ and the difference $a - b$ are called conjugates of each other.

Solution

Multiply the numerator and denominator of the left side of the identity by the conjugate of $1 + \cos x$, which is $1 - \cos x$.

$$\frac{\sin x}{1 + \cos x} = \frac{\sin x}{1 + \cos x} \cdot \frac{1 - \cos x}{1 - \cos x}$$
$$= \frac{\sin x(1 - \cos x)}{1 - \cos^2 x}$$
$$= \frac{\sin x(1 - \cos x)}{\sin^2 x} = \frac{1 - \cos x}{\sin x}$$

Try Exercise 34, page 223

In Example 5 we verify an identity by first rewriting the more complicated side so that it involves only sines and cosines. After further algebraic simplification we are able to establish that the equation is an identity.

EXAMPLE 5 Change to Sines and Cosines to Verify an Identity

Verify the identity $\dfrac{\sin x + \tan x}{1 + \cos x} = \tan x$.

Solution

The left side of the identity is more complicated than the right side. We will try to verify the identity by rewriting the left side so that it involves only sines and cosines.

$$\frac{\sin x + \tan x}{1 + \cos x} = \frac{\sin x + \dfrac{\sin x}{\cos x}}{1 + \cos x}$$

- Use the identity $\tan x = \dfrac{\sin x}{\cos x}$.

$$= \frac{\dfrac{\sin x \cos x}{\cos x} + \dfrac{\sin x}{\cos x}}{1 + \cos x}$$

- The common denominator for the terms in the numerator is $\cos x$. Rewrite the numerator so that each term has a denominator of $\cos x$.

$$= \frac{\dfrac{\sin x \cos x + \sin x}{\cos x}}{1 + \cos x}$$

- Add the terms in the numerator.

$$= \frac{\sin x \cos x + \sin x}{\cos x} \div \frac{1 + \cos x}{1}$$

- Rewrite the complex fraction as a division.

$$= \frac{\sin x \cos x + \sin x}{\cos x} \cdot \frac{1}{1 + \cos x}$$

- Invert and multiply.

$$= \frac{\sin x\,(\cos x + 1)}{\cos x} \cdot \frac{1}{1 + \cos x}$$

- Factor.

$$= \frac{\sin x \,\cancel{(1 + \cos x)}}{\cos x} \cdot \frac{1}{\cancel{1 + \cos x}}$$ • Simplify.

$$= \frac{\sin x}{\cos x}$$ • Use the identity $\dfrac{\sin x}{\cos x} = \tan x$.

$$= \tan x$$

We have rewritten the left side of the equation so that it is identical to the right side. Thus we have verified that the equation is an identity.

Try Exercise 44, page 224

In the previous examples we verified trigonometric identities by rewriting one side of the identity in equivalent forms until that side appeared identical to the other side. A second method is to work with each side *separately* to produce a trigonometric expression that is equivalent to both sides. In the following paragraphs we use this procedure to verify the identity

$$\frac{1 + \cos x}{1 - \cos x} = (\csc x + \cot x)^2$$

Working with the left side gives us

$$\frac{1 + \cos x}{1 - \cos x} = \frac{1 + \cos x}{1 - \cos x} \cdot \frac{1 + \cos x}{1 + \cos x} \qquad (1)$$ • Multiply both numerator and denominator by the conjugate of the denominator.

$$= \frac{1 + 2\cos x + \cos^2 x}{1 - \cos^2 x} \qquad (2)$$ • Multiply and rewrite as a single fraction.

$$= \frac{1 + 2\cos x + \cos^2 x}{\sin^2 x} \qquad (3)$$ • Use the identity $1 - \cos^2 x = \sin^2 x$.

When you reach a point at which you are unsure how to proceed, stop and see what results can be obtained by working with the right side.

$$(\csc x + \cot x)^2 = \csc^2 x + 2\csc x \cot x + \cot^2 x \qquad (4)$$ • Square the binomial.

$$= \frac{1}{\sin^2 x} + 2\left(\frac{1}{\sin x}\right)\left(\frac{\cos x}{\sin x}\right) + \frac{\cos^2 x}{\sin^2 x} \qquad (5)$$ • Rewrite using fundamental identities.

$$= \frac{1 + 2\cos x + \cos^2 x}{\sin^2 x} \qquad (6)$$ • Add the fractions.

At this point we have verified the identity because we have shown that each side is equal to the same trigonometric expression. It is worth noting that if we arranged the above equations in the order (1), (2), (3), (5), and (4), we would have a verification that starts with the left side of the original identity and produces the right side.

When we verify an identity, we are not allowed to perform the same mathematical operation on both sides of the equation because this can lead to an incorrect result. For instance, consider the equation $\tan x = -\tan x$, which is not an identity. If we square both sides of the equation, we produce the identity $\tan^2 x = \tan^2 x$. We can avoid the problem of converting a non-identity into an identity and vice versa by using only mathematical operations and procedures that are *reversible*. We can see that the operation of squaring is not a *reversible* operation because it is not possible to take square roots (reverse the squaring operation) and convert $\tan^2 x = \tan^2 x$ back into $\tan x = -\tan x$.

There is often more than one way to verify a trigonometric identity. Some of these approaches are shorter and considered more elegant than others. As a beginner, you should be content with finding a verification. After you have had lots of practice, you may decide to seek out different verifications and compare them to see which you think is most efficient.

Topics for Discussion

1. Explain why $\tan = \dfrac{\sin}{\cos}$ is not an identity.
2. Is $\cos|x| = |\cos x|$ an identity? Explain. What about $\cos|x| = \cos x$? Explain.
3. The identity $\sin^2 x + \cos^2 x = 1$ is one of the Pythagorean identities. What are the other two Pythagorean identities, and how are they derived?
4. The graph of $y = \sin\dfrac{x}{2}$ for $0 \le x \le 2\pi$ is shown on the left below. The graph of $y = \sqrt{\dfrac{1-\cos x}{2}}$ for $0 \le x \le 2\pi$ is shown on the right below.

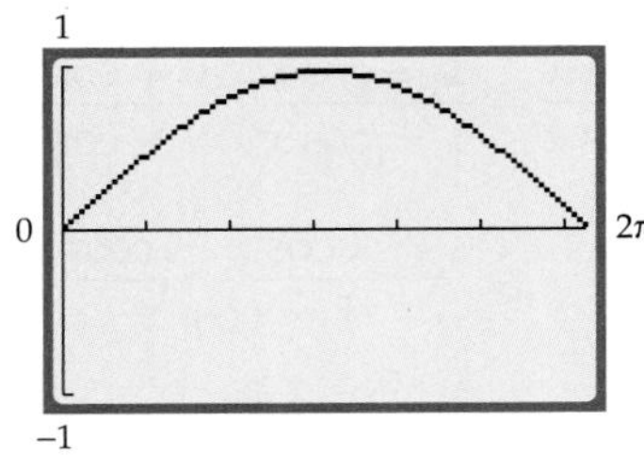

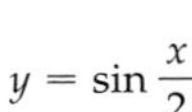

$y = \sin\dfrac{x}{2}$

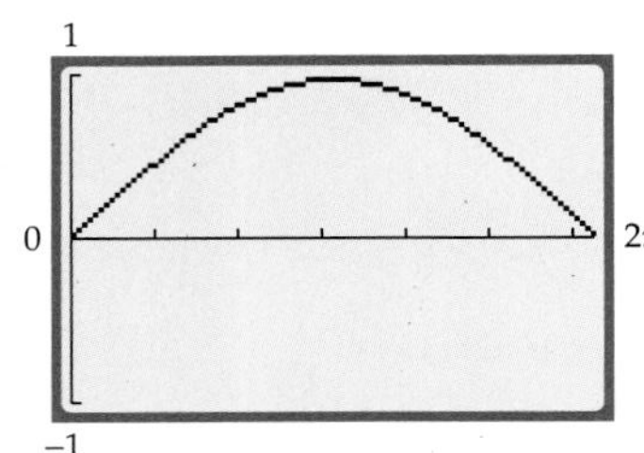

$y = \sqrt{\dfrac{1-\cos x}{2}}$

The graphs appear identical, but the equation

$$\sin\frac{x}{2} = \sqrt{\frac{1-\cos x}{2}}$$

is not an identity. Explain.

Exercise Set 3.1

In Exercises 1 to 56, verify each identity.

1. $\tan x \csc x \cos x = 1$
2. $\tan x \sec x \sin x = \tan^2 x$
3. $\dfrac{4\sin^2 x - 1}{2\sin x + 1} = 2\sin x - 1$
4. $\dfrac{\sin^2 x - 2\sin x + 1}{\sin x - 1} = \sin x - 1$
5. $(\sin x - \cos x)(\sin x + \cos x) = 1 - 2\cos^2 x$
6. $(\tan x)(1 - \cot x) = \tan x - 1$
7. $\dfrac{1}{\sin x} - \dfrac{1}{\cos x} = \dfrac{\cos x - \sin x}{\sin x \cos x}$
8. $\dfrac{1}{\sin x} + \dfrac{3}{\cos x} = \dfrac{\cos x + 3\sin x}{\sin x \cos x}$

9. $\dfrac{\cos x}{1 - \sin x} = \sec x + \tan x$

10. $\dfrac{\sin x}{1 - \cos x} = \csc x + \cot x$

11. $\dfrac{1 - \tan^4 x}{\sec^2 x} = 1 - \tan^2 x$

12. $\sin^4 x - \cos^4 x = \sin^2 x - \cos^2 x$

13. $\dfrac{1 + \tan^3 x}{1 + \tan x} = 1 - \tan x + \tan^2 x$

14. $\dfrac{\cos x \tan x - \sin x}{\cot x} = 0$

15. $\dfrac{\sin x - 2 + \dfrac{1}{\sin x}}{\sin x - \dfrac{1}{\sin x}} = \dfrac{\sin x - 1}{\sin x + 1}$

16. $\dfrac{\sin x}{1 - \cos x} - \dfrac{\sin x}{1 + \cos x} = 2 \cot x$

17. $(\sin x + \cos x)^2 = 1 + 2 \sin x \cos x$

18. $(\tan x + 1)^2 = \sec^2 x + 2 \tan x$

19. $\dfrac{\cos x}{1 + \sin x} = \sec x - \tan x$

20. $\dfrac{\sin x}{1 + \cos x} = \csc x - \cot x$

21. $\csc x = \dfrac{\cot x + \tan x}{\sec x}$

22. $\sec x = \dfrac{\cot x + \tan x}{\csc x}$

23. $\dfrac{\cos x \tan x + 2 \cos x - \tan x - 2}{\tan x + 2} = \cos x - 1$

24. $\dfrac{2 \sin x \cot x + \sin x - 4 \cot x - 2}{2 \cot x + 1} = \sin x - 2$

25. $\sec x - \tan x = \dfrac{1 - \sin x}{\cos x}$

26. $\cot x - \csc x = \dfrac{\cos x - 1}{\sin x}$

27. $\sin^2 x - \cos^2 x = 2 \sin^2 x - 1$

28. $\sin^2 x - \cos^2 x = 1 - 2 \cos^2 x$

29. $\dfrac{1}{\sin^2 x} + \dfrac{1}{\cos^2 x} = \csc^2 x \sec^2 x$

30. $\dfrac{1}{\tan^2 x} - \dfrac{1}{\cot^2 x} = \csc^2 x - \sec^2 x$

31. $\sec x - \cos x = \sin x \tan x$

32. $\tan x + \cot x = \sec x \csc x$

33. $\dfrac{\dfrac{1}{\sin x} + 1}{\dfrac{1}{\sin x} - 1} = \tan^2 x + 2 \tan x \sec x + \sec^2 x$

34. $\dfrac{\dfrac{1}{\sin x} + \dfrac{1}{\cos x}}{\dfrac{1}{\sin x} - \dfrac{1}{\cos x}} = \dfrac{\cos^2 x - \sin^2 x}{1 - 2 \cos x \sin x}$

35. $\sin^4 x - \cos^4 x = 2 \sin^2 x - 1$

36. $\sin^6 x + \cos^6 x = \sin^4 x - \sin^2 x \cos^2 x + \cos^4 x$

37. $\dfrac{1}{1 - \cos x} = \dfrac{1 + \cos x}{\sin^2 x}$

38. $1 + \sin x = \dfrac{\cos^2 x}{1 - \sin x}$

39. $\dfrac{\sin x}{1 - \sin x} - \dfrac{\cos x}{1 - \sin x} = \dfrac{1 - \cot x}{\csc x - 1}$

40. $\dfrac{\tan x}{1 + \tan x} - \dfrac{\cot x}{1 + \tan x} = 1 - \cot x$

41. $\dfrac{1}{1 + \cos x} - \dfrac{1}{1 - \cos x} = -2 \cot x \csc x$

42. $\dfrac{1}{1 - \sin x} - \dfrac{1}{1 + \sin x} = 2 \tan x \sec x$

43. $\dfrac{\dfrac{1}{\sin x} + \csc x}{\dfrac{1}{\sin x} - \sin x} = \dfrac{2}{\cos^2 x}$

44. $\dfrac{2\cot x}{\cot x + \tan x} = 2\cos^2 x$

45. $\dfrac{\cot x}{1 + \csc x} + \dfrac{1 + \csc x}{\cot x} = 2\sec x$

46. $\sec^2 x - \csc^2 x = \dfrac{\tan x - \cot x}{\sin x \cos x}$

47. $\sqrt{\dfrac{1 + \sin x}{1 - \sin x}} = \dfrac{1 + \sin x}{\cos x}, \quad \cos x > 0$

48. $\dfrac{\cos x + \cot x \sin x}{\cot x} = 2\sin x$

49. $\dfrac{\sin^3 x + \cos^3 x}{\sin x + \cos x} = 1 - \sin x \cos x$

50. $\dfrac{1 - \sin x}{1 + \sin x} - \dfrac{1 + \sin x}{1 - \sin x} = -4\sec x \tan x$

51. $\dfrac{\sec x - 1}{\sec x + 1} - \dfrac{\sec x + 1}{\sec x - 1} = -4\csc x \cot x$

52. $\dfrac{1}{1 - \cos x} - \dfrac{\cos x}{1 + \cos x} = 2\csc^2 x - 1$

53. $\dfrac{1 + \sin x}{\cos x} - \dfrac{\cos x}{1 - \sin x} = 0$

54. $(\sin x + \cos x + 1)^2 = 2(\sin x + 1)(\cos x + 1)$

55. $\dfrac{\sec x + \tan x}{\sec x - \tan x} = \dfrac{(\sin x + 1)^2}{\cos^2 x}$

56. $\dfrac{\sin^3 x - \cos^3 x}{\sin x + \cos x} = \dfrac{\csc^2 x - \cot x - 2\cos^2 x}{1 - \cot^2 x}$

In Exercises 57 to 64, compare the graphs of each side of the equation to predict whether the equation is an identity.

57. $\sin 2x = 2\sin x \cos x$

58. $\sin^2 x + \cos^2 x = 1$

59. $\sin x + \cos x = \sqrt{2}\sin\left(x + \dfrac{\pi}{4}\right)$

60. $\cos 2x = 2\cos^2 x - 1$

61. $\cos\left(x + \dfrac{\pi}{3}\right) = \cos x \cos\dfrac{\pi}{3} - \sin x \sin\dfrac{\pi}{3}$

62. $\cos\left(x - \dfrac{\pi}{4}\right) = \cos x \cos\dfrac{\pi}{4} + \sin x \sin\dfrac{\pi}{4}$

63. $\sin\left(x + \dfrac{\pi}{6}\right) = \sin x \sin\dfrac{\pi}{6} + \cos x \cos\dfrac{\pi}{6}$

64. $\sin\left(x - \dfrac{\pi}{3}\right) = \sin x \cos\dfrac{\pi}{3} + \cos x \sin\dfrac{\pi}{3}$

In Exercises 65 to 70, verify that the equation is *not* an identity by finding an *x* value for which the left side of the equation is not equal to the right side.

65. $(\sin x + \cos x)^2 = \sin^2 x + \cos^2 x$

66. $\tan 2x = 2\tan x$

67. $\cos(x + 30°) = \cos x + \cos 30°$

68. $\sqrt{1 - \sin^2 x} = \cos x$

69. $\tan^4 x - \sec^4 x = \tan^2 x + \sec^2 x$

70. $\sqrt{1 + \tan^2 x} = \sec x$

Connecting Concepts

In Exercises 71 to 76, verify the identity.

71. $\dfrac{1 - \sin x + \cos x}{1 + \sin x + \cos x} = \dfrac{\cos x}{\sin x + 1}$

72. $\dfrac{1 - \tan x + \sec x}{1 + \tan x - \sec x} = \dfrac{1 + \sec x}{\tan x}$

73. $\dfrac{2\sin^4 x + 2\sin^2 x \cos^2 x - 3\sin^2 x - 3\cos^2 x}{2\sin^2 x} = 1 - \dfrac{3}{2}\csc^2 x$

74. $\dfrac{4\tan x \sec^2 x - 4\tan x - \sec^2 x + 1}{4\tan^3 x - \tan^2 x} = 1$

75. $\dfrac{\sin x(\tan x + 1) - 2\tan x \cos x}{\sin x - \cos x} = \tan x$

76. $\dfrac{\sin^2 x \cos x + \cos^3 x - \sin^3 x \cos x - \sin x \cos^3 x}{1 - \sin^2 x} = \dfrac{\cos x}{1 + \sin x}$

77. Verify the identity $\sin^4 x + \cos^4 x = 1 - 2\sin^2 x \cos^2 x$ by completing the square of the left side of the identity.

78. Verify the identity $\tan^4 x + \sec^4 x = 1 + 2\tan^2 x \sec^2 x$ by completing the square of the left side of the identity.

»»» Projects »»»

1. GRADING A QUIZ Suppose that you are a teacher's assistant. You are to assist the teacher of a trigonometry class by grading a four-question quiz. Each question asks the student to find a trigonometric expression that models a given application. The teacher has prepared an answer key. These answers are shown in the next column. A student gives as answers the expressions shown in the far right column. Determine for which problems the student has given a correct response.

Answer Key	Student's Response
1. $\csc x \sec x$	1. $\cot x + \tan x$
2. $\cos^2 x$	2. $(1 + \sin x)(1 - \sin x)$
3. $\cos x \cot x$	3. $\csc x - \sec x$
4. $\csc x \cot x$	4. $\sin x(\cot x + \cot^3 x)$

Section 3.2

- Identities That Involve $(\alpha \pm \beta)$
- Cofunctions
- Additional Sum and Difference Identities
- Reduction Formulas

Sum, Difference, and Cofunction Identities

PREPARE FOR THIS SECTION

Prepare for this section by completing the following exercises. The answers can be found on page A16.

PS1. Compare $\cos(\alpha - \beta)$ and $\cos\alpha\cos\beta + \sin\alpha\sin\beta$ for $\alpha = \dfrac{\pi}{2}$ and $\beta = \dfrac{\pi}{6}$. [2.2]

PS2. Compare $\sin(\alpha + \beta)$ and $\sin\alpha\cos\beta + \cos\alpha\sin\beta$ for $\alpha = \dfrac{\pi}{2}$ and $\beta = \dfrac{\pi}{3}$. [2.2]

PS3. Compare $\sin(90° - \theta)$ and $\cos\theta$ for $\theta = 30°$, $\theta = 45°$, and $\theta = 120°$. [2.2]

PS4. Compare $\tan\left(\dfrac{\pi}{2} - \theta\right)$ and $\cot\theta$ for $\theta = \dfrac{\pi}{6}$, $\theta = \dfrac{\pi}{4}$, and $\theta = \dfrac{4\pi}{3}$. [2.2]

PS5. Compare $\tan(\alpha - \beta)$ and $\dfrac{\tan\alpha - \tan\beta}{1 + \tan\alpha\tan\beta}$ for $\alpha = \dfrac{\pi}{3}$ and $\beta = \dfrac{\pi}{6}$. [2.2]

PS6. Find the value of $\sin[(2k + 1)\pi]$, where k is any arbitrary integer. [2.2]

Identities That Involve $(\alpha \pm \beta)$

Each identity in Section 3.1 involved only one variable. We now consider identities that involve a trigonometric function of the sum or difference of two variables.

Sum and Difference Identities

$$\cos(\alpha - \beta) = \cos\alpha\cos\beta + \sin\alpha\sin\beta$$
$$\cos(\alpha + \beta) = \cos\alpha\cos\beta - \sin\alpha\sin\beta$$
$$\sin(\alpha - \beta) = \sin\alpha\cos\beta - \cos\alpha\sin\beta$$
$$\sin(\alpha + \beta) = \sin\alpha\cos\beta + \cos\alpha\sin\beta$$
$$\tan(\alpha + \beta) = \frac{\tan\alpha + \tan\beta}{1 - \tan\alpha\tan\beta}$$
$$\tan(\alpha - \beta) = \frac{\tan\alpha - \tan\beta}{1 + \tan\alpha\tan\beta}$$

Proof To establish the identity for $\cos(\alpha - \beta)$, we make use of the unit circle shown in **Figure 3.2**. The angles α and β are drawn in standard position, with OA and OB as the terminal sides of α and β, respectively. The coordinates of A are $(\cos\alpha, \sin\alpha)$, and the coordinates of B are $(\cos\beta, \sin\beta)$. The angle $(\alpha - \beta)$ is formed by the terminal sides of the angles α and β (angle AOB).

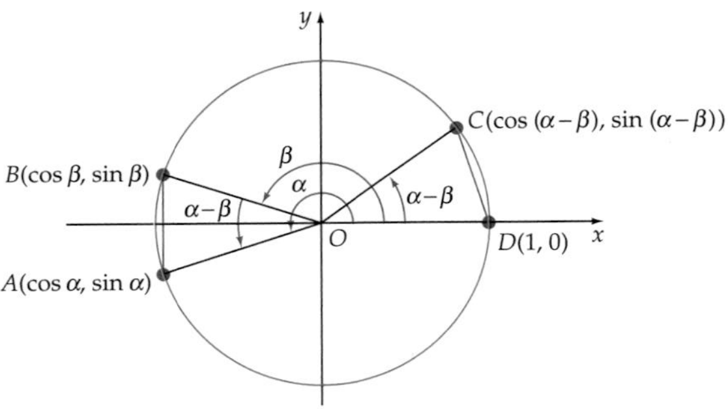

Figure 3.2

An angle equal in measure to angle $(\alpha - \beta)$ is placed in standard position in the same figure (angle COD). From geometry, if two central angles of a circle have the same measure, then their chords are also equal in measure. Thus the chords AB and CD are equal in length. Using the distance formula, we can calculate the lengths of the chords AB and CD.

$$d(A, B) = \sqrt{(\cos\alpha - \cos\beta)^2 + (\sin\alpha - \sin\beta)^2}$$
$$d(C, D) = \sqrt{[\cos(\alpha - \beta) - 1]^2 + [\sin(\alpha - \beta) - 0]^2}$$

Because $d(A, B) = d(C, D)$, we have

$$\sqrt{(\cos\alpha - \cos\beta)^2 + (\sin\alpha - \sin\beta)^2} = \sqrt{[\cos(\alpha - \beta) - 1]^2 + [\sin(\alpha - \beta)]^2}$$

Squaring each side of the equation and simplifying, we obtain

$$(\cos\alpha - \cos\beta)^2 + (\sin\alpha - \sin\beta)^2 = [\cos(\alpha - \beta) - 1]^2 + [\sin(\alpha - \beta)]^2$$
$$\cos^2\alpha - 2\cos\alpha\cos\beta + \cos^2\beta + \sin^2\alpha - 2\sin\alpha\sin\beta + \sin^2\beta$$
$$= \cos^2(\alpha - \beta) - 2\cos(\alpha - \beta) + 1 + \sin^2(\alpha - \beta)$$

$$\cos^2\alpha + \sin^2\alpha + \cos^2\beta + \sin^2\beta - 2\cos\alpha\cos\beta - 2\sin\alpha\sin\beta$$
$$= \cos^2(\alpha - \beta) + \sin^2(\alpha - \beta) + 1 - 2\cos(\alpha - \beta)$$

Simplifying by using $\sin^2\theta + \cos^2\theta = 1$, we have

$$2 - 2\sin\alpha\sin\beta - 2\cos\alpha\cos\beta = 2 - 2\cos(\alpha - \beta)$$

Solving for $\cos(\alpha - \beta)$ gives us

$$\cos(\alpha - \beta) = \cos\alpha\cos\beta + \sin\alpha\sin\beta$$ ◆

To derive an identity for $\cos(\alpha + \beta)$, write $\cos(\alpha + \beta)$ as $\cos[\alpha - (-\beta)]$.

$$\cos(\alpha + \beta) = \cos[\alpha - (-\beta)] = \cos\alpha\cos(-\beta) + \sin\alpha\sin(-\beta)$$

Recall that $\cos(-\beta) = \cos\beta$ and $\sin(-\beta) = -\sin\beta$. Substituting into the previous equation, we obtain the identity

$$\cos(\alpha + \beta) = \cos\alpha\cos\beta - \sin\alpha\sin\beta$$

EXAMPLE 1 ⟫ Evaluate a Trigonometric Expression

Use an identity to find the *exact* value of $\cos(60° - 45°)$.

Solution

Use the identity $\cos(\alpha - \beta) = \cos\alpha\cos\beta + \sin\alpha\sin\beta$ with $\alpha = 60°$ and $\beta = 45°$.

$$\cos(60° - 45°) = \cos 60° \cos 45° + \sin 60° \sin 45° \quad \text{• Substitute.}$$
$$= \left(\frac{1}{2}\right)\left(\frac{\sqrt{2}}{2}\right) + \left(\frac{\sqrt{3}}{2}\right)\left(\frac{\sqrt{2}}{2}\right) \quad \text{• Evaluate each factor.}$$
$$= \frac{\sqrt{2}}{4} + \frac{\sqrt{6}}{4} \quad \text{• Simplify.}$$
$$= \frac{\sqrt{2} + \sqrt{6}}{4}$$

⟫ *Try Exercise 4, page 233*

Cofunctions

Any pair of trigonometric functions f and g for which $f(x) = g(90° - x)$ and $g(x) = f(90° - x)$ are said to be **cofunctions.**

take note

To visualize the cofunction identities, consider the right triangle shown in the following figure.

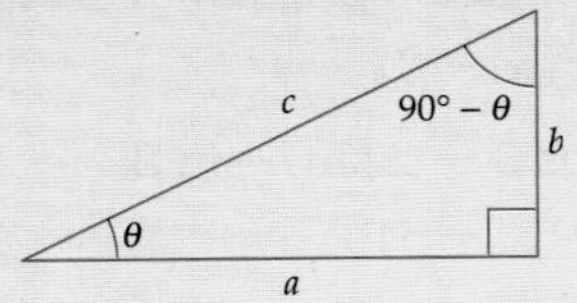

If θ is the degree measure of one of the acute angles, then the degree measure of the other acute angle is $(90° - \theta)$. Using the definitions of the trigonometric functions gives us

$$\sin\theta = \frac{b}{c} = \cos(90° - \theta)$$

$$\tan\theta = \frac{b}{a} = \cot(90° - \theta)$$

$$\sec\theta = \frac{c}{a} = \csc(90° - \theta)$$

These identities state that *the value of a trigonometric function of θ is equal to the cofunction of the complement of θ.*

Cofunction Identities

$$\sin(90° - \theta) = \cos\theta \qquad \cos(90° - \theta) = \sin\theta$$

$$\tan(90° - \theta) = \cot\theta \qquad \cot(90° - \theta) = \tan\theta$$

$$\sec(90° - \theta) = \csc\theta \qquad \csc(90° - \theta) = \sec\theta$$

If θ is in radian measure, replace 90° with $\frac{\pi}{2}$.

To verify that the sine function and the cosine function are cofunctions, we make use of the identity for $\cos(\alpha - \beta)$.

$$\begin{aligned}\cos(90° - \beta) &= \cos 90° \cos\beta + \sin 90° \sin\beta \\ &= 0 \cdot \cos\beta + 1 \cdot \sin\beta\end{aligned}$$

which gives

$$\cos(90° - \beta) = \sin\beta$$

Thus the sine of an angle is equal to the cosine of its complement. Using $\cos(90° - \beta) = \sin\beta$ with $\beta = 90° - \alpha$, we have

$$\cos\alpha = \cos[90° - (90° - \alpha)] = \sin(90° - \alpha)$$

Therefore,

$$\cos\alpha = \sin(90° - \alpha)$$

We can use the ratio identities to show that the tangent and cotangent functions are cofunctions.

$$\tan(90° - \theta) = \frac{\sin(90° - \theta)}{\cos(90° - \theta)} = \frac{\cos\theta}{\sin\theta} = \cot\theta$$

$$\cot(90° - \theta) = \frac{\cos(90° - \theta)}{\sin(90° - \theta)} = \frac{\sin\theta}{\cos\theta} = \tan\theta$$

The secant and cosecant functions are also cofunctions.

EXAMPLE 2 》 Write an Equivalent Expression

Use a cofunction identity to write an equivalent expression for sin 20°.

Solution

The value of a given trigonometric function of θ, measured in degrees, is equal to its cofunction of $90° - \theta$. Thus

$$\begin{aligned}\sin 20° &= \cos(90° - 20°) \\ &= \cos 70°\end{aligned}$$

》 *Try Exercise 20, page 233*

Additional Sum and Difference Identities

We can use the cofunction identities to verify the remaining sum and difference identities. To derive an identity for $\sin(\alpha + \beta)$, substitute $\alpha + \beta$ for θ in the cofunction identity $\sin\theta = \cos(90° - \theta)$.

$$\begin{aligned}
\sin\theta &= \cos(90° - \theta) \\
\sin(\alpha + \beta) &= \cos[90° - (\alpha + \beta)] && \bullet \text{ Replace } \theta \text{ with } \alpha + \beta. \\
&= \cos[(90° - \alpha) - \beta] && \bullet \text{ Rewrite as the difference of two angles.} \\
&= \cos(90° - \alpha)\cos\beta + \sin(90° - \alpha)\sin\beta \\
&= \sin\alpha\cos\beta + \cos\alpha\sin\beta
\end{aligned}$$

Therefore,

$$\sin(\alpha + \beta) = \sin\alpha\cos\beta + \cos\alpha\sin\beta$$

We also can derive an identity for $\sin(\alpha - \beta)$ by rewriting $(\alpha - \beta)$ as $[\alpha + (-\beta)]$.

$$\begin{aligned}
\sin(\alpha - \beta) &= \sin[\alpha + (-\beta)] \\
&= \sin\alpha\cos(-\beta) + \cos\alpha\sin(-\beta) \\
&= \sin\alpha\cos\beta - \cos\alpha\sin\beta && \bullet \cos(-\beta) = \cos\beta,\ \sin(-\beta) = -\sin\beta
\end{aligned}$$

Thus

$$\sin(\alpha - \beta) = \sin\alpha\cos\beta - \cos\alpha\sin\beta$$

The identity for $\tan(\alpha + \beta)$ is a result of the identity $\tan\theta = \dfrac{\sin\theta}{\cos\theta}$ and the identities for $\sin(\alpha + \beta)$ and $\cos(\alpha + \beta)$.

$$\begin{aligned}
\tan(\alpha + \beta) &= \frac{\sin(\alpha + \beta)}{\cos(\alpha + \beta)} = \frac{\sin\alpha\cos\beta + \cos\alpha\sin\beta}{\cos\alpha\cos\beta - \sin\alpha\sin\beta} \\
&= \frac{\dfrac{\sin\alpha\cos\beta}{\cos\alpha\cos\beta} + \dfrac{\cos\alpha\sin\beta}{\cos\alpha\cos\beta}}{\dfrac{\cos\alpha\cos\beta}{\cos\alpha\cos\beta} - \dfrac{\sin\alpha\sin\beta}{\cos\alpha\cos\beta}} && \bullet \text{ Multiply both the numerator and the denominator by } \frac{1}{\cos\alpha\cos\beta} \text{ and simplify.}
\end{aligned}$$

Therefore,

$$\tan(\alpha + \beta) = \frac{\tan\alpha + \tan\beta}{1 - \tan\alpha\tan\beta}$$

The tangent function is an odd function, so $\tan(-\theta) = -\tan\theta$. Rewriting $(\alpha - \beta)$ as $[\alpha + (-\beta)]$ enables us to derive an identity for $\tan(\alpha - \beta)$.

$$\tan(\alpha - \beta) = \tan[\alpha + (-\beta)] = \frac{\tan\alpha + \tan(-\beta)}{1 - \tan\alpha\tan(-\beta)}$$

Therefore,

$$\tan(\alpha - \beta) = \frac{\tan\alpha - \tan\beta}{1 + \tan\alpha\tan\beta}$$

The sum and difference identities can be used to simplify some trigonometric expressions.

EXAMPLE 3 » Simplify Trigonometric Expressions

Write each expression in terms of a single trigonometric function.

a. $\sin 5x \cos 3x - \cos 5x \sin 3x$ **b.** $\dfrac{\tan 4\alpha + \tan \alpha}{1 - \tan 4\alpha \tan \alpha}$

Solution

a. $\sin 5x \cos 3x - \cos 5x \sin 3x = \sin(5x - 3x) = \sin 2x$

b. $\dfrac{\tan 4\alpha + \tan \alpha}{1 - \tan 4\alpha \tan \alpha} = \tan(4\alpha + \alpha) = \tan 5\alpha$

» *Try Exercise 26, page 233*

EXAMPLE 4 » Evaluate a Trigonometric Function

Given $\tan \alpha = -\dfrac{4}{3}$ for α in Quadrant II and $\tan \beta = -\dfrac{5}{12}$ for β in Quadrant IV, find $\sin(\alpha + \beta)$.

Solution

See **Figure 3.3**. Because $\tan \alpha = \dfrac{y}{x} = -\dfrac{4}{3}$ and the terminal side of α is in Quadrant II, $P_1(-3, 4)$ is a point on the terminal side of α. Similarly, $P_2(12, -5)$ is a point on the terminal side of β.

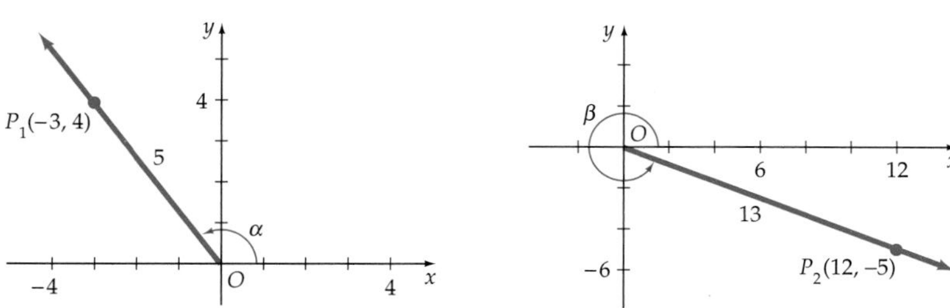

Figure 3.3

Using the Pythagorean Theorem, we find that the length of the line segment OP_1 is 5 and the length of OP_2 is 13.

$$\begin{aligned}\sin(\alpha + \beta) &= \sin \alpha \cos \beta + \cos \alpha \sin \beta \\ &= \frac{4}{5} \cdot \frac{12}{13} + \frac{-3}{5} \cdot \frac{-5}{13} = \frac{48}{65} + \frac{15}{65} = \frac{63}{65}\end{aligned}$$

» *Try Exercise 38, page 233*

EXAMPLE 5 ›› Verify an Identity

Verify the identity $\cos(\pi - \theta) = -\cos\theta$.

Solution

$$\begin{aligned}\cos(\pi - \theta) &= \cos\pi\cos\theta + \sin\pi\sin\theta \\ &= -1\cdot\cos\theta + 0\cdot\sin\theta \\ &= -\cos\theta\end{aligned}$$

• **Use the identity for $\cos(\alpha - \beta)$.**

›› *Try Exercise 50, page 234*

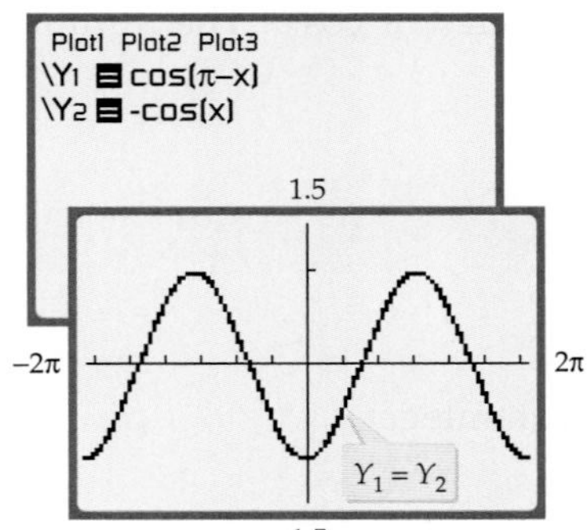

Figure 3.4

Figure 3.4 shows the graphs of $Y1 = \cos(\pi - \theta)$ and $Y2 = -\cos\theta$ on the same coordinate axes. The fact that the graphs appear to be identical supports the verification in Example 5.

EXAMPLE 6 ›› Verify an Identity

Verify the identity $\dfrac{\cos 4\theta}{\sin\theta} - \dfrac{\sin 4\theta}{\cos\theta} = \dfrac{\cos 5\theta}{\sin\theta\cos\theta}$.

Solution

Subtract the fractions on the left side of the equation.

$$\begin{aligned}\frac{\cos 4\theta}{\sin\theta} - \frac{\sin 4\theta}{\cos\theta} &= \frac{\cos 4\theta\cos\theta - \sin 4\theta\sin\theta}{\sin\theta\cos\theta} \\ &= \frac{\cos(4\theta + \theta)}{\sin\theta\cos\theta} \\ &= \frac{\cos 5\theta}{\sin\theta\cos\theta}\end{aligned}$$

• **Use the identity for $\cos(\alpha + \beta)$.**

›› *Try Exercise 62, page 234*

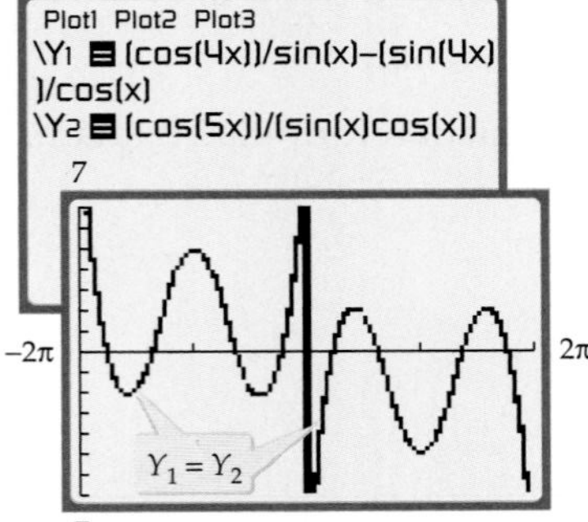

Figure 3.5

Figure 3.5 shows the graph of $Y1 = \dfrac{\cos 4\theta}{\sin\theta} - \dfrac{\sin 4\theta}{\cos\theta}$ and the graph of $Y2 = \dfrac{\cos 5\theta}{\sin\theta\cos\theta}$ on the same coordinate axes. The fact that the graphs appear to be identical supports the verification in Example 6.

Reduction Formulas

The sum or difference identities can be used to write expressions such as

$$\sin(\theta + k\pi) \qquad \sin(\theta + 2k\pi) \qquad \text{and} \qquad \cos[\theta + (2k + 1)\pi]$$

where k is an integer, as expressions involving only $\sin \theta$ or $\cos \theta$. The resulting formulas are called **reduction formulas.**

EXAMPLE 7 Find Reduction Formulas

Write as a function involving only $\sin \theta$.

$$\sin[\theta + (2k + 1)\pi], \quad \text{where } k \text{ is an integer}$$

Solution

Applying the identity $\sin(\alpha + \beta) = \sin \alpha \cos \beta + \cos \alpha \sin \beta$ yields

$$\sin[\theta + (2k + 1)\pi] = \sin \theta \cos[(2k + 1)\pi] + \cos \theta \sin[(2k + 1)\pi]$$

If k is an integer, then $2k + 1$ is an odd integer. The cosine of any odd multiple of π equals -1, and the sine of any odd multiple of π is 0. This gives us

$$\sin[\theta + (2k + 1)\pi] = (\sin \theta)(-1) + (\cos \theta)(0) = -\sin \theta$$

Thus $\sin[\theta + (2k + 1)\pi] = -\sin \theta$ for any integer k.

Try Exercise 76, page 235

QUESTION Is $\sin(\theta + 2k\pi) = \sin \theta$ a reduction formula?

Topics for Discussion

1. Does $\sin(\alpha + \beta) = \sin \alpha + \sin \beta$ for all values of α and β? If not, find nonzero values of α and β for which $\sin(\alpha + \beta) \neq \sin \alpha + \sin \beta$.
2. If k is an integer, then $2k + 1$ is an odd integer. Do you agree? Explain.
3. What are the trigonometric cofunction identities? Explain.
4. Is $\tan(\theta + k\pi) = \tan \theta$, where k is an integer, a reduction formula? Explain.

ANSWER Yes. $\sin (\theta + 2k\pi) = \sin \theta \cos (2k\pi) + \cos \theta \sin (2k\pi)$
$= (\sin \theta)(1) + (\cos \theta)(0) = \sin \theta.$

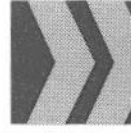

Exercise Set 3.2

In Exercises 1 to 18, find (if possible) the exact value of the expression.

1. $\sin(45° + 30°)$

2. $\sin(330° + 45°)$

3. $\cos(45° - 30°)$

4. $\cos(120° - 45°)$

5. $\tan(45° - 30°)$

6. $\tan(240° - 45°)$

7. $\sin\left(\frac{5\pi}{4} - \frac{\pi}{6}\right)$

8. $\sin\left(\frac{4\pi}{3} + \frac{\pi}{4}\right)$

9. $\cos\left(\frac{3\pi}{4} + \frac{\pi}{6}\right)$

10. $\cos\left(\frac{\pi}{4} - \frac{\pi}{3}\right)$

11. $\tan\left(\frac{\pi}{6} + \frac{\pi}{4}\right)$

12. $\tan\left(\frac{11\pi}{6} - \frac{\pi}{4}\right)$

13. $\cos 212° \cos 122° + \sin 212° \sin 122°$

14. $\sin 167° \cos 107° - \cos 167° \sin 107°$

15. $\sin\frac{5\pi}{12}\cos\frac{\pi}{4} - \cos\frac{5\pi}{12}\sin\frac{\pi}{4}$

16. $\cos\frac{\pi}{12}\cos\frac{\pi}{4} - \sin\frac{\pi}{12}\sin\frac{\pi}{4}$

17. $\dfrac{\tan\frac{7\pi}{12} - \tan\frac{\pi}{4}}{1 + \tan\frac{7\pi}{12}\tan\frac{\pi}{4}}$

18. $\dfrac{\tan\frac{\pi}{6} + \tan\frac{\pi}{3}}{1 - \tan\frac{\pi}{6}\tan\frac{\pi}{3}}$

In Exercises 19 to 24, use a cofunction identity to write an equivalent expression for the given value.

19. $\sin 42°$

20. $\cos 80°$

21. $\tan 15°$

22. $\cot 2°$

23. $\sec 25°$

24. $\csc 84°$

In Exercises 25 to 36, write each expression in terms of a single trigonometric function.

25. $\sin 7x \cos 2x - \cos 7x \sin 2x$

26. $\sin x \cos 3x + \cos x \sin 3x$

27. $\cos x \cos 2x + \sin x \sin 2x$

28. $\cos 4x \cos 2x - \sin 4x \sin 2x$

29. $\sin 7x \cos 3x - \cos 7x \sin 3x$

30. $\cos x \cos 5x - \sin x \sin 5x$

31. $\cos 4x \cos(-2x) - \sin 4x \sin(-2x)$

32. $\sin(-x) \cos 3x - \cos(-x) \sin 3x$

33. $\sin\frac{x}{3}\cos\frac{2x}{3} + \cos\frac{x}{3}\sin\frac{2x}{3}$

34. $\cos\frac{3x}{4}\cos\frac{x}{4} + \sin\frac{3x}{4}\sin\frac{x}{4}$

35. $\dfrac{\tan 3x + \tan 4x}{1 - \tan 3x \tan 4x}$

36. $\dfrac{\tan 2x - \tan 3x}{1 + \tan 2x \tan 3x}$

In Exercises 37 to 48, find the exact value of the given functions.

37. Given $\tan\alpha = -\frac{4}{3}$, α in Quadrant II, and $\tan\beta = \frac{15}{8}$, β in Quadrant III, find

a. $\sin(\alpha - \beta)$ **b.** $\cos(\alpha + \beta)$ **c.** $\tan(\alpha - \beta)$

38. Given $\tan\alpha = \frac{24}{7}$, α in Quadrant I, and $\sin\beta = -\frac{8}{17}$, β in Quadrant III, find

a. $\sin(\alpha + \beta)$ **b.** $\cos(\alpha + \beta)$ **c.** $\tan(\alpha - \beta)$

39. Given $\sin\alpha = \frac{3}{5}$, α in Quadrant I, and $\cos\beta = -\frac{5}{13}$, β in Quadrant II, find

a. $\sin(\alpha - \beta)$ **b.** $\cos(\alpha + \beta)$ **c.** $\tan(\alpha - \beta)$

40. Given $\sin\alpha = \frac{24}{25}$, α in Quadrant II, and $\cos\beta = -\frac{4}{5}$, β in Quadrant III, find

a. $\cos(\beta - \alpha)$ **b.** $\sin(\alpha + \beta)$ **c.** $\tan(\alpha + \beta)$

41. Given $\sin\alpha = -\frac{4}{5}$, α in Quadrant III, and $\cos\beta = -\frac{12}{13}$, β in Quadrant II, find

a. $\sin(\alpha - \beta)$ **b.** $\cos(\alpha + \beta)$ **c.** $\tan(\alpha + \beta)$

42. Given $\sin\alpha = -\frac{7}{25}$, α in Quadrant IV, and $\cos\beta = \frac{8}{17}$, β in Quadrant IV, find

a. $\sin(\alpha + \beta)$ **b.** $\cos(\alpha - \beta)$ **c.** $\tan(\alpha + \beta)$

43. Given $\cos\alpha = \frac{15}{17}$, α in Quadrant I, and $\sin\beta = -\frac{3}{5}$, β in Quadrant III, find

a. $\sin(\alpha + \beta)$ **b.** $\cos(\alpha - \beta)$ **c.** $\tan(\alpha - \beta)$

44. Given $\cos\alpha = -\frac{7}{25}$, α in Quadrant II, and $\sin\beta = -\frac{12}{13}$, β in Quadrant IV, find

a. $\sin(\alpha + \beta)$ **b.** $\cos(\alpha + \beta)$ **c.** $\tan(\alpha - \beta)$

45. Given $\cos\alpha = -\frac{3}{5}$, α in Quadrant III, and $\sin\beta = \frac{5}{13}$, β in Quadrant I, find

a. $\sin(\alpha - \beta)$ **b.** $\cos(\alpha + \beta)$ **c.** $\tan(\alpha + \beta)$

46. Given $\cos\alpha = \frac{8}{17}$, α in Quadrant IV, and $\sin\beta = -\frac{24}{25}$, β in Quadrant III, find

a. $\sin(\alpha - \beta)$ **b.** $\cos(\alpha + \beta)$ **c.** $\tan(\alpha + \beta)$

47. Given $\sin\alpha = \frac{3}{5}$, α in Quadrant I, and $\tan\beta = \frac{5}{12}$, β in Quadrant III, find

a. $\sin(\alpha + \beta)$ **b.** $\cos(\alpha - \beta)$ **c.** $\tan(\alpha - \beta)$

48. Given $\tan\alpha = \frac{15}{8}$, α in Quadrant I, and $\tan\beta = -\frac{7}{24}$, β in Quadrant IV, find

a. $\sin(\alpha - \beta)$ **b.** $\cos(\alpha - \beta)$ **c.** $\tan(\alpha + \beta)$

In Exercises 49 to 74, verify the identity.

49. $\cos\left(\frac{\pi}{2} - \theta\right) = \sin\theta$

50. $\cos(\theta + \pi) = -\cos\theta$

51. $\sin\left(\theta + \frac{\pi}{2}\right) = \cos\theta$

52. $\sin(\theta + \pi) = -\sin\theta$

53. $\tan\left(\theta + \frac{\pi}{4}\right) = \frac{\tan\theta + 1}{1 - \tan\theta}$

54. $\tan 2\theta = \frac{2\tan\theta}{1 - \tan^2\theta}$

55. $\cos\left(\frac{3\pi}{2} - \theta\right) = -\sin\theta$

56. $\sin\left(\frac{3\pi}{2} + \theta\right) = -\cos\theta$

57. $\cot\left(\frac{\pi}{2} - \theta\right) = \tan\theta$

58. $\cot(\pi + \theta) = \cot\theta$

59. $\csc(\pi - \theta) = \csc\theta$

60. $\sec\left(\frac{\pi}{2} - \theta\right) = \csc\theta$

61. $\sin 6x\cos 2x - \cos 6x\sin 2x = 2\sin 2x\cos 2x$

62. $\cos 5x\cos 3x + \sin 5x\sin 3x = \cos^2 x - \sin^2 x$

63. $\cos(\alpha + \beta) + \cos(\alpha - \beta) = 2\cos\alpha\cos\beta$

64. $\cos(\alpha - \beta) - \cos(\alpha + \beta) = 2\sin\alpha\sin\beta$

65. $\sin(\alpha + \beta) + \sin(\alpha - \beta) = 2\sin\alpha\cos\beta$

66. $\sin(\alpha - \beta) - \sin(\alpha + \beta) = -2\cos\alpha\sin\beta$

67. $\frac{\cos(\alpha - \beta)}{\sin(\alpha + \beta)} = \frac{\cot\alpha + \tan\beta}{1 + \cot\alpha\tan\beta}$

68. $\dfrac{\sin(\alpha + \beta)}{\sin(\alpha - \beta)} = \dfrac{1 + \cot \alpha \tan \beta}{1 - \cot \alpha \tan \beta}$

69. $\dfrac{\sin(x + h) - \sin x}{h} = \cos x \cdot \dfrac{\sin h}{h} + \sin x \cdot \dfrac{(\cos h - 1)}{h}$

70. $\dfrac{\cos(x + h) - \cos x}{h} = \cos x \cdot \dfrac{(\cos h - 1)}{h} - \sin x \cdot \dfrac{\sin h}{h}$

71. $\sin\left(\dfrac{\pi}{2} + \alpha - \beta\right) = \cos \alpha \cos \beta + \sin \alpha \sin \beta$

72. $\cos\left(\dfrac{\pi}{2} + \alpha + \beta\right) = -(\sin \alpha \cos \beta + \cos \alpha \sin \beta)$

73. $\sin 3x = 3 \sin x - 4 \sin^3 x$

74. $\cos 3x = 4 \cos^3 x - 3 \cos x$

In Exercises 75 to 80, write the given expression as a function that involves only sin θ, cos θ, or tan θ. (In Exercises 78, 79, and 80, assume k is an integer.)

75. $\cos(\theta + 3\pi)$

76. $\sin(\theta + 2\pi)$

77. $\tan(\theta + \pi)$

78. $\cos[\theta + (2k + 1)\pi]$

79. $\sin(\theta + 2k\pi)$

80. $\sin(\theta - k\pi)$

In Exercises 81 to 84, compare the graphs of each side of the equation to predict whether the equation is an identity.

81. $\sin\left(\dfrac{\pi}{2} - x\right) = \cos x$

82. $\cos(x + \pi) = -\cos x$

83. $\sin 7x \cos 2x - \cos 7x \sin 2x = \sin 5x$

84. $\sin 3x = 3 \sin x - 4 \sin^3 x$

Connecting Concepts

In Exercises 85 to 89, verify the identity.

85. $\sin(x - y) \cdot \sin(x + y) = \sin^2 x \cos^2 y - \cos^2 x \sin^2 y$

86. $\sin(x + y + z) = \sin x \cos y \cos z + \cos x \sin y \cos z + \cos x \cos y \sin z - \sin x \sin y \sin z$

87. $\cos(x + y + z) = \cos x \cos y \cos z - \sin x \sin y \cos z - \sin x \cos y \sin z - \cos x \sin y \sin z$

88. $\dfrac{\sin(x + y)}{\sin x \sin y} = \cot x + \cot y$

89. $\dfrac{\cos(x - y)}{\cos x \sin y} = \cot y + \tan x$

90. **Model Resistance** The drag (resistance) on a fish when it is swimming is two to three times the drag when it is gliding. To compensate for this, some fish swim in a saw-tooth pattern, as shown in the accompanying figure. The ratio of the amount of energy the fish expends when swimming upward at angle β and then gliding down at angle α to the energy it expends swimming horizontally is given by

$$E_R = \frac{k \sin \alpha + \sin \beta}{k \sin(\alpha + \beta)}$$

where k is a value such that $2 \le k \le 3$, and k depends on the assumptions we make about the amount of drag experienced by the fish. Find E_R for $k = 2$, $\alpha = 10°$, and $\beta = 20°$.

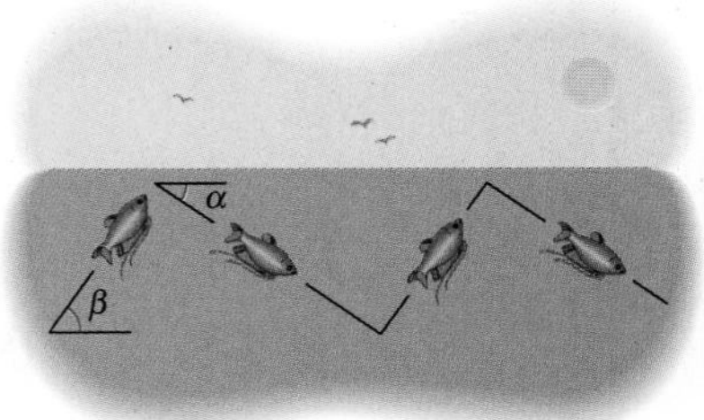

Projects

1. **Intersecting Lines** In the figure shown at the right, two nonvertical lines intersect in a plane. The slope of line l_1 is m_1 and the slope of line l_2 is m_2.

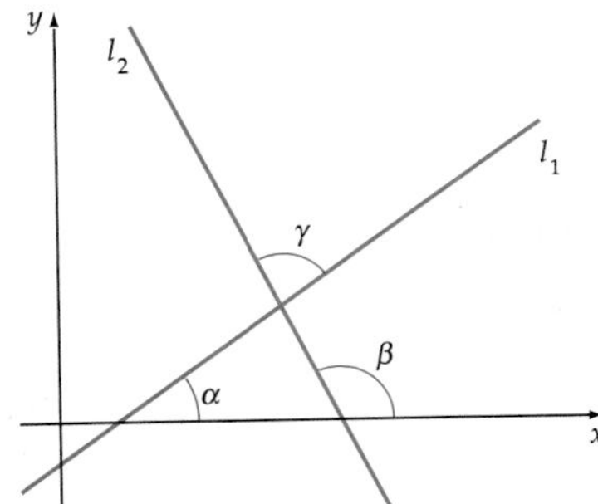

   a. Show that the tangent of the smallest positive angle γ between l_1 and l_2 is given by

$$\tan \gamma = \frac{m_2 - m_1}{1 + m_1 m_2}$$

   b. Two nonvertical lines intersect at the point (1,5). The measure of the smallest positive angle between the lines is $\gamma = 60°$. The first line is given by $y = 0.5x + 4.5$. What is the equation (in slope-intercept form) of the second line?

Section 3.3

- Double-Angle Identities
- Power-Reducing Identities
- Half-Angle Identities

Double- and Half-Angle Identities

PREPARE FOR THIS SECTION

Prepare for this section by completing the following exercises. The answers can be found on page A16.

PS1. Use the identity for $\sin(\alpha + \beta)$ to rewrite $\sin 2\alpha$. [3.2]

PS2. Use the identity for $\cos(\alpha + \beta)$ to rewrite $\cos 2\alpha$. [3.2]

PS3. Use the identity for $\tan(\alpha + \beta)$ to rewrite $\tan 2\alpha$. [3.2]

PS4. Compare $\tan \dfrac{\alpha}{2}$ and $\dfrac{\sin \alpha}{1 + \cos \alpha}$ for $\alpha = 60°$, $\alpha = 90°$, and $\alpha = 120°$. [2.2]

PS5. Verify that $\sin 2\alpha = 2 \sin \alpha$ is *not* an identity. [2.2]

PS6. Verify that $\cos \dfrac{\alpha}{2} = \dfrac{1}{2} \cos \alpha$ is *not* an identity. [2.2]

Integrating Technology

One way of showing that $\sin 2x \neq 2 \sin x$ is by graphing $y = \sin 2x$ and $y = 2 \sin x$ and observing that the graphs are not the same.

Double-Angle Identities

By using the sum identities, we can derive identities for $f(2\alpha)$, where f is a trigonometric function. These are called the *double-angle identities*. To find the sine of a double angle, substitute α for β in the identity for $\sin(\alpha + \beta)$.

$$\sin (\alpha + \beta) = \sin \alpha \cos \beta + \cos \alpha \sin \beta$$

$$\sin (\alpha + \alpha) = \sin \alpha \cos \alpha + \cos \alpha \sin \alpha \qquad \text{• Let } \beta = \alpha.$$

$$\sin 2\alpha = 2 \sin \alpha \cos \alpha$$

A double-angle identity for cosine is derived in a similar manner.

$$\cos(\alpha + \beta) = \cos\alpha\cos\beta - \sin\alpha\sin\beta$$

$$\cos(\alpha + \alpha) = \cos\alpha\cos\alpha - \sin\alpha\sin\alpha \qquad \text{• Let } \beta = \alpha.$$

$$\cos 2\alpha = \cos^2\alpha - \sin^2\alpha$$

There are two alternative forms of the double-angle identity for $\cos 2\alpha$. Using $\cos^2\alpha = 1 - \sin^2\alpha$, we can rewrite the identity for $\cos 2\alpha$ as follows:

$$\cos 2\alpha = \cos^2\alpha - \sin^2\alpha$$

$$\cos 2\alpha = (1 - \sin^2\alpha) - \sin^2\alpha \qquad \text{• } \cos^2\alpha = 1 - \sin^2\alpha$$

$$\cos 2\alpha = 1 - 2\sin^2\alpha$$

We also can rewrite $\cos 2\alpha$ as

$$\cos 2\alpha = \cos^2\alpha - \sin^2\alpha$$

$$\cos 2\alpha = \cos^2\alpha - (1 - \cos^2\alpha) \qquad \text{• } \sin^2\alpha = 1 - \cos^2\alpha$$

$$\cos 2\alpha = 2\cos^2\alpha - 1$$

The double-angle identity for the tangent function is derived from the identity for $\tan(\alpha + \beta)$ with $\beta = \alpha$.

$$\tan(\alpha + \beta) = \frac{\tan\alpha + \tan\beta}{1 - \tan\alpha\tan\beta}$$

$$\tan(\alpha + \alpha) = \frac{\tan\alpha + \tan\alpha}{1 - \tan\alpha\tan\alpha} \qquad \text{• Let } \beta = \alpha.$$

$$\tan 2\alpha = \frac{2\tan\alpha}{1 - \tan^2\alpha}$$

Here is a summary of the double-angle identities.

Double-Angle Identities

$$\sin 2\alpha = 2\sin\alpha\cos\alpha$$

$$\cos 2\alpha = \cos^2\alpha - \sin^2\alpha = 1 - 2\sin^2\alpha = 2\cos^2\alpha - 1$$

$$\tan 2\alpha = \frac{2\tan\alpha}{1 - \tan^2\alpha}$$

The double-angle identities are often used to write a trigonometric expression in terms of a single trigonometric function.

EXAMPLE 1 Simplify a Trigonometric Expression

Write $4\sin 5\theta\cos 5\theta$ in terms of a single trigonometric function.

Solution

$$4\sin 5\theta\cos 5\theta = 2(2\sin 5\theta\cos 5\theta)$$

$$= 2(\sin 10\theta) = 2\sin 10\theta$$

• Use $2\sin\alpha\cos\alpha = \sin 2\alpha$, with $\alpha = 5\theta$.

Try Exercise 2, page 243

QUESTION Does $\sin\theta\cos\theta = \frac{1}{2}\sin 2\theta$?

In Example 2, we use given information concerning the angle α to find the exact value of $\sin 2\alpha$.

EXAMPLE 2 Evaluate a Trigonometric Function

If $\sin\alpha = \frac{4}{5}$ and $0° < \alpha < 90°$, find the exact value of $\sin 2\alpha$.

Solution

Use the identity $\sin 2\alpha = 2\sin\alpha\cos\alpha$. Find $\cos\alpha$ by substituting for $\sin\alpha$ in $\sin^2\alpha + \cos^2\alpha = 1$ and solving for $\cos\alpha$.

$$\cos\alpha = \sqrt{1 - \sin^2\alpha} = \sqrt{1 - \left(\frac{4}{5}\right)^2} = \frac{3}{5}$$

• $\cos\alpha > 0$ if α is in Quadrant I.

Substitute the values of $\sin\alpha$ and $\cos\alpha$ in the double-angle formula for $\sin 2\alpha$.

$$\sin 2\alpha = 2\sin\alpha\cos\alpha = 2\left(\frac{4}{5}\right)\left(\frac{3}{5}\right) = \frac{24}{25}$$

Try Exercise 10, page 243

EXAMPLE 3 Verify an Identity

Verify the identity $\csc 2\alpha = \frac{1}{2}(\tan\alpha + \cot\alpha)$.

Solution

Work on the right-hand side of the equation.

$$\begin{aligned}\frac{1}{2}(\tan\alpha + \cot\alpha) &= \frac{1}{2}\left(\frac{\sin\alpha}{\cos\alpha} + \frac{\cos\alpha}{\sin\alpha}\right)\\ &= \frac{1}{2}\left(\frac{\sin^2\alpha + \cos^2\alpha}{\cos\alpha\sin\alpha}\right)\\ &= \frac{1}{2\cos\alpha\sin\alpha} = \frac{1}{\sin 2\alpha} = \csc 2\alpha\end{aligned}$$

Try Exercise 50, page 244

ANSWER Yes. $\sin\theta\cos\theta = \frac{2\sin\theta\cos\theta}{2} = \frac{\sin 2\theta}{2} = \frac{1}{2}\sin 2\theta$.

Power-Reducing Identities

The double-angle identities can be used to derive the following power-reducing identities. These identities can be used to write trigonometric expressions involving even powers of sine, cosine, and tangent in terms of the first power of a cosine function.

Power-Reducing Identities

$$\sin^2 \alpha = \frac{1 - \cos 2\alpha}{2} \qquad \cos^2 \alpha = \frac{1 + \cos 2\alpha}{2} \qquad \tan^2 \alpha = \frac{1 - \cos 2\alpha}{1 + \cos 2\alpha}$$

The first power-reducing identity is derived by solving the double-angle identity $\cos 2\alpha = 1 - 2\sin^2\alpha$ for $\sin^2\alpha$. The second identity is derived by solving the double-angle identity $\cos 2\alpha = 2\cos^2\alpha - 1$ for $\cos^2\alpha$. The identity for $\tan^2\alpha$ can be derived by using the ratio identity, as shown below.

$$\tan^2 \alpha = \frac{\sin^2 \alpha}{\cos^2 \alpha} = \frac{\dfrac{1 - \cos 2\alpha}{2}}{\dfrac{1 + \cos 2\alpha}{2}} = \frac{1 - \cos 2\alpha}{1 + \cos 2\alpha}$$

EXAMPLE 4 ⟫ Use Power-Reducing Identities

Write $\sin^4\alpha$ in terms of the first power of one or more cosine functions.

Solution

$$\sin^4\alpha = (\sin^2\alpha)^2$$

$$= \left(\frac{1 - \cos 2\alpha}{2}\right)^2 \qquad \text{• Power-reducing identity}$$

$$= \frac{1}{4}(1 - 2\cos 2\alpha + \cos^2 2\alpha) \qquad \text{• Square.}$$

$$= \frac{1}{4}\left(1 - 2\cos 2\alpha + \frac{1 + \cos 4\alpha}{2}\right) \qquad \text{• Power-reducing identity}$$

$$= \frac{1}{4}\left(\frac{2 - 4\cos 2\alpha + 1 + \cos 4\alpha}{2}\right) \qquad \text{• Simplify.}$$

$$= \frac{1}{8}(3 - 4\cos 2\alpha + \cos 4\alpha)$$

⟫ *Try Exercise 22, page 243*

Half-Angle Identities

The following identities, called *half-angle identities,* can be derived from the power-reducing identities by replacing α with $\frac{\alpha}{2}$ and taking the square root of each side. Two additional identities are given for $\tan\frac{\alpha}{2}$.

Half-Angle Identities

$$\sin\frac{\alpha}{2} = \pm\sqrt{\frac{1-\cos\alpha}{2}}$$

$$\cos\frac{\alpha}{2} = \pm\sqrt{\frac{1+\cos\alpha}{2}}$$

$$\tan\frac{\alpha}{2} = \pm\sqrt{\frac{1-\cos\alpha}{1+\cos\alpha}} = \frac{\sin\alpha}{1+\cos\alpha} = \frac{1-\cos\alpha}{\sin\alpha}$$

The choice of the plus or minus sign depends on the quadrant in which $\frac{\alpha}{2}$ lies.

In Example 5, we use a half-angle identity to find the exact value of a trigonometric function.

EXAMPLE 5 Evaluate a Trigonometric Function

Find the exact value of $\cos 105°$.

Solution

Because $105° = \frac{1}{2}(210°)$, we can find $\cos 105°$ by using the half-angle identity for $\cos\frac{\alpha}{2}$ with $\alpha = 210°$. The angle $\frac{\alpha}{2} = 105°$ lies in Quadrant II and the cosine function is negative in Quadrant II. Thus $\cos 105° < 0$, and we must select the minus sign that precedes the radical in $\cos\frac{\alpha}{2} = \pm\sqrt{\frac{1+\cos\alpha}{2}}$ to produce the correct result.

$$\cos 105° = -\sqrt{\frac{1+\cos 210°}{2}}$$

• Use the formula for $\cos\frac{\alpha}{2}$ with $\alpha = 210°$ and a minus sign in front of the radical.

$$= -\sqrt{\frac{1+\left(-\frac{\sqrt{3}}{2}\right)}{2}}$$

• $\cos 210° = -\frac{\sqrt{3}}{2}$

$$= -\sqrt{\frac{\frac{2}{2}-\frac{\sqrt{3}}{2}}{2}}$$

• Simplify the numerator of the radicand.

$$= -\sqrt{\left(\frac{2-\sqrt{3}}{2}\right)\cdot\frac{1}{2}}$$ • **Definition of division**

$$= -\sqrt{\frac{2-\sqrt{3}}{4}}$$ • **Simplify.**

$$= -\frac{\sqrt{2-\sqrt{3}}}{2}$$

›› *Try Exercise 28, page 243*

In Example 6, we use given information concerning an angle α to find the exact values of the cosine and tangent of $\frac{\alpha}{2}$.

EXAMPLE 6 ›› Evaluate Trigonometric Functions

If $\sin\alpha = -\frac{3}{5}$ and $180° < \alpha < 270°$, find the exact value of

a. $\cos\frac{\alpha}{2}$ **b.** $\tan\frac{\alpha}{2}$

Solution

To apply the half-angle identities, we need to find $\cos\alpha$. We can use the identity $\cos^2\alpha = 1 - \sin^2\alpha$ to find $\cos\alpha$, but first we need to determine the sign of $\cos\alpha$. Because $180° < \alpha < 270°$, we know that $\cos\alpha < 0$. Thus

$$\cos^2\alpha = 1 - \sin^2\alpha = 1 - \left(-\frac{3}{5}\right)^2 = \frac{16}{25} \quad\text{and}\quad \cos\alpha = -\sqrt{\frac{16}{25}} = -\frac{4}{5}$$

Multiply each part of $180° < \alpha < 270°$ by $\frac{1}{2}$ to obtain $90° < \frac{\alpha}{2} < 135°$.

Therefore, $\frac{\alpha}{2}$ lies in Quadrant II. Thus $\cos\frac{\alpha}{2} < 0$ and $\tan\frac{\alpha}{2} < 0$.

a. In the following work, we use the half-angle identity for $\cos\frac{\alpha}{2}$ with a minus sign in front of the radical.

$$\cos\frac{\alpha}{2} = -\sqrt{\frac{1+\cos\alpha}{2}} = -\sqrt{\frac{1+\left(-\frac{4}{5}\right)}{2}} = -\sqrt{\frac{1}{5}\cdot\frac{1}{2}} = -\sqrt{\frac{1}{10}} = -\frac{\sqrt{10}}{10}$$

b. We could use the half-angle identity $\tan\frac{\alpha}{2} = \pm\sqrt{\frac{1-\cos\alpha}{1+\cos\alpha}}$; however, the identity $\tan\frac{\alpha}{2} = \frac{\sin\alpha}{1+\cos\alpha}$ is simpler and easier to evaluate.

$$\tan\frac{\alpha}{2} = \frac{\sin\alpha}{1+\cos\alpha} = \frac{-\frac{3}{5}}{1+\left(-\frac{4}{5}\right)} = \left(-\frac{3}{5}\right)\div\frac{1}{5} = \left(-\frac{3}{5}\right)\cdot 5 = -3$$

›› *Try Exercise 38, page 243*

The power-reducing identities and half-angle identities can be used to verify other identities.

EXAMPLE 7 » Verify an Identity

Verify the identity $2 \csc x \cos^2 \frac{x}{2} = \frac{\sin x}{1 - \cos x}$.

Solution

Work on the left side of the identity.

$$2 \csc x \cos^2 \frac{x}{2} = 2 \csc x \left(\frac{1 + \cos x}{2}\right)$$ • $\cos^2 \frac{x}{2} = \frac{1 + \cos x}{2}$

$$= \frac{1 + \cos x}{\sin x}$$ • $\csc x = \frac{1}{\sin x}$

$$= \frac{1 + \cos x}{\sin x} \cdot \frac{1 - \cos x}{1 - \cos x}$$ • **Multiply the numerator and denominator by the conjugate of the numerator.**

$$= \frac{1 - \cos^2 x}{\sin x(1 - \cos x)}$$

$$= \frac{\sin^2 x}{\sin x(1 - \cos x)}$$ • $1 - \cos^2 x = \sin^2 x$

$$= \frac{\sin x}{1 - \cos x}$$ • **Simplify.**

» *Try Exercise 68, page 244*

Topics for Discussion

1. True or false: If $\sin \alpha = \sin \beta$, then $\alpha = \beta$. Why?
2. Because

$$\tan \frac{\alpha}{2} = \frac{\sin \alpha}{1 + \cos \alpha} \quad \text{and} \quad \tan \frac{\alpha}{2} = \frac{1 - \cos \alpha}{\sin \alpha}$$

are both identities, it follows that

$$\frac{\sin \alpha}{1 + \cos \alpha} = \frac{1 - \cos \alpha}{\sin \alpha}$$

is also an identity. Do you agree? Explain.

3. Is $\sin 10x = 2 \sin 5x \cos 5x$ an identity? Explain.
4. Is $\sin \frac{\alpha}{2} = \cos \frac{\alpha}{2}$ an identity? Explain.

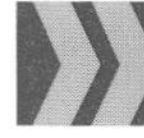

Exercise Set 3.3

In Exercises 1 to 8, write each trigonometric expression in terms of a single trigonometric function.

1. $2\sin 2\alpha \cos 2\alpha$
2. $2\sin 3\theta \cos 3\theta$
3. $1 - 2\sin^2 5\beta$
4. $2\cos^2 2\beta - 1$
5. $\cos^2 3\alpha - \sin^2 3\alpha$
6. $\cos^2 6\alpha - \sin^2 6\alpha$
7. $\dfrac{2\tan 3\alpha}{1 - \tan^2 3\alpha}$
8. $\dfrac{2\tan 4\theta}{1 - \tan^2 4\theta}$

In Exercises 9 to 18, find the exact values of sin 2α, cos 2α, and tan 2α given the following information.

9. $\cos\alpha = -\dfrac{4}{5}$ $\quad 90° < \alpha < 180°$
10. $\cos\alpha = \dfrac{24}{25}$ $\quad 270° < \alpha < 360°$
11. $\sin\alpha = \dfrac{8}{17}$ $\quad 90° < \alpha < 180°$
12. $\sin\alpha = -\dfrac{9}{41}$ $\quad 180° < \alpha < 270°$
13. $\tan\alpha = -\dfrac{24}{7}$ $\quad 270° < \alpha < 360°$
14. $\tan\alpha = \dfrac{4}{3}$ $\quad 0° < \alpha < 90°$
15. $\sin\alpha = \dfrac{15}{17}$ $\quad 0° < \alpha < 90°$
16. $\sin\alpha = -\dfrac{3}{5}$ $\quad 180° < \alpha < 270°$
17. $\cos\alpha = \dfrac{40}{41}$ $\quad 270° < \alpha < 360°$
18. $\cos\alpha = \dfrac{4}{5}$ $\quad 270° < \alpha < 360°$

In Exercises 19 to 24, use the power-reducing identities to write each trigonometric expression in terms of the first power of one or more cosine functions.

19. $6\cos^2 x$
20. $\sin^4 x \cos^4 x$
21. $\cos^4 x$
22. $\sin^2 x \cos^4 x$
23. $\sin^4 x \cos^2 x$
24. $\sin^6 x$

In Exercises 25 to 36, use the half-angle identities to find the exact value of each trigonometric expression.

25. $\sin 75°$
26. $\cos 105°$
27. $\tan 67.5°$
28. $\cos 165°$
29. $\cos 157.5°$
30. $\sin 112.5°$
31. $\sin 22.5°$
32. $\cos 67.5°$
33. $\sin\dfrac{7\pi}{8}$
34. $\cos\dfrac{5\pi}{8}$
35. $\cos\dfrac{5\pi}{12}$
36. $\sin\dfrac{3\pi}{8}$

In Exercises 37 to 44, find the exact values of the sine, cosine, and tangent of $\dfrac{\alpha}{2}$ given the following information.

37. $\sin\alpha = \dfrac{5}{13}$ $\quad 90° < \alpha < 180°$
38. $\sin\alpha = -\dfrac{7}{25}$ $\quad 180° < \alpha < 270°$
39. $\cos\alpha = -\dfrac{8}{17}$ $\quad 180° < \alpha < 270°$
40. $\cos\alpha = \dfrac{12}{13}$ $\quad 0° < \alpha < 90°$
41. $\tan\alpha = \dfrac{4}{3}$ $\quad 0° < \alpha < 90°$
42. $\tan\alpha = -\dfrac{8}{15}$ $\quad 90° < \alpha < 180°$
43. $\cos\alpha = \dfrac{24}{25}$ $\quad 270° < \alpha < 360°$
44. $\sin\alpha = -\dfrac{9}{41}$ $\quad 270° < \alpha < 360°$

In Exercises 45 to 90, verify the given identity.

45. $\sin 3x \cos 3x = \dfrac{1}{2}\sin 6x$
46. $\cos 8x = \cos^2 4x - \sin^2 4x$

47. $\sin^2 x + \cos 2x = \cos^2 x$

48. $\dfrac{\cos 2x}{\sin^2 x} = \cot^2 x - 1$

49. $\dfrac{1 + \cos 2x}{\sin 2x} = \cot x$

50. $\dfrac{1}{1 - \cos 2x} = \dfrac{1}{2}\csc^2 x$

51. $\dfrac{\sin 2x}{1 - \sin^2 x} = 2 \tan x$

52. $\dfrac{\cos^2 x - \sin^2 x}{2 \sin x \cos x} = \cot 2x$

53. $1 - \tan^2 x = \dfrac{\cos 2x}{\cos^2 x}$

54. $\tan 2x = \dfrac{2 \sin x \cos x}{\cos^2 x - \sin^2 x}$

55. $\sin 2x - \tan x = \tan x \cos 2x$

56. $\sin 2x - \cot x = -\cot x \cos 2x$

57. $\cos^4 x - \sin^4 x = \cos 2x$

58. $\sin 4x = 4 \sin x \cos^3 x - 4 \cos x \sin^3 x$

59. $\cos^2 x - 2 \sin^2 x \cos^2 x - \sin^2 x + 2 \sin^4 x = \cos^2 2x$

60. $2 \cos^4 x - \cos^2 x - 2 \sin^2 x \cos^2 x + \sin^2 x = \cos^2 2x$

61. $\cos 4x = 1 - 8 \cos^2 x + 8 \cos^4 x$

62. $\sin 4x = 4 \sin x \cos x - 8 \cos x \sin^3 x$

63. $\cos 3x - \cos x = 4 \cos^3 x - 4 \cos x$

64. $\sin 3x + \sin x = 4 \sin x - 4 \sin^3 x$

65. $\sin^3 x + \cos^3 x = (\sin x + \cos x)\left(1 - \dfrac{1}{2}\sin 2x\right)$

66. $\cos^3 x - \sin^3 x = (\cos x - \sin x)\left(1 + \dfrac{1}{2}\sin 2x\right)$

67. $\sin^2 \dfrac{x}{2} = \dfrac{\sec x - 1}{2 \sec x}$

68. $\cos^2 \dfrac{x}{2} = \dfrac{\sec x + 1}{2 \sec x}$

69. $\tan \dfrac{x}{2} = \csc x - \cot x$

70. $\tan \dfrac{x}{2} = \dfrac{\tan x}{\sec x + 1}$

71. $2 \sin \dfrac{x}{2} \cos \dfrac{x}{2} = \sin x$

72. $\cos^2 \dfrac{x}{2} - \sin^2 \dfrac{x}{2} = \cos x$

73. $\left(\cos \dfrac{x}{2} + \sin \dfrac{x}{2}\right)^2 = 1 + \sin x$

74. $\tan^2 \dfrac{x}{2} = \dfrac{\sec x - 1}{\sec x + 1}$

75. $\sin^2 \dfrac{x}{2} \sec x = \dfrac{1}{2}(\sec x - 1)$

76. $\cos^2 \dfrac{x}{2} \sec x = \dfrac{1}{2}(\sec x + 1)$

77. $\cos^2 \dfrac{x}{2} - \cos x = \sin^2 \dfrac{x}{2}$

78. $\sin^2 \dfrac{x}{2} + \cos x = \cos^2 \dfrac{x}{2}$

79. $\sin^2 \dfrac{x}{2} - \cos^2 \dfrac{x}{2} = -\cos x$

80. $\cos^2 \dfrac{x}{2} - \sin^2 \dfrac{x}{2} = \dfrac{1}{2} \csc x \sin 2x$

81. $\sin 2x - \cos x = (\cos x)(2 \sin x - 1)$

82. $\dfrac{\cos 2x}{\sin^2 x} = \csc^2 x - 2$

83. $\tan 2x = \dfrac{2}{\cot x - \tan x}$

84. $\dfrac{2 \cos 2x}{\sin 2x} = \cot x - \tan x$

85. $2 \tan \dfrac{x}{2} = \dfrac{\sin^2 x + 1 - \cos^2 x}{(\sin x)(1 + \cos x)}$

86. $\dfrac{1}{2} \csc^2 \dfrac{x}{2} = \csc^2 x + \cot x \csc x$

87. $\csc 2x = \dfrac{1}{2} \csc x \sec x$

88. $\sec 2x = \dfrac{\sec^2 x}{2 - \sec^2 x}$

89. $\cos \dfrac{x}{5} = 1 - 2 \sin^2 \dfrac{x}{10}$

90. $\sec^2 \dfrac{x}{2} = \dfrac{2}{1 + \cos x}$

91. MACH NUMBERS Ernst Mach (1838–1916) was an Austrian physicist who made a study of the motion of objects at high speeds. Today we often state the speed of aircraft in terms of a *Mach number*. A **Mach number** is the speed of an object divided by the speed of sound. For example, a plane flying at the speed of sound is said to have a speed M of Mach 1. Mach 2 is twice the speed of sound. An airplane that travels faster than the speed of sound creates a sonic boom. This sonic boom emanates from the airplane in the shape of a cone.

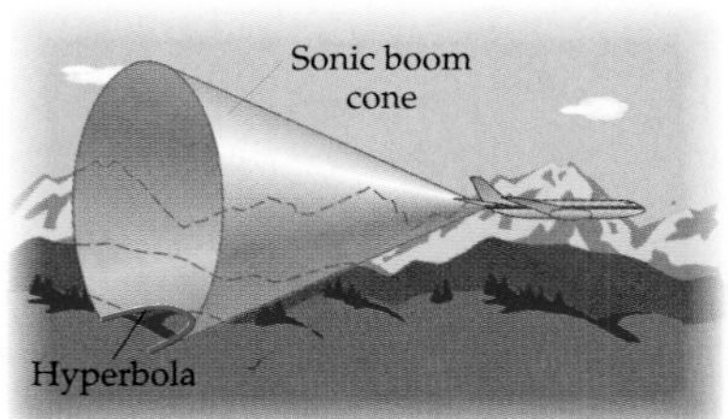

The following equation shows the relationship between the measure of the cone's vertex angle α and the Mach speed M of an aircraft that is flying faster than the speed of sound.

$$M \sin \frac{\alpha}{2} = 1$$

a. If $\alpha = \frac{\pi}{4}$, determine the Mach speed M of the airplane. State your answer as an *exact* value and as a decimal accurate to the nearest hundredth.

b. Solve $M \sin \frac{\alpha}{2} = 1$ for α.

c. Does the vertex angle α increase or decrease as the Mach number M increases?

»»» Connecting Concepts

In Exercises 92 to 95, compare the graphs of each side of the equation to predict whether the equation is an identity.

92. $\sin^2 x + \cos 2x = \cos^2 x$

93. $\frac{\sin 2x}{1 - \sin^2 x} = 2 \tan x$

94. $\sin \frac{x}{2} \cos \frac{x}{2} = \sin x$

95. $\left(\cos \frac{x}{2} + \sin \frac{x}{2}\right)^2 = 1 + \sin x$

In Exercises 96 to 98, verify the identity.

96. $\frac{\sin^3 x + \cos^3 x}{\sin x + \cos x} = 1 - \frac{1}{2} \sin 2x$

97. $\cos^4 x = \frac{1}{8} \cos 4x + \frac{1}{2} \cos 2x + \frac{3}{8}$

98. $\frac{\sin x - \sin 2x}{\cos x + \cos 2x} = -\tan \frac{x}{2}$

»»» Projects

1.  VISUAL INSIGHT Explain how the figure at the right can be used to verify the half-angle identity

$$\tan \frac{\theta}{2} = \frac{\sin \theta}{1 + \cos \theta}$$

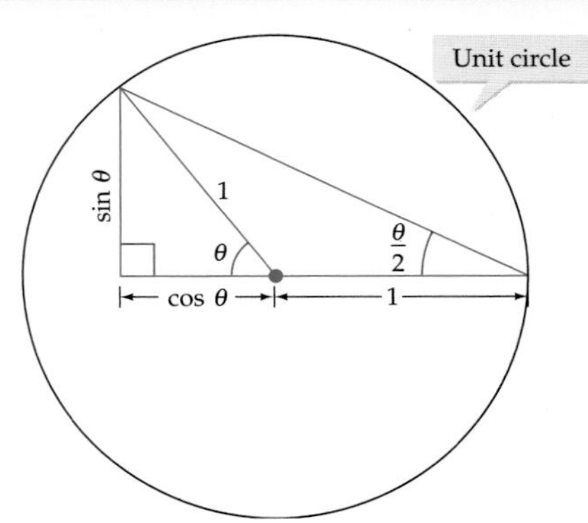

Section 3.4

- The Product-to-Sum Identities
- The Sum-to-Product Identities
- Functions of the Form $f(x) = a \sin x + b \cos x$

Identities Involving the Sum of Trigonometric Functions

PREPARE FOR THIS SECTION

Prepare for this section by completing the following exercises. The answers can be found on page A17.

PS1. Use sum and difference identities to rewrite $\frac{1}{2}[\sin(\alpha + \beta) + \sin(\alpha - \beta)]$. [3.2]

PS2. Use sum and difference identities to rewrite $\frac{1}{2}[\cos(\alpha + \beta) + \cos(\alpha - \beta)]$. [3.2]

PS3. Compare $\sin x - \sin y$ and $2 \cos \frac{x + y}{2} \sin \frac{x - y}{2}$ for $x = \pi$ and $y = \frac{\pi}{6}$. [2.4]

PS4. Use a sum identity to rewrite $\sqrt{2} \sin\left(x + \frac{\pi}{4}\right)$. [3.2]

PS5. Find a real number x and a real number y to verify that $\sin x - \sin y = \sin(x - y)$ is *not* an identity. [2.4]

PS6. Evaluate $\sqrt{a^2 + b^2}$ for $a = -1$ and $b = \sqrt{3}$.

The Product-to-Sum Identities

Some applications require that a product of trigonometric functions be written as a sum or difference of these functions. Other applications require that the sum or difference of trigonometric functions be represented as a product of these functions. The *product-to-sum identities* are particularly useful in these types of problems.

Product-to-Sum Identities

$$\sin \alpha \cos \beta = \frac{1}{2}[\sin(\alpha + \beta) + \sin(\alpha - \beta)]$$

$$\cos \alpha \sin \beta = \frac{1}{2}[\sin(\alpha + \beta) - \sin(\alpha - \beta)]$$

$$\cos \alpha \cos \beta = \frac{1}{2}[\cos(\alpha + \beta) + \cos(\alpha - \beta)]$$

$$\sin \alpha \sin \beta = \frac{1}{2}[\cos(\alpha - \beta) - \cos(\alpha + \beta)]$$

The product-to-sum identities can be derived by using the sum or difference identities. Adding the identities for $\sin(\alpha + \beta)$ and $\sin(\alpha - \beta)$, we have

$$\begin{aligned} \sin(\alpha + \beta) &= \sin\alpha\cos\beta + \cos\alpha\sin\beta \\ \underline{\sin(\alpha - \beta)} &= \underline{\sin\alpha\cos\beta - \cos\alpha\sin\beta} \\ \sin(\alpha + \beta) + \sin(\alpha - \beta) &= 2\sin\alpha\cos\beta \end{aligned}$$

• **Add the identities.**

Solving for $\sin\alpha\cos\beta$, we obtain the first product-to-sum identity:

$$\sin\alpha\cos\beta = \frac{1}{2}[\sin(\alpha + \beta) + \sin(\alpha - \beta)]$$

The identity for $\cos\alpha\sin\beta$ is obtained when $\sin(\alpha - \beta)$ is subtracted from $\sin(\alpha + \beta)$. The result is

$$\cos\alpha\sin\beta = \frac{1}{2}[\sin(\alpha + \beta) - \sin(\alpha - \beta)]$$

In like manner, the identities for $\cos(\alpha + \beta)$ and $\cos(\alpha - \beta)$ are used to derive the identities for $\cos\alpha\cos\beta$ and $\sin\alpha\sin\beta$.

The product-to-sum identities can be used to verify some identities.

EXAMPLE 1 ›› Verify an Identity

Verify the identity $\cos 2x \sin 5x = \frac{1}{2}(\sin 7x + \sin 3x)$.

Solution

$$\begin{aligned} \cos 2x \sin 5x &= \frac{1}{2}[\sin(2x + 5x) - \sin(2x - 5x)] \\ &= \frac{1}{2}[\sin 7x - \sin(-3x)] \\ &= \frac{1}{2}(\sin 7x + \sin 3x) \end{aligned}$$

• **Use the product-to-sum identity for $\cos\alpha\sin\beta$.**

• **$\sin(-3x) = -\sin 3x$**

›› *Try Exercise 36, page 252*

The Sum-to-Product Identities

The *sum-to-product identities* can be derived from the product-to-sum identities.

Sum-to-Product Identities

$$\sin x + \sin y = 2 \sin \frac{x + y}{2} \cos \frac{x - y}{2}$$

$$\cos x + \cos y = 2 \cos \frac{x + y}{2} \cos \frac{x - y}{2}$$

$$\sin x - \sin y = 2 \cos \frac{x + y}{2} \sin \frac{x - y}{2}$$

$$\cos x - \cos y = -2 \sin \frac{x + y}{2} \sin \frac{x - y}{2}$$

To derive the sum-to-product identity for $\sin x + \sin y$, we first let $x = \alpha + \beta$ and $y = \alpha - \beta$. Then

$$x + y = \alpha + \beta + \alpha - \beta \quad \text{and} \quad x - y = \alpha + \beta - (\alpha - \beta)$$

$$x + y = 2\alpha \qquad\qquad x - y = 2\beta$$

$$\alpha = \frac{x + y}{2} \qquad\qquad \beta = \frac{x - y}{2}$$

Substituting these expressions for α and β into the product-to-sum identity

$$\frac{1}{2}[\sin(\alpha + \beta) + \sin(\alpha - \beta)] = \sin \alpha \cos \beta$$

yields

$$\sin\left(\frac{x + y}{2} + \frac{x - y}{2}\right) + \sin\left(\frac{x + y}{2} - \frac{x - y}{2}\right) = 2 \sin \frac{x + y}{2} \cos \frac{x - y}{2}$$

Simplifying the left side, we have the sum-to-product identity.

$$\sin x + \sin y = 2 \sin \frac{x + y}{2} \cos \frac{x - y}{2}$$

The other three sum-to-product identities can be derived in a similar manner. The proofs of these identities are left as exercises.

EXAMPLE 2 » Write the Difference of Trigonometric Expressions as a Product

Write $\sin 4\theta - \sin \theta$ as the product of two functions.

Solution

$$\sin 4\theta - \sin \theta = 2 \cos \frac{4\theta + \theta}{2} \sin \frac{4\theta - \theta}{2} = 2 \cos \frac{5\theta}{2} \sin \frac{3\theta}{2}$$

» *Try Exercise 22, page 251*

? QUESTION Does $\cos 4\theta + \cos 2\theta = 2\cos 3\theta \cos\theta$?

EXAMPLE 3 ⟫ Verify a Sum-to-Product Identity

Verify the identity $\dfrac{\sin 6x + \sin 2x}{\sin 6x - \sin 2x} = \tan 4x \cot 2x$.

Solution

$$\frac{\sin 6x + \sin 2x}{\sin 6x - \sin 2x} = \frac{2\sin\dfrac{6x+2x}{2}\cos\dfrac{6x-2x}{2}}{2\cos\dfrac{6x+2x}{2}\sin\dfrac{6x-2x}{2}}$$

$$= \frac{\sin 4x \cos 2x}{\cos 4x \sin 2x}$$

$$= \tan 4x \cot 2x$$

⟫ *Try Exercise 44, page 252*

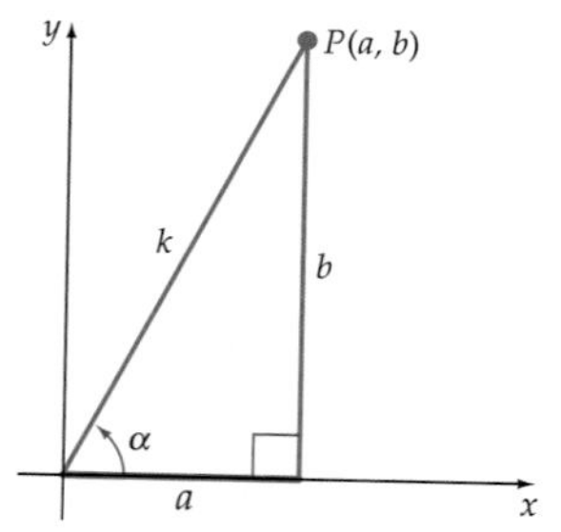

Figure 3.6

Functions of the Form $f(x) = a \sin x + b \cos x$

The function given by the equation $f(x) = a\sin x + b\cos x$ can be written in the form $f(x) = k\sin(x + \alpha)$. This form of the function is useful in graphing and engineering applications because the amplitude, period, and phase shift can be readily calculated.

Let $P(a, b)$ be a point on a coordinate plane, and let α represent an angle in standard position whose terminal side contains P. See **Figure 3.6.** To rewrite $y = a\sin x + b\cos x$, multiply and divide the expression $a\sin x + b\cos x$ by $\sqrt{a^2 + b^2}$.

$$a\sin x + b\cos x = \frac{\sqrt{a^2+b^2}}{\sqrt{a^2+b^2}}(a\sin x + b\cos x)$$

$$= \sqrt{a^2+b^2}\left(\frac{a}{\sqrt{a^2+b^2}}\sin x + \frac{b}{\sqrt{a^2+b^2}}\cos x\right) \qquad (1)$$

From the definition of the sine and cosine of an angle in standard position, let

$$k = \sqrt{a^2+b^2}, \quad \cos\alpha = \frac{a}{\sqrt{a^2+b^2}}, \quad \text{and} \quad \sin\alpha = \frac{b}{\sqrt{a^2+b^2}}$$

Substituting these expressions into Equation (1) yields

$$a\sin x + b\cos x = k(\cos\alpha \sin x + \sin\alpha \cos x)$$

Now, using the identity for the sine of the sum of two angles, we have

$$a\sin x + b\cos x = k\sin(x + \alpha)$$

Thus $a\sin x + b\cos x = k\sin(x+\alpha)$, where $k = \sqrt{a^2+b^2}$ and α is the angle for which $\sin\alpha = \dfrac{b}{\sqrt{a^2+b^2}}$ and $\cos\alpha = \dfrac{a}{\sqrt{a^2+b^2}}$.

? ANSWER Yes. $\cos 4\theta + \cos 2\theta = 2\cos\left(\dfrac{4\theta + 2\theta}{2}\right)\cos\left(\dfrac{4\theta - 2\theta}{2}\right) = 2\cos 3\theta \cos\theta$.

Functions of the Form $a \sin x + b \cos x$

$$a \sin x + b \cos x = k \sin(x + \alpha)$$

where $k = \sqrt{a^2 + b^2}$, $\sin \alpha = \dfrac{b}{\sqrt{a^2 + b^2}}$, and $\cos \alpha = \dfrac{a}{\sqrt{a^2 + b^2}}$

EXAMPLE 4 ⟫ Rewrite $a \sin x + b \cos x$

Rewrite $\sin x + \cos x$ in the form $k \sin(x + \alpha)$.

Solution

Comparing $\sin x + \cos x$ to $a \sin x + b \cos x$, $a = 1$ and $b = 1$. Thus $k = \sqrt{1^2 + 1^2} = \sqrt{2}$, $\sin \alpha = \dfrac{1}{\sqrt{2}}$, and $\cos \alpha = \dfrac{1}{\sqrt{2}}$. Thus $\alpha = \dfrac{\pi}{4}$.

$$\sin x + \cos x = k \sin(x + \alpha) = \sqrt{2} \sin\left(x + \frac{\pi}{4}\right)$$

⟫ *Try Exercise 62, page 252*

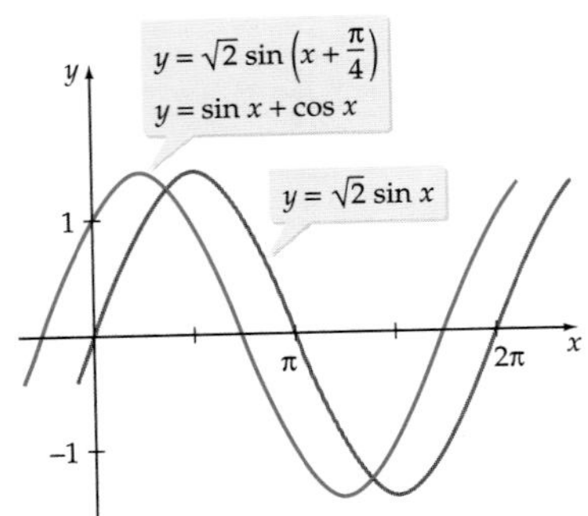

Figure 3.7

The graphs of $y = \sin x + \cos x$ and $y = \sqrt{2} \sin\left(x + \dfrac{\pi}{4}\right)$ are both the graph of $y = \sqrt{2} \sin x$ shifted $\dfrac{\pi}{4}$ units to the left. See **Figure 3.7**.

EXAMPLE 5 ⟫ Graph a Function of the Form $f(x) = a \sin x + b \cos x$

Graph: $f(x) = -\sin x + \sqrt{3} \cos x$

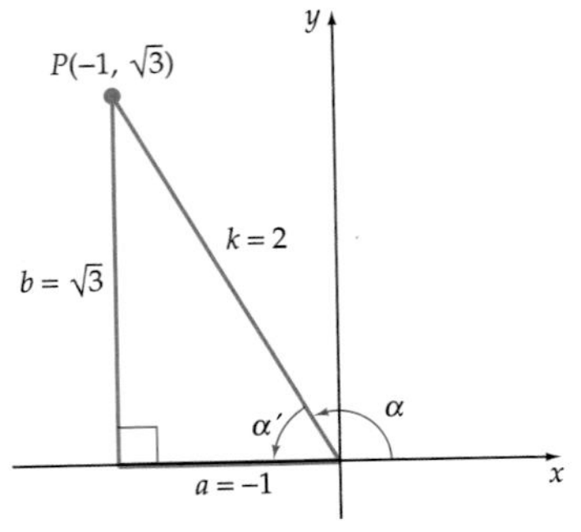

Figure 3.8

Solution

First, we write $f(x)$ as $k \sin(x + \alpha)$. Let $a = -1$ and $b = \sqrt{3}$; then $k = \sqrt{(-1)^2 + (\sqrt{3})^2} = 2$. The amplitude is 2. The point $P(-1, \sqrt{3})$ is in the second quadrant (see **Figure 3.8**). Let α be an angle in standard position with P on its terminal side. Let α' be the reference angle for α. Then

$$\sin \alpha' = \frac{\sqrt{3}}{2}$$

$$\alpha' = \frac{\pi}{3}$$

$$\alpha = \pi - \alpha' = \pi - \frac{\pi}{3} = \frac{2\pi}{3}$$

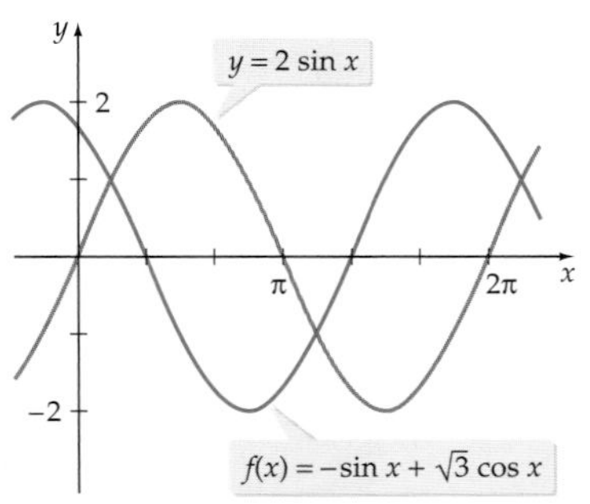

Figure 3.9

Substituting 2 for k and $\frac{2\pi}{3}$ for α in $y = k\sin(x + \alpha)$, we have

$$y = 2\sin\left(x + \frac{2\pi}{3}\right)$$

The phase shift is $-\frac{2\pi}{3}$. The graph of $f(x) = -\sin x + \sqrt{3}\cos x$ is the graph of $y = 2\sin x$ shifted $\frac{2\pi}{3}$ units to the left. See **Figure 3.9.**

Try Exercise 70, page 252

Topics for Discussion

1. The *exact* value of $\sin 75° \cos 15°$ is $\frac{2 + \sqrt{3}}{4}$. Do you agree? Explain.
2. Do you agree with the following work? Explain.

$$\cos 195° + \cos 105° = \cos(195° + 105°) = \cos 300° = -\frac{1}{2}$$

3. The graphs of $y_1 = \sin x + \cos x$ and $y_2 = \sqrt{2}\sin\left(x + \frac{\pi}{4}\right)$ are identical. Do you agree? Explain.
4. Explain how to find the amplitude of the graph of $y = a\sin x + b\cos x$.

Exercise Set 3.4

In Exercises 1 to 8, write each expression as the sum or difference of two functions.

1. $2\sin x \cos 2x$
2. $2\sin 4x \sin 2x$
3. $\cos 6x \sin 2x$
4. $\cos 3x \cos 5x$
5. $2\sin 5x \cos 3x$
6. $2\sin 2x \cos 6x$
7. $\sin x \sin 5x$
8. $\cos 3x \sin x$

In Exercises 9 to 16, find the exact value of each expression. Do not use a calculator.

9. $\cos 75° \cos 15°$
10. $\sin 105° \cos 15°$
11. $\cos 157.5° \sin 22.5°$
12. $\sin 195° \cos 15°$
13. $\sin\frac{13\pi}{12}\cos\frac{\pi}{12}$
14. $\sin\frac{11\pi}{12}\sin\frac{7\pi}{12}$
15. $\sin\frac{\pi}{12}\cos\frac{7\pi}{12}$
16. $\cos\frac{17\pi}{12}\sin\frac{7\pi}{12}$

In Exercises 17 to 32, write each expression as the product of two functions.

17. $\sin 4\theta + \sin 2\theta$
18. $\cos 5\theta - \cos 3\theta$
19. $\cos 3\theta + \cos \theta$
20. $\sin 7\theta - \sin 3\theta$
21. $\cos 6\theta - \cos 2\theta$
22. $\cos 3\theta + \cos 5\theta$
23. $\cos \theta + \cos 7\theta$
24. $\sin 3\theta + \sin 7\theta$

25. $\sin 5\theta + \sin 9\theta$

26. $\cos 5\theta - \cos \theta$

27. $\cos 2\theta - \cos \theta$

28. $\sin 2\theta + \sin 6\theta$

29. $\cos \frac{\theta}{2} - \cos \theta$

30. $\sin \frac{3\theta}{4} + \sin \frac{\theta}{2}$

31. $\sin \frac{\theta}{2} - \sin \frac{\theta}{3}$

32. $\cos \theta + \cos \frac{\theta}{2}$

In Exercises 33 to 48, verify the identity.

33. $2 \cos \alpha \cos \beta = \cos(\alpha + \beta) + \cos(\alpha - \beta)$

34. $2 \sin \alpha \sin \beta = \cos(\alpha - \beta) - \cos(\alpha + \beta)$

35. $2 \cos 3x \sin x = 2 \sin x \cos x - 8 \cos x \sin^3 x$

36. $\sin 5x \cos 3x = \sin 4x \cos 4x + \sin x \cos x$

37. $2 \cos 5x \cos 7x = \cos^2 6x - \sin^2 6x + 2 \cos^2 x - 1$

38. $\sin 3x \cos x = \sin x \cos x(3 - 4 \sin^2 x)$

39. $\sin 3x - \sin x = 2 \sin x - 4 \sin^3 x$

40. $\cos 5x - \cos 3x = -8 \sin^2 x(2 \cos^3 x - \cos x)$

41. $\sin 2x + \sin 4x = 2 \sin x \cos x(4 \cos^2 x - 1)$

42. $\cos 3x + \cos x = 4 \cos^3 x - 2 \cos x$

43. $\frac{\sin 3x - \sin x}{\cos 3x - \cos x} = -\cot 2x$

44. $\frac{\cos 5x - \cos 3x}{\sin 5x + \sin 3x} = -\tan x$

45. $\frac{\sin 5x + \sin 3x}{4 \sin x \cos^3 x - 4 \sin^3 x \cos x} = 2 \cos x$

46. $\frac{\cos 4x - \cos 2x}{\sin 2x - \sin 4x} = \tan 3x$

47. $\sin(x + y) \cos(x - y) = \sin x \cos x + \sin y \cos y$

48. $\sin(x + y) \sin(x - y) = \sin^2 x - \sin^2 y$

In Exercises 49 to 58, write the given equation in the form $y = k \sin (x + \alpha)$, where the measure of α is in degrees.

49. $y = -\sin x - \cos x$

50. $y = \sqrt{3} \sin x - \cos x$

51. $y = \frac{1}{2} \sin x - \frac{\sqrt{3}}{2} \cos x$

52. $y = \frac{\sqrt{3}}{2} \sin x - \frac{1}{2} \cos x$

53. $y = \frac{1}{2} \sin x - \frac{1}{2} \cos x$

54. $y = -\frac{\sqrt{3}}{2} \sin x - \frac{1}{2} \cos x$

55. $y = -3 \sin x + 3 \cos x$

56. $y = \frac{\sqrt{2}}{2} \sin x + \frac{\sqrt{2}}{2} \cos x$

57. $y = \pi \sin x - \pi \cos x$

58. $y = -0.4 \sin x + 0.4 \cos x$

In Exercises 59 to 66, write the given equation in the form $y = k \sin (x + \alpha)$, where the measure of α is in radians.

59. $y = -\sin x + \cos x$

60. $y = -\sqrt{3} \sin x - \cos x$

61. $y = \frac{\sqrt{3}}{2} \sin x + \frac{1}{2} \cos x$

62. $y = \sin x + \sqrt{3} \cos x$

63. $y = -10 \sin x + 10\sqrt{3} \cos x$

64. $y = 3 \sin x - 3\sqrt{3} \cos x$

65. $y = -5 \sin x + 5 \cos x$

66. $y = 3 \sin x - 3 \cos x$

In Exercises 67 to 76, graph one cycle of the function. Do not use a graphing calculator.

67. $y = -\sin x - \sqrt{3} \cos x$

68. $y = -\sqrt{3} \sin x + \cos x$

69. $y = 2 \sin x + 2 \cos x$

70. $y = \sin x + \sqrt{3} \cos x$

71. $y = -\sqrt{3} \sin x - \cos x$

72. $y = -\sin x + \cos x$

73. $y = -5 \sin x + 5\sqrt{3} \cos x$

74. $y = -\sqrt{2} \sin x + \sqrt{2} \cos x$

75. $y = 6\sqrt{3} \sin x - 6 \cos x$

76. $y = 5\sqrt{2} \sin x - 5\sqrt{2} \cos x$

TONES ON A TOUCH-TONE PHONE **In Exercises 77 and 78, use the following information about touch-tone phones. Every tone made on a touch-tone phone is produced by adding a pair of sounds. The following chart shows the sound fre-**

quencies used for each key on the telephone keypad. For example, the sound emitted by pressing 3 on the keypad is produced by adding a 1477-hertz (cycles per second) sound to a 697-hertz sound. An equation that models this tone is

$$p(t) = \sin(2\pi \cdot 1477t) + \sin(2\pi \cdot 697t)$$

where *p* is the pressure on the eardrum and *t* is the time in seconds.

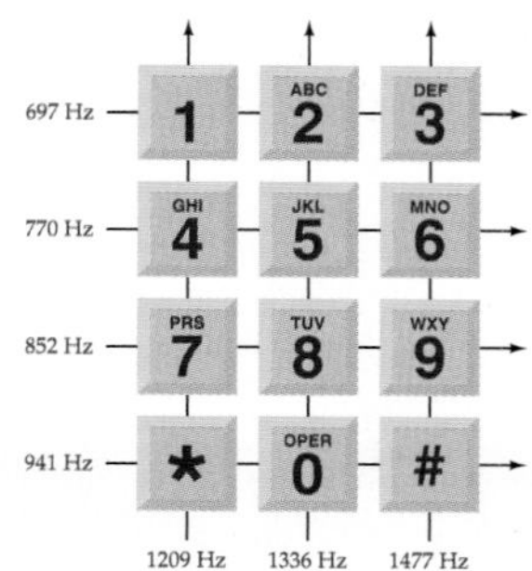

Source: Data in chart from http://www.howstuffworks.com/telephone2.htm

77. a. Write an equation of the form

$$p(t) = \sin(2\pi f_1 t) + \sin(2\pi f_2 t)$$

that models the tone produced by pressing the 5 key on a touch-tone phone.

b. Use a sum-to-product identity to write your equation from **a.** in the form

$$p(t) = A\sin(B\pi t)\sin(C\pi t)$$

c. When a sound of frequency f_1 is combined with a sound of frequency f_2, the combined sound has a frequency of $\frac{f_1 + f_2}{2}$. What is the frequency of the tone produced when the 5 key is pressed?

78. a. Write an equation of the form

$$p(t) = \sin(2\pi f_1 t) + \sin(2\pi f_2 t)$$

that models the tone produced by pressing the 8 key on a touch-tone phone.

b. Use a sum-to-product identity to write your equation from **a.** in the form

$$p(t) = A\sin(B\pi t)\sin(C\pi t)$$

c. What is the frequency of the tone produced when the 8 key is pressed? (*Hint:* See **c.** of Exercise 77.)

In Exercises 79 to 84, compare the graphs of each side of the equation to predict whether the equation is an identity.

79. $\sin 3x - \sin x = 2\sin x - 4\sin^3 x$

80. $\dfrac{\sin 3x - \sin x}{\cos 3x - \cos x} = -\dfrac{1}{\tan 2x}$

81. $-\sqrt{3}\sin x - \cos x = 2\sin\left(x - \dfrac{5\pi}{6}\right)$

82. $-\sqrt{3}\sin x + \cos x = 2\sin\left(x + \dfrac{5\pi}{6}\right)$

83. $\dfrac{1}{2}\sin x - \dfrac{\sqrt{3}}{2}\cos x = \sin\left(x - \dfrac{\pi}{3}\right)$

84. $\dfrac{\sqrt{3}}{2}\sin x + \dfrac{1}{2}\cos x = \sin\left(x + \dfrac{\pi}{6}\right)$

»»» Connecting Concepts »»»

85. Derive the sum-to-product identity

$$\cos x + \cos y = 2\cos\frac{x+y}{2}\cos\frac{x-y}{2}$$

86. Derive the product-to-sum identity

$$\sin x \sin y = \frac{1}{2}[\cos(x - y) - \cos(x + y)]$$

87. If $x + y = 180°$, show that $\sin x + \sin y = 2\sin x$.

88. If $x + y = 360°$, show that $\cos x + \cos y = 2\cos x$.

In Exercises 89 to 94, verify the identity.

89. $\sin 2x + \sin 4x + \sin 6x = 4\sin 3x\cos 2x\cos x$

90. $\sin 4x - \sin 2x + \sin 6x = 4\cos 3x\sin 2x\cos x$

91. $\dfrac{\cos 10x + \cos 8x}{\sin 10x - \sin 8x} = \cot x$

92. $\dfrac{\sin 10x + \sin 2x}{\cos 10x + \cos 2x} = \dfrac{2\tan 3x}{1 - \tan^2 3x}$

93. $\dfrac{\sin 2x + \sin 4x + \sin 6x}{\cos 2x + \cos 4x + \cos 6x} = \tan 4x$

94. $\dfrac{\sin 2x + \sin 6x}{\cos 6x - \cos 2x} = -\cot 2x$

95. Verify that $\cos^2 x - \sin^2 x = \cos 2x$ by using a product-to-sum identity.

96. Verify that $2 \sin x \cos x = \sin 2x$ by using a product-to-sum identity.

97. Verify that $a \sin x + b \cos x = k \cos(x - \alpha)$, where $k = \sqrt{a^2 + b^2}$ and $\tan \alpha = \dfrac{a}{b}$.

98. Verify that $a \sin cx + b \cos cx = k \sin(cx + \alpha)$, where $k = \sqrt{a^2 + b^2}$ and $\tan \alpha = \dfrac{b}{a}$.

»»» Projects »»

1. **BEATS** If two tuning forks that are close in frequency are struck at the same time, the sound we hear fluctuates between a loud tone and silence. These regular fluctuations are called **beats**. The loud periods occur when the sound waves reinforce (interfere constructively with) one another, and the silent periods occur when the waves interfere destructively with one another. If the frequencies of the tuning forks are 442 cycles per second and 440 cycles per second then the pressure produced on our eardrums by each tuning fork, respectively, is given by

$$\text{Y1} = \sin(2\pi \cdot 442x) \quad \text{and} \quad \text{Y2} = \sin(2\pi \cdot 440x)$$

where x is the time in seconds. The combined pressure produced when both tuning forks are struck simultaneously is modeled by

$$\text{Y3} = \text{Y1} + \text{Y2}$$

a. Graph Y3. Use a viewing window with Xmin = 0, Xmax = 1, Ymin = −2.1, Ymax = 2.1. To produce a graph that better illustrates the beats produced when both tuning forks are struck simultaneously, graph Y3, $\text{Y4} = 2\cos(2\pi x)$, and $\text{Y5} = -2\cos(2\pi x)$ together in the same window. What is the relationship between the graph of Y3 and the graphs of Y4 and Y5?

b. Use a sum-to-product identity to write Y3 as a product.

c. The rate of the beats produced by two sounds with the same intensity is the absolute value of the difference between their frequencies. Consider two tuning forks that are struck at the same time with the same force and held on a sounding board. The tuning forks have frequencies of 564 and 568 cycles per second, respectively. How many beats will be heard each second?

d. A piano tuner strikes a tuning fork and a key on a piano that is supposed to have the same frequency as the tuning fork. The piano tuner notices that the sound produced by the piano is lower than that produced by the tuning fork. The piano tuner also notes that the combined sound of the piano and the tuning fork has 2 beats per second. How much lower is the frequency of the piano than the frequency of the tuning fork?

Section 3.5

- Inverse Trigonometric Functions
- Composition of Trigonometric Functions and Their Inverses
- Graphs of Inverse Trigonometric Functions

Inverse Trigonometric Functions

PREPARE FOR THIS SECTION

Prepare for this section by completing the following exercises. The answers can be found on page A17.

PS1. What is a one-to-one function? [1.3]

PS2. State the horizontal line test. [1.3]

PS3. Find $f[g(x)]$ given that $f(x) = 2x + 4$ and $g(x) = \frac{1}{2}x - 2$. [1.5]

PS4. If f and f^{-1} are inverse functions, then determine $f[f^{-1}(x)]$ for any x in the domain of f^{-1}. [1.6]

PS5. If f and f^{-1} are inverse functions, then explain how the graph of f^{-1} is related to the graph of f. [1.6]

PS6. Use the horizontal line test to determine whether the graph of $y = \sin x$, where x is any real number, is a one-to-one function. [1.3/2.5]

Inverse Trigonometric Functions

Because the graph of $y = \sin x$ fails the horizontal line test, it is not the graph of a one-to-one function. Therefore, it does not have an inverse function. **Figure 3.10** shows the graph of $y = \sin x$ on the interval $-2\pi \le x \le 2\pi$ and the graph of the inverse relation $x = \sin y$. Note that the graph of $x = \sin y$ does not satisfy the vertical line test and therefore is not the graph of a function.

If the domain of the function $y = \sin x$ is restricted to $-\frac{\pi}{2} \le x \le \frac{\pi}{2}$, the graph of $y = \sin x$ satisfies the horizontal line test and therefore the function has an inverse function. The graphs of $y = \sin x$ for $-\frac{\pi}{2} \le x \le \frac{\pi}{2}$ and its inverse are shown in **Figure 3.11** on the next page.

Inverse Functions
See section 1.6.

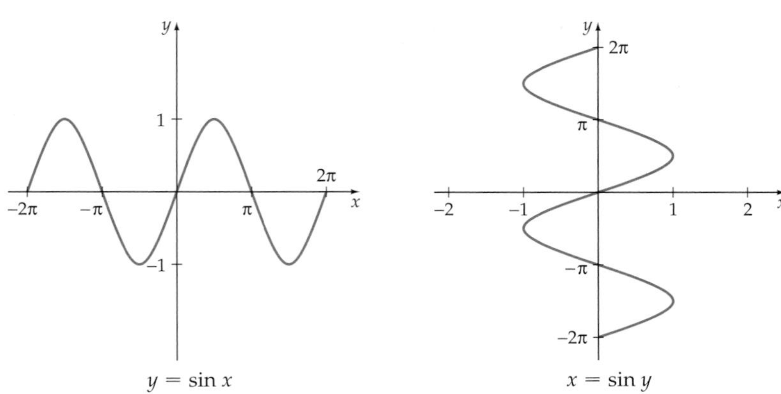

Figure 3.10

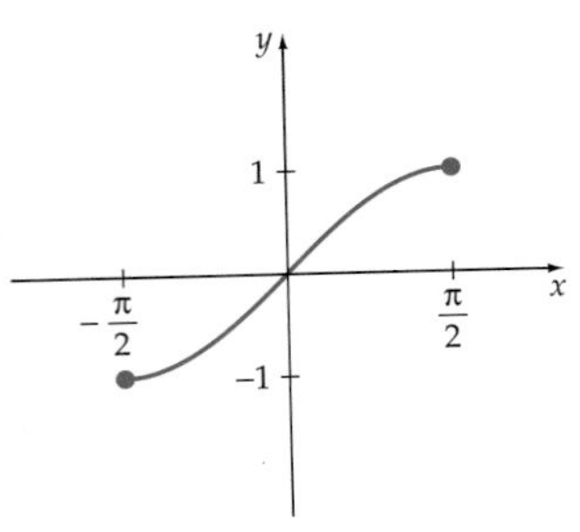

$y = \sin x$: $-\frac{\pi}{2} \le x \le \frac{\pi}{2}$

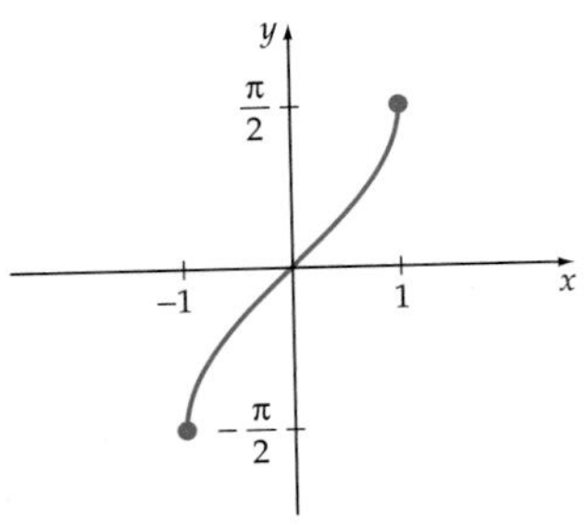

$y = \sin^{-1} x$: $-1 \le x \le 1$

Figure 3.11

> **take note**
>
> The −1 in $\sin^{-1} x$ is not an exponent. The −1 is used to denote the inverse function. To use −1 as an exponent for a sine function, enclose the function in parentheses.
>
> $$(\sin x)^{-1} = \frac{1}{\sin x} = \csc x$$
>
> $$\sin^{-1} x \ne \frac{1}{\sin x}$$

To find the inverse of the function defined by $y = \sin x$, with $-\frac{\pi}{2} \le x \le \frac{\pi}{2}$, interchange x and y. Then solve for y.

$y = \sin x$ • $-\frac{\pi}{2} \le x \le \frac{\pi}{2}$

$x = \sin y$ • **Interchange *x* and *y*.**

$y = ?$ • **Solve for *y*.**

Unfortunately, there is no algebraic solution for y. Thus we establish new notation and write

$$y = \sin^{-1} x$$

which is read "y is the inverse sine of x." Some textbooks use the notation $\arcsin x$ instead of $\sin^{-1} x$.

Definition of $\sin^{-1} x$

$$y = \sin^{-1} x \quad \text{if and only if} \quad x = \sin y$$

where $-1 \le x \le 1$ and $-\frac{\pi}{2} \le y \le \frac{\pi}{2}$.

It is convenient to think of the value of an inverse trigonometric function as an angle. For instance, if $y = \sin^{-1}\left(\frac{1}{2}\right)$, then y is the angle in the interval $\left[-\frac{\pi}{2}, \frac{\pi}{2}\right]$ whose sine is $\frac{1}{2}$. Thus $y = \frac{\pi}{6}$.

Because the graph of $y = \cos x$ fails the horizontal line test, it is not the graph of a one-to-one function. Therefore, it does not have an inverse function. **Figure 3.12** shows the graph of $y = \cos x$ on the interval $-2\pi \le x \le 2\pi$ and the graph of the inverse relation $x = \cos y$. Note that the graph of $x = \cos y$ does not satisfy the vertical line test and therefore is not the graph of a function.

If the domain of $y = \cos x$ is restricted to $0 \le x \le \pi$, the graph of $y = \cos x$ satisfies the horizontal line test and therefore is the graph of a one-to-one function. The graphs of $y = \cos x$ for $0 \le x \le \pi$ and its inverse are shown in **Figure 3.13.**

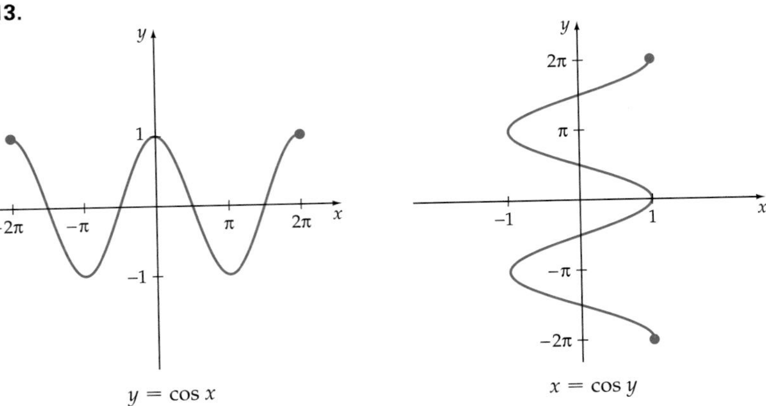

Figure 3.12

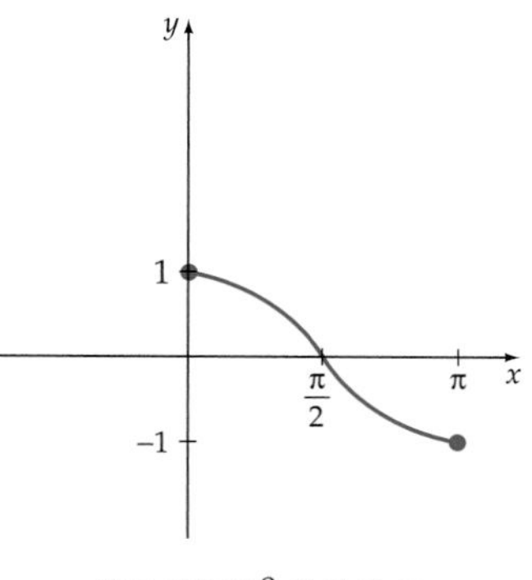

$y = \cos x$: $0 \le x \le \pi$

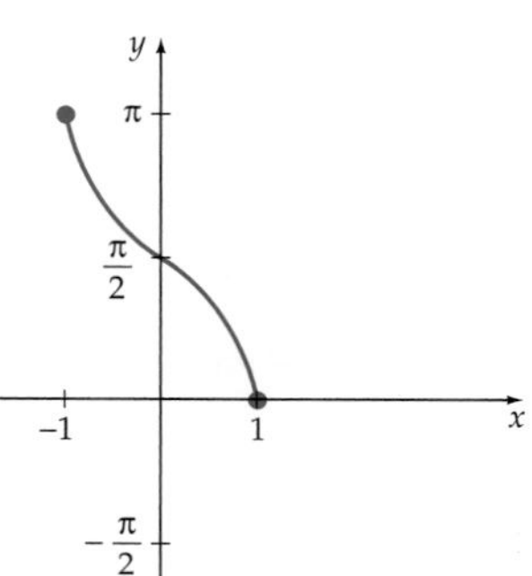

$y = \cos^{-1} x$: $-1 \le x \le 1$

Figure 3.13

To find the inverse of the function defined by $y = \cos x$, with $0 \le x \le \pi$, interchange x and y. Then solve for y.

$y = \cos x$ • $0 \le x \le \pi$

$x = \cos y$ • Interchange x and y.

$y = ?$ • Solve for y.

As in the case for the inverse sine function, there is no algebraic solution for y. Thus the notation for the inverse cosine function becomes $y = \cos^{-1} x$. We can write the following definition of the inverse cosine function.

Definition of $\cos^{-1} x$

$$y = \cos^{-1} x \quad \text{if and only if} \quad x = \cos y$$

where $-1 \le x \le 1$ and $0 \le y \le \pi$.

Because the graphs of $y = \tan x$, $y = \csc x$, $y = \sec x$, and $y = \cot x$ fail the horizontal line test, these functions are not one-to-one functions. Therefore, these functions do not have inverse functions. If the domains of all these functions are restricted in a certain way, however, the graphs satisfy the horizontal line test. Thus each of these functions has an inverse function over a restricted domain. **Table 3.2** on page 258 shows the restricted function and the inverse function for $\tan x$, $\csc x$, $\sec x$, and $\cot x$.

The choice of ranges for $y = \sec^{-1} x$ and $y = \csc^{-1} x$ is not universally accepted. For example, some calculus texts use $\left[0, \frac{\pi}{2}\right) \cup \left[\pi, \frac{3\pi}{2}\right)$ as the range of $y = \sec^{-1} x$. This definition has some advantages and some disadvantages that are explained in more advanced mathematics courses.

EXAMPLE 1 ❯❯ Evaluate Inverse Functions

Find the exact value of each inverse function.

a. $y = \tan^{-1} \frac{\sqrt{3}}{3}$ **b.** $y = \cos^{-1}\left(-\frac{\sqrt{2}}{2}\right)$

Solution

a. Because $y = \tan^{-1} \frac{\sqrt{3}}{3}$, y is the angle whose measure is in the interval $\left(-\frac{\pi}{2}, \frac{\pi}{2}\right)$, and $\tan y = \frac{\sqrt{3}}{3}$. Therefore, $y = \frac{\pi}{6}$.

b. Because $y = \cos^{-1}\left(-\frac{\sqrt{2}}{2}\right)$, y is the angle whose measure is in the interval $[0, \pi]$, and $\cos y = -\frac{\sqrt{2}}{2}$. Therefore, $y = \frac{3}{4}\pi$.

❯❯ *Try Exercise 2, page 265*

Table 3.2

	$y = \tan x$	$y = \tan^{-1} x$	$y = \csc x$	$y = \csc^{-1} x$
Domain	$-\frac{\pi}{2} < x < \frac{\pi}{2}$	$-\infty < x < \infty$	$-\frac{\pi}{2} \le x \le \frac{\pi}{2}, x \ne 0$	$x \le -1$ or $x \ge 1$
Range	$-\infty < y < \infty$	$-\frac{\pi}{2} < y < \frac{\pi}{2}$	$y \le -1$ or $y \ge 1$	$-\frac{\pi}{2} \le y \le \frac{\pi}{2}, y \ne 0$
Asymptotes	$x = -\frac{\pi}{2}, x = \frac{\pi}{2}$	$y = -\frac{\pi}{2}, y = \frac{\pi}{2}$	$x = 0$	$y = 0$
Graph	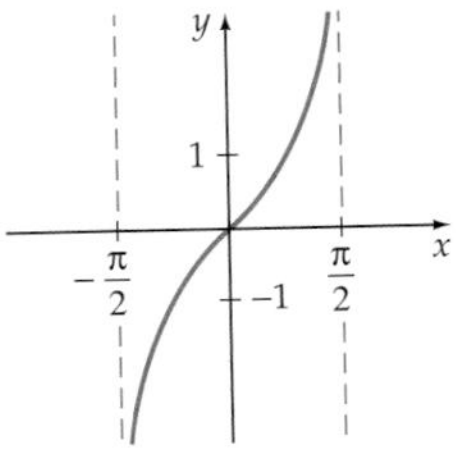	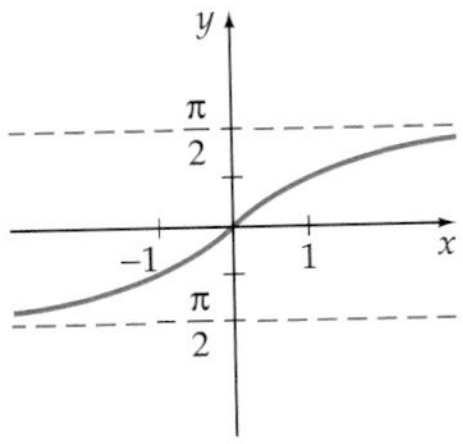	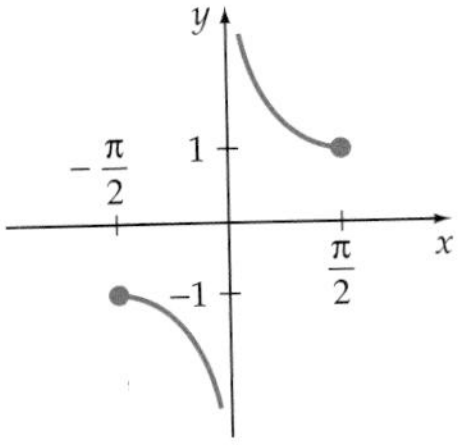	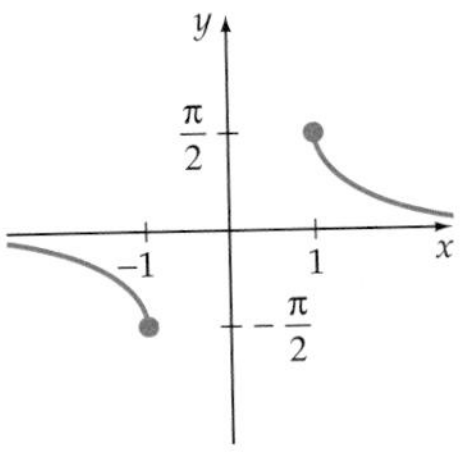

	$y = \sec x$	$y = \sec^{-1} x$	$y = \cot x$	$y = \cot^{-1} x$
Domain	$0 \le x \le \pi, x \ne \frac{\pi}{2}$	$x \le -1$ or $x \ge 1$	$0 < x < \pi$	$-\infty < x < \infty$
Range	$y \le -1$ or $y \ge 1$	$0 \le y \le \pi, y \ne \frac{\pi}{2}$	$-\infty < y < \infty$	$0 < y < \pi$
Asymptotes	$x = \frac{\pi}{2}$	$y = \frac{\pi}{2}$	$x = 0, x = \pi$	$y = 0, y = \pi$
Graph	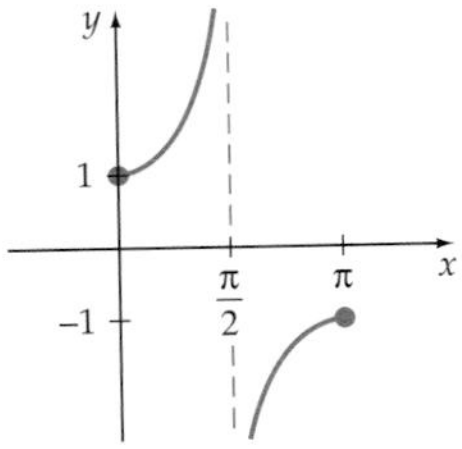	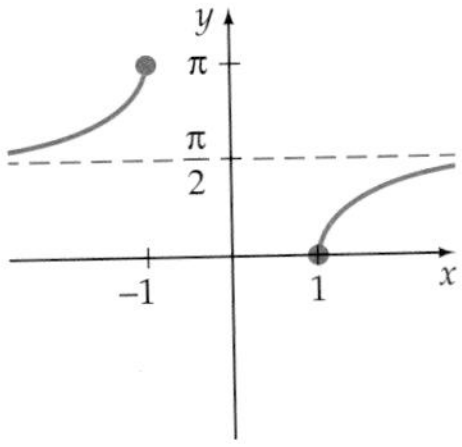	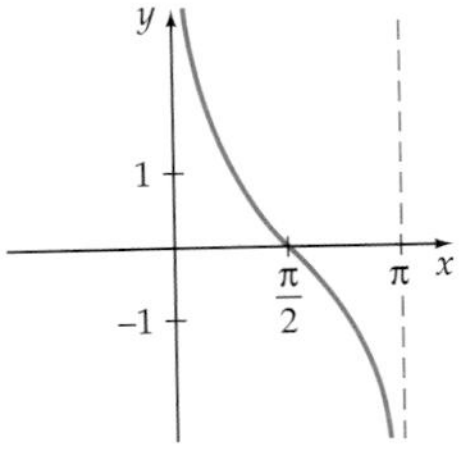	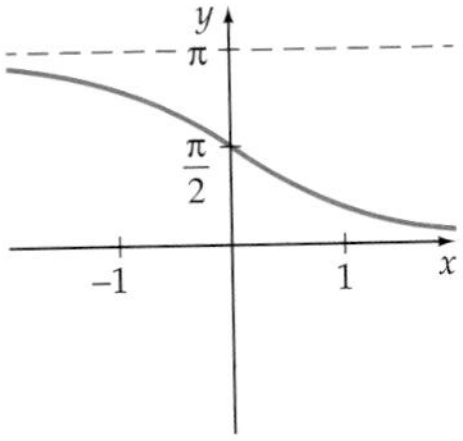

A calculator may not have keys for the inverse secant, cosecant, and cotangent functions. The following procedure shows an identity for the inverse cosecant function in terms of the inverse sine function. If we need to determine y, which is the angle whose cosecant is x, we can rewrite $y = \csc^{-1} x$ as follows:

$$y = \csc^{-1} x$$

• Domain: $x \le -1$ or $x \ge 1$
Range: $-\frac{\pi}{2} \le y \le \frac{\pi}{2}, y \ne 0$

$$\csc y = x$$

• Definition of inverse function

$$\frac{1}{\sin y} = x$$

• Substitute $\frac{1}{\sin y}$ for $\csc y$.

$$\sin y = \frac{1}{x} \qquad \text{• Solve for } \sin y.$$

$$y = \sin^{-1}\frac{1}{x} \qquad \text{• Write using inverse notation.}$$

$$\csc^{-1} x = \sin^{-1}\frac{1}{x} \qquad \text{• Replace } y \text{ with } \csc^{-1} x.$$

Thus $\csc^{-1} x$ is the same as $\sin^{-1}\frac{1}{x}$. There is a similar identity for $\sec^{-1} x$.

Identities for the Inverse Secant, Cosecant, and Cotangent Functions

If $x \le -1$ or $x \ge 1$, then

$$\csc^{-1} x = \sin^{-1}\frac{1}{x} \quad \text{and} \quad \sec^{-1} x = \cos^{-1}\frac{1}{x}$$

If x is a real number, then

$$\cot^{-1} x = \frac{\pi}{2} - \tan^{-1} x$$

Composition of Trigonometric Functions and Their Inverses

Recall that a function f and its inverse f^{-1} have the property that $f[f^{-1}(x)] = x$ for all x in the domain of f^{-1} and that $f^{-1}[f(x)] = x$ for all x in the domain of f. Applying this property to the functions $\sin x$, $\cos x$, and $\tan x$ and their inverse functions produces the following theorems.

TO REVIEW

Composition of Functions
See section 1.5.

Composition of Trigonometric Functions and Their Inverses

- If $-1 \le x \le 1$, then $\sin(\sin^{-1} x) = x$ and $\cos(\cos^{-1} x) = x$.
- If x is any real number, then $\tan(\tan^{-1} x) = x$.
- If $-\frac{\pi}{2} \le x \le \frac{\pi}{2}$, then $\sin^{-1}(\sin x) = x$.
- If $0 \le x \le \pi$, then $\cos^{-1}(\cos x) = x$.
- If $-\frac{\pi}{2} < x < \frac{\pi}{2}$, then $\tan^{-1}(\tan x) = x$.

In the next example we make use of some of the composition theorems to evaluate trigonometric expressions.

EXAMPLE 2 » Evaluate the Composition of a Function and Its Inverse

Find the exact value of each composition of functions.

a. $\sin(\sin^{-1} 0.357)$ b. $\cos^{-1}(\cos 3)$ c. $\tan[\tan^{-1}(-11.27)]$

d. $\sin(\sin^{-1} \pi)$ e. $\cos(\cos^{-1} 0.277)$ f. $\tan^{-1}\left(\tan \frac{4\pi}{3}\right)$

Solution

a. Because 0.357 is in the interval $[-1, 1]$, $\sin(\sin^{-1} 0.357) = 0.357$.

b. Because 3 is in the interval $[0, \pi]$, $\cos^{-1}(\cos 3) = 3$.

c. Because -11.27 is a real number, $\tan[\tan^{-1}(-11.27)] = -11.27$.

d. Because π is not in the domain of the inverse sine function, $\sin(\sin^{-1} \pi)$ is undefined.

e. Because 0.277 is in the interval $[-1, 1]$, $\cos(\cos^{-1} 0.277) = 0.277$.

f. $\frac{4\pi}{3}$ is not in the interval $\left(-\frac{\pi}{2}, \frac{\pi}{2}\right)$; however, the reference angle for $\theta = \frac{4\pi}{3}$ is $\theta' = \frac{\pi}{3}$. Thus $\tan^{-1}\left(\tan \frac{4\pi}{3}\right) = \tan^{-1}\left(\tan \frac{\pi}{3}\right)$. Because $\frac{\pi}{3}$ is in the interval $\left(-\frac{\pi}{2}, \frac{\pi}{2}\right)$, $\tan^{-1}\left(\tan \frac{\pi}{3}\right) = \frac{\pi}{3}$. Hence $\tan^{-1}\left(\tan \frac{4\pi}{3}\right) = \frac{\pi}{3}$.

» *Try Exercise 28, page 265*

QUESTION Is $\tan^{-1}(\tan x) = x$ an identity?

It is often easy to evaluate a trigonometric expression by referring to a sketch of a right triangle that satisfies given conditions. In Example 3 we make use of this technique.

EXAMPLE 3 » Evaluate a Trigonometric Expression

Find the exact value of $\sin\left(\cos^{-1} \frac{2}{5}\right)$.

ANSWER No. $\tan^{-1}(\tan x) = x$ only if $-\frac{\pi}{2} < x < \frac{\pi}{2}$.

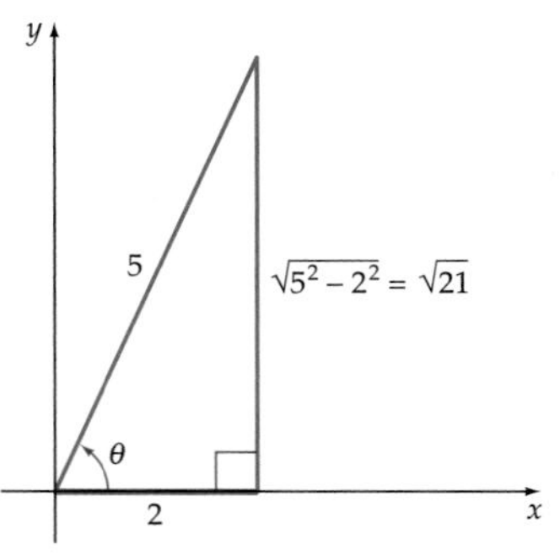

Figure 3.14

Solution

Let $\theta = \cos^{-1}\frac{2}{5}$, which implies that $\cos\theta = \frac{2}{5}$. Because $\cos\theta$ is positive, θ is a first-quadrant angle. We draw a right triangle with base 2 and hypotenuse 5 so that we can view θ, as shown in **Figure 3.14**. The height of the triangle is $\sqrt{5^2 - 2^2} = \sqrt{21}$. Our goal is to find $\sin\theta$, which by definition is $\frac{\text{opp}}{\text{hyp}} = \frac{\sqrt{21}}{5}$. Thus

$$\sin\left(\cos^{-1}\frac{2}{5}\right) = \sin(\theta) = \frac{\sqrt{21}}{5}$$

Try Exercise 50, page 265

In Example 4, we sketch two right triangles to evaluate the given expression.

EXAMPLE 4 Evaluate a Trigonometric Expression

Find the exact value of $\sin\left[\sin^{-1}\frac{3}{5} + \cos^{-1}\left(-\frac{5}{13}\right)\right]$.

Solution

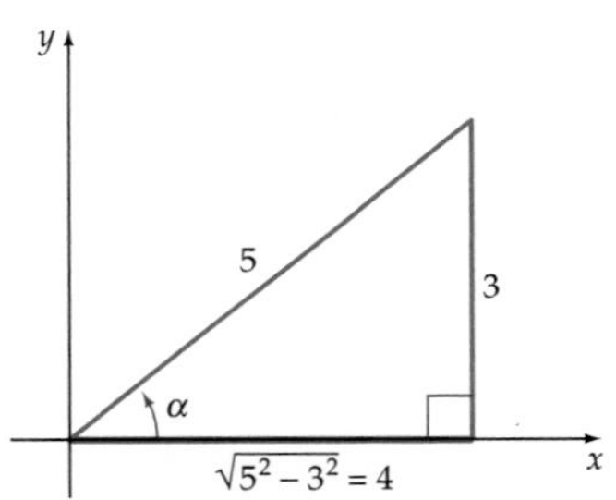

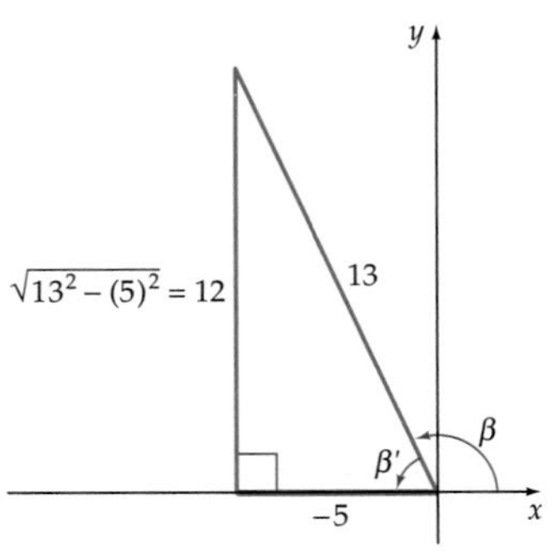

Figure 3.15

Let $\alpha = \sin^{-1}\frac{3}{5}$. Thus $\sin\alpha = \frac{3}{5}$. Let $\beta = \cos^{-1}\left(-\frac{5}{13}\right)$, which implies that $\cos\beta = -\frac{5}{13}$. Sketch angles α and β as shown in **Figure 3.15**. We wish to evaluate

$$\begin{aligned}\sin\left[\sin^{-1}\frac{3}{5} + \cos^{-1}\left(-\frac{5}{13}\right)\right] &= \sin(\alpha + \beta)\\ &= \sin\alpha\cos\beta + \cos\alpha\sin\beta \qquad (1)\end{aligned}$$

A close look at the triangles in **Figure 3.15** shows that

$$\cos\alpha = \frac{4}{5} \quad \text{and} \quad \sin\beta = \frac{12}{13}$$

Substituting in Equation (1) gives us our desired result.

$$\begin{aligned}\sin\left[\sin^{-1}\frac{3}{5} + \cos^{-1}\left(-\frac{5}{13}\right)\right] &= \sin\alpha\cos\beta + \cos\alpha\sin\beta\\ &= \left(\frac{3}{5}\right)\left(-\frac{5}{13}\right) + \left(\frac{4}{5}\right)\left(\frac{12}{13}\right) = \frac{33}{65}\end{aligned}$$

Try Exercise 56, page 266

In Example 5 we make use of the identity $\cos(\cos^{-1} x) = x$, where $-1 \le x \le 1$, to solve an equation.

EXAMPLE 5 Solve an Inverse Trigonometric Equation

Solve $\sin^{-1}\frac{3}{5} + \cos^{-1} x = \pi$.

Solution

Solve for $\cos^{-1} x$ and then take the cosine of both sides of the equation.

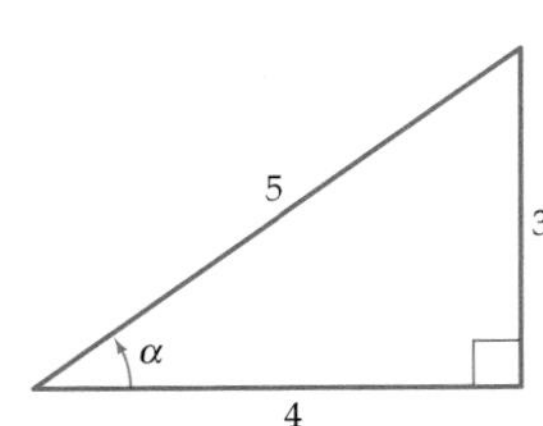

Figure 3.16

$$\sin^{-1}\frac{3}{5} + \cos^{-1} x = \pi$$

$$\cos^{-1} x = \pi - \sin^{-1}\frac{3}{5}$$

$$\cos(\cos^{-1} x) = \cos\left(\pi - \sin^{-1}\frac{3}{5}\right)$$

$$x = \cos(\pi - \alpha)$$

• Let $\alpha = \sin^{-1}\frac{3}{5}$. Note that α is the angle whose sine is $\frac{3}{5}$. (See **Figure 3.16.**)

$$= \cos \pi \cos \alpha + \sin \pi \sin \alpha$$

• Difference identity for cosine

$$= (-1)\cos \alpha + (0)\sin \alpha$$

$$= -\cos \alpha$$

$$= -\frac{4}{5}$$

• $\cos \alpha = \frac{4}{5}$ (See **Figure 3.16.**)

Try Exercise 66, page 266

EXAMPLE 6 Verify a Trigonometric Identity That Involves Inverses

Verify the identity $\sin^{-1} x + \cos^{-1} x = \frac{\pi}{2}$.

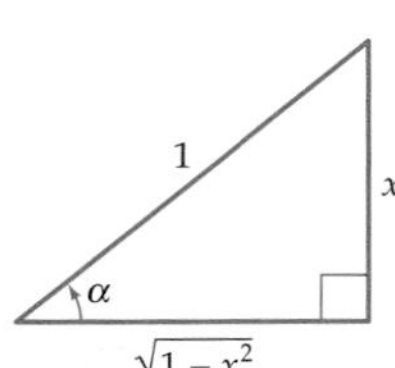

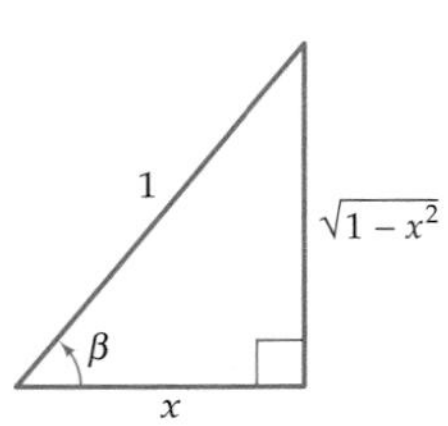

Figure 3.17

Solution

Let $\alpha = \sin^{-1} x$ and $\beta = \cos^{-1} x$. These equations imply that $\sin \alpha = x$ and $\cos \beta = x$. From the right triangles in **Figure 3.17,**

$$\cos \alpha = \sqrt{1 - x^2} \quad \text{and} \quad \sin \beta = \sqrt{1 - x^2}$$

Our goal is to show $\sin^{-1} x + \cos^{-1} x$ equals $\frac{\pi}{2}$.

$$\sin^{-1} x + \cos^{-1} x = \alpha + \beta$$

$$= \cos^{-1}[\cos(\alpha + \beta)]$$

• Because $0 \le \alpha + \beta \le \pi$, we can apply $\alpha + \beta = \cos^{-1}[\cos(\alpha + \beta)]$.

$$= \cos^{-1}[\cos\alpha\cos\beta - \sin\alpha\sin\beta]$$ • Addition identity for cosine

$$= \cos^{-1}\left[\left(\sqrt{1-x^2}\right)(x) - (x)\left(\sqrt{1-x^2}\right)\right]$$

$$= \cos^{-1} 0$$

$$= \frac{\pi}{2}$$

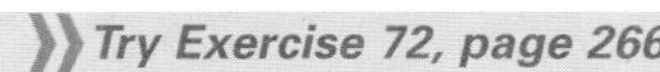
Try Exercise 72, page 266

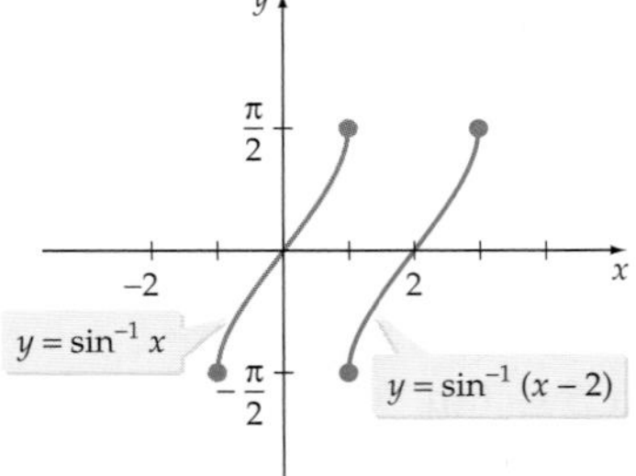

Figure 3.18

Graphs of Inverse Trigonometric Functions

The inverse trigonometric functions can be graphed by using the procedures of stretching, shrinking, and translation that were discussed earlier in the text. For instance, the graph of $y = \sin^{-1}(x-2)$ is a horizontal shift 2 units to the right of the graph of $y = \sin^{-1} x$, as shown in **Figure 3.18.**

EXAMPLE 7 Graph an Inverse Function

Graph: $y = \cos^{-1} x + 1$

Solution

Recall that the graph of $y = f(x) + c$ is a vertical translation of the graph of f. Because $c = 1$, a positive number, the graph of $y = \cos^{-1} x + 1$ is the graph of $y = \cos^{-1} x$ shifted 1 unit up. See **Figure 3.19.**

Try Exercise 76, page 266

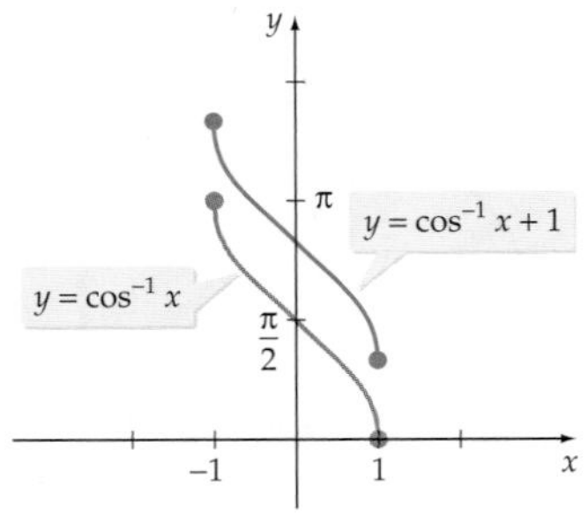

Figure 3.19

Integrating Technology

When you use a graphing utility to draw the graph of an inverse trigonometric function, use the properties of these functions to verify the correctness of your graph. For instance, the graph of $y = 3\sin^{-1} 0.5x$ is shown in **Figure 3.20.** The domain of $y = \sin^{-1} x$ is $-1 \le x \le 1$. Therefore, the domain of $y = 3\sin^{-1} 0.5x$ is $-1 \le 0.5x \le 1$ or, multiplying the inequality by 2, $-2 \le x \le 2$. This is consistent with the graph in **Figure 3.20.**

The range of $y = \sin^{-1} x$ is $-\frac{\pi}{2} \le y \le \frac{\pi}{2}$. Thus the range of $y = 3\sin^{-1} 0.5x$ is $-\frac{3\pi}{2} \le y \le \frac{3\pi}{2}$. This is also consistent with the graph. Verifying some of the properties of $y = \sin^{-1} x$ serves as a check that you have correctly entered the equation for the graph.

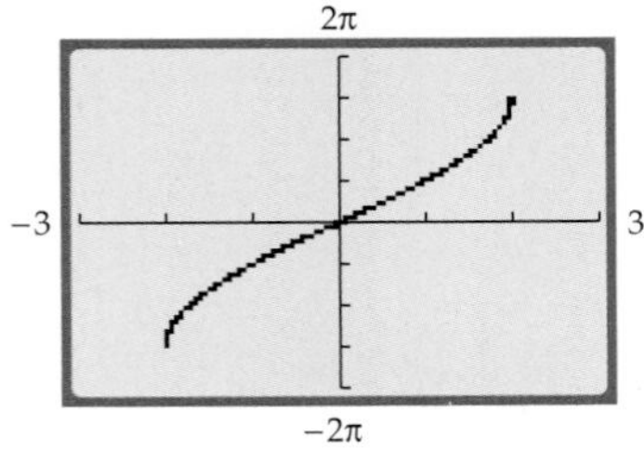

$y = 3\sin^{-1} 0.5x$

Figure 3.20

EXAMPLE 8 Solve an Application

A camera is placed on a deck of a pool as shown in **Figure 3.21.** A diver is 18 feet above the camera lens. The extended length of the diver is 8 feet.

a. Show that the angle θ subtended at the lens by the diver is

$$\theta = \tan^{-1}\frac{26}{x} - \tan^{-1}\frac{18}{x}$$

b. For what values of x will $\theta = 9°$?

c. What value of x maximizes θ?

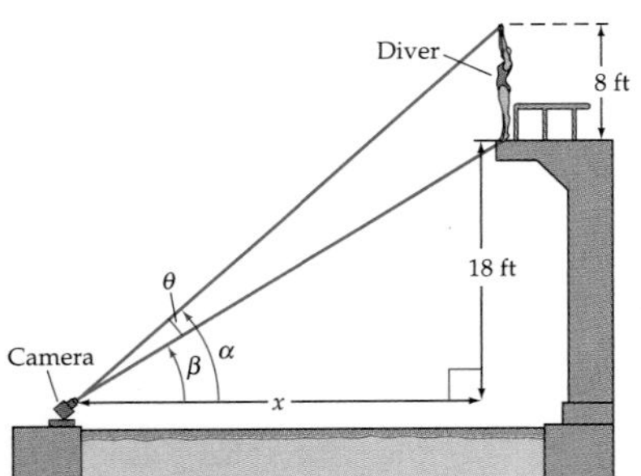

Figure 3.21

Solution

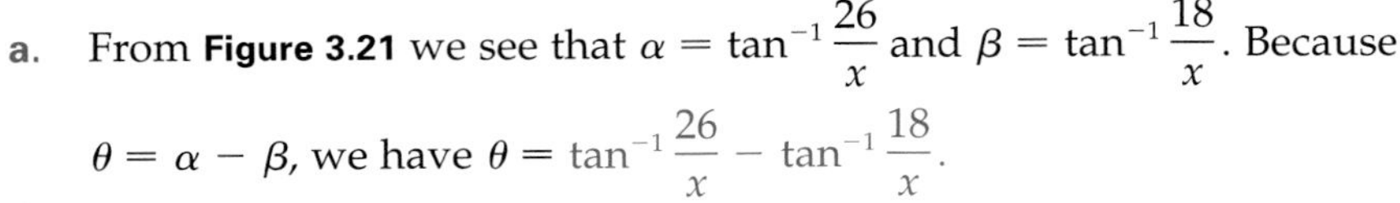

a. From **Figure 3.21** we see that $\alpha = \tan^{-1}\frac{26}{x}$ and $\beta = \tan^{-1}\frac{18}{x}$. Because $\theta = \alpha - \beta$, we have $\theta = \tan^{-1}\frac{26}{x} - \tan^{-1}\frac{18}{x}$.

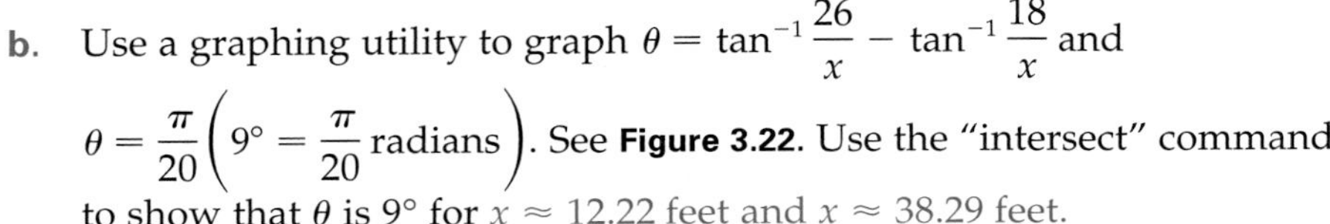

b. Use a graphing utility to graph $\theta = \tan^{-1}\frac{26}{x} - \tan^{-1}\frac{18}{x}$ and $\theta = \frac{\pi}{20}\left(9° = \frac{\pi}{20} \text{ radians}\right)$. See **Figure 3.22.** Use the "intersect" command to show that θ is 9° for $x \approx 12.22$ feet and $x \approx 38.29$ feet.

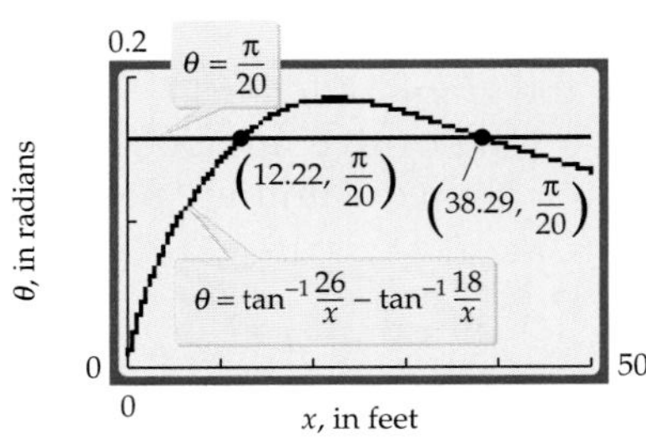

Figure 3.22

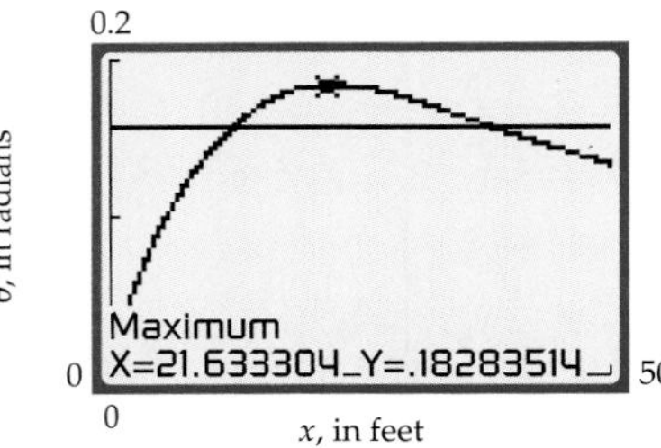

Figure 3.23

c. Use the "maximum" command to show that the maximum value of $\theta = \tan^{-1}\frac{26}{x} - \tan^{-1}\frac{18}{x}$ occurs when $x \approx 21.63$ feet. See **Figure 3.23.**

Try Exercise 84, page 266

Topics for Discussion

1. Is the equation

$$\tan^{-1} x = \frac{1}{\tan x}$$

true for all values of x, true for some values of x, or false for all values of x?

2. Are there real numbers x for which the following is true? Explain.

$$\sin(\sin^{-1} x) \neq \sin^{-1}(\sin x)$$

3. Explain how to find the value of $\sec^{-1} 3$ by using a calculator.

4. Explain how you can determine the range of $y = (2\cos^{-1} x) - 1$ using
   a. algebra b. a graph

Exercise Set 3.5

In Exercises 1 to 18, find the exact radian value.

1. $\sin^{-1} 1$
2. $\sin^{-1} \frac{\sqrt{2}}{2}$
3. $\cos^{-1}\left(-\frac{\sqrt{3}}{2}\right)$
4. $\cos^{-1}\left(-\frac{1}{2}\right)$
5. $\tan^{-1}(-1)$
6. $\tan^{-1} \sqrt{3}$
7. $\cot^{-1} \frac{\sqrt{3}}{3}$
8. $\cot^{-1} 1$
9. $\sec^{-1} 2$
10. $\sec^{-1} \frac{2\sqrt{3}}{3}$
11. $\csc^{-1}(-\sqrt{2})$
12. $\csc^{-1}(-2)$
13. $\sin^{-1}\left(-\frac{\sqrt{3}}{2}\right)$
14. $\sin^{-1} \frac{1}{2}$
15. $\cos^{-1}\left(-\frac{1}{2}\right)$
16. $\cos^{-1} \frac{\sqrt{3}}{2}$
17. $\tan^{-1} \frac{\sqrt{3}}{3}$
18. $\tan^{-1} 1$

In Exercises 19 to 22, use a calculator to approximate each function accurate to four decimal places.

19. a. $\sin^{-1}(0.8422)$ b. $\tan^{-1}(0.2385)$
20. a. $\cos^{-1}(-0.0356)$ b. $\tan^{-1}(3.7555)$
21. a. $\sec^{-1}(2.2500)$ b. $\cot^{-1}(3.4545)$
22. a. $\csc^{-1}(1.3465)$ b. $\cot^{-1}(0.1274)$

In Exercises 23 to 24, express θ as a function of x.

23.

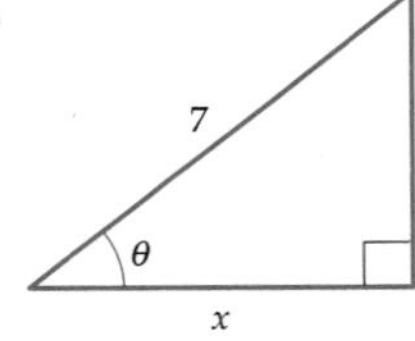

24.

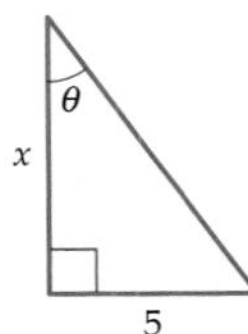

In Exercises 25 to 58, find the exact value of the given expression. If an exact value cannot be given, give the value to the nearest ten-thousandth.

25. $\cos\left(\cos^{-1} \frac{1}{2}\right)$
26. $\cos(\cos^{-1} 2)$
27. $\tan(\tan^{-1} 2)$
28. $\tan\left(\tan^{-1} \frac{1}{2}\right)$
29. $\sin\left(\tan^{-1} \frac{3}{4}\right)$
30. $\cos\left(\sin^{-1} \frac{5}{13}\right)$
31. $\tan\left(\sin^{-1} \frac{\sqrt{2}}{2}\right)$
32. $\sin\left[\cos^{-1}\left(-\frac{\sqrt{3}}{2}\right)\right]$
33. $\cos(\sec^{-1} 2)$
34. $\sin^{-1}(\sin 2)$
35. $\sin^{-1}\left(\sin \frac{\pi}{6}\right)$
36. $\sin^{-1}\left(\sin \frac{5\pi}{6}\right)$
37. $\cos^{-1}\left(\sin \frac{\pi}{4}\right)$
38. $\cos^{-1}\left(\cos \frac{5\pi}{4}\right)$
39. $\sin^{-1}\left(\tan \frac{\pi}{3}\right)$
40. $\cos^{-1}\left(\tan \frac{2\pi}{3}\right)$
41. $\tan^{-1}\left(\sin \frac{\pi}{6}\right)$
42. $\cot^{-1}\left(\cos \frac{2\pi}{3}\right)$
43. $\sin^{-1}\left[\cos\left(-\frac{2\pi}{3}\right)\right]$
44. $\cos^{-1}\left[\tan\left(-\frac{\pi}{3}\right)\right]$
45. $\tan\left(\sin^{-1} \frac{1}{2}\right)$
46. $\cot(\csc^{-1} 2)$
47. $\sec\left(\sin^{-1} \frac{1}{4}\right)$
48. $\csc\left(\cos^{-1} \frac{3}{4}\right)$
49. $\cos\left(\sin^{-1} \frac{7}{25}\right)$
50. $\tan\left(\cos^{-1} \frac{3}{5}\right)$

51. $\cos\left(2\sin^{-1}\dfrac{\sqrt{2}}{2}\right)$

52. $\tan\left(2\sin^{-1}\dfrac{\sqrt{3}}{2}\right)$

53. $\sin\left(2\sin^{-1}\dfrac{4}{5}\right)$

54. $\cos(2\tan^{-1}1)$

55. $\sin\left(\sin^{-1}\dfrac{2}{3} + \cos^{-1}\dfrac{1}{2}\right)$

56. $\cos\left(\sin^{-1}\dfrac{3}{4} + \cos^{-1}\dfrac{5}{13}\right)$

57. $\tan\left(\cos^{-1}\dfrac{1}{2} - \sin^{-1}\dfrac{3}{4}\right)$

58. $\sec\left(\cos^{-1}\dfrac{2}{3} + \sin^{-1}\dfrac{2}{3}\right)$

In Exercises 59 to 68, solve the equation for *x* algebraically.

59. $\sin^{-1}x = \cos^{-1}\dfrac{5}{13}$

60. $\tan^{-1}x = \sin^{-1}\dfrac{24}{25}$

61. $\sin^{-1}(x-1) = \dfrac{\pi}{2}$

62. $\cos^{-1}\left(x - \dfrac{1}{2}\right) = \dfrac{\pi}{3}$

63. $\tan^{-1}\left(x + \dfrac{\sqrt{2}}{2}\right) = \dfrac{\pi}{4}$

64. $\sin^{-1}(x-2) = -\dfrac{\pi}{6}$

65. $\sin^{-1}\dfrac{3}{5} + \cos^{-1}x = \dfrac{\pi}{4}$

66. $\sin^{-1}x + \cos^{-1}\dfrac{4}{5} = \dfrac{\pi}{6}$

67. $\sin^{-1}\dfrac{\sqrt{2}}{2} + \cos^{-1}x = \dfrac{2\pi}{3}$

68. $\cos^{-1}x + \sin^{-1}\dfrac{\sqrt{3}}{2} = \dfrac{\pi}{2}$

In Exercises 69 and 70, write each expression in terms of *x*.

69. $\tan(\cos^{-1}x)$

70. $\sin(\sec^{-1}x)$

In Exercises 71 to 74, verify the identity.

71. $\sin^{-1}x + \sin^{-1}(-x) = 0$

72. $\cos^{-1}x + \cos^{-1}(-x) = \pi$

73. $\tan^{-1}x + \tan^{-1}\dfrac{1}{x} = \dfrac{\pi}{2}, x > 0$

74. $\sec^{-1}\dfrac{1}{x} + \csc^{-1}\dfrac{1}{x} = \dfrac{\pi}{2}$

In Exercises 75 to 82, use stretching, shrinking, and translation procedures to graph each equation.

75. $y = \sin^{-1}x + 2$

76. $y = \cos^{-1}(x-1)$

77. $y = \sin^{-1}(x+1) - 2$

78. $y = \tan^{-1}(x-1) + 2$

79. $y = 2\cos^{-1}x$

80. $y = -2\tan^{-1}x$

81. $y = \tan^{-1}(x+1) - 2$

82. $y = \sin^{-1}(x-2) + 1$

83. **SATELLITE COVERAGE** A communications satellite orbits Earth at an altitude of a miles. The beam coverage provided by the satellite is shown in the figure below.

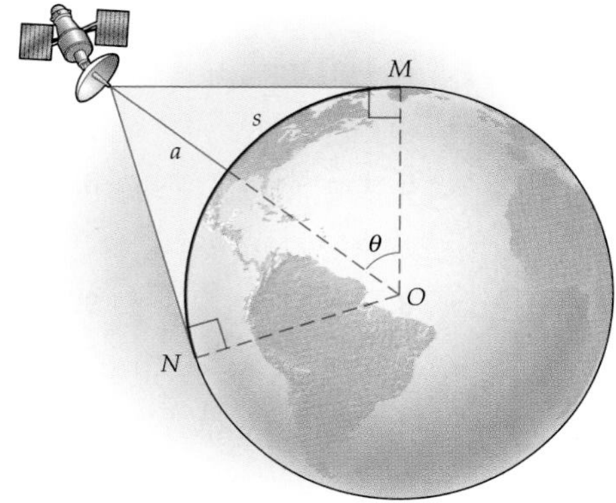

a. Find the distance s (length of arc MN) as a function of the altitude a of the satellite. Assume the radius of Earth is 3960 miles.

b. What altitude does the satellite need to attain to provide coverage over a distance of $s = 5500$ miles? Round to the nearest 10 miles.

84. **VOLUME IN A WATER TANK** The volume V of water (measured in cubic feet) in a horizontal cylindrical tank of radius 5 feet and length 12 feet is given by

$$V(x) = 12\left[25\cos^{-1}\left(\frac{5-x}{5}\right) - (5-x)\sqrt{10x - x^2}\right]$$

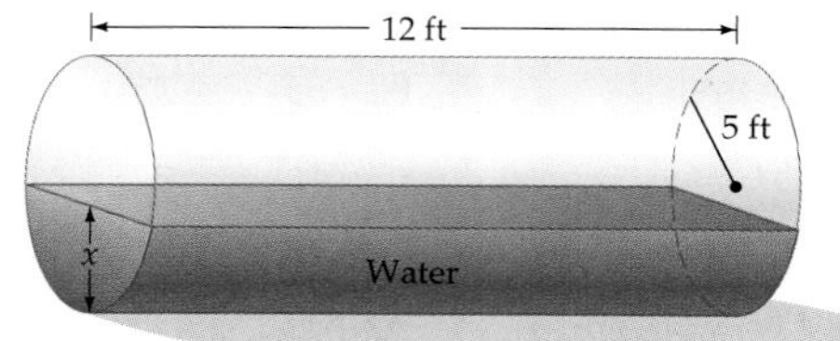

where x is the depth of the water in feet.

a. Graph V over its domain $0 \le x \le 10$.

b. Write a sentence that explains why the graph of V increases more rapidly when x increases from 4.9 feet to 5 feet than it does when x increases from 0.1 foot to 0.2 foot.

c. If $x = 4$ feet, find the volume (to the nearest 0.01 cubic foot) of the water in the tank.

d. Find the depth x (to the nearest 0.01 foot) if there are 288 cubic feet of water in the tank.

85. Graph $f(x) = \cos^{-1} x$ and $g(x) = \sin^{-1} \sqrt{1 - x^2}$ on the same coordinate axes. Does $f(x) = g(x)$ on the interval $[-1, 1]$?

86. Graph $y = \cos(\cos^{-1} x)$ on $[-1, 1]$. Graph $y = \cos^{-1}(\cos x)$ on $[-2\pi, 2\pi]$.

In Exercises 87 to 92, use a graphing utility to graph each equation.

87. $y = \csc^{-1} 2x$

88. $y = 0.5 \sec^{-1} \dfrac{x}{2}$

89. $y = \sec^{-1}(x - 1)$

90. $y = \sec^{-1}(x + \pi)$

91. $y = 2 \tan^{-1} 2x$

92. $y = \tan^{-1}(x - 1)$

Connecting Concepts

In Exercises 93 to 96, verify the identity.

93. $\cos(\sin^{-1} x) = \sqrt{1 - x^2}$

94. $\sec(\sin^{-1} x) = \dfrac{\sqrt{1 - x^2}}{1 - x^2}$

95. $\tan(\csc^{-1} x) = \dfrac{\sqrt{x^2 - 1}}{x^2 - 1}, x > 1$

96. $\sin(\cot^{-1} x) = \dfrac{\sqrt{x^2 + 1}}{x^2 + 1}$

In Exercises 97 to 100, solve for *y* in terms of *x*.

97. $5x = \tan^{-1} 3y$

98. $2x = \dfrac{1}{2} \sin^{-1} 2y$

99. $x - \dfrac{\pi}{3} = \cos^{-1}(y - 3)$

100. $x + \dfrac{\pi}{2} = \tan^{-1}(2y - 1)$

Projects

1. Visual Insight

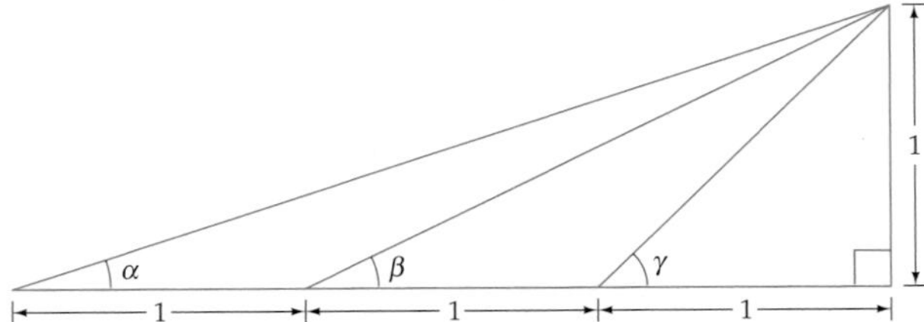

Explain how the figure above can be used to verify each identity.

a. $\tan^{-1} \dfrac{1}{3} + \tan^{-1} \dfrac{1}{2} = \dfrac{\pi}{4}$ [*Hint:* Start by using an identity to find the value of $\tan(\alpha + \beta)$.]

b. $\alpha + \beta = \gamma$

Section 3.6

- Solve Trigonometric Equations
- Model Sinusoidal Data

Trigonometric Equations

PREPARE FOR THIS SECTION

Prepare for this section by completing the following exercises. The answers can be found on page A18.

PS1. Use the quadratic formula to solve $3x^2 - 5x - 4 = 0$. [1.1]

PS2. Use a Pythagorean identity to write $\sin^2 x$ as a function involving $\cos^2 x$. [3.1]

PS3. Evaluate $\frac{\pi}{2} + 2k\pi$ for $k = 1, 2$, and 3.

PS4. Factor by grouping: $x^2 - \frac{\sqrt{3}}{2}x + x - \frac{\sqrt{3}}{2}$.

PS5. Graph the scatter plot for the following data. Use a viewing window with Xmin=0, Xmax=40, Ymin=0, and Ymax=100. [1.7]

x	3	7	11	15	19	23	27	31
y	14	55	90	99	80	44	8	4

PS6. Solve $2x^2 - 2x = 0$ by factoring. [1.1]

Solve Trigonometric Equations

Consider the equation $\sin x = \frac{1}{2}$. The graph of $y = \sin x$, along with the line $y = \frac{1}{2}$, is shown in **Figure 3.24.** The x values of the intersections of the two graphs are the solutions of $\sin x = \frac{1}{2}$. The solutions in the interval $0 \le x < 2\pi$ are $x = \frac{\pi}{6}$ and $\frac{5\pi}{6}$.

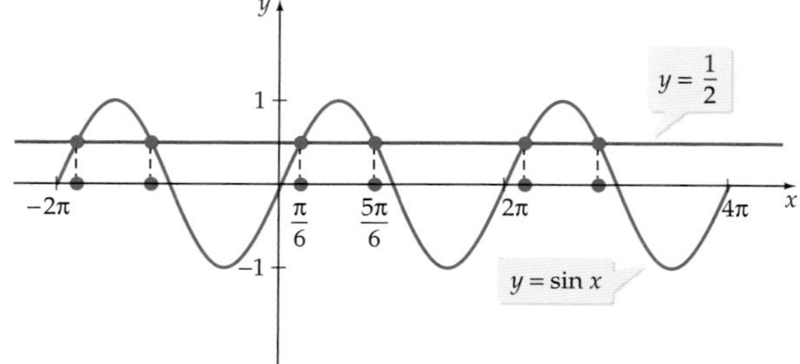

Figure 3.24

If we remove the restriction $0 \le x < 2\pi$, there are many more solutions. Because the sine function is periodic with a period of 2π, other solutions are obtained by adding $2k\pi$, k an integer, to either of the previous solutions. Thus the solutions of $\sin x = \frac{1}{2}$ are

$$x = \frac{\pi}{6} + 2k\pi, \quad k \text{ an integer}$$

$$x = \frac{5\pi}{6} + 2k\pi, \quad k \text{ an integer}$$

QUESTION How many solutions does the equation $\cos x = \frac{\sqrt{3}}{2}$ have on the interval $0 \le x < 2\pi$?

Algebraic methods and trigonometric identities are used frequently to find the solutions of trigonometric equations. Algebraic methods that are often employed include solving by factoring, solving by using the quadratic formula, and squaring each side of the equation.

EXAMPLE 1 ›› Solve a Trigonometric Equation by Factoring

Solve $2\sin^2 x \cos x - \cos x = 0$, where $0 \le x < 2\pi$.

ALGEBRAIC SOLUTION

$$2\sin^2 x \cos x - \cos x = 0$$

$$\cos x(2\sin^2 x - 1) = 0 \qquad \text{• Factor } \cos x \text{ from each term.}$$

$$\cos x = 0 \quad \text{or} \quad 2\sin^2 x - 1 = 0 \qquad \text{• Use the Principle of Zero Products.}$$

$$x = \frac{\pi}{2}, \frac{3\pi}{2} \qquad \sin^2 x = \frac{1}{2} \qquad \text{• Solve each equation for } x \text{ with } 0 \le x < 2\pi.$$

$$\sin x = \pm\frac{\sqrt{2}}{2}$$

$$x = \frac{\pi}{4}, \frac{3\pi}{4}, \frac{5\pi}{4}, \frac{7\pi}{4}$$

The solutions in the interval $0 \le x < 2\pi$ are

$$\frac{\pi}{4}, \frac{\pi}{2}, \frac{3\pi}{4}, \frac{5\pi}{4}, \frac{3\pi}{2}, \text{ and } \frac{7\pi}{4}$$

VISUALIZE THE SOLUTION

The solutions are the x-coordinates of the x-intercepts of the graph of $y = 2\sin^2 x \cos x - \cos x$ on the interval $[0, 2\pi)$.

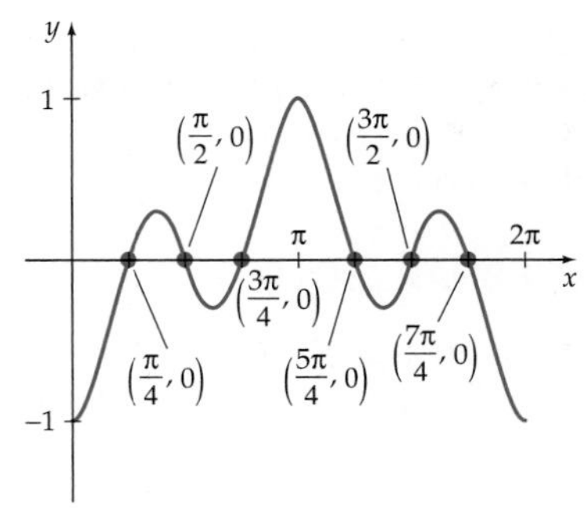

$y = 2\sin^2 x \cos x - \cos x$

›› Try Exercise 14, page 278

ANSWER Two

Squaring both sides of an equation may not produce an equivalent equation. Thus, when this method is used, the proposed solutions must be checked to eliminate any extraneous solutions.

EXAMPLE 2 Solve a Trigonometric Equation by Squaring Each Side of the Equation

Solve $\sin x + \cos x = 1$, where $0 \le x < 2\pi$.

ALGEBRAIC SOLUTION

$$\sin x + \cos x = 1 \qquad \text{• Solve for } \sin x.$$
$$\sin x = 1 - \cos x$$
$$\sin^2 x = (1 - \cos x)^2 \qquad \text{• Square each side.}$$
$$\sin^2 x = 1 - 2\cos x + \cos^2 x$$
$$1 - \cos^2 x = 1 - 2\cos x + \cos^2 x \qquad \text{• } \sin^2 x = 1 - \cos^2 x$$
$$2\cos^2 x - 2\cos x = 0$$
$$2\cos x(\cos x - 1) = 0 \qquad \text{• Factor.}$$
$$2\cos x = 0 \quad \text{or} \quad \cos x = 1$$
$$x = \frac{\pi}{2}, \frac{3\pi}{2} \qquad x = 0 \qquad \text{• Solve each equation for } x \text{ with } 0 \le x < 2\pi.$$

A check will show that 0 and $\frac{\pi}{2}$ are solutions but $\frac{3\pi}{2}$ is not a solution.

VISUALIZE THE SOLUTION

The solutions are the x-coordinates of the points of intersection of $y = \sin x + \cos x$ and $y = 1$ on the interval $[0, 2\pi)$.

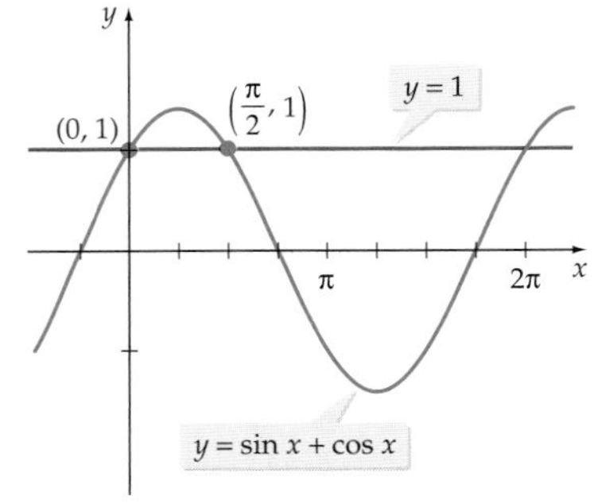

Try Exercise 52, page 278

EXAMPLE 3 Solve a Trigonometric Equation by Using the Quadratic Formula

Solve $3\cos^2 x - 5\cos x - 4 = 0$, where $0 \le x < 2\pi$.

ALGEBRAIC SOLUTION

The given equation is quadratic in form and cannot be factored easily. However, we can use the quadratic formula to solve for $\cos x$.

$$3\cos^2 x - 5\cos x - 4 = 0 \qquad \text{• } a = 3,\ b = -5,\ c = -4$$
$$\cos x = \frac{-(-5) \pm \sqrt{(-5)^2 - 4(3)(-4)}}{(2)(3)} = \frac{5 \pm \sqrt{73}}{6}$$

The equation $\cos x = \frac{5 + \sqrt{73}}{6}$ does not have a solution because $\frac{5 + \sqrt{73}}{6} > 2$, and for any x the maximum value of $\cos x$ is 1. Thus

VISUALIZE THE SOLUTION

The solutions are the x-coordinates of the x-intercepts of $y = 3\cos^2 x - 5\cos x - 4$ on the interval $[0, 2\pi)$. See the following figure.

$\cos x = \dfrac{5 - \sqrt{73}}{6}$, and because $\dfrac{5 - \sqrt{73}}{6}$ is a negative number (about -0.59), the equation $\cos x = \dfrac{5 - \sqrt{73}}{6}$ will have two solutions on the interval $[0, 2\pi)$. Thus

$$x = \cos^{-1}\left(\frac{5 - \sqrt{73}}{6}\right) \approx 2.2027 \quad \text{or}$$

$$x = 2\pi - \cos^{-1}\left(\frac{5 - \sqrt{73}}{6}\right) \approx 4.0805$$

To the nearest 0.0001, the solutions on $[0, 2\pi)$ are 2.2027 and 4.0805.

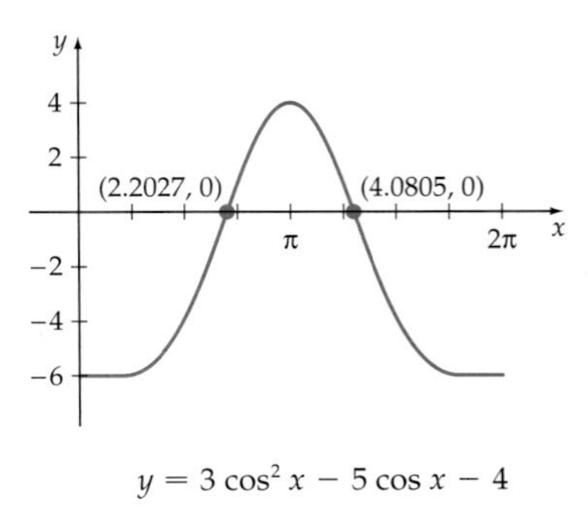

Try Exercise 56, page 278

When solving an equation that has multiple solutions, we must be sure we find all solutions of the equation for the given interval. For example, to find all solutions of $\sin 2x = \dfrac{1}{2}$, where $0 \le x < 2\pi$, we first solve for $2x$.

$$\sin 2x = \frac{1}{2}$$

$$2x = \frac{\pi}{6} + 2k\pi \quad \text{or} \quad 2x = \frac{5\pi}{6} + 2k\pi$$

• **k is an integer.**

Solving for x, we have $x = \dfrac{\pi}{12} + k\pi$ or $x = \dfrac{5\pi}{12} + k\pi$. Substituting integers for k, we obtain

$$k = 0: \quad x = \frac{\pi}{12} \quad \text{or} \quad x = \frac{5\pi}{12}$$

$$k = 1: \quad x = \frac{13\pi}{12} \quad \text{or} \quad x = \frac{17\pi}{12}$$

$$k = 2: \quad x = \frac{25\pi}{12} \quad \text{or} \quad x = \frac{29\pi}{12}$$

Note that for $k \ge 2$, $x \ge 2\pi$ and the solutions to $\sin 2x = \dfrac{1}{2}$ are not in the interval $0 \le x < 2\pi$. Also if $k < 0$, then $x < 0$ and no solutions in the interval $0 \le x < 2\pi$ are produced. Thus, for $0 \le x < 2\pi$, the solutions are $\dfrac{\pi}{12}$, $\dfrac{5\pi}{12}$, $\dfrac{13\pi}{12}$, and $\dfrac{17\pi}{12}$. See **Figure 3.25.**

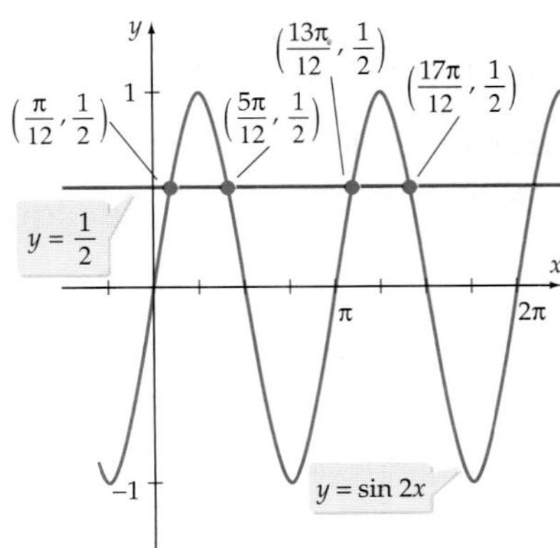

Figure 3.25

$$f_4(x) = x + \frac{x^3}{2 \cdot 3} + \frac{1 \cdot 3x^5}{2 \cdot 4 \cdot 5} + \frac{1 \cdot 3 \cdot 5x^7}{2 \cdot 4 \cdot 6 \cdot 7} + \frac{1 \cdot 3 \cdot 5 \cdot 7x^9}{2 \cdot 4 \cdot 6 \cdot 8 \cdot 9} \quad \text{where } -1 \le x \le 1$$

$$\vdots$$

$$f_n(x) = x + \frac{x^3}{2 \cdot 3} + \frac{1 \cdot 3x^5}{2 \cdot 4 \cdot 5} + \frac{1 \cdot 3 \cdot 5x^7}{2 \cdot 4 \cdot 6 \cdot 7} + \cdots + \frac{(2n)!\, x^{2n+1}}{(2^n n!)^2(2n+1)}$$

where $-1 \le x \le 1,\ n! = 1 \cdot 2 \cdot 3 \cdots (n-1)n$

and $(2n)! = 1 \cdot 2 \cdot 3 \cdots (2n-1)(2n)$

Use a graphing utility for the following exercises.

1. Graph $y = f_1(x)$, $y = f_2(x)$, $y = f_3(x)$, and $y = f_4(x)$ in the viewing window Xmin = −1, Xmax = 1, Ymin = −1.5708, Ymax = 1.5708.
2. Determine the values of x for which $f_3(x)$ and $\sin^{-1} x$ differ by less than 0.001. That is, determine the values of x for which $|f_3(x) - \sin^{-1} x| < 0.001$.
3. Determine the values of x for which $|f_4(x) - \sin^{-1} x| < 0.001$.
4. Write all seven terms of $f_6(x)$. Graph $y = f_6(x)$ and $y = \sin^{-1} x$ on the viewing window Xmin = −1, Xmax = 1, Ymin = $-\frac{\pi}{2}$, Ymax = $\frac{\pi}{2}$.
5. Write all seven terms of $f_6(1)$. What do you notice about the size of a term compared to that of the preceding term?
6. What is the largest-degree term in $f_{10}(x)$?

Chapter 3 Summary

3.1 Verification of Trigonometric Identities

- Trigonometric identities are verified by using algebraic methods and previously proved identities.
- The Fundamental Trigonometric Identities are given in Table 3.1, page 217.
- Guidelines for verifying trigonometric identities are given on pages 217 and 218.

3.2 Sum, Difference, and Cofunction Identities

- Sum and difference identities for the cosine function

$$\cos(\alpha - \beta) = \cos\alpha\cos\beta + \sin\alpha\sin\beta$$
$$\cos(\alpha + \beta) = \cos\alpha\cos\beta - \sin\alpha\sin\beta$$

- Sum and difference identities for the sine function

$$\sin(\alpha - \beta) = \sin\alpha\cos\beta - \cos\alpha\sin\beta$$
$$\sin(\alpha + \beta) = \sin\alpha\cos\beta + \cos\alpha\sin\beta$$

- Sum and difference identities for the tangent function

$$\tan(\alpha + \beta) = \frac{\tan\alpha + \tan\beta}{1 - \tan\alpha\tan\beta}$$
$$\tan(\alpha - \beta) = \frac{\tan\alpha - \tan\beta}{1 + \tan\alpha\tan\beta}$$

- Cofunction identities with θ in degrees

$$\sin(90° - \theta) = \cos\theta \qquad \cos(90° - \theta) = \sin\theta$$
$$\tan(90° - \theta) = \cot\theta \qquad \cot(90° - \theta) = \tan\theta$$
$$\sec(90° - \theta) = \csc\theta \qquad \csc(90° - \theta) = \sec\theta$$

If θ is in radian measure, replace 90° with $\frac{\pi}{2}$.

3.3 Double- and Half-Angle Identities

- Double-angle identities

$$\sin 2\alpha = 2 \sin \alpha \cos \alpha$$
$$\cos 2\alpha = \cos^2 \alpha - \sin^2 \alpha = 1 - 2\sin^2 \alpha = 2\cos^2 \alpha - 1$$
$$\tan 2\alpha = \frac{2 \tan \alpha}{1 - \tan^2 \alpha}$$

- Power-reducing identities

$$\sin^2 \alpha = \frac{1 - \cos 2\alpha}{2}$$
$$\cos^2 \alpha = \frac{1 + \cos 2\alpha}{2}$$
$$\tan^2 \alpha = \frac{1 - \cos 2\alpha}{1 + \cos 2\alpha}$$

- Half-angle identities

$$\sin \frac{\alpha}{2} = \pm\sqrt{\frac{1 - \cos \alpha}{2}}$$
$$\cos \frac{\alpha}{2} = \pm\sqrt{\frac{1 + \cos \alpha}{2}}$$
$$\tan \frac{\alpha}{2} = \pm\sqrt{\frac{1 - \cos \alpha}{1 + \cos \alpha}} = \frac{\sin \alpha}{1 + \cos \alpha} = \frac{1 - \cos \alpha}{\sin \alpha}$$

The choice of the plus or minus sign depends on the quadrant in which $\frac{\alpha}{2}$ lies.

3.4 Identities Involving the Sum of Trigonometric Functions

- Product-to-sum identities

$$\sin \alpha \cos \beta = \frac{1}{2}[\sin(\alpha + \beta) + \sin(\alpha - \beta)]$$
$$\cos \alpha \sin \beta = \frac{1}{2}[\sin(\alpha + \beta) - \sin(\alpha - \beta)]$$
$$\cos \alpha \cos \beta = \frac{1}{2}[\cos(\alpha + \beta) + \cos(\alpha - \beta)]$$
$$\sin \alpha \sin \beta = \frac{1}{2}[\cos(\alpha - \beta) - \cos(\alpha + \beta)]$$

- Sum-to-product identities

$$\sin x + \sin y = 2 \sin \frac{x + y}{2} \cos \frac{x - y}{2}$$
$$\cos x + \cos y = 2 \cos \frac{x + y}{2} \cos \frac{x - y}{2}$$
$$\sin x - \sin y = 2 \cos \frac{x + y}{2} \sin \frac{x - y}{2}$$
$$\cos x - \cos y = -2 \sin \frac{x + y}{2} \sin \frac{x - y}{2}$$

- For sums of the form $a \sin x + b \cos x$,

$$a \sin x + b \cos x = k \sin(x + \alpha)$$

where $k = \sqrt{a^2 + b^2}$, $\sin \alpha = \frac{b}{\sqrt{a^2 + b^2}}$, and $\cos \alpha = \frac{a}{\sqrt{a^2 + b^2}}$.

3.5 Inverse Trigonometric Functions

- The inverse of $y = \sin x$ is $y = \sin^{-1} x$, with $-1 \le x \le 1$ and $-\frac{\pi}{2} \le y \le \frac{\pi}{2}$.
- The inverse of $y = \cos x$ is $y = \cos^{-1} x$, with $-1 \le x \le 1$ and $0 \le y \le \pi$.
- The inverse of $y = \tan x$ is $y = \tan^{-1} x$, with $-\infty < x < \infty$ and $-\frac{\pi}{2} < y < \frac{\pi}{2}$.
- The inverse of $y = \cot x$ is $y = \cot^{-1} x$, with $-\infty < x < \infty$ and $0 < y < \pi$.
- The inverse of $y = \csc x$ is $y = \csc^{-1} x$, with $x \le -1$ or $x \ge 1$ and $-\frac{\pi}{2} \le y \le \frac{\pi}{2}$, $y \ne 0$.
- The inverse of $y = \sec x$ is $y = \sec^{-1} x$, with $x \le -1$ or $x \ge 1$ and $0 \le y \le \pi$, $y \ne \frac{\pi}{2}$.

3.6 Trigonometric Equations

- Algebraic methods and identities are used to solve trigonometric equations. Because the trigonometric functions are periodic, there may be an infinite number of solutions. If solutions cannot be found by algebraic methods, then we often use a graphing utility to find approximate solutions.

Chapter 3 Assessing Concepts

1. True or False: $\cos^2 \alpha = \dfrac{1 + \cos 2\alpha}{2}$ is an identity.

2. True or False: For all real numbers x, $\cos^{-1}(\cos x) = x$.

3. True or False: For all real numbers x, $\cos(\cos^{-1} x) = x$.

4. True or False: The graph of $y = \sin^{-1} x$ is symmetric with respect to the origin.

5. How many solutions does the equation $\sin 2x = 0.3$ have on the interval $0 \le x < 2\pi$?

6. What is the domain of $f(x) = \cos^{-1} 2x$?

7. What is the range of $y = \cos^{-1} x$?

8. Determine the exact value of $\sin^{-1}\left(\sin \dfrac{7\pi}{3}\right)$.

9. What is the amplitude of $f(x) = \sin x + \cos x$?

10. Determine the exact value of $\tan 75°$.

Chapter 3 Review Exercises

In Exercises 1 to 10, find the exact value.

1. $\cos(45° + 30°)$
2. $\tan(210° - 45°)$
3. $\sin\left(\dfrac{2\pi}{3} + \dfrac{\pi}{4}\right)$
4. $\sec\left(\dfrac{4\pi}{3} - \dfrac{\pi}{4}\right)$
5. $\sin(60° - 135°)$
6. $\cos\left(\dfrac{5\pi}{3} - \dfrac{7\pi}{4}\right)$
7. $\sin\left(22\dfrac{1}{2}\right)°$
8. $\cos 105°$
9. $\tan\left(67\dfrac{1}{2}\right)°$
10. $\sin 112.5°$

In Exercises 11 to 14, find the exact values of the given functions.

11. Given $\sin \alpha = \dfrac{1}{2}$ with $0° < \alpha < 90°$, and $\cos \beta = \dfrac{1}{2}$ with $270° < \beta < 360°$, find

    a. $\cos(\alpha - \beta)$ b. $\tan 2\alpha$ c. $\sin\left(\dfrac{\beta}{2}\right)$

12. Given $\sin \alpha = \dfrac{\sqrt{3}}{2}$ with $90° < \alpha < 180°$, and $\cos \beta = -\dfrac{1}{2}$ with $180° < \beta < 270°$, find

    a. $\sin(\alpha + \beta)$ b. $\sec 2\beta$ c. $\cos\left(\dfrac{\alpha}{2}\right)$

13. Given $\sin \alpha = -\dfrac{1}{2}$ with $270° < \alpha < 360°$, and $\cos \beta = -\dfrac{\sqrt{3}}{2}$ with $180° < \beta < 270°$, find

    a. $\sin(\alpha - \beta)$ b. $\tan 2\alpha$ c. $\cos\left(\dfrac{\beta}{2}\right)$

14. Given $\sin \alpha = \dfrac{\sqrt{2}}{2}$ with $0° < \alpha < 90°$, and $\cos \beta = \dfrac{\sqrt{3}}{2}$ with $270° < \beta < 360°$, find

    a. $\cos(\alpha - \beta)$ b. $\tan 2\beta$ c. $\sin 2\alpha$

In Exercises 15 to 20, write the given expression as a single trigonometric function.

15. $2 \sin 3x \cos 3x$
16. $\dfrac{\tan 2x + \tan x}{1 - \tan 2x \tan x}$
17. $\sin 4x \cos x - \cos 4x \sin x$
18. $\cos^2 2\theta - \sin^2 2\theta$
19. $\dfrac{\sin 2\theta}{\cos 2\theta}$
20. $\dfrac{1 - \cos 2\theta}{\sin 2\theta}$

In Exercises 21 to 24, write each expression as the product of two trigonometric functions.

21. $\cos 2\theta - \cos 4\theta$
22. $\sin 3\theta - \sin 5\theta$
23. $\sin 6\theta + \sin 2\theta$
24. $\sin 5\theta - \sin \theta$

In Exercises 25 to 42, verify the identity.

25. $\dfrac{1}{\sin x - 1} + \dfrac{1}{\sin x + 1} = -2 \tan x \sec x$

26. $\dfrac{\sin x}{1 - \cos x} = \csc x + \cot x, \quad 0 < x < \dfrac{\pi}{2}$

27. $\dfrac{1 + \sin x}{\cos^2 x} = \tan^2 x + 1 + \tan x \sec x$

28. $\cos^2 x - \sin^2 x - \sin 2x = \dfrac{\cos^2 2x - \sin^2 2x}{\cos 2x + \sin 2x}$

29. $\dfrac{1}{\cos x} - \cos x = \tan x \sin x$

30. $\sin(270° - \theta) - \cos(270° - \theta) = \sin \theta - \cos \theta$

31. $\sin\left(\dfrac{\pi}{4} - \alpha\right) = \dfrac{\sqrt{2}}{2}(\cos \alpha - \sin \alpha)$

32. $\sin(180° - \alpha + \beta) = \sin \alpha \cos \beta - \cos \alpha \sin \beta$

33. $\dfrac{\sin 4x - \sin 2x}{\cos 4x - \cos 2x} = -\cot 3x$

34. $2 \sin x \sin 3x = (1 - \cos 2x)(1 + 2 \cos 2x)$

35. $\sin x - \cos 2x = (2 \sin x - 1)(\sin x + 1)$

36. $\cos 4x = 1 - 8 \sin^2 x + 8 \sin^4 x$

37. $\tan 4x = \dfrac{4 \tan x - 4 \tan^3 x}{1 - 6 \tan^2 x + \tan^4 x}$

38. $\dfrac{\sin 2x - \sin x}{\cos 2x + \cos x} = \dfrac{1 - \cos x}{\sin x}$

39. $2 \cos 4x \sin 2x = 2 \sin 3x \cos 3x - 2 \sin x \cos x$

40. $2 \sin x \sin 2x = 4 \cos x \sin^2 x$

41. $\cos(x + y) \cos(x - y) = \cos^2 x + \cos^2 y - 1$

42. $\cos(x + y) \sin(x - y) = \sin x \cos x - \sin y \cos y$

In Exercises 43 to 46, evaluate each expression.

43. $\sec\left(\sin^{-1} \dfrac{12}{13}\right)$

44. $\cos\left(\sin^{-1} \dfrac{3}{5}\right)$

45. $\cos\left[\sin^{-1}\left(-\dfrac{3}{5}\right) + \cos^{-1} \dfrac{5}{13}\right]$

46. $\cos\left(2 \sin^{-1} \dfrac{3}{5}\right)$

In Exercises 47 and 48, solve each equation.

47. $2 \sin^{-1}(x - 1) = \dfrac{\pi}{3}$

48. $\sin^{-1} x + \cos^{-1} \dfrac{4}{5} = \dfrac{\pi}{2}$

In Exercises 49 and 50, find all solutions of each equation with $0° \le x < 360°$.

49. $4 \sin^2 x + 2\sqrt{3} \sin x - 2 \sin x - \sqrt{3} = 0$

50. $2 \sin x \cos x - \sqrt{2} \cos x - 2 \sin x + \sqrt{2} = 0$

In Exercises 51 and 52, solve the trigonometric equation where *x* is in radians. Round approximate solutions to four decimal places.

51. $3 \cos^2 x + \sin x = 1$

52. $\tan^2 x - 2 \tan x - 3 = 0$

In Exercises 53 and 54, solve each equation on $0 \le x < 2\pi$.

53. $\sin 3x \cos x - \cos 3x \sin x = \dfrac{1}{2}$

54. $\cos\left(2x - \dfrac{\pi}{3}\right) = -\dfrac{\sqrt{3}}{2}$

In Exercises 55 to 58, write the equation in the form $y = k \sin(x + \alpha)$, where the measure of α is in radians. Graph one period of each function.

55. $f(x) = \sqrt{3} \sin x + \cos x$

56. $f(x) = -2 \sin x - 2 \cos x$

57. $f(x) = -\sin x - \sqrt{3} \cos x$

58. $f(x) = \dfrac{\sqrt{3}}{2} \sin x - \dfrac{1}{2} \cos x$

In Exercises 59 to 62, graph each function.

59. $f(x) = 2 \cos^{-1} x$

60. $f(x) = \sin^{-1}(x - 1)$

61. $f(x) = \sin^{-1} \dfrac{x}{2}$

62. $f(x) = \sec^{-1} 2x$

63. **Sunrise Time** The table below shows the sunrise time for Flagstaff, Arizona, for selected days in 2009.

Date	Day Number	Sunrise Time (hours:minutes)
Jan. 1	1	7:35
Feb. 1	32	7:25
Mar. 1	60	6:56
April 1	91	6:13
May 1	121	5:35
June 1	152	5:13
July 1	182	5:16
Aug. 1	213	5:36
Sept. 1	244	5:59
Oct. 1	274	6:22
Nov. 1	305	6:48
Dec. 1	335	7:17

Source: The U.S. Naval Observatory. *Note:* The times are Mountain Standard Times.

a. Find the sine regression function that models the sunrise time, in hours, as a function of the day number. Let $x = 1$ represent January 1, 2009. Assume that the sunrise times have a period of 365.25 days.

b. Use the regression function to estimate the sunrise time (to the nearest minute) for April 14, 2009 ($x = 104$).

»»» Quantitative Reasoning: *Basketball and Trigonometric Equations* »»»

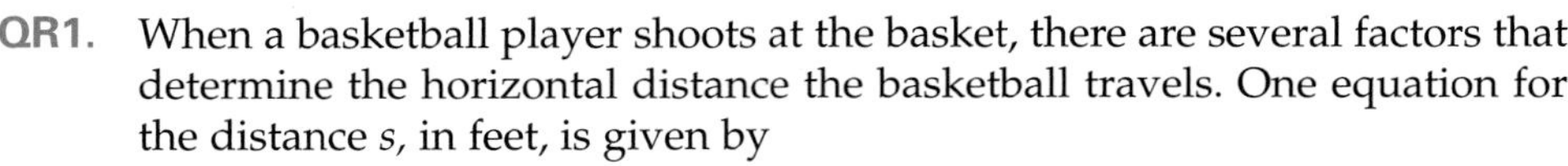

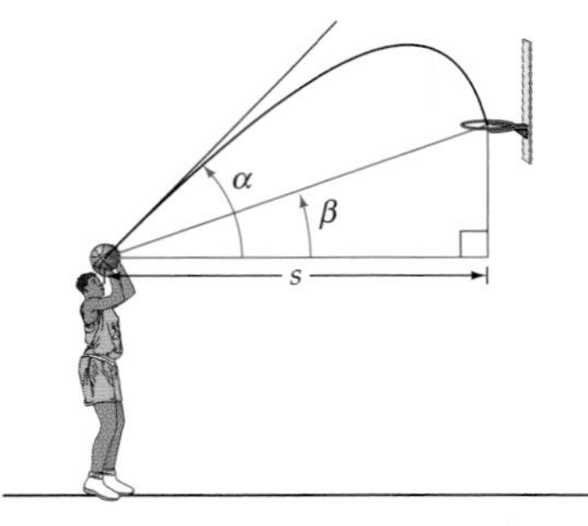

QR1. When a basketball player shoots at the basket, there are several factors that determine the horizontal distance the basketball travels. One equation for the distance s, in feet, is given by

$$s = \frac{v^2 \cos\alpha \sec\beta \sin(\alpha - \beta)}{16} \qquad (1)$$

where the angles α and β are as shown in the diagram at the left and v is the velocity in feet per second at which the ball leaves the player's hand.

a. Graph Equation (1) for $\beta = \frac{\pi}{18}$, $v = 28$ feet per second, and $\beta < \alpha < 1.35$. Use the graph to find the value of α that maximizes s for the given values.

b. Show that the value of α in **a.** is approximately $\frac{\pi}{4} + \frac{\beta}{2}$.

QR2. Verify the identity $\cos\alpha \sec\beta \sin(\alpha - \beta) = \cos^2\alpha(\tan\alpha - \tan\beta)$ and then use the identity to show that Equation (1) can be written as

$$v = 4\sec\alpha\sqrt{\frac{s}{\tan\alpha - \tan\beta}} \qquad (2)$$

Note: Equation (2) gives the velocity at which a ball must be launched for various launch angles to reach a basket s feet away.

a. Graph Equation (2) for $\beta = \frac{\pi}{18}$, $s = 15$ feet (the distance of a free throw), and $\beta < \alpha < 1.35$. Use the graph to find the value of α that maximizes v for the given values.

b. Show that the value of α in part **a.** is approximately $\frac{\pi}{4} + \frac{\beta}{2}$. *Note:* The significance of this result is that the ball must be launched at an angle of $\alpha = \frac{\pi}{4} + \frac{\beta}{2}$ to travel the required distance with the least effort.

Chapter 3 Test

1. Verify the identity $1 + \sin^2 x \sec^2 x = \sec^2 x$.

2. Verify the identity

$$\frac{1}{\sec x - \tan x} - \frac{1}{\sec x + \tan x} = 2\tan x$$

3. Verify the identity $\cos^3 x + \cos x \sin^2 x = \cos x$.

4. Verify the identity $\csc x - \cot x = \frac{1 - \cos x}{\sin x}$.

5. Find the exact value of $\sin 195°$.

6. Given $\sin\alpha = -\frac{3}{5}$, α in Quadrant III, and $\cos\beta = -\frac{\sqrt{2}}{2}$, β in Quadrant II, find $\sin(\alpha + \beta)$.

7. Verify the identity $\sin\left(\theta - \frac{3\pi}{2}\right) = \cos\theta$.

8. Write $\cos 6x \sin 3x + \sin 6x \cos 3x$ in terms of a single trigonometric function.

9. Find the exact value of $\cos 2\theta$ given that $\sin\theta = \frac{4}{5}$ and θ is in Quadrant II.

10. Verify the identity $\tan\frac{\theta}{2} + \frac{\cos\theta}{\sin\theta} = \csc\theta$.

11. Verify the identity $\sin^2 2x + 4\cos^4 x = 4\cos^2 x$.

12. Find the exact value of $\sin 15° \cos 75°$.

13. Write $y = -\frac{\sqrt{3}}{2}\sin x + \frac{1}{2}\cos x$ in the form $y = k\sin(x + \alpha)$, where α is measured in radians.

14. Use a calculator to approximate the radian measure of $\cos^{-1} 0.7644$ to the nearest thousandth.

15. Find the exact value of $\sin\left(\cos^{-1}\frac{12}{13}\right)$.

16. Graph $y = \sin^{-1}(x + 2)$.

17. Solve $3 \sin x - 2 = 0$, where $0° \le x < 360°$. (State solutions to the nearest tenth of a degree.)

18. Solve $\sin x \cos x - \frac{\sqrt{3}}{2} \sin x = 0$, where $0 \le x < 2\pi$.

19. Find the exact solutions of $\sin 2x + \sin x - 2 \cos x - 1 = 0$, where $0 \le x < 2\pi$.

20.  **Hours of Daylight** The table below shows the hours of daylight for Tampa, Florida, for selected days in 2009.

Date	Day Number	Hours of Daylight
Jan. 1	1	10:24
Feb. 1	32	10:53
Mar. 1	60	11:37
April 1	91	12:28
May 1	121	13:15
June 1	152	13:48
July 1	182	13:53
Aug. 1	213	13:26
Sept. 1	244	12:41
Oct. 1	274	11:52
Nov. 1	305	11:04
Dec. 1	335	10:30

Source: The U.S. Naval Observatory

a. Find the sine regression function that models the hours of daylight as a function of the day number. Let $x = 1$ represent January 1, 2009. Use 365.25 for the period of the data.

b. Use the regression function to estimate the hours of daylight (stated in hours and minutes, with the minutes rounded to the nearest minute) for Tampa on March 16, 2009 ($x = 75$).

Cumulative Review Exercises

1. Solve: $-2x + 1 < 7$

2. Explain how to use the graph of $y = f(x)$ to produce the graph of $y = f(x + 1) + 2$.

3. Explain how to use the graph of $y = f(x)$ to produce the graph of $y = -f(x)$.

4. Determine whether $f(x) = x - \sin x$ is an even function or an odd function.

5. Find the inverse of $f(x) = \frac{5x}{x - 1}$.

6. Convert 240° to radians.

7. Convert $\frac{5\pi}{3}$ to degrees.

8. Evaluate: $\sin \frac{\pi}{3}$

9. Evaluate: $\csc 60°$

10. Find $\tan \theta$, given that θ is an acute angle and $\sin \theta = \frac{2}{3}$.

11. Determine the sign of $\cot \theta$ given that $\pi < \theta < \dfrac{3\pi}{2}$.

12. What is the measure of the reference angle for the angle $\theta = 310°$?

13. What is the measure of the reference angle for the angle $\theta = \dfrac{5\pi}{3}$?

14. Find the x- and y-coordinates of the point defined by $W\left(\dfrac{\pi}{3}\right)$, where W is the wrapping function.

15. Find the amplitude, the period, and the phase shift for the graph of $y = 0.43 \sin\left(2x - \dfrac{\pi}{6}\right)$.

16. Evaluate: $\sin^{-1}\dfrac{1}{2}$

17. Use a calculator to evaluate $\cos^{-1}(-0.8)$. Round to the nearest thousandth.

18. Use interval notation to state the domain of $f(x) = \cos^{-1} x$.

19. Use interval notation to state the range of $f(x) = \tan^{-1} x$.

20. Find the exact solutions of $2\cos^2 x - 1 = -\sin x$, where $0 \le x < 2\pi$.

4 Applications of Trigonometry

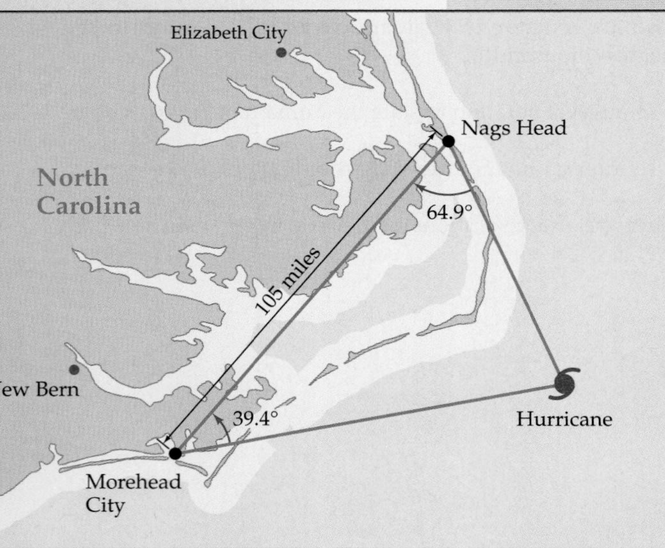

Trigonometry and Indirect Measurement

In Chapter 2 we used trigonometric functions to find the unknown length of a side of a given *right triangle*. In this chapter we develop theorems that can be used to find the length of a side or the measure of an angle of *any* triangle, even if it is not a right triangle. These theorems are often used in the areas of navigation, surveying, and building design. Meterologists use these theorems to estimate the distance from an approaching hurricane to cities in the projected path of the hurricane. For instance, in the diagram on the left, the distance from the hurricane to Nags Head can be determined using the Law of Sines, a theorem presented in this chapter.

See Exercises 30 and 31 on page 299 for additional applications that can be solved by using the Law of Sines.

Online Study Center
For online student resources, such as section quizzes, visit this website: college.hmco.com/info/aufmannCAT

Section 4.1

- The Law of Sines
- The Ambiguous Case (SSA)
- Applications of the Law of Sines

The Law of Sines

The Law of Sines

Solving a triangle involves finding the lengths of all sides and the measures of all angles in the triangle. In this section and the next we develop formulas for solving an **oblique triangle,** which is a triangle that does not contain a right angle. The *Law of Sines* can be used to solve oblique triangles in which either two angles and a side or two sides and an angle opposite one of the sides are known. In **Figure 4.1,** altitude CD is drawn from C. The length of the altitude is h. Triangles ACD and BCD are right triangles.

Using the definition of the sine of an angle of a right triangle, we have from **Figure 4.1**

$$\sin B = \frac{h}{a} \qquad \sin A = \frac{h}{b}$$

$$h = a \sin B \quad (1) \qquad h = b \sin A \quad (2)$$

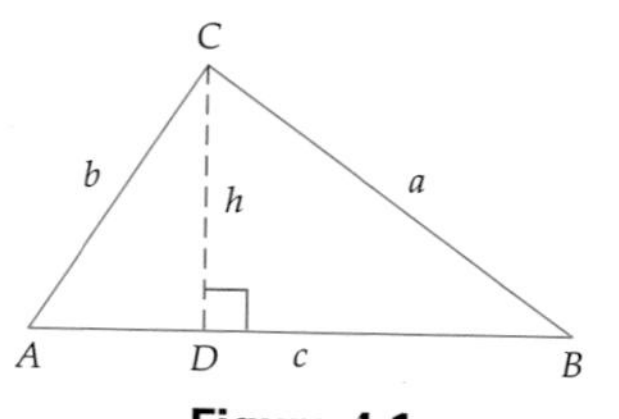

Figure 4.1

Equating the values of h in Equations (1) and (2), we obtain

$$a \sin B = b \sin A$$

Dividing each side of the equation by $\sin A \sin B$, we obtain

$$\frac{a}{\sin A} = \frac{b}{\sin B}$$

Similarly, when an altitude is drawn to a different side, the following formulas result:

$$\frac{c}{\sin C} = \frac{b}{\sin B} \quad \text{and} \quad \frac{c}{\sin C} = \frac{a}{\sin A}$$

take note

The Law of Sines may also be written as

$$\frac{\sin A}{a} = \frac{\sin B}{b} = \frac{\sin C}{c}$$

The Law of Sines

If A, B, and C are the measures of the angles of a triangle and a, b, and c are the lengths of the sides opposite those angles, then

$$\frac{a}{\sin A} = \frac{b}{\sin B} = \frac{c}{\sin C}$$

EXAMPLE 1 ⟫ Solve a Triangle Using the Law of Sines

Solve triangle ABC if $A = 42°$, $B = 63°$, and $c = 18$ centimeters.

Continued ▶

Solution

Find C by using the fact that the sum of the interior angles of a triangle is 180°.

$$A + B + C = 180^\circ$$
$$42^\circ + 63^\circ + C = 180^\circ$$
$$C = 75^\circ$$

take note

We have used the rounding conventions stated on page 140 to determine the number of significant digits to be used for a and b.

Use the Law of Sines to find a.

$$\frac{a}{\sin A} = \frac{c}{\sin C}$$

$$\frac{a}{\sin 42^\circ} = \frac{18}{\sin 75^\circ} \qquad \bullet\ A = 42^\circ,\ c = 18,\ C = 75^\circ$$

$$a = \frac{18 \sin 42^\circ}{\sin 75^\circ} \approx 12 \text{ centimeters}$$

Use the Law of Sines again, this time to find b.

$$\frac{b}{\sin B} = \frac{c}{\sin C}$$

$$\frac{b}{\sin 63^\circ} = \frac{18}{\sin 75^\circ} \qquad \bullet\ B = 63^\circ,\ c = 18,\ C = 75^\circ$$

$$b = \frac{18 \sin 63^\circ}{\sin 75^\circ} \approx 17 \text{ centimeters}$$

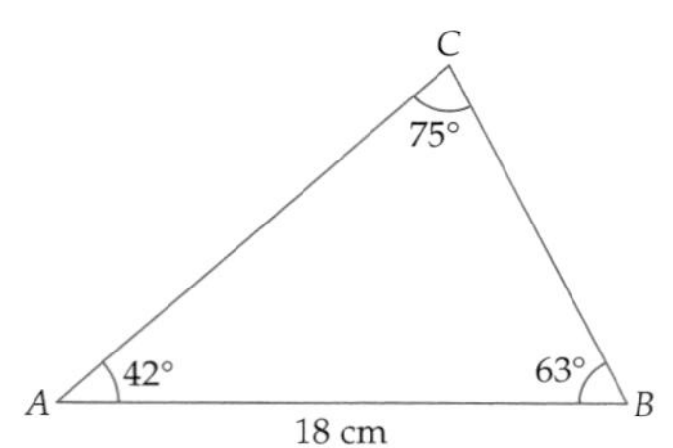

Figure 4.2

The solution is $C = 75^\circ$, $a \approx 12$ centimeters, and $b \approx 17$ centimeters. A scale drawing can be used to see if these results are reasonable. See **Figure 4.2**.

Try Exercise 4, page 298

The Ambiguous Case (SSA)

When you are given two sides of a triangle and an angle opposite one of them, you may find that the triangle is not unique. Some information may result in two triangles, and some may result in no triangle at all. It is because of this that the case of knowing two sides and an angle opposite one of them (SSA) is called the *ambiguous case* of the Law of Sines.

Suppose that sides a and c and the nonincluded angle A of a triangle are known and we are asked to solve triangle ABC. The relationships among h, the height of the triangle, a (the side opposite $\angle A$), and c determine whether there are no, one, or two triangles.

Case 1 First consider the case in which $\angle A$ is an acute angle (see **Figure 4.3**). There are four possible situations.

1. $a < h$; there is no possible triangle.
2. $a = h$; there is one triangle, a right triangle.
3. $h < a < c$; there are two possible triangles.
4. $a \geq c$; there is one triangle, which is not a right triangle.

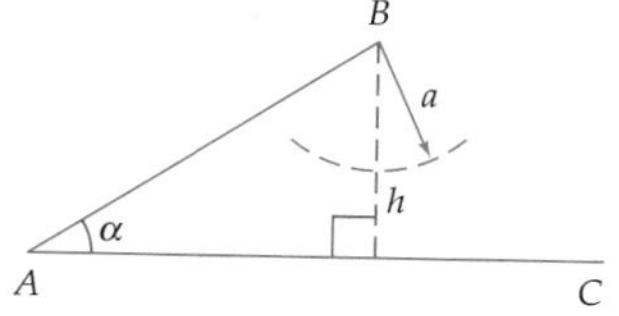

1. $a < h$; no triangle

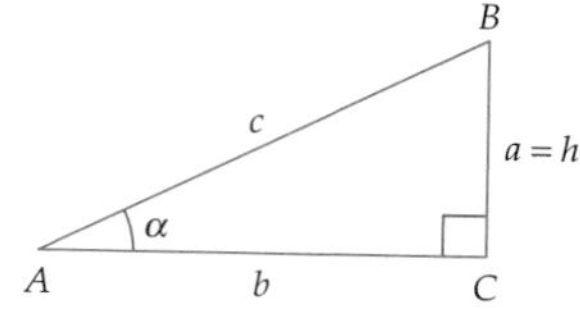

2. $a = h$; one triangle

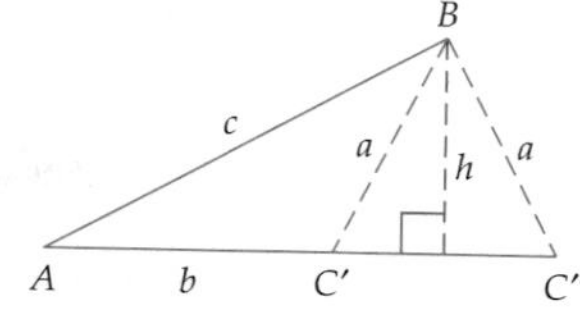

3. $h < a < c$; two triangles

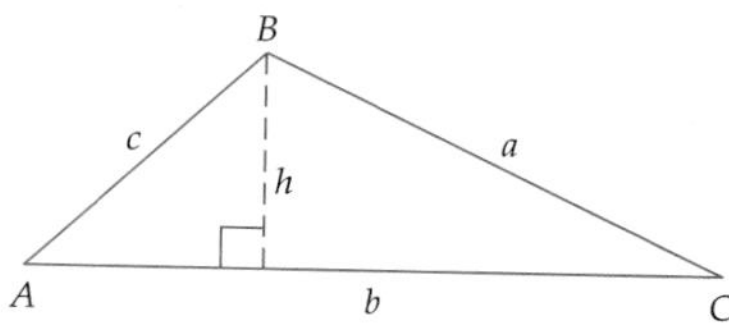

4. $a \geq c$; one triangle

Figure 4.3
Case 1: A is an acute angle.

Case 2 Now consider the case in which $\angle A$ is an obtuse angle (see **Figure 4.4**). Here, there are two possible situations.

1. $a \leq c$; there is no triangle.
2. $a > c$; there is one triangle.

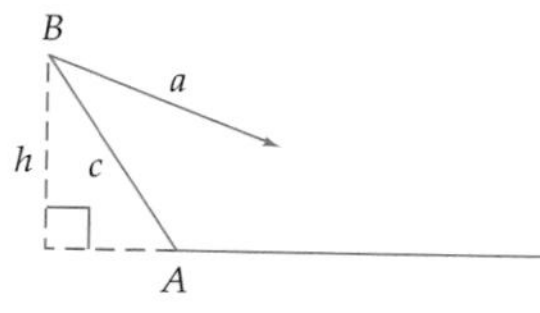

1. $a \leq c$; no triangle

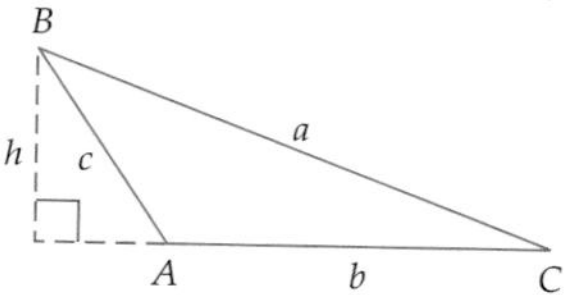

2. $a > c$; one triangle

Figure 4.4
Case 2: A is an obtuse angle.

take note

When you solve for an angle of a triangle by using the Law of Sines, be aware that the value of the inverse sine function will give the measure of an acute angle. If the situation is the ambiguous case (SSA), you must consider a second, obtuse angle by using the supplement of the angle. You can use a scale drawing to see whether your results are reasonable.

EXAMPLE 2 ›› Solve a Triangle Using the Law of Sines (SSA)

a. Find A, given triangle ABC with $B = 32°$, $a = 42$, and $b = 30$.

b. Find C, given triangle ABC with $A = 57°$, $a = 15$ feet, and $c = 20$ feet.

Solution

a. $$\frac{b}{\sin B} = \frac{a}{\sin A}$$

$$\frac{30}{\sin 32°} = \frac{42}{\sin A}$$ • $B = 32°$, $a = 42$, $b = 30$

$$\sin A = \frac{42 \sin 32°}{30} \approx 0.7419$$

$$A \approx 48° \text{ or } 132°$$ • The two angles with measure between 0° and 180° that have a sine of 0.7419 are approximately 48° and 132°.

To check that $A \approx 132°$ is a valid result, add 132° to the measure of the given angle B (32°). Because $132° + 32° < 180°$, we know that $A \approx 132°$ is a valid result. Thus angle $A \approx 48°$ or $A \approx 132°$ ($\angle BAC$ in **Figure 4.5**).

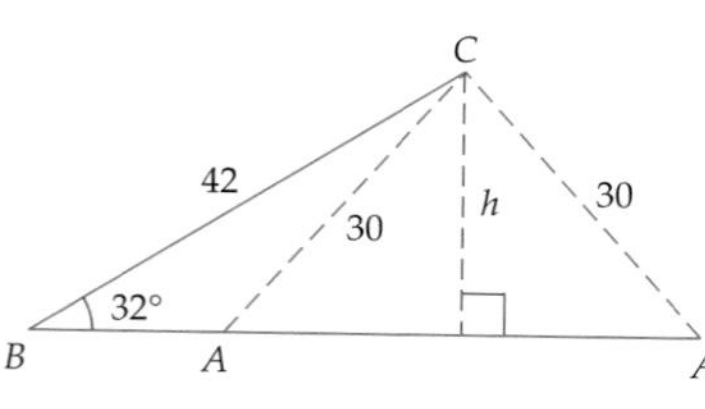

Figure 4.5

Continued ▶

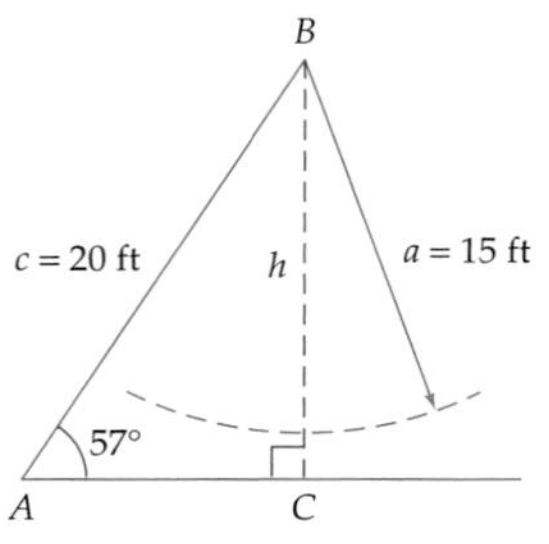

Figure 4.6

b. $$\frac{a}{\sin A} = \frac{c}{\sin C}$$

$$\frac{15}{\sin 57^\circ} = \frac{20}{\sin C}$$ • $A = 57^\circ$, $a = 15$, $c = 20$

$$\sin C = \frac{20 \sin 57^\circ}{15} \approx 1.1182$$

Because 1.1182 is not in the range of the sine function, there is no solution of the equation. Thus there is no triangle for these values of A, a, and c. See **Figure 4.6.**

Try Exercise 20, page 298

Applications of the Law of Sines

EXAMPLE 3 Solve an Application Using the Law of Sines

A radio antenna 85 feet high is located on top of an office building. At a distance AD from the base of the building, the angle of elevation to the top of the antenna is 26°, and the angle of elevation to the bottom of the antenna is 16°. Find the height of the building.

Solution

Sketch a diagram. See **Figure 4.7**. Find the measure of angle B and the measure of angle β.

$$B = 90^\circ - 26^\circ = 64^\circ$$
$$\beta = 26^\circ - 16^\circ = 10^\circ$$

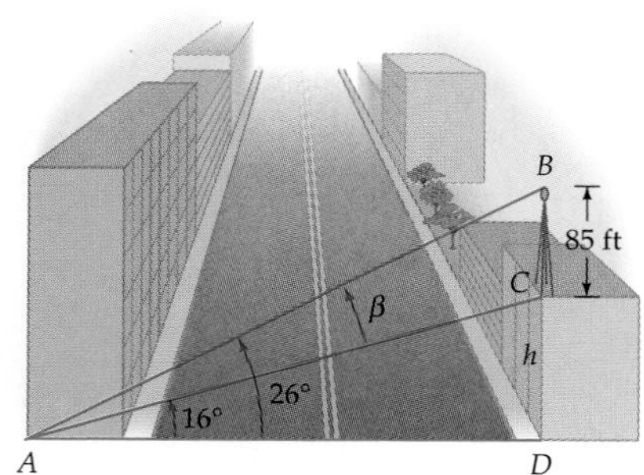

Figure 4.7

Because we know the length BC and the measure of β, we can use triangle ABC and the Law of Sines to find length AC.

$$\frac{BC}{\sin \beta} = \frac{AC}{\sin B}$$

$$\frac{85}{\sin 10^\circ} = \frac{AC}{\sin 64^\circ}$$ • $BC = 85$, $\beta = 10^\circ$, $B = 64^\circ$

$$AC = \frac{85 \sin 64^\circ}{\sin 10^\circ}$$

Having found AC, we can now find the height of the building.

$$\sin 16^\circ = \frac{h}{AC}$$
$$h = AC \sin 16^\circ$$
$$= \frac{85 \sin 64^\circ}{\sin 10^\circ} \sin 16^\circ \approx 121 \text{ feet}$$ • Substitute for AC.

The height of the building, to two significant digits, is 120 feet.

Try Exercise 30, page 299

take note

In Example 3 we rounded the height of the building to two significant digits to comply with the rounding conventions given on page 140.

In navigation and surveying problems, there are two commonly used methods for specifying direction. The angular direction in which a craft is pointed is called the **heading.** Heading is expressed in terms of an angle measured clockwise from north. **Figure 4.8** shows a heading of 65° and a heading of 285°.

The angular direction used to locate one object in relation to another object is called the **bearing.** Bearing is expressed in terms of the acute angle formed by a north–south line and the line of direction. **Figure 4.9** shows a bearing of N38°W and a bearing of S15°E.

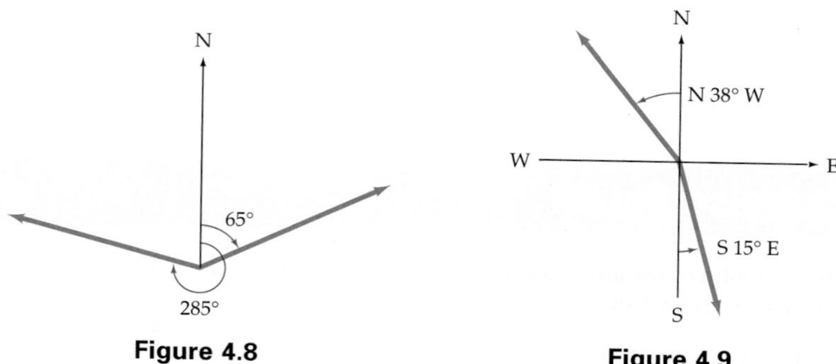

Figure 4.8 **Figure 4.9**

? QUESTION Can a bearing of N50°E be written as N310°W?

EXAMPLE 4 » Solve an Application

A ship with a heading of 330° first sighted a lighthouse at a bearing of N65°E. After traveling 8.5 miles, the ship observed the lighthouse at a bearing of S50°E. Find the distance from the ship to the lighthouse when the first sighting was made.

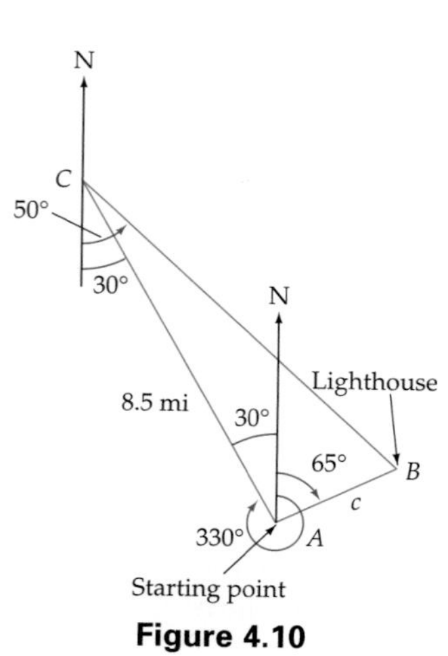

Figure 4.10

Solution

Use the given information to draw a diagram. See **Figure 4.10** which shows that the measure of $\angle CAB = 65° + 30° = 95°$, the measure of $\angle BCA = 50° - 30° = 20°$, and $B = 180° - 95° - 20° = 65°$. Use triangle ABC and the Law of Sines to find c.

$$\frac{b}{\sin B} = \frac{c}{\sin C}$$

$$\frac{8.5}{\sin 65°} = \frac{c}{\sin 20°} \qquad \text{• } b = 8.5,\ B = 65°,\ C = 20°$$

$$c = \frac{8.5 \sin 20°}{\sin 65°} \approx 3.2$$

The lighthouse was 3.2 miles (to two significant digits) from the ship when the first sighting was made.

» *Try Exercise 40, page 300*

? ANSWER No. A bearing is always expressed using an acute angle.

Topics for Discussion

1. Is it possible to solve a triangle if the only given information consists of the measures of the three angles of the triangle? Explain.
2. Explain why it is not possible (in general) to use the Law of Sines to solve a triangle for which we are given only the lengths of all the sides.
3. Draw a triangle with dimensions $A = 30°$, $c = 3$ inches, and $a = 2.5$ inches. Is your answer unique? That is, can more than one triangle with the given dimensions be drawn?

Exercise Set 4.1

In Exercises 1 to 44, round answers according to the rounding conventions on page 140.

In Exercises 1 to 14, solve the triangles.

1. $A = 42°, B = 61°, a = 12$
2. $B = 25°, C = 125°, b = 5.0$
3. $A = 110°, C = 32°, b = 12$

4. $B = 28°, C = 78°, c = 44$
5. $A = 132°, a = 22, b = 16$
6. $B = 82.0°, b = 6.0, c = 3.0$
7. $A = 22.5°, B = 112.4°, a = 16.3$
8. $A = 21.5°, B = 104.2°, c = 57.4$
9. $A = 82.0°, B = 65.4°, b = 36.5$
10. $B = 54.8°, C = 72.6°, a = 14.4$
11. $A = 33.8°, C = 98.5°, c = 102$
12. $B = 36.9°, C = 69.2°, a = 166$
13. $C = 114.2°, c = 87.2, b = 12.1$
14. $A = 54.32°, a = 24.42, c = 16.92$

In Exercises 15 to 28, solve the triangles that exist.

15. $A = 37°, c = 40, a = 28$
16. $B = 32°, c = 14, b = 9.0$
17. $C = 65°, b = 10, c = 8.0$
18. $A = 42°, a = 12, c = 18$
19. $A = 30°, a = 1.0, b = 2.4$
20. $B = 22.6°, b = 5.55, a = 13.8$
21. $A = 14.8°, c = 6.35, a = 4.80$
22. $C = 37.9°, b = 3.50, c = 2.84$
23. $C = 47.2°, a = 8.25, c = 5.80$
24. $B = 52.7°, b = 12.3, c = 16.3$
25. $B = 117.32°, b = 67.25, a = 15.05$
26. $A = 49.22°, a = 16.92, c = 24.62$
27. $A = 20.5°, a = 10.3, c = 14.1$
28. $B = 41.2°, a = 31.5, b = 21.6$
29. **Hurricane Watch** A satellite weather map shows a hurricane off the coast of North Carolina. Use the information in the map to find the distance from the hurricane to Nags Head.

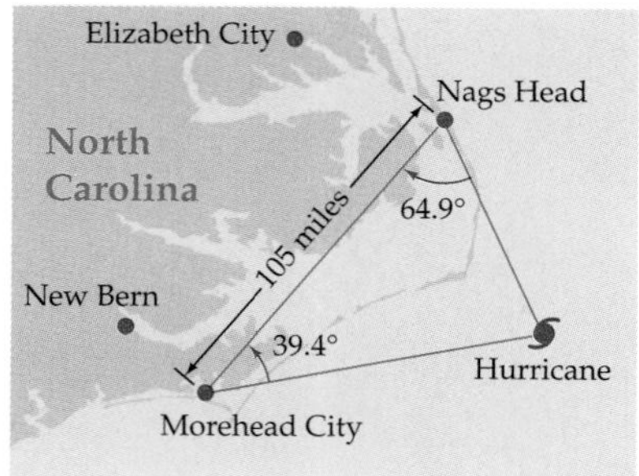

30. Naval Maneuvers The distance between an aircraft carrier and a Navy destroyer is 7620 feet. The angle of elevation from the destroyer to a helicopter is 77.2°, and the angle of elevation from the aircraft carrier to the helicopter is 59.0°. The helicopter is in the same vertical plane as the two ships, as shown in the following figure. Use this data to determine the distance x from the helicopter to the aircraft carrier.

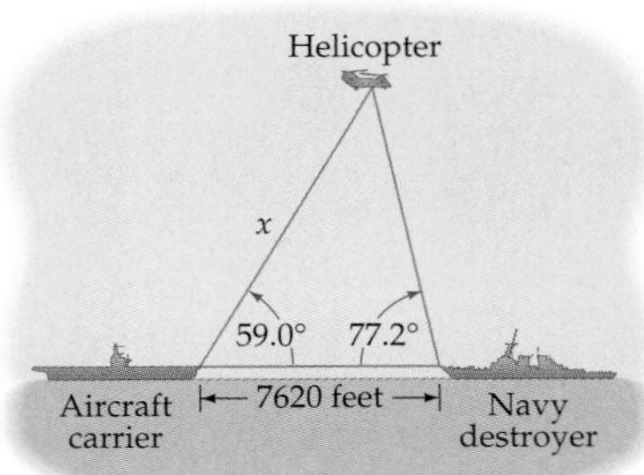

31. Choosing a Golf Strategy The following diagram shows two ways to play a golf hole. One is to hit the ball down the fairway on your first shot and then hit an approach shot to the green on your second shot. A second way is to hit directly toward the pin. Due to the water hazard, this is a more risky strategy. The distance AB is 165 yards, BC is 155 yards, and angle $A = 42.0°$. Find the distance AC from the tee directly to the pin. Assume that angle B is an obtuse angle.

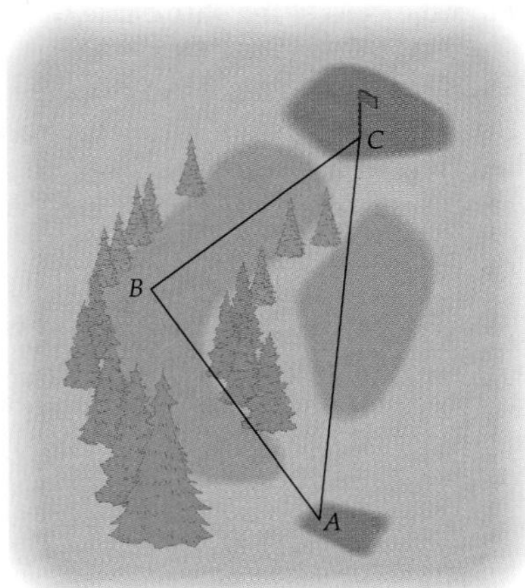

32. Driving Distance A golfer drives a golf ball from the tee at point A to point B, as shown in the following diagram. The distance AC from the tee directly to the pin is 365 yards. Angle A measures 11.2° and angle C measures 22.9°.

a. Find the distance AB that the golfer drove the ball.

b. Find the distance BC from the present position of the ball to the pin.

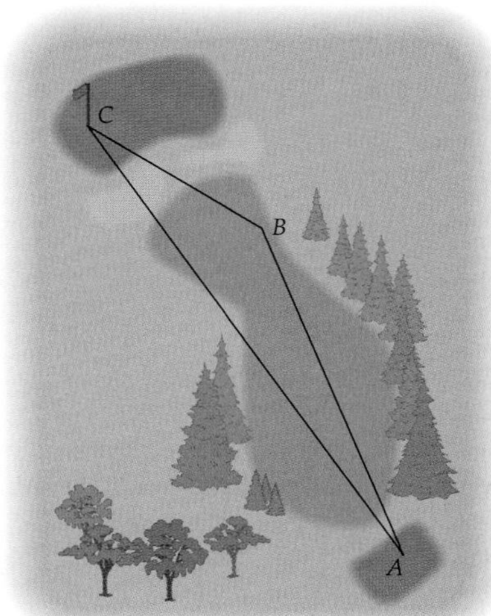

33. Distance to a Hot Air Balloon The angle of elevation to a balloon from one observer is 67°, and the angle of elevation from another observer, 220 feet away, is 31°. If the balloon is in the same vertical plane as the two observers and between them, find the distance of the balloon from the first observer.

34. Height of a Space Shuttle Use the Law of Sines to solve Example 6 on page 141. Compare this method with the method used in Example 6. Which method do you prefer? Explain.

35. Runway Replacement An airport runway is 3550 feet long and has an incline of 3.0°. The airport planning committee plans to replace this runway with a new runway, as shown in the following figure. The new runway will be inclined at an angle of 2.2°. What will be the length of the new runway?

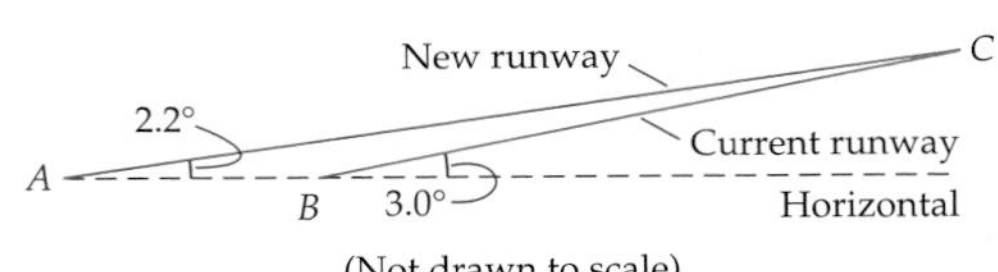

(Not drawn to scale)

36. **HEIGHT OF A KITE** Two observers, in the same vertical plane as a kite and at a distance of 30 feet apart, observe the kite at angles of 62° and 78°, as shown in the following diagram. Find the height of the kite.

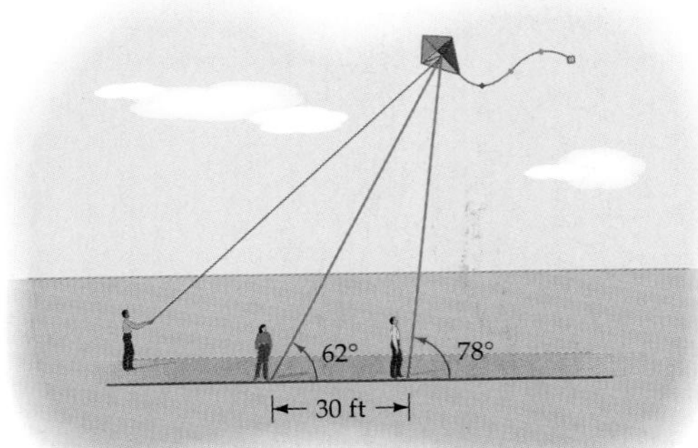

37. **LENGTH OF A GUY WIRE** A telephone pole 35 feet high is situated on an 11° slope from the horizontal. The measure of angle CAB is 21°. Find the length of the guy wire AC.

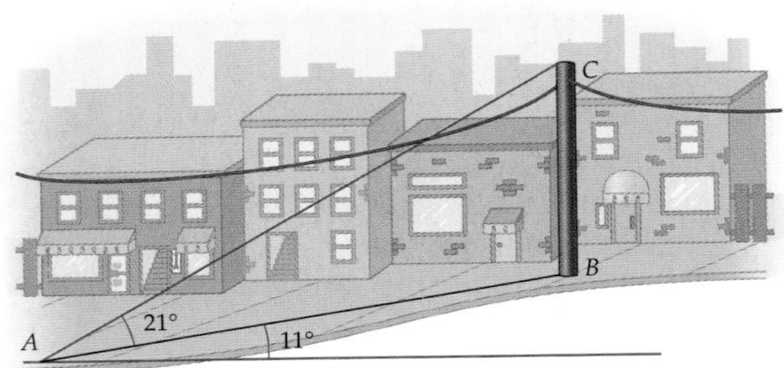

38. **DIMENSIONS OF A PLOT OF LAND** Three roads intersect in such a way as to form a triangular piece of land. See the accompanying figure. Find the lengths of the other two sides of the land.

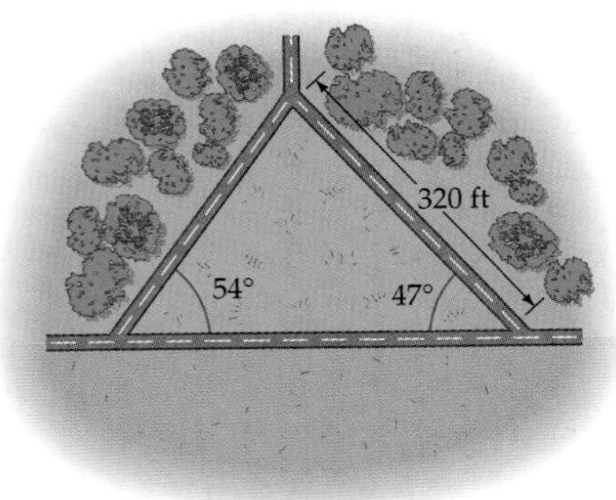

39. **HEIGHT OF A HILL** A surveying team determines the height of a hill by placing a 12-foot pole at the top of the hill and measuring the angles of elevation to the bottom and the top of the pole. They find the angles of elevation to be as shown in the following figure. Find the height of the hill.

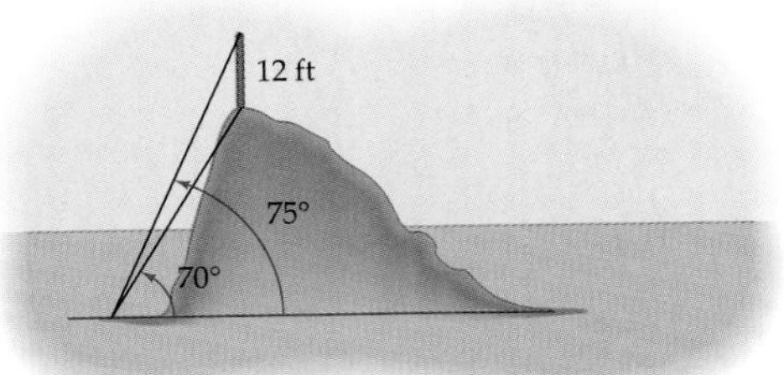

40. **DISTANCE TO A FIRE** Two fire lookouts are located on mountains 20 miles apart. Lookout B is at a bearing of S65°E from lookout A. A fire was sighted at a bearing of N50°E from A and at a bearing of N8°E from B. Find the distance of the fire from lookout A.

41. **DISTANCE TO A LIGHTHOUSE** A navigator on a ship sights a lighthouse at a bearing of N36°E. After traveling 8.0 miles at a heading of 332°, the ship sights the lighthouse at a bearing of S82°E. How far is the ship from the lighthouse at the second sighting?

42. **MINIMUM DISTANCE** The navigator on a ship traveling due east at 8 mph sights a lighthouse at a bearing of S55°E. One hour later the lighthouse is sighted at a bearing of S25°W. Find the closest the ship came to the lighthouse.

43. **DISTANCE BETWEEN AIRPORTS** An airplane flew 450 miles at a bearing of N65°E from airport A to airport B. The plane then flew at a bearing of S38°E to airport C. Find the distance from A to C if the bearing from airport A to airport C is S60°E.

44. **LENGTH OF A BRACE** A 12-foot solar panel is to be installed on a roof with a 15° pitch. Find the length of the vertical brace d if the panel must be installed to make a 40° angle with the horizontal.

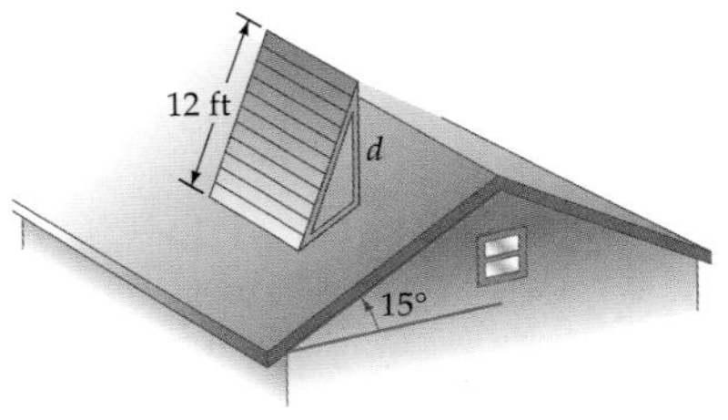

Connecting Concepts

45. DISTANCES BETWEEN HOUSES House B is located at a bearing of N67°E from house A. House C is 300 meters from house A at a bearing of S68°E. House B is located at a bearing of N11°W from house C. Find the distance from house A to house B.

46. Show that for any triangle ABC, $\dfrac{a-b}{b} = \dfrac{\sin A - \sin B}{\sin B}$.

47. Show that for any triangle ABC, $\dfrac{a+b}{b} = \dfrac{\sin A + \sin B}{\sin B}$.

48. Show that for any triangle ABC, $\dfrac{a-b}{a+b} = \dfrac{\sin A - \sin B}{\sin A + \sin B}$.

49. MAXIMUM LENGTH OF A ROD The longest rod that can be carried horizontally around a corner from a hall 3 meters wide into one that is 5 meters wide is the minimum of the length L of the black dashed line shown in the figure below.

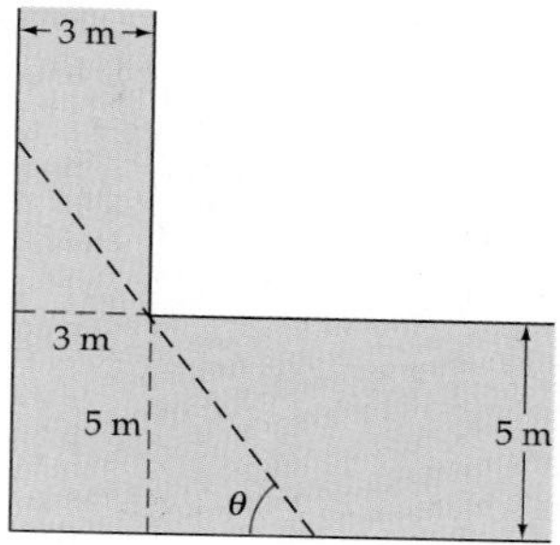

Use similar triangles to show that the length L is a function of the angle θ, given by

$$L(\theta) = \frac{5}{\sin \theta} + \frac{3}{\cos \theta}$$

Graph L and estimate the minimum value of L. Round to the nearest hundredth of a meter.

Projects

1. FERMAT'S PRINCIPLE AND SNELL'S LAW State Fermat's Principle and Snell's Law. The refractive index of the glass in a ring is found to be 1.82. The refractive index of a particular diamond in a ring is 2.38. Use Snell's Law to explain what this means in terms of the reflective properties of the glass and the diamond.

Section 4.2

- The Law of Cosines
- Area of a Triangle
- Heron's Formula

The Law of Cosines and Area

PREPARE FOR THIS SECTION

Prepare for this section by completing the following exercises. The answers can be found on page A20.

PS1. Evaluate $\sqrt{a^2 + b^2 - 2ab\cos C}$ for $a = 10.0$, $b = 15.0$, and $C = 110.0°$. Round your result to the nearest tenth. [2.3]

PS2. Find the area of a triangle with a base of 6 inches and a height of 8.5 inches.

PS3. Solve $c^2 = a^2 + b^2 - 2ab\cos C$ for C. [3.5]

PS4. The **semiperimeter** of a triangle is defined as one-half the perimeter of the triangle. Find the semiperimeter of a triangle with sides of 6 meters, 9 meters, and 10 meters.

PS5. Evaluate $\sqrt{s(s-a)(s-b)(s-c)}$ for $a = 3$, $b = 4$, $c = 5$, and $s = \dfrac{a+b+c}{2}$.

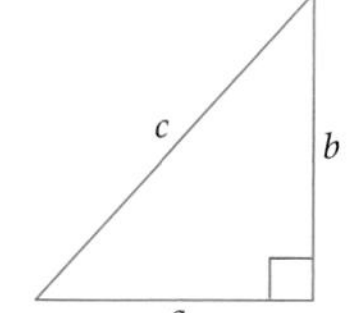

PS6. State a relationship between the lengths a, b, and c in the triangle shown at the right. [1.2]

The Law of Cosines

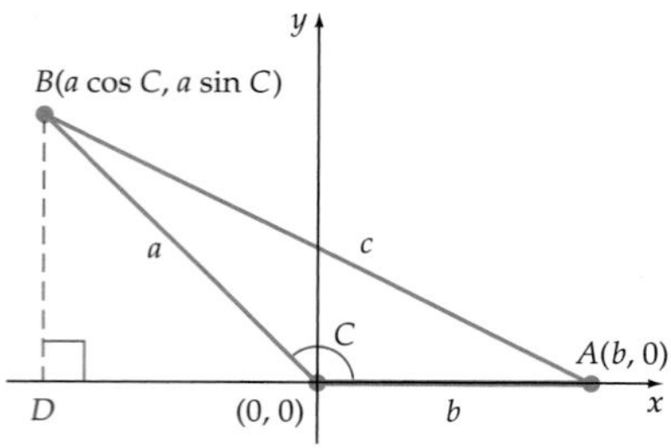

Figure 4.11

The *Law of Cosines* can be used to solve triangles in which two sides and the included angle (SAS) are known or in which three sides (SSS) are known. Consider the triangle in **Figure 4.11**. The height BD is drawn from B perpendicular to the x-axis. The triangle BDA is a right triangle, and the coordinates of B are $(a\cos C, a\sin C)$. The coordinates of A are $(b, 0)$. Using the distance formula, we can find the distance c.

$$c = \sqrt{(a\cos C - b)^2 + (a\sin C - 0)^2}$$
$$c^2 = a^2\cos^2 C - 2ab\cos C + b^2 + a^2\sin^2 C$$
$$c^2 = a^2(\cos^2 C + \sin^2 C) + b^2 - 2ab\cos C$$
$$c^2 = a^2 + b^2 - 2ab\cos C$$

The Law of Cosines

If A, B, and C are the measures of the angles of a triangle and a, b, and c are the lengths of the sides opposite these angles, then

$$c^2 = a^2 + b^2 - 2ab\cos C$$
$$a^2 = b^2 + c^2 - 2bc\cos A$$
$$b^2 = a^2 + c^2 - 2ac\cos B$$

EXAMPLE 1 » Use the Law of Cosines (SAS)

In triangle ABC, $B = 110.0°$, $a = 10.0$ centimeters, and $c = 15.0$ centimeters. See **Figure 4.12**. Find b.

Figure 4.12

Solution

The Law of Cosines can be used because two sides and the included angle are known.

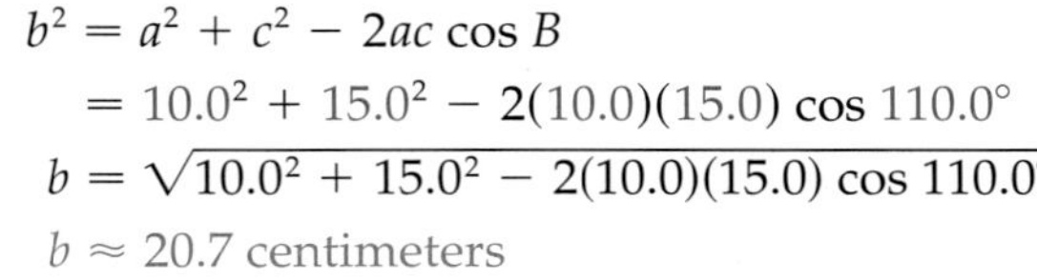

$$b^2 = a^2 + c^2 - 2ac \cos B$$
$$= 10.0^2 + 15.0^2 - 2(10.0)(15.0) \cos 110.0°$$
$$b = \sqrt{10.0^2 + 15.0^2 - 2(10.0)(15.0) \cos 110.0°}$$
$$b \approx 20.7 \text{ centimeters}$$

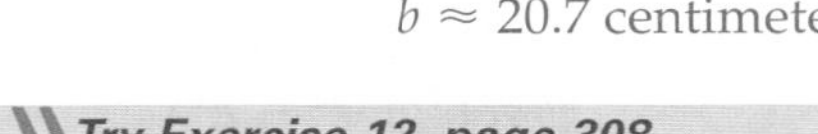

» *Try Exercise 12, page 308*

In the next example we know the length of each side, but we do not know the measure of any of the angles.

EXAMPLE 2 » Use the Law of Cosines (SSS)

In triangle ABC, $a = 32$ feet, $b = 20$ feet, and $c = 40$ feet. Find B. This is the SSS case.

Solution

$$b^2 = a^2 + c^2 - 2ac \cos B$$

$$\cos B = \frac{a^2 + c^2 - b^2}{2ac}$$ • **Solve for cos *B*.**

$$= \frac{32^2 + 40^2 - 20^2}{2(32)(40)}$$ • **Substitute for *a*, *b*, and *c*.**

$$B = \cos^{-1}\left(\frac{32^2 + 40^2 - 20^2}{2(32)(40)}\right)$$ • **Solve for angle *B*.**

$$B \approx 30°$$ • **To the nearest degree**

» *Try Exercise 18, page 308*

EXAMPLE 3 » Solve an Application Using the Law of Cosines

A boat sailed 3.0 miles at a heading of 78° and then turned to a heading of 138° and sailed another 4.3 miles. Find the distance and the bearing of the boat from the starting point.

Continued ►

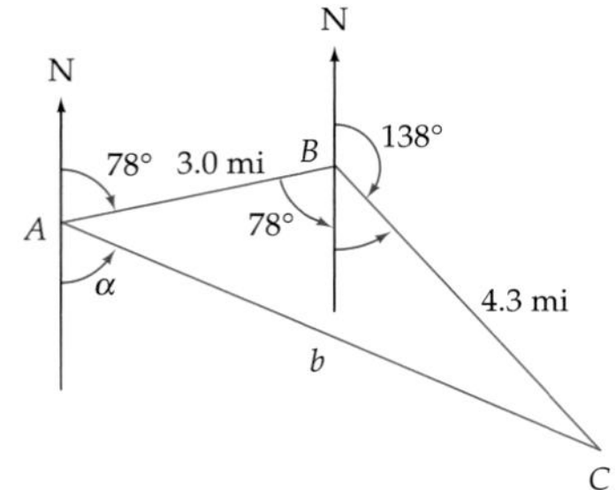

Figure 4.13

Solution

Sketch a diagram (see **Figure 4.13**). First find the measure of angle B in triangle ABC.

$$B = 78° + (180° - 138°) = 120°$$

Use the Law of Cosines first to find b and then to find A.

$$b^2 = a^2 + c^2 - 2ac \cos B$$

$$= 4.3^2 + 3.0^2 - 2(4.3)(3.0) \cos 120°$$ • **Substitute for *a*, *c*, and *B*.**

$$b = \sqrt{4.3^2 + 3.0^2 - 2(4.3)(3.0) \cos 120°}$$

$$b \approx 6.4 \text{ miles}$$

$$\cos A = \frac{b^2 + c^2 - a^2}{2bc}$$ • **Solve the Law of Cosines for cos *A*.**

$$A = \cos^{-1}\left(\frac{b^2 + c^2 - a^2}{2bc}\right) \approx \cos^{-1}\left(\frac{6.4^2 + 3.0^2 - 4.3^2}{(2)(6.4)(3.0)}\right) \approx 35°$$

The bearing of the present position of the boat from the starting point A can be determined by calculating the measure of angle α in **Figure 4.13.**

$$\alpha \approx 180° - (78° + 35°) = 67°$$

The distance is approximately 6.4 miles, and the bearing (to the nearest degree) is S67°E.

take note

The measure of angle A in Example 3 can also be determined by using the Law of Sines.

»Try Exercise 52, page 310

There are five different cases that we may encounter when solving an oblique triangle. Each case is listed below under the law that can be used to solve the triangle.

Choosing Between the Law of Sines and the Law of Cosines

Apply the Law of Sines to solve an oblique triangle for each of the following cases.

ASA The measures of two angles of the triangle and the length of the included side are known.

AAS The measures of two angles of the triangle and the length of a side opposite one of these angles are known.

SSA The lengths of two sides of the triangle and the measure of an angle opposite one of these sides are known. This case is called the ambiguous case. It may yield one solution, two solutions, or no solution.

Apply the Law of Cosines to solve an oblique triangle for each of the following cases.

SSS The lengths of all three sides of the triangle are known. After finding the measure of an angle, you can complete your solution by using the Law of Sines.

SAS The lengths of two sides of the triangle and the measure of the included angle are known. After finding the measure of the third side, you can complete your solution by using the Law of Sines.

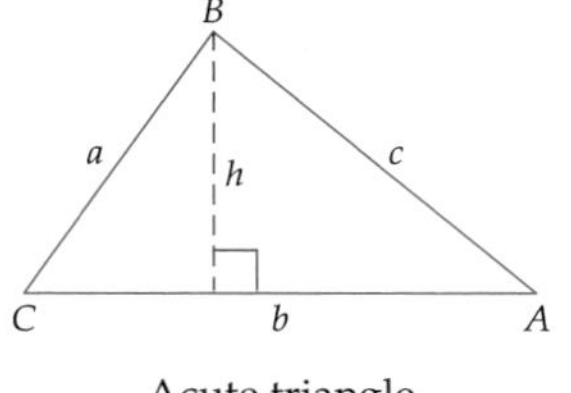

Acute triangle

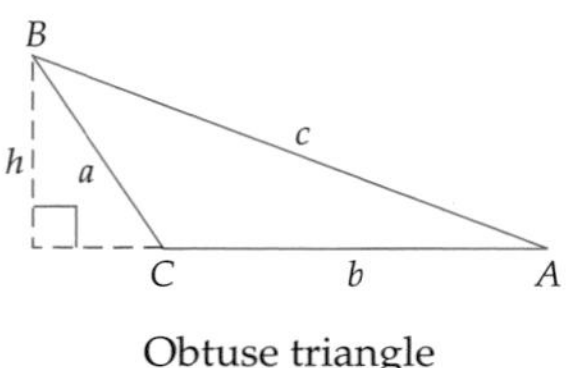

Obtuse triangle

Figure 4.14

QUESTION In triangle ABC, $A = 40°$, $C = 60°$, and $b = 114$. Should you use the Law of Sines or the Law of Cosines to solve this triangle?

Area of a Triangle

The formula $A = \frac{1}{2}bh$ can be used to find the area of a triangle when the base and height are given. In this section we will find the areas of triangles when the height is not given. We will use K for the area of a triangle because the letter A is often used to represent the measure of an angle.

Consider the areas of the acute and obtuse triangles in **Figure 4.14.**

Height of each triangle: $h = c \sin A$

Area of each triangle: $K = \frac{1}{2}bh$

$K = \frac{1}{2}bc \sin A$ • **Substitute for h.**

Thus we have established the following theorem.

take note

Because each formula requires two sides and the included angle, it is necessary to learn only one formula.

Area of a Triangle

The area K of triangle ABC is one-half the product of the lengths of any two sides and the sine of the included angle. Thus

$$K = \frac{1}{2}bc \sin A \qquad K = \frac{1}{2}ab \sin C \qquad K = \frac{1}{2}ac \sin B$$

EXAMPLE 4 Find the Area of a Triangle

Given angle $A = 62°$, $b = 12$ meters, and $c = 5.0$ meters, find the area of triangle ABC.

Solution

In **Figure 4.15,** two sides and the included angle of the triangle are given. Using the formula for area, we have

$$K = \frac{1}{2}bc \sin A = \frac{1}{2}(12)(5.0)(\sin 62°) \approx 26 \text{ square meters}$$

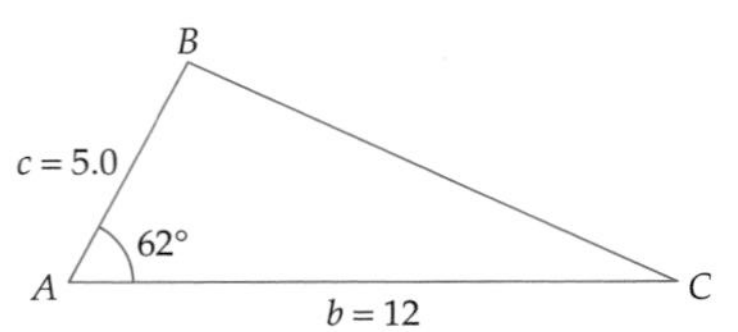

Figure 4.15

Try Exercise 30, page 309

ANSWER Because the measure of two angles and the length of the included side are given, the triangle can be solved by using the Law of Sines.

When two angles and an included side are given, the Law of Sines is used to derive a formula for the area of a triangle. First, solve for c in the Law of Sines.

$$\frac{c}{\sin C} = \frac{b}{\sin B}$$
$$c = \frac{b \sin C}{\sin B}$$

Substitute for c in the formula $K = \frac{1}{2}bc \sin A$.

$$K = \frac{1}{2}bc \sin A = \frac{1}{2}b\left(\frac{b \sin C}{\sin B}\right) \sin A$$
$$K = \frac{b^2 \sin C \sin A}{2 \sin B}$$

In like manner, the following two alternative formulas can be derived for the area of a triangle.

$$K = \frac{a^2 \sin B \sin C}{2 \sin A} \quad \text{and} \quad K = \frac{c^2 \sin A \sin B}{2 \sin C}$$

EXAMPLE 5 》 Find the Area of a Triangle

Given $A = 32°$, $C = 77°$, and $a = 14$ inches, find the area of triangle ABC.

Solution

To use the above area formula, we need to know two angles and the included side. Therefore, we need to determine the measure of angle B.

$$B = 180° - 32° - 77° = 71°$$

Thus

$$K = \frac{a^2 \sin B \sin C}{2 \sin A} = \frac{14^2 \sin 71° \sin 77°}{2 \sin 32°} \approx 170 \text{ square inches}$$

》 *Try Exercise 32, page 309*

Math Matters

Recent findings indicate that Heron's formula for finding the area of a triangle was first discovered by Archimedes. However, the formula is called Heron's formula in honor of the geometer Heron of Alexandria (A.D. 50), who gave an ingenious proof of the theorem in his work *Metrica.* Because Heron of Alexandria was also known as Hero, some texts refer to Heron's formula as Hero's formula.

Heron's Formula

The Law of Cosines can be used to derive *Heron's formula* for the area of a triangle in which three sides of the triangle are given.

Heron's Formula for Finding the Area of a Triangle

If a, b, and c are the lengths of the sides of a triangle, then the area K of the triangle is

$$K = \sqrt{s(s-a)(s-b)(s-c)}, \quad \text{where } s = \frac{1}{2}(a + b + c)$$

Because s is one-half the perimeter of the triangle, it is called the **semiperimeter.**

EXAMPLE 6 Find an Area by Heron's Formula

Find, to two significant digits, the area of the triangle with $a = 7.0$ meters, $b = 15$ meters, and $c = 12$ meters.

Solution

Calculate the semiperimeter s.

$$s = \frac{a + b + c}{2} = \frac{7.0 + 15 + 12}{2} = 17$$

Use Heron's formula.

$$\begin{aligned} K &= \sqrt{s(s - a)(s - b)(s - c)} \\ &= \sqrt{17(17 - 7.0)(17 - 15)(17 - 12)} \\ &= \sqrt{1700} \approx 41 \text{ square meters} \end{aligned}$$

Try Exercise 40, page 309

EXAMPLE 7 Use Heron's Formula to Solve an Application

The original portion of the Luxor Hotel in Las Vegas has the shape of a square pyramid. Each face of the pyramid is an isosceles triangle with a base of 646 feet and sides of length 576 feet. Assuming that the glass on the exterior of the Luxor Hotel costs \$35 per square foot, determine the cost of the glass, to the nearest \$10,000, for one of the triangular faces of the hotel.

Solution

The lengths (in feet) of the sides of a triangular face are $a = 646$, $b = 576$, and $c = 576$.

$$s = \frac{a + b + c}{2} = \frac{646 + 576 + 576}{2} = 899 \text{ feet}$$

$$\begin{aligned} K &= \sqrt{s(s - a)(s - b)(s - c)} \\ &= \sqrt{899(899 - 646)(899 - 576)(899 - 576)} \\ &= \sqrt{23{,}729{,}318{,}063} \\ &\approx 154{,}043 \text{ square feet} \end{aligned}$$

The cost C of the glass is the product of the cost per square foot and the area.

$$C \approx 35 \cdot 154{,}043 = 5{,}391{,}505$$

The approximate cost of the glass for one face of the Luxor Hotel is \$5,390,000.

Try Exercise 60, page 310

The pyramid portion of the Luxor Hotel in Las Vegas, Nevada

Topics for Discussion

1. Explain why there is no triangle that has sides of lengths $a = 2$ inches, $b = 11$ inches, and $c = 3$ inches.
2. The Pythagorean Theorem is a special case of the Law of Cosines. Explain.
3. To solve a triangle in which the lengths of the three sides are given (SSS), a mathematics professor recommends the following procedure.

   (i) Use the Law of Cosines to find the measure of the largest angle.

   (ii) Use the Law of Sines to find the measure of a second angle.

   (iii) Find the measure of the third angle by using the formula $A + B + C = 180°$.

   Explain why this procedure is easier than using the Law of Cosines to find the measure of all three angles.
4. Explain why the Law of Cosines cannot be used to solve a triangle in which you are given the measures of two angles and the length of the included side (ASA).

Exercise Set 4.2

In Exercises 1 to 52, round answers according to the rounding conventions on page 140.

In Exercises 1 to 14, find the third side of the triangle.

1. $a = 12, b = 18, C = 44°$

2. $b = 30, c = 24, A = 120°$

3. $a = 120, c = 180, B = 56°$

4. $a = 400, b = 620, C = 116°$

5. $b = 60, c = 84, A = 13°$

6. $a = 122, c = 144, B = 48°$

7. $a = 9.0, b = 7.0, C = 72°$

8. $b = 12, c = 22, A = 55°$

9. $a = 4.6, b = 7.2, C = 124°$

10. $b = 12.3, c = 14.5, A = 6.5°$

11. $a = 25.9, c = 33.4, B = 84.0°$

12. $a = 14.2, b = 9.30, C = 9.20°$

13. $a = 122, c = 55.9, B = 44.2°$

14. $b = 444.8, c = 389.6, A = 78.44°$

In Exercises 15 to 24, given three sides of a triangle, find the specified angle.

15. $a = 25, b = 32, c = 40$; find A.

16. $a = 60, b = 88, c = 120$; find B.

17. $a = 8.0, b = 9.0, c = 12$; find C.

18. $a = 108, b = 132, c = 160$; find A.

19. $a = 80.0, b = 92.0, c = 124$; find B.

20. $a = 166, b = 124, c = 139$; find B.

21. $a = 1025, b = 625.0, c = 1420$; find C.

22. $a = 4.7, b = 3.2, c = 5.9$; find A.

23. $a = 32.5, b = 40.1, c = 29.6$; find B.

24. $a = 112.4, b = 96.80, c = 129.2$; find C.

In Exercises 25 to 28, solve the triangle.

25. $A = 39.4°, b = 15.5, c = 17.2$

26. $C = 98.4°, a = 141, b = 92.3$

27. $a = 83.6, b = 144, c = 98.1$

28. $a = 25.4, b = 36.3, c = 38.2$

In Exercises 29 to 40, find the area of the given triangle. Round each area to the same number of significant digits given for each of the given sides.

29. $A = 105°, b = 12, c = 24$

30. $B = 127°, a = 32, c = 25$

31. $A = 42°, B = 76°, c = 12$

32. $B = 102°, C = 27°, a = 8.5$

33. $a = 16, b = 12, c = 14$

34. $a = 32, b = 24, c = 36$

35. $B = 54.3°, a = 22.4, b = 26.9$

36. $C = 18.2°, b = 13.4, a = 9.84$

37. $A = 116°, B = 34°, c = 8.5$

38. $B = 42.8°, C = 76.3°, c = 17.9$

39. $a = 3.6, b = 4.2, c = 4.8$

40. $a = 10.2, b = 13.3, c = 15.4$

41. Distance Between Airports A plane leaves airport A and travels 560 miles to airport B at a bearing of N32°E. The plane leaves airport B and travels to airport C 320 miles away at a bearing of S72°E. Find the distance from airport A to airport C.

42. Length of a Street A developer owns a triangular lot at the intersection of two streets. The streets meet at an angle of 72°, and the lot has 300 feet of frontage along one street and 416 feet of frontage along the other street. Find the length of the third side of the lot.

43. Baseball In a baseball game, a batter hits a ground ball 26 feet in the direction of the pitcher's mound. See the figure at the top of the next column. The pitcher runs forward and reaches for the ball. At that moment, how far is the ball from first base? (*Note:* A baseball infield is a square that measures 90 feet on each side.)

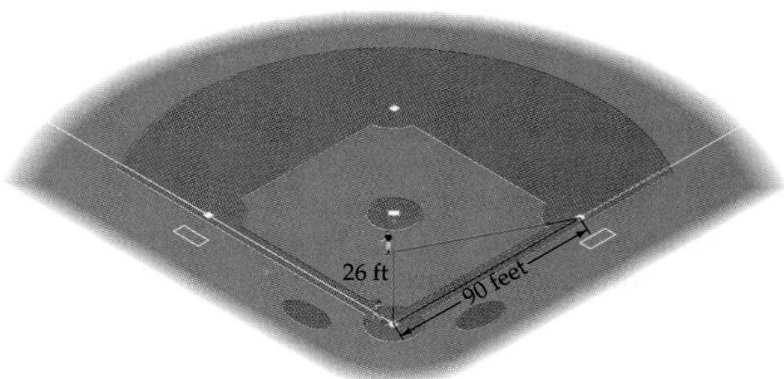

44. B-2 Bomber The leading edge of each wing of the B-2 Stealth Bomber measures 105.6 feet in length. The angle between the wing's leading edges ($\angle ABC$) is 109.05°. What is the wing span (the distance from A to C) of the B-2 Bomber?

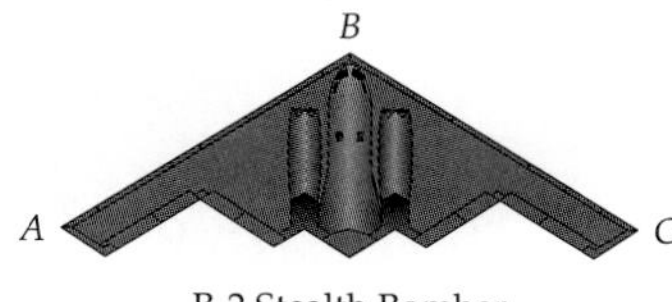

B-2 Stealth Bomber

45. Angle Between the Diagonals of a Box The rectangular box in the figure measures 6.50 feet by 3.25 feet by 4.75 feet. Find the measure of the angle θ that is formed by the union of the diagonal shown on the front of the box and the diagonal shown on the right side of the box.

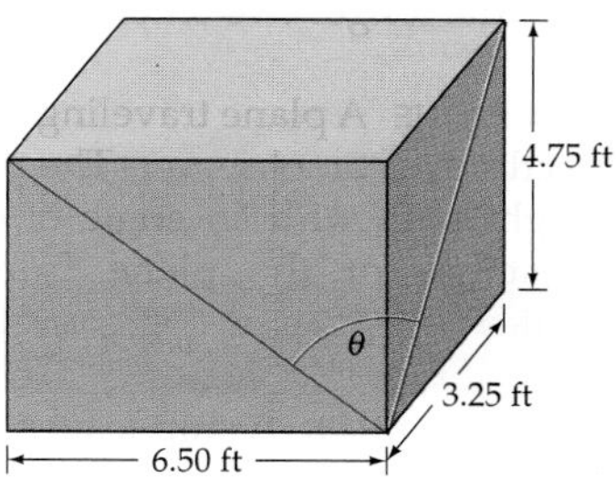

46. Submarine Rescue Mission Use the distances shown in the following figure to determine the depth of the submarine below the surface of the water. Assume that the line segment between the surface ships is directly above the submarine.

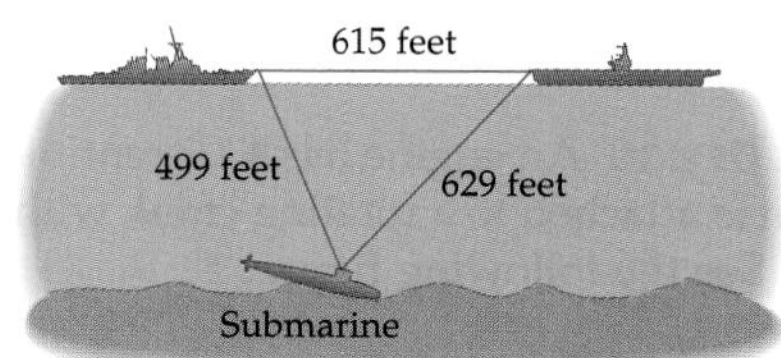

Given any nonzero vector **v**, we can obtain a unit vector in the direction of **v** by dividing each component of **v** by the magnitude of **v**, $\|\mathbf{v}\|$.

EXAMPLE 3 ›› Find a Unit Vector

Find a unit vector **u** in the direction of $\mathbf{v} = \langle -4, 2 \rangle$.

Solution

Find the magnitude of **v**.

$$\|\mathbf{v}\| = \sqrt{(-4)^2 + 2^2} = \sqrt{16 + 4} = \sqrt{20} = 2\sqrt{5}$$

Divide each component of **v** by $\|\mathbf{v}\|$.

$$\mathbf{u} = \left\langle \frac{-4}{2\sqrt{5}}, \frac{2}{2\sqrt{5}} \right\rangle = \left\langle \frac{-2}{\sqrt{5}}, \frac{1}{\sqrt{5}} \right\rangle = \left\langle -\frac{2\sqrt{5}}{5}, \frac{\sqrt{5}}{5} \right\rangle.$$

A unit vector in the direction of **v** is **u**.

›› Try Exercise 12, page 325

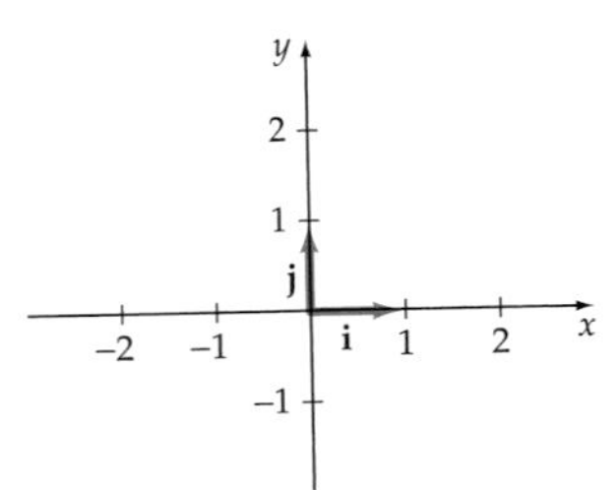

Figure 4.27

Two unit vectors, one parallel to the *x*-axis and one parallel to the *y*-axis, are of special importance. See **Figure 4.27.**

Definitions of Unit Vectors i and j

$$\mathbf{i} = \langle 1, 0 \rangle \qquad \mathbf{j} = \langle 0, 1 \rangle$$

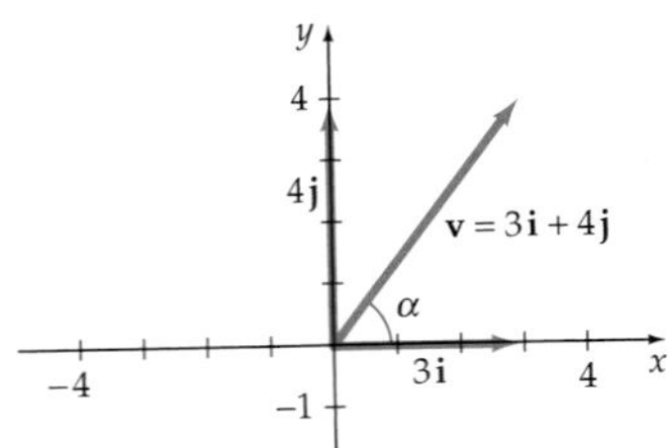

Figure 4.28

The vector $\mathbf{v} = \langle 3, 4 \rangle$ can be written in terms of the unit vectors **i** and **j** as shown in **Figure 4.28.**

$$\begin{aligned} \langle 3, 4 \rangle &= \langle 3, 0 \rangle + \langle 0, 4 \rangle && \text{• Definition of vector addition} \\ &= 3\langle 1, 0 \rangle + 4\langle 0, 1 \rangle && \text{• Definition of scalar multiplication of a vector} \\ &= 3\mathbf{i} + 4\mathbf{j} && \text{• Definition of i and j} \end{aligned}$$

By means of scalar multiplication and addition of vectors, any vector can be expressed in terms of the unit vectors **i** and **j**. Let $\mathbf{v} = \langle a_1, a_2 \rangle$. Then

$$\mathbf{v} = \langle a_1, a_2 \rangle = a_1\langle 1, 0 \rangle + a_2\langle 0, 1 \rangle = a_1\mathbf{i} + a_2\mathbf{j}$$

This gives the following result.

Representation of a Vector in Terms of i and j

If **v** is a vector and $\mathbf{v} = \langle a_1, a_2 \rangle$, then $\mathbf{v} = a_1\mathbf{i} + a_2\mathbf{j}$.

The definitions for addition and scalar multiplication of vectors can be restated in terms of **i** and **j**. If $\mathbf{v} = a_1\mathbf{i} + a_2\mathbf{j}$ and $\mathbf{w} = b_1\mathbf{i} + b_2\mathbf{j}$, then

$$\mathbf{v} + \mathbf{w} = (a_1\mathbf{i} + a_2\mathbf{j}) + (b_1\mathbf{i} + b_2\mathbf{j}) = (a_1 + b_1)\mathbf{i} + (a_2 + b_2)\mathbf{j}$$
$$k\mathbf{v} = k(a_1\mathbf{i} + a_2\mathbf{j}) = ka_1\mathbf{i} + ka_2\mathbf{j}$$

EXAMPLE 4 ⟩⟩ Operate on Vectors Written in Terms of i and j

Given $\mathbf{v} = 3\mathbf{i} - 4\mathbf{j}$ and $\mathbf{w} = 5\mathbf{i} + 3\mathbf{j}$, find $3\mathbf{v} - 2\mathbf{w}$.

Solution

$$\begin{aligned} 3\mathbf{v} - 2\mathbf{w} &= 3(3\mathbf{i} - 4\mathbf{j}) - 2(5\mathbf{i} + 3\mathbf{j}) \\ &= (9\mathbf{i} - 12\mathbf{j}) - (10\mathbf{i} + 6\mathbf{j}) \\ &= (9 - 10)\mathbf{i} + (-12 - 6)\mathbf{j} \\ &= -\mathbf{i} - 18\mathbf{j} \end{aligned}$$

⟩⟩ *Try Exercise 30, page 325*

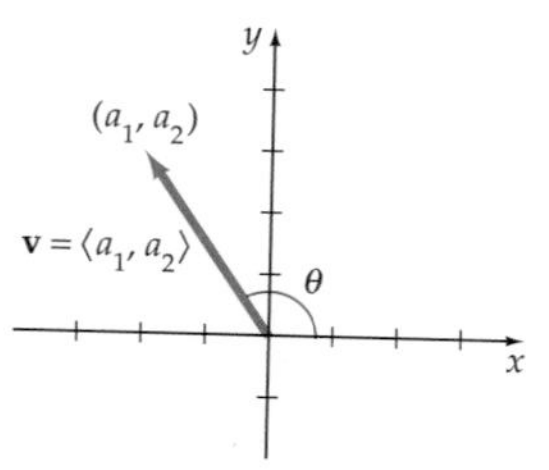

Figure 4.29

The components a_1 and a_2 of the vector $\mathbf{v} = \langle a_1, a_2 \rangle$ can be expressed in terms of the magnitude of $\mathbf{v}$ and the direction angle of $\mathbf{v}$ (the angle that $\mathbf{v}$ makes with the positive x-axis). Consider the vector $\mathbf{v}$ in **Figure 4.29**. Then

$$\|\mathbf{v}\| = \sqrt{(a_1)^2 + (a_2)^2}$$

From the definitions of sine and cosine, we have

$$\cos\theta = \frac{a_1}{\|\mathbf{v}\|} \quad \text{and} \quad \sin\theta = \frac{a_2}{\|\mathbf{v}\|}$$

Rewriting the last two equations, we find that the components of $\mathbf{v}$ are

$$a_1 = \|\mathbf{v}\|\cos\theta \quad \text{and} \quad a_2 = \|\mathbf{v}\|\sin\theta$$

Definitions of Horizontal and Vertical Components of a Vector

Let $\mathbf{v} = \langle a_1, a_2 \rangle$, where $\mathbf{v} \neq \mathbf{0}$, the zero vector. Then

$$a_1 = \|\mathbf{v}\|\cos\theta \quad \text{and} \quad a_2 = \|\mathbf{v}\|\sin\theta$$

where θ is the angle between the positive x-axis and $\mathbf{v}$.

The **horizontal component** of $\mathbf{v}$ is $\|\mathbf{v}\|\cos\theta$. The **vertical component** of $\mathbf{v}$ is $\|\mathbf{v}\|\sin\theta$.

? QUESTION Is $\mathbf{u} = \cos\theta\mathbf{i} + \sin\theta\mathbf{j}$ a unit vector?

Any nonzero vector can be written in terms of its horizontal and vertical components. Let $\mathbf{v} = a_1\mathbf{i} + a_2\mathbf{j}$. Then

$$\begin{aligned} \mathbf{v} &= a_1\mathbf{i} + a_2\mathbf{j} \\ &= (\|\mathbf{v}\|\cos\theta)\mathbf{i} + (\|\mathbf{v}\|\sin\theta)\mathbf{j} \\ &= \|\mathbf{v}\|(\cos\theta\mathbf{i} + \sin\theta\mathbf{j}) \end{aligned}$$

where $\|\mathbf{v}\|$ is the magnitude of $\mathbf{v}$ and the vector $\cos\theta\mathbf{i} + \sin\theta\mathbf{j}$ is a unit vector. The last equation shows that any vector $\mathbf{v}$ can be written as the product of its magnitude and a unit vector in the direction of $\mathbf{v}$.

? ANSWER Yes, because $\|\cos\theta\mathbf{i} + \sin\theta\mathbf{j}\| = \sqrt{\cos^2\theta + \sin^2\theta} = \sqrt{1} = 1$.

EXAMPLE 5 ⟩⟩ Find the Horizontal and Vertical Components of a Vector

Find, to the nearest tenth, the horizontal and vertical components of a vector **v** of magnitude 10 meters with direction angle 228°. Write the vector in the form $a_1\mathbf{i} + a_2\mathbf{j}$.

Solution

$a_1 = 10 \cos 228° \approx -6.7$

$a_2 = 10 \sin 228° \approx -7.4$

The approximate horizontal and vertical components are −6.7 and −7.4, respectively.

$$\mathbf{v} \approx -6.7\mathbf{i} - 7.4\mathbf{j}$$

⟩⟩ *Try Exercise 40, page 325*

Applications of Vectors

take note

Ground speed is the magnitude of the resultant of the plane's velocity vector and the wind velocity vector.

Vectors are used to solve applied problems in which forces are acting simultaneously on an object. For instance, an airplane flying in a wind is being acted upon by the force of its engine, $\mathbf{F}_1$, and the force of the wind, $\mathbf{F}_2$. The combined effect of these two forces is given by the resultant $\mathbf{F}_1 + \mathbf{F}_2$. In the following example, the **airspeed** of the plane is the speed at which the plane would be moving if there were no wind. The **actual velocity** of the plane is the plane's velocity with respect to the ground. The magnitude of the plane's actual velocity is called its **ground speed.**

EXAMPLE 6 ⟩⟩ Solve an Application Involving Airspeed

An airplane is traveling with an airspeed of 320 mph and a heading of 62°. A wind of 42 mph is blowing at a heading of 125°. Find the ground speed and the course of the airplane.

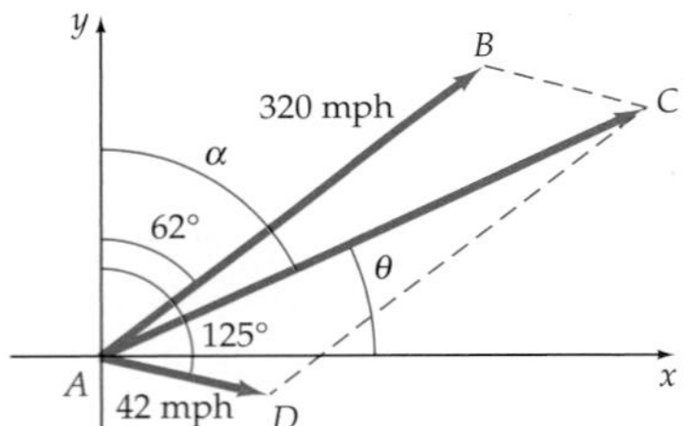

Figure 4.30

Solution

Sketch a diagram similar to **Figure 4.30** showing the relevant vectors. **AB** represents the heading and the airspeed, **AD** represents the wind velocity, and **AC** represents the course and the ground speed. By vector addition, **AC** = **AB** + **AD**. From the figure,

$$\begin{aligned}
\mathbf{AB} &= 320(\cos 28°\mathbf{i} + \sin 28°\mathbf{j}) \\
\mathbf{AD} &= 42[\cos(-35°)\mathbf{i} + \sin(-35°)\mathbf{j}] \\
\mathbf{AC} &= 320(\cos 28°\mathbf{i} + \sin 28°\mathbf{j}) + 42[\cos(-35°)\mathbf{i} + \sin(-35°)\mathbf{j}] \\
&\approx (282.5\mathbf{i} + 150.2\mathbf{j}) + (34.4\mathbf{i} - 24.1\mathbf{j}) \\
&= 316.9\mathbf{i} + 126.1\mathbf{j}
\end{aligned}$$

AC is the course of the plane. The ground speed is $\|\mathbf{AC}\|$. The heading is $\alpha = 90° - \theta$.

$$\|\mathbf{AC}\| = \sqrt{(316.9)^2 + (126.1)^2} \approx 340$$

$$\alpha = 90° - \theta = 90° - \tan^{-1}\left(\frac{126.1}{316.9}\right) \approx 68°$$

The ground speed is approximately 340 mph at a heading of 68°.

⟩⟩ *Try Exercise 44, page 325*

We can add two vectors to produce a resultant vector, and we can also find two vectors whose sum is a given vector. The process of finding two vectors whose sum is a given vector is called **resolving the vector.** Many applied problems can be analyzed by resolving a vector into two vectors that are *perpendicular* to each other. For instance, **Figure 4.31** shows a car on a ramp. The force of gravity on the car is shown by the downward vector **w**, which has been resolved into the vectors $\mathbf{F}_1$ and $\mathbf{F}_2$. The vector $\mathbf{F}_1$ is parallel to the ramp and represents the force pushing the car down the ramp. The vector $\mathbf{F}_2$ is perpendicular to the ramp and represents the force the car exerts against the ramp. These two perpendicular forces $\mathbf{F}_1$ and $\mathbf{F}_2$ are **vector components** of **w**. That is, $\mathbf{w} = \mathbf{F}_1 + \mathbf{F}_2$. The force needed to keep the car from rolling down the ramp is $-\mathbf{F}_1$. The angle formed by **w** and $\mathbf{F}_2$ is complementary to the 75.0° angle. Thus the angle formed by **w** and $\mathbf{F}_2$ has a measure of 15.0°, which is the same as the angular measure of the incline of the ramp.

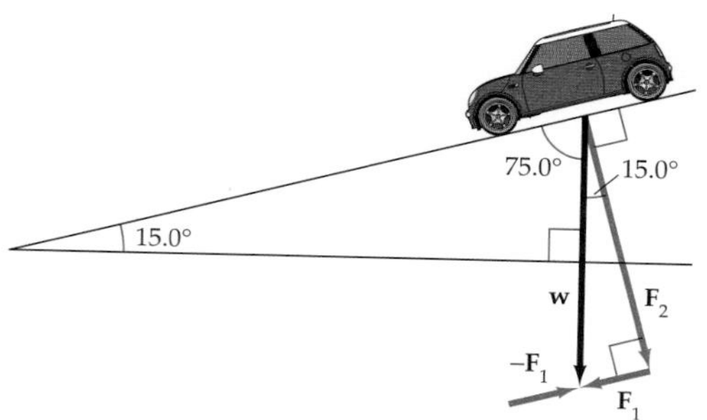

Figure 4.31

EXAMPLE 7 ⟫ Solve an Application Involving Force

The car shown in **Figure 4.31** has a weight of 2855 pounds.

a. Find the magnitude of the force needed to keep the car from rolling down the ramp. Round to the nearest pound.

b. Find the magnitude of the force the car exerts against the ramp. Round to the nearest pound.

Solution

a. The force needed to keep the car from rolling down the ramp is represented by $-\mathbf{F}_1$ in **Figure 4.31.** The magnitude of $-\mathbf{F}_1$ equals the magnitude of $\mathbf{F}_1$. The triangle formed by **w**, $\mathbf{F}_1$, and $\mathbf{F}_2$ is a right triangle in which $\sin 15.0° = \dfrac{\|\mathbf{F}_1\|}{\|\mathbf{w}\|}$. Use this equation to find $\|\mathbf{F}_1\|$.

$$\sin 15.0° = \frac{\|\mathbf{F}_1\|}{\|\mathbf{w}\|}$$

$$\sin 15.0° = \frac{\|\mathbf{F}_1\|}{2855} \qquad \text{• } \|\mathbf{w}\| = 2855$$

$$\|\mathbf{F}_1\| = 2855 \sin 15.0° \qquad \text{• Solve for } \|\mathbf{F}_1\|.$$

$$\approx 739$$

Approximately 739 pounds of force is needed to keep the car from rolling down the ramp.

Continued ▶

b. The vector $\mathbf{F}_2$ in **Figure 4.31** is the force the car exerts against the ramp. The triangle formed by $\mathbf{w}$, $\mathbf{F}_1$, and $\mathbf{F}_2$ is a right triangle in which $\cos 15.0° = \frac{\|\mathbf{F}_2\|}{\|\mathbf{w}\|}$. Use this equation to find $\|\mathbf{F}_2\|$.

$$\cos 15.0° = \frac{\|\mathbf{F}_2\|}{\|\mathbf{w}\|}$$

$$\cos 15.0° = \frac{\|\mathbf{F}_2\|}{2855} \qquad \bullet\ \|\mathbf{w}\| = 2855$$

$$\|\mathbf{F}_2\| = 2855 \cos 15.0° \qquad \bullet\ \text{Solve for } \|\mathbf{F}_2\|.$$

$$\approx 2758$$

The magnitude of the force the car exerts against the ramp is approximately 2758 pounds.

Try Exercise 48, Page 326

Dot Product

We have considered the product of a real number (scalar) and a vector. We now turn our attention to the product of two vectors. Finding the *dot product* of two vectors is one way to multiply a vector by a vector. The dot product of two vectors is a real number and *not* a vector. The dot product is also called the *inner product* or the *scalar product*. This product is useful in engineering and physics.

Definition of Dot Product

Given $\mathbf{v} = \langle a, b\rangle$ and $\mathbf{w} = \langle c, d\rangle$, the **dot product** of $\mathbf{v}$ and $\mathbf{w}$ is given by

$$\mathbf{v} \cdot \mathbf{w} = ac + bd$$

EXAMPLE 8 Find the Dot Product of Two Vectors

Find the dot product of $\mathbf{v} = \langle 6, -2\rangle$ and $\mathbf{w} = \langle -3, 4\rangle$.

Solution

$\mathbf{v} \cdot \mathbf{w} = 6(-3) + (-2)4 = -18 - 8 = -26$

Try Exercise 60, page 326

QUESTION Is the dot product of two vectors a vector or a real number?

If the vectors in Example 8 were given in terms of the vectors $\mathbf{i}$ and $\mathbf{j}$, then $\mathbf{v} = 6\mathbf{i} - 2\mathbf{j}$ and $\mathbf{w} = -3\mathbf{i} + 4\mathbf{j}$. In this case,

$$\mathbf{v} \cdot \mathbf{w} = (6\mathbf{i} - 2\mathbf{j}) \cdot (-3\mathbf{i} + 4\mathbf{j}) = 6(-3) + (-2)4 = -26$$

ANSWER A real number

Properties of the Dot Product

In the following properties, **u**, **v**, and **w** are vectors and a is a scalar.

1. $\mathbf{v} \cdot \mathbf{w} = \mathbf{w} \cdot \mathbf{v}$
2. $\mathbf{u} \cdot (\mathbf{v} + \mathbf{w}) = \mathbf{u} \cdot \mathbf{v} + \mathbf{u} \cdot \mathbf{w}$
3. $a(\mathbf{u} \cdot \mathbf{v}) = (a\mathbf{u}) \cdot \mathbf{v} = \mathbf{u} \cdot (a\mathbf{v})$
4. $\mathbf{v} \cdot \mathbf{v} = \|\mathbf{v}\|^2$
5. $\mathbf{0} \cdot \mathbf{v} = 0$
6. $\mathbf{i} \cdot \mathbf{i} = \mathbf{j} \cdot \mathbf{j} = 1$
7. $\mathbf{i} \cdot \mathbf{j} = \mathbf{j} \cdot \mathbf{i} = 0$

The proofs of these properties follow from the definition of dot product. Here is the proof of the fourth property. Let $\mathbf{v} = a\mathbf{i} + b\mathbf{j}$.

$$\mathbf{v} \cdot \mathbf{v} = (a\mathbf{i} + b\mathbf{j}) \cdot (a\mathbf{i} + b\mathbf{j}) = a^2 + b^2 = \|\mathbf{v}\|^2$$

Rewriting the fourth property of the dot product yields an alternative way of expressing the magnitude of a vector.

Magnitude of a Vector in Terms of the Dot Product

If $\mathbf{v} = \langle a, b \rangle$, then $\|\mathbf{v}\| = \sqrt{\mathbf{v} \cdot \mathbf{v}}$.

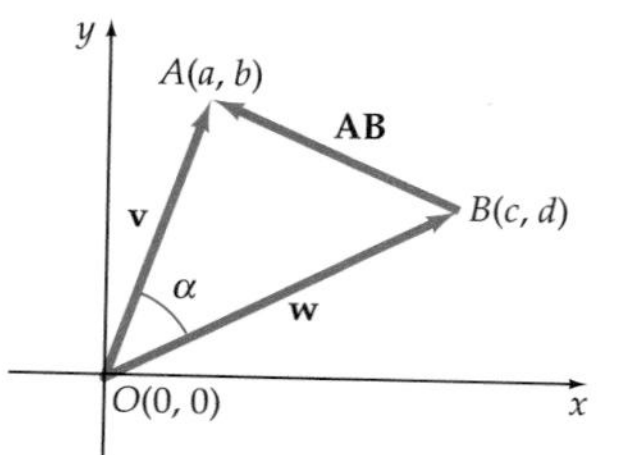

Figure 4.32

The Law of Cosines can be used to derive an alternative formula for the dot product. Consider the vectors $\mathbf{v} = \langle a, b \rangle$ and $\mathbf{w} = \langle c, d \rangle$ as shown in **Figure 4.32.** Using the Law of Cosines for triangle OAB, we have

$$\|\mathbf{AB}\|^2 = \|\mathbf{v}\|^2 + \|\mathbf{w}\|^2 - 2\|\mathbf{v}\|\,\|\mathbf{w}\| \cos \alpha$$

By the distance formula, $\|\mathbf{AB}\|^2 = (a - c)^2 + (b - d)^2$, $\|\mathbf{v}\|^2 = a^2 + b^2$, and $\|\mathbf{w}\|^2 = c^2 + d^2$. Thus

$$\begin{aligned}
(a - c)^2 + (b - d)^2 &= (a^2 + b^2) + (c^2 + d^2) - 2\|\mathbf{v}\|\,\|\mathbf{w}\|\cos \alpha \\
a^2 - 2ac + c^2 + b^2 - 2bd + d^2 &= a^2 + b^2 + c^2 + d^2 - 2\|\mathbf{v}\|\,\|\mathbf{w}\| \cos \alpha \\
-2ac - 2bd &= -2\|\mathbf{v}\|\,\|\mathbf{w}\| \cos \alpha \\
ac + bd &= \|\mathbf{v}\|\,\|\mathbf{w}\| \cos \alpha \\
\mathbf{v} \cdot \mathbf{w} &= \|\mathbf{v}\|\,\|\mathbf{w}\| \cos \alpha
\end{aligned}$$

• $\mathbf{v} \cdot \mathbf{w} = ac + bd$

Alternative Formula for the Dot Product

If **v** and **w** are two nonzero vectors and α is the smallest nonnegative angle between **v** and **w**, then $\mathbf{v} \cdot \mathbf{w} = \|\mathbf{v}\|\,\|\mathbf{w}\| \cos \alpha$.

Solving the alternative formula for the dot product for $\cos \alpha$, we have a formula for the cosine of the angle between two vectors.

Angle Between Two Vectors

If $\mathbf{v}$ and $\mathbf{w}$ are two nonzero vectors and α is the smallest nonnegative angle between $\mathbf{v}$ and $\mathbf{w}$, then $\cos\alpha = \dfrac{\mathbf{v}\cdot\mathbf{w}}{\|\mathbf{v}\|\,\|\mathbf{w}\|}$ and $\alpha = \cos^{-1}\left(\dfrac{\mathbf{v}\cdot\mathbf{w}}{\|\mathbf{v}\|\,\|\mathbf{w}\|}\right)$.

EXAMPLE 9 Find the Angle Between Two Vectors

Find, to the nearest tenth of a degree, the measure of the smallest nonnegative angle between the vectors $\mathbf{v} = 2\mathbf{i} - 3\mathbf{j}$ and $\mathbf{w} = -\mathbf{i} + 5\mathbf{j}$, as shown in **Figure 4.33.**

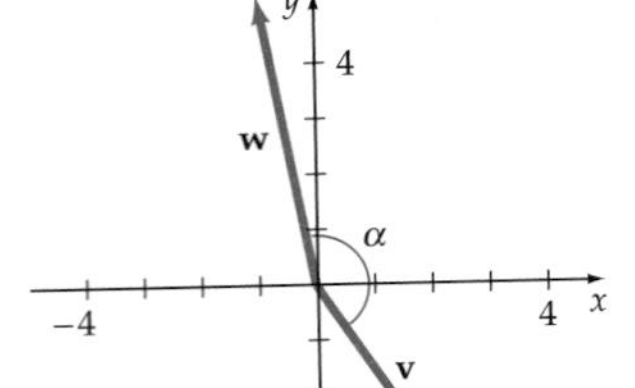

Figure 4.33

Solution

Use the equation for the angle between two vectors.

$$\cos\alpha = \frac{\mathbf{v}\cdot\mathbf{w}}{\|\mathbf{v}\|\|\mathbf{w}\|} = \frac{(2\mathbf{i} - 3\mathbf{j})\cdot(-\mathbf{i} + 5\mathbf{j})}{(\sqrt{2^2 + (-3)^2})(\sqrt{(-1)^2 + 5^2})}$$

$$= \frac{-2 - 15}{\sqrt{13}\,\sqrt{26}} = \frac{-17}{\sqrt{338}}$$

$$\alpha = \cos^{-1}\left(\frac{-17}{\sqrt{338}}\right) \approx 157.6^\circ$$

The measure of the smallest nonnegative angle between the two vectors is approximately 157.6°.

Try Exercise 70, page 326

Scalar Projection

Let $\mathbf{v} = \langle a_1, a_2\rangle$ and $\mathbf{w} = \langle b_1, b_2\rangle$ be two nonzero vectors, and let α be the angle between the vectors. Two possible configurations, one for which α is an acute angle and one for which α is an obtuse angle, are shown in **Figure 4.34**. In each case, a right triangle is formed by drawing a line segment from the terminal point of $\mathbf{v}$ to a line through $\mathbf{w}$.

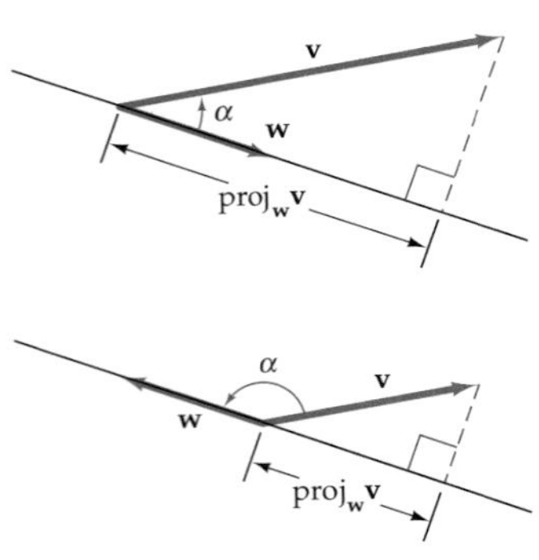

Figure 4.34

Definition of the Scalar Projection of v onto w

If $\mathbf{v}$ and $\mathbf{w}$ are two nonzero vectors and α is the angle between $\mathbf{v}$ and $\mathbf{w}$, then the scalar projection of $\mathbf{v}$ onto $\mathbf{w}$, $\text{proj}_{\mathbf{w}}\mathbf{v}$, is given by

$$\text{proj}_{\mathbf{w}}\mathbf{v} = \|\mathbf{v}\| \cos\alpha$$

To derive an alternate formula for $\text{proj}_{\mathbf{w}}\mathbf{v}$, consider the dot product $\mathbf{v}\cdot\mathbf{w} = \|\mathbf{v}\|\,\|\mathbf{w}\| \cos\alpha$. Solving for $\|\mathbf{v}\| \cos\alpha$, which is $\text{proj}_{\mathbf{w}}\mathbf{v}$, we have

$$\text{proj}_{\mathbf{w}}\mathbf{v} = \frac{\mathbf{v}\cdot\mathbf{w}}{\|\mathbf{w}\|}$$

When the angle α between the two vectors is an acute angle, $\text{proj}_{\mathbf{w}}\mathbf{v}$ is positive. When α is an obtuse angle, $\text{proj}_{\mathbf{w}}\mathbf{v}$ is negative.

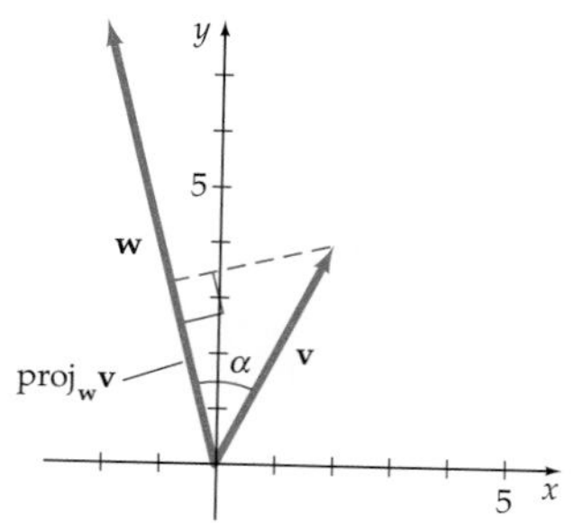

Figure 4.35

EXAMPLE 10 ⟫ Find the Projection of v onto w

Given $\mathbf{v} = 2\mathbf{i} + 4\mathbf{j}$ and $\mathbf{w} = -2\mathbf{i} + 8\mathbf{j}$ as shown in **Figure 4.35,** find $\text{proj}_{\mathbf{w}}\mathbf{v}$.

Solution

Use the equation $\text{proj}_{\mathbf{w}}\mathbf{v} = \dfrac{\mathbf{v} \cdot \mathbf{w}}{\|\mathbf{w}\|}$.

$$\text{proj}_{\mathbf{w}}\,\mathbf{v} = \frac{(2\mathbf{i} + 4\mathbf{j}) \cdot (-2\mathbf{i} + 8\mathbf{j})}{\sqrt{(-2)^2 + 8^2}} = \frac{28}{\sqrt{68}} = \frac{14\sqrt{17}}{17} \approx 3.4$$

⟫ *Try Exercise 72, page 326*

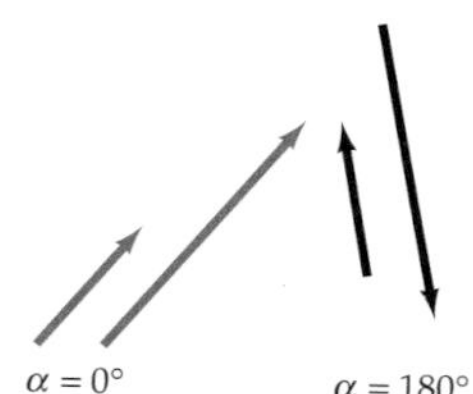

Figure 4.36

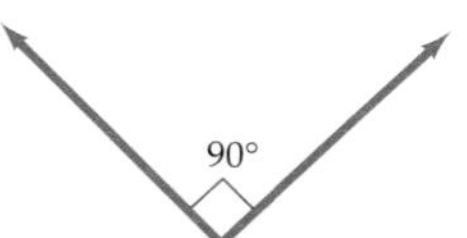

Figure 4.37

Parallel and Perpendicular Vectors

Two vectors are *parallel* when the angle α between the vectors is 0° or 180°, as shown in **Figure 4.36.** When the angle α is 0°, the vectors point in the same direction; the vectors point in opposite directions when α is 180°.

Let $\mathbf{v} = a_1\mathbf{i} + b_1\mathbf{j}$, let c be a real number, and let $\mathbf{w} = c\mathbf{v}$. Because $\mathbf{w}$ is a constant multiple of $\mathbf{v}$, $\mathbf{w}$ and $\mathbf{v}$ are parallel vectors. When $c > 0$, the vectors point in the same direction. When $c < 0$, the vectors point in opposite directions.

Two vectors are *perpendicular* when the angle between the vectors is 90°. See **Figure 4.37.** Perpendicular vectors are referred to as **orthogonal vectors.** If $\mathbf{v}$ and $\mathbf{w}$ are two nonzero orthogonal vectors, then from the formula for the angle between two vectors and the fact that $\cos\alpha = 0$, we have

$$0 = \frac{\mathbf{v} \cdot \mathbf{w}}{\|\mathbf{v}\|\,\|\mathbf{w}\|}$$

If a fraction equals zero, the numerator must be zero. Thus, for orthogonal vectors $\mathbf{v}$ and $\mathbf{w}$, $\mathbf{v} \cdot \mathbf{w} = 0$. This gives the following result.

Condition for Perpendicular Vectors

Two nonzero vectors $\mathbf{v}$ and $\mathbf{w}$ are orthogonal if and only if $\mathbf{v} \cdot \mathbf{w} = 0$.

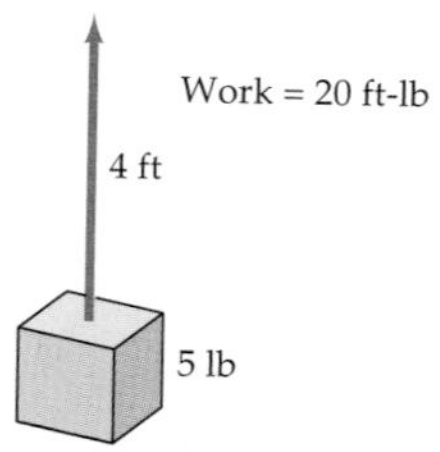

Work = 20 ft-lb
4 ft
5 lb

Figure 4.38

Work: An Application of the Dot Product

When a 5-pound force is used to lift a box from the ground a distance of 4 feet, *work* is done. The amount of **work** is the product of the force on the box and the distance the box is moved. In this case the work is 20 foot-pounds. When the box is lifted, the force and the displacement vector (the direction in which and the distance the box was moved) are in the same direction. (See **Figure 4.38.**)

Now consider a sled being pulled by a child along the ground by a rope attached to the sled, as shown in **Figure 4.39.** The force vector (along the rope) is *not* in the same direction as the displacement vector (parallel to the ground). In this case the dot product is used to determine the work done by the force.

Definition of Work

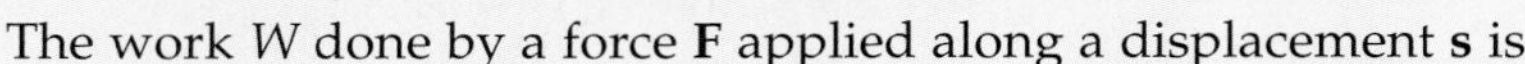

The work W done by a force $\mathbf{F}$ applied along a displacement $\mathbf{s}$ is

$$W = \mathbf{F} \cdot \mathbf{s} = \|\mathbf{F}\|\,\|\mathbf{s}\| \cos \alpha$$

where α is the angle between $\mathbf{F}$ and $\mathbf{s}$.

Figure 4.39

In the case of the child pulling the sled a horizontal distance of 7 feet, the work done is

$$\begin{aligned} W &= \mathbf{F} \cdot \mathbf{s} \\ &= \|\mathbf{F}\|\,\|\mathbf{s}\| \cos \alpha && \bullet\ \alpha \text{ is the angle between } \mathbf{F} \text{ and } \mathbf{s}. \\ &= (25)(7) \cos 37^\circ \approx 140 \text{ foot-pounds} \end{aligned}$$

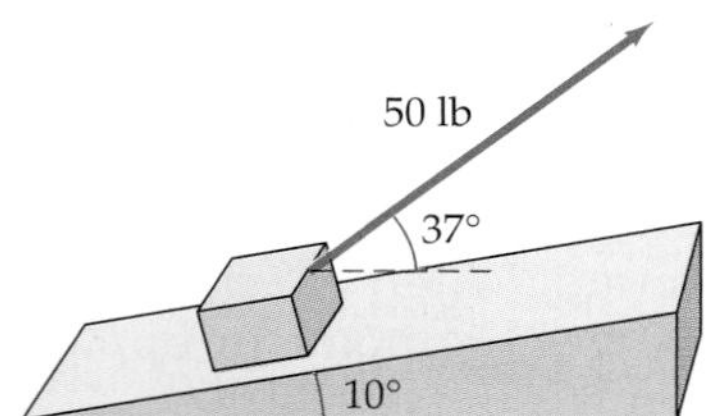

Figure 4.40

EXAMPLE 11 Solve a Work Problem

A force of 50 pounds on a rope is used to drag a box up a ramp that is inclined 10°. If the rope makes an angle of 37° with the ground, find the work done in moving the box 15 feet along the ramp. See **Figure 4.40.**

Solution

In the formula $W = \|\mathbf{F}\|\,\|\mathbf{s}\| \cos \alpha$, α is the angle between the force and the displacement. Thus $\alpha = 37^\circ - 10^\circ = 27^\circ$. The work done is

$$W = \|\mathbf{F}\|\,\|\mathbf{s}\| \cos \alpha = 50 \cdot 15 \cdot \cos 27^\circ \approx 670 \text{ foot-pounds}$$

Try Exercise 80, page 326

Topics for Discussion

1. Is the dot product of two vectors a vector or a scalar? Explain.
2. Is the projection of **v** onto **w** a vector or a scalar? Explain.
3. Is the nonzero vector $\langle a, b \rangle$ perpendicular to the vector $\langle -b, a \rangle$? Explain.
4. Consider the nonzero vector $\mathbf{u} = \langle a, b \rangle$ and the vector

$$\mathbf{v} = \left\langle \frac{a}{\sqrt{a^2 + b^2}}, \frac{b}{\sqrt{a^2 + b^2}} \right\rangle$$

   a. Are the vectors parallel? Explain.
   b. Which one of the vectors is a unit vector?
   c. Which vector has the larger magnitude? Explain.

Exercise Set 4.3

In Exercises 1 to 10, find the components of the vector with the initial point P_1 and terminal point P_2. Use these components to write a vector that is equivalent to $\mathbf{P_1 P_2}$.

1.

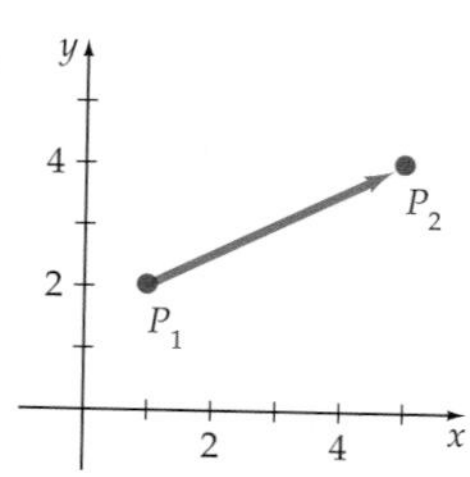

2.

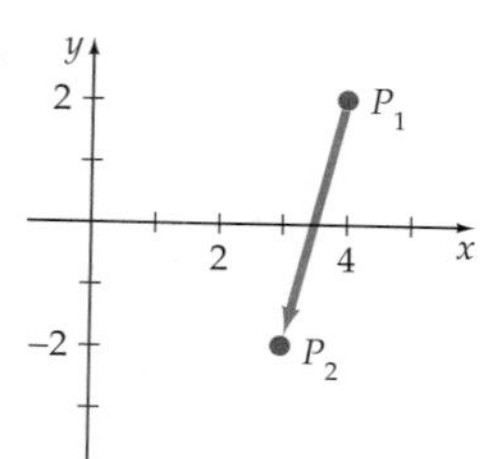

3.

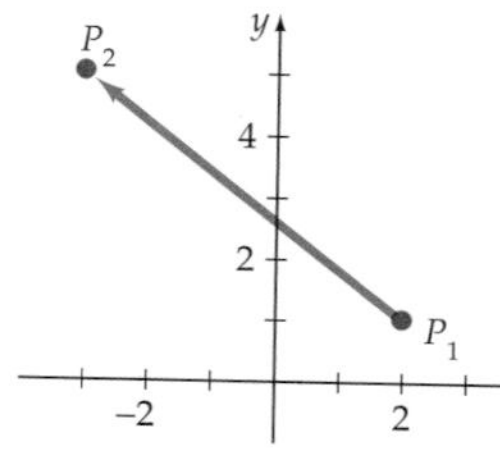

4.

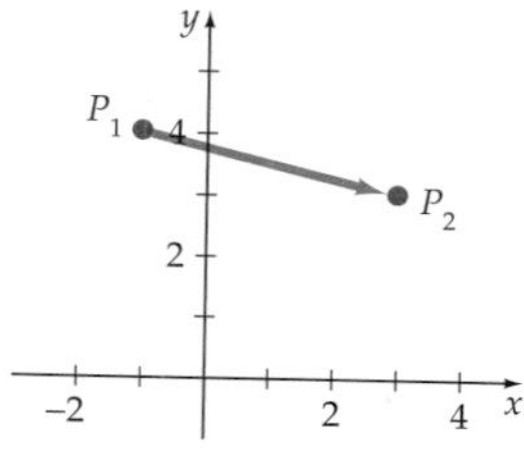

5. $P_1(-3, 0)$; $P_2(4, -1)$

6. $P_1(5, -1)$; $P_2(3, 1)$

7. $P_1(4, 2)$; $P_2(-3, -3)$

8. $P_1(0, -3)$; $P_2(0, 4)$

9. $P_1(2, -5)$; $P_2(2, 3)$

10. $P_1(3, -2)$; $P_2(3, 0)$

In Exercises 11 to 18, find the magnitude and direction of each vector. Find the unit vector in the direction of the given vector.

11. $\mathbf{v} = \langle -3, 4 \rangle$

12. $\mathbf{v} = \langle 6, 10 \rangle$

13. $\mathbf{v} = \langle 20, -40 \rangle$

14. $\mathbf{v} = \langle -50, 30 \rangle$

15. $\mathbf{v} = 2\mathbf{i} - 4\mathbf{j}$

16. $\mathbf{v} = -5\mathbf{i} + 6\mathbf{j}$

17. $\mathbf{v} = 42\mathbf{i} - 18\mathbf{j}$

18. $\mathbf{v} = -22\mathbf{i} - 32\mathbf{j}$

In Exercises 19 to 27, perform the indicated operations, where $\mathbf{u} = \langle -2, 4 \rangle$ and $\mathbf{v} = \langle -3, -2 \rangle$.

19. $3\mathbf{u}$

20. $-4\mathbf{v}$

21. $2\mathbf{u} - \mathbf{v}$

22. $4\mathbf{v} - 2\mathbf{u}$

23. $\frac{2}{3}\mathbf{u} + \frac{1}{6}\mathbf{v}$

24. $\frac{3}{4}\mathbf{u} - 2\mathbf{v}$

25. $\|\mathbf{u}\|$

26. $\|\mathbf{v} + 2\mathbf{u}\|$

27. $\|3\mathbf{u} - 4\mathbf{v}\|$

In Exercises 28 to 36, perform the indicated operations, where $\mathbf{u} = 3\mathbf{i} - 2\mathbf{j}$ and $\mathbf{v} = -2\mathbf{i} + 3\mathbf{j}$.

28. $-2\mathbf{u}$

29. $4\mathbf{v}$

30. $3\mathbf{u} + 2\mathbf{v}$

31. $6\mathbf{u} + 2\mathbf{v}$

32. $\frac{1}{2}\mathbf{u} - \frac{3}{4}\mathbf{v}$

33. $\frac{2}{3}\mathbf{v} + \frac{3}{4}\mathbf{u}$

34. $\|\mathbf{v}\|$

35. $\|\mathbf{u} - 2\mathbf{v}\|$

36. $\|2\mathbf{v} + 3\mathbf{u}\|$

In Exercises 37 to 40, find the horizontal and vertical components of each vector. Round to the nearest tenth. Write an equivalent vector in the form $\mathbf{v} = a_1\mathbf{i} + a_2\mathbf{j}$.

37. Magnitude $= 5$, direction angle $= 27°$

38. Magnitude $= 4$, direction angle $= 127°$

39. Magnitude $= 4$, direction angle $= \frac{\pi}{4}$

40. Magnitude $= 2$, direction angle $= \frac{8\pi}{7}$

41. **Ground Speed of a Plane** A plane is flying at an airspeed of 340 mph at a heading of 124°. A wind of 45 mph is blowing from the west. Find the ground speed of the plane.

42. **Heading of a Boat** A person who can row 2.6 mph in still water wants to row due east across a river. The river is flowing from the north at a rate of 0.8 mph. Determine the heading of the boat required for the boat to travel due east across the river.

43. **Ground Speed and Course of a Plane** A pilot is flying at a heading of 96° at 225 mph. A 50-mph wind is blowing from the southwest at a heading of 37°. Find the ground speed and course of the plane.

44. **Course of a Boat** The captain of a boat is steering at a heading of 327° at 18 mph. The current is flowing at 4 mph at a heading of 60°. Find the course (to the nearest degree) of the boat.

45. **Magnitude of a Force** Find the magnitude of the force necessary to keep a 3000-pound car from sliding down a ramp inclined at an angle of 5.6°. Round to the nearest pound.

46. **ANGLE OF A RAMP** A 120-pound force keeps an 800-pound object from sliding down an inclined ramp. Find the angle of the ramp. Round to the nearest tenth of a degree.

47. **MAGNITUDE OF A FORCE** A 345-pound box is placed on a ramp that is inclined 22.4°.

  a. Find the magnitude of the force needed to keep the box from sliding down the ramp. Ignore the effects of friction. Round to the nearest pound.

  b. Find the magnitude of the force the box exerts against the ramp. Round to the nearest pound.

48. **MAGNITUDE OF A FORCE** A motorcycle that weighs 811 pounds is placed on a ramp that is inclined 31.8°.

  a. Find the magnitude of the force needed to keep the motorcycle from rolling down the ramp. Round to the nearest pound.

  b. Find the magnitude of the force the motorcycle exerts against the ramp. Round to the nearest pound.

The forces $\mathbf{F}_1, \mathbf{F}_2, \mathbf{F}_3, \ldots, \mathbf{F}_n$ acting on an object are in *equilibrium* provided the resultant of all the forces is the zero vector:

$$\mathbf{F}_1 + \mathbf{F}_2 + \mathbf{F}_3 + \cdots + \mathbf{F}_n = \mathbf{0}$$

In Exercises 49 to 53, determine whether the given forces are in equilibrium. If the forces are not in equilibrium, determine an additional force that would bring the forces into equilibrium.

49. $\mathbf{F}_1 = \langle 18.2, 13.1 \rangle$, $\mathbf{F}_2 = \langle -12.4, 3.8 \rangle$, $\mathbf{F}_3 = \langle -5.8, -16.9 \rangle$

50. $\mathbf{F}_1 = \langle -4.6, 5.3 \rangle$, $\mathbf{F}_2 = \langle 6.2, 4.9 \rangle$, $\mathbf{F}_3 = \langle -1.6, -10.2 \rangle$

51. $\mathbf{F}_1 = 155\mathbf{i} - 257\mathbf{j}$, $\mathbf{F}_2 = -124\mathbf{i} + 149\mathbf{j}$, $\mathbf{F}_3 = -31\mathbf{i} + 98\mathbf{j}$

52. $\mathbf{F}_1 = 23.5\mathbf{i} + 18.9\mathbf{j}$, $\mathbf{F}_2 = -18.7\mathbf{i} + 2.5\mathbf{j}$, $\mathbf{F}_3 = -5.6\mathbf{i} - 15.6\mathbf{j}$

53. $\mathbf{F}_1 = 189.3\mathbf{i} + 235.7\mathbf{j}$, $\mathbf{F}_2 = 45.8\mathbf{i} - 205.6\mathbf{j}$, $\mathbf{F}_3 = -175.2\mathbf{i} - 37.7\mathbf{j}$, $\mathbf{F}_4 = -59.9\mathbf{i} + 7.6\mathbf{j}$

54. The cranes in the following figure are holding a 9450-pound steel girder in midair. The forces $\mathbf{F}_1$ and $\mathbf{F}_2$, along with the force $\mathbf{F}_3 = \langle 0, -9450 \rangle$, are in equilibrium as defined above Exercise 49. Find the magnitude of $\mathbf{F}_2$, given that $\|\mathbf{F}_1\| = 6223$ pounds. Round to the nearest 10 pounds.

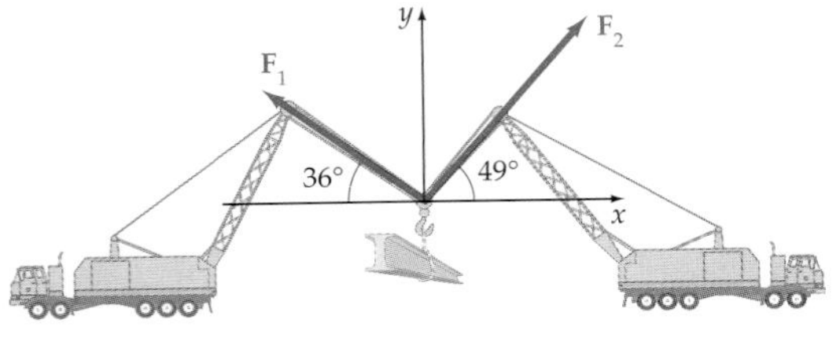

In Exercises 55 to 62, find the dot product of the vectors.

55. $\mathbf{v} = \langle 3, -2 \rangle$; $\mathbf{w} = \langle 1, 3 \rangle$

56. $\mathbf{v} = \langle 2, 4 \rangle$; $\mathbf{w} = \langle 0, 2 \rangle$

57. $\mathbf{v} = \langle 4, 1 \rangle$; $\mathbf{w} = \langle -1, 4 \rangle$

58. $\mathbf{v} = \langle 2, -3 \rangle$; $\mathbf{w} = \langle 3, 2 \rangle$

59. $\mathbf{v} = \mathbf{i} + 2\mathbf{j}$; $\mathbf{w} = -\mathbf{i} + \mathbf{j}$

60. $\mathbf{v} = 5\mathbf{i} + 3\mathbf{j}$; $\mathbf{w} = 4\mathbf{i} - 2\mathbf{j}$

61. $\mathbf{v} = 6\mathbf{i} - 4\mathbf{j}$; $\mathbf{w} = -2\mathbf{i} - 3\mathbf{j}$

62. $\mathbf{v} = -4\mathbf{i} + 2\mathbf{j}$; $\mathbf{w} = -2\mathbf{i} - 4\mathbf{j}$

In Exercises 63 to 70, find the measure of the angle between the two vectors. State which pairs of vectors are orthogonal. Round approximate measures to the nearest tenth of a degree.

63. $\mathbf{v} = \langle 2, -1 \rangle$; $\mathbf{w} = \langle 3, 4 \rangle$

64. $\mathbf{v} = \langle 1, -5 \rangle$; $\mathbf{w} = \langle -2, 3 \rangle$

65. $\mathbf{v} = \langle 0, 3 \rangle$; $\mathbf{w} = \langle 2, 2 \rangle$

66. $\mathbf{v} = \langle -1, 7 \rangle$; $\mathbf{w} = \langle 3, -2 \rangle$

67. $\mathbf{v} = 5\mathbf{i} - 2\mathbf{j}$; $\mathbf{w} = 2\mathbf{i} + 5\mathbf{j}$

68. $\mathbf{v} = 8\mathbf{i} + \mathbf{j}$; $\mathbf{w} = -\mathbf{i} + 8\mathbf{j}$

69. $\mathbf{v} = 5\mathbf{i} + 2\mathbf{j}$; $\mathbf{w} = -5\mathbf{i} - 2\mathbf{j}$

70. $\mathbf{v} = 3\mathbf{i} - 4\mathbf{j}$; $\mathbf{w} = 6\mathbf{i} - 12\mathbf{j}$

In Exercises 71 to 78, find $\text{proj}_{\mathbf{w}}\mathbf{v}$.

71. $\mathbf{v} = \langle 6, 7 \rangle$; $\mathbf{w} = \langle 3, 4 \rangle$

72. $\mathbf{v} = \langle -7, 5 \rangle$; $\mathbf{w} = \langle -4, 1 \rangle$

73. $\mathbf{v} = \langle -3, 4 \rangle$; $\mathbf{w} = \langle 2, 5 \rangle$

74. $\mathbf{v} = \langle 2, 4 \rangle$; $\mathbf{w} = \langle -1, 5 \rangle$

75. $\mathbf{v} = 2\mathbf{i} + \mathbf{j}$; $\mathbf{w} = 6\mathbf{i} + 3\mathbf{j}$

76. $\mathbf{v} = 5\mathbf{i} + 2\mathbf{j}$; $\mathbf{w} = -5\mathbf{i} - 2\mathbf{j}$

77. $\mathbf{v} = 3\mathbf{i} - 4\mathbf{j}$; $\mathbf{w} = -6\mathbf{i} + 12\mathbf{j}$

78. $\mathbf{v} = 2\mathbf{i} + 2\mathbf{j}$; $\mathbf{w} = -4\mathbf{i} - 2\mathbf{j}$

79. **WORK** A 150-pound box is dragged 15 feet along a level floor. Find the work done if a force of 75 pounds with a direction angle of 32° is used. Round to the nearest foot-pound.

80. **WORK** A 100-pound force is pulling a sled loaded with bricks that weighs 400 pounds. The force is at an angle of 42° with the displacement. Find the work done in moving the sled 25 feet.

81. **WORK** A rope is being used to pull a box up a ramp that is inclined at 15°. The rope exerts a force of 75 pounds on the box, and it makes an angle of 30° with the plane of the ramp. Find the work done in moving the box 12 feet.

82. **WORK** A dock worker exerts a force on a box sliding down the ramp of a truck. The ramp makes an angle of 48° with the road, and the worker exerts a 50-pound force parallel to the road. Find the work done in sliding the box 6 feet.

Connecting Concepts

83. For $\mathbf{u} = \langle -1, 1\rangle$, $\mathbf{v} = \langle 2, 3\rangle$, and $\mathbf{w} = \langle 5, 5\rangle$, find the sum of the three vectors geometrically by using the triangle method of adding vectors.

84. For $\mathbf{u} = \langle 1, 2\rangle$, $\mathbf{v} = \langle 3, -2\rangle$, and $\mathbf{w} = \langle -1, 4\rangle$, find $\mathbf{u} + \mathbf{v} - \mathbf{w}$ geometrically by using the triangle method of adding vectors.

85. Find a vector that has initial point $(3, -1)$ and is equivalent to $\mathbf{v} = 2\mathbf{i} - 3\mathbf{j}$.

86. Find a vector that has initial point $(-2, 4)$ and is equivalent to $\mathbf{v} = \langle -1, 3\rangle$.

87. If $\mathbf{v} = 2\mathbf{i} - 5\mathbf{j}$ and $\mathbf{w} = 5\mathbf{i} + 2\mathbf{j}$ have the same initial point, is $\mathbf{v}$ perpendicular to $\mathbf{w}$? Why or why not?

88. If $\mathbf{v} = \langle 5, 6\rangle$ and $\mathbf{w} = \langle 6, 5\rangle$ have the same initial point, is $\mathbf{v}$ perpendicular to $\mathbf{w}$? Why or why not?

89. Let $\mathbf{v} = \langle -2, 7\rangle$. Find a vector perpendicular to $\mathbf{v}$.

90. Let $\mathbf{w} = 4\mathbf{i} + \mathbf{j}$. Find a vector perpendicular to $\mathbf{w}$.

91. Is the dot product an associative operation? That is, given any nonzero vectors $\mathbf{u}$, $\mathbf{v}$, and $\mathbf{w}$, does

$$(\mathbf{u} \cdot \mathbf{v}) \cdot \mathbf{w} = \mathbf{u} \cdot (\mathbf{v} \cdot \mathbf{w})?$$

92. Prove that $\mathbf{v} \cdot \mathbf{w} = \mathbf{w} \cdot \mathbf{v}$.

93. Prove that $c(\mathbf{v} \cdot \mathbf{w}) = (c\mathbf{v}) \cdot \mathbf{w}$.

94. Show that the dot product of two nonzero vectors is positive if the angle between the vectors is an acute angle and negative if the angle between the two vectors is an obtuse angle.

95. **Comparison of Work Done** Consider the following two situations. (1) A rope is being used to pull a box up a ramp inclined at an angle α. The rope exerts a force $\mathbf{F}$ on the box, and the rope makes an angle θ with the ramp. The box is pulled s feet. (2) A rope is being used to pull the same box along a level floor. The rope exerts the same force $\mathbf{F}$ on the box. The box is pulled the same s feet. In which case is more work done?

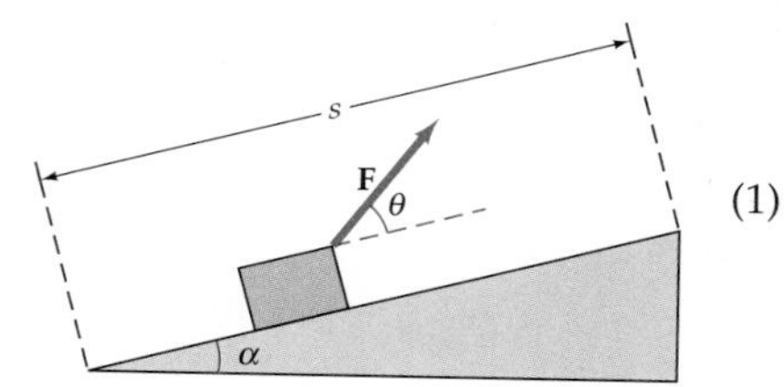

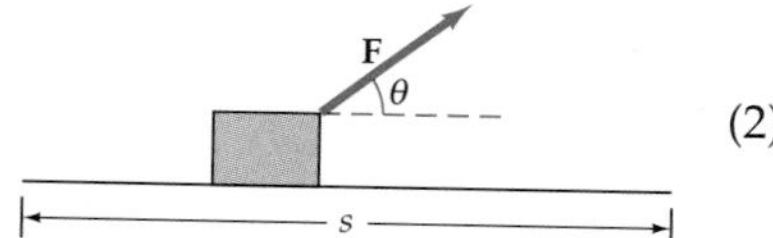

Projects

1. **Same Direction or Opposite Directions** Let $\mathbf{v} = c\mathbf{w}$, where c is a nonzero real number and $\mathbf{w}$ is a nonzero vector. Show that $\dfrac{\mathbf{v} \cdot \mathbf{w}}{\|\mathbf{v}\|\,\|\mathbf{w}\|} = \pm 1$ and that the result is 1 when $c > 0$ and -1 when $c < 0$.

2. **The Law of Cosines and Vectors** Prove that

$$\|\mathbf{v} - \mathbf{w}\|^2 = \|\mathbf{v}\|^2 + \|\mathbf{w}\|^2 - 2\mathbf{v} \cdot \mathbf{w}$$

3. **Projection Relationships** Determine the relationship between the nonzero vectors $\mathbf{v}$ and $\mathbf{w}$ if

a. $\text{proj}_{\mathbf{w}}\mathbf{v} = 0$ **b.** $\text{proj}_{\mathbf{w}}\mathbf{v} = \|\mathbf{v}\|$

Exploring Concepts with Technology

Optimal Branching of Arteries

The physiologist Jean Louis Poiseuille (1799–1869) developed several laws concerning the flow of blood. One of his laws states that the resistance R of a blood vessel of length l and radius r is given by

$$R = k\frac{l}{r^4} \tag{1}$$

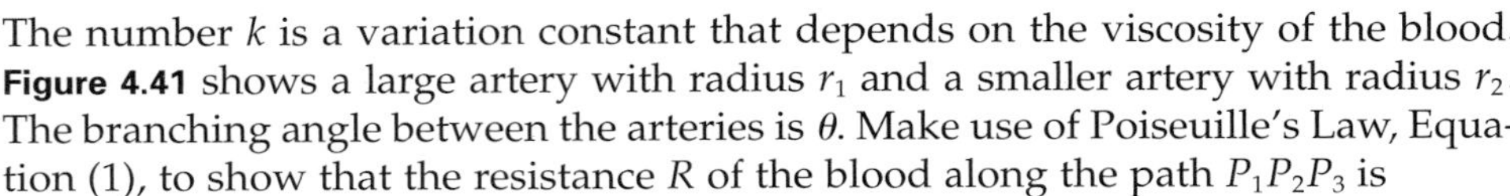

The number k is a variation constant that depends on the viscosity of the blood. **Figure 4.41** shows a large artery with radius r_1 and a smaller artery with radius r_2. The branching angle between the arteries is θ. Make use of Poiseuille's Law, Equation (1), to show that the resistance R of the blood along the path $P_1P_2P_3$ is

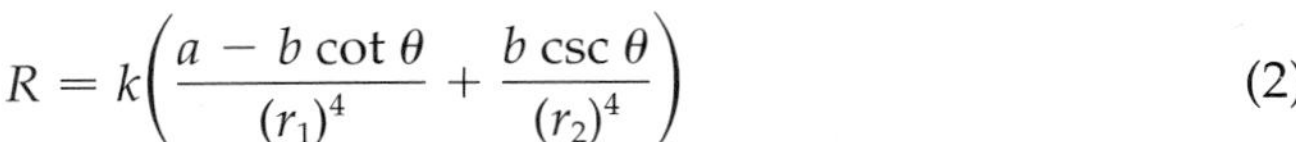

$$R = k\left(\frac{a - b\cot\theta}{(r_1)^4} + \frac{b\csc\theta}{(r_2)^4}\right) \tag{2}$$

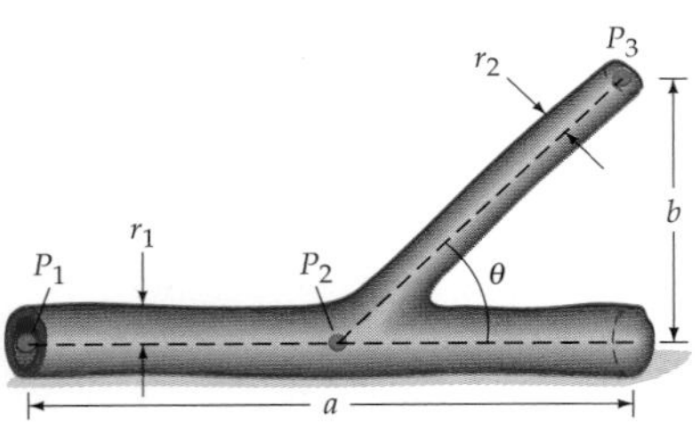

Figure 4.41

Use a graphing utility to graph R with $k = 0.0563$, $a = 8$ centimeters, $b = 4$ centimeters, $r_1 = 0.4$ centimeter, and $r_2 = \frac{3}{4}r_1 = 0.3$ centimeter. Then estimate (to the nearest degree) the angle θ that minimizes R. By using calculus, it can be demonstrated that R is minimized when

$$\cos\theta = \left(\frac{r_2}{r_1}\right)^4 \tag{3}$$

This equation is remarkable because it is much simpler than Equation (2) and because it does not involve the distance a or b. Solve Equation (3) for θ, with $r_2 = \frac{3}{4}r_1$. How does this value of θ compare with the value of θ you obtained by graphing?

Chapter 4 Summary

4.1 The Law of Sines

- The following Law of Sines is used to solve triangles when two angles and a side are given or when two sides and an angle opposite one of them are given.

$$\frac{a}{\sin A} = \frac{b}{\sin B} = \frac{c}{\sin C}$$

4.2 The Law of Cosines and Area

- The Law of Cosines, $a^2 = b^2 + c^2 - 2bc\cos A$, is used to solve general triangles when two sides and the included angle or three sides of the triangle are given.
- The area K of triangle ABC is

$$K = \frac{1}{2}bc\sin A = \frac{b^2\sin C\sin A}{2\sin B}$$

- The area of a triangle for which three sides are given (Heron's formula) is

$$K = \sqrt{s(s-a)(s-b)(s-c)}, \quad \text{where } s = \frac{1}{2}(a + b + c)$$

4.3 Vectors

- A vector is a directed line segment. The length of the line segment is the magnitude of the vector, and the direction of the vector is measured by an angle. The angle between a vector and the positive x-axis is called the direction angle of the vector. Two vectors are equivalent if they have the same magnitude and the same direction. The resultant of two or more vectors is the sum of the vectors.

- Vectors can be added by the parallelogram method, the triangle method, or addition of the x- and y-components.
- If $\mathbf{v} = \langle a, b \rangle$ and k is a real number, then $k\mathbf{v} = \langle ka, kb \rangle$.
- The dot product of $\mathbf{v} = \langle a, b \rangle$ and $\mathbf{w} = \langle c, d \rangle$ is given by

$$\mathbf{v} \cdot \mathbf{w} = ac + bd$$

- If $\mathbf{v}$ and $\mathbf{w}$ are two nonzero vectors and α is the smallest nonnegative angle between $\mathbf{v}$ and $\mathbf{w}$, then $\cos \alpha = \dfrac{\mathbf{v} \cdot \mathbf{w}}{\|\mathbf{v}\| \|\mathbf{w}\|}$.

Chapter 4 Assessing Concepts

1. What is an oblique triangle?
2. In triangle ABC, $a = 4.5$, $b = 6.2$, and $C = 107°$. Which law, the Law of Sines or the Law of Cosines, can be used to find c?
3. Which of the following cases,

   ASA, AAS, SSA, SSS, or SAS

   is known as the ambiguous case of the Law of Sines?
4. In Heron's formula, what does the variable s represent?
5. Is the dot product of two vectors a vector or a scalar?
6. Let $\mathbf{v}$ and $\mathbf{w}$ be nonzero vectors. Is $\text{proj}_{\mathbf{w}}\mathbf{v}$ a vector or a scalar?
7. True or false: The vector $\left\langle \frac{12}{13}, -\frac{5}{13} \right\rangle$ is a unit vector.
8. True or false: $\mathbf{i} \cdot \mathbf{j} = 0$.
9. True or false: The Law of Sines can be used to solve any triangle, given two angles and any side.
10. True or false: If two nonzero vectors are orthogonal, then their dot product is 0.

Chapter 4 Review Exercises

In Exercises 1 to 10, solve each triangle.

1. $A = 37°, b = 14, C = 92°$
2. $B = 77.4°, c = 11.8, C = 94.0°$
3. $a = 12, b = 15, c = 20$
4. $a = 24, b = 32, c = 28$
5. $a = 18, b = 22, C = 35°$
6. $b = 102, c = 150, A = 82°$
7. $A = 105°, a = 8, c = 10$
8. $C = 55°, c = 80, b = 110$
9. $A = 55°, B = 80°, c = 25$
10. $B = 25°, C = 40°, c = 40$

In Exercises 11 to 18, find the area of each triangle. Round each area accurate to two significant digits.

11. $a = 24, b = 30, c = 36$
12. $a = 9.0, b = 7.0, c = 12$
13. $a = 60, b = 44, C = 44°$
14. $b = 8.0, c = 12, A = 75°$
15. $b = 50, c = 75, C = 15°$
16. $b = 18, a = 25, A = 68°$
17. $A = 110°, a = 32, b = 15$
18. $A = 45°, c = 22, b = 18$

In Exercises 19 and 20, find the components of each vector with the given initial and terminal points. Write an equivalent vector in terms of its components.

19. $P_1(-2, 4); P_2(3, 7)$
20. $P_1(-4, 0); P_2(-3, 6)$

In Exercises 21 to 24, find the magnitude and direction angle of each vector.

21. $\mathbf{v} = \langle -4, 2 \rangle$
22. $\mathbf{v} = \langle 6, -3 \rangle$
23. $\mathbf{u} = -2\mathbf{i} + 3\mathbf{j}$
24. $\mathbf{u} = -4\mathbf{i} - 7\mathbf{j}$

In Exercises 25 to 28, find a unit vector in the direction of the given vector.

25. $\mathbf{w} = \langle -8, 5 \rangle$

26. $\mathbf{w} = \langle 7, -12 \rangle$

27. $\mathbf{v} = 5\mathbf{i} + \mathbf{j}$

28. $\mathbf{v} = 3\mathbf{i} - 5\mathbf{j}$

In Exercises 29 and 30, perform the indicated operation, where $\mathbf{u} = \langle 3, 2 \rangle$ and $\mathbf{v} = \langle -4, -1 \rangle$.

29. $\mathbf{v} - \mathbf{u}$

30. $2\mathbf{u} - 3\mathbf{v}$

In Exercises 31 and 32, perform the indicated operation, where $\mathbf{u} = 10\mathbf{i} + 6\mathbf{j}$ and $\mathbf{v} = 8\mathbf{i} - 5\mathbf{j}$.

31. $-\mathbf{u} + \frac{1}{2}\mathbf{v}$

32. $\frac{2}{3}\mathbf{v} - \frac{3}{4}\mathbf{u}$

33. **Ground Speed of a Plane** A plane is flying at an airspeed of 400 mph at a heading of 204°. A wind of 45 mph is blowing from the east. Find the ground speed of the plane.

34. **Angle of a Ramp** A 40-pound force keeps a 320-pound object from sliding down an inclined ramp. Find the angle of the ramp.

In Exercises 35 to 38, find the dot product of the vectors.

35. $\mathbf{u} = \langle 3, 7 \rangle$; $\mathbf{v} = \langle -1, 3 \rangle$

36. $\mathbf{v} = \langle -8, 5 \rangle$; $\mathbf{u} = \langle 2, -1 \rangle$

37. $\mathbf{v} = -4\mathbf{i} - \mathbf{j}$; $\mathbf{u} = 2\mathbf{i} + \mathbf{j}$

38. $\mathbf{u} = -3\mathbf{i} + 7\mathbf{j}$; $\mathbf{v} = -2\mathbf{i} + 2\mathbf{j}$

In Exercises 39 to 42, find the angle between the vectors. Round to the nearest degree.

39. $\mathbf{u} = \langle 7, -4 \rangle$; $\mathbf{v} = \langle 2, 3 \rangle$

40. $\mathbf{v} = \langle -5, 2 \rangle$; $\mathbf{u} = \langle 2, -4 \rangle$

41. $\mathbf{v} = 6\mathbf{i} - 11\mathbf{j}$; $\mathbf{u} = 2\mathbf{i} + 4\mathbf{j}$

42. $\mathbf{u} = \mathbf{i} - 5\mathbf{j}$; $\mathbf{v} = \mathbf{i} + 5\mathbf{j}$

In Exercises 43 and 44, find $\text{proj}_{\mathbf{w}}\mathbf{v}$.

43. $\mathbf{v} = \langle -2, 5 \rangle$; $\mathbf{w} = \langle 5, 4 \rangle$

44. $\mathbf{v} = 4\mathbf{i} - 7\mathbf{j}$; $\mathbf{w} = -2\mathbf{i} - 5\mathbf{j}$

45. **Work** A 120-pound box is dragged 14 feet along a level floor. Find the work done if a force of 60 pounds with a direction angle of 38° is used. Round to the nearest foot-pound.

»»» Quantitative Reasoning: *Trigonometry and Great Circle Routes* »»»

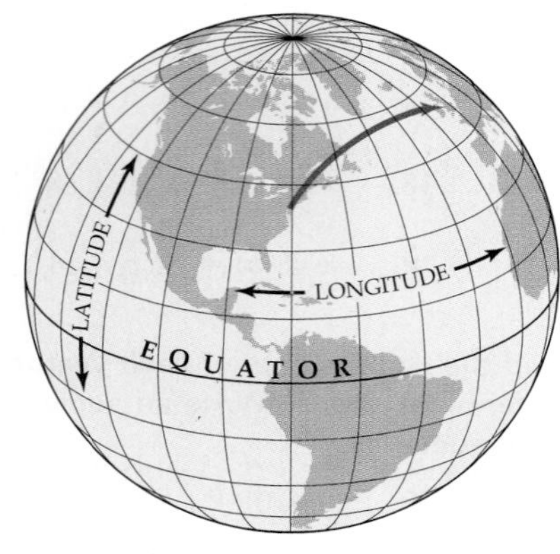

Great circle route from New York's John F. Kennedy airport to London's Heathrow airport.

A **great circle** is a circle on a sphere's surface whose center is the same as the center of the sphere. A great circle is a path on the sphere with the smallest curvature, and hence an arc of a great circle is the shortest path between two given points on the sphere. The distance between any two points on a sphere is called the **great circle distance** for the given points.

Great circle routes are often used by pilots when winds are not a significant factor. A formula that is used to estimate the great circle distance d, measured in radians, between two airports is given by

$$d = \cos^{-1}[\cos(\text{lat1})\cos(\text{lat2})\cos(\text{lon1}-\text{lon2}) + \sin(\text{lat1})\sin(\text{lat2})] \quad (1)$$

where lat1 is the latitude of airport 1, lon1 is the longitude of airport 1, lat2 is the latitude of airport 2, and lon2 is the longitude of airport 2.

For example, the coordinates of New York's John F. Kennedy airport (JFK) are 40°39′ N and 73°47′ W. The coordinates of London's Heathrow airport (LHR) are 51°29′ N and 0°27′ W. Converting each of these coordinates to radians gives

$$(\text{lat1}, \text{lon1}) \approx (0.709476, -1.287756)$$
$$(\text{lat2}, \text{lon2}) \approx (0.898554, -0.007854)$$

Note: North latitudes and east longitudes are represented by positive radian values, whereas south latitudes and west longitudes are represented by negative radian values. Substitute the radian values into Formula (1) to produce $d \approx 0.869496$ radian. Because the earth is nearly spherical with a radius of about 3960 miles, we can convert d to miles by multiplying by 3960.

$$d \approx 0.869496 \times 3960 \approx 3443$$

Rounded to the nearest 10 miles, the great circle distance from JFK to LHR is approximately 3440 miles.

To fly a great circle route, the pilot must adjusts his or her heading several times during the flight. The following formulas can be used to find the *initial heading* (h1) a pilot should use to fly a great circle route from airport 1 to airport 2.

If sin(lon2 – lon1) > 0, then

$$\text{h1} = \cos^{-1}\left[\frac{\sin(\text{lat2}) - \sin(\text{lat1})\cos(d)}{\sin(d)\cos(\text{lat1})}\right] \tag{2}$$

Otherwise,

$$\text{h1} = 2\pi - \cos^{-1}\left[\frac{\sin(\text{lat2}) - \sin(\text{lat1})\cos(d)}{\sin(d)\cos(\text{lat1})}\right] \tag{3}$$

In Formulas (2) and (3), each of the values lat1, lat2, and d must be in radians. Also, these formulas are not valid if airport 1 is located at the North or South Pole.

To find the initial heading required to fly the great circle route from JFK to LHR, we first note that

$$\sin(\text{lon2} - \text{lon1}) = \sin[-0.007854 - (-1.287756)] \approx 0.957988$$

Because $\sin(\text{lon2} - \text{lon1}) > 0$, we use Formula (2) to find the initial heading. Substitute 0.709476 for lat1, 0.898554 for lat2, and 0.869496 for d to produce h1 $\approx$ 0.896040 radian. Converting to degrees gives 51° as the initial heading from JFK to LHR, to the nearest degree.

QR1. Find the great circle distance, in miles, between Orlando International airport (KMCO) (latitude = 0.496187 radian, longitude = −1.419110 radians) and Los Angeles International airport (LAX) (latitude = 0.592409 radian, longitude = −2.066611 radians). Use 3960 miles as the radius of Earth. Round to the nearest 10 miles.

QR2. Find the initial heading needed to fly the great circle route from KMCO to LAX. Round to the nearest degree.

QR3. Find the great circle distance, in miles, between JFK and Denver International airport (KDEN) (latitude = 0.695717 radian, longitude = −1.826892 radians). Use 3960 miles as the radius of Earth. Round to the nearest 10 miles.

QR4. Find the initial heading required to fly the great circle route from KDEN to JFK. Round to the nearest degree.

Chapter 4 Test

1. Solve triangle ABC if $A = 70°$, $C = 16°$, and $c = 14$.

2. Find B in triangle ABC if $A = 140°$, $b = 13$, and $a = 45$.

3. In triangle ABC, $C = 42°$, $a = 20$, and $b = 12$. Find side c.

4. In triangle ABC, $a = 32$, $b = 24$, and $c = 18$. Find angle B.

In Exercises 5 to 7, round your answers to two significant digits.

5. Given angle $C = 110°$, side $a = 7.0$, and side $b = 12$, find the area of triangle ABC.

6. Given angle $B = 42°$, angle $C = 75°$, and side $b = 12$, find the area of triangle ABC.

7. Given side $a = 17$, side $b = 55$, and side $c = 42$, find the area of triangle ABC.

8. Given $\mathbf{v} = -2\mathbf{i} + 3\mathbf{j}$, find $\|\mathbf{v}\|$.

9. A vector has a magnitude of 12 and direction 220°. Write an equivalent vector in the form $\mathbf{v} = a_1\mathbf{i} + a_2\mathbf{j}$. Round a_1 and a_2 to four significant digits.

10. Find $3\mathbf{u} - 5\mathbf{v}$ given the vectors $\mathbf{u} = 2\mathbf{i} - 3\mathbf{j}$ and $\mathbf{v} = 5\mathbf{i} + 4\mathbf{j}$.

11. Find the dot product of $\mathbf{u} = -2\mathbf{i} + 3\mathbf{j}$ and $\mathbf{v} = 5\mathbf{i} + 3\mathbf{j}$.

12. Find the smallest positive angle, to the nearest degree, between the vectors $\mathbf{u} = \langle 3, 5\rangle$ and $\mathbf{v} = \langle -6, 2\rangle$.

13. One ship leaves a port at 1:00 P.M. traveling at 12 mph at a heading of 65°. At 2:00 P.M. another ship leaves the port traveling at 18 mph at a heading of 142°. Find the distance between the ships at 3:00 P.M.

14. Two fire lookouts are located 12 miles apart. Lookout A is at a bearing of N32°W from lookout B. A fire was sighted at a bearing of S82°E from A and N72°E from B. Find the distance of the fire from lookout B.

15. A triangular commercial piece of real estate is priced at \$8.50 per square foot. Find the cost, to the nearest \$100, of the lot, which measures 112 feet by 165 feet by 140 feet.

Cumulative Review Exercises

1. Find the distance between $P_1(-3, 4)$ and $P_2(4, -1)$.

2. Given $f(x) = \cos x$ and $g(x) = \sin x$, find $(f + g)(x)$.

3. Given $f(x) = \sec x$ and $g(x) = \cos x$, find $(f \circ g)(x)$.

4. Given $f(x) = \frac{1}{2}x - 3$, find $f^{-1}(x)$.

5. How is the graph of $F(x) = f(x - 2) + 3$ related to the graph of $y = f(x)$?

6. For the right triangle shown at the right, find a.

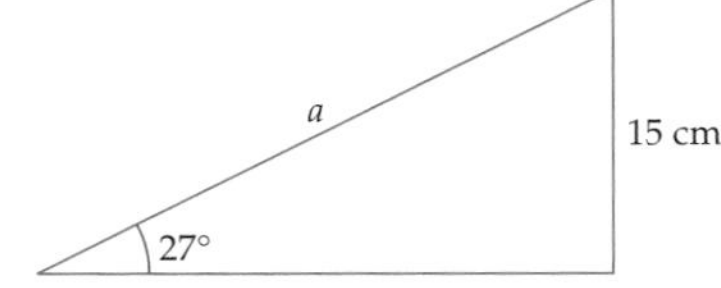

7. Graph $y = 3 \sin \pi x$.

8. Graph $y = \frac{1}{4}\tan 2x$.

9. Graph $y = 2 \sin(\pi x) + 1$.

10. Find the amplitude, period, and phase shift of the graph of $y = 3 \sin\left(\frac{1}{3}x - \frac{\pi}{2}\right)$.

11. Find the amplitude, period, and phase shift of the graph of $y = \sin x + \cos x$.

12. Find c for the triangle at the right.

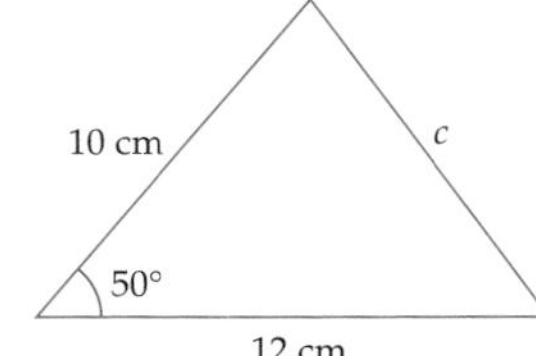

13. Verify the identity $\dfrac{1}{\cos x} - \cos x = \sin x \tan x$.

14. Evaluate $\sin^{-1}\left(\sin\left(\dfrac{2\pi}{3}\right)\right)$.

15. Evaluate $\tan\left(\cos^{-1}\left(\dfrac{12}{13}\right)\right)$.

16. Solve $\sin x \tan x - \dfrac{1}{2}\tan x = 0$ for $0 \le x < 2\pi$.

17. Find the magnitude and the positive direction angle for the vector $\langle 4, -3\rangle$. Round the angle to the nearest tenth of a degree.

18. Find the angle between the vectors $\mathbf{v} = \langle 1, 2\rangle$ and $\mathbf{w} = \langle -2, 3\rangle$. Round to the nearest tenth of a degree.

19. **Heading of a Boat** A person who can row at 3 mph in still water wants to row due west across a river. This river is flowing north to south at a rate of 1 mph. Determine the heading of the boat that is required to travel due west across the river.

20. **Ground Speed and Course of a Plane** An airplane is traveling with an airspeed of 515 mph at a heading of 54.0°. A wind of 150 mph is blowing at a heading of 120.0°. Find the ground speed and the course of the plane.

5 Complex Numbers

A fractal image produced on a computer. As you zoom in on a region, additonal details of the fractal are displayed.

Fractals

A fractal is a geometric figure that reveals greater detail and complexity as it is magnified over and over. Traditional Euclidean figures, such as a circle, a parabola, and a hyperbola, appear simpler as they are magnified. For instance, if you use a computer to repeatedly zoom in on a region of one of these figures, the figure starts to appear more and more like a straight line. However, as you repeatedly zoom in on a region of a fractal, more and more complex details of the figure are displayed.

The word *fractal* was first used by the mathematician Benoit B. Mandelbrot (1924–). Earlier mathematicians such as Cantor, Lebesgue, Julia, Koch, Hausdorff, Peano, Bolzano, and Sierpinski developed many concepts regarding fractals; however, these ideas were largely neglected until Mandelbrot publicized *The Fractal Geometry of Nature* in 1977.

Many fractals are generated using complex numbers and a feedback process. The process starts by substituting a complex number into a complex function to produce an output. The output is substituted back into the function and the process is repeated. The Exploring Concepts with Technology on page 356 and the Quantitative Reasoning feature on page 359 explain the details of how a fractal called the *Mandelbrot set* is generated and how you can use a graphing calculator to produce its graph.

Fractal images can be produced using complex numbers and simple algebraic procedures.

Online Study Center
For online student resources, such as section quizzes, visit this website: college.hmco.com/info/aufmannCAT6e

Complex Numbers

Section 5.1

- Introduction to Complex Numbers
- Addition and Subtraction of Complex Numbers
- Multiplication of Complex Numbers
- Division of Complex Numbers
- Powers of i

Math Matters

It may seem strange to just invent new numbers, but that is how mathematics evolves. For instance, negative numbers were not an accepted part of mathematics until well into the thirteenth century. In fact, these numbers often were referred to as "fictitious numbers."

In the seventeenth century, Rene Descartes called square roots of negative numbers "imaginary numbers," an unfortunate choice of words, and started using the letter i to denote these numbers. These numbers were subjected to the same skepticism as negative numbers.

It is important to understand that these numbers are not *imaginary* in the dictionary sense of the word. This misleading word is similar to the situation of negative numbers being called *fictitious.*

If you think of a number line, then the numbers to the right of zero are positive numbers and the numbers to the left of zero are negative numbers. One way to think of an imaginary number is to visualize it as *up* or *down* from zero. See the Project on page 342 for more information on this topic.

Introduction to Complex Numbers

Recall that $\sqrt{9} = 3$ because $3^2 = 9$. Now consider the expression $\sqrt{-9}$. To find $\sqrt{-9}$, we need to find a number c such that $c^2 = -9$. However, the square of any real number c (except zero) is a *positive* number. Consequently, we must expand our concept of number to include numbers whose squares are negative numbers.

Around the seventeenth century, a new number, called an *imaginary number,* was defined so that a negative number would have a square root. The letter i was chosen to represent the number whose square is -1.

Definition of i

The **imaginary unit,** designated by the letter i, is the number such that $i^2 = -1$.

The principal square root of a negative number is defined in terms of i.

Definition of an Imaginary Number

If a is a positive real number, then $\sqrt{-a} = i\sqrt{a}$. The number $i\sqrt{a}$ is called an **imaginary number.**

Example

$\sqrt{-36} = i\sqrt{36} = 6i$ $\quad$ $\sqrt{-18} = i\sqrt{18} = 3i\sqrt{2}$

$\sqrt{-23} = i\sqrt{23}$ $\quad$ $\sqrt{-1} = i\sqrt{1} = i$

It is customary to write i in front of a radical sign, as we did for $i\sqrt{23}$, to avoid confusing $\sqrt{a}\,i$ with $\sqrt{ai}$.

Definition of a Complex Number

A **complex number** is a number of the form $a + bi$, where a and b are real numbers and $i = \sqrt{-1}$. The number a is the **real part** of $a + bi$, and b is the **imaginary part.**

Example

$-3 + 5i$ • Real part: -3; imaginary part: 5

$2 - 6i$ • Real part: 2; imaginary part: -6

5 • Real part: 5; imaginary part: 0

$7i$ • Real part: 0; imaginary part: 7

Note from these examples that a real number is a complex number whose imaginary part is zero, and an imaginary number is a complex number whose real part is zero, and whose imaginary part is not zero.

Math Matters

The imaginary unit i is important in the field of electrical engineering. However, because the letter i is used by engineers as the symbol for electric current, these engineers use j for the complex unit.

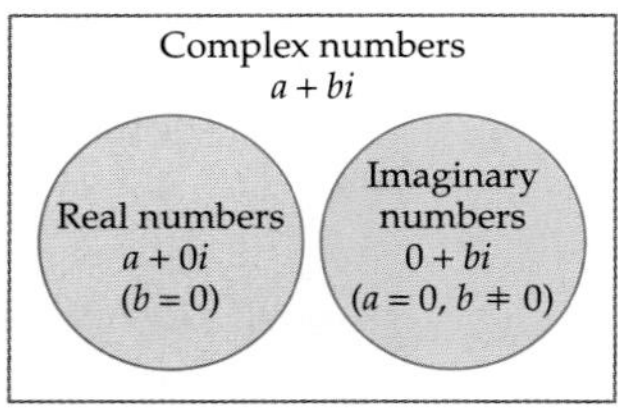

QUESTION What are the real part and imaginary part of $3 - 5i$?

Note from the diagram at the left that the set of real numbers is a subset of the complex numbers, and the set of imaginary numbers is a separate subset of the complex numbers. The set of real numbers and the set of imaginary numbers are disjoint sets.

Example 1 illustrates how to write a complex number in the **standard form** $a + bi$.

EXAMPLE 1 Write a Complex Number in Standard Form

Write $7 + \sqrt{-45}$ in the form $a + bi$.

Solution

$$\begin{aligned} 7 + \sqrt{-45} &= 7 + i\sqrt{45} \\ &= 7 + i\sqrt{9} \cdot \sqrt{5} \\ &= 7 + 3i\sqrt{5} \end{aligned}$$

Try Exercise 8, page 340

Addition and Subtraction of Complex Numbers

All the standard arithmetic operations that are applied to real numbers can be applied to complex numbers.

Definition of Addition and Subtraction of Complex Numbers

If $a + bi$ and $c + di$ are complex numbers, then

Addition $(a + bi) + (c + di) = (a + c) + (b + d)i$

Subtraction $(a + bi) - (c + di) = (a - c) + (b - d)i$

Basically, these definitions say that to add two complex numbers, add the real parts and add the imaginary parts. To subtract two complex numbers, subtract the real parts and subtract the imaginary parts.

EXAMPLE 2 Add or Subtract Complex Numbers

Simplify.

a. $(7 - 2i) + (-2 + 4i)$ b. $(-9 + 4i) - (2 - 6i)$

Solution

a. $(7 - 2i) + (-2 + 4i) = (7 + (-2)) + (-2 + 4)i = 5 + 2i$

b. $(-9 + 4i) - (2 - 6i) = (-9 - 2) + (4 - (-6))i = -11 + 10i$

Try Exercise 18, page 340

ANSWER Real part: 3; imaginary part: -5

Multiplication of Complex Numbers

When multiplying complex numbers, the term i^2 is frequently a part of the product. Recall that $i^2 = -1$. Therefore,

$$3i(5i) = 15i^2 = 15(-1) = -15$$
$$-2i(6i) = -12i^2 = -12(-1) = 12$$
$$4i(3 - 2i) = 12i - 8i^2 = 12i - 8(-1) = 8 + 12i$$

> **take note**
>
> Recall that the definition of the product of radical expressions requires that the radicand be a positive number. Therefore, when multiplying expressions containing negative radicands, we first must rewrite the expression using *i* and a positive radicand.

When multiplying square roots of negative numbers, first rewrite the radical expressions using i. For instance,

$$\sqrt{-6} \cdot \sqrt{-24} = i\sqrt{6} \cdot i\sqrt{24} \qquad \text{• } \sqrt{-6} = i\sqrt{6},\ \sqrt{-24} = i\sqrt{24}$$
$$= i^2\sqrt{144} = -1 \cdot 12$$
$$= -12$$

Note from this example that it would have been incorrect to multiply the radicands of the two radical expressions. To illustrate:

$$\sqrt{-6} \cdot \sqrt{-24} \neq \sqrt{(-6)(-24)}$$

? QUESTION What is the product of $\sqrt{-2}$ and $\sqrt{-8}$?

To multiply two complex numbers, we use the following definition.

Definition of Multiplication of Complex Numbers

If $a + bi$ and $c + di$ are complex numbers, then

$$(a + bi)(c + di) = (ac - bd) + (ad + bc)i$$

Because every complex number can be written as a sum of two terms, it is natural to perform multiplication on complex numbers in a manner consistent with the operation defined on binomials and the definition $i^2 = -1$. By using this analogy, you can multiply complex numbers without memorizing the definition.

EXAMPLE 3 ❯❯ Multiply Complex Numbers

Simplify. **a.** $(3 - 4i)(2 + 5i)$ **b.** $(2 + \sqrt{-3})(4 - 5\sqrt{-3})$

Solution

a. $(3 - 4i)(2 + 5i) = 6 + 15i - 8i - 20i^2$

$$= 6 + 15i - 8i - 20(-1) \qquad \text{• Replace } i^2 \text{ by } -1.$$
$$= 6 + 15i - 8i + 20 \qquad \text{• Simplify.}$$
$$= 26 + 7i$$

Continued ▶

? ANSWER $\sqrt{-2} \cdot \sqrt{-8} = i\sqrt{2} \cdot i\sqrt{8} = i^2\sqrt{16} = -1 \cdot 4 = -4$

> **Integrating Technology**
>
> Some graphing calculators can be used to perform operations on complex numbers. Here are some typical screens for a TI-83/TI-83 Plus/TI-84 Plus graphing calculator.
>
> Press MODE. Use the down arrow key to highlight $a + bi$.
>
> 
>
>
> Press ENTER 2nd [QUIT].
>
> The following screen shows two examples of computations on complex numbers. To enter an i, use 2nd [i], which is located above the decimal point key.
>
> (3–4i)(2+5i)
> 26+7i
> (16–11i)/(5+2i)
> 2–3i

b. $$\begin{aligned}(2+\sqrt{-3})(4-5\sqrt{-3}) &= (2+i\sqrt{3})(4-5i\sqrt{3})\\ &= 8-10i\sqrt{3}+4i\sqrt{3}-5i^2(3)\\ &= 8-10i\sqrt{3}+4i\sqrt{3}-5(-1)(3)\\ &= 8-10i\sqrt{3}+4i\sqrt{3}+15 = 23-6i\sqrt{3}\end{aligned}$$

Try Exercise 34, page 340

Division of Complex Numbers

Recall that the number $\frac{3}{\sqrt{2}}$ is not in simplest form because there is a radical expression in the denominator. Similarly, $\frac{3}{i}$ is not in simplest form because $i = \sqrt{-1}$. To write this expression in simplest form, multiply the numerator and denominator by i.

$$\frac{3}{i}\cdot\frac{i}{i} = \frac{3i}{i^2} = \frac{3i}{-1} = -3i$$

Here is another example.

$$\frac{3-6i}{2i} = \frac{3-6i}{2i}\cdot\frac{i}{i} = \frac{3i-6i^2}{2i^2} = \frac{3i-6(-1)}{2(-1)} = \frac{3i+6}{-2} = -3-\frac{3}{2}i$$

Recall that to simplify the quotient $\frac{2+\sqrt{3}}{5+2\sqrt{3}}$, we multiply the numerator and denominator by the conjugate of $5+2\sqrt{3}$, which is $5-2\sqrt{3}$. In a similar manner, to find the quotient of two complex numbers, we multiply the numerator and denominator by the conjugate of the denominator.

The complex numbers $a + bi$ and $a - bi$ are called **complex conjugates** or **conjugates** of each other. The conjugate of the complex number z is denoted by $\overline{z}$. For instance,

$$\overline{2+5i} = 2-5i \qquad \text{and} \qquad \overline{3-4i} = 3+4i$$

Consider the product of a complex number and its conjugate. For instance,

$$\begin{aligned}(2+5i)(2-5i) &= 4-10i+10i-25i^2\\ &= 4-25(-1) = 4+25\\ &= 29\end{aligned}$$

Note that the product is a *real* number. This is always true.

Product of Complex Conjugates

The product of a complex number and its conjugate is a real number. That is, $(a+bi)(a-bi) = a^2+b^2$.

Example

$(5+3i)(5-3i) = 5^2+3^2 = 25+9 = 34$

The next example shows how the quotient of two complex numbers is determined by using conjugates.

EXAMPLE 4 Divide Complex Numbers

Simplify: $\dfrac{16 - 11i}{5 + 2i}$

Solution

$$\begin{aligned}\frac{16 - 11i}{5 + 2i} &= \frac{16 - 11i}{5 + 2i} \cdot \frac{5 - 2i}{5 - 2i} \\ &= \frac{80 - 32i - 55i + 22i^2}{5^2 + 2^2} \\ &= \frac{80 - 32i - 55i + 22(-1)}{25 + 4} \\ &= \frac{80 - 87i - 22}{29} \\ &= \frac{58 - 87i}{29} \\ &= \frac{29(2 - 3i)}{29} = 2 - 3i\end{aligned}$$

• **Multiply numerator and denominator by the conjugate of the denominator.**

Try Exercise 48, page 341

Powers of *i*

The following powers of i illustrate a pattern:

$$\begin{aligned} i^1 &= i & i^5 &= i^4 \cdot i = 1 \cdot i = i \\ i^2 &= -1 & i^6 &= i^4 \cdot i^2 = 1(-1) = -1 \\ i^3 &= i^2 \cdot i = (-1)i = -i & i^7 &= i^4 \cdot i^3 = 1(-i) = -i \\ i^4 &= i^2 \cdot i^2 = (-1)(-1) = 1 & i^8 &= (i^4)^2 = 1^2 = 1 \end{aligned}$$

Because $i^4 = 1$, $(i^4)^n = 1^n = 1$ for any integer n. Thus it is possible to evaluate powers of i by factoring out powers of i^4, as shown in the following:

$$i^{27} = (i^4)^6 \cdot i^3 = 1^6 \cdot i^3 = 1 \cdot (-i) = -i$$

The following theorem can also be used to evaluate powers of i.

Powers of *i*

If n is a positive integer, then $i^n = i^r$, where r is the remainder of the division of n by 4.

EXAMPLE 5 Evaluate a Power of i

Evaluate: i^{153}

Solution

Use the powers of i theorem.

$$i^{153} = i^1 = i$$ • Remainder of $153 \div 4$ is 1.

Try Exercise 60, page 341

Topics for Discussion

1. What is an imaginary number? What is a complex number?
2. How are the real numbers related to the complex numbers?
3. Is zero a complex number?
4. What is the conjugate of a complex number?
5. If a and b are real numbers and $ab = 0$, then $a = 0$ or $b = 0$. Is the same true for complex numbers? That is, if u and v are complex numbers and $uv = 0$, must one of the complex numbers be zero?

Exercise Set 5.1

In Exercises 1 to 10, write the complex number in standard form.

1. $\sqrt{-81}$
2. $\sqrt{-64}$
3. $\sqrt{-98}$
4. $\sqrt{-27}$
5. $\sqrt{16} + \sqrt{-81}$
6. $\sqrt{25} + \sqrt{-9}$
7. $5 + \sqrt{-49}$
8. $6 - \sqrt{-1}$
9. $8 - \sqrt{-18}$
10. $11 + \sqrt{-48}$

In Exercises 11 to 36, simplify and write the complex number in standard form.

11. $(5 + 2i) + (6 - 7i)$
12. $(4 - 8i) + (5 + 3i)$
13. $(-2 - 4i) - (5 - 8i)$
14. $(3 - 5i) - (8 - 2i)$
15. $(1 - 3i) + (7 - 2i)$
16. $(2 - 6i) + (4 - 7i)$
17. $(-3 - 5i) - (7 - 5i)$
18. $(5 - 3i) - (2 + 9i)$
19. $8i - (2 - 8i)$
20. $3 - (4 - 5i)$
21. $5i \cdot 8i$
22. $(-3i)(2i)$
23. $\sqrt{-50} \cdot \sqrt{-2}$
24. $\sqrt{-12} \cdot \sqrt{-27}$
25. $3(2 + 5i) - 2(3 - 2i)$
26. $3i(2 + 5i) + 2i(3 - 4i)$
27. $(4 + 2i)(3 - 4i)$
28. $(6 + 5i)(2 - 5i)$
29. $(-3 - 4i)(2 + 7i)$
30. $(-5 - i)(2 + 3i)$
31. $(4 - 5i)(4 + 5i)$
32. $(3 + 7i)(3 - 7i)$
33. $(3 + \sqrt{-4})(2 - \sqrt{-9})$
34. $(5 + 2\sqrt{-16})(1 - \sqrt{-25})$
35. $(3 + 2\sqrt{-18})(2 + 2\sqrt{-50})$
36. $(5 - 3\sqrt{-48})(2 - 4\sqrt{-27})$

In Exercises 37 to 54, write each expression as a complex number in standard form.

37. $\frac{6}{i}$

38. $\frac{-8}{2i}$

39. $\frac{6+3i}{i}$

40. $\frac{4-8i}{4i}$

41. $\frac{1}{7+2i}$

42. $\frac{5}{3+4i}$

43. $\frac{2i}{1+i}$

44. $\frac{5i}{2-3i}$

45. $\frac{5-i}{4+5i}$

46. $\frac{4+i}{3+5i}$

47. $\frac{3+2i}{3-2i}$

48. $\frac{8-i}{2+3i}$

49. $\frac{-7+26i}{4+3i}$

50. $\frac{-4-39i}{5-2i}$

51. $(3-5i)^2$

52. $(2+4i)^2$

53. $(1+2i)^3$

54. $(2-i)^3$

In Exercises 55 to 62, evaluate the power of *i*.

55. i^{15}

56. i^{66}

57. $-i^{40}$

58. $-i^{51}$

59. $\frac{1}{i^{25}}$

60. $\frac{1}{i^{83}}$

61. i^{-34}

62. i^{-52}

FRACTAL GEOMETRY **Many fractal images are generated by substituting an initial value into a complex function, calculating the output, and then using the output as the next value to substitute into the function. The process is then repeated indefinitely. This recycling of outputs is called iteration, and each output is called an iterate. In Exercises 63 to 66, we will use the letter z to symbolize a complex number. The initial value is called the *seed* and is denoted by z_0. Successive iterates are denoted by $z_1, z_2, z_3, \ldots$.**

63. Let $f(z) = z + 4 + 3i$. Begin with the intial value $z_0 = -2 + i$. Determine $z_1, z_2, \ldots, z_5$.

64. Let $f(z) = 2iz$. Begin with the intial value $z_0 = 1 + 3i$. Determine $z_1, z_2, \ldots, z_5$.

65. Let $f(z) = iz$. Begin with the intial value $z_0 = 1 - i$. Determine z_1, z_2, z_3, and z_4. Find a pattern and use it to predict z_5, z_6, z_7, and z_8.

66. Let $f(z) = z^2$. Begin with the intial value $z_0 = 0.5i$. Determine z_1, z_2, and z_3.

In Exercises 67 to 72, evaluate $\frac{-b+\sqrt{b^2-4ac}}{2a}$ for the given values of *a*, *b*, and *c*. Write your answer as a complex number in standard form.

67. $a = 3, b = -3, c = 3$

68. $a = 2, b = 4, c = 4$

69. $a = 2, b = 6, c = 6$

70. $a = 2, b = 1, c = 3$

71. $a = 4, b = -4, c = 2$

72. $a = 3, b = -2, c = 4$

Connecting Concepts

The property that the product of conjugates of the form $(a+bi)(a-bi)$ is equal to a^2+b^2 can be used to factor the sum of two perfect squares over the set of complex numbers. For example, $x^2+y^2=(x+yi)(x-yi)$. In Exercises 73 to 78, factor the binomial over the set of complex numbers.

73. $x^2 + 16$

74. $x^2 + 9$

75. $z^2 + 25$

76. $z^2 + 64$

77. $4x^2 + 81$

78. $9x^2 + 1$

79. Show that if $x = 1 + 2i$, then $x^2 - 2x + 5 = 0$.

80. Show that if $x = 1 - 2i$, then $x^2 - 2x + 5 = 0$.

81. When we think of the cube root of 8, $\sqrt[3]{8}$, we normally mean the *real* cube root of 8 and write $\sqrt[3]{8} = 2$. However, there are two other cube roots of 8 that are complex numbers. Verify that $-1 + i\sqrt{3}$ and $-1 - i\sqrt{3}$ are cube roots of 8 by showing that $\left(-1 + i\sqrt{3}\right)^3 = 8$ and $\left(-1 - i\sqrt{3}\right)^3 = 8$.

82. It is possible to find the square root of a complex number. Verify that $\sqrt{i} = \frac{\sqrt{2}}{2}(1 + i)$ by showing that

$$\left[\frac{\sqrt{2}}{2}(1+i)\right]^2 = i.$$

83. Simplify $i + i^2 + i^3 + i^4 + \cdots + i^{28}$.

84. Simplify $i + i^2 + i^3 + i^4 + \cdots + i^{100}$.

»»» Projects »»»

ARGAND DIAGRAM Just as we can graph a real number on a real number line, we can graph a complex number. This is accomplished by using one number line for the real part of the complex number and one number line for the imaginary part of the complex number. These two number lines are drawn perpendicular to each other and pass through their respective origins, as shown below.

The result is called the *complex plane* or an *Argand diagram* after Jean-Robert Argand (1768–1822), an accountant and amateur mathematician. Although he is given credit for this representation of complex numbers, Caspar Wessel (1745–1818) actually conceived the idea before Argand.

To graph the complex number $3 + 4i$, start at 3 on the real axis. Now move 4 units up (for positive numbers move up; for negative numbers move down) and place a dot at that point, as shown in the diagram. The graphs of several other complex numbers are also shown.

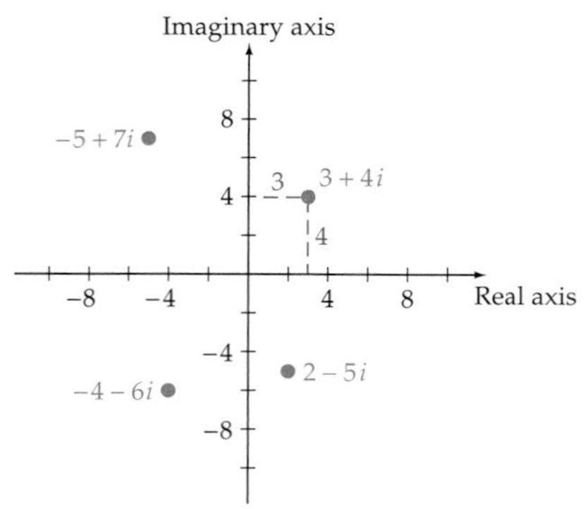

In Exercises 1 to 8, graph the complex number.

1. $2 + 5i$ **2.** $4 - 3i$ **3.** $-2 + 6i$ **4.** $-3 - 5i$

5. 4 **6.** $-2i$ **7.** $3i$ **8.** -5

The absolute value of a complex number is given by $|a + bi| = \sqrt{a^2 + b^2}$. In Exercises 9 to 12, find the absolute value of the complex number.

9. $2 + 5i$ **10.** $4 - 3i$ **11.** $-2 + 6i$ **12.** $-3 - 5i$

13. The additive inverse of $a + bi$ is $-a - bi$. Show that the absolute value of a complex number and the absolute value of its additive inverse are equal.

14. A *real* number and its additive inverse are the same distance from zero but on opposite sides of zero on a real number line. Describe the relationship between the graphs of a complex number and its additive inverse.

Section 5.2

- Trigonometric Form of a Complex Number
- The Product and Quotient of Complex Numbers Written in Trigonometric Form

Trigonometric Form of Complex Numbers

PREPARE FOR THIS SECTION

Prepare for this section by completing the following exercises. The answers can be found on page A22.

PS1. Simplify: $(1 + i)(2 + i)$ [5.1]

PS2. Simplify: $\frac{2 + i}{3 - i}$ [5.1]

PS3. What is the conjugate of $2 + 3i$? [5.1]

PS4. What is the conjugate of $3 - 5i$? [5.1]

PS5. Use the quadratic formula to find the solutions of $x^2 + x = -1$. [1.1/5.1]

PS6. Solve: $x^2 + 9 = 0$ [1.1/5.1]

Trigonometric Form of a Complex Number

Real numbers are graphed as points on a number line. Complex numbers can be graphed in a coordinate plane called the **complex plane.** The horizontal axis of the complex plane is called the **real axis;** the vertical axis is called the **imaginary axis.**

A complex number written in the form $z = a + bi$ is written in **standard form** or **rectangular form.** The graph of $a + bi$ is associated with the point $P(a, b)$ in the complex plane. **Figure 5.1** shows the graphs of several complex numbers.

The length of the line segment from the origin to the point $(-3, 4)$ in the complex plane is the *absolute value* of $z = -3 + 4i$. See **Figure 5.2.** From the Pythagorean Theorem, the absolute value of $z = -3 + 4i$ is

$$\sqrt{(-3)^2 + 4^2} = \sqrt{25} = 5$$

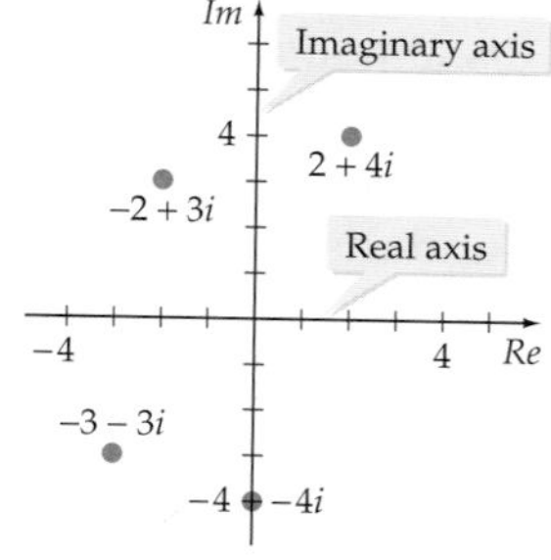

Figure 5.1

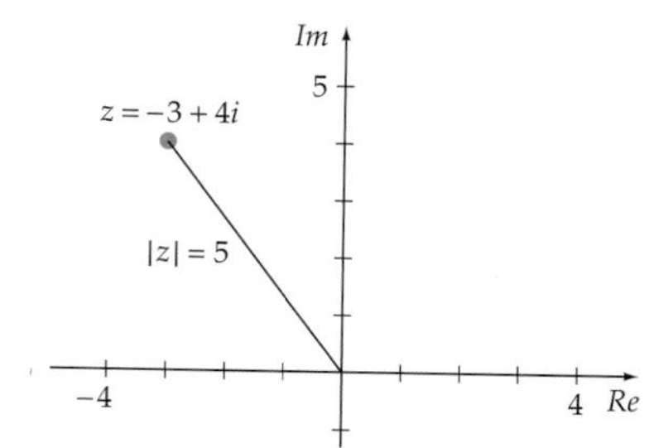

Figure 5.2

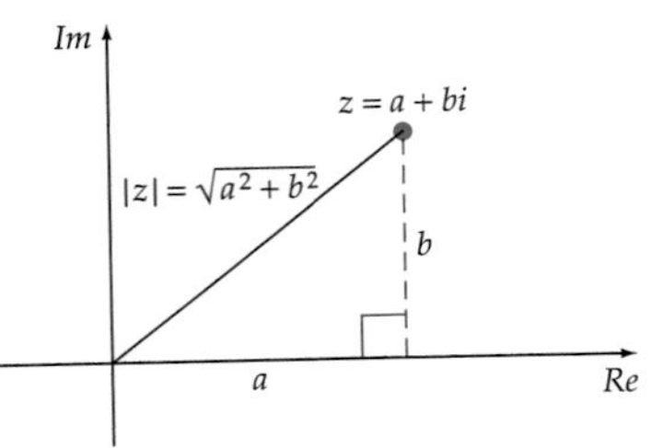

Figure 5.3

Definition of the Absolute Value of a Complex Number

The absolute value of the complex number $z = a + bi$, denoted by $|z|$, is

$$|z| = |a + bi| = \sqrt{a^2 + b^2}$$

Thus $|z|$ is the distance from the origin to z (see **Figure 5.3**).

QUESTION The conjugate of $a + bi$ is $a - bi$. Does $|a + bi| = |a - bi|$?

A complex number $z = a + bi$ can be written in terms of trigonometric functions. Consider the complex number graphed in **Figure 5.4**. We can write a and b in terms of the sine and the cosine.

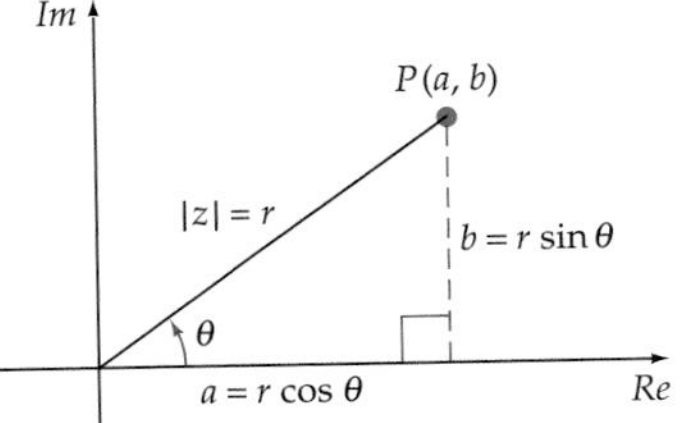

Figure 5.4

$$\cos \theta = \frac{a}{r} \qquad \sin \theta = \frac{b}{r}$$

$$a = r \cos \theta \qquad b = r \sin \theta$$

where $r = |z| = \sqrt{a^2 + b^2}$. Substituting for a and b in $z = a + bi$, we obtain

$$z = r \cos \theta + ir \sin \theta = r(\cos \theta + i \sin \theta)$$

The expression $z = r(\cos \theta + i \sin \theta)$ is known as the **trigonometric form** of a complex number. The trigonometric form of a complex number is also called the **polar form** of the complex number. The notation $\cos \theta + i \sin \theta$ is often abbreviated as $\operatorname{cis} \theta$ using the c from $\cos \theta$, the imaginary unit i, and the s from $\sin \theta$.

Trigonometric Form of a Complex Number

The complex number $z = a + bi$ can be written in trigonometric form as

$$z = r(\cos \theta + i \sin \theta) = r \operatorname{cis} \theta$$

where $a = r \cos \theta$, $b = r \sin \theta$, $r = \sqrt{a^2 + b^2}$, and $\tan \theta = \dfrac{b}{a}$.

In this text we will often write the trigonometric form of a complex number in its abbreviated form $z = r \operatorname{cis} \theta$. The value of r is called the **modulus** of the complex number z, and the angle θ is called the **argument** of the complex number z. The modulus r and the argument θ of a complex number $z = a + bi$ are given by

$$r = \sqrt{a^2 + b^2} \quad \text{and} \quad \cos \theta = \frac{a}{r}, \quad \sin \theta = \frac{b}{r}$$

ANSWER Yes, because $\sqrt{a^2 + b^2} = \sqrt{a^2 + (-b)^2}$.

We also can write $\alpha = \tan^{-1}\left|\frac{b}{a}\right|$, where α is the reference angle for θ. Due to the periodic nature of the sine and cosine functions, the trigonometric form of a complex number is not unique. Because $\cos\theta = \cos(\theta + 2k\pi)$ and $\sin\theta = \sin(\theta + 2k\pi)$, where k is an integer, the complex numbers $r \operatorname{cis}\theta$ and $r\operatorname{cis}(\theta + 2k\pi)$ are equal. For example, $2\operatorname{cis}\frac{\pi}{6} = 2\operatorname{cis}\left(\frac{\pi}{6} + 2\pi\right)$.

EXAMPLE 1 ›› Write a Complex Number in Trigonometric Form

Write $z = -2 - 2i$ in trigonometric form.

Solution

Find the modulus and the argument of z. Then substitute these values in the trigonometric form of z.

$$r = \sqrt{(-2)^2 + (-2)^2} = \sqrt{8} = 2\sqrt{2}$$

To determine θ, we first determine α. See **Figure 5.5**.

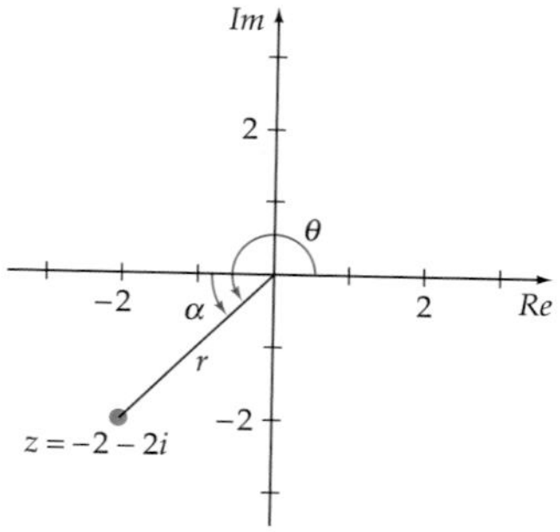

Figure 5.5

$$\alpha = \tan^{-1}\left|\frac{b}{a}\right|$$ • **α is the reference angle of angle θ.**

$$\alpha = \tan^{-1}\left|\frac{-2}{-2}\right| = \tan^{-1}1 = 45^\circ$$ • **$a = -2$ and $b = -2$**

$$\theta = 180^\circ + 45^\circ = 225^\circ$$ • **Because z is in the third quadrant, $180^\circ < \theta < 270^\circ$.**

The trigonometric form is

$$z = r\operatorname{cis}\theta = 2\sqrt{2}\operatorname{cis}225^\circ$$ • **$r = 2\sqrt{2}$, $\theta = 225^\circ$**

›› *Try Exercise 12, page 349*

EXAMPLE 2 ›› Write a Complex Number in Standard Form

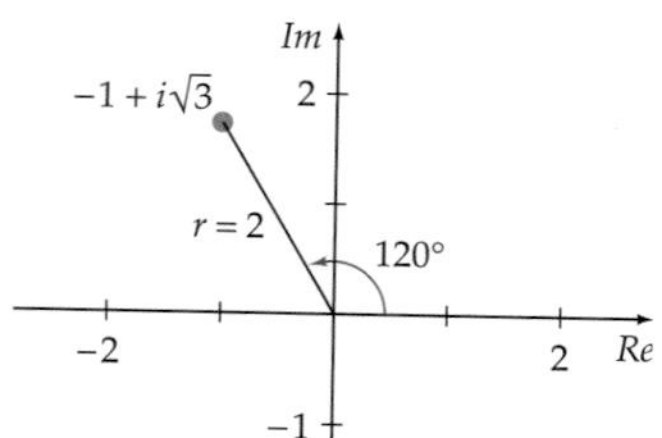

Figure 5.6

Write $z = 2\operatorname{cis}120^\circ$ in standard form.

Solution

Write z in the form $r(\cos\theta + i\sin\theta)$ and then evaluate $\cos\theta$ and $\sin\theta$. See **Figure 5.6**.

$$z = 2\operatorname{cis}120^\circ = 2(\cos 120^\circ + i\sin 120^\circ) = 2\left(-\frac{1}{2} + \frac{\sqrt{3}}{2}i\right) = -1 + i\sqrt{3}$$

›› *Try Exercise 30, page 349*

The Product and Quotient of Complex Numbers Written in Trigonometric Form

Let z_1 and z_2 be two complex numbers written in trigonometric form. The product of z_1 and z_2 can be found by using trigonometric identities. If $z_1 = r_1(\cos\theta_1 + i\sin\theta_1)$ and $z_2 = r_2(\cos\theta_2 + i\sin\theta_2)$, then

$$\begin{aligned} z_1z_2 &= r_1(\cos\theta_1 + i\sin\theta_1)\cdot r_2(\cos\theta_2 + i\sin\theta_2) \\ &= r_1r_2(\cos\theta_1\cos\theta_2 + i\cos\theta_1\sin\theta_2 + i\sin\theta_1\cos\theta_2 + i^2\sin\theta_1\sin\theta_2) \\ &= r_1r_2[(\cos\theta_1\cos\theta_2 - \sin\theta_1\sin\theta_2) + i(\sin\theta_1\cos\theta_2 + \cos\theta_1\sin\theta_2)] \\ &= r_1r_2[\cos(\theta_1+\theta_2) + i\sin(\theta_1+\theta_2)] \end{aligned}$$

• Identities for $\cos(\theta_1 + \theta_2)$ and $\sin(\theta_1 + \theta_2)$

Thus, using cis notation, we have $z_1z_2 = r_1r_2\operatorname{cis}(\theta_1+\theta_2)$.

The above result is called the **Product Property of Complex Numbers.** It states that the modulus for the product of two complex numbers in trigonometric form is the product of the moduli of the two numbers, and the argument of the product is the sum of the arguments of the two numbers.

EXAMPLE 3 ›› Find the Product of Two Complex Numbers

Find the product of $z_1 = -1 + i\sqrt{3}$ and $z_2 = -\sqrt{3} + i$ by using the trigonometric forms of the complex numbers. Write the answer in standard form.

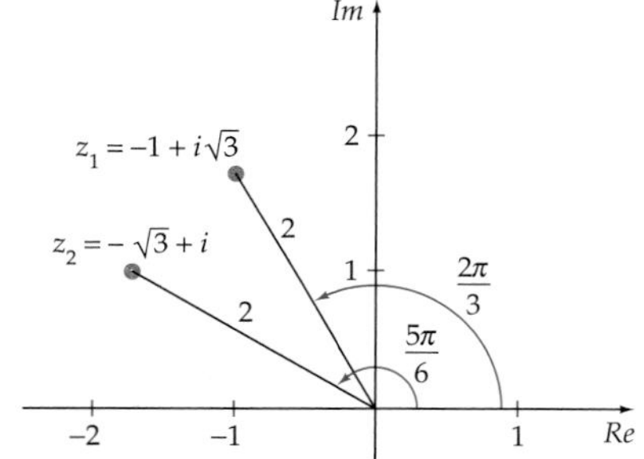

Figure 5.7

Solution

Write each complex number in trigonometric form. Then use the Product Property of Complex Numbers. See **Figure 5.7**.

$$z_1 = -1 + i\sqrt{3} = 2\operatorname{cis}\frac{2\pi}{3}$$ • $r_1 = 2,\ \theta_1 = \frac{2\pi}{3}$

$$z_2 = -\sqrt{3} + i = 2\operatorname{cis}\frac{5\pi}{6}$$ • $r_2 = 2,\ \theta_2 = \frac{5\pi}{6}$

$$\begin{aligned} z_1z_2 &= 2\operatorname{cis}\frac{2\pi}{3}\cdot 2\operatorname{cis}\frac{5\pi}{6} \\ &= 4\operatorname{cis}\left(\frac{2\pi}{3} + \frac{5\pi}{6}\right) \\ &= 4\operatorname{cis}\frac{3\pi}{2} \\ &= 4\left(\cos\frac{3\pi}{2} + i\sin\frac{3\pi}{2}\right) \\ &= 4(0 - i) = -4i \end{aligned}$$

• The Product Property of Complex Numbers

• Simplify.

›› Try Exercise 56, page 349

Similarly, the quotient of z_1 and z_2 can be found by using trigonometric identities. If $z_1 = r_1(\cos\theta_1 + i\sin\theta_1)$ and $z_2 = r_2(\cos\theta_2 + i\sin\theta_2)$, then

$$\frac{z_1}{z_2} = \frac{r_1(\cos\theta_1 + i\sin\theta_1)}{r_2(\cos\theta_2 + i\sin\theta_2)}$$

$$= \frac{r_1(\cos\theta_1 + i\sin\theta_1)(\cos\theta_2 - i\sin\theta_2)}{r_2(\cos\theta_2 + i\sin\theta_2)(\cos\theta_2 - i\sin\theta_2)}$$

$$= \frac{r_1(\cos\theta_1\cos\theta_2 - i\cos\theta_1\sin\theta_2 + i\sin\theta_1\cos\theta_2 - i^2\sin\theta_1\sin\theta_2)}{r_2(\cos^2\theta_2 - i^2\sin^2\theta_2)}$$

$$= \frac{r_1[(\cos\theta_1\cos\theta_2 + \sin\theta_1\sin\theta_2) + i(\sin\theta_1\cos\theta_2 - \cos\theta_1\sin\theta_2)]}{r_2(\cos^2\theta_2 + \sin^2\theta_2)}$$

$$= \frac{r_1}{r_2}[\cos(\theta_1 - \theta_2) + i\sin(\theta_1 - \theta_2)]$$

• **Identities for $\cos(\theta_1 - \theta_2)$, $\sin(\theta_1 - \theta_2)$, and $\cos^2\theta_2 + \sin^2\theta_2$**

Thus, using cis notation, we have $\frac{z_1}{z_2} = \frac{r_1}{r_2}\operatorname{cis}(\theta_1 - \theta_2)$.

The above result is called the **Quotient Property of Complex Numbers.** It states that the modulus for the quotient of two complex numbers in trigonometric form is the quotient of the moduli of the two numbers, and the argument of the quotient is the difference of the arguments of the two numbers.

EXAMPLE 4 » Find the Quotient of Two Complex Numbers

Find the quotient of $z_1 = -1 + i$ and $z_2 = \sqrt{3} - i$ by using the trigonometric forms of the complex numbers. Write the answer in standard form. Round approximate constants to the nearest thousandth.

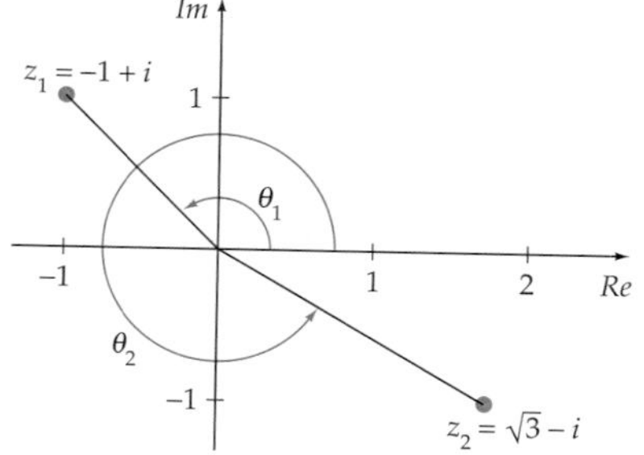

Figure 5.8

Solution

Write the numbers in trigonometric form. Then use the Quotient Property of Complex Numbers. See **Figure 5.8.**

$z_1 = -1 + i = \sqrt{2}\operatorname{cis}135^\circ$ • $r_1 = \sqrt{2}$, $\theta_1 = 135^\circ$

$z_2 = \sqrt{3} - i = 2\operatorname{cis}330^\circ$ • $r_2 = 2$, $\theta_2 = 330^\circ$

$$\frac{z_1}{z_2} = \frac{-1 + i}{\sqrt{3} - i} = \frac{\sqrt{2}\operatorname{cis}135^\circ}{2\operatorname{cis}330^\circ}$$

$$= \frac{\sqrt{2}}{2}\operatorname{cis}(-195^\circ)$$ • **The Quotient Property of Complex Numbers**

$$= \frac{\sqrt{2}}{2}[\cos(-195^\circ) + i\sin(-195^\circ)]$$ • **Simplify.**

$$= \frac{\sqrt{2}}{2}(\cos 195^\circ - i\sin 195^\circ)$$ • $\cos(-x) = \cos x$, $\sin(-x) = -\sin x$

$$\approx -0.683 + 0.183i$$

» Try Exercise 60, page 349

Here is a summary of the product and quotient theorems.

Product and Quotient Properties of Complex Numbers Written in Trigonometric Form

Product Property of Complex Numbers

Let $z_1 = r_1(\cos\theta_1 + i\sin\theta_1)$ and $z_2 = r_2(\cos\theta_2 + i\sin\theta_2)$. Then

$$z_1z_2 = r_1r_2[\cos(\theta_1 + \theta_2) + i\sin(\theta_1 + \theta_2)]$$

$$z_1z_2 = r_1r_2\operatorname{cis}(\theta_1 + \theta_2)$$ • **Using cis notation**

Quotient Property of Complex Numbers

Let $z_1 = r_1(\cos\theta_1 + i\sin\theta_1)$ and $z_2 = r_2(\cos\theta_2 + i\sin\theta_2)$. Then

$$\frac{z_1}{z_2} = \frac{r_1}{r_2}[\cos(\theta_1 - \theta_2) + i\sin(\theta_1 - \theta_2)]$$

$$\frac{z_1}{z_2} = \frac{r_1}{r_2}\operatorname{cis}(\theta_1 - \theta_2)$$ • **Using cis notation**

Topics for Discussion

1. Describe an algebraic procedure that can be used to verify that the absolute value of $a + bi$ is equal to the absolute value of $-a + bi$.
2. Describe the graph of all complex numbers with an absolute value of 5.
3. Describe two different methods that can be used to simplify $\frac{4}{i}$.
4. The complex numbers z_1 and z_2 both have an absolute value of 1. What is the absolute value of the product z_1z_2? Explain.
5. Explain how to use the Product Property of Complex Numbers to prove that the product of a complex number and its conjugate is a real number.

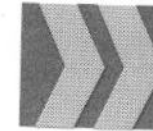

Exercise Set 5.2

In Exercises 1 to 8, graph each complex number. Find the absolute value of each complex number.

1. $z = -2 - 2i$
2. $z = 4 - 4i$
3. $z = \sqrt{3} - i$
4. $z = 1 + i\sqrt{3}$
5. $z = -2i$
6. $z = -5$
7. $z = 3 - 5i$
8. $z = -5 - 4i$

In Exercises 9 to 20, write each complex number in trigonometric form.

9. $z = 1 - i$
10. $z = -4 - 4i$
11. $z = \sqrt{3} - i$
12. $z = 1 + i\sqrt{3}$
13. $z = 3i$
14. $z = -2i$
15. $z = -5$
16. $z = 3$
17. $z = -8 + 8i\sqrt{3}$
18. $z = -2\sqrt{2} + 2i\sqrt{2}$
19. $z = -2 - 2i\sqrt{3}$
20. $z = \sqrt{2} - i\sqrt{2}$

In Exercises 21 to 38, write each complex number in standard form.

21. $z = 2(\cos 45° + i \sin 45°)$
22. $z = 3(\cos 240° + i \sin 240°)$
23. $z = (\cos 315° + i \sin 315°)$
24. $z = 5(\cos 120° + i \sin 120°)$
25. $z = 6 \text{ cis } 135°$
26. $z = \text{cis } 315°$
27. $z = 8 \text{ cis } 0°$
28. $z = 5 \text{ cis } 90°$
29. $z = 2\left(\cos \frac{5\pi}{6} + i \sin \frac{5\pi}{6}\right)$
30. $z = 4\left(\cos \frac{5\pi}{3} + i \sin \frac{5\pi}{3}\right)$
31. $z = 3\left(\cos \frac{3\pi}{2} + i \sin \frac{3\pi}{2}\right)$
32. $z = 5(\cos \pi + i \sin \pi)$
33. $z = 8 \text{ cis } \frac{3\pi}{4}$
34. $z = 9 \text{ cis } \frac{4\pi}{3}$
35. $z = 9 \text{ cis } \frac{11\pi}{6}$
36. $z = \text{cis } \frac{3\pi}{2}$
37. $z = 2 \text{ cis } 2$
38. $z = 5 \text{ cis } 4$

In Exercises 39 to 46, multiply the complex numbers. Write the answer in trigonometric form.

39. $2 \text{ cis } 30° \cdot 3 \text{ cis } 225°$
40. $4 \text{ cis } 120° \cdot 6 \text{ cis } 315°$
41. $3(\cos 122° + i \sin 122°) \cdot 4(\cos 213° + i \sin 213°)$
42. $8(\cos 88° + i \sin 88°) \cdot 12(\cos 112° + i \sin 112°)$
43. $5\left(\cos \frac{2\pi}{3} + i \sin \frac{2\pi}{3}\right) \cdot 2\left(\cos \frac{2\pi}{5} + i \sin \frac{2\pi}{5}\right)$
44. $5 \text{ cis } \frac{11\pi}{12} \cdot 3 \text{ cis } \frac{4\pi}{3}$
45. $4 \text{ cis } 2.4 \cdot 6 \text{ cis } 4.1$
46. $7 \text{ cis } 0.88 \cdot 5 \text{ cis } 1.32$

In Exercises 47 to 54, divide the complex numbers. Write the answer in standard form. Round approximate constants to the nearest thousandth.

47. $\frac{32 \text{ cis } 30°}{4 \text{ cis } 150°}$
48. $\frac{15 \text{ cis } 240°}{3 \text{ cis } 135°}$
49. $\frac{27(\cos 315° + i \sin 315°)}{9(\cos 225° + i \sin 225°)}$
50. $\frac{9(\cos 25° + i \sin 25°)}{3(\cos 175° + i \sin 175°)}$
51. $\frac{12 \text{ cis } \frac{2\pi}{3}}{4 \text{ cis } \frac{11\pi}{6}}$
52. $\frac{10 \text{ cis } \frac{\pi}{3}}{5 \text{ cis } \frac{\pi}{4}}$
53. $\frac{25(\cos 3.5 + i \sin 3.5)}{5(\cos 1.5 + i \sin 1.5)}$
54. $\frac{18(\cos 0.56 + i \sin 0.56)}{6(\cos 1.22 + i \sin 1.22)}$

In Exercises 55 to 62, perform the indicated operation in trigonometric form. Write the solution in standard form. Round approximate constants to the nearest ten-thousandth.

55. $(1 - i\sqrt{3})(1 + i)$
56. $(\sqrt{3} - i)(1 + i\sqrt{3})$
57. $(3 - 3i)(1 + i)$
58. $(2 + 2i)(\sqrt{3} - i)$
59. $\frac{1 + i\sqrt{3}}{1 - i\sqrt{3}}$
60. $\frac{1 + i}{1 - i}$
61. $\frac{\sqrt{2} - i\sqrt{2}}{1 + i}$
62. $\frac{1 + i\sqrt{3}}{4 - 4i}$

Connecting Concepts

In Exercises 63 to 68, perform the indicated operation in trigonometric form. Write the solution in standard form.

63. $(\sqrt{3} - i)(2 + 2i)(2 - 2i\sqrt{3})$

64. $(1 - i)(1 + i\sqrt{3})(\sqrt{3} - i)$

65. $\dfrac{\sqrt{3} + i\sqrt{3}}{(1 - i\sqrt{3})(2 - 2i)}$

66. $\dfrac{(2 - 2i\sqrt{3})(1 - i\sqrt{3})}{4\sqrt{3} + 4i}$

67. $(1 - 3i)(2 + 3i)(4 + 5i)$

68. $\dfrac{(2 - 5i)(1 - 6i)}{3 + 4i}$

In Exercises 69 and 70, the notation $\bar{z}$ is used to represent the conjugate of the complex number z.

69. Use the trigonometric forms of z and $\bar{z}$ to find $z \cdot \bar{z}$.

70. Use the trigonometric forms of z and $\bar{z}$ to find $\dfrac{z}{\bar{z}}$.

Projects

1. **A Geometric Interpretation** Multiplying a real number by a number greater than 1 increases the magnitude of that number. For example, multiplying 2 by 3 triples the magnitude of 2. Explain the effect of multiplying a real number by i, the imaginary unit. Now multiply a complex number by i and note the effect. The use of a complex plane may be helpful in your explanation.

Section 5.3

- De Moivre's Theorem
- De Moivre's Theorem for Finding Roots

De Moivre's Theorem

PREPARE FOR THIS SECTION

Prepare for this section by completing the following exercises. The answers can be found on page A22.

PS1. Find $\left(\dfrac{\sqrt{2}}{2} + \dfrac{\sqrt{2}}{2}i\right)^2$. [5.1]

PS2. Find the real root of $x^3 - 8 = 0$.

PS3. Find the real root of $x^5 - 243 = 0$.

PS4. Write $2 + 2i$ in trigonometric form. [5.2]

PS5. Write $2(\cos 150° + i \sin 150°)$ in standard form. [5.2]

PS6. Find the absolute value of $\dfrac{\sqrt{2}}{2} - \dfrac{\sqrt{2}}{2}i$. [5.2]

De Moivre's Theorem

De Moivre's Theorem is a procedure for finding powers and roots of complex numbers when the complex numbers are expressed in trigonometric form. This theorem can be illustrated by repeated multiplication of a complex number.

Let $z = r \operatorname{cis} \theta$. Then z^2 can be written as

$$z \cdot z = r \operatorname{cis} \theta \cdot r \operatorname{cis} \theta$$

$$z^2 = r^2 \operatorname{cis} 2\theta$$

Math Matters

Abraham de Moivre (1667–1754) was a French mathematician who fled to England during the expulsion of the Huguenots in 1685.

De Moivre made important contributions to probability theory and analytic geometry. In 1718 he published *The Doctrine of Chance,* in which he developed the theory of annuities, mortality statistics, and the concept of statistical independence. In 1730 de Moivre stated the theorem shown above Example 1, which we now call De Moivre's Theorem. This theorem is significant because it provides a connection between trigonometry and mathematical analysis.

Although de Moivre was well respected in the mathematical community and was elected to the Royal Society of England, he never was able to secure a university teaching position. His income came mainly from tutoring, and he died in poverty.

De Moivre is often remembered for predicting the day of his own death. At one point in his life, he noticed that he was sleeping a few minutes longer each night. Thus he calculated that he would die when he needed 24 hours of sleep. As it turned out, his calculation was correct.

The product $z^2 \cdot z$ is

$$z^2 \cdot z = r^2 \operatorname{cis} 2\theta \cdot r \operatorname{cis} \theta$$
$$z^3 = r^3 \operatorname{cis} 3\theta$$

If we continue this process, the results suggest a formula known as De Moivre's Theorem for the nth power of a complex number.

De Moivre's Theorem

If $z = r \operatorname{cis} \theta$ and n is a positive integer, then

$$z^n = r^n \operatorname{cis} n\theta$$

EXAMPLE 1 ›› Find the Power of a Complex Number

Find $(2 \operatorname{cis} 30°)^5$. Write the answer in standard form.

Solution

By De Moivre's Theorem,

$$\begin{aligned}(2 \operatorname{cis} 30°)^5 &= 2^5 \operatorname{cis}(5 \cdot 30°)\\ &= 2^5[\cos(5 \cdot 30°) + i \sin(5 \cdot 30°)]\\ &= 32(\cos 150° + i \sin 150°)\\ &= 32\left(-\frac{\sqrt{3}}{2} + \frac{1}{2}i\right) = -16\sqrt{3} + 16i\end{aligned}$$

›› Try Exercise 6, page 354

EXAMPLE 2 ›› Use De Moivre's Theorem

Find $(1 + i)^8$ using De Moivre's Theorem. Write the answer in standard form.

Solution

Convert $1 + i$ to trigonometric form and then use De Moivre's Theorem.

$$\begin{aligned}(1 + i)^8 &= (\sqrt{2} \operatorname{cis} 45°)^8 = (\sqrt{2})^8 \operatorname{cis} 8(45°) = 16 \operatorname{cis} 360°\\ &= 16(\cos 360° + i \sin 360°) = 16(1 + 0i) = 16\end{aligned}$$

›› Try Exercise 16, page 354

? QUESTION Is $(1 + i)^4$ a real number?

? ANSWER Yes. $(1 + i)^4 = (\sqrt{2} \operatorname{cis} 45°)^4 = (\sqrt{2})^4 \operatorname{cis} 180° = -4$.

De Moivre's Theorem for Finding Roots

De Moivre's Theorem can be used to find the nth roots of any number.

De Moivre's Theorem for Finding Roots

If $z = r \operatorname{cis} \theta$ is a complex number, then there exist n distinct nth roots of z given by

$$w_k = r^{1/n} \operatorname{cis} \frac{\theta + 360°k}{n} \quad \text{for } k = 0, 1, 2, \ldots, n-1, \text{ and } n \geq 1$$

EXAMPLE 3 Find Cube Roots by De Moivre's Theorem

Find the three cube roots of 27.

ALGEBRAIC SOLUTION

Write 27 in trigonometric form: $27 = 27 \operatorname{cis} 0°$. Then, from De Moivre's Theorem for finding roots, the cube roots of 27 are

$$w_k = 27^{1/3} \operatorname{cis} \frac{0° + 360°k}{3} \quad \text{for } k = 0, 1, 2$$

Substitute for k to find the three cube roots of 27.

$$\begin{aligned} w_0 &= 27^{1/3} \operatorname{cis} 0° & \bullet\ k = 0;\ \frac{0° + 360°(0)}{3} = 0° \\ &= 3(\cos 0° + i \sin 0°) \\ &= 3 \end{aligned}$$

$$\begin{aligned} w_1 &= 27^{1/3} \operatorname{cis} 120° & \bullet\ k = 1;\ \frac{0° + 360°(1)}{3} = 120° \\ &= 3(\cos 120° + i \sin 120°) \\ &= -\frac{3}{2} + \frac{3\sqrt{3}}{2}i \end{aligned}$$

$$\begin{aligned} w_2 &= 27^{1/3} \operatorname{cis} 240° & \bullet\ k = 2;\ \frac{0° + 360°(2)}{3} = 240° \\ &= 3(\cos 240° + i \sin 240°) \\ &= -\frac{3}{2} - \frac{3\sqrt{3}}{2}i \end{aligned}$$

For $k = 3$, $\frac{0° + 1080°}{3} = 360°$. The angles start repeating; thus there are only three cube roots of 27. The three cube roots are graphed in **Figure 5.9**.

VISUALIZE THE SOLUTION

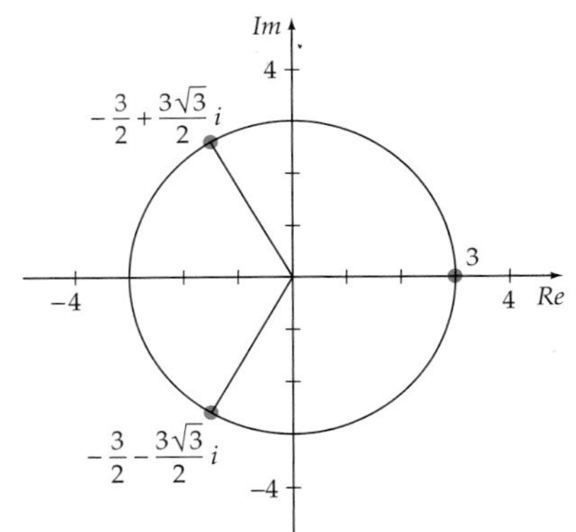

Figure 5.9

Note that the arguments of the three cube roots of 27 are 0°, 120°, and 240° and that $|w_0| = |w_1| = |w_2| = 3$. In geometric terms, this means that the three cube roots of 27 are equally spaced on a circle centered at the origin with a radius of 3.

Try Exercise 28, page 355

EXAMPLE 4 Find the Fifth Roots of a Complex Number

Find the fifth roots of $z = 1 + i\sqrt{3}$.

ALGEBRAIC SOLUTION

Write z in trigonometric form: $z = r \text{ cis } \theta$.

$$r = \sqrt{1^2 + (\sqrt{3})^2} = 2$$

$$z = 2 \text{ cis } 60° \qquad \bullet\ \theta = \tan^{-1}\frac{\sqrt{3}}{1} = 60°$$

From De Moivre's Theorem, the modulus of each root is $\sqrt[5]{2}$, and the arguments are determined by $\frac{60° + 360°k}{5}$, $k = 0, 1, 2, 3, 4$.

$$w_k = \sqrt[5]{2} \text{ cis } \frac{60° + 360°k}{5} \qquad \bullet\ k = 0, 1, 2, 3, 4$$

Substitute for k to find the five fifth roots of z.

$$w_0 = \sqrt[5]{2} \text{ cis } 12° \qquad \bullet\ k = 0; \frac{60° + 360°(0)}{5} = 12°$$

$$w_1 = \sqrt[5]{2} \text{ cis } 84° \qquad \bullet\ k = 1; \frac{60° + 360°(1)}{5} = 84°$$

$$w_2 = \sqrt[5]{2} \text{ cis } 156° \qquad \bullet\ k = 2; \frac{60° + 360°(2)}{5} = 156°$$

$$w_3 = \sqrt[5]{2} \text{ cis } 228° \qquad \bullet\ k = 3; \frac{60° + 360°(3)}{5} = 228°$$

$$w_4 = \sqrt[5]{2} \text{ cis } 300° \qquad \bullet\ k = 4; \frac{60° + 360°(4)}{5} = 300°$$

VISUALIZE THE SOLUTION

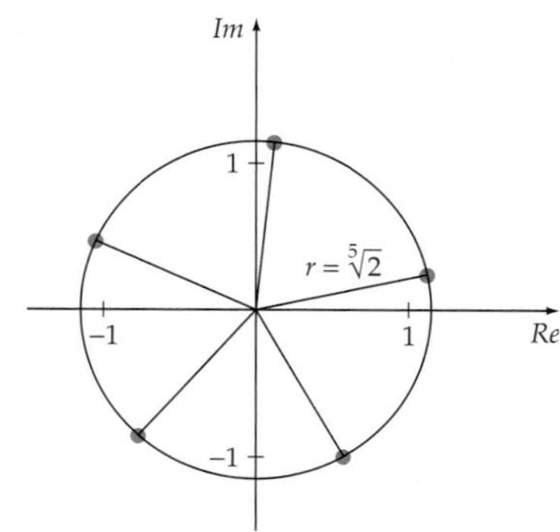

Figure 5.10

The five fifth roots of $1 + i\sqrt{3}$ are graphed in **Figure 5.10.** Note that the roots are equally spaced on a circle with center (0, 0) and a radius of $\sqrt[5]{2} \approx 1.15$.

Try Exercise 30, page 355

Keep the following properties in mind as you compute the n distinct nth roots of the complex number z.

Properties of the nth Roots of z

Geometric Property

All nth roots of z are equally spaced on a circle with center (0, 0) and a radius of $|z|^{1/n}$.

Absolute Value Properties

1. If $|z| = 1$, then each nth root of z has an absolute value of 1.
2. If $|z| > 1$, then each nth root of z has an absolute value of $|z|^{1/n}$, where $|z|^{1/n}$ is greater than 1 but less than $|z|$.
3. If $|z| < 1$, then each nth root of z has an absolute value of $|z|^{1/n}$, where $|z|^{1/n}$ is less than 1 but greater than $|z|$.

Integrating Technology

A web applet is available to show the nth roots of a complex number. This applet, Nth Roots, can be found on our website at college.hmco.com/info/aufmannCAT

Argument Property

Given that the argument of z is θ, then the argument of w_0 is $\frac{\theta}{n}$ and the arguments of the remaining nth roots can be determined by adding multiples of $\frac{360°}{n}$ (or $\frac{2\pi}{n}$ if you are using radians) to $\frac{\theta}{n}$.

QUESTION Are all fourth roots of 1 equally spaced on a circle with center (0, 0) and radius 1?

Topics for Discussion

1. How many solutions are there for $z = (a + bi)^8$? How many solutions are there for $z^8 = a + bi$? Explain.
2. To solve $z^4 = 16$, a student first observes that $z = 2$ is one solution. The other solutions are equally spaced on a circle with center (0, 0) and radius 2, so the student reasons that $z = 2i$, $z = -2$, and $z = -2i$ are the other three solutions. Do you agree? Explain.
3. If $|z| = 1$, then the n solutions of $w^n = z$ all have an absolute value of 1. Do you agree? Explain.
4. If z is a solution of $z^2 = c + di$, then the conjugate of z is also a solution. Do you agree? Explain.

ANSWER Yes.

Exercise Set 5.3

In Exercises 1 to 16, find the indicated power. Write the answer in standard form.

1. $[2(\cos 30° + i \sin 30°)]^8$
2. $(\cos 240° + i \sin 240°)^{12}$
3. $[2(\cos 240° + i \sin 240°)]^5$
4. $[2(\cos 45° + i \sin 45°)]^{10}$
5. $(2 \text{ cis } 225°)^5$
6. $(2 \text{ cis } 330°)^4$
7. $\left(2 \text{ cis } \frac{2\pi}{3}\right)^6$
8. $\left(4 \text{ cis } \frac{5\pi}{6}\right)^3$
9. $(1 - i)^{10}$
10. $(1 + i\sqrt{3})^8$
11. $(1 + i)^4$
12. $(2 - 2i\sqrt{3})^3$
13. $(2 + 2i)^7$
14. $(2\sqrt{3} - 2i)^5$
15. $\left(\frac{\sqrt{2}}{2} + i\frac{\sqrt{2}}{2}\right)^6$
16. $\left(-\frac{\sqrt{2}}{2} + i\frac{\sqrt{2}}{2}\right)^{12}$

In Exercises 17 to 30, find all of the indicated roots. Write all answers in standard form. Round approximate constants to the nearest thousandth.

17. The two square roots of 9
18. The two square roots of 16
19. The six sixth roots of 64
20. The five fifth roots of 32

21. The five fifth roots of -1

22. The four fourth roots of -16

23. The three cube roots of 1

24. The three cube roots of i

25. The four fourth roots of $1 + i$

26. The five fifth roots of $-1 + i$

27. The three cube roots of $2 - 2i\sqrt{3}$

28. The three cube roots of $-2 + 2i\sqrt{3}$

29. The two square roots of $-16 + 16i\sqrt{3}$

30. The two square roots of $-1 + i\sqrt{3}$

In Exercises 31 to 42, find all roots of the equation. Write the answers in trigonometric form.

31. $x^3 + 8 = 0$

32. $x^5 - 32 = 0$

33. $x^4 + i = 0$

34. $x^3 - 2i = 0$

35. $x^3 - 27 = 0$

36. $x^5 + 32i = 0$

37. $x^4 + 81 = 0$

38. $x^3 - 64i = 0$

39. $x^4 - (1 - i\sqrt{3}) = 0$

40. $x^3 + (2\sqrt{3} - 2i) = 0$

41. $x^3 + (1 + i\sqrt{3}) = 0$

42. $x^6 - (4 - 4i) = 0$

Connecting Concepts

43. Show that the conjugate of $z = r(\cos\theta + i\sin\theta)$ is equal to $\bar{z} = r(\cos\theta - i\sin\theta)$.

44. Show that if $z = r(\cos\theta + i\sin\theta)$, then

$$z^{-1} = r^{-1}(\cos\theta - i\sin\theta)$$

45. Show that if $z = r(\cos\theta + i\sin\theta)$, then

$$z^{-2} = r^{-2}(\cos 2\theta - i\sin 2\theta)$$

Note that Exercises 44 and 45 suggest that if $z = r(\cos\theta + i\sin\theta)$, then, for positive integers n,

$$z^{-n} = r^{-n}(\cos n\theta - i\sin n\theta)$$

46. Use the above equation to find z^{-4} for $z = 1 - i\sqrt{3}$.

47. **Sum of the nth Roots of 1** Make a conjecture about the *sum* of the nth roots of 1 for any natural number $n \geq 2$. (*Hint*: Experiment by finding the sum of the two square roots of 1, the sum of the three cube roots of 1, the sum of the four fourth roots of 1, the sum of the five fifth roots of 1, and the sum of the six sixth roots of 1.)

48. **Product of the nth Roots of 1** Make a conjecture about the *product* of the nth roots of 1 for any natural number $n \geq 2$. (*Hint*: Experiment by finding the product of the two square roots of 1, the product of the three cube roots of 1, the product of the four fourth roots of 1, the product of the five fifth roots of 1, and the product of the six sixth roots of 1.)

Projects

1. **Verify Identities** Raise $(\cos\theta + i\sin\theta)$ to the second power by using De Moivre's Theorem. Now square $(\cos\theta + i\sin\theta)$ as a binomial. Equate the real and imaginary parts of the two complex numbers and show that

   a. $\cos 2\theta = \cos^2\theta - \sin^2\theta$

   b. $\sin 2\theta = 2\sin\theta\cos\theta$

2. **Discover Identities** Raise $(\cos\theta + i\sin\theta)$ to the fourth power by using De Moivre's Theorem. Now find the fourth power of the binomial $(\cos\theta + i\sin\theta)$ by multiplying. Equate the real and imaginary parts of the two complex numbers.

   a. What identity have you discovered for $\cos 4\theta$?

   b. What identity have you discovered for $\sin 4\theta$?

Exploring Concepts with Technology

The Mandelbrot Iteration Procedure

The following procedure is called the **Mandelbrot iteration procedure**.

Mandelbrot Iteration Procedure

Choose a complex number z_0.

1. Square z_0 and add the reult to z_0.
2. Square the preceding result and add it to z_0.
3. Repeat step 2.

Applying the Mandelbrot iteration procedure to a complex number z_0 generates a sequence of numbers $z_1, z_2, z_3, \ldots$. The numbers $z_1, z_2, z_3, \ldots$ are called **iterates**. In this case, the iterates are generated using the complex function $f(z) = z^2 + z_0$. The number z_0 is referred to as the *seed* or *initial value*. If you change the seed, a different sequence of iterates is generated. Some seeds generate iterates that grow without bound; some seeds generate iterates that approach a constant; and some seeds generate iterates that are cyclic. For instance:

- Let $z_0 = 1$. Then

  $$z_1 = 1^2 + 1 = 2,\ z_2 = 2^2 + 1 = 5,\ z_3 = 5^2 + 1 = 26,\ z_4 = 26^2 + 1 = 677$$

  In this case, the iterates grow larger and larger.

- Let $z_0 = -1$. Then

  $$z_1 = (-1)^2 + (-1) = 0,\ z_2 = 0^2 + (-1) = -1,\ z_3 = (-1)^2 + (-1) = 0$$

  In this case, the iterates cycle: $0, -1, 0, -1, 0, \ldots$.

- Let $z_0 = 0.25$. Then

  $$z_1 = 0.25^2 + 0.25 = 0.3125,\ z_2 = 0.3125^2 + 0.25 = 0.34765625,$$
  $$z_3 = 0.34765625^2 + 0.25 \approx 0.3708648682$$

  In this case, subsequent iterates can be calculated easily using a graphing calculator and the method shown below.

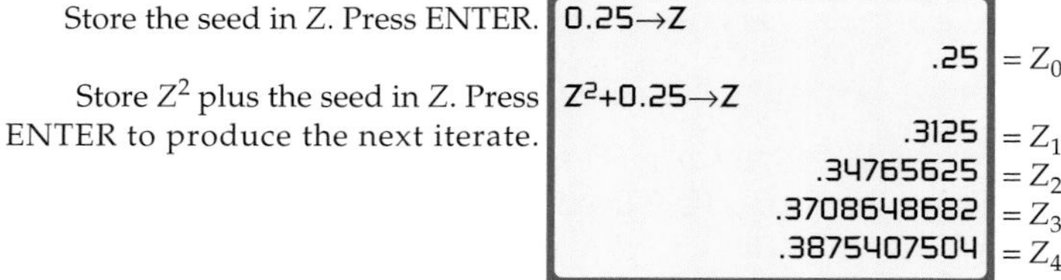

1. Use a graphing calculator to continue the above Mandelbrot iteration procedure, with $z_0 = 0.25$, to find z_5, z_{10}, z_{100}, and z_{200}. It can be shown that the sequence of iterates is approaching a constant. What constant do you think the sequence of iterates is approaching?

2. Let $z_0 = i$. Use the Mandelbrot iterartion procedure to find z_1, z_2, z_3, and z_4. What happens as the iteration procedure progresses?

The black region in **Figure 5.11** is called the **Mandelbrot set**. The Mandelbrot set consists of all complex numbers for which the absolute value of each of their Mandelbrot iteration procedure iterates $z_1, z_2, z_3, \ldots$ is less than or equal to 2.

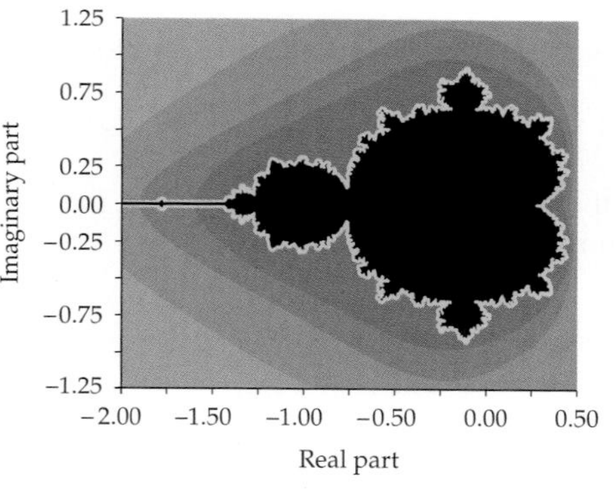

Figure 5.11

The Mandelbrot set has been called the most complex object in the realm of mathematics. There are many Internet sites that provide additional information on the Mandelbrot set. Visit some of these sites and use the computer programs that are provided to graph the Mandelbrot set and zoom in on different regions of the graph to see some of its complex structure.

Chapter 5 Summary

5.1 Complex Numbers

- The number i, called the *imaginary unit*, is the number such that $i^2 = -1$.
- If a is positive real number, then $\sqrt{-a} = i\sqrt{a}$. The number $i\sqrt{a}$ is called an *imaginary number*.
- A complex number is a number of the form $a + bi$, where a and b are real numbers and $i = \sqrt{-1}$. The number a is the *real part* of $a + bi$, and b is the *imaginary part*.
- The complex number $a + bi$ and $a - bi$ are called *complex conjugates* or *conjugates* of each other.
- **Operations on Complex Numbers**

$$(a + bi) + (c + di) = (a + c) + (b + d)i$$

$$(a + bi) - (c + di) = (a - c) + (b - d)i$$

$$(a + bi)(c + di) = (ac - bd) + (ad + bc)i$$

$$\frac{a + bi}{c + di} = \frac{a + bi}{c + di} \cdot \frac{c - di}{c - di}$$

- **Multiply the numerator and denominator by the conjugate of the denominator.**

5.2 Trignometric Form of Complex Numbers

- The complex number $z = a + bi$ can be written in trigonometric form as

$$z = r(\cos\theta + i\sin\theta) = r\,\mathrm{cis}\,\theta$$

where $a = r\cos\theta$, $b = r\sin\theta$, $r = \sqrt{a^2 + b^2}$, and $\tan\theta = \dfrac{b}{a}$

- If $z_1 = r_1(\cos\theta_1 + i\sin\theta_1)$ and $z_2 = r_2(\cos\theta_2 + i\sin\theta_2)$, then

$$z_1 z_2 = r_1 r_2\,\mathrm{cis}(\theta_1 + \theta_2) \quad \text{and} \quad \frac{z_1}{z_2} = \frac{r_1}{r_2}\mathrm{cis}(\theta_1 - \theta_2)$$

5.3 De Moivre's Theorem

- **De Moivre's Theorem**

If $z = r\,\mathrm{cis}\,\theta$ and n is a positive integer, then

$$z^n = r^n\,\mathrm{cis}\,n\theta$$

- If $z = r\,\mathrm{cis}\,\theta$, then the n distinct roots of z are given by

$$w_k = r^{1/n}\,\mathrm{cis}\,\frac{\theta + 360°k}{n} \quad \text{for } k = 0, 1, 2, \ldots, n - 1$$

Chapter 5 Assessing Concepts

1. True or false: The real number 7 is also a complex number.

2. True or false: The product of a complex number z and its conjugate $\bar{z}$ is a real number.

3. True or false: $z = \cos 45° + i \sin 45°$ is a square root of i.

4. True or false: $\sqrt{-4} \cdot \sqrt{-9} = 6$

5. If $|z| = 1$, describe the geometric relationship among the graphs of the fourth roots of z.

6. How many solutions exist for the equation $z = (1 + i)^5$?

7. How many solutions exist for the equation $z^5 = (1 + i)$?

8. Does $|a + bi| = \sqrt{a^2 + (bi)^2}$?

9. What is the conjugate of $-3 - 5i$?

10. What is the modulus of $2\left(\cos \frac{\pi}{3} + i \sin \frac{\pi}{3}\right)$?

Chapter 5 Review Exercises

In Exercises 1 to 4, write the complex number in standard form and give its conjugate.

1. $3 - \sqrt{-64}$
2. $\sqrt{-4} + 6$
3. $-2 + \sqrt{-5}$
4. $-5 - \sqrt{-27}$

In Exercises 5 to 20, simplify and write the complex number in standard form.

5. $(\sqrt{-4})(\sqrt{-4})$
6. $(-\sqrt{-27})(\sqrt{-3})$
7. $(3 + 7i) + (2 - 5i)$
8. $(3 - 4i) + (-6 + 8i)$
9. $(6 - 8i) - (9 - 11i)$
10. $(-3 - 5i) - (2 + 10i)$
11. $(5 + 3i)(2 - 5i)$
12. $(-2 - 3i)(-4 + 7i)$
13. $\dfrac{-2i}{3 - 4i}$
14. $\dfrac{4 + i}{7 - 2i}$
15. $i(2i) - (1 + i)^2$
16. $(2 - i)^3$
17. $(3 + \sqrt{-4}) - (-3 - \sqrt{-16})$
18. $(-2 + \sqrt{-9}) + (-3 - \sqrt{-81})$
19. $(2 - \sqrt{-3})(2 + \sqrt{-3})$
20. $(3 - \sqrt{-5})(2 + \sqrt{-5})$

In Exercises 21 to 24, simplify and write each complex number as i, $-i$, 1, or -1.

21. i^{27}
22. i^{105}
23. $\dfrac{i}{i^{17}}$
24. i^{62}

In Exercises 25 to 28, find the absolute value of each complex number.

25. $|-8i|$
26. $|2 - 3i|$
27. $|-4 + 5i|$
28. $|-1 - i|$

In Exercises 29 to 32, write the complex number in trigonometric form.

29. $z = 2 - 2i$
30. $z = -\sqrt{3} + i$
31. $z = -3 + 2i$
32. $z = 4 - i$

In Exercises 33 to 36, write the complex number in standard form. Round approximate constants to the nearest thousandth.

33. $z = 5(\cos 315° + i \sin 315°)$
34. $z = 6\left(\cos \dfrac{4\pi}{3} + i \sin \dfrac{4\pi}{3}\right)$
35. $z = 2(\cos 2 + i \sin 2)$
36. $z = 3(\cos 115° + i \sin 115°)$

In Exercises 37 to 42, multiply the complex numbers. Write the answer in standard form.

37. $3(\cos 225^\circ + i \sin 225^\circ) \cdot 10(\cos 45^\circ + i \sin 45^\circ)$

38. $5(\cos 162^\circ + i \sin 162^\circ) \cdot 2(\cos 63^\circ + i \sin 63^\circ)$

39. $3(\cos 12^\circ + i \sin 12^\circ) \cdot 4(\cos 126^\circ + i \sin 126^\circ)$

40. $(\cos 23^\circ + i \sin 23^\circ) \cdot 4(\cos 233^\circ + i \sin 233^\circ)$

41. $3(\cos 1.8 + i \sin 1.8) \cdot 5(\cos 2.5 + i \sin 2.5)$

42. $6(\cos 3.1 + i \sin 3.1) \cdot 5(\cos 4.3 + i \sin 4.3)$

In Exercises 43 to 48, divide the complex numbers. Write the answer in trigonometric form.

43. $\dfrac{6(\cos 50^\circ + i \sin 50^\circ)}{2(\cos 150^\circ + i \sin 150^\circ)}$

44. $\dfrac{30(\cos 165^\circ + i \sin 165^\circ)}{10(\cos 55^\circ + i \sin 55^\circ)}$

45. $\dfrac{40(\cos 66^\circ + i \sin 66^\circ)}{8(\cos 125^\circ + i \sin 125^\circ)}$

46. $\dfrac{2(\cos 150^\circ + i \sin 150^\circ)}{\sqrt{2}(\cos 200^\circ + i \sin 200^\circ)}$

47. $\dfrac{10(\cos 3.7 + i \sin 3.7)}{6(\cos 1.8 + i \sin 1.8)}$

48. $\dfrac{4(\cos 1.2 + i \sin 1.2)}{8(\cos 5.2 + i \sin 5.2)}$

In Exercises 49 to 54, find the indicated power. Write the answer in standard form.

49. $[3(\cos 45^\circ + i \sin 45^\circ)]^5$

50. $\left[\cos\left(\dfrac{11\pi}{8}\right) + i \sin\left(\dfrac{11\pi}{8}\right)\right]^8$

51. $(1 - i\sqrt{3})^7$

52. $(-2 - 2i)^{10}$

53. $(\sqrt{2} - i\sqrt{2})^5$

54. $(3 - 4i)^5$

In Exercises 55 to 60, find the indicated roots. Write the answer in trigonometric form.

55. The three cube roots of $27i$

56. The four fourth roots of $8i$

57. The four fourth roots of 256

58. The five fifth roots of $-16\sqrt{2} - 16\sqrt{2}i$

59. The four fourth roots of 81

60. The three cube roots of -125

»»» Quantitative Reasoning: *Graphing the Mandelbrot Set* »»»

```
PROGRAM: MANBROT
Disp " THIS PROGRAM "
Disp " GRAPHS THE TOP "
Disp " HALF OF THE "
Disp " MANDELBROT SET "
Disp " "
Disp " PRESS ENTER "
Disp " TO START "
Pause
24→I
AxesOff:ClrDraw:FnOff
-2→Xmin:1.03→Xmax
0→Ymin:2→Ymax
For(A,-2,.4,.032)
For(B,0,√(4-A²),.032)
Pt-On(A,B)
(A+Bi)→S
S→Z
For (C,0,I)
(Z²+S)→Z
If abs(Z)>2
Then
Pt-Off(A,B)
I→C
End:End:End:End
StorePic 2
```

The T1-83/T1-83 Plus/T1-84 Plus program at the left graphs the top half of the Mandelbrot set. The bottom half of the Mandelbrot set is a reflection across the x-axis of the top half. See **Figure 5.11** on page 357. The program uses the Mandelbrot iteration procedure, presented in the Exploring Concepts with Technology on page 356, to generate the iterates $z_1, z_2, z_3, \ldots$. In theory, a complex number z_0 is defined to be an element of the Mandelbrot set if and only if the absolute value of each of its iterates is less than or equal to 2. In reality, a calculator or computer program cannot check an infinite number of iterates, but a fairly accurate graph can be produced by checking the first 24 iterates. If the program finds that the absolute values of the first 24 iterates of z_0 are all less than or equal to 2, then the program assumes that z_0 is an element of the Mandelbrot set and plots a point at z_0 in the complex plane. If the program finds an iterate whose absolute value is greater than 2, then the program does not plot a point at z_0. After the graph is produced, it is stored in memory, location `Pic 2`. You can recall the graph by pressing `RecallPic 2`. The `RecallPic` instruction is found in the `DRAW STO` menu.

QR1. Use the MANBROT program to graph the top half of the Mandelbrot set. The program requires about 25 minutes to complete the graph. The program step 24 → I sets the number of iterates to check. You can make the program run faster by using a natural number less than 24, but the graph will be less accurate. The graph will be slightly more accurate if you use a natural number larger than 24, but the program will run slower.

QR2. Examine the graph from Exercise QR1 or **Figure 5.11** to determine which of the following complex numbers are elements of the Mandelbrot set. The WINDOW settings for the program are Xmin = −2, Xmax = 1.03, Ymin = 0, and Ymax = 2.

$$-0.25 + 0.25i, -1 + 0.1i, -0.75 + 0.75i, 0.1 + 0.2i$$

QR3. Examine the iterates of -2 to verify that -2 is an element of the Mandelbrot set.

QR4. What is the first iterate of $2i$? Is $2i$ an element of the Mandelbrot set?

QR5. Which step in the program applies the Mandelbrot iteration procedure?

QR6. Search the Internet for additional information on the Mandelbrot set. Describe two properties of the Mandelbrot set that you find the most interesting.

Chapter 5 Test

1. Write $6 + \sqrt{-9}$ in the form $a + bi$.

2. Simplify: $\sqrt{-18}$

3. Simplify $(3 + \sqrt{-4}) + (7 - \sqrt{-9})$. Write the answer in standard form.

4. Simplify $(-1 + \sqrt{-25}) - (8 - \sqrt{-16})$. Write the answer in standard form.

5. Simplify: $(\sqrt{-12})(\sqrt{-3})$

6. Simplify: i^{263}

7. Simplify: $(3 + 7i) - (-2 - 9i)$

8. Simplify: $(-6 - 9i)(4 + 3i)$

9. Simplify: $(3 - 5i)(-3 + 5i)$

10. Simplify: $\dfrac{4 - 5i}{i}$

11. Simplify: $\dfrac{2 - 7i}{4 + 3i}$

12. Simplify: $\dfrac{6 + 2i}{1 - i}$

13. Find the absolute value of $3 - 5i$.

14. Write $3 - 3i$ in trigonometric form.

15. Write $-6i$ in trigonometric form.

16. Write $4(\cos 120° + i \sin 120°)$ in standard form.

17. Write $5(\cos 225° + i \sin 225°)$ in standard form.

18. Simplify $3(\cos 28° + i \sin 28°) \cdot 4(\cos 17° + i \sin 17°)$. Write the answer in standard form.

19. Simplify $5(\cos 115° + i \sin 115°) \cdot 4(\cos 10° + i \sin 10°)$. Write the answer in standard form.

20. Simplify $\dfrac{24(\cos 258° + i \sin 258°)}{6(\cos 78° + i \sin 78°)}$. Write the answer in standard form.

21. Simplify $\dfrac{18(\cos 50° + i \sin 50°)}{3(\cos 140° + i \sin 140°)}$. Write the answer in standard form.

22. Simplify $(2 - 2i\sqrt{3})^{12}$. Write the answer in standard form.

In Exercises 23 to 25, write the indicated roots and solutions in trigonometric form.

23. Find the six sixth roots of 64.

24. Find the three cube roots of $-1 + i\sqrt{3}$.

25. Find the five solutions of $z^5 + 32 = 0$.

Cumulative Review Exercises

1. Solve $x^2 - x - 6 \le 0$. Write the answer using interval notation.

2. What is the domain of $f(x) = \dfrac{x^2}{x^2 - 4}$?

3. Find c in the domain of $f(x) = \dfrac{x}{x + 1}$ such that $f(c) = 2$.

4. Given $f(x) = \sin 3x$ and $g(x) = \dfrac{x^2 - 1}{3}$, find $(f \circ g)(x)$.

5. Given $f(x) = \dfrac{x}{x - 1}$, find $f^{-1}(3)$.

6. Convert $\dfrac{3\pi}{2}$ radians to degrees.

7. Find the length of the hypotenuse a for the right triangle shown at the right.

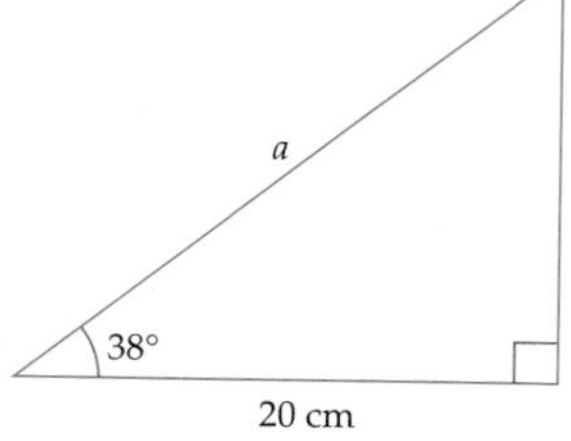

8. If t is any real number, what are the values of a and b for the inequality $a \le \sin t \le b$?

9. Graph $y = 3 \sin \pi x$.

10. Graph $y = \dfrac{1}{2} \tan \dfrac{\pi x}{4}$.

11. Verify the identity $\dfrac{\sin x}{1 + \cos x} = \csc x - \cot x$.

12. Express $\sin 2x \cos 3x - \sin 3x \cos 2x$ in terms of the sine function.

13. Given $\sin \alpha = \dfrac{4}{5}$ in Quadrant I and $\cos \beta = \dfrac{12}{13}$ in Quadrant IV, find $\cos(\alpha + \beta)$.

14. Find the exact value of $\sin\left[\sin^{-1}\left(\dfrac{3}{5}\right) + \cos^{-1}\left(-\dfrac{5}{13}\right)\right]$.

15. Solve $\sin 2x = \sqrt{3} \sin x$ for $0 \le x < 2\pi$.

16. For the triangle at the right, find the length of side c. Round to the nearest ten.

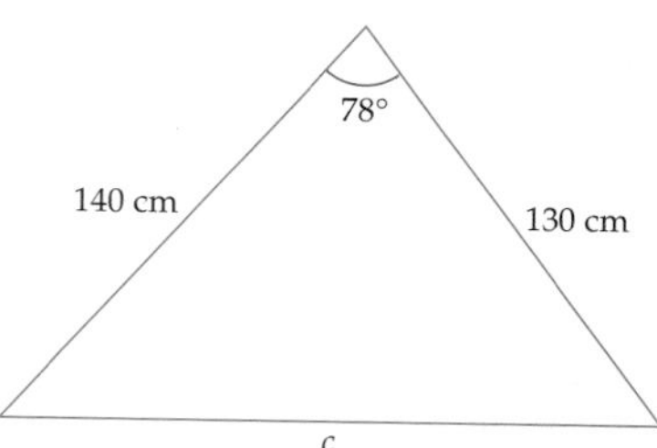

17. Find the angle between the vectors $\mathbf{v} = 3\mathbf{i} + 2\mathbf{j}$ and $\mathbf{w} = 5\mathbf{i} - 3\mathbf{j}$. Round to the nearest tenth of a degree.

18. Work A force of 100 pounds on a rope is used to drag a box up a ramp that is inclined 15°. If the rope makes an angle of 30° with the ground, find the work done in moving the box 15 feet along the ramp. Round to the nearest foot-pound.

19. Write the complex number $2 + 2i$ in trigonometric form.

20. Find the three cube roots of -27. Write the roots in standard form.

6 Topics in Analytic Geometry

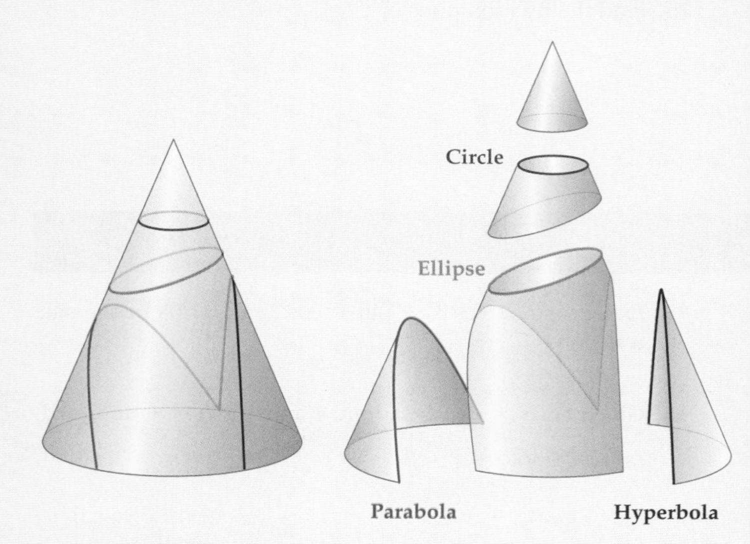

Conic Sections and Their Applications

The study of the geometric figures called *conic sections* is one of the topics of this chapter. Each of these figures is formed by the intersection of a plane and a cone.[1]

The ancient Greeks were the first to study the conic sections. Their study was motivated by the many new and interesting mathematical concepts they were able to discover, without regard to finding or producing practical applications. The ancient Greeks would be surprised to learn that their study of the conic sections helped produce a body of knowledge with many practical applications in several diverse fields, including astronomy, architecture, engineering, and satellite communications. Exercise 55 on page 386 illustrates a medical application of conic sections, and Exercise 47 on page 373 illustrates an application of conic sections in the design of a ski with a parabolic sidecut.

Online Study Center For online student resources, such as section quizzes, visit this website: college.hmco.com/info/aufmannCAT

[1]Only one branch of the hyperbola in the figure is displayed; however, all hyperbolas have two branches that are formed by the intersection of a plane and a double-napped cone as shown in Figure 6.1 on page 363.

Parabolas

Section 6.1

- Parabolas with Vertex at (0, 0)
- Parabolas with Vertex at (h, k)
- Applications

The graph of a circle, an ellipse, a parabola, or a hyperbola can be formed by the intersection of a plane and a cone. Hence these figures are referred to as conic sections. See **Figure 6.1.**

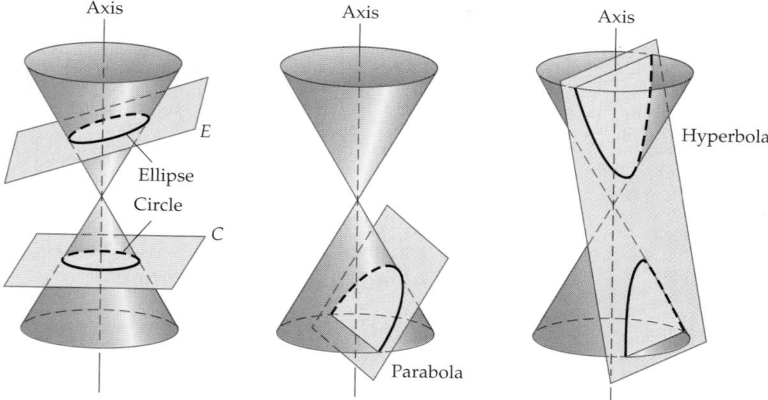

Figure 6.1
Cones intersected by planes

A plane perpendicular to the axis of the cone intersects the cone in a circle (plane C). The plane E, tilted so that it is not perpendicular to the axis, intersects the cone in an ellipse. When the plane is parallel to a line on the surface of the cone, the plane intersects the cone in a parabola. When the plane intersects both portions of the cone, a hyperbola is formed.

take note

If the intersection of a plane and a cone is a point, a line, or two intersecting lines, then the intersection is called a *degenerate conic section.*

Math Matters

Appollonius (262–200 B.C.) wrote an eight-volume treatise entitled *On Conic Sections* in which he derived the formulas for all the conic sections. He was the first to use the words *parabola, ellipse,* and *hyperbola.*

Parabolas with Vertex at (0, 0)

In addition to the geometric description of a conic section just given, a conic section can be defined as a set of points. This method uses specified conditions about a curve to determine which points in the coordinate system are points of the graph. For example, a parabola can be defined by the following set of points.

Definition of a Parabola

A **parabola** is the set of points in a plane that are equidistant from a fixed line (the **directrix**) and a fixed point (the **focus**) not on the directrix.

The line that passes through the focus and is perpendicular to the directrix is called the **axis of symmetry** of the parabola. The midpoint of the line segment between the focus and directrix on the axis of symmetry is the **vertex** of the parabola, as shown in **Figure 6.2.**

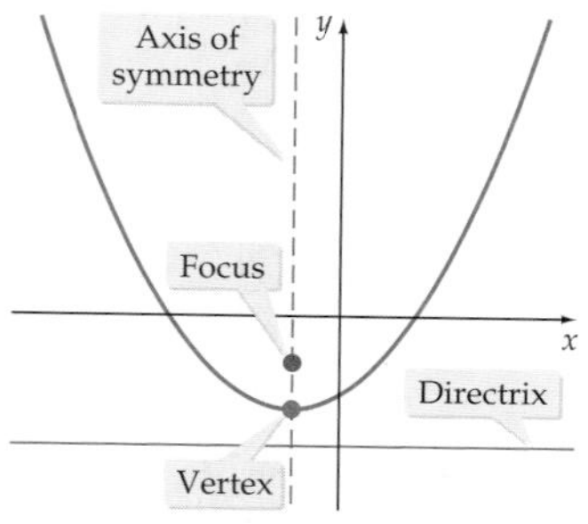

Figure 6.2

Using this definition of a parabola, we can determine an equation of a parabola. Suppose that the coordinates of the vertex of a parabola are $V(0, 0)$ and the axis of symmetry is the y-axis. The equation of the directrix is $y = -p, p > 0$. The focus lies on the axis of symmetry and is the same distance from the vertex as the vertex is from the directrix. Thus the coordinates of the focus are $F(0, p)$, as shown in **Figure 6.3.**

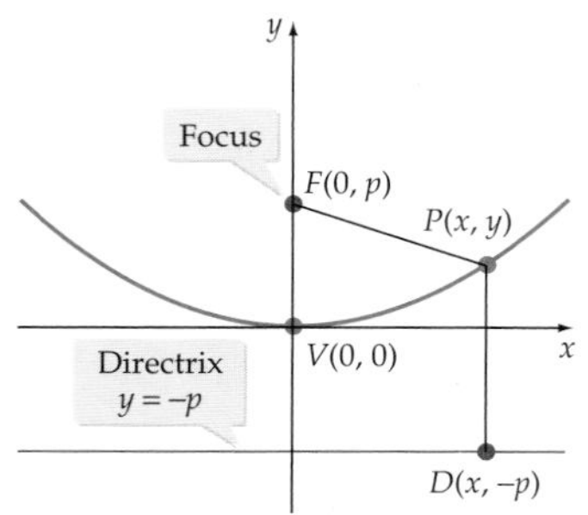

Figure 6.3

TO REVIEW

Axis of Symmetry
See page 54.

Let $P(x, y)$ be any point P on the parabola. Then, using the distance formula and the fact that the distance between any point P on the parabola and the focus is equal to the distance from the point P to the directrix, we can write the equation

$$d(P, F) = d(P, D)$$

By the distance formula,

$$\sqrt{(x - 0)^2 + (y - p)^2} = y + p$$

Now, squaring each side and simplifying, we get

$$\left(\sqrt{(x - 0)^2 + (y - p)^2}\right)^2 = (y + p)^2$$
$$x^2 + y^2 - 2py + p^2 = y^2 + 2py + p^2$$
$$x^2 = 4py$$

This is the standard form of the equation of a parabola with vertex at the origin and the y-axis as its axis of symmetry. The standard form of the equation of a parabola with vertex at the origin and the x-axis as its axis of symmetry is derived in a similar manner.

take note

The tests for y-axis and x-axis symmetry can be used to verify these statements and provide connections to earlier topics on symmetry.

Standard Forms of the Equation of a Parabola with Vertex at the Origin

Axis of Symmetry Is the y-Axis

The standard form of the equation of a parabola with vertex $(0, 0)$ and the y-axis as its axis of symmetry is

$$x^2 = 4py$$

The focus is $(0, p)$, and the equation of the directrix is $y = -p$. If $p > 0$, the graph of the parabola opens up. See **Figure 6.4a.** If $p < 0$, the graph of the parabola opens down. See **Figure 6.4b.**

Axis of Symmetry Is the x-Axis

The standard form of the equation of a parabola with vertex $(0, 0)$ and the x-axis as its axis of symmetry is

$$y^2 = 4px$$

The focus is $(p, 0)$, and the equation of the directrix is $x = -p$. If $p > 0$, the graph of the parabola opens to the right. See **Figure 6.4c.** If $p < 0$, the graph of the parabola opens to the left. See **Figure 6.4d.**

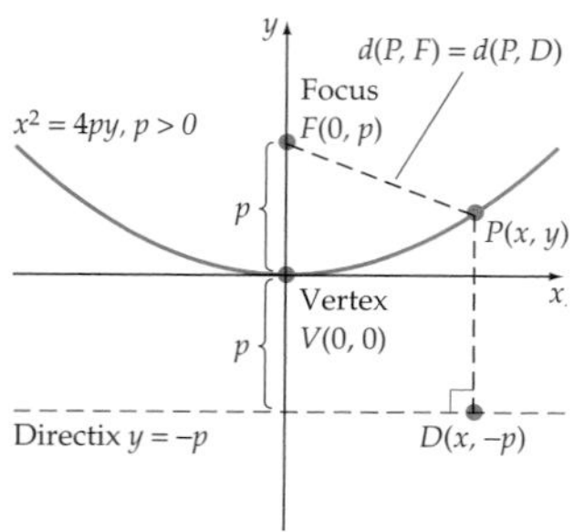

a. The graph of $x^2 = 4py$ with $p > 0$

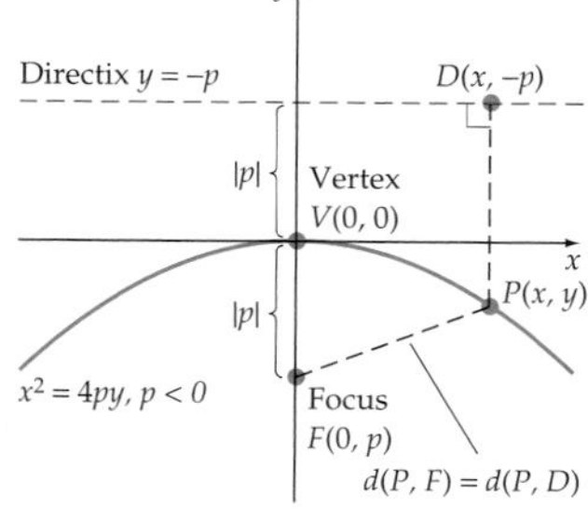

b. The graph of $x^2 = 4py$ with $p < 0$

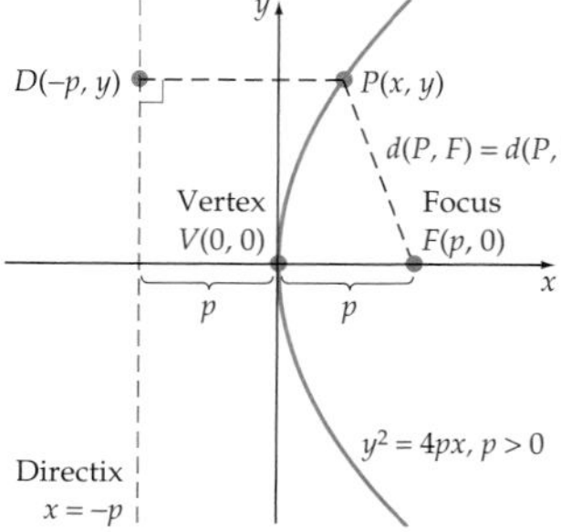

c. The graph of $y^2 = 4px$ with $p > 0$

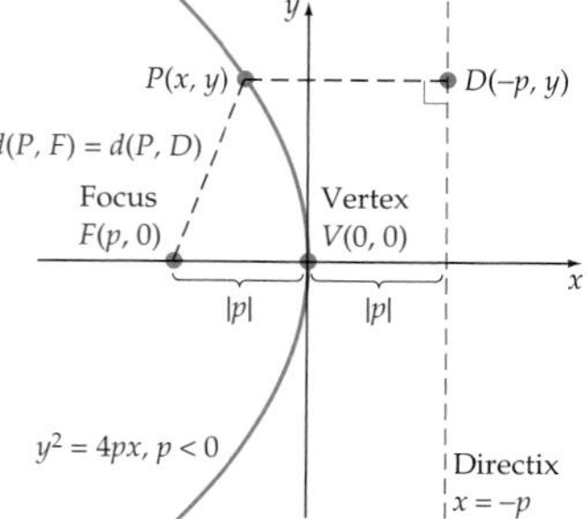

d. The graph of $y^2 = 4px$ with $p < 0$

Figure 6.4

❓ **QUESTION** Does the graph of $y^2 = -4x$ open up, down, to the left, or to the right?

EXAMPLE 1 ⟫ Find the Focus and Directrix of a Parabola

Find the focus and directrix of the parabola given by the equation $y = -\frac{1}{2}x^2$.

Solution

Because the x term is squared, the standard form of the equation is $x^2 = 4py$.

$$y = -\frac{1}{2}x^2$$

$$x^2 = -2y$$

• **Write the given equation in standard form.**

Comparing this equation with $x^2 = 4py$ gives

$$4p = -2$$

$$p = -\frac{1}{2}$$

Because p is negative, the parabola opens down, and the focus is below the vertex (0, 0), as shown in **Figure 6.5**. The coordinates of the focus are $\left(0, -\frac{1}{2}\right)$. The equation of the directrix is $y = \frac{1}{2}$.

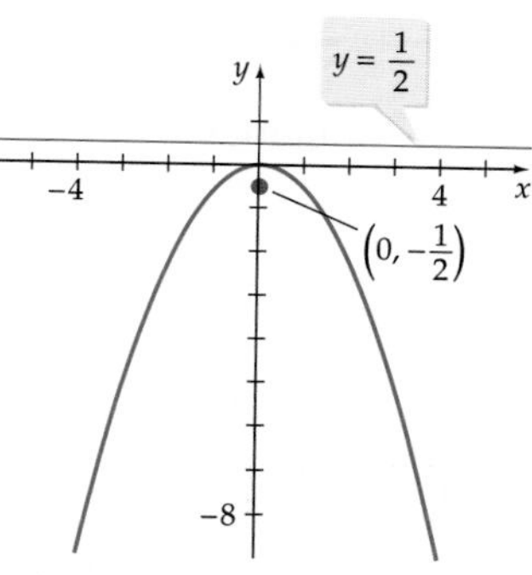

Figure 6.5

⟫ *Try Exercise 6, page 371*

EXAMPLE 2 ⟫ Find the Equation of a Parabola in Standard Form

Find the equation in standard form of the parabola with vertex at the origin and focus at (−2, 0).

Solution

Because the vertex is (0, 0) and the focus is at (−2, 0), $p = -2$. The graph of the parabola opens toward the focus, so in this case the parabola opens to

Continued ▶

❓ **ANSWER** The graph opens to the left.

the left. The equation in standard form of the parabola that opens to the left is $y^2 = 4px$. Substitute -2 for p in this equation and simplify.

$$y^2 = 4(-2)x = -8x$$

The equation of the parabola is $y^2 = -8x$.

Try Exercise 30, page 371

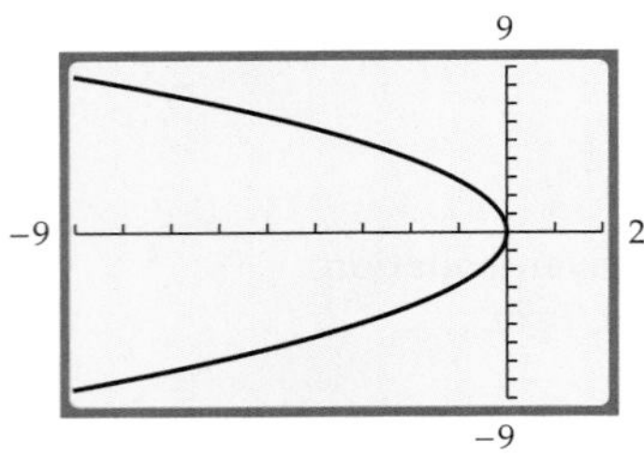

Figure 6.6

Integrating Technology

The graph of $y^2 = -8x$ is shown in **Figure 6.6**. Note that the graph is not the graph of a function. To graph $y^2 = -8x$ with a graphing utility, we first solve for y to produce $y = \pm\sqrt{-8x}$. From this equation we can see that for any $x < 0$, there are two values of y. For example, when $x = -2$,

$$y = \pm\sqrt{(-8)(-2)} = \pm\sqrt{16} = \pm 4$$

The graph of $y^2 = -8x$ in **Figure 6.6** was constructed by graphing both Y1 $= \sqrt{-8x}$ and Y2 $= -\sqrt{-8x}$ in the same window.

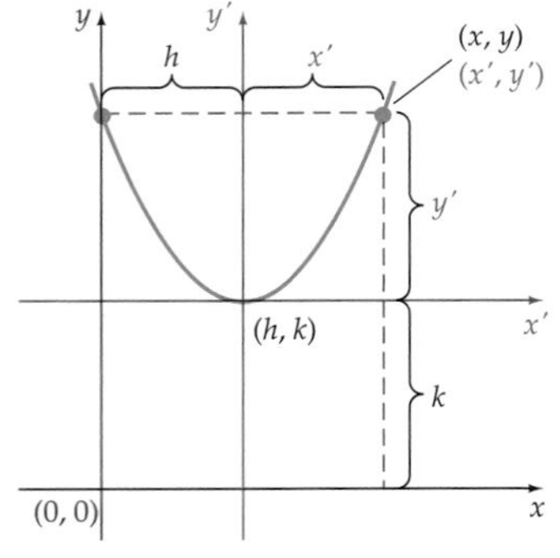

Figure 6.7

Parabolas with Vertex at (h, k)

The equation of a parabola with a vertical or horizontal axis of symmetry and with vertex at a point (h, k) can be found by using the translations discussed previously. Consider a coordinate system with coordinate axes labeled x' and y' placed so that its origin is at (h, k) of the xy-coordinate system.

The relationship between an ordered pair in the $x'y'$-coordinate system and in the xy-coordinate system is given by the **transformation equations**

$$\begin{aligned} x' &= x - h \\ y' &= y - k \end{aligned} \qquad (1)$$

Now consider a parabola with vertex at (h, k), as shown in **Figure 6.7**. Create a new coordinate system with axes labeled x' and y' and with its origin at (h, k). The equation of a parabola in the $x'y'$-coordinate system is

$$(x')^2 = 4py' \qquad (2)$$

Using the transformation Equations (1), we can substitute the expressions for x' and y' into Equation (2). The standard form of the equation of a parabola with vertex (h, k) and a vertical axis of symmetry is

$$(x - h)^2 = 4p(y - k)$$

Similarly, we can derive the standard form of the equation of a parabola with vertex (h, k) and a horizontal axis of symmetry.

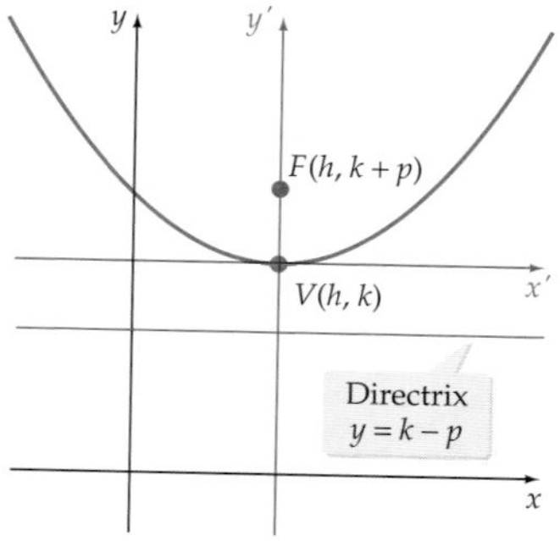

Figure 6.8

Standard Forms of the Equation of a Parabola with Vertex at (h, k)

Vertical Axis of Symmetry

The standard form of the equation of a parabola with vertex (h, k) and a vertical axis of symmetry is

$$(x - h)^2 = 4p(y - k)$$

The focus is $(h, k + p)$, and the equation of the directrix is $y = k - p$. If $p > 0$, the parabola opens up. See **Figure 6.8.** If $p < 0$, the parabola opens down.

Horizontal Axis of Symmetry

The standard form of the equation of a parabola with vertex (h, k) and a horizontal axis of symmetry is

$$(y - k)^2 = 4p(x - h)$$

The focus is $(h + p, k)$, and the equation of the directrix is $x = h - p$. If $p > 0$, the parabola opens to the right. If $p < 0$, the parabola opens to the left.

In Example 3 we complete the square to find the standard form of a parabola, and then use the standard form to determine the vertex, focus, and directrix of the parabola.

TO REVIEW

Completing the Square

See page 26.

EXAMPLE 3 》 Find the Focus and Directrix of a Parabola

Find the equation of the directrix and the coordinates of the vertex and focus of the parabola given by the equation $3x + 2y^2 + 8y - 4 = 0$.

Solution

Rewrite the equation so that the y terms are on one side of the equation, and then complete the square on y.

$$3x + 2y^2 + 8y - 4 = 0$$
$$2y^2 + 8y = -3x + 4$$
$$2(y^2 + 4y) = -3x + 4$$
$$2(y^2 + 4y + 4) = -3x + 4 + 8$$

• **Complete the square. Note that $2 \cdot 4 = 8$ is added to each side.**

$$2(y + 2)^2 = -3(x - 4)$$

• **Simplify and factor.**

$$(y + 2)^2 = -\frac{3}{2}(x - 4)$$

• **Write the equation in standard form.**

Comparing this equation to $(y - k)^2 = 4p(x - h)$, we have a parabola that opens to the left with vertex $(4, -2)$ and $4p = -\dfrac{3}{2}$. Thus $p = -\dfrac{3}{8}$.

The coordinates of the focus are

$$\left(4 + \left(-\frac{3}{8}\right), -2\right) = \left(\frac{29}{8}, -2\right)$$

Continued ▶

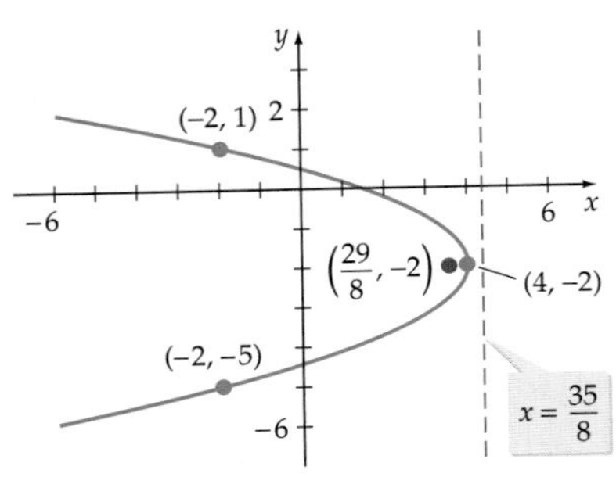

Figure 6.9

The equation of the directrix is

$$x = 4 - \left(-\frac{3}{8}\right) = \frac{35}{8}$$

Choosing some values for y and finding the corresponding values for x, we plot a few points. Because the line $y = -2$ is the axis of symmetry, for each point on one side of the axis of symmetry there is a corresponding point on the other side. Two points are $(-2, 1)$ and $(-2, -5)$. See **Figure 6.9.**

Try Exercise 22, page 371

EXAMPLE 4 Find the Equation in Standard Form of a Parabola

Find the equation in standard form of the parabola with directrix $x = -1$ and focus $(3, 2)$.

Solution

The vertex is the midpoint of the line segment joining the focus $(3, 2)$ and the point $(-1, 2)$ on the directrix.

$$(h, k) = \left(\frac{-1 + 3}{2}, \frac{2 + 2}{2}\right) = (1, 2)$$

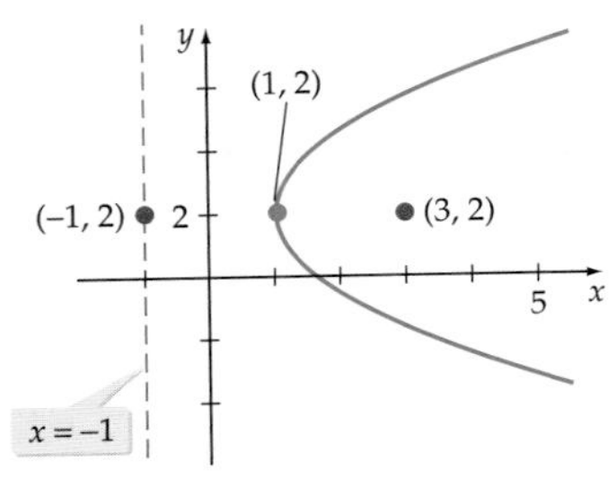

Figure 6.10

The standard form of the equation is $(y - k)^2 = 4p(x - h)$. The distance from the vertex to the focus is 2. Thus $4p = 4(2) = 8$, and the equation of the parabola in standard form is $(y - 2)^2 = 8(x - 1)$. See **Figure 6.10.**

Try Exercise 32, page 371

Applications

A principle of physics states that when light is reflected from a point P on a surface, the angle of incidence (that of the incoming ray) equals the angle of reflection (that of the outgoing ray). See **Figure 6.11.** This principle applied to parabolas has some useful consequences.

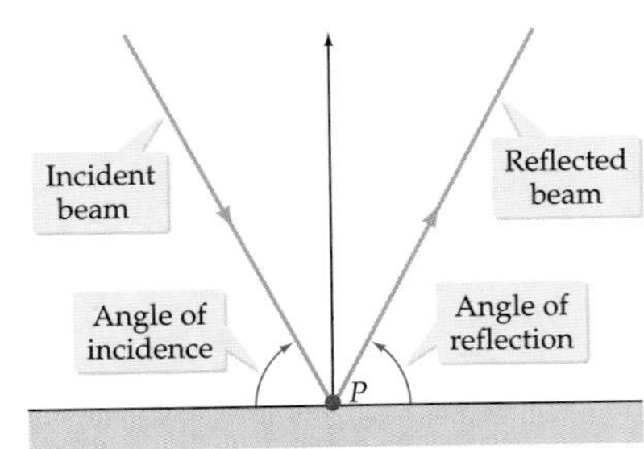

Figure 6.11

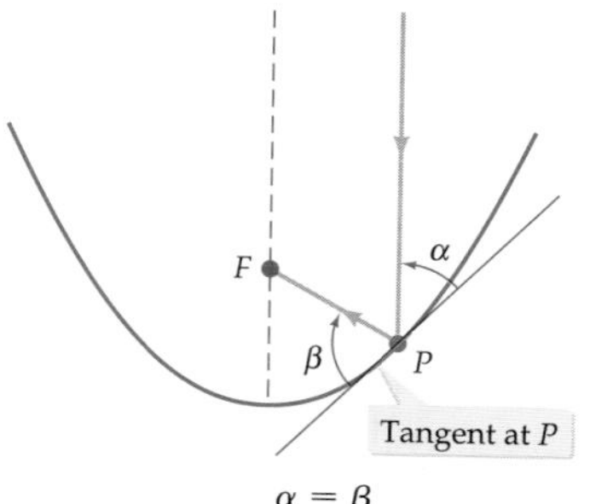

$\alpha = \beta$

Figure 6.12

Reflective Property of a Parabola

The line tangent to a parabola at a point P makes equal angles with the line through P and parallel to the axis of symmetry and the line through P and the focus of the parabola (see **Figure 6.12**).

A cross section of the reflecting mirror of a telescope has the shape of a parabola. The incoming parallel rays of light are reflected from the surface of the mirror to the eyepiece. See **Figure 6.13.**

Flashlights and car headlights also make use of this reflective property. The light bulb is positioned at the focus of the parabolic reflector, which causes the reflected light to be reflected outward in parallel rays. See **Figure 6.14.**

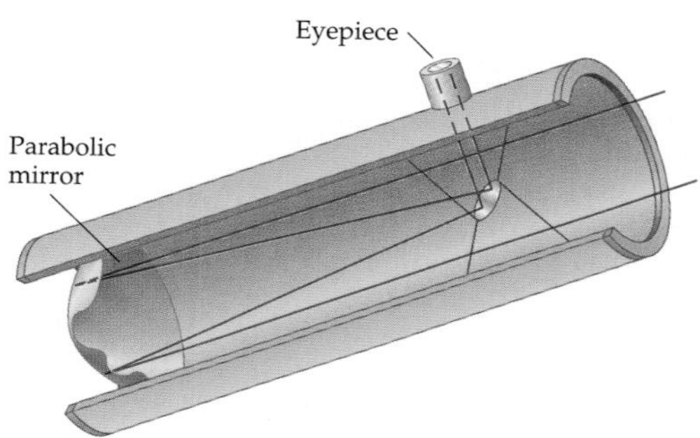

Figure 6.13

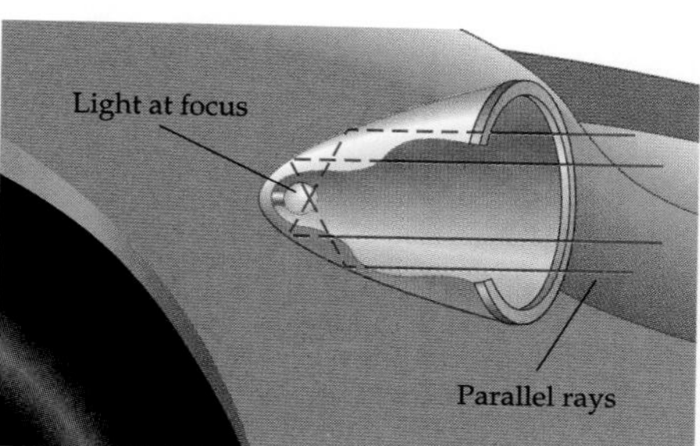

Figure 6.14

When a parabola is revolved about its axis, it produces a three-dimensional surface called a **paraboloid**. The **focus of a paraboloid** is the same as the focus of the parabola that was revolved to generate the paraboloid. The **vertex of a paraboloid** is the same as the vertex of the parabola that was revolved to generate the paraboloid. In Example 5 we find the focus of a satellite dish that has the shape of a paraboloid.

EXAMPLE 5 » Find the Focus of a Satellite Dish

A satellite dish has the shape of a paraboloid. The signals that it receives are reflected to a receiver that is located at the focus of the paraboloid. If the dish is 8 feet across at its opening and 1.25 feet deep at its center, determine the location of its focus.

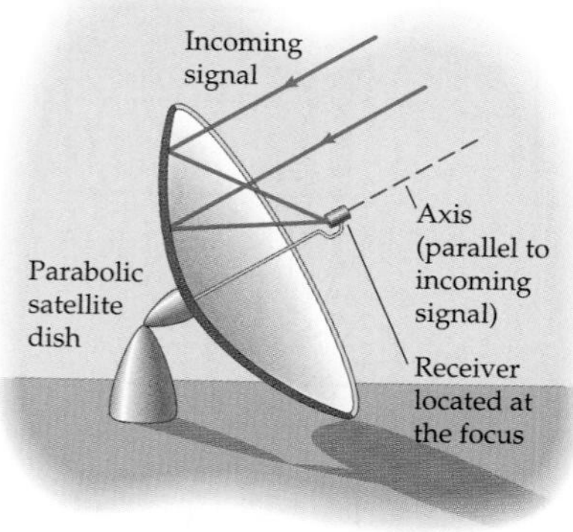

Figure 6.15

Solution

Figure 6.15 shows that a cross section of the paraboloid along its axis of symmetry is a parabola. **Figure 6.16** shows this cross section placed in a rectangular coordinate system with the vertex of the parabola at $(0, 0)$ and the axis of symmetry of the parabola on the y-axis. The parabola has an equation of the form

$$4py = x^2$$

Continued ▶

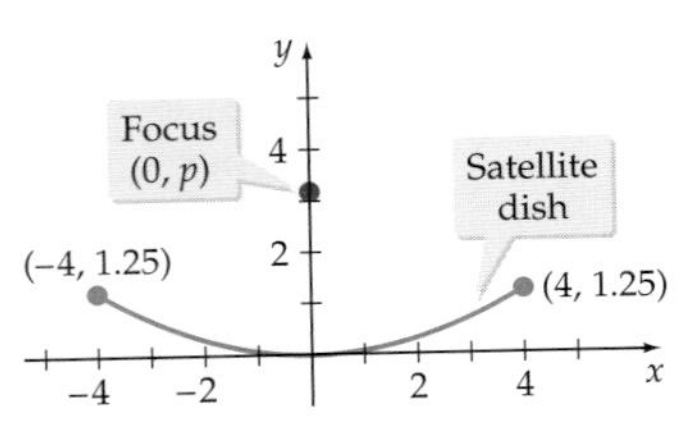

Figure 6.16

Because the parabola contains the point (4, 1.25), this equation is satisfied by the substitutions $x = 4$ and $y = 1.25$. Thus we have

$$4p(1.25) = 4^2$$
$$5p = 16$$
$$p = \frac{16}{5}$$

The focus of the satellite dish is on the axis of symmetry of the dish, and it is $3\frac{1}{5}$ feet above the vertex of the dish. See **Figure 6.16.**

Try Exercise 40, page 372

Topics for Discussion

1. Do the graphs of the parabola given by $y = x^2$ and the vertical line given by $x = 10{,}000$ intersect? Explain.
2. A student claims that the focus of the parabola given by $y = 8x^2$ is at (0, 2) because $4p = 8$ implies that $p = 2$. Explain the error in the student's reasoning.
3. The vertex of a parabola is always halfway between its focus and its directrix. Do you agree? Explain.
4. A tutor claims that the graph of $(x - h)^2 = 4p(y - k)$ has a y-intercept of $\left(0, \frac{h^2}{4p} + k\right)$. Explain why the tutor is correct.

Exercise Set 6.1

1. Examine the following four equations and the graphs labeled **i, ii, iii,** and **iv.** Determine which graph is the graph of each equation.

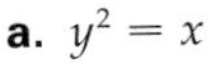

a. $y^2 = x$

b. $x^2 = 4y$

c. $x^2 = -\frac{1}{2}y$

d. $y^2 = -12x$

i.

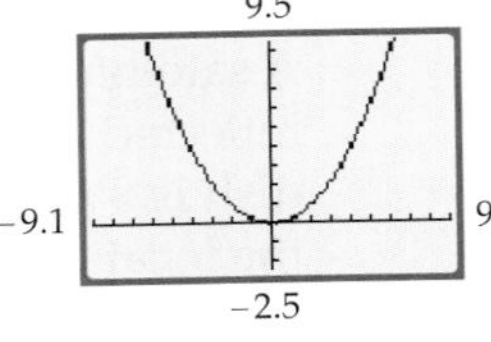

ii.

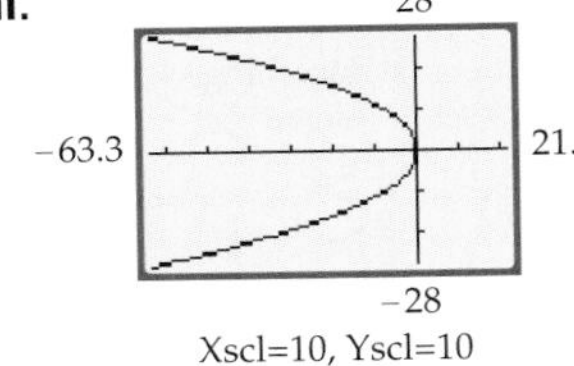

Xscl=10, Yscl=10

iii.

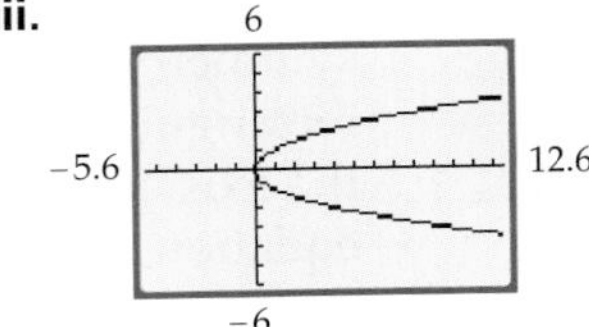

iv.

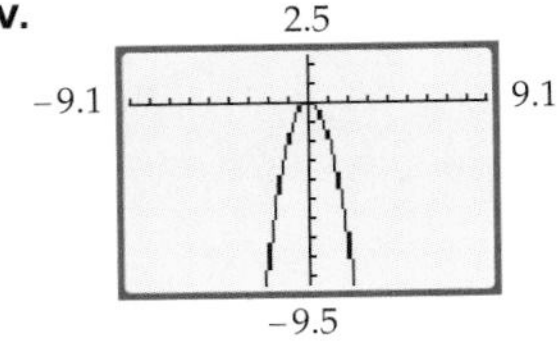

2. Examine the following four equations and the graphs labeled **i, ii, iii,** and **iv.** Determine which graph is the graph of each equation.

a. $(y - 3)^2 = x - 2$

b. $(y - 3)^2 = -2(x - 2)$

c. $\frac{1}{2}(y - 3) = (x - 2)^2$

d. $-2(y - 3) = (x - 2)^2$

i.

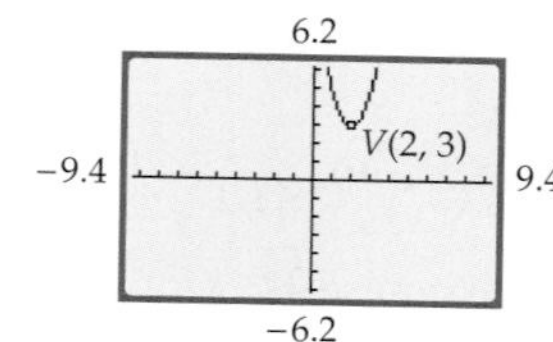

ii.

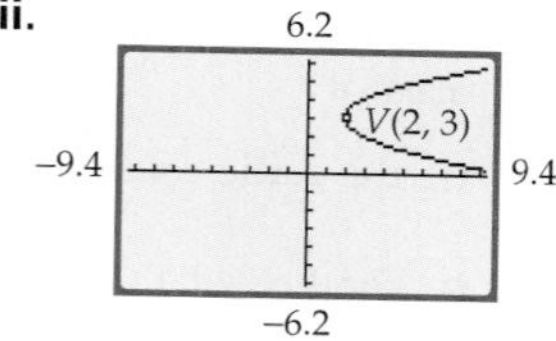

iii.

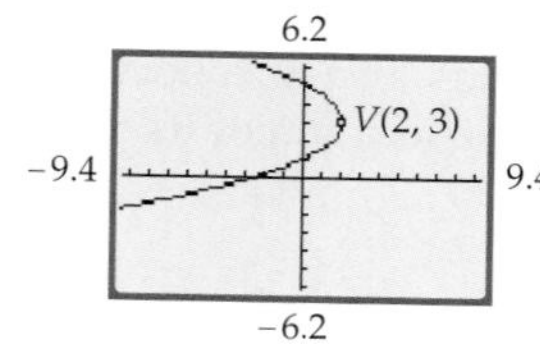

iv.

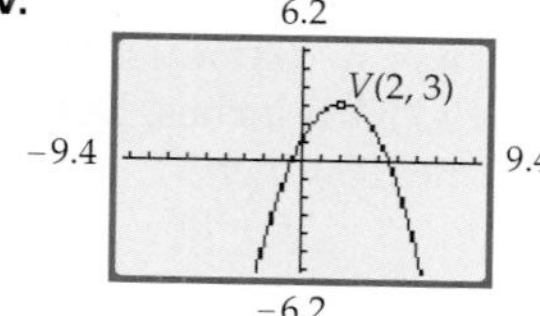

In Exercises 3 to 28, find the vertex, focus, and directrix of the parabola given by each equation. Sketch the graph.

3. $x^2 = -4y$

4. $2y^2 = x$

5. $y^2 = \frac{1}{3}x$

6. $x^2 = -\frac{1}{4}y$

7. $(x - 2)^2 = 8(y + 3)$

8. $(y + 1)^2 = 6(x - 1)$

9. $(y + 4)^2 = -4(x - 2)$

10. $(x - 3)^2 = -(y + 2)$

11. $(y - 1)^2 = 2x + 8$

12. $(x + 2)^2 = 3y - 6$

13. $(2x - 4)^2 = 8y - 16$

14. $(3x + 6)^2 = 18y - 36$

15. $x^2 + 8x - y + 6 = 0$

16. $x^2 - 6x + y + 10 = 0$

17. $x + y^2 - 3y + 4 = 0$

18. $x - y^2 - 4y + 9 = 0$

19. $2x - y^2 - 6y + 1 = 0$

20. $3x + y^2 + 8y + 4 = 0$

21. $x^2 + 3x + 3y - 1 = 0$

22. $x^2 + 5x - 4y - 1 = 0$

23. $2x^2 - 8x - 4y + 3 = 0$

24. $6x - 3y^2 - 12y + 4 = 0$

25. $2x + 4y^2 + 8y - 5 = 0$

26. $4x^2 - 12x + 12y + 7 = 0$

27. $3x^2 - 6x - 9y + 4 = 0$

28. $2x - 3y^2 + 9y + 5 = 0$

29. Find the equation in standard form of the parabola with vertex at the origin and focus $(0, -4)$.

30. Find the equation in standard form of the parabola with vertex at the origin and focus $(5, 0)$.

31. Find the equation in standard form of the parabola with vertex at $(-1, 2)$ and focus $(-1, 3)$.

32. Find the equation in standard form of the parabola with vertex at $(2, -3)$ and focus $(0, -3)$.

33. Find the equation in standard form of the parabola with focus $(3, -3)$ and directrix $y = -5$.

34. Find the equation in standard form of the parabola with focus $(-2, 4)$ and directrix $x = 4$.

35. Find the equation in standard form of the parabola that has vertex $(-4, 1)$, has its axis of symmetry parallel to the y-axis, and passes through the point $(-2, 2)$.

36. Find the equation in standard form of the parabola that has vertex $(3, -5)$, has its axis of symmetry parallel to the x-axis, and passes through the point $(4, 3)$.

37. **Structural Defects** Ultrasound is used as a nondestructive method of determining whether a support beam for a structure has an internal fracture. In one scanning procedure, if the resulting image is a parabola, engineers know that there is a structural defect. Suppose that a scan produced an image whose equation is

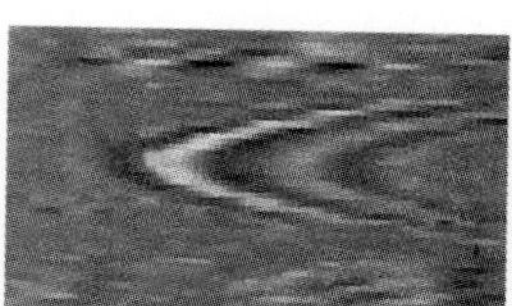

$$x = -0.325y^2 + 13y + 120$$

Determine the vertex and focus of the graph of this parabola.

38. **Fountain Design** A fountain in a shopping mall has two parabolic arcs of water intersecting as shown below. The equation of one parabola is $y = -0.25x^2 + 2x$ and the equation of the second parabola is

$$y = -0.25x^2 + 4.5x - 16.25$$

How high above the base of the fountain do the parabolas intersect? All dimensions are in feet.

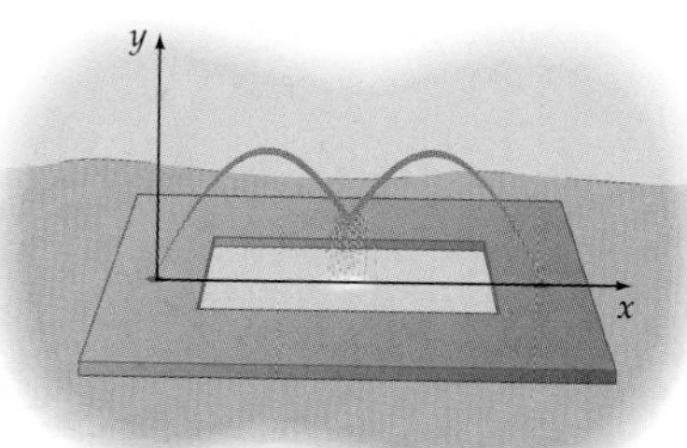

39. **Satellite Dish** A satellite dish has the shape of a paraboloid. The signals that it receives are reflected to a receiver that is located at the focus of the paraboloid. If the dish is 8 feet across at its opening and 1 foot deep at its vertex, determine the location (distance above the vertex of the dish) of its focus.

40. **Radio Telescopes** The antenna of a radio telescope is a paraboloid measuring 81 feet across with a depth of 16 feet. Determine, to the nearest tenth of a foot, the distance from the vertex to the focus of this antenna.

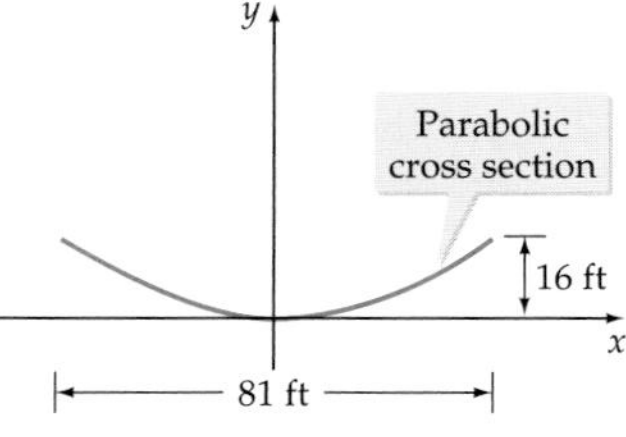

41. **Capturing Sound** During televised football games, a parabolic microphone is used to capture sounds. The shield of the microphone is a paraboloid with a diameter of 18.75 inches and a depth of 3.66 inches. To pick up the sounds, a microphone is placed at the focus of the paraboloid. How far (to the nearest tenth of an inch) from the vertex of the paraboloid should the microphone be placed?

42. **The Lovell Telescope** The Lovell Telescope is a radio telescope located at the Jodrell Bank Observatory in Cheshire, England. The dish of the telescope has the shape of a paraboloid with a diameter of 250 feet. The distance from the vertex of the dish to its focus is 75 feet.

a. Find an equation of a cross section of the paraboloid that passes through the vertex of the paraboloid. Assume that the dish has its vertex at $(0, 0)$ and a vertical axis of symmetry.

b. Find the depth of the dish. Round to the nearest foot.

43. The surface area of a paraboloid with radius r and depth d is given by $S = \frac{\pi r}{6d^2}[(r^2 + 4d^2)^{3/2} - r^3]$. Approximate (to the nearest 100 square feet) the surface area of

a. the radio telescope in Exercise 40.

b. the Lovell Telescope (see Exercise 42).

44. **The Hale Telescope** The parabolic mirror in the Hale telescope at the Palomar Observatory in southern California has a diameter of 200 inches and a concave depth of 3.75375 inches. Determine the location of its focus (to the nearest inch).

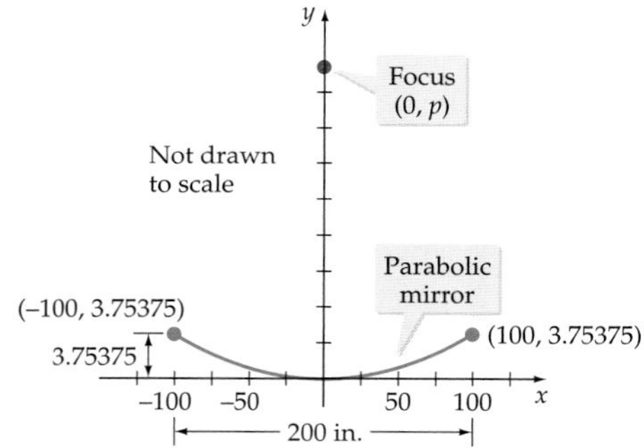

Cross Section of the Mirror in the Hale Telescope

45. **The Lick Telescope** The parabolic mirror in the Lick telescope at the Lick Observatory on Mount Hamilton has a diameter of 120 inches and a focal length of 600 inches. (*Note:* The **focal length** of a parabolic mirror is the distance from the vertex of the mirror to the mirror's focus.) In the construction of the mirror, workers ground the mirror as shown in the following diagram. Determine the dimension a, which is the concave depth of the mirror.

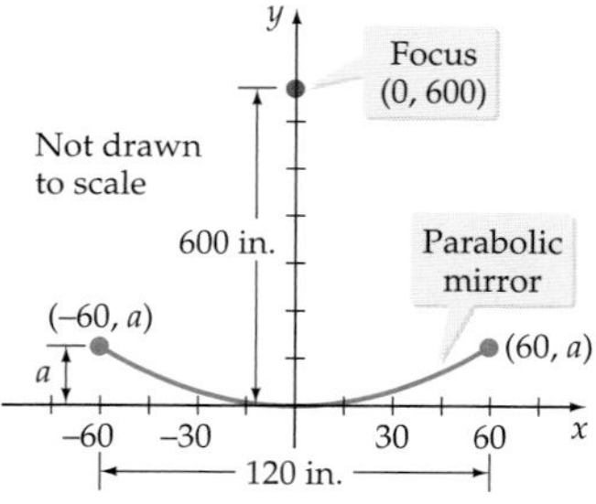

Cross Section of the Mirror in the Lick Telescope

46. **Headlight Design** A light source is to be placed on the axis of symmetry of the parabolic reflector shown in the figure below. How far to the right of the vertex point should the light source be located if the designer wishes the reflected light rays to form a beam of parallel rays?

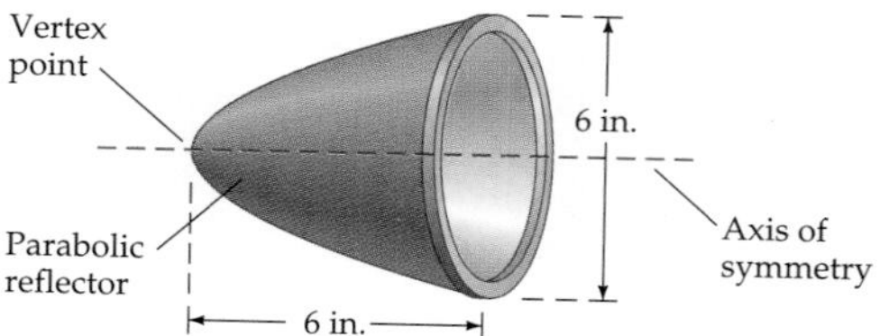

47. **Ski Design** Many contemporary skis have parabolic sidecuts that allow a skier to carve tighter turns than are possible with traditional skis. In the following diagram, x is the directed distance to the right or left of the y-axis (with x and y measured in millimeters). The x-axis is on the center horizontal axis of the ski. The vertex of the parabolic sidecut is $V(0, 32)$.

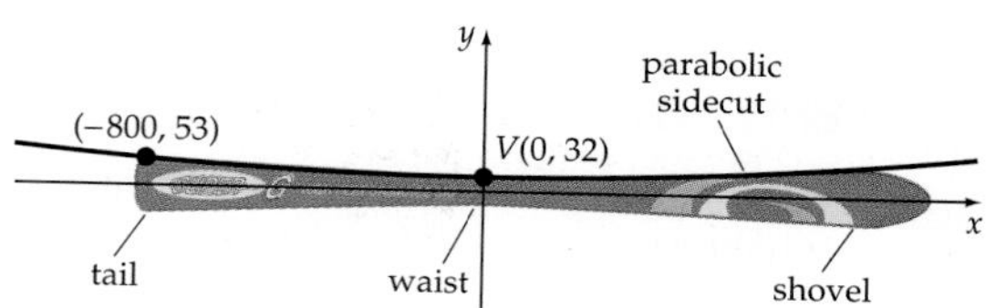

a. Find the equation in standard form of the parabolic sidecut.

b. How wide is the ski at its shovel (the widest point near the front of the ski), where $x = 900$? Round to the nearest millimeter.

48. The only information we have about a particular parabola is that $(2, 3)$ and $(-2, 3)$ are points on the parabola. Explain why it is not possible to find the equation of this particular parabola using just this information.

Connecting Concepts

In Exercises 49 to 51, use the following definition of latus rectum: The line segment that has endpoints on a parabola, passes through the focus of the parabola, and is perpendicular to the axis of symmetry is called the *latus rectum* of the parabola.

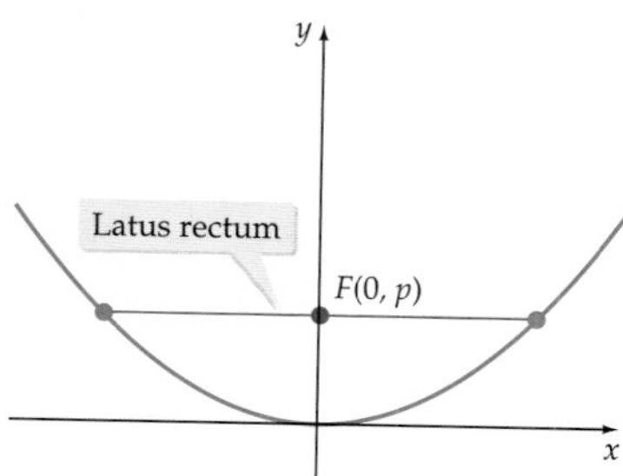

49. Find the length of the latus rectum for the parabola $x^2 = 4y$.

50. Find the length of the latus rectum for the parabola $y^2 = -8x$.

51. Find the length of the latus rectum for any parabola in terms of $|p|$, the distance from the vertex of the parabola to its focus.

The result of Exercise 51 can be stated as the following theorem: Two points on a parabola will be $2|p|$ units on each side of the axis of symmetry on the line through the focus and perpendicular to that axis.

52. Use the theorem to sketch a graph of the parabola given by the equation $(x - 3)^2 = 2(y + 1)$.

53. Use the theorem to sketch a graph of the parabola given by the equation $(y + 4)^2 = -(x - 1)$.

54. By using the definition of a parabola, find the equation in standard form of the parabola with $V(0, 0)$, $F(-c, 0)$, and directrix $x = c$.

55. Sketch a graph of $4(y - 2) = x|x| - 1$.

56. Find the equation of the directrix of the parabola with vertex at the origin and focus at the point $(1, 1)$.

57. Find the equation of the parabola with vertex at the origin and focus at the point $(1, 1)$. (*Hint:* You will need the answer to Exercise 56 and the definition of a parabola.)

Projects

1. 3-D OPTICAL ILLUSION In the photo below, a strawberry appears to float above a black disc.

Mirage by InnovaToys
Photo source: http://innovatoys.com/c/OPT

If you try to touch the strawberry, you will discover that the strawberry is not located where it appears. Search the Internet to discover how parabolic mirrors are used to create this illusion. Write a few sentences, along with a diagram that explains how the illusion is accomplished.

Section 6.2

- Ellipses with Center at (0, 0)
- Ellipses with Center at (*h*, *k*)
- Eccentricity of an Ellipse
- Applications

Ellipses

PREPARE FOR THIS SECTION

Prepare for this section by completing the following exercises. The answers can be found on page A25.

PS1. Find the midpoint and the length of the line segment between $P_1(5, 1)$ and $P_2(-1, 5)$. [1.2]

PS2. Solve: $x^2 + 6x - 16 = 0$ [1.1]

PS3. Solve: $x^2 - 2x = 2$ [1.1]

PS4. Complete the square of $x^2 - 8x$ and write the result as the square of a binomial. [1.2]

PS5. Solve $(x - 2)^2 + y^2 = 4$ for y. [1.2]

PS6. Graph: $(x - 2)^2 + (y + 3)^2 = 16$ [1.2]

An ellipse is another of the conic sections formed when a plane intersects a right circular cone. If β is the angle at which the plane intersects the axis of the cone and

take note

If a plane intersects a cone at the vertex of the cone so that the resulting figure is a point, the point is called a *degenerate ellipse*. See the accompanying figure.

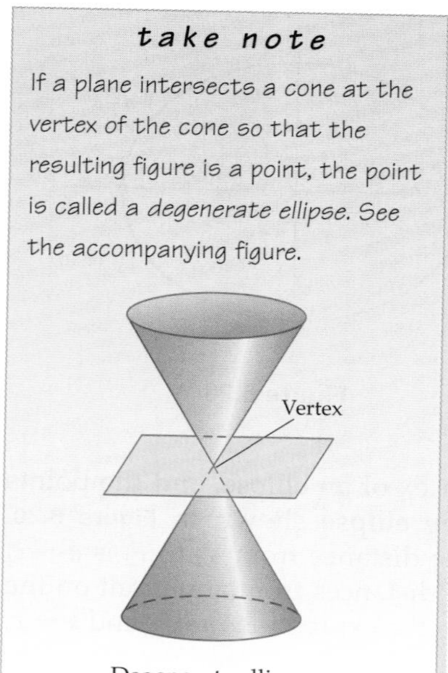

Degenerate ellipse

α is the angle shown in **Figure 6.17**, an ellipse is formed when $\alpha < \beta < 90°$. If $\beta = 90°$, then a circle is formed.

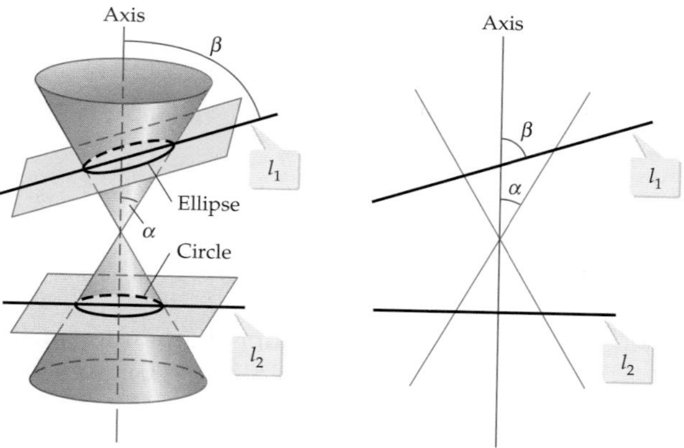

Figure 6.17

As was the case for a parabola, there is a definition for an ellipse in terms of a certain set of points in the plane.

Definition of an Ellipse

An **ellipse** is the set of all points in the plane the sum of whose distances from two fixed points (**foci**) is a positive constant.

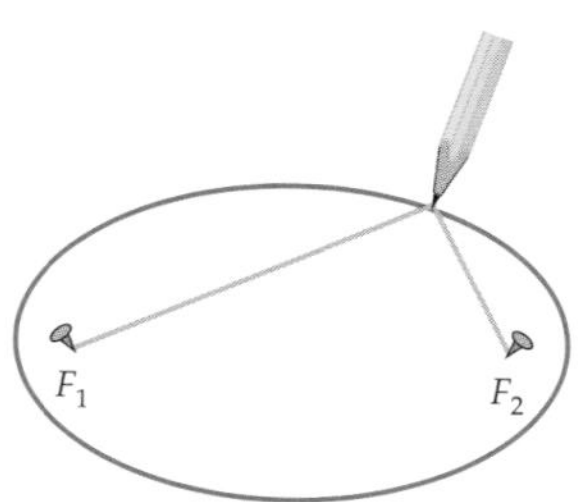

Figure 6.18

Equipped only with a piece of string and two tacks, we can use this definition to draw an ellipse (see **Figure 6.18**). Tack the ends of the string to the foci, and trace a curve with a pencil held tight against the string. The resulting curve is an ellipse. The positive constant mentioned in the definition of an ellipse is the length of the string.

Ellipses with Center at (0, 0)

The graph of an ellipse has two axes of symmetry (see **Figure 6.19**). The longer axis is called the **major axis.** The foci of the ellipse are on the major axis. The shorter axis is called the **minor axis.** It is customary to denote the length of the major axis by $2a$ and the length of the minor axis by $2b$. The **center** of the ellipse is the midpoint of the major axis. The endpoints of the major axis are the **vertices** (plural of *vertex*) of the ellipse.

A **semimajor axis** of an ellipse is a line segment that connects the center point of the ellipse with a vertex. Its length is half the length of the major axis. A **semiminor axis** of an ellipse is a line segment that lies on the minor axis and connects the center point with a point on the ellipse. Its length is half the length of the minor axis.

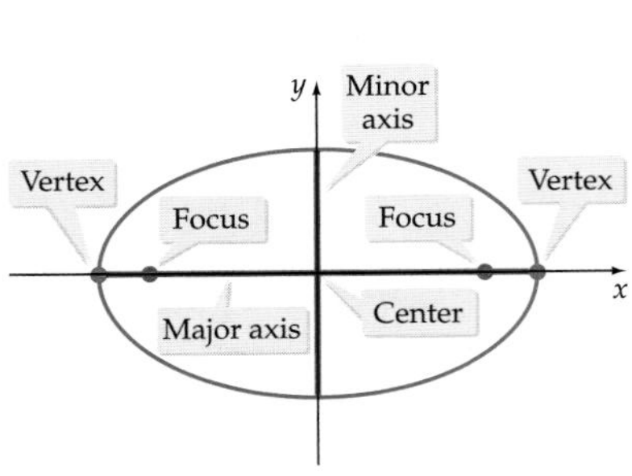

Figure 6.19

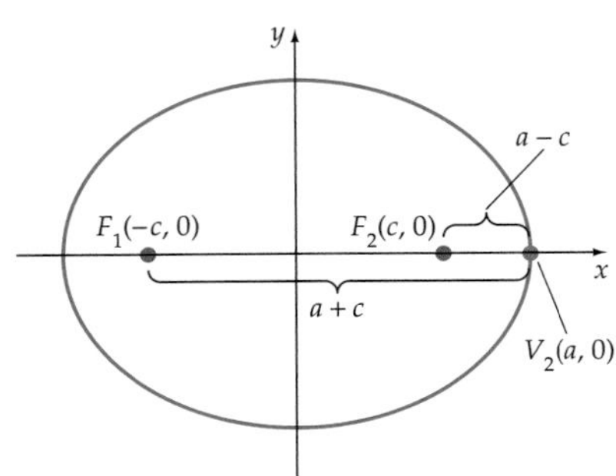

Figure 6.20

Consider the point $V_2(a, 0)$, which is one vertex of an ellipse, and the points $F_2(c, 0)$ and $F_1(-c, 0)$, which are the foci of the ellipse shown in **Figure 6.20**. The distance from V_2 to F_1 is $a + c$. Similarly, the distance from V_2 to F_2 is $a - c$. From the definition of an ellipse, the sum of the distances from any point on the ellipse to the foci is a positive constant. By adding the expressions $a + c$ and $a - c$, we have

$$(a + c) + (a - c) = 2a$$

Thus the positive constant referred to in the definition of an ellipse is $2a$, the length of the major axis.

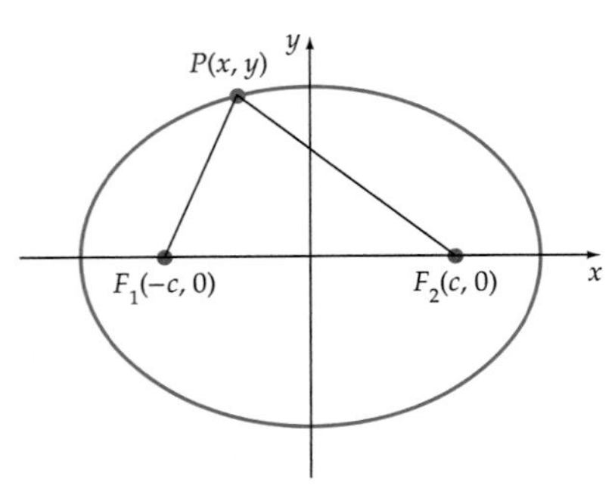

Figure 6.21

Now let $P(x, y)$ be any point on the ellipse (see **Figure 6.21**). By using the definition of an ellipse, we have

$$d(P, F_1) + d(P, F_2) = 2a$$
$$\sqrt{(x + c)^2 + y^2} + \sqrt{(x - c)^2 + y^2} = 2a$$

Subtract the second radical from each side of the equation, and then square each side.

$$\left[\sqrt{(x + c)^2 + y^2}\right]^2 = \left[2a - \sqrt{(x - c)^2 + y^2}\right]^2$$
$$(x + c)^2 + y^2 = 4a^2 - 4a\sqrt{(x - c)^2 + y^2} + (x - c)^2 + y^2$$
$$x^2 + 2cx + c^2 + y^2 = 4a^2 - 4a\sqrt{(x - c)^2 + y^2} + x^2 - 2cx + c^2 + y^2$$
$$4cx - 4a^2 = -4a\sqrt{(x - c)^2 + y^2}$$
$$[-cx + a^2]^2 = \left[a\sqrt{(x - c)^2 + y^2}\right]^2$$

• **Divide by −4, and then square each side.**

$$c^2x^2 - 2cxa^2 + a^4 = a^2x^2 - 2cxa^2 + a^2c^2 + a^2y^2$$
$$-a^2x^2 + c^2x^2 - a^2y^2 = -a^4 + a^2c^2$$

• **Rewrite with x and y terms on the left side.**

$$-(a^2 - c^2)x^2 - a^2y^2 = -a^2(a^2 - c^2)$$

• **Factor.**

$$-b^2x^2 - a^2y^2 = -a^2b^2$$

• **Let $b^2 = a^2 - c^2$.**

$$\frac{x^2}{a^2} + \frac{y^2}{b^2} = 1$$

• **Divide each side by $-a^2b^2$. The result is an equation of an ellipse with center at (0, 0).**

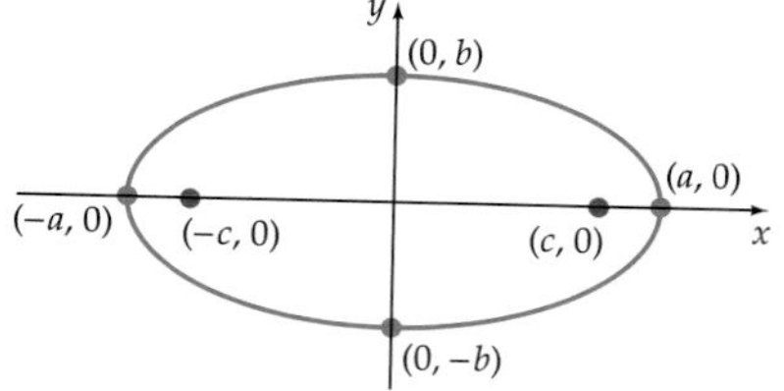

a. Major axis on x-axis

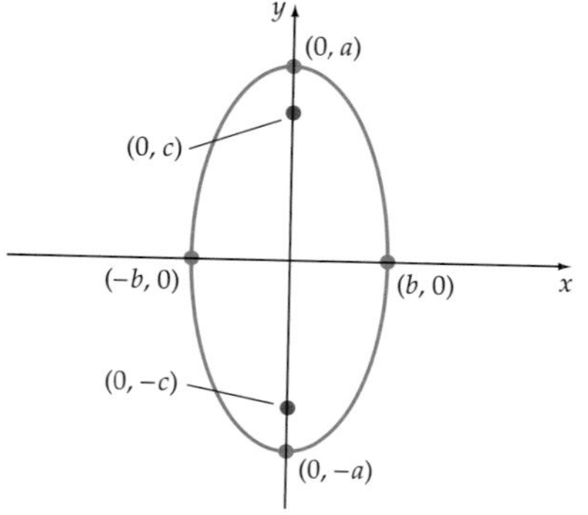

b. Major axis on y-axis

Figure 6.22

Standard Forms of the Equation of an Ellipse with Center at the Origin

Major Axis on the x-Axis

The standard form of the equation of an ellipse with center at the origin and major axis on the x-axis (see **Figure 6.22a**) is given by

$$\frac{x^2}{a^2} + \frac{y^2}{b^2} = 1, \quad a > b$$

The length of the major axis is $2a$. The length of the minor axis is $2b$. The coordinates of the vertices are $(a, 0)$ and $(-a, 0)$, and the coordinates of the foci are $(c, 0)$ and $(-c, 0)$, where $c^2 = a^2 - b^2$.

Major Axis on the y-Axis

The standard form of the equation of an ellipse with center at the origin and major axis on the y-axis (see **Figure 6.22b**) is given by

$$\frac{x^2}{b^2} + \frac{y^2}{a^2} = 1, \quad a > b$$

The length of the major axis is $2a$. The length of the minor axis is $2b$. The coordinates of the vertices are $(0, a)$ and $(0, -a)$, and the coordinates of the foci are $(0, c)$ and $(0, -c)$, where $c^2 = a^2 - b^2$.

? QUESTION For the graph of $\frac{x^2}{16} + \frac{y^2}{25} = 1$, is the major axis on the x-axis or the y-axis?

EXAMPLE 1 ⟫ Find the Vertices and Foci of an Ellipse

Find the vertices and foci of the ellipse given by the equation $\frac{x^2}{25} + \frac{y^2}{49} = 1$. Sketch the graph.

Solution

Because the y^2 term has the larger denominator, the major axis is on the y-axis.

$$a^2 = 49 \qquad b^2 = 25 \qquad c^2 = a^2 - b^2$$
$$a = 7 \qquad b = 5 \qquad = 49 - 25 = 24$$
$$c = \sqrt{24} = 2\sqrt{6}$$

Continued ▶

? ANSWER Because $25 > 16$, the major axis is on the y-axis.

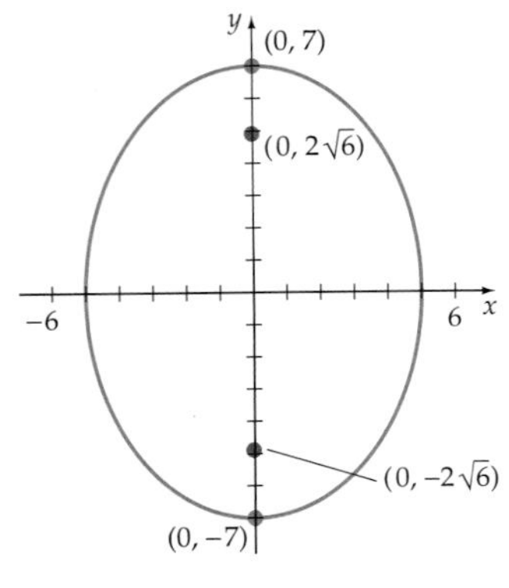

$\frac{x^2}{25} + \frac{y^2}{49} = 1$

Figure 6.23

The vertices are $(0, 7)$ and $(0, -7)$. The foci are $(0, 2\sqrt{6})$ and $(0, -2\sqrt{6})$. See **Figure 6.23.**

Try Exercise 22, page 385

An ellipse with foci $(3, 0)$ and $(-3, 0)$ and major axis of length 10 is shown in **Figure 6.24.** To find the equation of the ellipse in standard form, we must find a^2 and b^2. Because the foci are on the major axis, the major axis is on the x-axis. The length of the major axis is $2a$. Thus $2a = 10$. Solving for a, we have $a = 5$ and $a^2 = 25$.

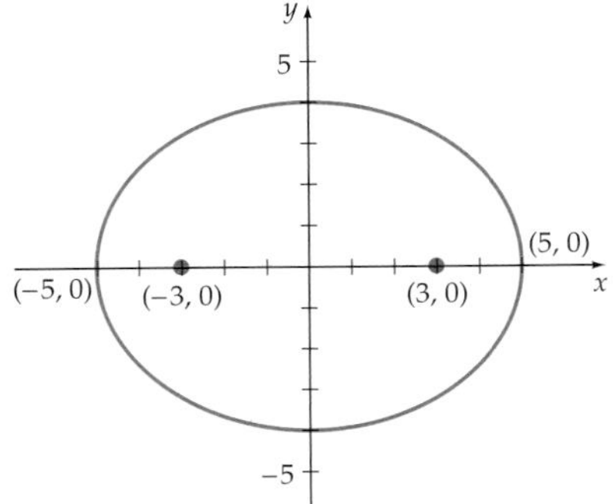

$\frac{x^2}{25} + \frac{y^2}{16} = 1$

Figure 6.24

Because the foci are $(3, 0)$ and $(-3, 0)$ and the center of the ellipse is the midpoint between the two foci, the distance from the center of the ellipse to a focus is 3. Therefore, $c = 3$. To find b^2, use the equation

$$c^2 = a^2 - b^2$$
$$9 = 25 - b^2$$
$$b^2 = 16$$

The equation of the ellipse in standard form is $\frac{x^2}{25} + \frac{y^2}{16} = 1$.

Ellipses with Center at (h, k)

The equation of an ellipse with center at (h, k) and with a horizontal or vertical major axis can be found by using a translation of coordinates. On a coordinate system with axes labeled x' and y', the standard form of the equation of an ellipse with center at the origin of the $x'y'$-coordinate system is

$$\frac{(x')^2}{a^2} + \frac{(y')^2}{b^2} = 1$$

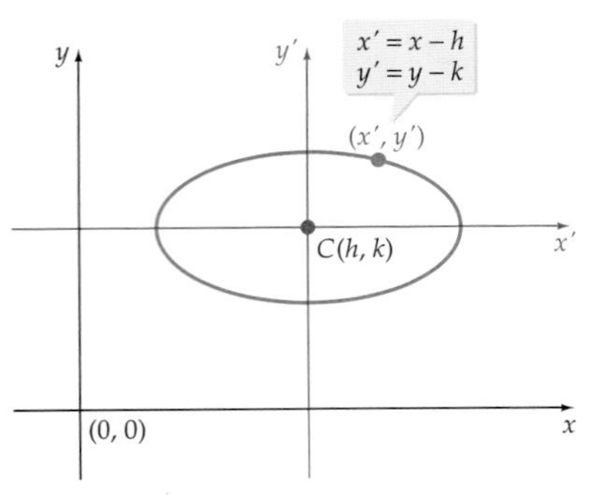

Figure 6.25

Now place the origin of the $x'y'$-coordinate system at (h, k) in an xy-coordinate system. See **Figure 6.25.**

The relationship between an ordered pair in the $x'y'$-coordinate system and one in the xy-coordinate system is given by the transformation equations

$$x' = x - h$$
$$y' = y - k$$

Substitute the expressions for x' and y' into the equation of an ellipse. The equation of the ellipse with center at (h, k) is

$$\frac{(x - h)^2}{a^2} + \frac{(y - k)^2}{b^2} = 1$$

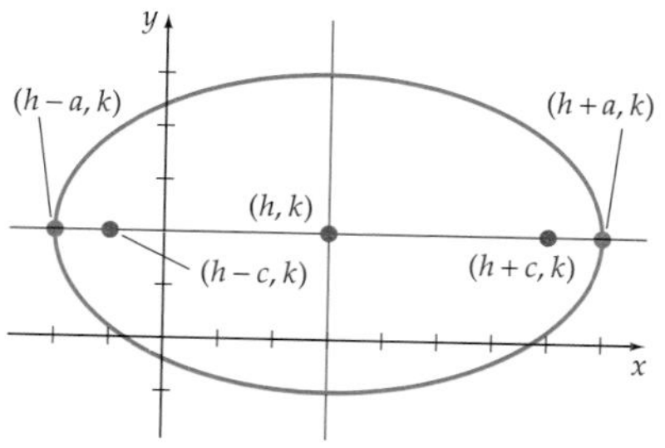

a. Major axis parallel to x-axis

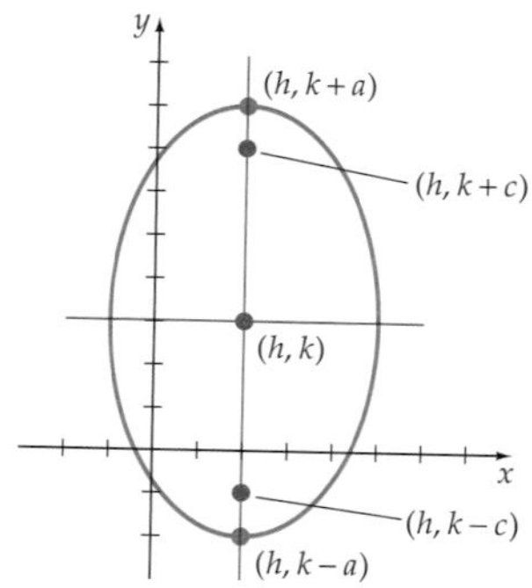

b. Major axis parallel to y-axis

Figure 6.26

Standard Forms of the Equation of an Ellipse with Center at (h, k)

Major Axis Parallel to the x-Axis

The standard form of the equation of an ellipse with center at (h, k) and major axis parallel to the x-axis (see **Figure 6.26a**) is given by

$$\frac{(x-h)^2}{a^2} + \frac{(y-k)^2}{b^2} = 1, \quad a > b$$

The length of the major axis is $2a$. The length of the minor axis is $2b$. The coordinates of the vertices are $(h + a, k)$ and $(h - a, k)$, and the coordinates of the foci are $(h + c, k)$ and $(h - c, k)$, where $c^2 = a^2 - b^2$.

Major Axis Parallel to the y-Axis

The standard form of the equation of an ellipse with center at (h, k) and major axis parallel to the y-axis (see **Figure 6.26b**) is given by

$$\frac{(x-h)^2}{b^2} + \frac{(y-k)^2}{a^2} = 1, \quad a > b$$

The length of the major axis is $2a$. The length of the minor axis is $2b$. The coordinates of the vertices are $(h, k + a)$ and $(h, k - a)$, and the coordinates of the foci are $(h, k + c)$ and $(h, k - c)$, where $c^2 = a^2 - b^2$.

EXAMPLE 2 ❯❯ Find the Center, Vertices, and Foci of an Ellipse

Find the center, vertices, and foci of the ellipse $4x^2 + 9y^2 - 8x + 36y + 4 = 0$. Sketch the graph.

Solution

Write the equation of the ellipse in standard form by completing the square.

$$4x^2 + 9y^2 - 8x + 36y + 4 = 0$$

$$4x^2 - 8x + 9y^2 + 36y = -4 \qquad \text{• Rearrange terms.}$$

$$4(x^2 - 2x) + 9(y^2 + 4y) = -4 \qquad \text{• Factor.}$$

$$4(x^2 - 2x + 1) + 9(y^2 + 4y + 4) = -4 + 4 + 36 \qquad \text{• Complete the square.}$$

$$4(x-1)^2 + 9(y+2)^2 = 36 \qquad \text{• Factor.}$$

$$\frac{(x-1)^2}{9} + \frac{(y+2)^2}{4} = 1 \qquad \text{• Divide each side by 36.}$$

From the equation of the ellipse in standard form, the coordinates of the center of the ellipse are $(1, -2)$. Because the larger denominator is 9, the major axis is parallel to the x-axis and $a^2 = 9$. Thus $a = 3$. The vertices are $(4, -2)$ and $(-2, -2)$.

y
x
$V_2(-2, -2)$
$C(1, -2)$
$V_1(4, -2)$
$F_2(1 - \sqrt{5}, -2)$
$F_1(1 + \sqrt{5}, -2)$

$$\frac{(x-1)^2}{9} + \frac{(y+2)^2}{4} = 1$$

Figure 6.27

Continued ►

To find the coordinates of the foci, we find c.

$$c^2 = a^2 - b^2 = 9 - 4 = 5$$
$$c = \sqrt{5}$$

The foci are $(1 + \sqrt{5}, -2)$ and $(1 - \sqrt{5}, -2)$. See **Figure 6.27.**

Try Exercise 28, page 385

Integrating Technology

A graphing utility can be used to graph an ellipse. For instance, consider the equation $4x^2 + 9y^2 - 8x + 36y + 4 = 0$ from Example 2. Rewrite the equation as

$$9y^2 + 36y + (4x^2 - 8x + 4) = 0$$

In this form, the equation is a quadratic equation in terms of the variable y with

$$A = 9, B = 36, \text{ and } C = 4x^2 - 8x + 4$$

TO REVIEW

Quadratic Formula
See page 5.

Apply the quadratic formula to produce

$$y = \frac{-36 \pm \sqrt{1296 - 36(4x^2 - 8x + 4)}}{18}$$

The graph of $Y_1 = \dfrac{-36 + \sqrt{1296 - 36(4x^2 - 8x + 4)}}{18}$ is the part of the ellipse on or above the line $y = -2$ (see **Figure 6.28**).

The graph of $Y_2 = \dfrac{-36 - \sqrt{1296 - 36(4x^2 - 8x + 4)}}{18}$ is the part of the ellipse on or below the line $y = -2$, as shown in **Figure 6.28.**

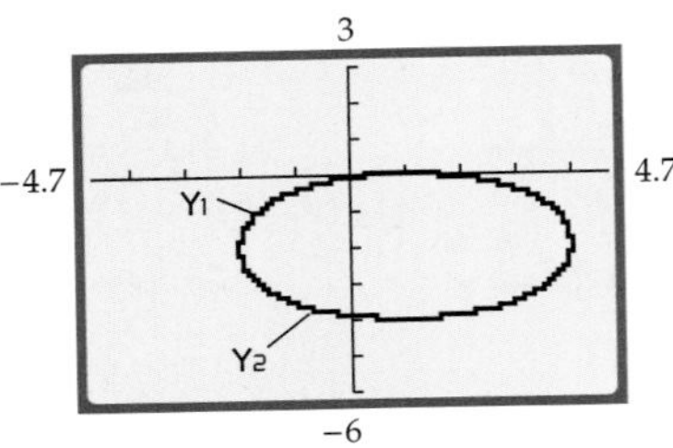

Figure 6.28

One advantage of this graphing procedure is that it does not require us to write the given equation in standard form. A disadvantage of the graphing procedure is that it does not indicate where the foci of the ellipse are located.

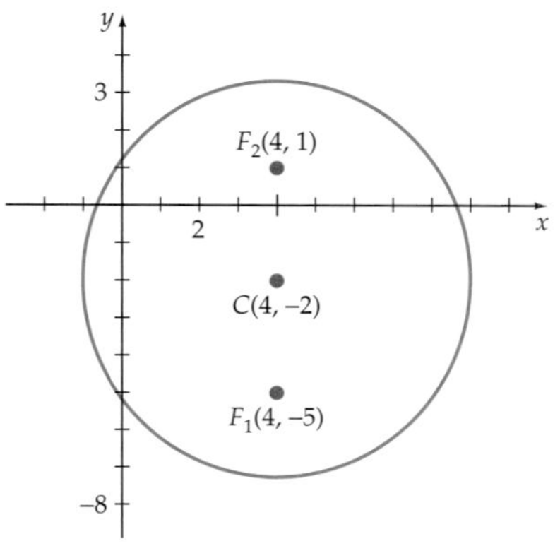

Figure 6.29

EXAMPLE 3 Find the Equation of an Ellipse

Find the standard form of the equation of the ellipse with center at $(4, -2)$, foci $F_2(4, 1)$ and $F_1(4, -5)$, and minor axis of length 10, as shown in **Figure 6.29.**

Solution

Because the foci are on the major axis, the major axis is parallel to the y-axis. The distance from the center of the ellipse to a focus is c. The distance between the center $(4, -2)$ and the focus $(4, 1)$ is 3. Therefore, $c = 3$.

The length of the minor axis is $2b$. Thus $2b = 10$ and $b = 5$.

To find a^2, use the equation $c^2 = a^2 - b^2$.

$$9 = a^2 - 25$$
$$a^2 = 34$$

Thus the equation in standard form is

$$\frac{(x-4)^2}{25} + \frac{(y+2)^2}{34} = 1$$

Try Exercise 44, page 386

Eccentricity of an Ellipse

take note

Eccentric literally means "out of the center." Eccentricity is a measure of how much an ellipse is unlike a set of points the same distance from the center. The higher the eccentricity, the more unlike a circle the ellipse is, and therefore the longer and thinner it is. A circle is also a conic section. Its standard form is given on page 25.

The graph of an ellipse can be very long and thin, or it can be much like a circle. The **eccentricity** of an ellipse is a measure of its "roundness."

Eccentricity (e) of an Ellipse

The eccentricity e of an ellipse is the ratio of c to a, where c is the distance from the center to a focus and a is one-half the length of the major axis. (See **Figure 6.30.**) That is,

$$e = \frac{c}{a}$$

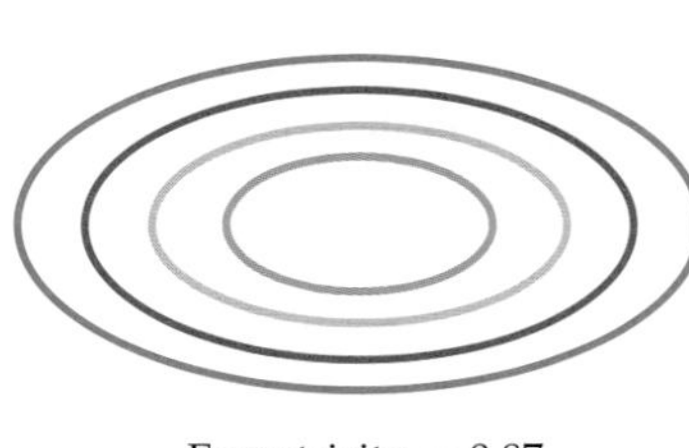

Eccentricity = 0.87

Figure 6.30

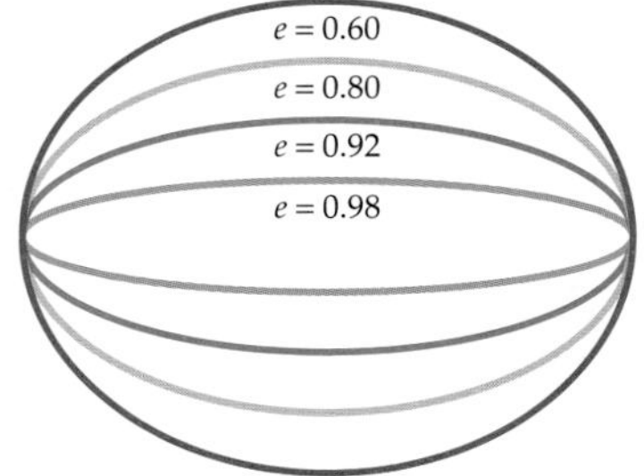

Figure 6.31

Because $c < a$, for an ellipse, $0 < e < 1$. When $e \approx 0$, the graph is almost a circle. When $e \approx 1$, the graph is long and thin. See **Figure 6.31.**

EXAMPLE 4 Find the Eccentricity of an Ellipse

Find the eccentricity of the ellipse given by $8x^2 + 9y^2 = 18$.

Solution

First, write the equation of the ellipse in standard form. Divide each side of the equation by 18.

$$\frac{8x^2}{18} + \frac{9y^2}{18} = 1$$

$$\frac{4x^2}{9} + \frac{y^2}{2} = 1$$

$$\frac{x^2}{9/4} + \frac{y^2}{2} = 1 \qquad \bullet\ \frac{4}{9} = \frac{1}{9/4}$$

The last step is necessary because the standard form of the equation has coefficients of 1 in the numerator. Thus

$$a^2 = \frac{9}{4} \quad \text{and} \quad a = \frac{3}{2}$$

Use the equation $c^2 = a^2 - b^2$ to find c.

$$c^2 = \frac{9}{4} - 2 = \frac{1}{4} \quad \text{and} \quad c = \sqrt{\frac{1}{4}} = \frac{1}{2}$$

Now find the eccentricity.

$$e = \frac{c}{a} = \frac{1/2}{3/2} = \frac{1}{3}$$

The eccentricity of the ellipse is $\frac{1}{3}$.

Try Exercise 50, page 386

Table 6.1

Planet	Eccentricity
Mercury	0.206
Venus	0.007
Earth	0.017
Mars	0.093
Jupiter	0.049
Saturn	0.051
Uranus	0.046
Neptune	0.005

Applications

The planets travel around the sun in elliptical orbits. The sun is located at a focus of the orbit. The eccentricities of the orbits for the planets in our solar system are given in **Table 6.1**.

QUESTION Which planet has the most nearly circular orbit?

ANSWER Neptune has the smallest eccentricity, so it is the planet with the most nearly circular orbit.

The terms *perihelion* and *aphelion* are used to denote the position of a planet in its orbit around the sun. The perihelion is the point nearest the sun; the aphelion is the point farthest from the sun. See **Figure 6.32**. The length of the semimajor axis of a planet's elliptical orbit is called the *mean distance* of the planet from the sun.

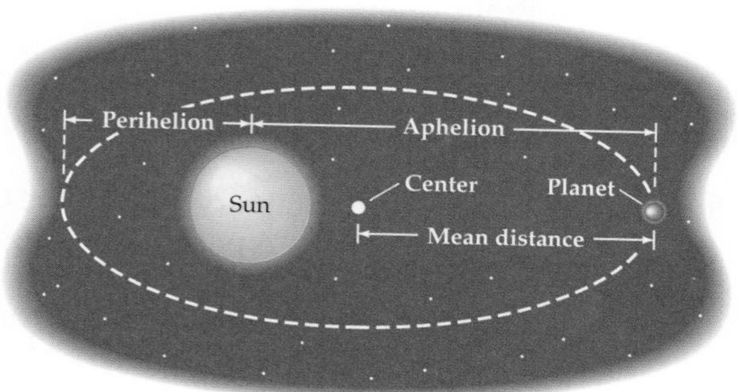

Figure 6.32

EXAMPLE 5 Determine an Equation for the Orbit of Earth

Earth has a mean distance of 93 million miles and a perihelion distance of 91.5 million miles. Find an equation for Earth's orbit.

Solution

A mean distance of 93 million miles implies that the length of the semimajor axis of the orbit is $a = 93$ million miles. Earth's aphelion distance is the length of the major axis less the length of the perihelion distance. Thus

$$\text{Aphelion distance} = 2(93) - 91.5 = 94.5 \text{ million miles}$$

The distance c from the sun to the center of Earth's orbit is

$$c = \text{aphelion distance} - 93 = 94.5 - 93 = 1.5 \text{ million miles}$$

The length b of the semiminor axis of the orbit is

$$b = \sqrt{a^2 - c^2} = \sqrt{93^2 - 1.5^2} = \sqrt{8646.75}$$

An equation of Earth's orbit is

$$\frac{x^2}{93^2} + \frac{y^2}{8646.75} = 1$$

Try Exercise 58, page 387

Sound waves, although different from light waves, have a similar reflective property. When sound is reflected from a point P on a surface, the angle of incidence equals the angle of reflection. Applying this principle to a room with an elliptical ceiling results in what are called whispering galleries. These galleries are based on the following theorem.

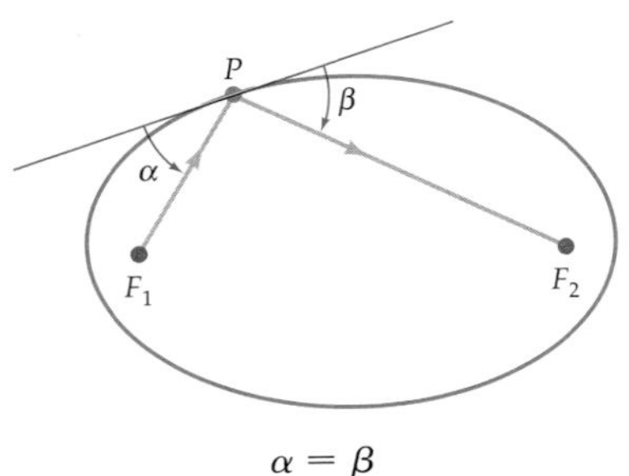

$\alpha = \beta$

Figure 6.33

Reflective Property of an Ellipse

The lines from the foci to a point on an ellipse make equal angles with the tangent line at that point. See **Figure 6.33.**

The reflective property of an ellipse can be used to show that sound waves, or light waves, that emanate from one focus of an ellipse will be reflected to the other focus.

The Statuary Hall in the Capitol Building in Washington, D.C., is a whispering gallery. A person standing at one focus of the elliptical ceiling can whisper and be heard by a person standing at the other focus. John Quincy Adams, while a member of the House of Representatives, was aware of this acoustical phenomenon. He situated his desk at a focus of the elliptical ceiling, which allowed him to eavesdrop on the conversations of his political adversaries, who were located near the other focus.

EXAMPLE 6 ›› Locate the Foci of a Whispering Gallery

A room 88 feet long is constructed to be a whispering gallery. The room has an elliptical ceiling, as shown in **Figure 6.34**. If the maximum height of the ceiling is 22 feet, determine where the foci are located.

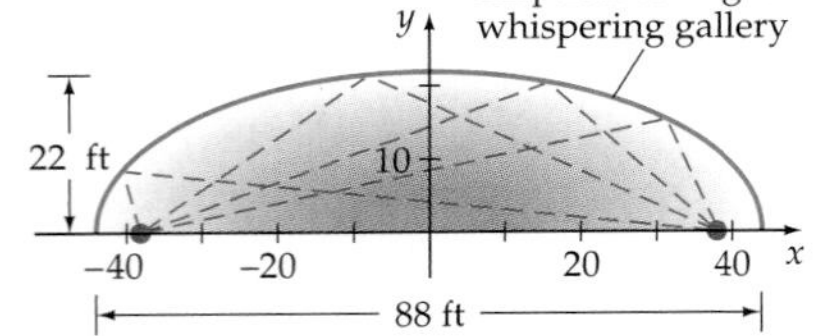

Figure 6.34

Solution

The length a of the semimajor axis of the elliptical ceiling is 44 feet. The height b of the semiminor axis is 22 feet. Thus

$$c^2 = a^2 - b^2$$
$$c^2 = 44^2 - 22^2$$
$$c = \sqrt{44^2 - 22^2} \approx 38.1 \text{ feet}$$

The foci are located about 38.1 feet from the center of the elliptical ceiling, along its major axis.

›› *Try Exercise 60, page 387*

Topics for Discussion

1. In every ellipse, the length of the semimajor axis a is greater than the length of the semiminor axis b and greater than the distance c from a focus to the center of the ellipse. Do you agree? Explain.
2. How many vertices does an ellipse have?
3. Every ellipse has two y-intercepts. Do you agree? Explain.
4. Explain why the eccentricity of every ellipse is a number between 0 and 1.

Exercise Set 6.2

1. Examine the following four equations and the graphs labeled **i, ii, iii,** and **iv.** Determine which graph is the graph of each equation.

a. $\dfrac{(x+1)^2}{9} + \dfrac{(y-2)^2}{1} = 1$ **b.** $\dfrac{x^2}{16} + \dfrac{y^2}{4} = 1$

c. $\dfrac{x^2}{4} + \dfrac{y^2}{9} = 1$ **d.** $\dfrac{(x+1)^2}{9} + \dfrac{(y-2)^2}{16} = 1$

i.

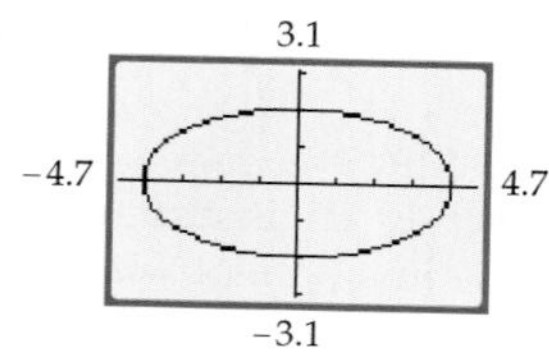

ii.

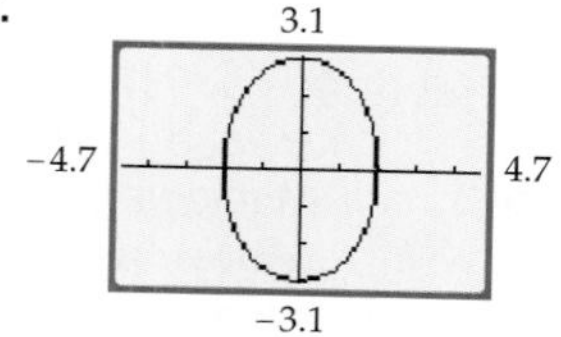

iii.

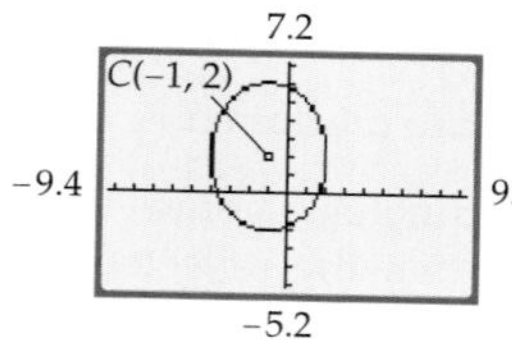

iv.

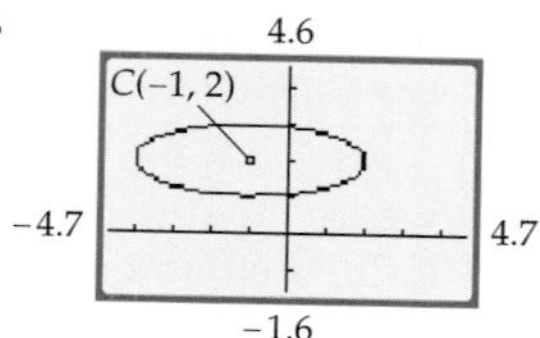

2. Examine the following four equations and the graphs labeled **i, ii, iii,** and **iv.** Determine which graph is the graph of each equation.

a. $\dfrac{(x-3)^2}{1} + \dfrac{(y-2)^2}{4} = 1$ **b.** $\dfrac{x^2}{9} + \dfrac{y^2}{4} = 1$

c. $\dfrac{(x-3)^2}{25} + \dfrac{(y-2)^2}{16} = 1$ **d.** $\dfrac{x^2}{4} + \dfrac{y^2}{16} = 1$

i.

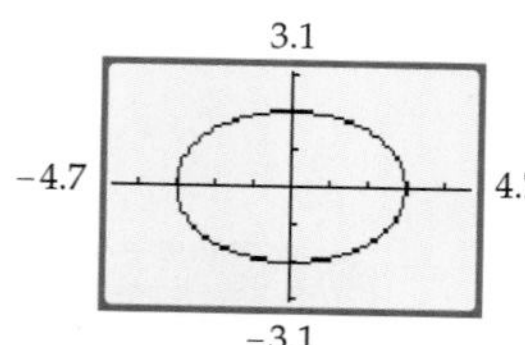

ii.

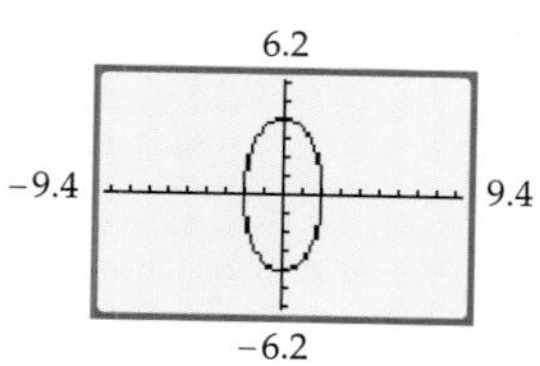

iii.

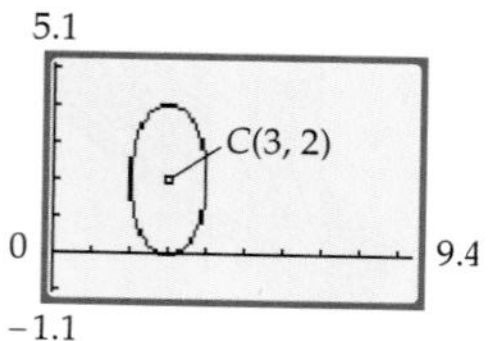

iv.

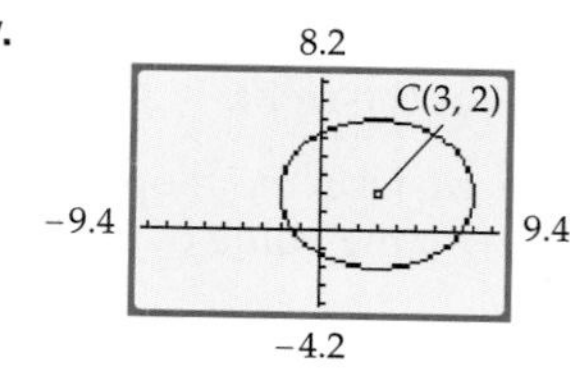

In Exercises 3 to 34, find the center, vertices, and foci of the ellipse given by each equation. Sketch the graph.

3. $\dfrac{x^2}{16} + \dfrac{y^2}{25} = 1$

4. $\dfrac{x^2}{49} + \dfrac{y^2}{36} = 1$

5. $\dfrac{x^2}{9} + \dfrac{y^2}{4} = 1$

6. $\dfrac{x^2}{64} + \dfrac{y^2}{25} = 1$

7. $\dfrac{x^2}{9} + \dfrac{y^2}{7} = 1$

8. $\dfrac{x^2}{5} + \dfrac{y^2}{4} = 1$

9. $\dfrac{4x^2}{9} + \dfrac{y^2}{16} = 1$

10. $\dfrac{x^2}{9} + \dfrac{9y^2}{16} = 1$

11. $\dfrac{(x-3)^2}{25} + \dfrac{(y+2)^2}{16} = 1$

12. $\dfrac{(x+3)^2}{9} + \dfrac{(y+1)^2}{16} = 1$

13. $\dfrac{(x+2)^2}{9} + \dfrac{y^2}{25} = 1$

14. $\dfrac{x^2}{25} + \dfrac{(y-2)^2}{81} = 1$

15. $\dfrac{(x-1)^2}{21} + \dfrac{(y-3)^2}{4} = 1$

16. $\dfrac{(x+5)^2}{9} + \dfrac{(y-3)^2}{7} = 1$

17. $\dfrac{9(x-1)^2}{16} + \dfrac{(y+1)^2}{9} = 1$

18. $\dfrac{(x+6)^2}{25} + \dfrac{25y^2}{144} = 1$

19. $3x^2 + 4y^2 = 12$

20. $5x^2 + 4y^2 = 20$

21. $25x^2 + 16y^2 = 400$

22. $25x^2 + 12y^2 = 300$

23. $64x^2 + 25y^2 = 400$

24. $9x^2 + 64y^2 = 144$

25. $4x^2 + y^2 - 24x - 8y + 48 = 0$

26. $x^2 + 9y^2 + 6x - 36y + 36 = 0$

27. $5x^2 + 9y^2 - 20x + 54y + 56 = 0$

28. $9x^2 + 16y^2 + 36x - 16y - 104 = 0$

29. $16x^2 + 9y^2 - 64x - 80 = 0$

30. $16x^2 + 9y^2 + 36y - 108 = 0$

31. $25x^2 + 16y^2 + 50x - 32y - 359 = 0$

32. $16x^2 + 9y^2 - 64x - 54y + 1 = 0$

33. $8x^2 + 25y^2 - 48x + 50y + 47 = 0$

34. $4x^2 + 9y^2 + 24x + 18y + 44 = 0$

In Exercises 35 to 46, find the equation in standard form of each ellipse, given the information provided.

35. Center $(0, 0)$, major axis of length 10, foci at $(4, 0)$ and $(-4, 0)$

36. Center $(0, 0)$, minor axis of length 6, foci at $(0, 4)$ and $(0, -4)$

37. Vertices $(6, 0)$, $(-6, 0)$; ellipse passes through $(0, -4)$ and $(0, 4)$

38. Vertices $(7, 0)$, $(-7, 0)$; ellipse passes through $(0, 5)$ and $(0, -5)$

39. Major axis of length 12 on the x-axis, center at $(0, 0)$; ellipse passes through $(2, -3)$

40. Major axis of length 8, center at $(0, 0)$; ellipse passes through $(-2, 2)$

41. Center $(-2, 4)$, vertices $(-6, 4)$ and $(2, 4)$, foci at $(-5, 4)$ and $(1, 4)$

42. Center $(0, 3)$, minor axis of length 4, foci at $(0, 0)$ and $(0, 6)$

43. Center $(2, 4)$, major axis parallel to the y-axis and of length 10; ellipse passes through the point $(3, 3)$

44. Center $(-4, 1)$, minor axis parallel to the y-axis and of length 8; ellipse passes through the point $(0, 4)$

45. Vertices $(5, 6)$ and $(5, -4)$, foci at $(5, 4)$ and $(5, -2)$

46. Vertices $(-7, -1)$ and $(5, -1)$, foci at $(-5, -1)$ and $(3, -1)$

In Exercises 47 to 54, use the eccentricity of each ellipse to find its equation in standard form.

47. Eccentricity $\frac{2}{5}$, major axis on the x-axis and of length 10, center at $(0, 0)$

48. Eccentricity $\frac{3}{4}$, foci at $(9, 0)$ and $(-9, 0)$

49. Foci at $(0, -4)$ and $(0, 4)$, eccentricity $\frac{2}{3}$

50. Foci at $(0, -3)$ and $(0, 3)$, eccentricity $\frac{1}{4}$

51. Eccentricity $\frac{2}{5}$, foci at $(-1, 3)$ and $(3, 3)$

52. Eccentricity $\frac{1}{4}$, foci at $(-2, 4)$ and $(-2, -2)$

53. Eccentricity $\frac{2}{3}$, major axis of length 24 on the y-axis, center at $(0, 0)$

54. Eccentricity $\frac{3}{5}$, major axis of length 15 on the x-axis, center at $(0, 0)$

55. **MEDICINE** A *lithotripter* is an instrument used to remove a kidney stone in a patient without having to do surgery. A high-frequency sound wave is emitted from a source that is located at the focus of an ellipse. The patient is placed so that the kidney stone is located at the other focus of the ellipse. If the equation of the ellipse is $\frac{(x - 11)^2}{484} + \frac{y^2}{64} = 1$ (x and y are measured in centimeters), where, to the nearest centimeter, should the patient's kidney stone be placed so that the reflected sound hits the kidney stone?

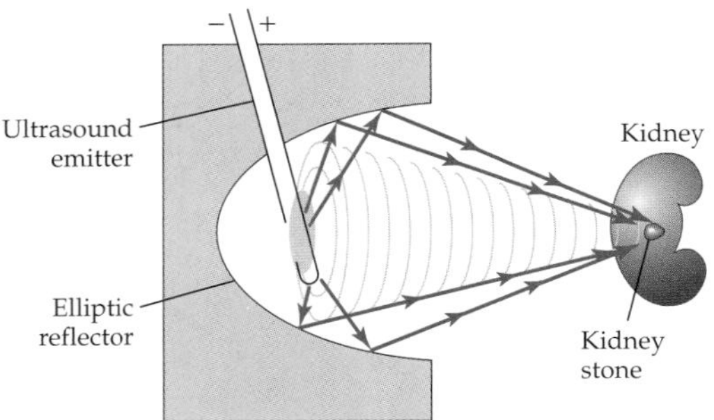

56. **CONSTRUCTION** A circular vent pipe is placed on a roof that has a slope of $\frac{4}{5}$, as shown in the figure at the right.

a. Use the slope to find the value of h.

b. The intersection of the vent pipe and the roof

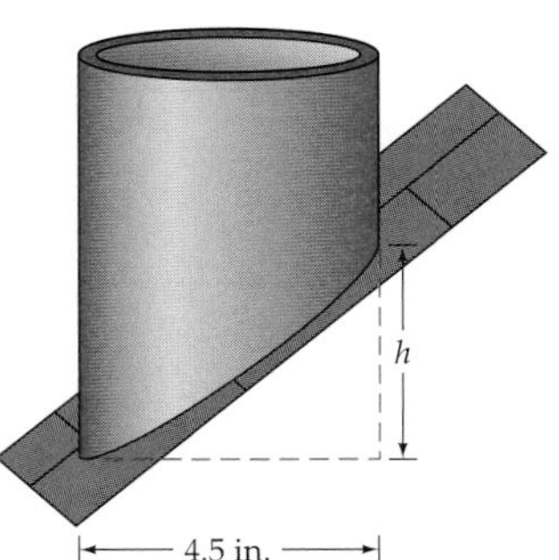

is an ellipse. To the nearest hundredth of an inch, what are the lengths of the major and minor axes?

c. Find an equation of the ellipse that should be cut from the roof so that the pipe will fit.

57. **The Orbit of Saturn** The distance from Saturn to the sun at Saturn's aphelion is 934.34 million miles, and the distance from Saturn to the sun at its perihelion is 835.14 million miles. Find an equation for the orbit of Saturn.

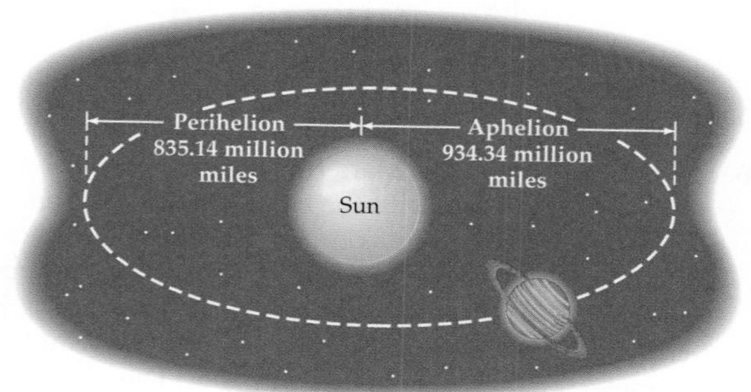

58. **The Orbit of Venus** Venus has a mean distance from the sun of 67.08 million miles, and the distance from Venus to the sun at its aphelion is 67.58 million miles. Find an equation for the orbit of Venus.

59. **Whispering Gallery** An architect wishes to design a large room that will be a whispering gallery. The ceiling of the room has a cross section that is an ellipse, as shown in the following figure.

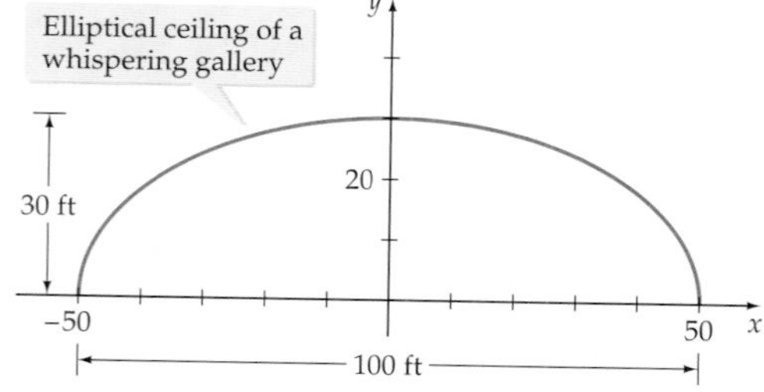

How far to the right and left of center are the foci located?

60. **Whispering Gallery** An architect wishes to design a large room 100 feet long that will be a whispering gallery. The ceiling of the room has a cross section that is an ellipse, as shown in the following figure.

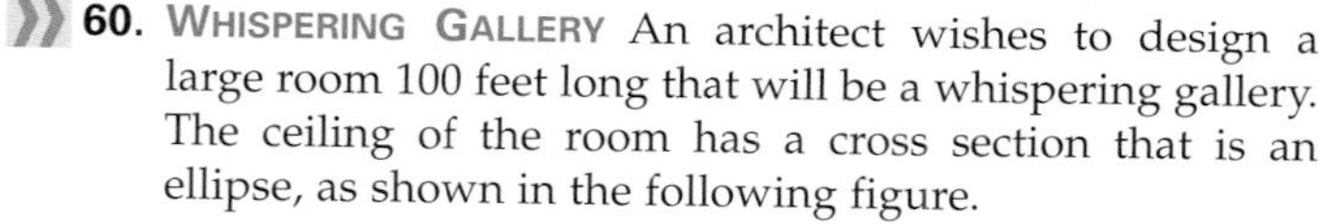

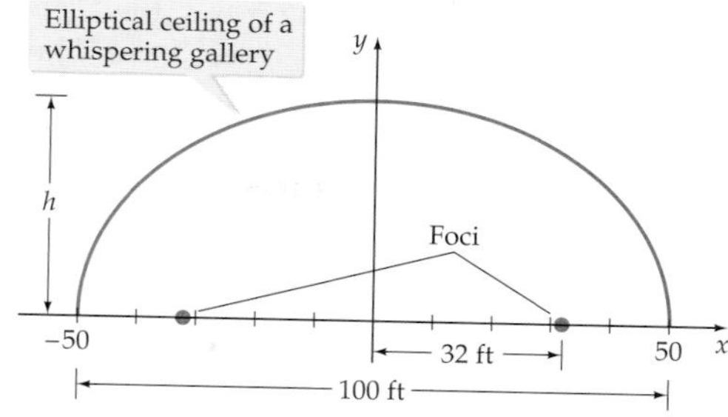

If the foci are to be located 32 feet to the right and the left of center, find the height h of the elliptical ceiling (to the nearest tenth of a foot).

61. **Halley's Comet** Find the equation of the path of Halley's comet in astronomical units by letting one focus (the sun) be at the origin and letting the other focus be on the positive x-axis. The length of the major axis of the orbit of Halley's comet is approximately 36 astronomical units (36 AU), and the length of the minor axis is 9 AU ($1 \text{ AU} = 92{,}960{,}000$ miles).

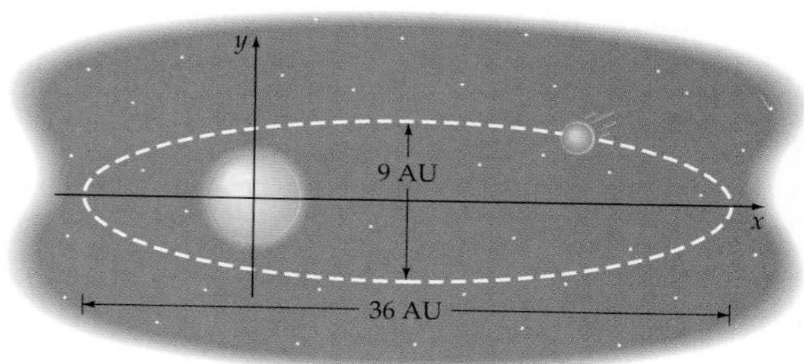

62. **Elliptical Pool Table** A pool table in the shape of an ellipse has only one pocket, which is located at a focus of the ellipse. A cue ball is placed at the other focus of the ellipse. Striking the cue ball firmly in any direction causes it to go into the pocket (assuming no side or back spin is introduced to the motion of the cue ball). Explain why this happens.

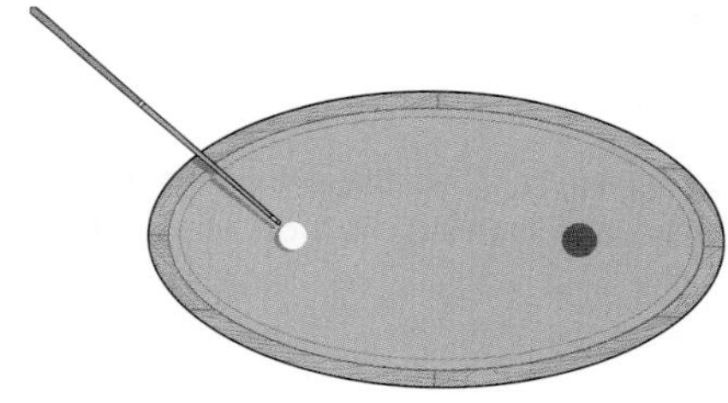

63. BRIDGE CLEARANCE During the 1960s, the London Bridge was dismantled and shipped to Lake Havasu City, Arizona, where it was reconstructed. Use the following diagram of an arch of the bridge to determine the vertical distance h from the water to the arch 55 feet to the right of point O, which is the center of the semielliptical arch. The distance from O to the water is 1 foot. Round to the nearest foot.

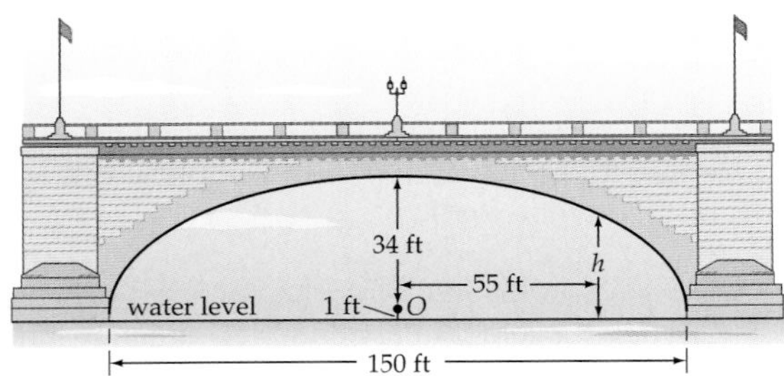

64. ELLIPTICAL GEARS The figure below shows two elliptical gears. Search the Internet for an animation that demonstrates the motion of an elliptical gear that is driven by another elliptical gear rotating at a constant speed. Explain what happens to the gear on the left as the gear on the right rotates once at a constant angular speed.

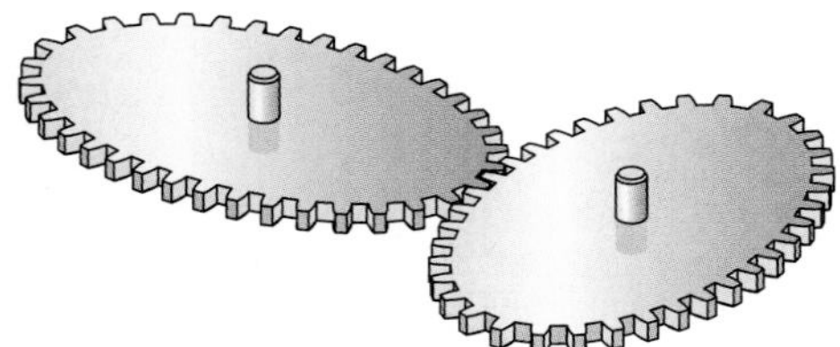

Source: http://www.stirlingsouth.com/richard/develop/engines_under_development.htm

65. CONSTRUCTION A carpenter needs to cut a semielliptical form from a 3-foot by 8-foot sheet of plywood, as shown in the following diagram.

a. Where should the carpenter place the push pins?

b. How long is the string that connects the push pins?

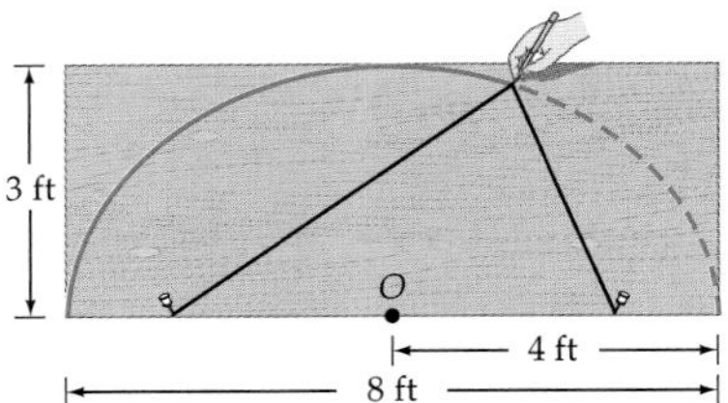

66. ORBIT OF MARS Mars travels around the sun in an elliptical orbit with the sun located at a focus of the orbit. The orbit has a major axis of 3.04 AU and a minor axis of 2.99 AU. (1 AU is 1 astronomical unit, or approximately 92,960,000 miles, the average distance of Earth from the sun.) Estimate, to the nearest million miles, the perimeter of the orbit of Mars. [*Note:* The approximate perimeter of an ellipse with semimajor axis a and semiminor axis b is $P = \pi\sqrt{2(a^2 + b^2)}$.]

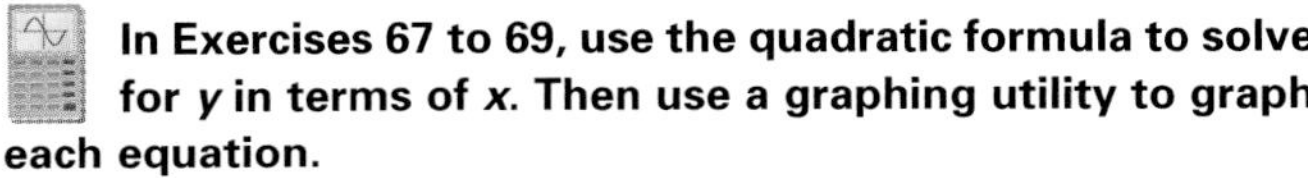

In Exercises 67 to 69, use the quadratic formula to solve for *y* in terms of *x*. Then use a graphing utility to graph each equation.

67. $16x^2 + 9y^2 + 36y - 108 = 0$

68. $8x^2 + 25y^2 - 48x + 50y + 47 = 0$

69. $4x^2 + 9y^2 + 24x + 18y + 44 = 0$

Connecting Concepts

70. Explain why the graph of $4x^2 + 9y - 16x - 2 = 0$ is or is not an ellipse. Sketch the graph of this equation.

In Exercises 71 to 74, find the equation in standard form of each ellipse by using the definition of an ellipse.

71. Find the equation of the ellipse with foci at $(-3, 0)$ and $(3, 0)$ that passes through the point $\left(3, \frac{9}{2}\right)$.

72. Find the equation of the ellipse with foci at $(0, 4)$ and $(0, -4)$ that passes through the point $\left(\frac{9}{5}, 4\right)$.

73. Find the equation of the ellipse with foci at $(-1, 2)$ and $(3, 2)$ that passes through the point $(3, 5)$.

74. Find the equation of the ellipse with foci at $(-1, 1)$ and $(-1, 7)$ that passes through the point $\left(\frac{3}{4}, 1\right)$.

In Exercises 75 and 76, find the latus rectum of the given ellipse. A line segment with endpoints on the ellipse that is perpendicular to the major axis and passes through a focus is a *latus rectum* of the ellipse.

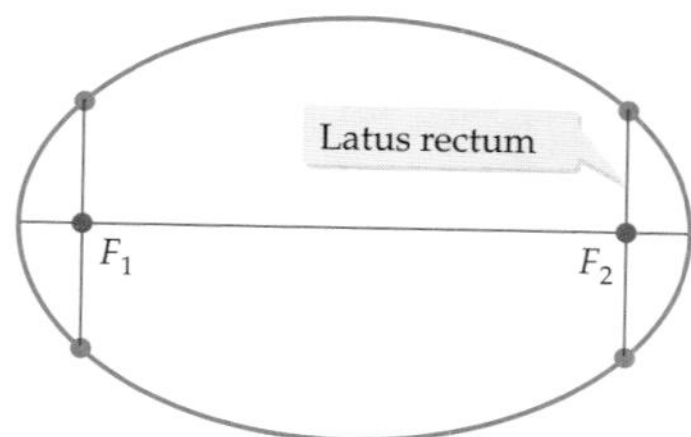

75. Find the length of a latus rectum of the ellipse given by
$$\frac{(x-1)^2}{9}+\frac{(y+1)^2}{16}=1$$

76. Find the length of a latus rectum of the ellipse given by
$$9x^2+16y^2-36x+96y+36=0$$

77. Show that for any ellipse, the length of a latus rectum is $\frac{2b^2}{a}$.

78. Use the definition of an ellipse to find the equation of an ellipse with center at (0, 0) and foci at $(0, c)$ and $(0, -c)$.

Projects

1. **Kepler's Laws** The German astronomer Johannes Kepler (1571–1630) derived three laws that describe how the planets orbit the sun. Write an essay that includes biographical information about Kepler and a statement of Kepler's Laws. In addition, use Kepler's Laws to answer the following questions.

a. Where is a planet located in its orbit around the sun when it achieves its greatest velocity?

b. What is the period of Mars if it has a mean distance from the sun of 1.52 astronomical units? (*Hint:* Use Earth as a reference with a period of 1 year and a mean distance from the sun of 1 astronomical unit.)

2. **Neptune** The position of the planet Neptune was discovered by using celestial mechanics and mathematics. Write an essay that tells how, when, and by whom Neptune was discovered.

3. **Graph the Colosseum** Some of the Colosseum scenes in the movie *Gladiator* (Universal Studios, 2000) were computer-generated.

a. You can create a simple but accurate scale image of the exterior of the Colosseum by using a computer and the mathematics software program *Maple*. Open a new *Maple* worksheet and enter the following two commands.

```
with(plots);
plots[implicitplot3d]((x^2)/(307.5^2)+(y^2)/(255^2)=
1, x= –310..310, y= –260..260, z=0..157,
scaling=CONSTRAINED, style=PATCHNOGRID,
axes=FRAMED);
```

Execute each of the commands by placing the cursor in a command and pressing the ENTER key. After execution of the second command, a three-dimensional graph will appear. Click and drag on the graph to rotate the image.

b. A graphing calculator can also be used to generate a simple "graph" of the exterior of the Colosseum. Here is a procedure for the TI-83/TI-83 Plus/TI-84 Plus graphing calculator.

Enter the following in the WINDOW menu.

Xmin=–4.7 Xmax=4.7 Xscl=1 Ymin=–4 Ymax=9 Yscl=1

Enter the following equations in the Y= menu.

$Y_1=\sqrt{(9-X^2)}$ $Y_2=Y_1+4$ $Y_3=-Y_1$ $Y_4=Y_3+4$

Press: QUIT (2nd MODE)

Enter: Shade(Y_3,Y_4) and press ENTER.
Note: "Shade(" is in the DRAW menu.

Explain why this "Colosseum graph" appears to be constructed with ellipses even though the functions entered in the Y= menu are the equations of semicircles.

Section 6.3

- Hyperbolas with Center at (0, 0)
- Hyperbolas with Center at (h, k)
- Eccentricity of a Hyperbola
- Applications

Hyperbolas

PREPARE FOR THIS SECTION

Prepare for this section by completing the following exercises. The answers can be found on page A27.

PS1. Find the midpoint and the length of the line segment between $P_1(4, -3)$ and $P_2(-2, 1)$. [1.2]

PS2. Solve: $(x - 1)(x + 3) = 5$ [1.1]

PS3. Simplify: $\frac{4}{\sqrt{8}}$

PS4. Complete the square of $4x^2 + 24x$ and write the result as the square of a binomial. [1.2]

PS5. Solve $\frac{x^2}{4} - \frac{y^2}{9} = 1$ for y. [1.2]

PS6. Graph: $\frac{(x-2)^2}{16} + \frac{(y+3)^2}{9} = 1$ [6.2]

---

take note

If a plane intersects a cone along the axis of the cone, the resulting curve is two intersecting straight lines. This is the *degenerate* form of a hyperbola. See the accompanying figure.

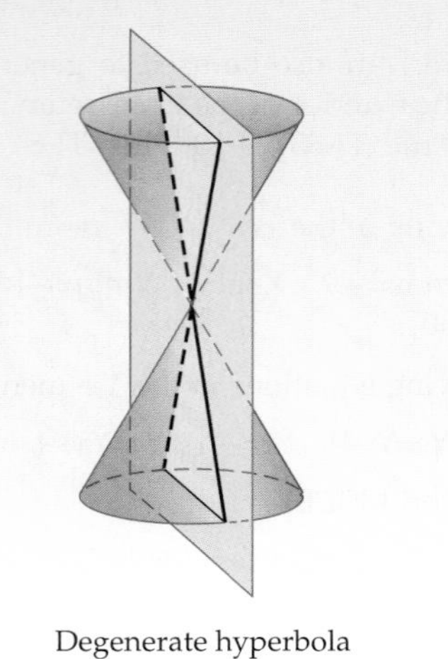

Degenerate hyperbola

A hyperbola is a conic section formed when a plane intersects a right circular cone at a certain angle. If β is the angle at which the plane intersects the axis of the cone and α is the angle shown in **Figure 6.35,** a hyperbola is formed when $0° < \beta < \alpha$ or when the plane is parallel to the axis of the cone.

As with the other conic sections, there is a definition of a hyperbola in terms of a certain set of points in the plane.

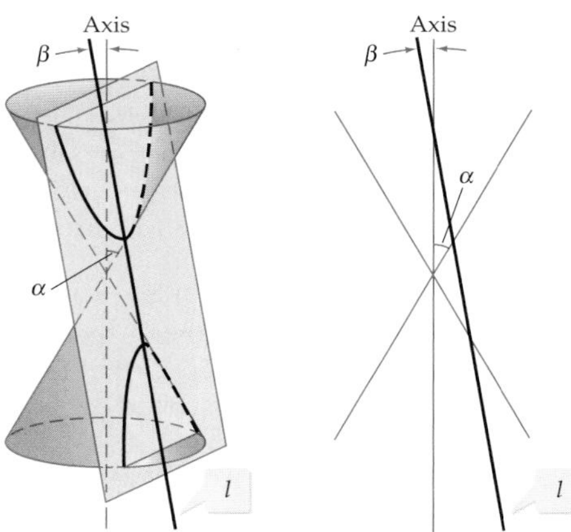

Figure 6.35

Definition of a Hyperbola

A **hyperbola** is the set of all points in the plane the difference between whose distances from two fixed points (**foci**) is a positive constant.

This definition differs from that of an ellipse in that the ellipse was defined in terms of the *sum* of two distances, whereas the hyperbola is defined in terms of the *difference* of two distances.

Hyperbolas with Center at (0, 0)

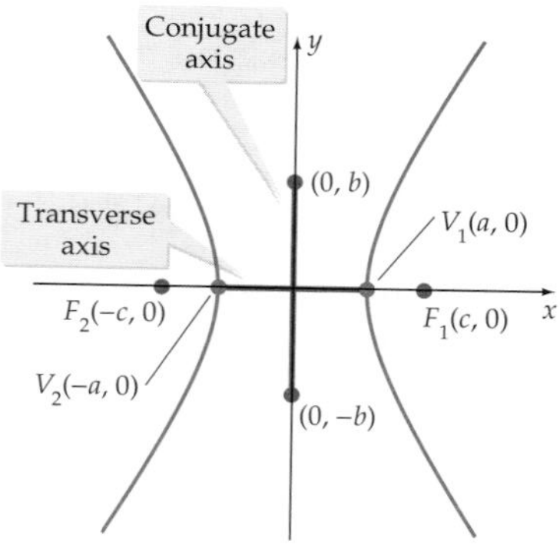

Figure 6.36

The **transverse axis** of a hyperbola shown in **Figure 6.36** is the line segment joining the intercepts. The midpoint of the transverse axis is called the **center** of the hyperbola. The **conjugate axis** is a line segment that passes through the center of the hyperbola and is perpendicular to the transverse axis.

The length of the transverse axis is customarily represented as $2a$, and the distance between the two foci is represented as $2c$. The length of the conjugate axis is represented as $2b$.

The **vertices** of a hyperbola are the points where the hyperbola intersects the transverse axis.

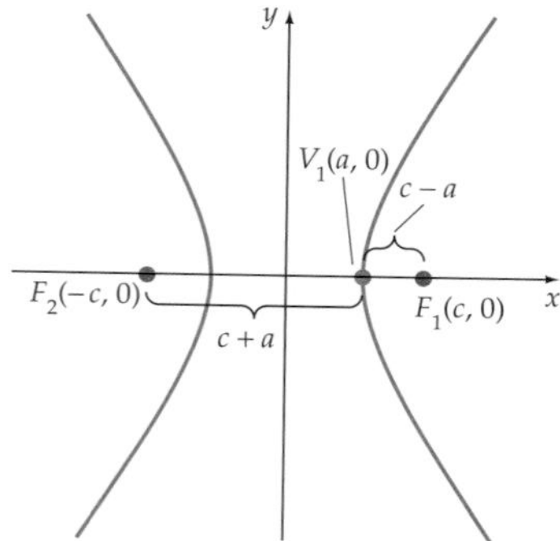

Figure 6.37

To determine the positive constant stated in the definition of a hyperbola, consider the point $V_1(a, 0)$, which is one vertex of a hyperbola, and the points $F_1(c, 0)$ and $F_2(-c, 0)$, which are the foci of the hyperbola (see **Figure 6.37**). The difference between the distance from $V_1(a, 0)$ to $F_1(c, 0)$, $c - a$, and the distance from $V_1(a, 0)$ to $F_2(-c, 0)$, $c + a$, must be a constant. By subtracting these distances, we find

$$|(c - a) - (c + a)| = |-2a| = 2a$$

Thus the constant is $2a$, and it is the length of the transverse axis. The absolute value is used to ensure that the distance is a positive number.

Standard Forms of the Equation of a Hyperbola with Center at the Origin

Transverse Axis on the x-Axis

The standard form of the equation of a hyperbola with center at the origin and transverse axis on the x-axis (see **Figure 6.38a**) is given by

$$\frac{x^2}{a^2} - \frac{y^2}{b^2} = 1$$

The coordinates of the vertices are $(a, 0)$ and $(-a, 0)$, and the coordinates of the foci are $(c, 0)$ and $(-c, 0)$, where $c^2 = a^2 + b^2$.

Transverse Axis on the y-Axis

The standard form of the equation of a hyperbola with center at the origin and transverse axis on the y-axis (see **Figure 6.38b**) is given by

$$\frac{y^2}{a^2} - \frac{x^2}{b^2} = 1$$

The coordinates of the vertices are $(0, a)$ and $(0, -a)$, and the coordinates of the foci are $(0, c)$ and $(0, -c)$, where $c^2 = a^2 + b^2$.

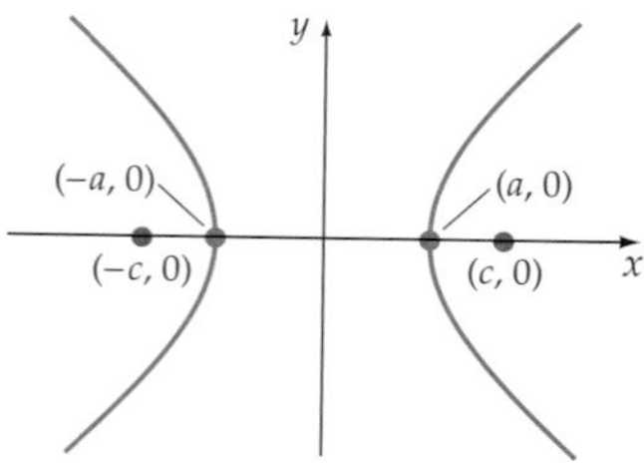

a. Transverse axis on the x-axis

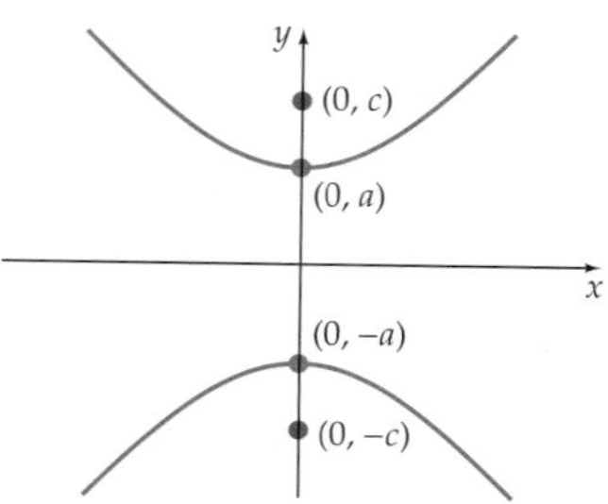

b. Transverse axis on the y-axis

Figure 6.38

By looking at the equations, it is possible to determine the location of the transverse axis by finding which term in the equation is positive. When the x^2 term is positive, the transverse axis is on the x-axis. When the y^2 term is positive, the transverse axis is on the y-axis.

QUESTION For the graph of $\frac{y^2}{9} - \frac{x^2}{4} = 1$, is the transverse axis on the x-axis or the y-axis?

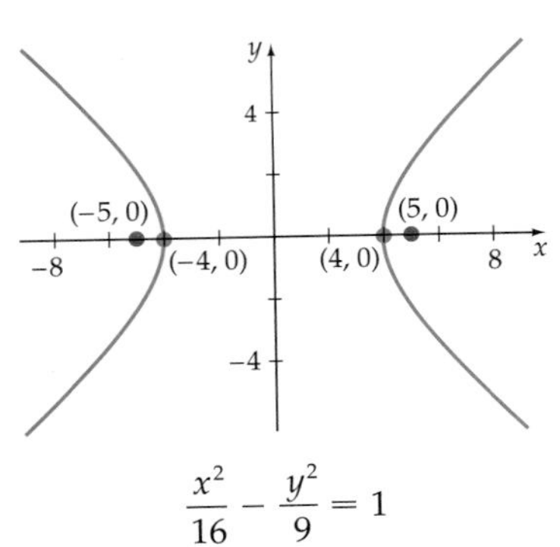

$\frac{x^2}{16} - \frac{y^2}{9} = 1$

Figure 6.39

Consider the hyperbola given by the equation $\frac{x^2}{16} - \frac{y^2}{9} = 1$. Because the x^2 term is positive, the transverse axis is on the x-axis, $a^2 = 16$, and thus $a = 4$. The vertices are $(4, 0)$ and $(-4, 0)$. To find the foci, we determine c.

$$c^2 = a^2 + b^2 = 16 + 9 = 25$$
$$c = \sqrt{25} = 5$$

The foci are $(5, 0)$ and $(-5, 0)$. The graph is shown in **Figure 6.39.**

Each hyperbola has two asymptotes that pass through the center of the hyperbola. The asymptotes of the hyperbola are a useful guide to sketching the graph of the hyperbola.

Asymptotes of a Hyperbola with Center at the Origin

The **asymptotes** of the hyperbola defined by $\frac{x^2}{a^2} - \frac{y^2}{b^2} = 1$ are given by the equations $y = \frac{b}{a}x$ and $y = -\frac{b}{a}x$ (see **Figure 6.40a**).

The asymptotes of the hyperbola defined by $\frac{y^2}{a^2} - \frac{x^2}{b^2} = 1$ are given by the equations $y = \frac{a}{b}x$ and $y = -\frac{a}{b}x$ (see **Figure 6.40b**).

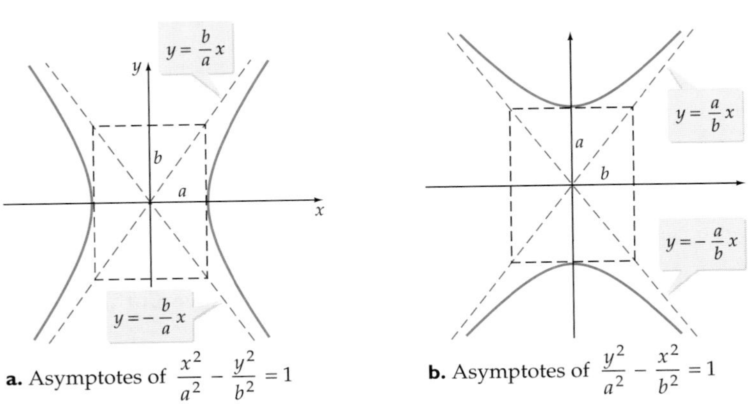

a. Asymptotes of $\frac{x^2}{a^2} - \frac{y^2}{b^2} = 1$ **b.** Asymptotes of $\frac{y^2}{a^2} - \frac{x^2}{b^2} = 1$

Figure 6.40

One method for remembering the equations of the asymptotes is to write the equation of a hyperbola in standard form, then replace 1 by 0 and solve for y.

$$\frac{x^2}{a^2} - \frac{y^2}{b^2} = 0 \quad \text{so} \quad y^2 = \frac{b^2}{a^2}x^2, \text{ or } y = \pm\frac{b}{a}x$$

$$\frac{y^2}{a^2} - \frac{x^2}{b^2} = 0 \quad \text{so} \quad y^2 = \frac{a^2}{b^2}x^2, \text{ or } y = \pm\frac{a}{b}x$$

ANSWER Because the y^2-term is positive, the transverse axis is on the y-axis.

EXAMPLE 1 Find the Vertices, Foci, and Asymptotes of a Hyperbola

Find the vertices, foci, and asymptotes of the hyperbola given by the equation $\frac{y^2}{9} - \frac{x^2}{4} = 1$. Sketch the graph.

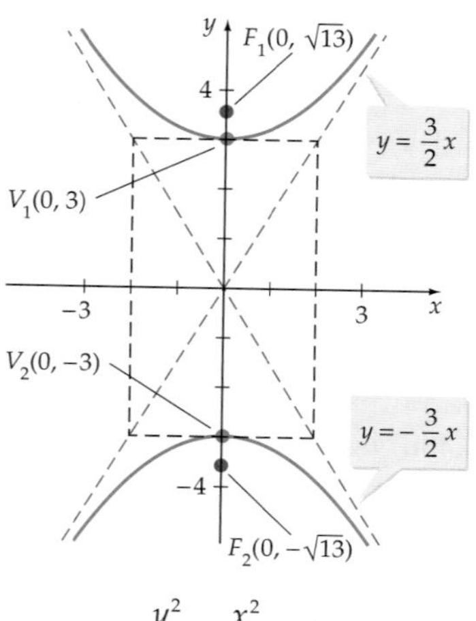

$$\frac{y^2}{9} - \frac{x^2}{4} = 1$$

Figure 6.41

Solution

Because the y^2 term is positive, the transverse axis is on the y-axis. We know that $a^2 = 9$; thus $a = 3$. The vertices are $V_1(0, 3)$ and $V_2(0, -3)$.

$$c^2 = a^2 + b^2 = 9 + 4$$
$$c = \sqrt{13}$$

The foci are $F_1(0, \sqrt{13})$ and $F_2(0, -\sqrt{13})$.

Because $a = 3$ and $b = 2$ ($b^2 = 4$), the equations of the asymptotes are $y = \frac{3}{2}x$ and $y = -\frac{3}{2}x$.

To sketch the graph, we draw a rectangle that has its center at the origin and has dimensions equal to the lengths of the transverse and conjugate axes. The asymptotes are extensions of the diagonals of the rectangle. See **Figure 6.41**.

Try Exercise 6, page 399

Hyperbolas with Center at (h, k)

Using a translation of coordinates similar to that used for ellipses, we can write the equation of a hyperbola with center at the point (h, k). Given coordinate axes labeled x' and y', an equation of a hyperbola with center at the origin is

$$\frac{(x')^2}{a^2} - \frac{(y')^2}{b^2} = 1 \qquad (1)$$

Now place the origin of this coordinate system at the point (h, k) of the xy-coordinate system, as shown in **Figure 6.42**. The relationship between an ordered pair in the $x'y'$-coordinate system and one in the xy-coordinate system is given by the transformation equations

$$x' = x - h$$
$$y' = y - k$$

Substitute the expressions for x' and y' into Equation (1). The equation of a hyperbola with center at (h, k) is

$$\frac{(x-h)^2}{a^2} - \frac{(y-k)^2}{b^2} = 1$$

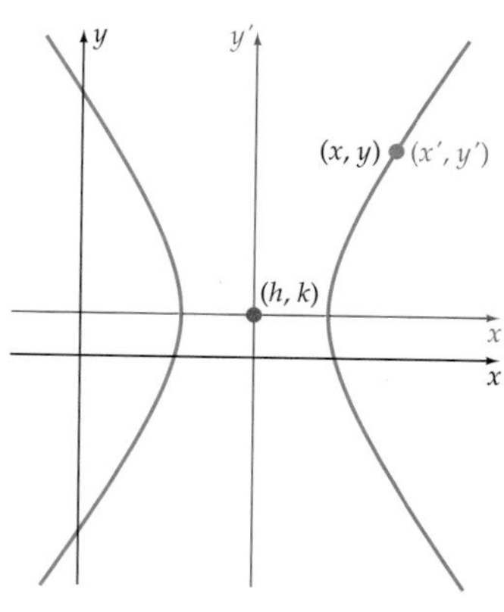

Figure 6.42

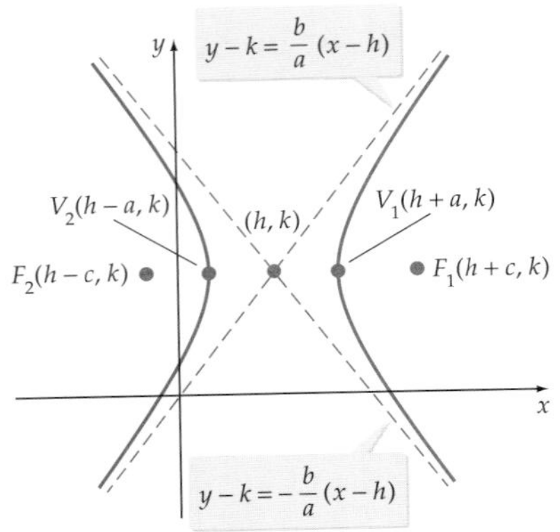

a. Transverse axis parallel to the x-axis

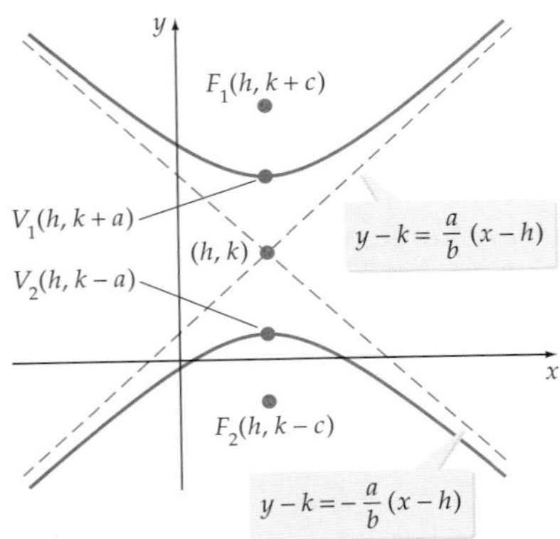

b. Transverse axis parallel to the y-axis

Figure 6.43

Standard Forms of the Equation of a Hyperbola with Center at (h, k)

Transverse Axis Parallel to the x-Axis

The standard form of the equation of a hyperbola with center at (h, k) and transverse axis parallel to the x-axis (see **Figure 6.43a**) is given by

$$\frac{(x - h)^2}{a^2} - \frac{(y - k)^2}{b^2} = 1$$

The coordinates of the vertices are $V_1(h + a, k)$ and $V_2(h - a, k)$. The coordinates of the foci are $F_1(h + c, k)$ and $F_2(h - c, k)$, where $c^2 = a^2 + b^2$. The equations of the asymptotes are $y - k = \pm\frac{b}{a}(x - h)$.

Transverse Axis Parallel to the y-Axis

The standard form of the equation of a hyperbola with center at (h, k) and transverse axis parallel to the y-axis (see **Figure 6.43b**) is given by

$$\frac{(y - k)^2}{a^2} - \frac{(x - h)^2}{b^2} = 1$$

The coordinates of the vertices are $V_1(h, k + a)$ and $V_2(h, k - a)$. The coordinates of the foci are $F_1(h, k + c)$ and $F_2(h, k - c)$, where $c^2 = a^2 + b^2$.

The equations of the asymptotes are $y - k = \pm\frac{a}{b}(x - h)$.

EXAMPLE 2 ⟫ Find the Center, Vertices, Foci, and Asymptotes of a Hyperbola

Find the center, vertices, foci, and asymptotes of the hyperbola given by the equation $4x^2 - 9y^2 - 16x + 54y - 29 = 0$. Sketch the graph.

Solution

Write the equation of the hyperbola in standard form by completing the square.

$$4x^2 - 9y^2 - 16x + 54y - 29 = 0$$

$$4x^2 - 16x - 9y^2 + 54y = 29$$ • Rearrange terms.

$$4(x^2 - 4x) - 9(y^2 - 6y) = 29$$ • Factor.

$$4(x^2 - 4x + 4) - 9(y^2 - 6y + 9) = 29 + 16 - 81$$ • Complete the square.

$$4(x - 2)^2 - 9(y - 3)^2 = -36$$ • Factor.

$$\frac{(y - 3)^2}{4} - \frac{(x - 2)^2}{9} = 1$$ • Divide each side by −36.

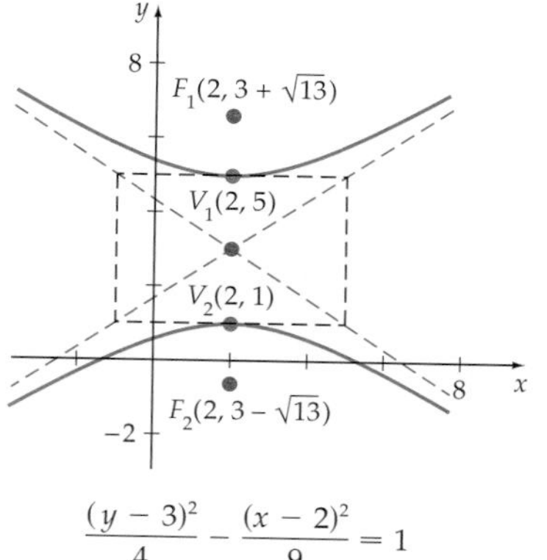

Figure 6.44

The coordinates of the center are (2, 3). Because the term containing $(y - 3)^2$ is positive, the transverse axis is parallel to the y-axis. We know that $a^2 = 4$; thus $a = 2$. The vertices are (2, 5) and (2, 1). See **Figure 6.44.** To find the coordinates of the foci, we find c.

$$c^2 = a^2 + b^2 = 4 + 9$$
$$c = \sqrt{13}$$

The foci are $\left(2, 3 + \sqrt{13}\right)$ and $\left(2, 3 - \sqrt{13}\right)$. We know that $b^2 = 9$; thus $b = 3$. The equations of the asymptotes are $y - 3 = \pm\left(\frac{2}{3}\right)(x - 2)$, which simplifies to

$$y = \frac{2}{3}x + \frac{5}{3} \qquad \text{and} \qquad y = -\frac{2}{3}x + \frac{13}{3}$$

Try Exercise 28, page 399

Integrating Technology

A graphing utility can be used to graph a hyperbola. For instance, consider the equation $4x^2 - 9y^2 - 16x + 54y - 29 = 0$ from Example 2. Rewrite the equation as

$$-9y^2 + 54y + (4x^2 - 16x - 29) = 0$$

In this form, the equation is a quadratic equation in terms of the variable y with

$$A = -9, B = 54, \text{ and } C = 4x^2 - 16x - 29$$

Apply the quadratic formula to produce

$$y = \frac{-54 \pm \sqrt{2916 + 36(4x^2 - 16x - 29)}}{-18}$$

The graph of $Y1 = \dfrac{-54 + \sqrt{2916 + 36(4x^2 - 16x - 29)}}{-18}$ is the upper branch of the hyperbola (see **Figure 6.45**).

The graph of $Y2 = \dfrac{-54 - \sqrt{2916 + 36(4x^2 - 16x - 29)}}{-18}$ is the lower branch of the hyperbola, as shown in **Figure 6.45.**

One advantage of this graphing procedure is that it does not require us to write the given equation in standard form. A disadvantage of the graphing procedure is that it does not indicate where the foci of the hyperbola are located.

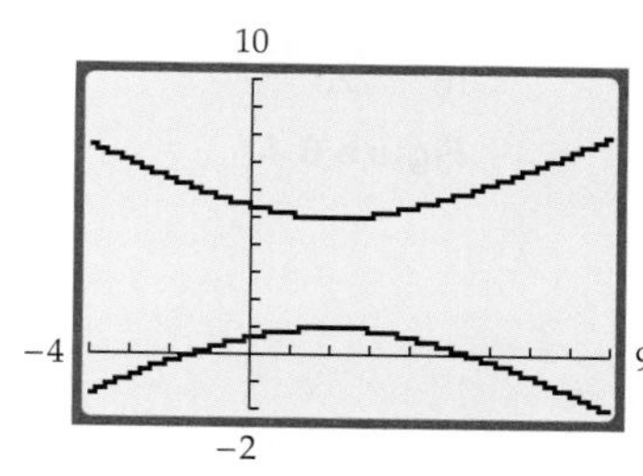

Figure 6.45

Eccentricity of a Hyperbola

The graph of a hyperbola can be very wide or very narrow. The **eccentricity** of a hyperbola is a measure of its "wideness."

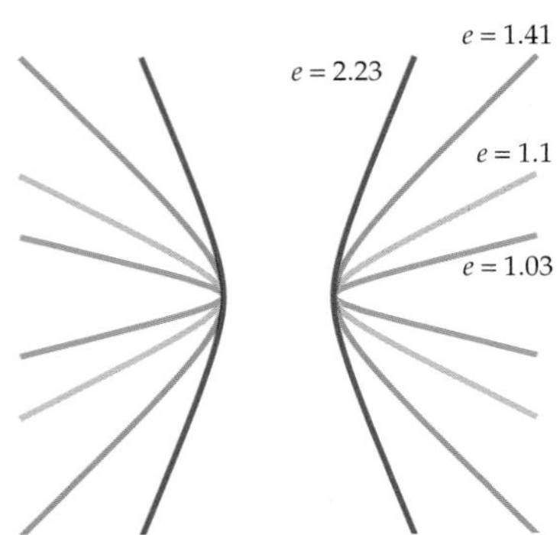

Figure 6.46

Definition of the Eccentricity (*e*) of a Hyperbola

The eccentricity e of a hyperbola is the ratio of c to a, where c is the distance from the center to a focus and a is half the length of the transverse axis.

$$e = \frac{c}{a}$$

For a hyperbola, $c > a$ and therefore $e > 1$. As the eccentricity of the hyperbola increases, the graph becomes wider and wider, as shown in **Figure 6.46.**

EXAMPLE 3 ⟫ Find the Equation of a Hyperbola Given Its Eccentricity

Find the standard form of the equation of the hyperbola that has eccentricity $\frac{3}{2}$, center at the origin, and a focus at (6, 0).

Solution

Because the focus is located at (6, 0) and the center is at the origin, $c = 6$. An extension of the transverse axis contains the foci, so the transverse axis is on the x-axis.

$$e = \frac{3}{2} = \frac{c}{a}$$

$$\frac{3}{2} = \frac{6}{a} \qquad \text{• Substitute 6 for } c.$$

$$a = 4 \qquad \text{• Solve for } a.$$

To find b^2, use the equation $c^2 = a^2 + b^2$ and the values for c and a.

$$c^2 = a^2 + b^2$$
$$36 = 16 + b^2$$
$$b^2 = 20$$

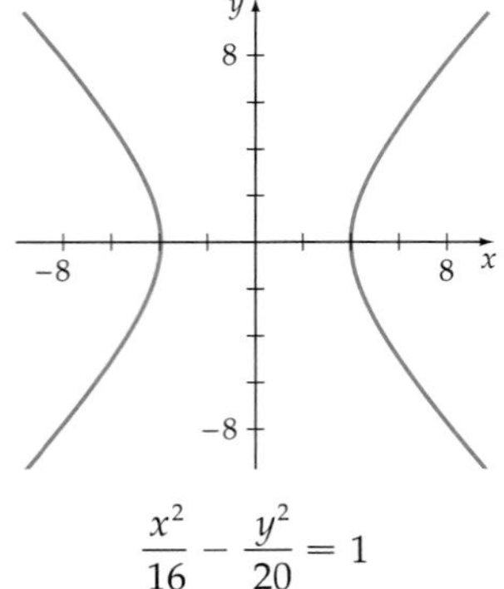

$\frac{x^2}{16} - \frac{y^2}{20} = 1$

Figure 6.47

The equation of the hyperbola is $\frac{x^2}{16} - \frac{y^2}{20} = 1$. See **Figure 6.47.**

⟫ *Try Exercise 50, page 400*

Applications

Orbits of Comets In Section 6.2 we noted that the orbits of the planets are elliptical. Some comets have elliptical orbits also, the most notable being Halley's comet, whose eccentricity is 0.97. See **Figure 6.48.**

Math Matters

Caroline Herschel (1750–1848) became interested in mathematics and astronomy after her brother William discovered the planet Uranus. She was the first woman to receive credit for the discovery of a comet. In fact, between 1786 and 1797, she discovered eight comets. In 1828 she completed a catalog of over 2000 nebulae, a feat for which the Royal Astronomical Society of England presented her with its prestigious gold medal.

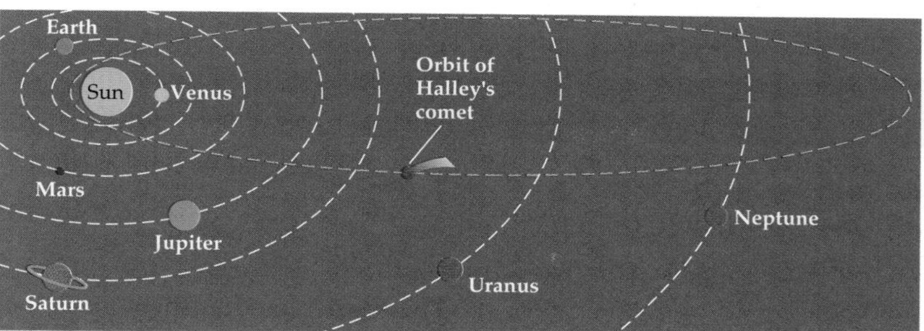

Not drawn to scale.

Figure 6.48

Other comets have hyperbolic orbits with the sun at a focus. These comets pass by the sun only once. The velocity of a comet determines whether its orbit is elliptical or hyperbolic.

Hyperbolas as an Aid to Navigation Consider two radio transmitters, T_1 and T_2, placed some distance apart. A ship with electronic equipment measures the difference between the times it takes signals from the transmitters to reach the ship. Because the difference between the times is proportional to the difference between the distances of the ship from the transmitters, the ship must be located on the hyperbola with foci at the two transmitters.

Using a third transmitter, T_3, we can find a second hyperbola with foci T_2 and T_3. The ship lies on the intersection of the two hyperbolas, as shown in **Figure 6.49.**

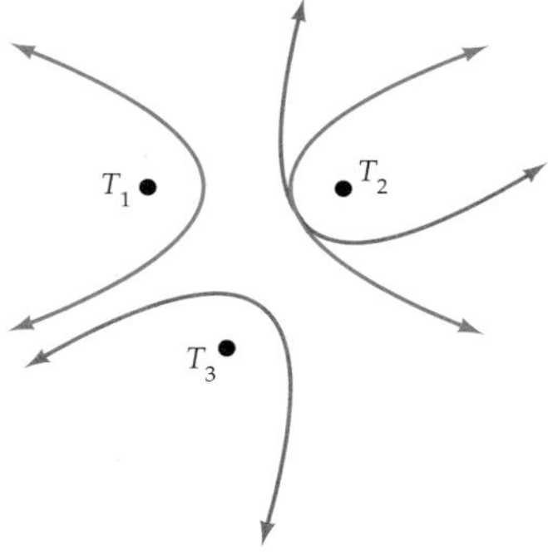

Figure 6.49

Figure 6.50

EXAMPLE 4 ⟫ Determine the Position of a Ship

Two radio transmitters are positioned along a coastline, 500 miles apart. See **Figure 6.50.** Using a LORAN (LOng RAnge Navigation) system, a ship determines that a radio signal from transmitter T_1 reaches the ship 1600 microseconds before it receives a simultaneous signal from transmitter T_2.

a. Find an equation of a hyperbola (with foci located at T_1 and T_2) on which the ship lies. See **Figure 6.50.** (Assume the radio signals travel at 0.186 mile per microsecond.)

b. If the ship is directly north of transmitter T_1, determine how far (to the nearest mile) the ship is from the transmitter.

Continued ▸

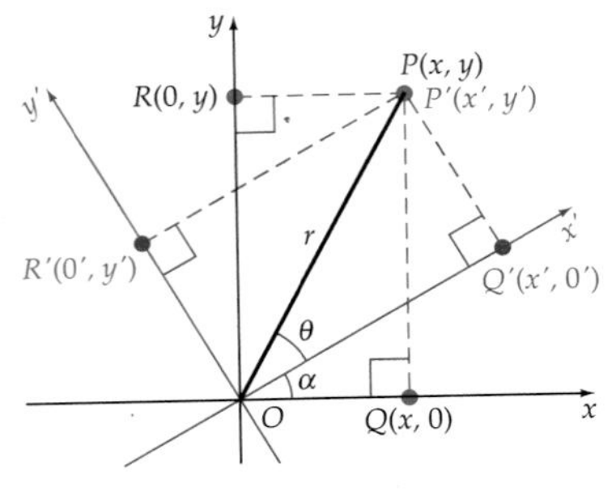

Figure 6.52

A **rotation of axes** is a rotation of the x- and y-axes about the origin to another position denoted by x' and y'. We denote the measure of the **angle of rotation** by α.

Let P be some point in the plane, and let r represent the distance of P from the origin. The coordinates of P relative to the xy-coordinate system and the $x'y'$-coordinate system are $P(x, y)$ and $P(x', y')$, respectively.

Let $Q(x, 0)$ and $R(0, y)$ be the projections of P onto the x- and the y-axis and let $Q'(x', 0')$ and $R'(0', y')$ be the projections of P onto the x'- and the y'-axis. (See **Figure 6.52**.) The angle between the x'-axis and OP is denoted by θ. We can express the coordinates of P in each coordinate system in terms of α and θ.

$$x = r\cos(\theta + \alpha) \qquad x' = r\cos\theta$$
$$y = r\sin(\theta + \alpha) \qquad y' = r\sin\theta$$

Applying the addition formulas for $\cos(\theta + \alpha)$ and $\sin(\theta + \alpha)$, we get

$$x = r\cos(\theta + \alpha) = r\cos\theta\cos\alpha - r\sin\theta\sin\alpha$$
$$y = r\sin(\theta + \alpha) = r\sin\theta\cos\alpha + r\cos\theta\sin\alpha$$

Now, substituting x' for $r\cos\theta$ and y' for $r\sin\theta$ into these equations yields

$$x = x'\cos\alpha - y'\sin\alpha \qquad \bullet\ x' = r\cos\theta,\ y' = r\sin\alpha$$
$$y = y'\cos\alpha + x'\sin\alpha$$

This proves the equations labeled (1) of the following theorem.

Rotation-of-Axes Formulas

Suppose that an xy-coordinate system and an $x'y'$-coordinate system have the same origin and that α is the angle between the positive x-axis and the positive x'-axis. If the coordinates of a point P are (x, y) in one system and (x', y') in the rotated system, then

$$\left.\begin{aligned} x &= x'\cos\alpha - y'\sin\alpha \\ y &= y'\cos\alpha + x'\sin\alpha \end{aligned}\right\} \quad (1) \qquad \left.\begin{aligned} x' &= x\cos\alpha + y\sin\alpha \\ y' &= y\cos\alpha - x\sin\alpha \end{aligned}\right\} \quad (2)$$

The derivations of the formulas for x' and y' are left as an exercise.

As we have noted, the appearance of the Bxy $(B \neq 0)$ term in the general second-degree equation indicates that the graph of the conic has been rotated. The angle through which the axes have been rotated can be determined from the following theorem.

Rotation Theorem for Conics

Let $Ax^2 + Bxy + Cy^2 + Dx + Ey + F = 0$, $B \neq 0$, be the equation of a conic in an xy-coordinate system, and let α be an angle of rotation such that

$$\cot 2\alpha = \frac{A - C}{B}, \quad 0^\circ < 2\alpha < 180^\circ \qquad (3)$$

Then the equation of the conic in the rotated coordinate system will be

$$A'(x')^2 + C'(y')^2 + D'x' + E'y' + F' = 0$$

where $0° < 2\alpha < 180°$ and

$$A' = A\cos^2\alpha + B\cos\alpha\sin\alpha + C\sin^2\alpha \qquad (4)$$
$$C' = A\sin^2\alpha - B\cos\alpha\sin\alpha + C\cos^2\alpha \qquad (5)$$
$$D' = D\cos\alpha + E\sin\alpha \qquad (6)$$
$$E' = -D\sin\alpha + E\cos\alpha \qquad (7)$$
$$F' = F \qquad (8)$$

Equation (3) has an infinite number of solutions. Any rotation through an angle α that is a solution of $\cot 2\alpha = \dfrac{A - C}{B}$ will eliminate the xy term. However, the equations of the conic in the rotated coordinate systems may differ, depending on the value of α that is used to produce these transformed equations. To avoid any confusion, we will always choose α to be the acute angle that satisfies Equation (3).

QUESTION For $x^2 - 4xy + 9y^2 + 6x - 8y - 20 = 0$, what is the value of $\cot 2\alpha$?

EXAMPLE 1 » Use the Rotation Theorem to Sketch a Conic

Sketch the graph of $7x^2 - 6\sqrt{3}xy + 13y^2 - 16 = 0$.

Solution

We are given

$$A = 7, \quad B = -6\sqrt{3}, \quad C = 13, \quad D = 0, \quad E = 0, \quad \text{and} \quad F = -16$$

The angle of rotation α can be determined by solving

$$\cot 2\alpha = \frac{A - C}{B} = \frac{7 - 13}{-6\sqrt{3}} = \frac{-6}{-6\sqrt{3}} = \frac{1}{\sqrt{3}} = \frac{\sqrt{3}}{3}$$

This gives us $2\alpha = 60°$, or $\alpha = 30°$. Because $\alpha = 30°$, we have

$$\sin\alpha = \frac{1}{2} \quad \text{and} \quad \cos\alpha = \frac{\sqrt{3}}{2}$$

We determine the coefficients A', C', D', E', and F' by using Equations (4) to (8).

$$A' = 7\left(\frac{\sqrt{3}}{2}\right)^2 + (-6\sqrt{3})\left(\frac{\sqrt{3}}{2}\right)\left(\frac{1}{2}\right) + 13\left(\frac{1}{2}\right)^2 = 4$$
$$C' = 7\left(\frac{1}{2}\right)^2 - (-6\sqrt{3})\left(\frac{\sqrt{3}}{2}\right)\left(\frac{1}{2}\right) + 13\left(\frac{\sqrt{3}}{2}\right)^2 = 16$$

Continued ►

ANSWER $\cot 2\alpha = \dfrac{1 - 9}{-4} = 2$

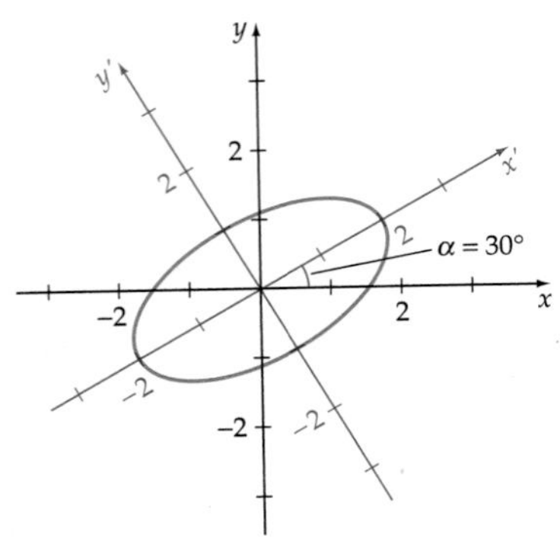

$7x^2 - 6\sqrt{3}xy + 13y^2 - 16 = 0$

Figure 6.53

$$D' = 0\left(\frac{\sqrt{3}}{2}\right)^2 + 0\left(\frac{1}{2}\right) = 0$$

$$E' = -0\left(\frac{1}{2}\right) + 0\left(\frac{\sqrt{3}}{2}\right) = 0$$

$$F' = F = -16$$

The equation of the conic in the $x'y'$-plane is $4(x')^2 + 16(y')^2 - 16 = 0$ or

$$\frac{(x')^2}{2^2} + \frac{(y')^2}{1^2} = 1$$

This is the equation of an ellipse that is centered at the origin of an $x'y'$-coordinate system. The ellipse has a semimajor axis $a = 2$ and a semiminor axis $b = 1$. See **Figure 6.53**.

Try Exercise 10, page 410

In Example 1, the angle of rotation α was 30°, which is a special angle. In the next example, we demonstrate a technique that is often used when the angle of rotation is not a special angle.

EXAMPLE 2 Use the Rotation Theorem to Sketch a Conic

Sketch the graph of $32x^2 - 48xy + 18y^2 - 15x - 20y = 0$.

Solution

We are given

$$A = 32, \quad B = -48, \quad C = 18, \quad D = -15, \quad E = -20, \quad \text{and} \quad F = 0$$

Therefore,

$$\cot 2\alpha = \frac{A - C}{B} = \frac{32 - 18}{-48} = -\frac{7}{24}$$

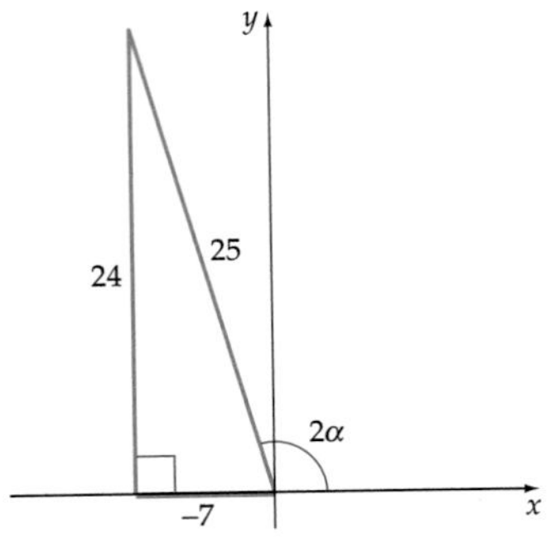

Figure 6.54

Figure 6.54 shows an angle 2α for which $\cot 2\alpha = -\frac{7}{24}$. From **Figure 6.54** we conclude that $\cos 2\alpha = -\frac{7}{25}$. The half-angle identities can be used to determine $\sin\alpha$ and $\cos\alpha$.

$$\sin\alpha = \sqrt{\frac{1 - (-7/25)}{2}} = \frac{4}{5} \quad \text{and} \quad \cos\alpha = \sqrt{\frac{1 + (-7/25)}{2}} = \frac{3}{5}$$

A calculator can be used to determine that $\alpha \approx 53.1°$.

Equations (4) to (8) give us

$$A' = 32\left(\frac{3}{5}\right)^2 + (-48)\left(\frac{3}{5}\right)\left(\frac{4}{5}\right) + 18\left(\frac{4}{5}\right)^2 = 0$$

$$C' = 32\left(\frac{4}{5}\right)^2 - (-48)\left(\frac{3}{5}\right)\left(\frac{4}{5}\right) + 18\left(\frac{3}{5}\right)^2 = 50$$

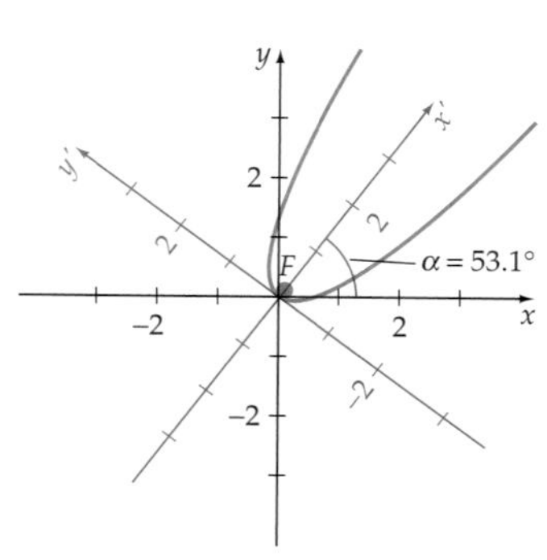

$32x^2 - 48xy + 18y^2 - 15x - 20y = 0$

Figure 6.55

$$D' = (-15)\left(\frac{3}{5}\right) + (-20)\left(\frac{4}{5}\right) = -25$$

$$E' = -(-15)\left(\frac{4}{5}\right) + (-20)\left(\frac{3}{5}\right) = 0$$

$$F' = F = 0$$

The equation of the conic in the $x'y'$-plane is $50(y')^2 - 25x' = 0$, or

$$(y')^2 = \frac{1}{2}x'$$

This is the equation of a parabola. Because $4p = \frac{1}{2}$, we know $p = \frac{1}{8}$, and the focus of the parabola is at $\left(\frac{1}{8}, 0\right)$ on the x'-axis. See **Figure 6.55.**

Try Exercise 20, page 410

The Conic Identification Theorem

The following theorem provides us with a procedure that can be used to identify the type of conic that will be produced by graphing an equation that is in the form of the general second-degree equation in two variables.

Conic Identification Theorem

The graph of

$$Ax^2 + Bxy + Cy^2 + Dx + Ey + F = 0$$

is either a conic or a degenerate conic. If the graph is a conic, then the graph can be identified by its *discriminant* $B^2 - 4AC$. The graph is

- an ellipse or a circle, provided $B^2 - 4AC < 0$.
- a parabola, provided $B^2 - 4AC = 0$.
- a hyperbola, provided $B^2 - 4AC > 0$.

EXAMPLE 3 Identify Conic Sections

Each of the following equations has a graph that is a nondegenerate conic. Compute $B^2 - 4AC$ to identify the type of conic given by each equation.

a. $2x^2 - 4xy + 2y^2 - 6x - 10 = 0$

b. $-2xy + 11 = 0$

c. $3x^2 + 5xy + 4y^2 - 8x + 10y + 6 = 0$

d. $xy - 3y^2 + 2 = 0$

Continued ►

Exercise Set 6.4

In Exercises 1 to 8, find the acute angle of rotation α that eliminates the xy term. State approximate solutions to the nearest 0.1°.

1. $xy = 3$

2. $5x^2 - 3xy - 5y^2 - 1 = 0$

3. $9x^2 - 24xy + 16y^2 - 320x - 240y = 0$

4. $x^2 + 4xy + 4y^2 - 6x - 5 = 0$

5. $5x^2 - 6\sqrt{3}xy - 11y^2 + 4x - 3y + 2 = 0$

6. $5x^2 + 4xy + 8y^2 - 6x + 3y - 12 = 0$

7. $2x^2 + xy + y^2 - 4 = 0$

8. $-2x^2 + \sqrt{3}xy - 3y^2 + 2x + 6y + 36 = 0$

In Exercises 9 to 20, find the acute angle of rotation α that eliminates the xy term. Then find an equation in x' and y'-coordinates. Graph the equation.

9. $xy = 4$

10. $xy = -10$

11. $6x^2 - 6xy + 14y^2 - 45 = 0$

12. $11x^2 - 10\sqrt{3}xy + y^2 - 20 = 0$

13. $x^2 + 4xy - 2y^2 - 1 = 0$

14. $9x^2 - 24xy + 16y^2 + 100 = 0$

15. $3x^2 + 2\sqrt{3}xy + y^2 + 2x - 2\sqrt{3}y + 16 = 0$

16. $x^2 + 2xy + y^2 + 2\sqrt{2}x - 2\sqrt{2}y = 0$

17. $9x^2 - 24xy + 16y^2 - 40x - 30y + 100 = 0$

18. $24x^2 + 16\sqrt{3}xy + 8y^2 - x + \sqrt{3}y - 8 = 0$

19. $6x^2 + 24xy - y^2 - 12x + 26y + 11 = 0$

20. $x^2 + 4xy + 4y^2 - 2\sqrt{5}x + \sqrt{5}y = 0$

In Exercises 21 to 26, use a graphing utility to graph each equation.

21. $6x^2 - xy + 2y^2 + 4x - 12y + 7 = 0$

22. $5x^2 - 2xy + 10y^2 - 6x - 9y - 20 = 0$

23. $x^2 - 6xy + y^2 - 2x - 5y + 4 = 0$

24. $2x^2 - 10xy + 3y^2 - x - 8y - 7 = 0$

25. $3x^2 - 6xy + 3y^2 + 10x - 8y - 2 = 0$

26. $2x^2 - 8xy + 8y^2 + 20x - 24y - 3 = 0$

27. Find the equations of the asymptotes, relative to an xy-coordinate system, for the hyperbola defined by the equation in Exercise 13. Assume that the xy-coordinate system has the same origin as the $x'y'$-coordinate system.

28. Find the coordinates of the foci and the equation of the directrix, relative to an xy-coordinate system, for the parabola defined by the equation in Exercise 16. Assume that the xy-coordinate system has the same origin as the $x'y'$-coordinate system.

29. Find the coordinates of the foci, relative to an xy-coordinate system, for the ellipse defined by the equation in Exercise 11. Assume that the xy-coordinate system has the same origin as the $x'y'$-coordinate system.

In Exercises 30 to 40, use the Conic Identification Theorem to identify the graph of each equation as a parabola, an ellipse (or a circle), or a hyperbola.

30. $xy = 4$

31. $x^2 + xy - y^2 - 40 = 0$

32. $11x^2 - 10\sqrt{3}xy + y^2 - 20 = 0$

33. $3x^2 + 2\sqrt{3}xy + y^2 - 3x + 2y + 20 = 0$

34. $9x^2 - 24xy + 16y^2 + 8x - 12y - 20 = 0$

35. $4x^2 - 4xy + y^2 - 12y + 20 = 0$

36. $5x^2 + 4xy + 8y^2 - 6x + 3y - 12 = 0$

37. $5x^2 - 6\sqrt{3}xy - 11y^2 + 4x - 3y + 2 = 0$

38. $6x^2 - 6xy + 14y^2 - 14x + 12y - 60 = 0$

39. $6x^2 + 2\sqrt{3}xy + 5y^2 - 3x + 2y - 20 = 0$

40. $5x^2 - 2\sqrt{3}xy + 3y^2 - x + y - 12 = 0$

Connecting Concepts

41. By using the rotation-of-axes equations, show that for every choice of α, the equation $x^2 + y^2 = r^2$ becomes $(x')^2 + (y')^2 = r^2$.

42. The vertices of a hyperbola are $(1, 1)$ and $(-1, -1)$. The foci are $(\sqrt{2}, \sqrt{2})$ and $(-\sqrt{2}, -\sqrt{2})$. Find an equation of the hyperbola.

43. The vertices on the major axis of an ellipse are the points $(2, 4)$ and $(-2, -4)$. The foci are the points $(\sqrt{2}, 2\sqrt{2})$ and $(-\sqrt{2}, -2\sqrt{2})$. Find an equation of the ellipse.

44. The vertex of a parabola is the origin, and the focus is the point $(1, 3)$. Find an equation of the parabola.

45. **An Invariant Theorem** Let

$$Ax^2 + Bxy + Cy^2 + Dx + Ey + F = 0$$

be an equation of a conic in an xy-coordinate system. Let the equation of the conic in the rotated $x'y'$-coordinate system be

$$A'(x')^2 + B'x'y' + C'(y')^2 + D'x' + E'y' + F' = 0$$

Show that

$$A' + C' = A + C$$

46. **An Invariant Theorem** Let

$$Ax^2 + Bxy + Cy^2 + Dx + Ey + F = 0$$

be an equation of a conic in an xy-coordinate system. Let the equation of the conic in the rotated $x'y'$-coordinate system be

$$A'(x')^2 + B'x'y' + C'(y')^2 + D'x' + E'y' + F' = 0$$

Show that

$$(B')^2 - 4A'C' = B^2 - 4AC$$

47. Use the result of Exercise 46 to verify the Conic Identification Theorem, stated on page 407.

48. Derive the equations labeled (2), page 404, of the rotation-of-axes formulas.

Projects

1. **Use the Invariant Theorems** The results of Exercises 45 and 46 illustrate that when the rotation theorem is used to transform equations of the form

$$Ax^2 + Bxy + Cy^2 - F = 0$$

to the form

$$A'(x')^2 + C'(y')^2 - F = 0$$

the following relationships hold:

$$A + C = A' + C' \quad \text{and} \quad B^2 - 4AC = (B')^2 - 4A'C'$$

Use these equations to transform

$$10x^2 + 24xy + 17y^2 - 26 = 0$$

to the form

$$A'(x')^2 + C'(y')^2 - 26 = 0$$

without applying the rotation theorem.

Section 6.5

- The Polar Coordinate System
- Graphs of Equations in a Polar Coordinate System
- Transformations Between Rectangular and Polar Coordinates
- Write Polar Coordinate Equations as Rectangular Equations and Rectangular Coordinate Equations as Polar Equations

Introduction to Polar Coordinates

PREPARE FOR THIS SECTION

Prepare for this section by completing the following exercises. The answers can be found on page A29.

PS1. Is $\sin x$ an even or odd function? [2.5]

PS2. Is $\cos x$ an even or odd function? [2.5]

PS3. Solve $\tan \alpha = -\sqrt{3}$ for $0 < \alpha < 2\pi$. [3.6]

PS4. If $\sin \alpha = -\dfrac{\sqrt{3}}{2}$ and $\cos \alpha = -\dfrac{1}{2}$, with $0° \le \alpha < 360°$, find α. [3.6]

PS5. Write $(r \cos \theta)^2 + (r \sin \theta)^2$ in simplest form. [3.1]

PS6. For the graph at the right, find the coordinates of point A. Round each coordinate to the nearest tenth. [2.2]

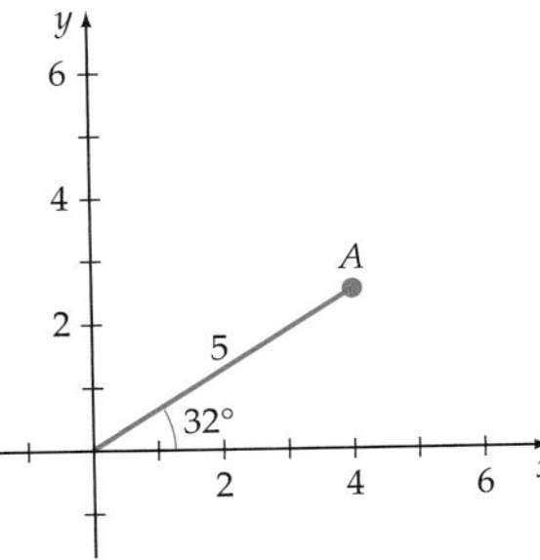

The Polar Coordinate System

Until now, we have used a *rectangular coordinate system* to locate a point in the coordinate plane. An alternative method is to use a *polar coordinate system*, wherein a point is located by giving a distance from a fixed point and an angle from some fixed direction.

A **polar coordinate system** is formed by drawing a horizontal ray. The ray is called the **polar axis**, and the initial point of the ray is called the **pole**. A point $P(r, \theta)$ in the plane is located by specifying a distance r from the pole and an angle θ measured from the polar axis to the line segment OP. The angle can be measured in degrees or radians. See **Figure 6.59**.

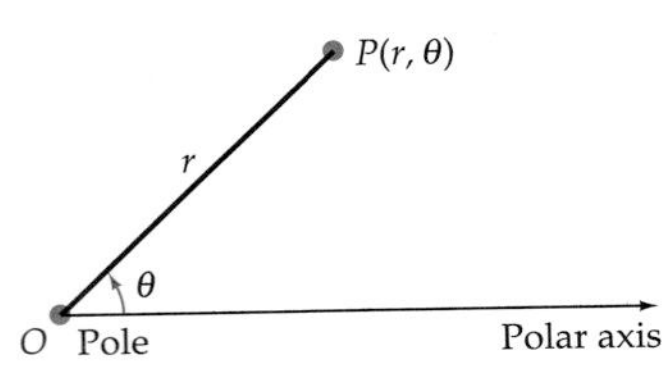

Figure 6.59

The coordinates of the pole are $(0, \theta)$, where θ is an arbitrary angle. Positive angles are measured counterclockwise from the polar axis. Negative angles are measured clockwise from the axis. Positive values of r are measured along the ray that makes an angle of θ from the polar axis. Negative values of r are measured along the ray that makes an angle of $\theta + 180°$ from the polar axis. See **Figures 6.60** and **6.61.**

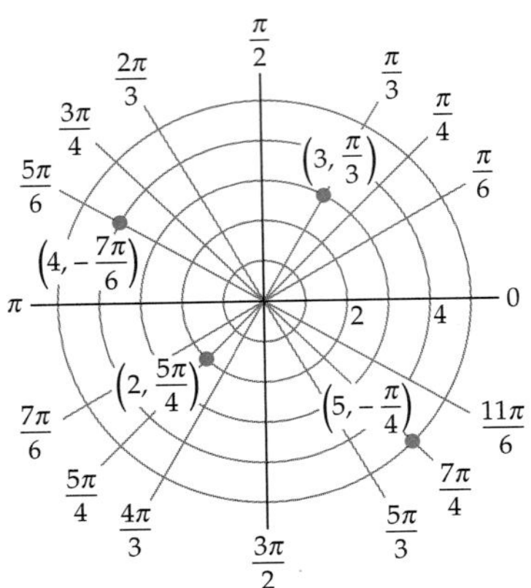

Figure 6.60

In a rectangular coordinate system, there is a one-to-one correspondence between the points in the plane and the ordered pairs (x, y). This is not true for a polar coordinate system. For polar coordinates, the relationship is one-to-many. Infinitely many ordered-pair descriptions correspond to each point $P(r, \theta)$ in a polar coordinate system.

For example, consider a point whose coordinates are $P(3, 45°)$. Because there are 360° in one complete revolution around a circle, the point P also could be written as $(3, 405°)$, as $(3, 765°)$, as $(3, 1125°)$, and generally as $(3, 45° + n \cdot 360°)$, where n is an integer. It is also possible to describe the point $P(3, 45°)$ by $(-3, 225°)$, $(-3, -135°)$, and $(3, -315°)$, to name just a few options.

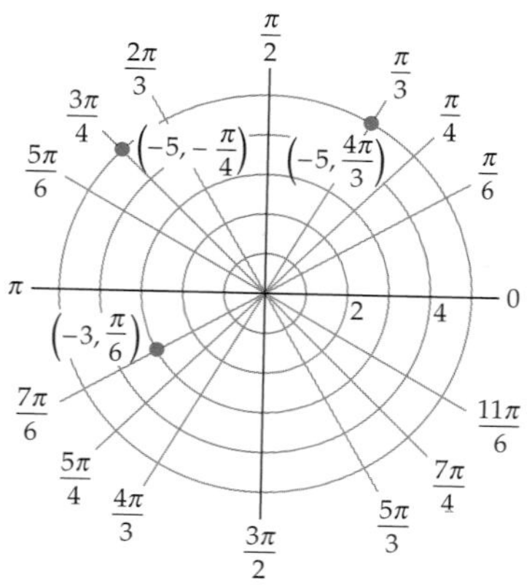

Figure 6.61

The relationship between an ordered pair and a point is not one-to-many. That is, given an ordered pair (r, θ), there is exactly one point in the plane that corresponds to that ordered pair.

Graphs of Equations in a Polar Coordinate System

A **polar equation** is an equation in r and θ. A **solution** to a polar equation is an ordered pair (r, θ) that satisfies the equation. The **graph** of a polar equation is the set of all points whose ordered pairs are solutions of the equation.

The graph of the polar equation $\theta = \frac{\pi}{6}$ is a line. Because θ is independent of r, θ is $\frac{\pi}{6}$ radian from the polar axis for all values of r. The graph is a line that makes an angle of $\frac{\pi}{6}$ radian (30°) from the polar axis. See **Figure 6.62**.

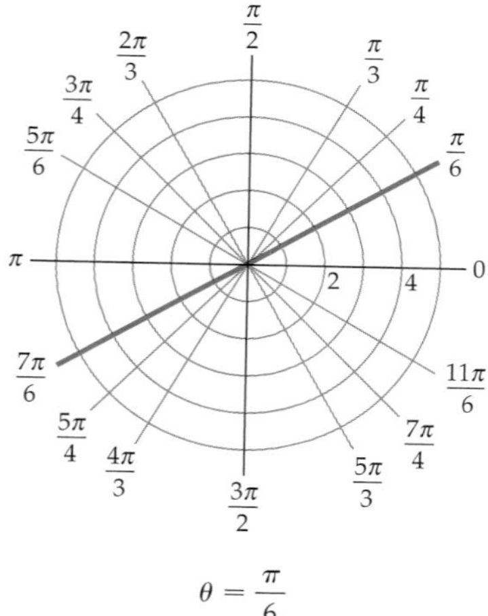

$\theta = \frac{\pi}{6}$

Figure 6.62

Polar Equations of a Line

The graph of $\theta = \alpha$ is a line through the pole at an angle of α from the polar axis. See **Figure 6.63a.**

The graph of $r \sin \theta = a$ is a horizontal line passing through the point $\left(a, \frac{\pi}{2}\right)$. See **Figure 6.63b.**

The graph of $r \cos \theta = a$ is a vertical line passing through the point $(a, 0)$. See **Figure 6.63c.**

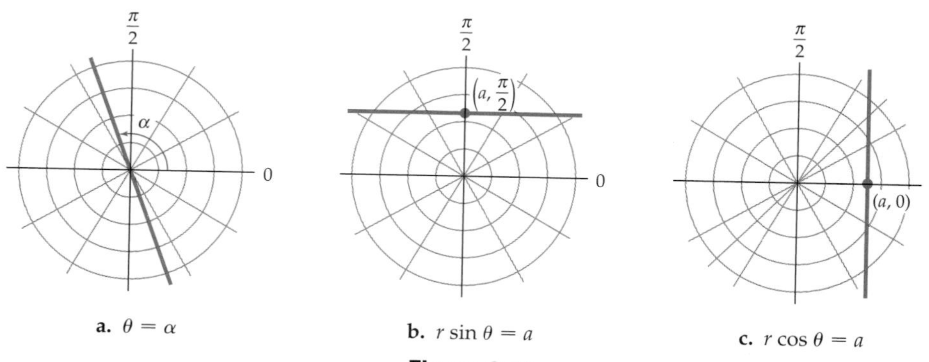

a. $\theta = \alpha$ **b.** $r \sin \theta = a$ **c.** $r \cos \theta = a$

Figure 6.63

Figure 6.64 is the graph of the polar equation $r = 2$. Because r is independent of θ, r is 2 units from the pole for all values of θ. The graph is a circle of radius 2 with center at the pole.

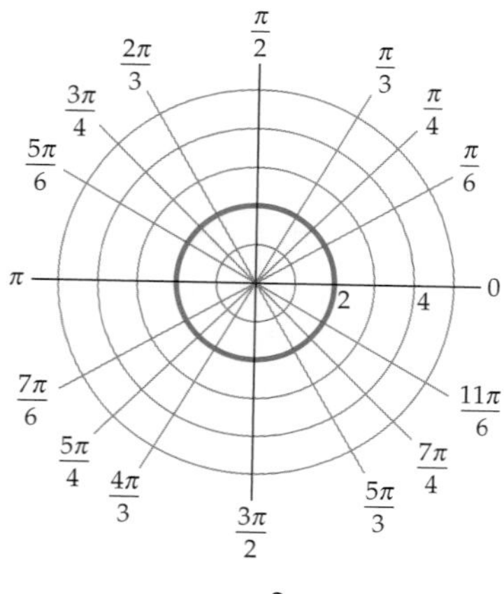

$r = 2$

Figure 6.64

The Graph of $r = a$

The graph of $r = a$ is a circle with center at the pole and radius a.

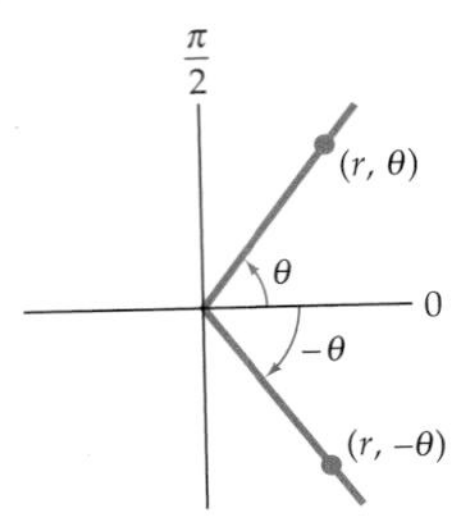

Symmetry with respect to the line $\theta = 0$

Figure 6.65

Suppose that whenever the ordered pair (r, θ) lies on the graph of a polar equation, $(r, -\theta)$ also lies on the graph. From **Figure 6.65,** the graph will have symmetry with respect to the line $\theta = 0$. Thus one test for symmetry is to replace θ by $-\theta$ in the polar equation. If the resulting equation is equivalent to the original equation, the graph is symmetric with respect to the line $\theta = 0$.

Table 6.2 shows the types of symmetry and their associated tests. For each type, if the recommended substitution results in an equivalent equation, the graph will have the indicated symmetry. **Figure 6.66** illustrates the tests for symmetry with respect to the line $\theta = \frac{\pi}{2}$ and for symmetry with respect to the pole.

Table 6.2 Tests for Symmetry

Substitution	Symmetry with respect to
$-\theta$ for θ	The line $\theta = 0$
$\pi - \theta$ for θ, $-r$ for r	The line $\theta = 0$
$\pi - \theta$ for θ	The line $\theta = \frac{\pi}{2}$
$-\theta$ for θ, $-r$ for r	The line $\theta = \frac{\pi}{2}$
$-r$ for r	The pole
$\pi + \theta$ for θ	The pole

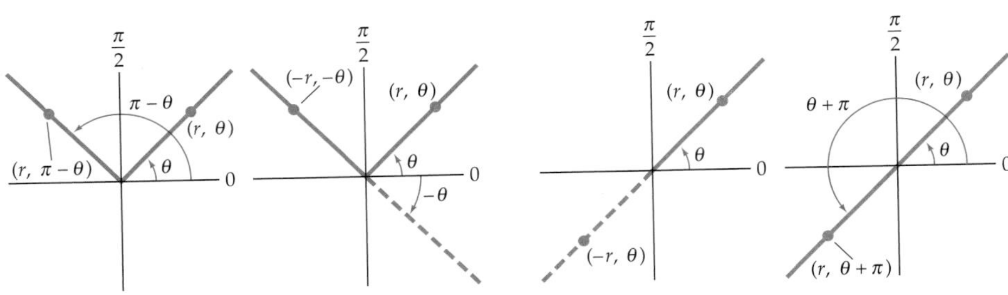

Symmetry with respect to the line $\theta = \frac{\pi}{2}$

Symmetry with respect to the pole

Figure 6.66

The graph of a polar equation may have a type of symmetry even though a test for that symmetry fails. For example, as we will see later, the graph of $r = \sin 2\theta$ is symmetric with respect to the line $\theta = 0$. However, using the symmetry test of substituting $-\theta$ for θ, we have

$$\sin 2(-\theta) = -\sin 2\theta = -r \neq r$$

Thus this test fails to show symmetry with respect to the line $\theta = 0$. The symmetry test of substituting $\pi - \theta$ for θ and $-r$ for r establishes symmetry with respect to the line $\theta = 0$.

EXAMPLE 1 Graph a Polar Equation

Show that the graph of $r = 4\cos\theta$ is symmetric with respect to the line $\theta = 0$. Graph the equation.

Solution

Test for symmetry with respect to the line $\theta = 0$. Replace θ by $-\theta$.

$$r = 4\cos(-\theta) = 4\cos\theta \qquad \bullet\ \cos(-\theta) = \cos\theta$$

Because replacing θ by $-\theta$ results in the original equation $r = 4\cos\theta$, the graph is symmetric with respect to the line $\theta = 0$.

To graph the equation, begin choosing various values of θ and finding the corresponding values of r. However, before doing so, consider two further observations that will reduce the number of θ-values you must choose.

First, because the cosine function is a periodic function with period 2π, it is only necessary to choose θ-values between 0 and 2π (0° and 360°). Second, when $\frac{\pi}{2} < \theta < \frac{3\pi}{2}$, $\cos\theta$ is negative, which means that any θ between these values will produce a negative r. Thus the point will be in the first or fourth quadrant. That is, we need consider only angles θ in the first or fourth quadrants. However, because the graph is symmetric with respect to the line $\theta = 0$, it is only necessary to choose values of θ between 0 and $\frac{\pi}{2}$.

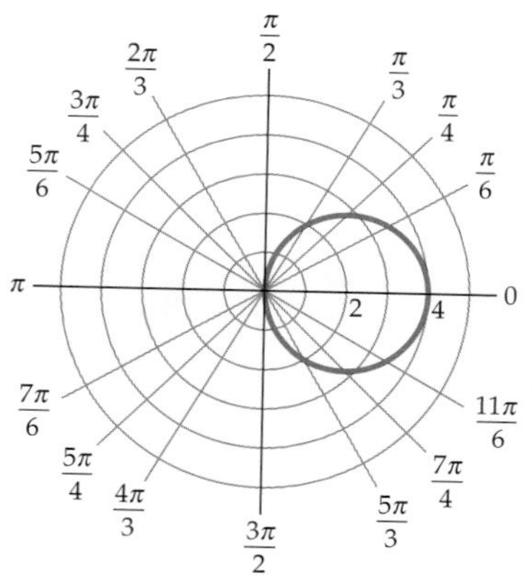

$r = 4\cos\theta$

Figure 6.67

By symmetry (applies to the negative θ columns)

θ	0	$\frac{\pi}{6}$	$\frac{\pi}{4}$	$\frac{\pi}{3}$	$\frac{\pi}{2}$	$-\frac{\pi}{6}$	$-\frac{\pi}{4}$	$-\frac{\pi}{3}$	$-\frac{\pi}{2}$
r	4.0	3.5	2.8	2.0	0.0	3.5	2.8	2.0	0.0

The graph of $r = 4\cos\theta$ is a circle with center at (2, 0). See **Figure 6.67.**

Try Exercise 14, page 423

Polar Equations of a Circle

The graph of the equation $r = a$ is a circle with center at the pole and radius a. See **Figure 6.68a.**

The graph of the equation $r = a\cos\theta$ is a circle that is symmetric with respect to the line $\theta = 0$. See **Figure 6.68b.**

The graph of $r = a\sin\theta$ is a circle that is symmetric with respect to the line $\theta = \frac{\pi}{2}$. See **Figure 6.68c.**

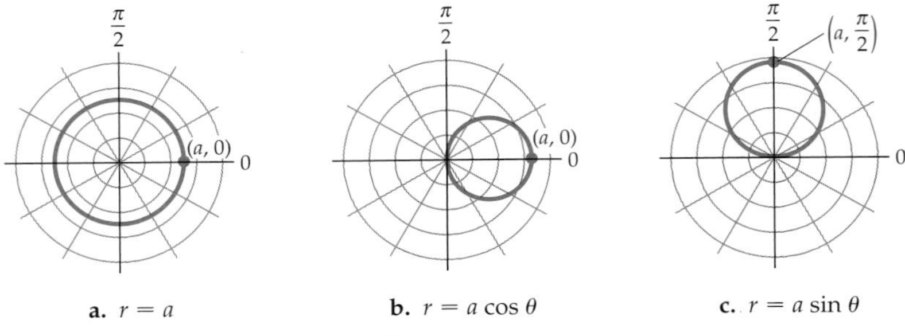

a. $r = a$ **b.** $r = a \cos \theta$ **c.** $r = a \sin \theta$

Figure 6.68

Just as there are specifically named curves in an xy-coordinate system (such as parabola and ellipse), there are named curves in an $r\theta$-coordinate system. Two of the many types are the *limaçon* and the *rose curve.*

Polar Equations of a Limaçon

The graph of the equation $r = a + b \cos \theta$ is a **limaçon** that is symmetric with respect to the line $\theta = 0$.

The graph of the equation $r = a + b \sin \theta$ is a limaçon that is symmetric with respect to the line $\theta = \frac{\pi}{2}$.

In the special case where $|a| = |b|$, the graph is called a **cardioid.**

The graph of $r = a + b \cos \theta$ is shown in **Figure 6.69** for various values of a and b.

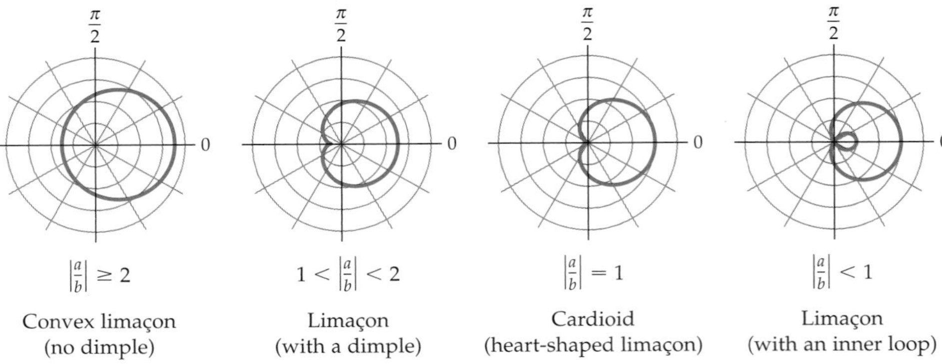

$\left|\frac{a}{b}\right| \geq 2$ Convex limaçon (no dimple)

$1 < \left|\frac{a}{b}\right| < 2$ Limaçon (with a dimple)

$\left|\frac{a}{b}\right| = 1$ Cardioid (heart-shaped limaçon)

$\left|\frac{a}{b}\right| < 1$ Limaçon (with an inner loop)

Figure 6.69

EXAMPLE 2 » Sketch the Graph of a Limaçon

Sketch the graph of $r = 2 - 2 \sin \theta$.

Solution

From the general equation of a limaçon $r = a + b \sin \theta$ with $|a| = |b|$ $(|2| = |-2|)$, the graph of $r = 2 - 2 \sin \theta$ is a cardioid that is symmetric with respect to the line $\theta = \frac{\pi}{2}$.

Because we know that the graph is heart-shaped, we can sketch the graph by finding r for a few values of θ. When $\theta = 0$, $r = 2$. When $\theta = \frac{\pi}{2}$, $r = 0$. When $\theta = \pi$, $r = 2$. When $\theta = \frac{3\pi}{2}$, $r = 4$. Sketching a heart-shaped curve through the four points

$$(2, 0), \quad \left(0, \frac{\pi}{2}\right), \quad (2, \pi), \quad \text{and} \quad \left(4, \frac{3\pi}{2}\right)$$

produces the cardioid in **Figure 6.70.**

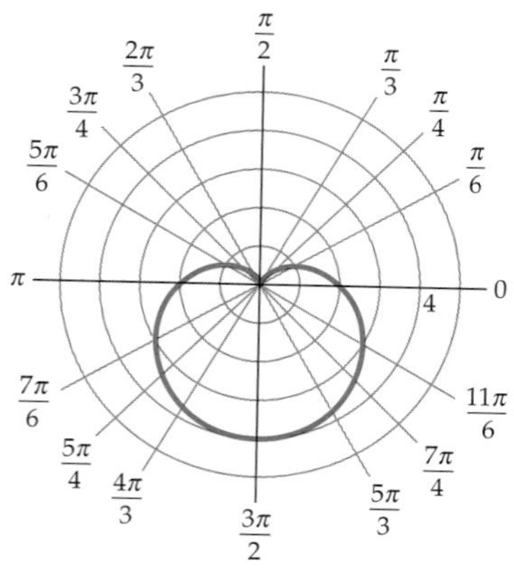

$r = 2 - 2 \sin \theta$

Figure 6.70

» *Try Exercise 20, page 424*

Example 3 shows how to use a graphing utility to construct a polar graph.

EXAMPLE 3 » Use a Graphing Utility to Graph a Polar Equation

Use a graphing utility to graph $r = 3 - 2 \cos \theta$.

Solution

From the general equation of a limaçon $r = a + b \cos \theta$, with $a = 3$ and $b = -2$, we know that the graph will be a limaçon with a dimple. The graph will be symmetric with respect to the line $\theta = 0$.

Use polar mode with angle measure in radians. Enter the equation $r = 3 - 2 \cos \theta$ in the polar function editing menu. The graph in **Figure 6.71** was produced with a TI-84 by using the following window:

θmin=0	Xmin=-6	Ymin=-4
θmax=2π	Xmax=6	Ymax=4
θstep=0.1	Xscl=1	Yscl=1

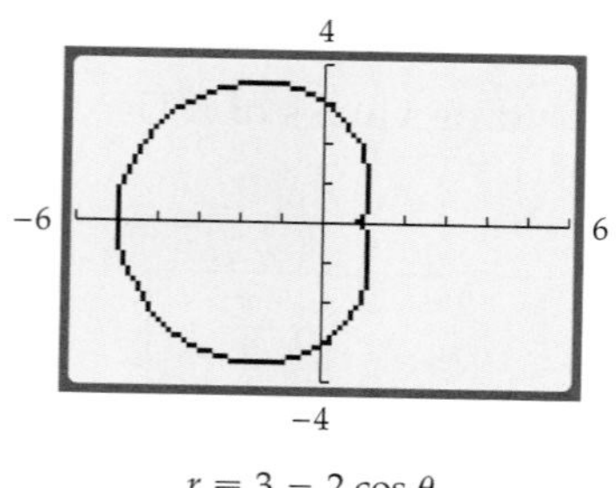

$r = 3 - 2 \cos \theta$

Figure 6.71

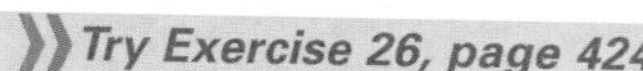

» *Try Exercise 26, page 424*

When using a graphing utility in polar mode, choose the value of θstep carefully. If θstep is set too small, the graphing utility may require an excessively long period of time to complete the graph. If θstep is set too large, the resulting graph may give only a very rough approximation of the actual graph.

Polar Equations of Rose Curves

The graphs of the equations $r = a\cos n\theta$ and $r = a\sin n\theta$ are **rose curves.** When n is an even number, the number of petals is $2n$. See **Figure 6.72a.** When n is an odd number, the number of petals is n. See **Figure 6.72b.**

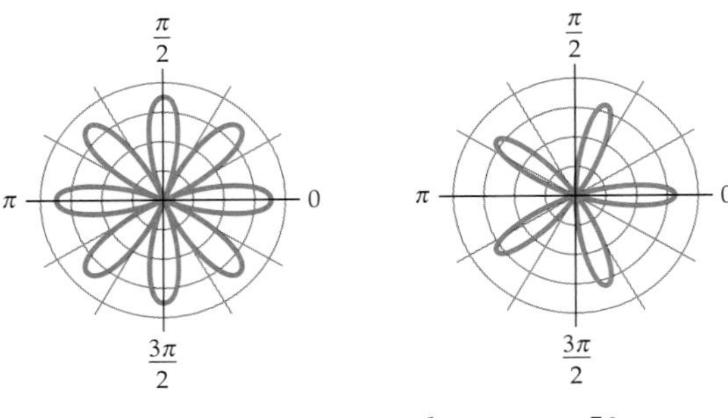

a. $r = a\cos 4\theta$
$n = 4$ is even, $2n = 8$ petals

b. $r = a\cos 5\theta$
$n = 5$ is odd, 5 petals

Figure 6.72

? QUESTION How many petals are in the graph of
a. $r = 4\cos 3\theta$? **b.** $r = 5\sin 2\theta$?

EXAMPLE 4 ›› Sketch the Graph of a Rose Curve

Sketch the graph of $r = 2\sin 3\theta$.

Solution

From the general equation of a rose curve $r = a\sin n\theta$, with $a = 2$ and $n = 3$, the graph of $r = 2\sin 3\theta$ is a rose curve that is symmetric with respect to the line $\theta = \frac{\pi}{2}$. Because n is an odd number ($n = 3$), there will be three petals in the graph.

Choose some values for θ and find the corresponding values of r. Use symmetry to sketch the graph. See **Figure 6.73.**

θ	0	$\frac{\pi}{18}$	$\frac{\pi}{6}$	$\frac{5\pi}{18}$	$\frac{\pi}{3}$	$\frac{7\pi}{18}$	$\frac{\pi}{2}$
r	0.0	1.0	2.0	1.0	0.0	−1.0	−2.0

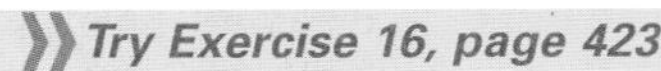
›› Try Exercise 16, page 423

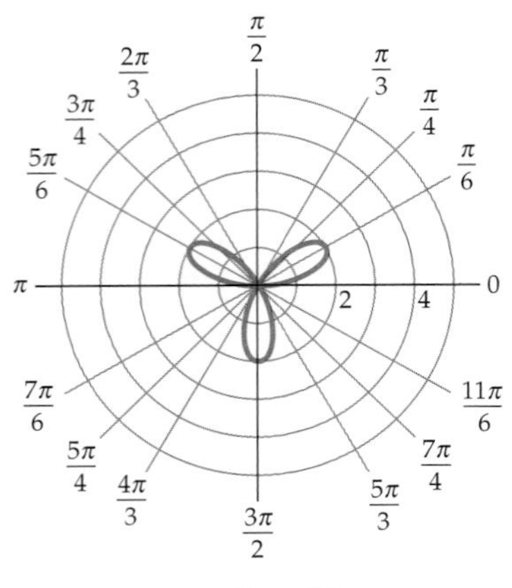

$r = 2\sin 3\theta$

Figure 6.73

? ANSWER **a.** Because 3 is an odd number, there are three petals in the graph.
b. Because 2 is an even number, there are $2(2) = 4$ petals in the graph.

EXAMPLE 5 » Use a Graphing Utility to Graph a Rose Curve

Use a graphing utility to graph $r = 4 \cos 2\theta$.

Solution

From the general equation of a rose curve $r = a \cos n\theta$, with $a = 4$ and $n = 2$, we know that the graph will be a rose curve with $2n = 4$ petals. The very tip of each petal will be $a = 4$ units away from the pole. Our symmetry tests also indicate that the graph is symmetric with respect to the line $\theta = 0$, the line $\theta = \frac{\pi}{2}$, and the pole.

Use polar mode with angle measure in radians. Enter the equation $r = 4 \cos 2\theta$ in the polar function editing menu. The graph in **Figure 6.74** was produced with a TI-84 by using the following window:

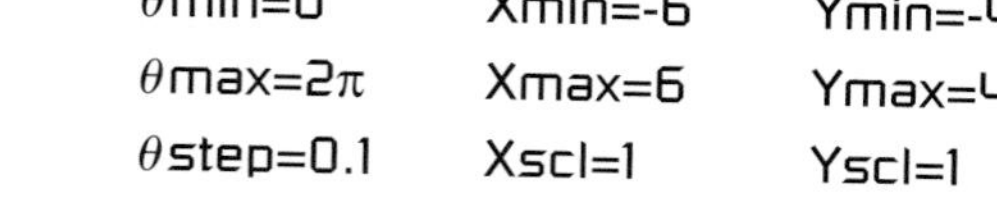

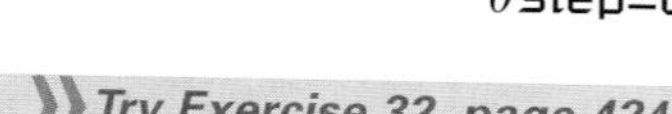

θmin=0	Xmin=-6	Ymin=-4
θmax=2π	Xmax=6	Ymax=4
θstep=0.1	Xscl=1	Yscl=1

» Try Exercise 32, page 424

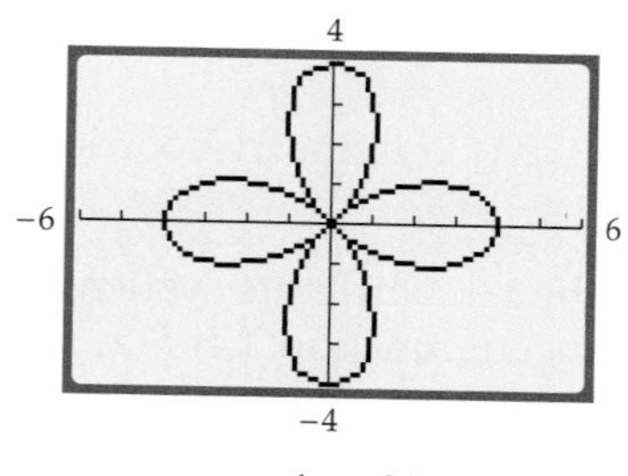

$r = 4 \cos 2\theta$

Figure 6.74

Transformations Between Rectangular and Polar Coordinates

A transformation between coordinate systems is a set of equations that relate the coordinates of a point in one system with the coordinates of the point in a second system. By superimposing a rectangular coordinate system on a polar system, we can derive the set of transformation equations.

Construct a polar coordinate system and a rectangular system such that the pole coincides with the origin and the polar axis coincides with the positive x-axis. Let a point P have coordinates (x, y) in one system and (r, θ) in the other $(r > 0)$.

From the definitions of $\sin \theta$ and $\cos \theta$, we have

$$\frac{x}{r} = \cos \theta \quad \text{or} \quad x = r \cos \theta$$

$$\frac{y}{r} = \sin \theta \quad \text{or} \quad y = r \sin \theta$$

It can be shown that these equations are also true when $r < 0$.

Thus, given the point (r, θ) in a polar coordinate system (see **Figure 6.75**), the coordinates of the point in the xy-coordinate system are given by

$$x = r \cos \theta \qquad y = r \sin \theta$$

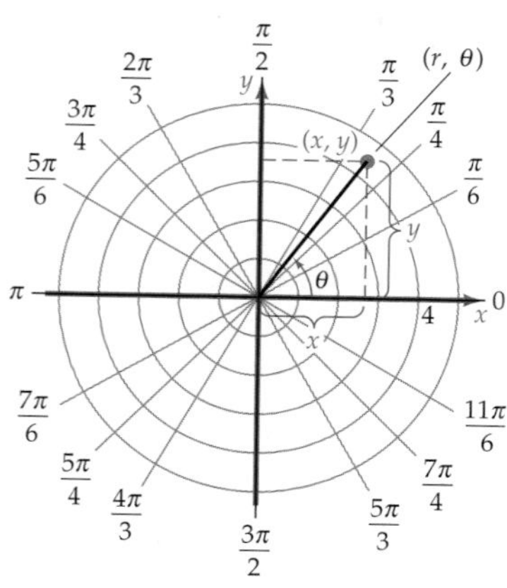

Figure 6.75

For example, to find the point in the xy-coordinate system that corresponds to the point $\left(4, \frac{2\pi}{3}\right)$ in the $r\theta$-coordinate system, substitute 4 for r and $\frac{2\pi}{3}$ for θ into the equations and simplify.

$$x = 4\cos\frac{2\pi}{3} = 4\left(-\frac{1}{2}\right) = -2$$

$$y = 4\sin\frac{2\pi}{3} = 4\left(\frac{\sqrt{3}}{2}\right) = 2\sqrt{3}$$

The point $\left(4, \frac{2\pi}{3}\right)$ in the $r\theta$-coordinate system is $(-2, 2\sqrt{3})$ in the xy-coordinate system.

To find the polar coordinates of a given point in the xy-coordinate system, use the Pythagorean Theorem and the definition of the tangent function. Let $P(x, y)$ be a point in the plane, and let r be the distance from the origin to the point P. Then $r = \sqrt{x^2 + y^2}$.

From the definition of the tangent function of an angle in a right triangle,

$$\tan\theta = \frac{y}{x}$$

Thus θ is the angle whose tangent is $\frac{y}{x}$. The quadrant for θ depends on the sign of x and the sign of y.

The equations of transformation between a polar and a rectangular coordinate system are summarized as follows:

Transformations Between Polar and Rectangular Coordinates

Given the point (r, θ) in the polar coordinate system, the transformation equations to change from polar to rectangular coordinates are

$$x = r\cos\theta \qquad y = r\sin\theta$$

Given the point (x, y) in the rectangular coordinate system, the transformation equations to change from rectangular to polar coordinates are

$$r = \sqrt{x^2 + y^2} \qquad \tan\theta = \frac{y}{x}, \quad x \neq 0$$

where $r \geq 0$, $0 \leq \theta < 2\pi$, and θ is chosen so that the point lies in the appropriate quadrant. If $x = 0$, then $\theta = \frac{\pi}{2}$ or $\theta = \frac{3\pi}{2}$.

EXAMPLE 6 ⟫ Transform from Polar to Rectangular Coordinates

Find the rectangular coordinates of the points whose polar coordinates are: **a.** $\left(6, \frac{3\pi}{4}\right)$ **b.** $(-4, 30°)$

Solution

Use the transformation equations $x = r\cos\theta$ and $y = r\sin\theta$.

a. $x = 6\cos\frac{3\pi}{4} = -3\sqrt{2}$ $\quad y = 6\sin\frac{3\pi}{4} = 3\sqrt{2}$

The rectangular coordinates of $\left(6, \frac{3\pi}{4}\right)$ are $\left(-3\sqrt{2}, 3\sqrt{2}\right)$.

b. $x = -4\cos 30° = -2\sqrt{3}$ $\quad y = -4\sin 30° = -2$

The rectangular coordinates of $(-4, 30°)$ are $\left(-2\sqrt{3}, -2\right)$.

⟫ *Try Exercise 44, page 424*

EXAMPLE 7 ⟫ Transform from Rectangular to Polar Coordinates

Find the polar coordinates of the point whose rectangular coordinates are $\left(-2, -2\sqrt{3}\right)$.

Solution

Use the transformation equations $r = \sqrt{x^2 + y^2}$ and $\tan\theta = \frac{y}{x}$.

$$r = \sqrt{(-2)^2 + (-2\sqrt{3})^2} = \sqrt{4 + 12} = \sqrt{16} = 4$$

$$\tan\theta = \frac{-2\sqrt{3}}{-2} = \sqrt{3}$$

From this and the fact that $\left(-2, -2\sqrt{3}\right)$ lies in the third quadrant, $\theta = \frac{4\pi}{3}$.

The polar coordinates of $\left(-2, -2\sqrt{3}\right)$ are $\left(4, \frac{4\pi}{3}\right)$.

⟫ *Try Exercise 48, page 424*

Write Polar Coordinate Equations as Rectangular Equations and Rectangular Coordinate Equations as Polar Equations

Using the transformation equations, it is possible to write a polar coordinate equation in rectangular form or a rectangular coordinate equation in polar form.

EXAMPLE 8 » Write a Polar Coordinate Equation in Rectangular Form

Find a rectangular form of the equation $r^2 \cos 2\theta = 3$.

Solution

$$\begin{aligned} r^2 \cos 2\theta &= 3 \\ r^2(1 - 2\sin^2\theta) &= 3 && \bullet \cos 2\theta = 1 - 2\sin^2\theta \\ r^2 - 2r^2\sin^2\theta &= 3 \\ r^2 - 2(r\sin\theta)^2 &= 3 \\ x^2 + y^2 - 2y^2 &= 3 && \bullet r^2 = x^2 + y^2;\ \sin\theta = \frac{y}{r} \\ x^2 - y^2 &= 3 \end{aligned}$$

A rectangular form of $r^2 \cos 2\theta = 3$ is $x^2 - y^2 = 3$.

» *Try Exercise 58, page 424*

Sometimes the procedures of squaring each side of a polar equation or multiplying each side of a polar equation by r can be used to write the polar equation in rectangular form. In Example 9 we multiply each side of a polar equation by r to convert the equation from polar to rectangular form.

EXAMPLE 9 » Write a Polar Coordinate Equation in Rectangular Form

Find a rectangular form of the equation $r = 8 \cos \theta$.

Solution

$$\begin{aligned} r &= 8\cos\theta \\ r^2 &= 8r\cos\theta && \bullet \text{ Multiply each side by } r. \\ x^2 + y^2 &= 8x && \bullet \text{ Use the equations } r^2 = x^2 + y^2 \text{ and } x = r\cos\theta. \\ (x^2 - 8x) + y^2 &= 0 && \bullet \text{ Subtract } 8x \text{ from each side.} \\ (x^2 - 8x + 16) + y^2 &= 16 && \bullet \text{ Complete the square in } x. \\ (x - 4)^2 + y^2 &= 4^2 && \bullet \text{ Write in standard form.} \end{aligned}$$

A rectangular form of $r = 8 \cos \theta$ is $(x - 4)^2 + y^2 = 4^2$. The graph of each of these equations is a circle with center (4, 0) and radius 4.

» *Try Exercise 50, page 424*

take note

Squaring each side of the transformation equation $r = \sqrt{x^2 + y^2}$ produces $r^2 = x^2 + y^2$. This equation is often used to write polar coordinate equations in rectangular form.

In Example 10, we use the transformation equations $x = r \cos \theta$ and $y = r \sin \theta$ to convert a rectangular coordinate equation to polar form.

EXAMPLE 10 Write a Rectangular Coordinate Equation in Polar Form

Find a polar form of the equation $x^2 + y^2 - 2x = 3$.

Solution

$$x^2 + y^2 - 2x = 3$$

$$(r\cos\theta)^2 + (r\sin\theta)^2 - 2r\cos\theta = 3$$ • Use the transformation equations $x = r\cos\theta$ and $y = r\sin\theta$.

$$r^2(\cos^2\theta + \sin^2\theta) - 2r\cos\theta = 3$$ • Simplify.

$$r^2 - 2r\cos\theta = 3$$

A polar form of $x^2 + y^2 - 2x = 3$ is $r^2 - 2r\cos\theta = 3$.

Try Exercise 70, page 424

Topics for Discussion

1. In what quadrant is the point $(-2, 150°)$ located?
2. To explain why the graph of $\theta = \dfrac{\pi}{6}$ is a line, a tutor rewrites the equation in the form $\theta = \dfrac{\pi}{6} + 0 \cdot r$. In this form, regardless of the value of r, $\theta = \dfrac{\pi}{6}$. Use an analogous approach to explain why the graph of $r = a$ is a circle.
3. Two students use a graphing calculator to graph the polar equation $r = 2\sin\theta$. One graph appears to be a circle, and the other graph appears to be an ellipse. Both graphs are correct. Explain.
4. Is the graph of $r^2 = 6\cos 2\theta$ a rose curve with four petals? Explain your answer.

Exercise Set 6.5

In Exercises 1 to 8, plot the point on a polar coordinate system.

1. $(2, 60°)$
2. $(3, -90°)$
3. $(1, 315°)$
4. $(2, 400°)$
5. $\left(-2, \dfrac{\pi}{4}\right)$
6. $\left(4, \dfrac{7\pi}{6}\right)$
7. $\left(-3, \dfrac{5\pi}{3}\right)$
8. $(-3, \pi)$

In Exercises 9 to 24, sketch the graph of each polar equation.

9. $r = 3$
10. $r = 5$
11. $\theta = 2$
12. $\theta = -\dfrac{\pi}{3}$
13. $r = 6\cos\theta$
14. $r = 4\sin\theta$
15. $r = 4\cos 2\theta$
16. $r = 5\cos 3\theta$

17. $r = 2 \sin 5\theta$

18. $r = 3 \cos 5\theta$

19. $r = 2 - 3 \sin \theta$

20. $r = 2 - 2 \cos \theta$

21. $r = 4 + 3 \sin \theta$

22. $r = 2 + 4 \sin \theta$

23. $r = 2[1 + 1.5 \sin(-\theta)]$

24. $r = 4(1 - \sin \theta)$

In Exercises 25 to 40, use a graphing utility to graph each equation.

25. $r = 3 + 3 \cos \theta$

26. $r = 4 - 4 \sin \theta$

27. $r = 4 \cos 3\theta$

28. $r = 2 \sin 4\theta$

29. $r = 3 \sec \theta$

30. $r = 4 \csc \theta$

31. $r = -5 \csc \theta$

32. $r = -4 \sec \theta$

33. $r = 4 \sin(3.5\theta)$

34. $r = 6 \cos(2.25\theta)$

35. $r = \theta, 0 \le \theta \le 6\pi$

36. $r = -\theta, 0 \le \theta \le 6\pi$

37. $r = 2^{\theta}, 0 \le \theta \le 2\pi$

38. $r = \dfrac{1}{\theta}, 0 \le \theta \le 4\pi$

39. $r = \dfrac{6 \cos 7\theta + 2 \cos 3\theta}{\cos \theta}$

40. $r = \dfrac{4 \cos 3\theta + \cos 5\theta}{\cos \theta}$

In Exercises 41 to 48, transform the given coordinates to the indicated ordered pair. Round approximate angle measures to the nearest tenth of a degree.

41. $(1, -\sqrt{3})$ to (r, θ)

42. $(-2\sqrt{3}, 2)$ to (r, θ)

43. $\left(-3, \dfrac{2\pi}{3}\right)$ to (x, y)

44. $\left(2, -\dfrac{\pi}{3}\right)$ to (x, y)

45. $\left(0, -\dfrac{\pi}{2}\right)$ to (x, y)

46. $\left(3, \dfrac{5\pi}{6}\right)$ to (x, y)

47. $(3, 4)$ to (r, θ)

48. $(12, -5)$ to (r, θ)

In Exercises 49 to 62, find a rectangular form of each of the equations.

49. $r = 3 \cos \theta$

50. $r = 2 \sin \theta$

51. $r = 3 \sec \theta$

52. $r = 4 \csc \theta$

53. $r = 4$

54. $\theta = \dfrac{\pi}{4}$

55. $\theta = \dfrac{\pi}{6}$

56. $r \cos \theta = -4$

57. $r = \tan \theta$

58. $r = \cot \theta$

59. $r = \dfrac{2}{1 + \cos \theta}$

60. $r = \dfrac{2}{1 - \sin \theta}$

61. $r(\sin \theta - 2 \cos \theta) = 6$

62. $r(2 \cos \theta + \sin \theta) = 3$

In Exercises 63 to 74, find a polar form of each of the equations.

63. $y = 2$

64. $x = -4$

65. $y = \sqrt{3}x$

66. $y = x^2$

67. $x = 3$

68. $xy = 4$

69. $x^2 + y^2 = 4$

70. $2x - 3y = 6$

71. $x^2 = 8y$

72. $y^2 = 4y$

73. $x^2 - y^2 = 25$

74. $x^2 + 4y^2 = 16$

In Exercises 75 to 82, use a graphing utility to graph each equation.

75. $r = 3 \cos\left(\theta + \dfrac{\pi}{4}\right)$

76. $r = 2 \sin\left(\theta - \dfrac{\pi}{6}\right)$

77. $r = 2 \sin\left(2\theta - \dfrac{\pi}{3}\right)$

78. $r = 3 \cos\left(2\theta + \dfrac{\pi}{4}\right)$

79. $r = 2 + 2 \sin\left(\theta - \dfrac{\pi}{6}\right)$

80. $r = 3 - 2 \cos\left(\theta + \dfrac{\pi}{3}\right)$

81. $r = 1 + 3 \cos\left(\theta + \dfrac{\pi}{3}\right)$

82. $r = 2 - 4 \sin\left(\theta - \dfrac{\pi}{4}\right)$

Connecting Concepts

83. Explain why the graph of $r^2 = \cos^2\theta$ and the graph of $r = \cos\theta$ are not the same.

84. Explain why the graph of $r = \cos 2\theta$ and the graph of $r = 2\cos^2\theta - 1$ are identical.

In Exercises 85 to 92, use a graphing utility to graph each equation.

85. $r^2 = 4\cos 2\theta$ (lemniscate)

86. $r^2 = -2\sin 2\theta$ (lemniscate)

87. $r = 2(1 + \sec\theta)$ (conchoid)

88. $r = 2\cos 2\theta \sec\theta$ (strophoid)

89. $r\theta = 2$ (spiral)

90. $r = 2\sin\theta\cos^2 2\theta$ (bifolium)

91. $r = |\theta|$

92. $r = \ln\theta$

93. The graph of

$$r = 1.5^{\sin\theta} - 2.5\cos 4\theta + \sin^7\frac{\theta}{15}$$

is a *butterfly curve* similar to the one shown below.

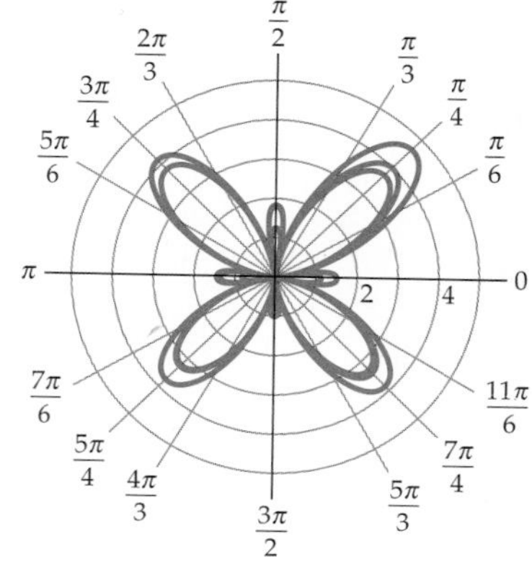

Use a graphing utility to graph the butterfly curve for

a. $0 \le \theta \le 5\pi$ **b.** $0 \le \theta \le 20\pi$

For additional information on butterfly curves, read "The Butterfly Curve" by Temple H. Fay, *The American Mathematical Monthly,* vol. 96, no. 5 (May 1989), p. 442.

Projects

1. A Polar Distance Formula Let $P_1(r_1, \theta_1)$ and $P_2(r_2, \theta_2)$ be two distinct points in the $r\theta$-plane.

a. Verify that the distance d between the points is

$$d = \sqrt{(r_1)^2 + (r_2)^2 - 2r_1r_2\cos(\theta_2 - \theta_1)}$$

b. Use the above formula to find the distance (to the nearest hundredth) between $(3, 60°)$ and $(5, 170°)$.

c. Does the formula

$$d = \sqrt{(r_1)^2 + (r_2)^2 - 2r_1r_2\cos(\theta_1 - \theta_2)}$$

also produce the correct distance between P_1 and P_2? Explain.

2. Another Polar Form for a Circle

a. Verify that the graph of the polar equation $r = a\sin\theta + b\cos\theta$ is a circle. Assume that a and b are not both 0.

b. What are the center (in rectangular coordinates) and the radius of the circle?

Section 6.6

- Polar Equations of the Conics
- Graph a Conic Given in Polar Form
- Write the Polar Equation of a Conic

Polar Equations of the Conics

PREPARE FOR THIS SECTION

Prepare for this section by completing the following exercises. The answers can be found on page A31.

PS1. Find the eccentricity of the graph of $\frac{x^2}{25} + \frac{y^2}{16} = 1$. [6.2]

PS2. What is the equation of the directrix of the graph of $y^2 = 4x$? [6.1]

PS3. Solve $y = 2(1 + yx)$ for y. [1.1]

PS4. For the function $f(x) = \frac{3}{1 + \sin x}$, what is the smallest positive value of x that must be excluded from the domain of f? [1.3/2.2]

PS5. Let e be the eccentricity of a hyperbola. Which of the following statements is true: $e = 0$, $0 < e < 1$, $e = 1$, or $e > 1$? [6.3]

PS6. Write $\frac{4 \sec x}{2 \sec x - 1}$ in terms of $\cos x$. [3.1]

Polar Equations of the Conics

The definition of a parabola was given in terms of a point (the focus) and a line (the directrix). The definitions of both an ellipse and a hyperbola were given in terms of two points (the foci). It is possible to define each conic in terms of a point and a line.

Calculus Connection

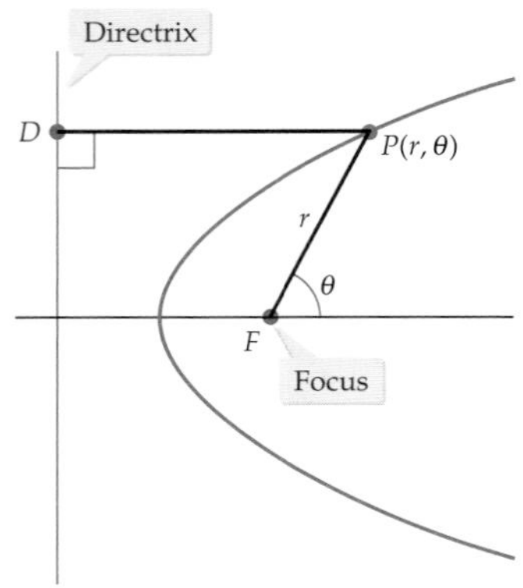

Figure 6.76

Focus-Directrix Definitions of the Conics

Let F be a fixed point and D a fixed line in a plane. Consider the set of all points P such that $\frac{d(P, F)}{d(P, D)} = e$, where e is a constant. The graph is a parabola for $e = 1$, an ellipse for $0 < e < 1$, and a hyperbola for $e > 1$. See **Figure 6.76.**

The fixed point is a focus of the conic, and the fixed line is a directrix. The constant e is the eccentricity of the conic. Using this definition, we can derive the polar equations of the conics.

Standard Forms of the Polar Equations of the Conics

Let the pole be a focus of a conic section of eccentricity e, with directrix d units from the focus. Then the equation of the conic is given by one of the following:

$$r = \frac{ed}{1 + e\cos\theta} \quad (1)$$

Vertical directrix to the right of the pole

$$r = \frac{ed}{1 - e\cos\theta} \quad (2)$$

Vertical directrix to the left of the pole

$$r = \frac{ed}{1 + e\sin\theta} \quad (3)$$

Horizontal directrix above the pole

$$r = \frac{ed}{1 - e\sin\theta} \quad (4)$$

Horizontal directrix below the pole

When the equation involves $\cos\theta$, the line $\theta = 0$ is an axis of symmetry.

When the equation involves $\sin\theta$, the line $\theta = \frac{\pi}{2}$ is an axis of symmetry.

Graphs of examples are shown in **Figure 6.77**.

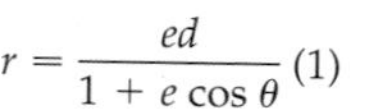

$r = \frac{ed}{1 + e\cos\theta}$ (1)

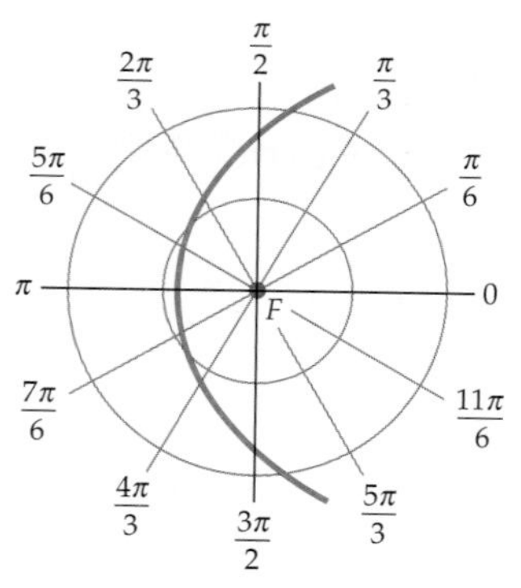

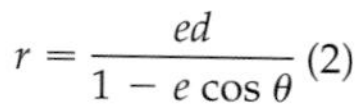

$r = \frac{ed}{1 - e\cos\theta}$ (2)

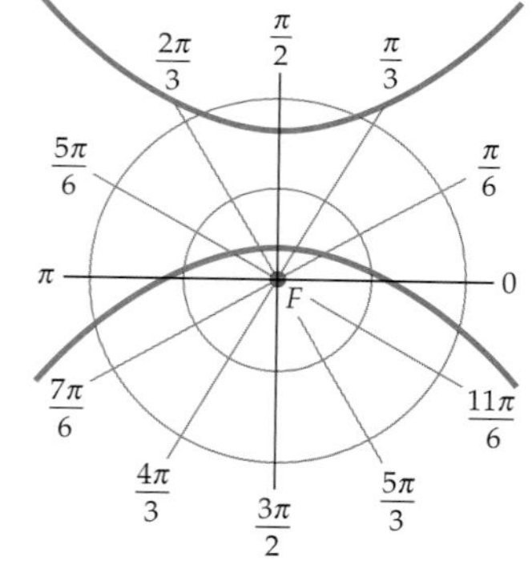

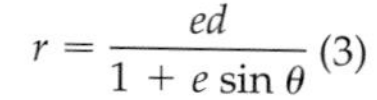

$r = \frac{ed}{1 + e\sin\theta}$ (3)

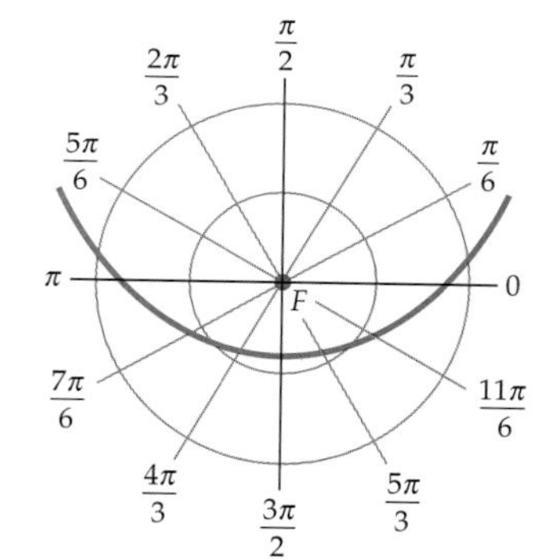

$r = \frac{ed}{1 - e\sin\theta}$ (4)

Figure 6.77

QUESTION Is the graph of $r = \frac{4}{5 - 3\sin\theta}$ a parabola, an ellipse, or a hyperbola?

ANSWER The eccentricity is $\frac{3}{5}$, which is less than 1. The graph is an ellipse.

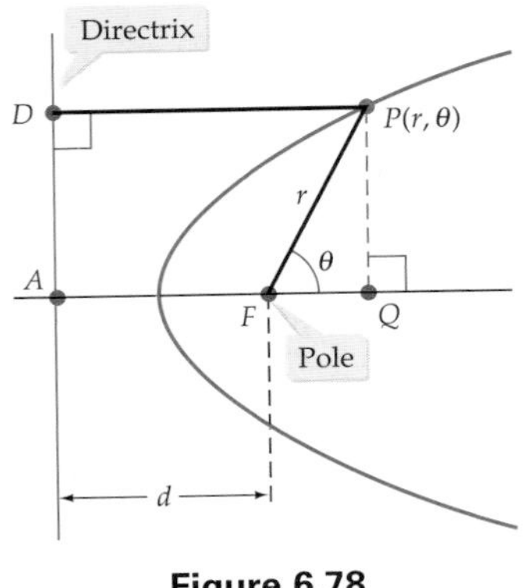

Figure 6.78

We will derive Equation (2). Let $P(r, \theta)$ be any point on a conic section. Then, by definition,

$$\frac{d(P, F)}{d(P, D)} = e \quad \text{or} \quad d(P, F) = e \cdot d(P, D)$$

From **Figure 6.78**, $d(P, F) = r$ and $d(P, D) = d(A, Q)$. But note that

$$d(A, Q) = d(A, F) + d(F, Q) = d + r\cos\theta$$

Thus

$$r = e(d + r\cos\theta) \qquad \text{• } d(P, F) = e \cdot d(P, D)$$

$$= ed + er\cos\theta$$

$$r - er\cos\theta = ed \qquad \text{• Subtract } er\cos\theta \text{ from each side.}$$

$$r = \frac{ed}{1 - e\cos\theta} \qquad \text{• Solve for } r.$$

The remaining equations can be derived in a similar manner.

Graph a Conic Given in Polar Form

EXAMPLE 1 » Sketch the Graph of a Hyperbola Given in Polar Form

Describe and sketch the graph of $r = \dfrac{8}{2 - 3\sin\theta}$.

Solution

Write the equation in standard form by dividing the numerator and denominator by 2, the constant term in the denominator.

$$r = \frac{4}{1 - \frac{3}{2}\sin\theta}$$

Because e is the coefficient of $\sin\theta$ and $e = \frac{3}{2} > 1$, the graph is a hyperbola with a focus at the pole. Because the equation contains the expression $\sin\theta$, the transverse axis is on the line $\theta = \frac{\pi}{2}$.

To find the vertices, choose θ equal to $\frac{\pi}{2}$ and $\frac{3\pi}{2}$. The corresponding values of r are -8 and $\frac{8}{5}$. The vertices are $\left(-8, \frac{\pi}{2}\right)$ and $\left(\frac{8}{5}, \frac{3\pi}{2}\right)$. By choosing θ equal to 0 and π, we can determine the points $(4, 0)$ and $(4, \pi)$ on

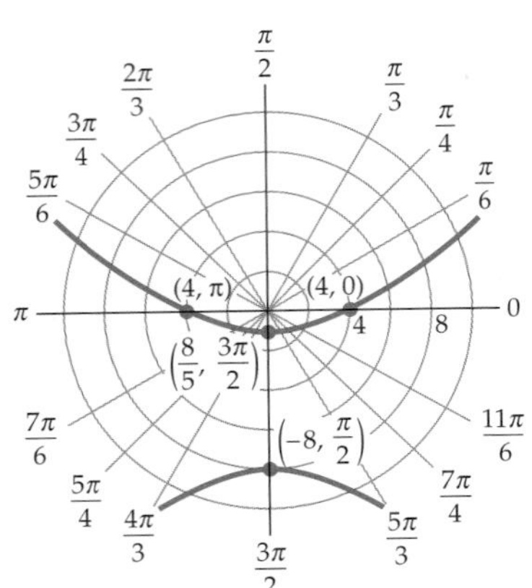

Figure 6.79

the upper branch of the hyperbola. The lower branch can be determined by symmetry.

Plot some points (r, θ) for additional values of θ and corresponding values of r. See **Figure 6.79.**

» *Try Exercise 2, page 431*

EXAMPLE 2 » Sketch the Graph of an Ellipse Given in Polar Form

Describe and sketch the graph of $r = \dfrac{4}{2 + \cos\theta}$.

Solution

Write the equation in standard form by dividing the numerator and denominator by 2, which is the constant term in the denominator.

$$r = \frac{2}{1 + \frac{1}{2}\cos\theta}$$

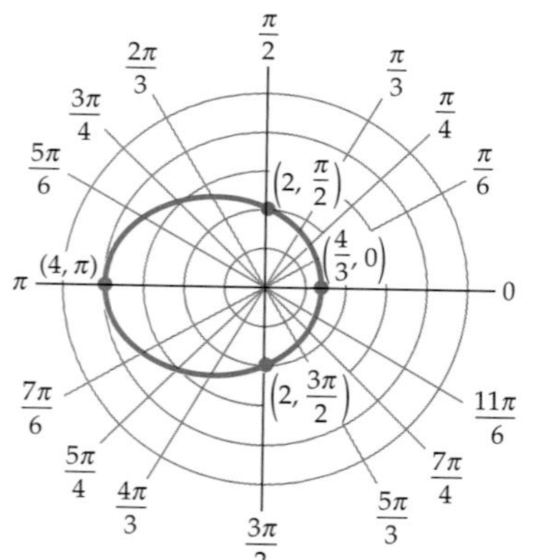

Figure 6.80

Thus $e = \frac{1}{2}$ and the graph is an ellipse with a focus at the pole. Because the equation contains the expression $\cos\theta$, the major axis is on the polar axis.

To find the vertices, choose θ equal to 0 and π. The corresponding values for r are $\frac{4}{3}$ and 4. The vertices on the major axis are $\left(\frac{4}{3}, 0\right)$ and $(4, \pi)$. Plot some points (r, θ) for additional values of θ and the corresponding values of r. Two possible points are $\left(2, \frac{\pi}{2}\right)$ and $\left(2, \frac{3\pi}{2}\right)$. See the graph of the ellipse in **Figure 6.80.**

» *Try Exercise 4, page 431*

Write the Polar Equation of a Conic

EXAMPLE 3 » Find the Equation of a Conic in Polar Form

Find the polar equation of the parabola, shown in **Figure 6.81** on page 430, with vertex at $\left(2, \frac{\pi}{2}\right)$ and focus at the pole.

Continued ▶

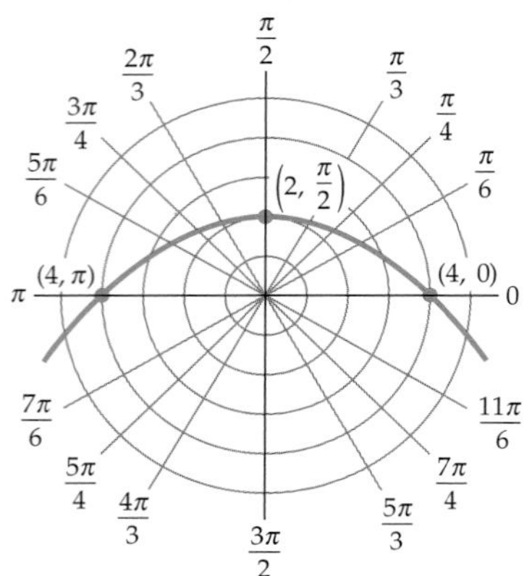

Figure 6.81

Solution

Because the vertex is on the line $\theta = \dfrac{\pi}{2}$ and the focus is at the pole, the axis of symmetry is the line $\theta = \dfrac{\pi}{2}$. Thus the equation of the parabola must involve $\sin\theta$. The parabola has a horizontal directrix above the pole, so the equation has the form

$$r = \frac{ed}{1 + e\sin\theta}$$

The distance from the vertex to the focus is 2, so the distance from the focus to the directrix is 4. Because the graph of the equation is a parabola, the eccentricity is 1. The equation is

$$r = \frac{(1)(4)}{1 + (1)\sin\theta} \qquad \bullet\ e = 1,\ d = 4$$

$$r = \frac{4}{1 + \sin\theta}$$

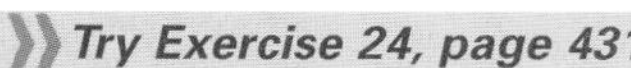
Try Exercise 24, page 431

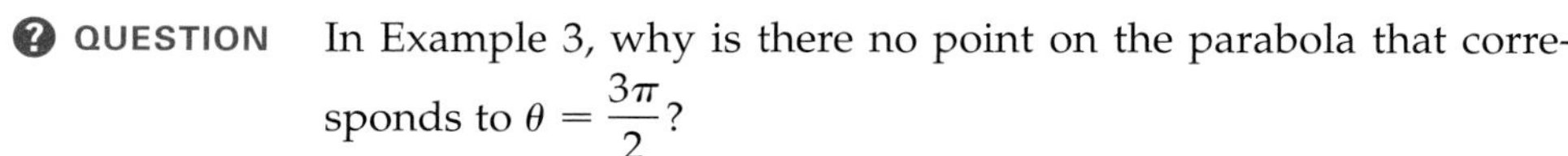
QUESTION In Example 3, why is there no point on the parabola that corresponds to $\theta = \dfrac{3\pi}{2}$?

Topics for Discussion

1. Is the graph of $r = \dfrac{12}{2 + \cos\theta}$ a parabola? Explain.
2. The graph of $r = \dfrac{2\sec\theta}{2\sec\theta + 1}$ is an ellipse except for the fact that it has two holes. Where are the holes located?
3. Are there two different ellipses that have a focus at the pole and a vertex at $\left(1, \dfrac{\pi}{2}\right)$? Explain.
4. Does the parabola given by $r = \dfrac{6}{1 + \sin\theta}$ have a horizontal axis of symmetry or a vertical axis of symmetry? Explain.

ANSWER When $\theta = \dfrac{3\pi}{2}$, $\sin\theta = -1$. Thus $1 + \sin\theta = 0$, and $r = \dfrac{4}{1 + \sin\theta}$ is undefined.

Exercise Set 6.6

In Exercises 1 to 14, describe and sketch the graph of each equation.

1. $r = \dfrac{12}{3 - 6\cos\theta}$

2. $r = \dfrac{8}{2 - 4\cos\theta}$

3. $r = \dfrac{8}{4 + 3\sin\theta}$

4. $r = \dfrac{6}{3 + 2\cos\theta}$

5. $r = \dfrac{9}{3 - 3\sin\theta}$

6. $r = \dfrac{5}{2 - 2\sin\theta}$

7. $r = \dfrac{10}{5 + 6\cos\theta}$

8. $r = \dfrac{8}{2 + 4\cos\theta}$

9. $r = \dfrac{4\sec\theta}{2\sec\theta - 1}$

10. $r = \dfrac{3\sec\theta}{2\sec\theta + 2}$

11. $r = \dfrac{12\csc\theta}{6\csc\theta - 2}$

12. $r = \dfrac{3\csc\theta}{2\csc\theta + 2}$

13. $r = \dfrac{3}{\cos\theta - 1}$

14. $r = \dfrac{2}{\sin\theta + 2}$

In Exercises 15 to 20, find a rectangular equation for each graph in Exercises 1 to 6.

In Exercises 21 to 28, find a polar equation of the conic with focus at the pole and the given eccentricity and directrix.

21. $e = 2, r\cos\theta = -1$

22. $e = \dfrac{3}{2}, r\sin\theta = 1$

23. $e = 1, r\sin\theta = 2$

24. $e = 1, r\cos\theta = -2$

25. $e = \dfrac{2}{3}, r\sin\theta = -4$

26. $e = \dfrac{1}{2}, r\cos\theta = 2$

27. $e = \dfrac{3}{2}, r = 2\sec\theta$

28. $e = \dfrac{3}{4}, r = 2\csc\theta$

29. Find the polar equation of the parabola with a focus at the pole and vertex $(2, \pi)$.

30. Find the polar equation of the ellipse with a focus at the pole, vertex at $(4, 0)$, and eccentricity $\dfrac{1}{2}$.

31. Find the polar equation of the hyperbola with a focus at the pole, vertex at $\left(1, \dfrac{3\pi}{2}\right)$, and eccentricity 2.

32. Find the polar equation of the ellipse with a focus at the pole, vertex at $\left(2, \dfrac{3\pi}{2}\right)$, and eccentricity $\dfrac{2}{3}$.

In Exercises 33 to 40, use a graphing utility to graph each equation. Write a sentence that explains how to obtain the graph from the graph of *r* as given in the exercise listed to the right of each equation.

33. $r = \dfrac{12}{3 - 6\cos\left(\theta - \dfrac{\pi}{6}\right)}$ (Compare with Exercise 1.)

34. $r = \dfrac{8}{2 - 4\cos\left(\theta - \dfrac{\pi}{2}\right)}$ (Compare with Exercise 2.)

35. $r = \dfrac{8}{4 + 3\sin(\theta - \pi)}$ (Compare with Exercise 3.)

36. $r = \dfrac{6}{3 + 2\cos\left(\theta - \dfrac{\pi}{3}\right)}$ (Compare with Exercise 4.)

37. $r = \dfrac{9}{3 - 3\sin\left(\theta + \dfrac{\pi}{6}\right)}$ (Compare with Exercise 5.)

38. $r = \dfrac{5}{2 - 2\sin\left(\theta + \dfrac{\pi}{2}\right)}$ (Compare with Exercise 6.)

39. $r = \dfrac{10}{5 + 6\cos(\theta + \pi)}$ (Compare with Exercise 7.)

40. $r = \dfrac{8}{2 + 4\cos\left(\theta + \dfrac{\pi}{3}\right)}$ (Compare with Exercise 8.)

Connecting Concepts

In Exercises 41 to 46, use a graphing utility to graph each equation.

41. $r = \dfrac{3}{3 - \sec\theta}$

42. $r = \dfrac{5}{4 - 2\csc\theta}$

43. $r = \dfrac{3}{1 + 2\csc\theta}$

44. $r = \dfrac{4}{1 + 3\sec\theta}$

45. $r = 4\sin\sqrt{2}\theta, 0 \le \theta \le 12\pi$

46. $r = 4\cos\sqrt{3}\theta, 0 \le \theta \le 8\pi$

47. Let $P(r, \theta)$ satisfy the equation $r = \dfrac{ed}{1 - e\cos\theta}$. Show that $\dfrac{d(P, F)}{d(P, D)} = e$.

48. Show that the equation of a conic with a focus at the pole and directrix $r\sin\theta = d$ is given by $r = \dfrac{ed}{1 + e\sin\theta}$.

Projects

1. Polar Equation of a Line Verify that the polar equation of a line that is d units from the pole is given by

$$r = \frac{d}{\cos(\theta - \theta_p)}$$

where θ_p is the angle from the polar axis to a line segment that passes through the pole and is perpendicular to the line.

2. Polar Equation of a Circle That Passes Through the Pole Verify that the polar equation of a circle with center (a, θ_c) that passes through the pole is given by

$$r = 2a\cos(\theta - \theta_c)$$

where a is the length of the radius and θ_c is the angle from the polar axis to the radius that connects the center of the circle with the pole.

Section 6.7

- Parametric Equations
- Eliminate the Parameter of a Pair of Parametric Equations
- Time as a Parameter
- The Brachistochrone Problem
- Parametric Equations and Projectile Motion

Parametric Equations

PREPARE FOR THIS SECTION

Prepare for this section by completing the following exercises. The answers can be found on page A31.

PS1. Complete the square of $y^2 + 3y$ and write the result as the square of a binomial. [1.1]

PS2. If $x = 2t + 1$ and $y = x^2$, write y in terms of t.

PS3. Identify the graph of $\left(\dfrac{x-2}{3^2}\right)^2 + \left(\dfrac{y-3}{2^2}\right)^2 = 1$. [6.2]

PS4. Let $x = \sin t$ and $y = \cos t$. What is the value of $x^2 + y^2$? [3.1]

PS5. If $x = t + 1$ and $y = 4 - t$, what is the ordered pair (x, y) that corresponds to $t = 5$?

PS6. What are the domain and range of $f(t) = 3\cos 2t$? Write the answers using interval notation. [2.5]

Parametric Equations

The graph of a function is a graph for which no vertical line can intersect the graph more than once. For a graph that is not the graph of a function (an ellipse or a hyperbola, for example), it is frequently useful to describe the graph by *parametric equations.*

Curve and Parametric Equations

Let t be a number in an interval I. A **curve** is a set of ordered pairs (x, y), where

$$x = f(t), \qquad y = g(t) \quad \text{for } t \in I$$

The variable t is called a **parameter,** and the equations $x = f(t)$ and $y = g(t)$ are **parametric equations.**

For instance,

$$x = 2t - 1, \qquad y = 4t + 1 \quad \text{for } t \in (-\infty, \infty)$$

is an example of a pair of parametric equations. By choosing arbitrary values of t, ordered pairs (x, y) can be created, as shown in the table below.

t	$x = 2t - 1$	$y = 4t + 1$	(x, y)
-2	-5	-7	$(-5, -7)$
0	-1	1	$(-1, 1)$
$\frac{1}{2}$	0	3	$(0, 3)$
2	3	9	$(3, 9)$

By plotting the points and drawing a curve through the points, a graph of the parametric equations is produced. See **Figure 6.82.**

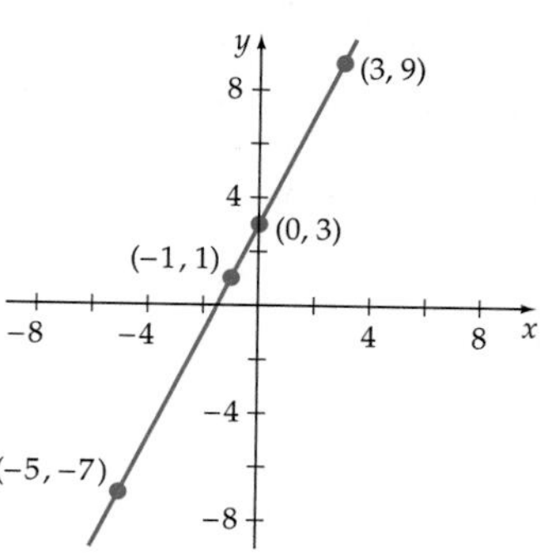

Figure 6.82

QUESTION If $x = t^2 + 1$ and $y = 3 - t$, what ordered pair corresponds to $t = -3$?

ANSWER (10, 6)

EXAMPLE 1 ⟫ Sketch the Graph of a Curve Given in Parametric Form

Sketch the graph of the curve given by the parametric equations

$$x = t^2 + t, \qquad y = t - 1 \quad \text{for } t \in R$$

Solution

Begin by making a table of values of t and the corresponding values of x and y. Five values of t were arbitrarily chosen for the table that follows. Many more values might be necessary to determine an accurate graph.

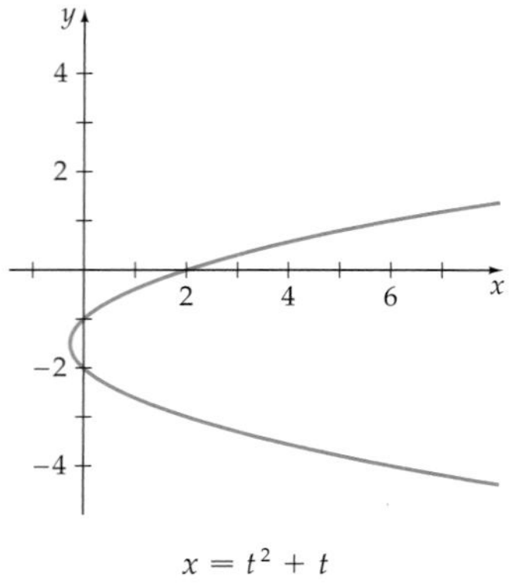

$x = t^2 + t$
$y = t - 1$

Figure 6.83

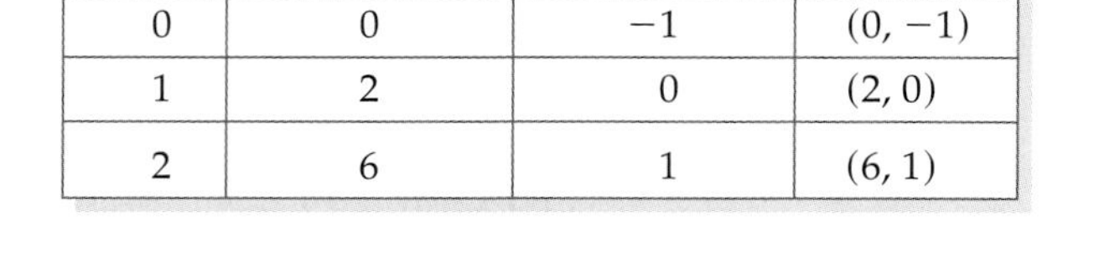

t	$x = t^2 + t$	$y = t - 1$	(x, y)
-2	2	-3	$(2, -3)$
-1	0	-2	$(0, -2)$
0	0	-1	$(0, -1)$
1	2	0	$(2, 0)$
2	6	1	$(6, 1)$

Graph the ordered pairs (x, y) and then draw a smooth curve through the points. See **Figure 6.83**.

⟫ *Try Exercise 6, page 440*

Eliminate the Parameter of a Pair of Parametric Equations

It may not be clear from Example 1 and the corresponding graph that the curve is a parabola. By **eliminating the parameter,** we can write one equation in x and y that is equivalent to the two parametric equations.

To eliminate the parameter, solve $y = t - 1$ for t.

$$y = t - 1 \qquad \text{or} \qquad t = y + 1$$

Substitute $y + 1$ for t in $x = t^2 + t$ and then simplify.

$$x = (y + 1)^2 + (y + 1)$$

$$x = y^2 + 3y + 2 \qquad \text{• The equation of a parabola}$$

Complete the square and write the equation in standard form.

$$\left(x + \frac{1}{4}\right) = \left(y + \frac{3}{2}\right)^2 \qquad \text{• This is the equation of a parabola vertex at } \left(-\frac{1}{4}, -\frac{3}{2}\right).$$

EXAMPLE 2 Eliminate the Parameter and Sketch the Graph of a Curve

Eliminate the parameter and sketch the curve of the parametric equations

$$x = \sin t, \qquad y = \cos t \quad \text{for } 0 \le t \le 2\pi$$

Solution

The process of eliminating the parameter sometimes involves trigonometric identities. To eliminate the parameter for the equations, square each side of each equation and then add.

$$\begin{aligned} x^2 &= \sin^2 t \\ y^2 &= \cos^2 t \\ x^2 + y^2 &= \sin^2 t + \cos^2 t \end{aligned}$$

Thus, using the trigonometric identity $\sin^2 t + \cos^2 t = 1$, we get

$$x^2 + y^2 = 1$$

This is the equation of a circle with center $(0, 0)$ and radius equal to 1. See **Figure 6.84.**

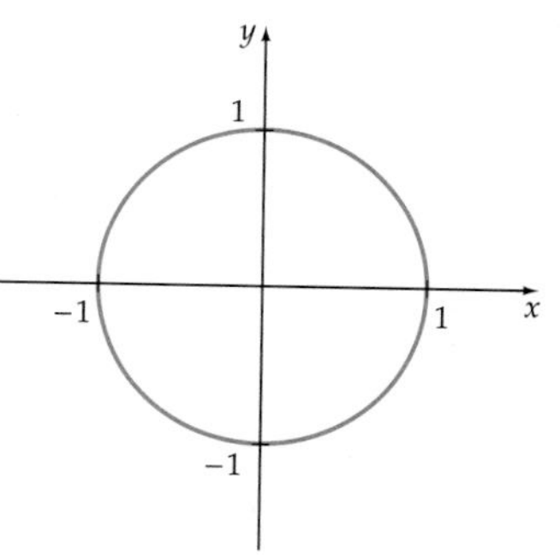

Figure 6.84

Try Exercise 12, page 440

A parametric representation of a curve is not unique. That is, it is possible that a curve may be given by many different pairs of parametric equations. We will demonstrate this by using the equation of a line and providing two different parametric representations of the line.

Consider a line with slope m passing through the point (x_1, y_1). By the point-slope formula, the equation of the line is

$$y - y_1 = m(x - x_1)$$

Let $t = x - x_1$. Then $y - y_1 = mt$. A parametric representation is

$$x = x_1 + t, \qquad y = y_1 + mt \quad \text{for } t \text{ a real number} \tag{1}$$

Let $x - x_1 = \cot t$. Then $y - y_1 = m \cot t$. A parametric representation is

$$x = x_1 + \cot t, \qquad y = y_1 + m \cot t \quad \text{for } 0 < t < \pi \tag{2}$$

It can be verified that Equations (1) and (2) represent the original line.

Example 3 illustrates that the domain of the parameter t can be used to determine the domain and range of the curve defined by the parametric equations.

EXAMPLE 3 ›› Sketch the Graph of a Curve Given by Parametric Equations

Eliminate the parameter and sketch the graph of the curve that is given by the parametric equations

$$x = 2 + 3\cos t, \qquad y = 3 + 2\sin t \quad \text{for } 0 \le t \le \pi$$

Solution

Solve each equation for its trigonometric function.

$$\frac{x-2}{3} = \cos t \qquad \frac{y-3}{2} = \sin t \tag{3}$$

Using the trigonometric identity $\cos^2 t + \sin^2 t = 1$, we have

$$\cos^2 t + \sin^2 t = \left(\frac{x-2}{3}\right)^2 + \left(\frac{y-3}{2}\right)^2 = 1$$

$$\frac{(x-2)^2}{9} + \frac{(y-3)^2}{4} = 1$$

This is the equation of an ellipse with center at (2, 3) and major axis parallel to the x-axis. However, because $0 \le t \le \pi$, it follows that $-1 \le \cos t \le 1$ and $0 \le \sin t \le 1$. Therefore, we have

$$-1 \le \frac{x-2}{3} \le 1 \qquad 0 \le \frac{y-3}{2} \le 1$$ • Using Equations (3)

Solving these inequalities for x and y yields

$$-1 \le x \le 5 \qquad \text{and} \qquad 3 \le y \le 5$$

Because the values of y are between 3 and 5, the graph of the parametric equations is only the top half of the ellipse. See **Figure 6.85**.

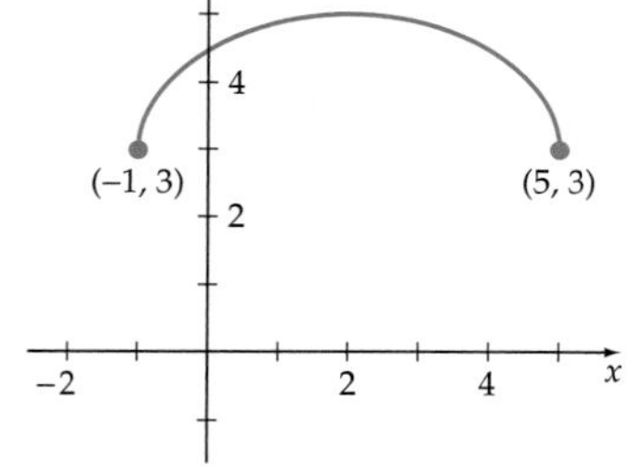

Figure 6.85

›› *Try Exercise 14, page 440*

Time as a Parameter

Parametric equations are often used to show the movement of a point or object as it travels along a curve. For instance, consider the parametric equations

$$x = t^2, \qquad y = t + 1 \quad \text{for } -2 \le t \le 3$$

For any given value of t in the interval $[-2, 3]$, we can evaluate $x = t^2$ and $y = t + 1$ to determine a point (x, y) on the curve defined by the equations. See the following table and **Figure 6.86**.

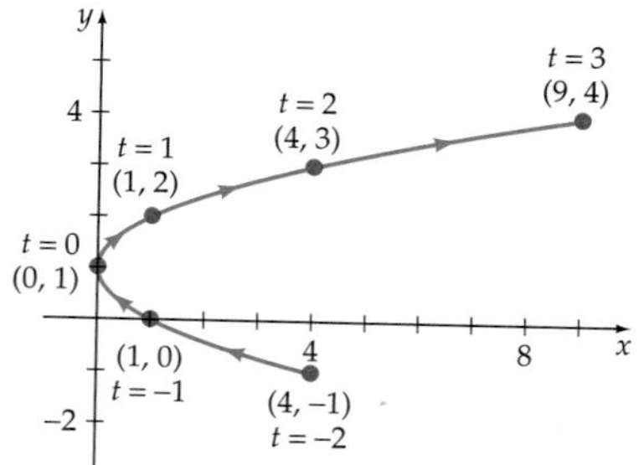

Figure 6.86

t	$x = t^2$	$y = t + 1$	(x, y)
−2	4	−1	(4, −1)
−1	1	0	(1, 0)
0	0	1	(0, 1)
1	1	2	(1, 2)
2	4	3	(4, 3)
3	9	4	(9, 4)

As t takes on values from -2 to 3, the curve defined by the parametric equations is traced in a particular direction. The direction in which the curve is traced—by increasing values of the parameter t—is referred to as its **orientation.** In **Figure 6.86** the arrowheads show the orientation of the curve.

In Example 4 we let the parameter t represent time. Then the parametric equations $x = f(t)$ and $y = g(t)$ indicate the x- and y-coordinates of a moving point as a function of t.

EXAMPLE 4 Describe the Motion of a Point

A point moves in a plane such that its position $P(x, y)$ at time t is

$$x = \sin t, \qquad y = \cos t \quad \text{for } 0 \le t \le 2\pi$$

Describe the motion of the point.

Solution

In Example 2 we determined that the graph of $x = \sin t, y = \cos t$ for $0 \le t \le 2\pi$ is a circle with center (0, 0) and radius 1. When $t = 0$, the point P is at $(\sin 0, \cos 0) = (0, 1)$. When $t = \frac{\pi}{2}$, P is at (1, 0); when $t = \pi$, P is at $(0, -1)$; when $t = \frac{3\pi}{2}$, P is at $(-1, 0)$; and when $t = 2\pi$, P is back to its starting position (0, 1). Thus P starts at the point (0, 1) and rotates clockwise around the circle with center (0, 0) and radius 1 as the time t increases from 0 to 2π. See **Figure 6.87.**

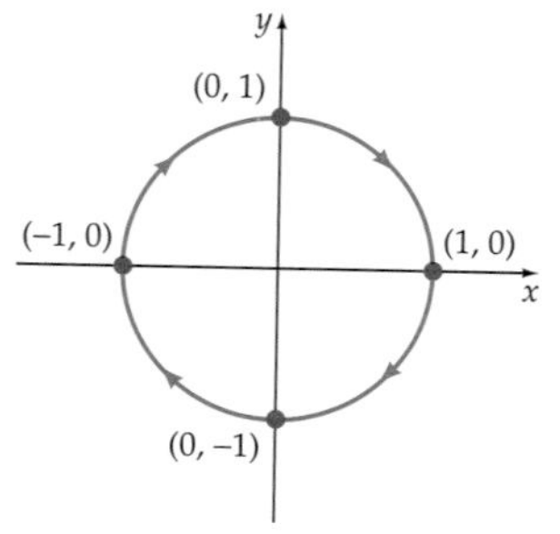

Figure 6.87

Try Exercise 22, page 440

The Brachistochrone Problem

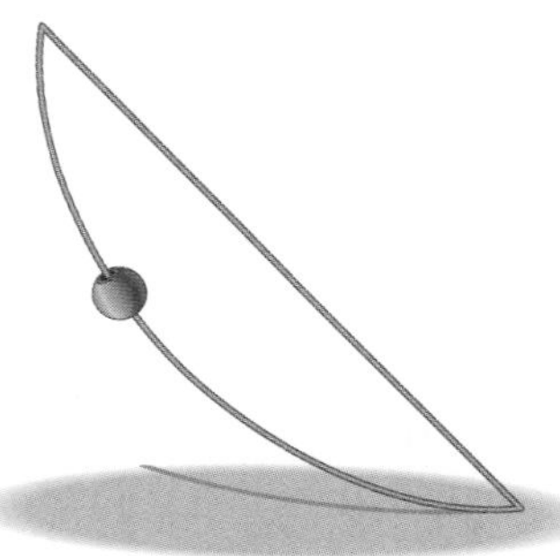
Figure 6.88

One famous problem, involving a bead traveling down a frictionless wire, was posed in 1696 by the mathematician Johann Bernoulli. The problem was to determine the shape of a wire a bead could slide down such that the distance between two points was traveled in the shortest time. Problems that involve "shortest time" are called *brachistochrone problems.* They are very important in physics and form the basis for much of the classical theory of light propagation.

The answer to Bernoulli's problem is an arc of an inverted cycloid. See **Figure 6.88.** A **cycloid** is formed by letting a circle of radius a roll on a straight line without slipping. See **Figure 6.89.** The curve traced by a point on the circumference of the circle is a cycloid. To find an equation for this curve, begin by placing a circle tangent to the x-axis with a point P on the circle and at the origin of a rectangular coordinate system.

Roll the circle along the x-axis. After the radius of the circle has rotated through an angle θ, the coordinates of the point $P(x, y)$ can be given by

$$x = h - a\sin\theta, \qquad y = k - a\cos\theta \tag{4}$$

where $C(h, k)$ is the current center of the circle.

Because the radius of the circle is a, $k = a$. See **Figure 6.89**. Because the circle rolls without slipping, the arc length subtended by θ equals h. Thus $h = a\theta$. Substituting for h and k in Equations (4), we have, after factoring,

$$x = a(\theta - \sin\theta), \qquad y = a(1 - \cos\theta) \quad \text{for } \theta \geq 0$$

See **Figure 6.90**.

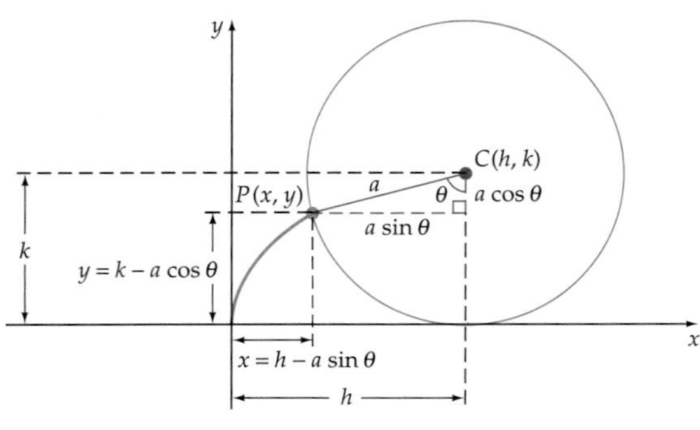

Figure 6.89

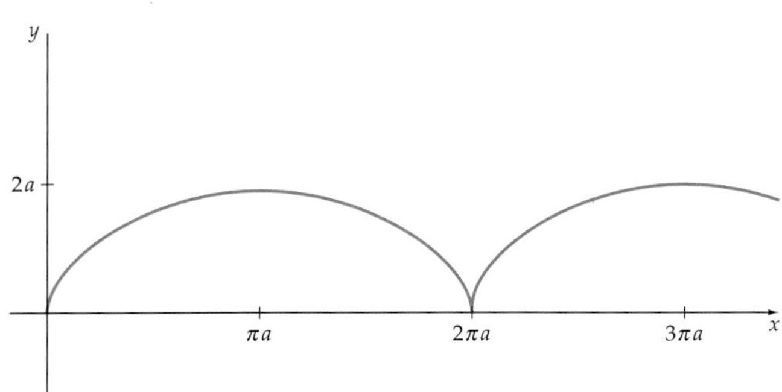

$x = a(\theta - \sin\theta), y = a(1 - \cos\theta)$

Figure 6.90
A cycloid

EXAMPLE 5 Graph a Cycloid

Use a graphing utility to graph the cycloid given by

$$x = 4(\theta - \sin\theta), \qquad y = 4(1 - \cos\theta) \quad \text{for } 0 \leq \theta \leq 4\pi$$

Solution

Although θ is the parameter in the above equations, many graphing utilities, such as the TI-83/TI-83 Plus/TI-84 Plus, use T as the parameter for parametric equations. Thus to graph the equations for $0 \leq \theta \leq 4\pi$, we use Tmin = 0 and Tmax = 4π, as shown below. Use radian mode and parametric mode to produce the graph in **Figure 6.91**.

Tmin=0	Xmin=-6	Ymin=-4
Tmax=4π	Xmax=16π	Ymax=10
Tstep=0.5	Xscl=2π	Yscl=1

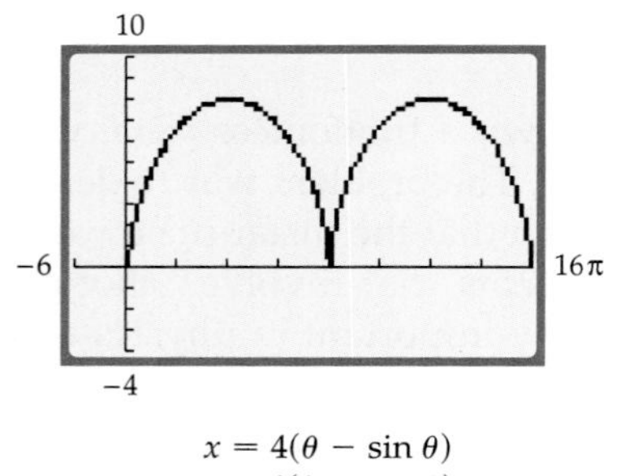

$x = 4(\theta - \sin\theta)$
$y = 4(1 - \cos\theta)$

Figure 6.91

Try Exercise 30, page 441

Parametric Equations and Projectile Motion

The path of a projectile (assume air resistance is negligible) that is launched at an angle θ from the horizon with an initial velocity of v_0 feet per second is given by the parametric equations

$$x = (v_0 \cos \theta)t, \qquad y = -16t^2 + (v_0 \sin \theta)t$$

where t is the time in seconds since the projectile was launched.

EXAMPLE 6 Graph the Path of a Projectile

Use a graphing utility to graph the path of a projectile that is launched at an angle of $\theta = 32°$ with an initial velocity of 144 feet per second. Use the graph to determine (to the nearest foot) the maximum height of the projectile and the range of the projectile. Assume the ground is level.

Solution

Use degree mode and parametric mode. Graph the parametric equations

$$x = (144 \cos 32°)t, \qquad y = -16t^2 + (144 \sin 32°)t \quad \text{for } 0 \le t \le 5$$

to produce the graph in **Figure 6.92**. Use the TRACE feature to determine that the maximum height of 91 feet is attained when $t \approx 2.38$ seconds and that the projectile strikes the ground about 581 feet downrange when $t \approx 4.76$ seconds.

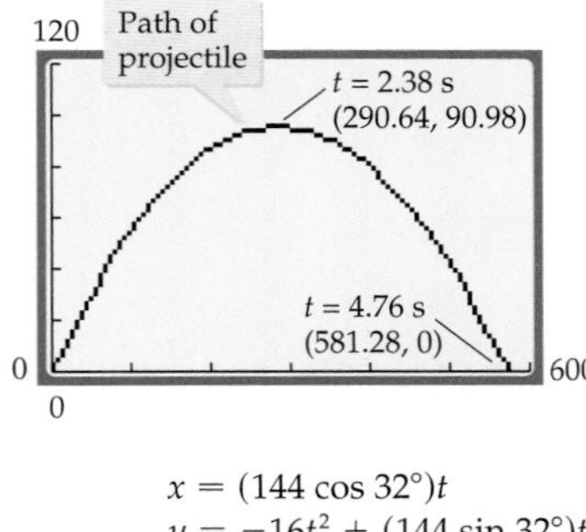

$x = (144 \cos 32°)t$
$y = -16t^2 + (144 \sin 32°)t$

Figure 6.92

In **Figure 6.92,** the angle of launch does not appear to be 32° because 1 foot on the x-axis is smaller than 1 foot on the y-axis.

Try Exercise 34, page 441

Topics for Discussion

1. It is always possible to eliminate the parameter of a pair of parametric equations. Do you agree? Explain.
2. The line $y = 3x + 5$ has more than one parametric representation. Do you agree? Explain.

3. Parametric equations are used only to graph functions. Do you agree? Explain.

4. Every function $y = f(x)$ can be written in parametric form by letting $x = t$ and $y = f(t)$. Do you agree? Explain.

Exercise Set 6.7

In Exercises 1 to 10, graph the parametric equations by plotting several points.

1. $x = 2t, y = -t$, for $t \in R$

2. $x = -3t, y = 6t$, for $t \in R$

3. $x = -t, y = t^2 - 1$, for $t \in R$

4. $x = 2t, y = 2t^2 - t + 1$, for $t \in R$

5. $x = t^2, y = t^3$, for $t \in R$

6. $x = t^2 + 1, y = t^2 - 1$, for $t \in R$

7. $x = 2 \cos t, y = 3 \sin t$, for $0 \le t < 2\pi$

8. $x = 1 - \sin t, y = 1 + \cos t$, for $0 \le t < 2\pi$

9. $x = 2^t, y = 2^{t+1}$, for $t \in R$

10. $x = t - 1, y = \sqrt{t}$, for $0 \le t \le 9$

In Exercises 11 to 18, eliminate the parameter and graph the equation.

11. $x = \sec t, y = \tan t$, for $-\frac{\pi}{2} < t < \frac{\pi}{2}$

12. $x = 3 + 2 \cos t, y = -1 - 3 \sin t$, for $0 \le t < 2\pi$

13. $x = 2 - t^2, y = 3 + 2t^2$, for $t \in R$

14. $x = 1 + t^2, y = 2 - t^2$, for $t \in R$

15. $x = \cos^3 t, y = \sin^3 t$, for $0 \le t < 2\pi$

16. $x = 3 \sin t, y = \cos t$, for $0 \le t \le \frac{\pi}{2}$

17. $x = \sqrt{t+1}, y = t$, for $t \ge -1$

18. $x = \sqrt{t}, y = 2t - 1$, for $t \ge 0$

In Exercises 19 to 24, the parameter *t* represents time and the parametric equations *x* = *f*(*t*) and *y* = *g*(*t*) indicate the *x*- and *y*-coordinates of a moving point *P* as a function of *t*. Describe the motion of the point as *t* increases.

19. $x = 2 + 3 \cos t, y = 3 + 2 \sin t$, for $0 \le t \le \pi$

20. $x = \sin t, y = -\cos t$, for $0 \le t \le \frac{3\pi}{2}$

21. $x = 2t - 1, y = t + 1$, for $0 \le t \le 3$

22. $x = t + 1, y = \sqrt{t}$, for $0 \le t \le 4$

23. $x = \tan\left(\frac{\pi}{4} - t\right), y = \sec\left(\frac{\pi}{4} - t\right)$, for $0 \le t \le \frac{\pi}{2}$

24. $x = 1 - t, y = t^2$, for $0 \le t \le 2$

25. Eliminate the parameter for the curves

$$C_1: \quad x = 2 + t^2, \quad y = 1 - 2t^2$$

and

$$C_2: \quad x = 2 + t, \quad y = 1 - 2t$$

and then discuss the differences between their graphs.

26. Eliminate the parameter for the curves

$$C_1: \quad x = \sec^2 t, \quad y = \tan^2 t$$

and

$$C_2: \quad x = 1 + t^2, \quad y = t^2$$

for $0 \le t < \frac{\pi}{2}$, and then discuss the differences between their graphs.

27. Sketch the graph of

$$x = \sin t, \qquad y = \csc t \quad \text{for } 0 < t \le \frac{\pi}{2}$$

Sketch another graph for the same pair of equations but choose the domain of t to be $\pi < t \le \frac{3\pi}{2}$.

28. Discuss the differences between the graphs of

$$C_1:\quad x = \cos t,\quad y = \cos^2 t$$

and
$$C_2:\quad x = \sin t,\quad y = \sin^2 t$$

for $0 \le t \le \pi$.

29. Use a graphing utility to graph the cycloid $x = 2(t - \sin t)$, $y = 2(1 - \cos t)$ for $0 \le t < 2\pi$.

30. Use a graphing utility to graph the cycloid $x = 3(t - \sin t)$, $y = 3(1 - \cos t)$ for $0 \le t \le 12\pi$.

In Exercises 31 and 32, the parameter t represents time and the parametric equations $x = f(t)$ and $y = g(t)$ indicate the x- and y-coordinates of a moving object as a function of t.

31. SIMULATE UNIFORM MOTION A Corvette is traveling east at 65 miles per hour. A Hummer is traveling north at 60 miles per hour. Both vehicles are heading toward the same intersection. At time $t = 0$ hours, the Corvette is 6 miles from the intersection and the Hummer is 5 miles from the intersection.

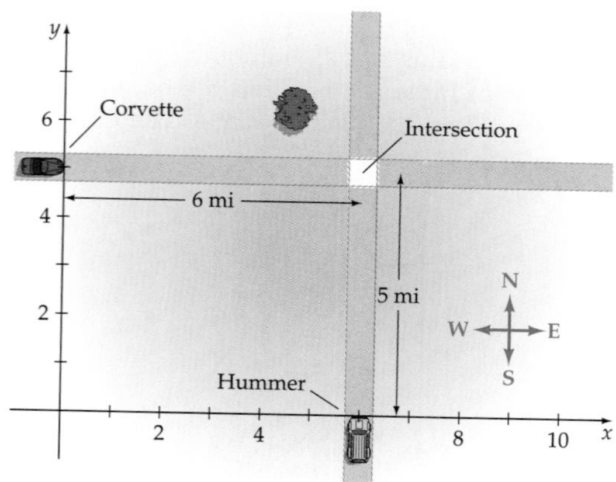

a. The location of the Corvette is described by the equations $x = 65t$, $y = 5$, for $t \ge 0$. Find a pair of parametric equations that describes the location of the Hummer at the time t, where t is measured in hours.

b. Graph the parametric equations of the vehicles to produce a simulation. In the MODE menu, select SIMUL so that the motion of both vehicles will be shown simultaneously. Use the following window settings.

Tmin=0	Xmin=–0.6	Ymin=0
Tmax=0.11	Xmax=10	Ymax=7
Tstep=0.0005	Xscl=1	Yscl=1

Which vehicle was the first to reach the intersection?

32. SIMULATE UNIFORM MOTION A Learjet is flying west at 420 miles per hour. A twin engine Piper Seneca is flying north at 235 miles per hour. Both Planes are flying at the same altitude. At time $t = 0$ hours, the Learjet is 800 miles from the intersection point of the flight paths and the Piper Seneca is 415 miles from the intersection point.

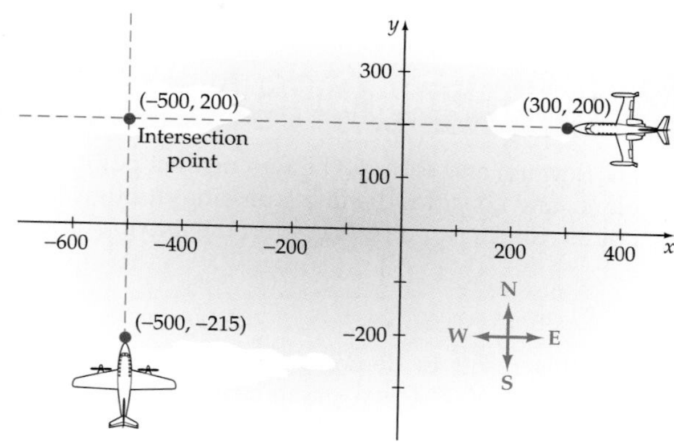

a. The location of the Piper Seneca is described by $x = -500$, $y = -215 + 235t$, $t \ge 0$. Find a pair of parametric equations that describes the location of the Learjet at time t, where t is measured in hours.

b. Graph the parametric equations of the planes to produce a simulation. In the MODE menu, select SIMUL so that the motion of both planes will be shown simultaneously. Use the following window settings.

Tmin=0	Xmin=–600	Ymin=–330
TMAX=2	Xmax=400	Ymax=330
Tstep=0.005	Xscl=100	Yscl=100

Which plane was the first to reach the intersection point?

In Exercises 33 to 36, graph the path of the projectile that is launched at an angle of θ with the horizon with an initial velocity of v_0. In each exercise, use the graph to determine the maximum height and the range of the projectile (to the nearest foot). Also state the time t at which the projectile reaches its maximum height and the time it hits the ground. Assume that the ground is level and the only force acting on the projectile is gravity.

33. $\theta = 55°$, $v_0 = 210$ feet per second

34. $\theta = 35°$, $v_0 = 195$ feet per second

35. $\theta = 42°$, $v_0 = 315$ feet per second

36. $\theta = 52°$, $v_0 = 315$ feet per second

Parametric equations of the form $x = a \sin \alpha t$, $y = b \cos \beta t$, for $t \geq 0$, are encountered in electrical circuit theory. The graphs of these equations are called *Lissajous figures.* In Exercises 37 to 40, use Tstep = 0.05 and $0 \leq t \leq 2\pi$.

37. Graph: $x = 5 \sin 2t$, $y = 5 \cos t$

38. Graph: $x = 5 \sin 3t$, $y = 5 \cos 2t$

39. Graph: $x = 4 \sin 2t$, $y = 4 \cos 3t$

40. Graph: $x = 5 \sin 10t$, $y = 5 \cos 9t$

Connecting Concepts

41. Let $P_1(x_1, y_1)$ and $P_2(x_2, y_2)$ be two distinct points in the plane, and consider the line L passing through those points. Choose a point $P(x, y)$ on the line L. Show that

$$\frac{x - x_1}{x_2 - x_1} = \frac{y - y_1}{y_2 - y_1}$$

Use this result to demonstrate that $x = (x_2 - x_1)t + x_1$, $y = (y_2 - y_1)t + y_1$ is a parametric representation of the line through the two points.

42. Show that $x = h + a \sin t$, $y = k + b \cos t$, for $a > 0$, $b > 0$, and $0 \leq t < 2\pi$, are parametric equations for an ellipse with center at (h, k).

43. Suppose a string, held taut, is unwound from the circumference of a circle of radius a. The path traced by the end of the string is called the *involute* of a circle. Find parametric equations for the involute of a circle.

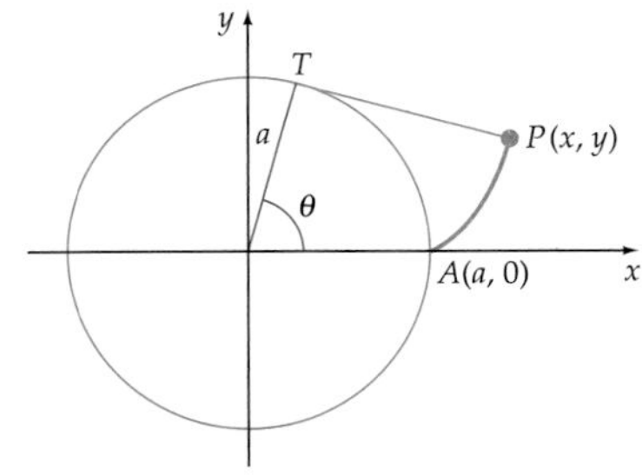

44. A circle of radius a rolls without slipping on the outside of a circle of radius $b > a$. Find the parametric equations of a point P on the smaller circle. The curve is called an *epicycloid.*

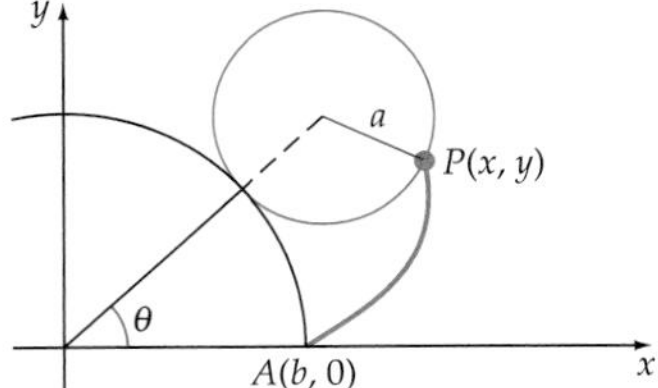

45. A circle of radius a rolls without slipping on the inside of a circle of radius $b > a$. Find the parametric equations of a point P on the smaller circle. The curve is called a *hypocycloid.*

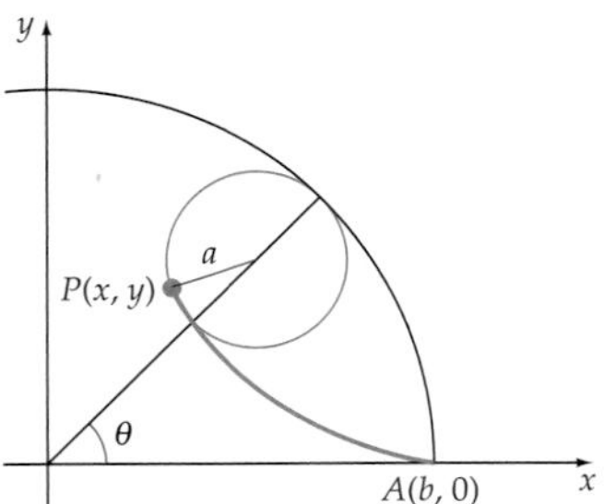

Projects

1. PARAMETRIC EQUATIONS IN AN *XYZ*-COORDINATE SYSTEM

a. Graph the three-dimensional curve given by

$$x = 3 \cos t, \quad y = 3 \sin t, \quad z = 0.5t$$

b. Graph the three-dimensional curve given by

$$x = 3 \cos t, \quad y = 6 \sin t, \quad z = 0.5t$$

c. What is the main difference between these curves?

d. What name is given to curves of this type?

Exploring Concepts with Technology

Using a Graphing Calculator to Find the *n*th Roots of *z*

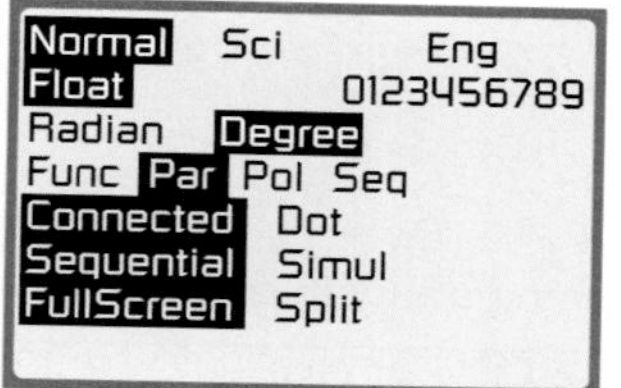

Figure 6.93

In Chapter 5 we used De Moivre's Theorem to find the nth roots of a number. The parametric feature of a graphing calculator can also be used to find and display the nth roots of $z = r(\cos\theta + i\sin\theta)$. Here is the procedure for a TI-83/TI-83 Plus/TI-84 Plus graphing calculator. Put the calculator in parametric and degree mode. See **Figure 6.93**. To find the nth roots of $z = r(\cos\theta + i\sin\theta)$, enter in the Y= menu

X1T=r^(1/n)cos(θ/n+T) and Y1T=r^(1/n)sin(θ/n+T)

In the WINDOW menu, set Tmin=0, Tmax=360, and Tstep=360/n. Set Xmin, Xmax, Ymin, and Ymax to appropriate values that will allow the roots to be seen in the graph window. Press GRAPH to display a polygon. The x- and y-coordinates of each vertex of the polygon represent a root of z in the rectangular form $x + yi$. Here is a specific example that illustrates this procedure.

Example Find the fourth roots of $z = 16i$.

In trigonometric form, $z = 16(\cos 90° + i\sin 90°)$. Thus, in this example, $r = 16$, $\theta = 90°$, and $n = 4$. In the Y= menu, enter

X1T=16^(1/4)cos(90/4+T) and Y1T=16^(1/4)sin(90/4+T)

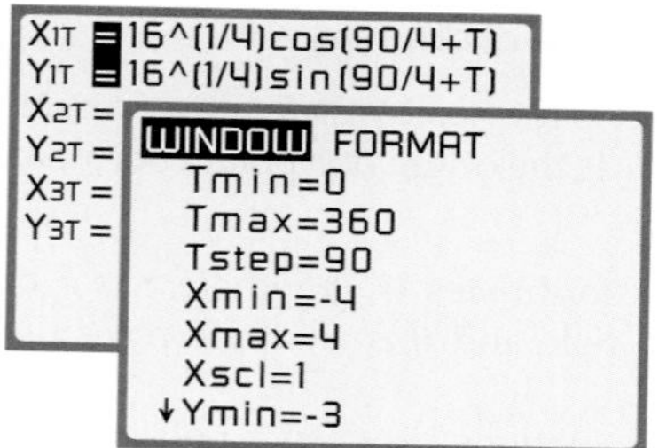

Figure 6.94

In the WINDOW menu, set

Tmin=0	Xmin=-4	Ymin=-3
Tmax=360	Xmax=4	Ymax=3
Tstep=360/4	Xscl=1	Yscl=1

See **Figure 6.94**. Press GRAPH to produce the quadrilateral in **Figure 6.95**. Use TRACE and the arrow key ▶ to move to each of the vertices of the quadrilateral. **Figure 6.95** shows that one of the roots of $z = 16i$ is $1.8477591 + 0.76536686i$. Continue to press the arrow key ▶ to find the other three roots, which are

$$-0.7653669 + 1.8477591i, -1.847759 - 0.7653669i, \text{ and } 0.76536686 - 1.847759i$$

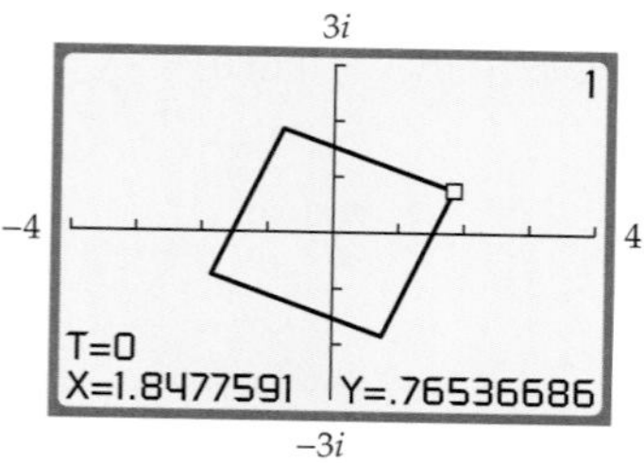

The vertices of the quadrilateral represent the fourth roots of $16i$ in the complex plane.

Figure 6.95

Use a graphing calculator to estimate, in rectangular form, each of the following.

1. The cube roots of -27
2. The fifth roots of $32i$
3. The fourth roots of $\sqrt{8} + \sqrt{8}i$
4. The sixth roots of $-64i$

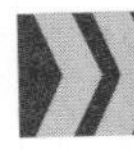

Chapter 6 Summary

6.1 Parabolas

- The equations of a parabola with vertex at (h, k) and axis of symmetry parallel to a coordinate axis are given by

$(x - h)^2 = 4p(y - k)$; focus $(h, k + p)$; directrix $y = k - p$

$(y - k)^2 = 4p(x - h)$; focus $(h + p, k)$; directrix $x = h - p$

6.2 Ellipses

- The equations of an ellipse with center at (h, k) and major axis parallel to a coordinate axis are given by

$$\frac{(x-h)^2}{a^2} + \frac{(y-k)^2}{b^2} = 1;\ \text{foci } (h \pm c, k);\ \text{vertices } (h \pm a, k)$$

$$\frac{(x-h)^2}{b^2} + \frac{(y-k)^2}{a^2} = 1;\ \text{foci } (h, k \pm c);\ \text{vertices } (h, k \pm a)$$

For each equation, $a > b$ and $c^2 = a^2 - b^2$.

- The eccentricity e of an ellipse is given by $e = \dfrac{c}{a}$.

6.3 Hyperbolas

- The equations of a hyperbola with center at (h, k) and transverse axis parallel to a coordinate axis are given by

$$\frac{(x-h)^2}{a^2} - \frac{(y-k)^2}{b^2} = 1;\ \text{foci } (h \pm c, k);\ \text{vertices } (h \pm a, k)$$

$$\frac{(y-k)^2}{a^2} - \frac{(x-h)^2}{b^2} = 1;\ \text{foci } (h, k \pm c);\ \text{vertices } (h, k \pm a)$$

For each equation, $c^2 = a^2 + b^2$.

- The eccentricity e of a hyperbola is given by $e = \dfrac{c}{a}$.

6.4 Rotation of Axes

- The rotation-of-axes formulas are

$$\begin{cases} x = x'\cos\alpha - y'\sin\alpha \\ y = y'\cos\alpha + x'\sin\alpha \end{cases} \qquad \begin{cases} x' = x\cos\alpha + y\sin\alpha \\ y' = y\cos\alpha - x\sin\alpha \end{cases}$$

- To eliminate the xy term from the general quadratic equation, rotate the coordinate axes through an angle α, where

$$\cot 2\alpha = \frac{A - C}{B}, \quad B \neq 0, \quad 0° < 2\alpha < 180°$$

- The graph of $Ax^2 + Bxy + Cy^2 + Dx + Ey + F = 0$ is either a conic or a degenerate conic. If the graph is a conic, then the graph can be identified by its *discriminant* $B^2 - 4AC$. The graph is

an ellipse or a circle, provided $B^2 - 4AC < 0$.
a parabola, provided $B^2 - 4AC = 0$.
a hyperbola, provided $B^2 - 4AC > 0$.

- The graph of $Ax^2 + Bxy + Cy^2 + Dx + Ey + F = 0$ can be constructed by using a graphing utility to graph both

$$y_1 = \frac{-(Bx + E) + \sqrt{(Bx + E)^2 - 4C(Ax^2 + Dx + F)}}{2C}$$

and

$$y_2 = \frac{-(Bx + E) - \sqrt{(Bx + E)^2 - 4C(Ax^2 + Dx + F)}}{2C}$$

6.5 Introduction to Polar Coordinates

- A polar coordinate system is formed by drawing a horizontal ray (*polar axis*). The *pole* is the origin of a polar coordinate system.
- A point is specified by coordinates (r, θ), where r is a directed distance from the pole and θ is an angle measured from the polar axis.

  The transformation equations between a polar coordinate system and a rectangular coordinate system are

  Polar to rectangular: $x = r\cos\theta \qquad y = r\sin\theta$

  Rectangular to polar: $r = \sqrt{x^2 + y^2} \qquad \tan\theta = \dfrac{y}{x}, x \neq 0$

- The polar equations of:

lines are on page 413.
circles are on page 415.
limaçons are on page 416.
rose curves are on page 418.

6.6 Polar Equations of the Conics

- The polar equations of the conics are given by

$$r = \frac{ed}{1 \pm e\cos\theta} \quad \text{or} \quad r = \frac{ed}{1 \pm e\sin\theta}$$

where e is the eccentricity and d is the distance of the directrix from the focus.

When

$0 < e < 1$, the graph is an ellipse.
$e = 1$, the graph is a parabola.
$e > 1$, the graph is a hyperbola.

6.7 Parametric Equations

- Let t be a number in an interval I. A *curve* is a set of ordered pairs (x, y), where

$$x = f(t), \qquad y = g(t) \quad \text{for } t \in I$$

The variable t is called a *parameter*, and the pair of equations are *parametric equations*.

- To *eliminate the parameter* is to find an equation in x and y that has the same graph as the given parametric equations.
- Parametric equations can be used to show the movement of a point as it travels along a curve.
- A cycloid is traced by a point on a circle as the circle rolls on a line. Parametric equations of a cycloid are given on page 438.
- The path of a projectile (assume air resistance is negligible) that is launched at an angle θ from the horizon with an initial velocity of v_0 feet per second is given by

$$x = (v_0 \cos \theta)t, \qquad y = -16t^2 + (v_0 \sin \theta)t$$

where t is the time in seconds since the projectile was launched.

Chapter 6 Assessing Concepts

In Exercises 1 to 12, match an equation to each description. A letter may be used more than once. Some letters may not be needed.

1. Equation, in rectangular form, of an ellipse with major axis of length 10 and minor axis of length 6 _____
2. Equation, in rectangular form, of a hyperbola with center (4, 3) _____
3. Equation, in rectangular form, of a parabola with vertex (2, −3) _____
4. Equation, in rectangular form, of an ellipse with eccentricity $\frac{1}{2}$ _____
5. Equation, in rectangular form, of a hyperbola with eccentricity of $\frac{\sqrt{7}}{2}$ _____
6. Equation, in rectangular form, of a parabola with focus (3, 2) _____
7. Equation, in rectangular form, of a parabola with vertex (2, 3) _____
8. Polar equation of an ellipse with major axis on the polar axis _____
9. Polar equation of a hyperbola with transverse axis on the polar axis _____
10. Polar equation of a parabola with axis of symmetry on the polar axis _____
11. Parametric equations of an ellipse with major axis parallel to the x-axis _____
12. Parametric equations of an ellipse with center (2, 3) _____

a. $3x^2 - 4y^2 = 12$

b. $\dfrac{(x-4)^2}{4^2} - \dfrac{(y-3)^2}{5^2} = 1$

c. $3x^2 + 4y^2 = 12$

d. $\dfrac{x^2}{25} + \dfrac{y^2}{9} = 1$

e. $-x + 2y^2 + 12y + 20 = 0$

f. $-8x + y^2 - 4y + 12 = 0$

g. $3x + y^2 - 6y + 3 = 0$

h. $r = \dfrac{5}{\frac{3}{5} + 2\cos\theta}$

i. $r = \dfrac{6}{2 + \cos\theta}$

j. $r = \dfrac{1}{\frac{2}{3}\cos\theta - \frac{2}{3}}$

k. $x = 2 + 4\cos t,\ y = 3 + 2\sin t$, for $0 \le t < 2\pi$

l. $x = -1 + 3\sin t,\ y = 3 + 8\cos t$, for $0 \le t < 2\pi$

Chapter 6 Review Exercises

In Exercises 1 to 12, if the equation is that of an ellipse or a hyperbola, find the center, vertices, and foci. For hyperbolas, find the equations of the asymptotes. If the equation is that of a parabola, find the vertex, the focus, and the equation of the directrix. Graph each equation.

1. $x^2 - y^2 = 4$
2. $y^2 = 16x$
3. $x^2 + 4y^2 - 6x + 8y - 3 = 0$
4. $3x^2 - 4y^2 + 12x - 24y - 36 = 0$
5. $3x - 4y^2 + 8y + 2 = 0$
6. $3x + 2y^2 - 4y - 7 = 0$
7. $9x^2 + 4y^2 + 36x - 8y + 4 = 0$
8. $11x^2 - 25y^2 - 44x - 50y - 256 = 0$
9. $4x^2 - 9y^2 - 8x + 12y - 144 = 0$
10. $9x^2 + 16y^2 + 36x - 16y - 104 = 0$
11. $4x^2 + 28x + 32y + 81 = 0$
12. $x^2 - 6x - 9y + 27 = 0$

In Exercises 13 to 20, find the equation of the conic that satisfies the given conditions.

13. Ellipse with vertices at (7, 3) and (−3, 3); length of minor axis is 8.
14. Hyperbola with vertices at (4, 1) and (−2, 1) and eccentricity $\frac{4}{3}$
15. Hyperbola with foci (−5, 2) and (1, 2); length of transverse axis is 4.
16. Parabola with focus (2, −3) and directrix $x = 6$
17. Parabola with vertex (0, −2) and passing through the point (3, 4)
18. Ellipse with eccentricity $\frac{2}{3}$ and foci (−4, −1) and (0, −1)
19. Hyperbola with vertices (±6, 0) and asymptotes whose equations are $y = \pm\frac{1}{9}x$.
20. Parabola passing through the points (1, 0), (2, 1), and (0, 1) with axis of symmetry parallel to the y-axis.

In Exercises 21 to 24, write the equation without an *xy* term. Name the graph of the equation.

21. $11x^2 - 6xy + 19y^2 - 40 = 0$
22. $3x^2 + 6xy + 3y^2 - 4x + 5y - 12 = 0$
23. $x^2 + 2\sqrt{3}xy + 3y^2 + 8\sqrt{3}x - 8y + 32 = 0$
24. $xy - x - y - 1 = 0$

In Exercises 25 to 34, graph each polar equation.

25. $r = 4 \cos 3\theta$
26. $r = 1 + \cos \theta$
27. $r = 2(1 - 2 \sin \theta)$
28. $r = 4 \sin 4\theta$
29. $r = 5 \sin \theta$
30. $r = 3 \sec \theta$
31. $r = 4 \csc \theta$
32. $r = 4 \cos \theta$
33. $r = 3 + 2 \cos \theta$
34. $r = 4 + 2 \sin \theta$

In Exercises 35 to 38, find a polar form of each equation.

35. $y^2 = 16x$
36. $x^2 + y^2 + 4x + 3y = 0$
37. $3x - 2y = 6$
38. $xy = 4$

In Exercises 39 to 42, find a rectangular form of each equation.

39. $r = \dfrac{4}{1 - \cos \theta}$
40. $r = 3 \cos \theta - 4 \sin \theta$
41. $r^2 = \cos 2\theta$
42. $\theta = 1$

In Exercises 43 to 46, graph the conic given by each polar equation.

43. $r = \dfrac{4}{3 - 6 \sin \theta}$
44. $r = \dfrac{2}{1 + \cos \theta}$
45. $r = \dfrac{2}{2 - \cos \theta}$
46. $r = \dfrac{6}{4 + 3 \sin \theta}$

In Exercises 47 to 53, eliminate the parameter and graph the curve given by the parametric equations.

47. $x = 4t - 2, y = 3t + 1$, for $t \in R$

48. $x = 1 - t^2, y = 3 - 2t^2$, for $t \in R$

49. $x = 4 \sin t, y = 3 \cos t$, for $0 \le t < 2\pi$

50. $x = \sec t, y = 4 \tan t$, for $-\frac{\pi}{2} < t < \frac{\pi}{2}$

51. $x = \frac{1}{t}, y = -\frac{2}{t}$, for $t > 0$

52. $x = 1 + \cos t, y = 2 - \sin t$, for $0 \le t < 2\pi$

53. $x = \sqrt{t}, y = 2^{-t}$, for $t \ge 0$

54. Use a graphing utility to graph the cycloid given by

$$x = 3(t - \sin t), \qquad y = 3(1 - \cos t)$$

for $0 \le t \le 18\pi$.

55. Use a graphing utility to graph the conic given by

$$x^2 + 4xy + 2y^2 - 2x + 5y + 1 = 0$$

56. Use a graphing utility to graph

$$r = \frac{6}{3 + \sin\left(\theta + \frac{\pi}{4}\right)}$$

57. **PHYSICS** The path of a projectile (assume air resistance is negligible) that is launched at an angle θ from the horizon with an initial velocity of v_0 feet per second is given by the parametric equations

$$x = (v_0 \cos \theta)t, \qquad y = -16t^2 + (v_0 \sin \theta)t$$

where t is the time in seconds since the projectile was launched. Use a graphing utility to graph the path of a projectile that is launched at an angle of 33° with an initial velocity of 245 feet per second. Use the graph to determine the maximum height of the projectile to the nearest foot.

»»» Quantitative Reasoning: *The Mathematics of a Rotary Engine* »»»

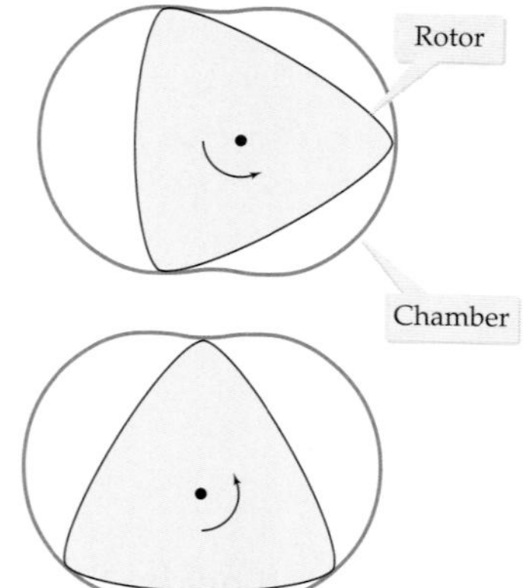

Figure 6.96

During the 1950s, Felix Wankel (1902–1988) developed a rotary engine. The engine has several remarkable properties.

- A *triangular rotor*[2] (see **Figure 6.96**) rotates inside a chamber in a manner such that all three vertices of the rotor maintain constant contact with the chamber.
- As the rotor rotates, the centroid (center) of the rotor moves on a circular path.
- Every time the rotor makes one complete revolution, the centroid makes three complete revolutions around its circular path.

The graph of the parametric equations

$$x = 3 \cos T + 0.5 \cos(3T), \qquad y = 3 \sin T + 0.5 \sin(3T), \text{ for } 0 \le T \le 2\pi \qquad (1)$$

where T is the counterclockwise angle of rotation as measured from the positive x-axis, yields the curve shown in **Figure 6.97** on page 448. The curve has the same shape as the chamber of a rotary engine.

The following TI-83/TI-83 Plus/TI-84 Plus program WANKEL simulates the motion of the rotor in a rotary engine.

[2]The actual rotor used in a rotary engine is a modified equilateral triangle with curved sides, as shown in Figure 6.96.

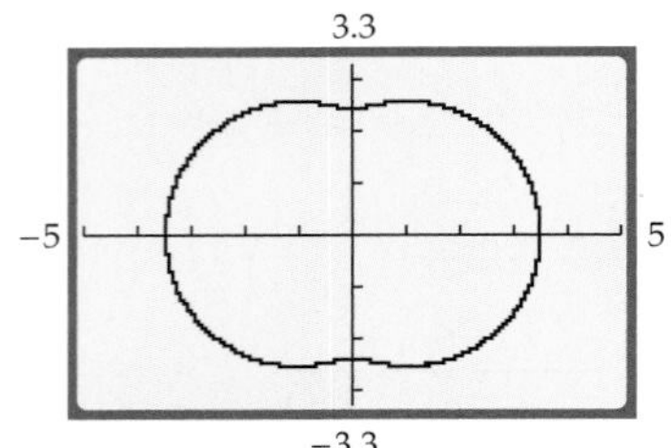

Figure 6.97

```
PROGRAM:WANKEL
ClrDraw:Radian:FnOff
-5→Xmin:5→Xmax:1→Xscl
-3.3→Ymin:3.3→Ymax:1→Yscl
For(T,0,2π/3,π/40)
Line(3cos(T)+.5cos(3T),
  3sin(T)+.5sin(3T),
  3cos(T+2π/3)+.5cos(3T),
  3sin(T+2π/3)+.5sin(3T))
Line(3cos(T+2π/3)+.5cos(3T),
  3sin(T+2π/3)+.5sin(3T),
  3cos(T+4π/3)+.5cos(3T),
  3sin(T+4π/3)+.5sin(3T))
Line(3cos(T+4π/3)+.5cos(3T),
  3sin(T+4π/3)+.5sin(3T),
  3cos(T+2π)+.5cos(3T),
  3sin(T+2π)+.5sin(3T))
Pt-On(.5cos(3T),.5sin(3T))
End
```

QR1. Run the WANKEL program. How does the path that is traced out by the vertices of the rotor compare with the graph in **Figure 6.97**?

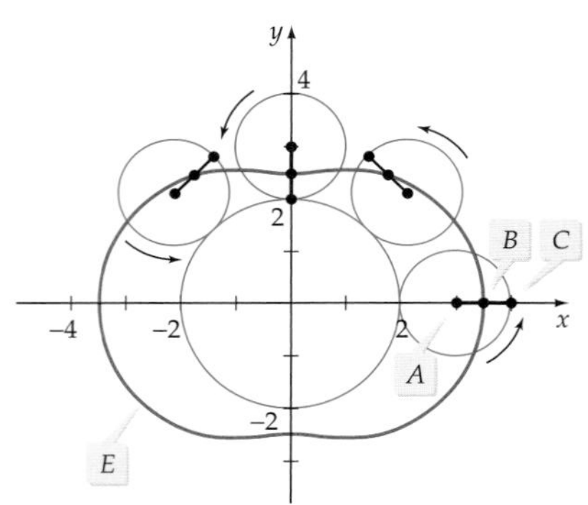

Figure 6.98

Felix Wankel designed the shape of the chamber of his rotary engine using trial and error. He did not know that mathematicians had made a study of this figure many years before by graphing the path that is traced out by a point on the radius of a circle that rolls around another circle. For example, **Figure 6.98** shows a circle rolling around a larger circle. Point B, the midpoint of radius AC of the small circle, generates curve E as the small circle is rolled around the larger circle. Curve E is called an *epitrochoid.*

The following TI-83/TI-83 Plus/TI-84 Plus program EPIT simulates the process of generating an epitrochoid by rolling a small circle around a larger circle and plotting the trace of the midpoint of a radius of the smaller circle (point B in **Figure 6.98**).

```
PROGRAM:EPIT
Radian:FnOff
ClrDraw
-1→Xmin:6→Xmax
1→Xscl
-.33→Ymin:4.3→Ymax
1→Yscl
Circle(0,0,2)
For(K,0,π/2,π/30)
Circle(3cos(K),3sin(K),1)
Circle(3cos(K)+.5cos(3K),
  3sin(K)+.5sin(3K),.09)
End
```

QR2. Run the EPIT program. How does the path that is traced out by the midpoint of the radius of the small circle compare with the graph in **Figure 6.98,** in Quadrant I?

QR3. Construct a two-dimensional model of a rotary engine. Demonstrate to your classmates how the rotor revolves inside the epitrochoid chamber. One method for constructing the model is to use thick cardboard from which you cut out an epitrochoid. To generate the epitrochoid, use a computer to print a large graph of the parametric equations given in (1).

Cut out an equilateral triangle from the cardboard to form the rotor. What is the length, to the nearest tenth of a unit, of each side of the triangle? (*Hint:* The length of each side of the triangle equals the distance between the points on the epitrochoid (1) at $T = 0$ and $T = \frac{2\pi}{3}$.)

For additional information on rotary engines and epitrochoids, visit website http://auto.howstuffworks.com/rotary-engine.htm and http://www.math.dartmouth.edu/~dlittle/java/SpiroGraph/.

Chapter 6 Test

1. Find the vertex, focus, and directrix of the parabola given by the equation $y = \frac{1}{8}x^2$.

2. Graph: $\frac{x^2}{16} + \frac{y^2}{1} = 1$

3. Find the vertices and foci of the ellipse given by the equation $25x^2 - 150x + 9y^2 + 18y + 9 = 0$.

4. Find the equation in standard form of the ellipse with center $(0, -3)$, foci $(-6, -3)$ and $(6, -3)$, and minor axis of length 6.

5. Graph: $\frac{y^2}{25} - \frac{x^2}{16} = 1$

6. Find the vertices, foci, and asymptotes of the hyperbola given by the equation $\frac{x^2}{36} - \frac{y^2}{64} = 1$.

7. Graph: $16y^2 + 32y - 4x^2 - 24x = 84$

8. For the equation $x^2 - 4xy - 5y^2 + 3x - 5y - 20 = 0$, determine what acute angle of rotation (to the nearest 0.01°) would eliminate the xy term.

9. Determine whether the graph of the following equation is the graph of a parabola, an ellipse, or a hyperbola.

$$8x^2 + 5xy + 2y^2 - 10x + 5y + 4 = 0$$

10. $P(1, -\sqrt{3})$ are the coordinates of a point in an xy-coordinate system. Find the polar coordinates of P.

11. Graph: $r = 4 \cos \theta$

12. Graph: $r = 3(1 - \sin \theta)$

13. Graph: $r = 2 \sin 4\theta$

14. Find the rectangular coordinates of the point whose polar coordinates are $\left(5, \frac{7\pi}{3}\right)$.

15. Find the rectangular form of $r - r\cos\theta = 4$.

16. Write $r = \frac{4}{1 + \sin\theta}$ as an equation in rectangular coordinates.

17. Eliminate the parameter and graph the curve given by the parametric equations $x = t - 3$, $y = 2t^2$.

18. Eliminate the parameter and graph the curve given by the parametric equations $x = 4\sin\theta$, $y = \cos\theta + 2$, where $0 \le \theta < 2\pi$.

19. Use a graphing utility to graph the cycloid given by

$$x = 2(t - \sin t), \qquad y = 2(1 - \cos t)$$

for $0 \le t \le 12\pi$.

20. The path of a projectile that is launched at an angle of 30° from the horizon with an initial velocity of 128 feet per second is given by

$$x = (128 \cos 30°)t, \qquad y = -16t^2 + (128 \sin 30°)t$$

where t is the time in seconds after the projectile is launched. Use a graphing utility to determine how far (to the nearest foot) the projectile will travel downrange if the ground is level.

Cumulative Review Exercises

1. Solve: $x^2 + 4x + 6 = 0$

2. Is the graph of $f(x) = x^3 - 4x$ symmetric with respect to the x-axis, the y-axis, or the origin, or does it exhibit none of these symmetries?

3. Given $f(x) = \sin x$ and $g(x) = 3x - 2$, find $(g \circ f)(x)$.

4. Convert 240° to radians.

5. An electric cart has 10-inch-radius wheels. What is the linear speed in miles per hour of this cart when the wheels are rotating at 3 radians per second? Round to the nearest mile per hour.

6. Given $\sin t = -\frac{\sqrt{3}}{2}$, $\frac{3\pi}{2} < t < 2\pi$, find $\tan t$.

7. Find the measure of a for the right triangle shown at the right. Round to the nearest centimeter.

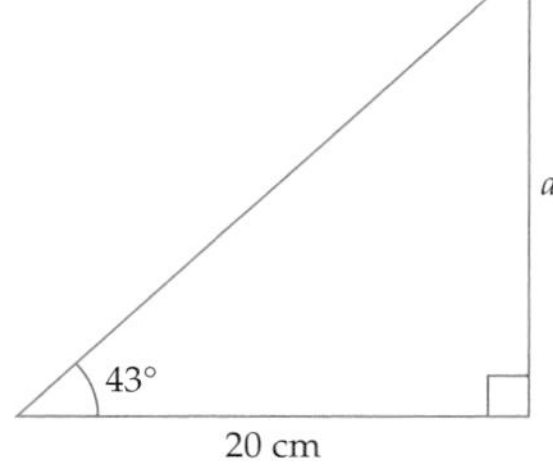

8. What are the amplitude and period of $y = \frac{1}{2}\cos\left(\frac{\pi x}{3}\right)$?

9. What is the period of $y = 2\tan\left(\frac{\pi x}{3}\right)$?

10. Verify the identity $\frac{\sin x}{1 - \cos x} = \csc x + \cot x$.

11. Given $\sin \alpha = \frac{3}{5}$ in Quadrant II and $\cos \beta = -\frac{5}{13}$ in Quadrant III, find $\sin(\alpha + \beta)$.

12. Find the exact value of $\sin\left(\cos^{-1}\frac{1}{5}\right)$.

13. Solve: $\cos^{-1} x = \sin^{-1}\frac{12}{13}$

14. To find the distance across a ravine, a surveying team locates points A and B on one side of the ravine and point C on the other side of the ravine. The distance between A and B is 155 feet. The measure of angle CAB is 71°, and the measure of angle CBA is 80°. Find the distance across the ravine. Round to the nearest foot.

15. The lengths of the sides of a triangular piece of wood are 2.5 feet, 4 feet, and 3.6 feet. Find the angle between the two longer sides of the triangle. Round to the nearest degree.

16. The magnitude of vector **v** is 30 and the direction angle is 145°. Write the vector in the form $\mathbf{v} = a\mathbf{i} + b\mathbf{j}$, where a and b are rounded to the nearest tenth.

17. Are the vectors $\mathbf{v} = 3\mathbf{i} + 2\mathbf{j}$ and $\mathbf{w} = 5\mathbf{i} - 7\mathbf{j}$ orthogonal?

18. Write the complex number $z = -2 + 2i\sqrt{3}$ in polar form.

19. Sketch the graph of $r = 3\sin 2\theta$.

20. Write x in terms of y by eliminating the parameter from the parametric equations $x = 2t - 1$, $y = 4t^2 + 1$.

7 Exponential and Logarithmic Functions

Modeling Data with an Exponential Function

The average annual salary of a basketball player in the National Basketball Association (NBA) has increased from \$12,000 a year in 1957 to over \$4.5 million a year in 2004.

The following bar graph shows the average annual NBA salary for selected years from 1990 to 2002.

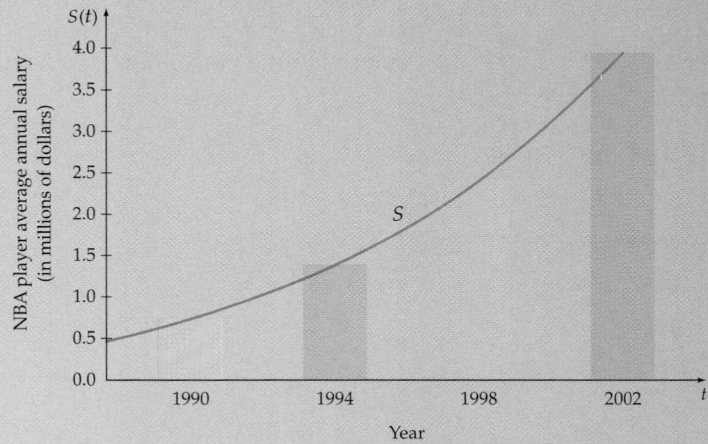

Source: www.sportsfansofamerica.com and the NBA Players Association

The function $S(t) = 0.7739(1.1485)^t$ closely models the average annual salary of an NBA player for the years 1990 ($t = 0$) to 2002 ($t = 12$). This function was determined by using the data from the bar graph and exponential regression, which is one of the topics of Section 7.6. See Exercise 23, page 531 for another application in which an exponential function is used to model data.

Online Study Center
For online student resources, such as section quizzes, visit this website: college.hmco.com/info/aufmannCAT

Section 7.1

- Exponential Functions
- Graphs of Exponential Functions
- The Natural Exponential Function

Exponential Functions and Their Applications

Exponential Functions

In 1965, Gordon Moore, one of the cofounders of Intel Corporation, observed that the maximum number of transistors that could be placed on a microprocessor seemed to be doubling every 18 to 24 months. This observation is known as Moore's Law. **Table 7.1** below shows how the maximum number of transistors on various Intel processors has changed over time. (*Source:* Intel Museum home page.)

The curve that approximately passes through the points is a mathematical model of the data. See **Figure 7.1.** The model is based on an *exponential* function.

Table 7.1

Year	1971	1979	1983	1985	1990	1993	1995	1998	2000	2004
Number of transistors per microprocessor (in thousands)	2.3	31	110	280	1200	3100	5500	14,000	42,000	592,000

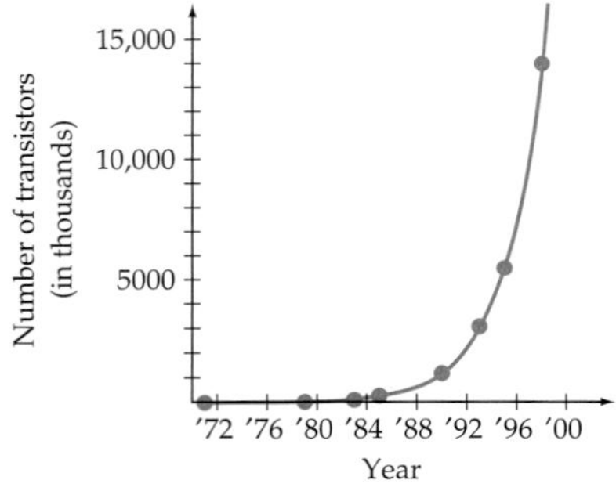

Figure 7.1
Moore's Law

When light enters water, the intensity of the light decreases with the depth of the water. The graph in **Figure 7.2** shows a model, for Lake Michigan, of the decrease in the percentage of available light as the depth of the water increases.

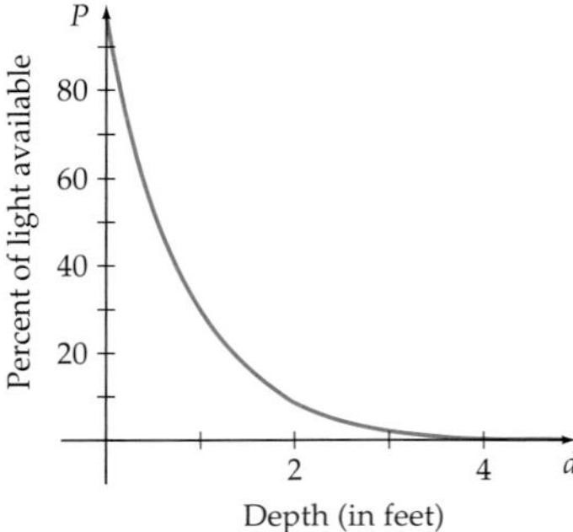

Figure 7.2

This model is also based on an exponential function.

Definition of an Exponential Function

The **exponential function with base b** is defined by

$$f(x) = b^x$$

where $b > 0$, $b \neq 1$, and x is a real number.

The base b of $f(x) = b^x$ is required to be positive. If the base were a negative number, the value of the function would be a complex number for some values of x. For instance, if $b = -4$ and $x = \frac{1}{2}$, then $f\left(\frac{1}{2}\right) = (-4)^{1/2} = 2i$. To avoid complex number values of a function, the base of any exponential function must be a positive number. Also, b is defined such that $b \neq 1$ because $f(x) = 1^x = 1$ is a constant function.

In the following examples we evaluate $f(x) = 2^x$ at $x = 3$ and $x = -2$.

$$f(3) = 2^3 = 8 \qquad f(-2) = 2^{-2} = \frac{1}{2^2} = \frac{1}{4}$$

To evaluate the exponential function $f(x) = 2^x$ at an irrational number such as $x = \sqrt{2}$, we use a rational approximation of $\sqrt{2}$, such as 1.4142, and a calculator to obtain an approximation of the function. For instance, if $f(x) = 2^x$, then $f(\sqrt{2}) = 2^{\sqrt{2}} \approx 2^{1.4142} \approx 2.6651$.

EXAMPLE 1 Evaluate an Exponential Function

Evaluate $f(x) = 3^x$ at $x = 2$, $x = -4$, and $x = \pi$.

Solution

$$f(2) = 3^2 = 9$$

$$f(-4) = 3^{-4} = \frac{1}{3^4} = \frac{1}{81}$$

$$f(\pi) = 3^{\pi} \approx 3^{3.1415927} \approx 31.54428$$ • Evaluate with the aid of a calculator.

Try Exercise 2, page 461

Graphs of Exponential Functions

The graph of $f(x) = 2^x$ is shown in **Figure 7.3**. The coordinates of some of the points on the curve are given in **Table 7.2**.

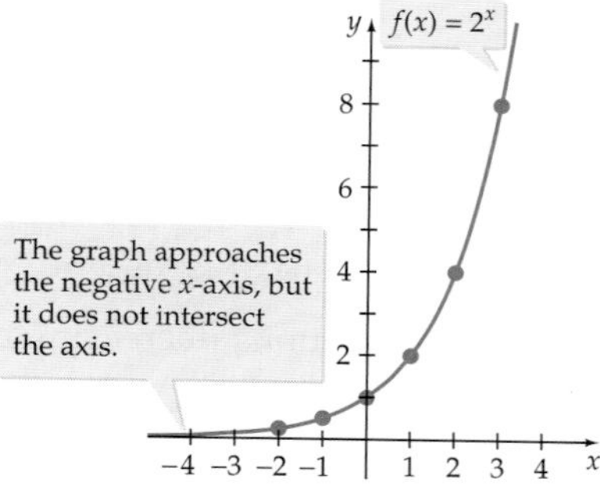

Figure 7.3

Table 7.2

x	$y = f(x) = 2^x$	(x, y)
−2	$f(-2) = 2^{-2} = \frac{1}{4}$	$\left(-2, \frac{1}{4}\right)$
−1	$f(-1) = 2^{-1} = \frac{1}{2}$	$\left(-1, \frac{1}{2}\right)$
0	$f(0) = 2^0 = 1$	(0, 1)
1	$f(1) = 2^1 = 2$	(1, 2)
2	$f(2) = 2^2 = 4$	(2, 4)
3	$f(3) = 2^3 = 8$	(3, 8)

Note the following properties of the graph of the exponential function $f(x) = 2^x$.

- The y-intercept is $(0, 1)$.
- The graph passes through $(1, 2)$.
- As x decreases without bound (that is, as $x \to -\infty$), $f(x) \to 0$.
- The graph is a smooth, continuous increasing curve.

Now consider the graph of an exponential function for which the base is between 0 and 1. The graph of $f(x) = \left(\frac{1}{2}\right)^x$ is shown in **Figure 7.4**. The coordinates of some of the points on the curve are given in **Table 7.3**.

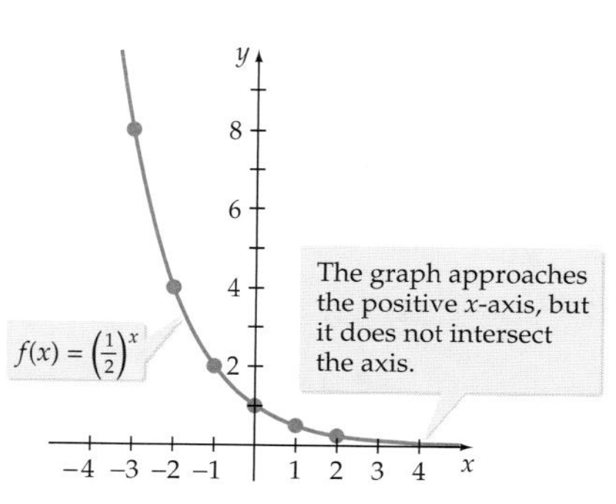

Figure 7.4

Table 7.3

x	$y = f(x) = \left(\frac{1}{2}\right)^x$	(x, y)
-3	$f(-3) = \left(\frac{1}{2}\right)^{-3} = 8$	$(-3, 8)$
-2	$f(-2) = \left(\frac{1}{2}\right)^{-2} = 4$	$(-2, 4)$
-1	$f(-1) = \left(\frac{1}{2}\right)^{-1} = 2$	$(-1, 2)$
0	$f(0) = \left(\frac{1}{2}\right)^{0} = 1$	$(0, 1)$
1	$f(1) = \left(\frac{1}{2}\right)^{1} = \frac{1}{2}$	$\left(1, \frac{1}{2}\right)$
2	$f(2) = \left(\frac{1}{2}\right)^{2} = \frac{1}{4}$	$\left(2, \frac{1}{4}\right)$

Note the following properties of the graph of $f(x) = \left(\frac{1}{2}\right)^x$.

- The y-intercept is $(0, 1)$.
- The graph passes through $\left(1, \frac{1}{2}\right)$.
- As x increases without bound (that is, as $x \to \infty$), $f(x) \to 0$.
- The graph is a smooth, continuous decreasing curve.

The basic properties of exponential functions are provided in the following summary.

Properties of $f(x) = b^x$

For positive real numbers b, $b \neq 1$, the exponential function defined by $f(x) = b^x$ has the following properties:

1. The function f is a one-to-one function. It has the set of real numbers as its domain and the set of positive real numbers as its range.
2. The graph of f is a smooth, continuous curve with a y-intercept of $(0, 1)$, and the graph passes through $(1, b)$.
3. If $b > 1$, f is an increasing function and the graph of f is asymptotic to the negative x-axis. [As $x \to \infty, f(x) \to \infty$, and as $x \to -\infty, f(x) \to 0$.] See **Figure 7.5a.**
4. If $0 < b < 1$, f is a decreasing function and the graph of f is asymptotic to the positive x-axis. [As $x \to -\infty, f(x) \to \infty$, and as $x \to \infty, f(x) \to 0$.] See **Figure 7.5b.**

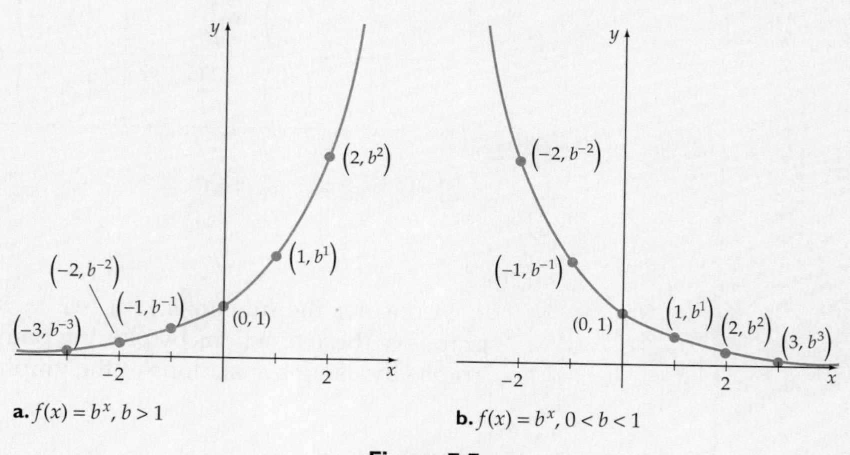

a. $f(x) = b^x, b > 1$

b. $f(x) = b^x, 0 < b < 1$

Figure 7.5

QUESTION What is the x-intercept of the graph of $f(x) = \left(\frac{1}{3}\right)^x$?

EXAMPLE 2 Graph an Exponential Function

Graph: $g(x) = \left(\frac{3}{4}\right)^x$

Continued ▶

ANSWER The graph does not have an x-intercept. As x increases without bound, the graph approaches the x-axis, but it does not intersect the x-axis.

Solution

Because the base $\frac{3}{4}$ is less than 1, we know that the graph of g is a decreasing function that is asymptotic to the positive x-axis. The y-intercept of the graph is the point $(0, 1)$, and the graph also passes through $\left(1, \frac{3}{4}\right)$. Plot a few additional points (see **Table 7.4**), and then draw a smooth curve through the points as in **Figure 7.6**.

Table 7.4

x	$y = g(x) = \left(\frac{3}{4}\right)^x$	(x, y)
-3	$\left(\frac{3}{4}\right)^{-3} = \frac{64}{27}$	$\left(-3, \frac{64}{27}\right)$
-2	$\left(\frac{3}{4}\right)^{-2} = \frac{16}{9}$	$\left(-2, \frac{16}{9}\right)$
-1	$\left(\frac{3}{4}\right)^{-1} = \frac{4}{3}$	$\left(-1, \frac{4}{3}\right)$
2	$\left(\frac{3}{4}\right)^{2} = \frac{9}{16}$	$\left(2, \frac{9}{16}\right)$
3	$\left(\frac{3}{4}\right)^{3} = \frac{27}{64}$	$\left(3, \frac{27}{64}\right)$

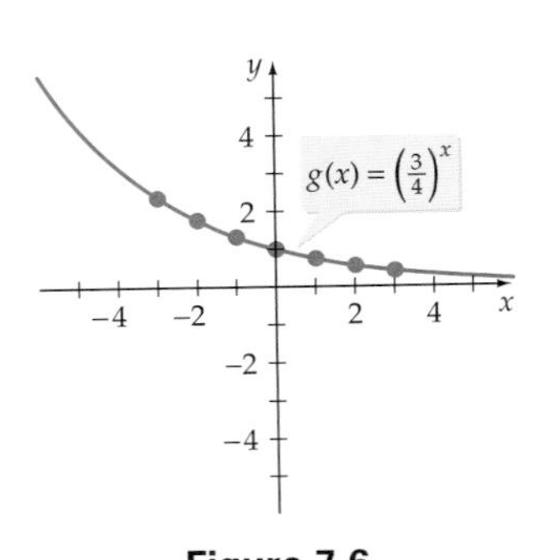

Figure 7.6

Try Exercise 22, page 462

Consider the functions $F(x) = 2^x - 3$ and $G(x) = 2^{x-3}$. You can construct the graphs of these functions by plotting points; however, it is easier to construct their graphs by using translations of the graph of $f(x) = 2^x$, as shown in Example 3.

EXAMPLE 3 Use a Translation to Produce a Graph

a. Explain how to use the graph of $f(x) = 2^x$ to produce the graph of $F(x) = 2^x - 3$.

b. Explain how to use the graph of $f(x) = 2^x$ to produce the graph of $G(x) = 2^{x-3}$.

Solution

a. $F(x) = 2^x - 3 = f(x) - 3$. The graph of F is a vertical translation of f down 3 units, as shown in **Figure 7.7**.

b. $G(x) = 2^{x-3} = f(x - 3)$. The graph of G is a horizontal translation of f to the right 3 units, as shown in **Figure 7.8**.

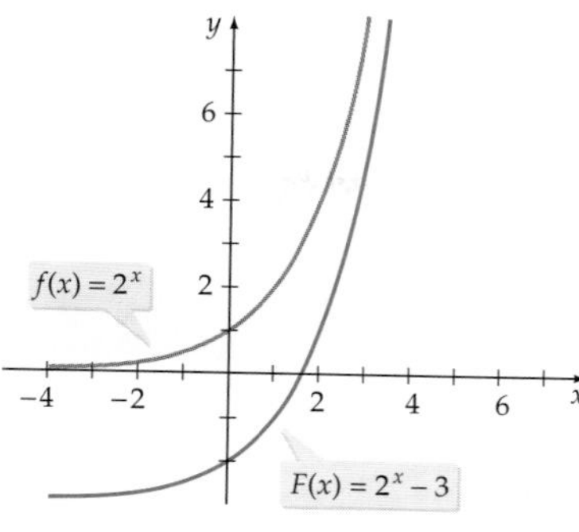

Figure 7.7

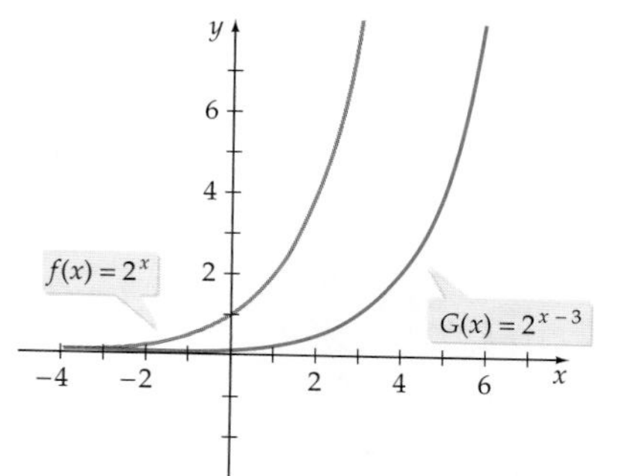

Figure 7.8

Try Exercise 28, page 462

The graphs of some functions can be constructed by stretching, compressing, or reflecting the graph of an exponential function.

EXAMPLE 4 Use Stretching or Reflecting Procedures to Produce a Graph

a. Explain how to use the graph of $f(x) = 2^x$ to produce the graph of $M(x) = 2(2^x)$.

b. Explain how to use the graph of $f(x) = 2^x$ to produce the graph of $N(x) = 2^{-x}$.

Solution

a. $M(x) = 2(2^x) = 2f(x)$. The graph of M is a vertical stretching of f away from the x-axis by a factor of 2, as shown in **Figure 7.9**. (*Note:* If (x, y) is a point on the graph of $f(x) = 2^x$, then $(x, 2y)$ is a point on the graph of M.)

b. $N(x) = 2^{-x} = f(-x)$. The graph of N is the graph of f reflected across the y-axis, as shown in **Figure 7.10**. (*Note:* If (x, y) is a point on the graph of $f(x) = 2^x$, then $(-x, y)$ is a point on the graph of N.)

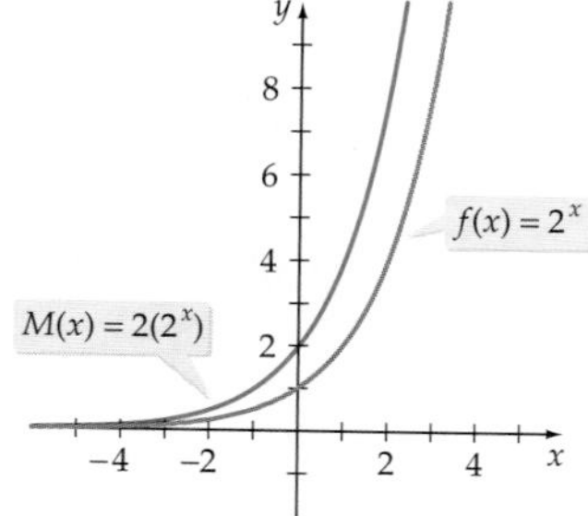

Figure 7.9

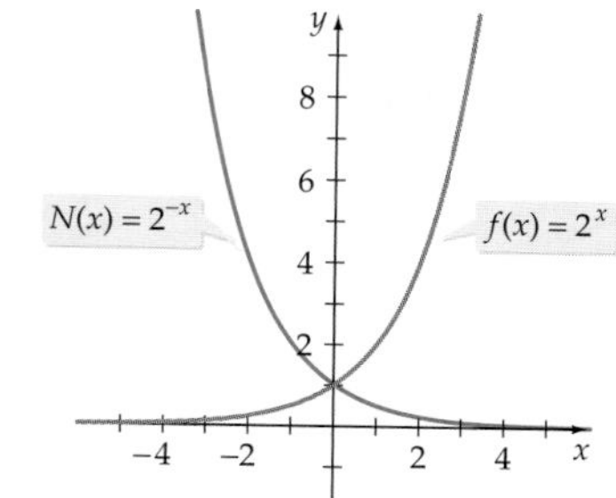

Figure 7.10

Try Exercise 30, page 462

Math Matters

Leonhard Euler (1707–1783)

Some mathematicians consider Euler to be the greatest mathematician of all time. He certainly was the most prolific writer of mathematics of all time. He made substantial contributions in the areas of number theory, geometry, calculus, differential equations, differential geometry, topology, complex variables, and analysis, to name but a few. Euler was the first to introduce many of the mathematical notations that we use today. For instance, he introduced the symbol *i* for the square root of −1, the symbol π for pi, the functional notation $f(x)$, and the letter *e* for the base of the natural exponential function. Euler's computational skills were truly amazing. The mathematician François Arago remarked, "Euler calculated without apparent effort, as men breathe, or as eagles sustain themselves in the wind."

The Natural Exponential Function

The irrational number π is often used in applications that involve circles. Another irrational number, denoted by the letter e, is useful in many applications that involve growth or decay.

Definition of *e*

The **number *e*** is defined as the number that

$$\left(1 + \frac{1}{n}\right)^n$$

approaches as n increases without bound.

The letter e was chosen in honor of the Swiss mathematician Leonhard Euler. He was able to compute the value of e to several decimal places by evaluating $\left(1 + \frac{1}{n}\right)^n$ for large values of n, as shown in **Table 7.5.**

Table 7.5

Value of n	Value of $\left(1 + \frac{1}{n}\right)^n$
1	2
10	2.59374246
100	2.704813829
1000	2.716923932
10,000	2.718145927
100,000	2.718268237
1,000,000	2.718280469
10,000,000	2.718281693

The value of e accurate to eight decimal places is 2.71828183.

Definition of the Natural Exponential Function

For all real numbers x, the function defined by

$$f(x) = e^x$$

is called the **natural exponential function.**

A calculator can be used to evaluate e^x for specific values of x. For instance,

$$e^2 \approx 7.389056, \quad e^{3.5} \approx 33.115452, \quad \text{and} \quad e^{-1.4} \approx 0.246597$$

On a TI-83/TI-83 Plus/TI-84 Plus calculator the e^x function is located above the LN key.

Integrating Technology

The graph of $f(x) = e^x$ below was produced on a TI-83/TI-83 Plus/ TI-84 Plus graphing calculator by entering e^x in the Y= menu.

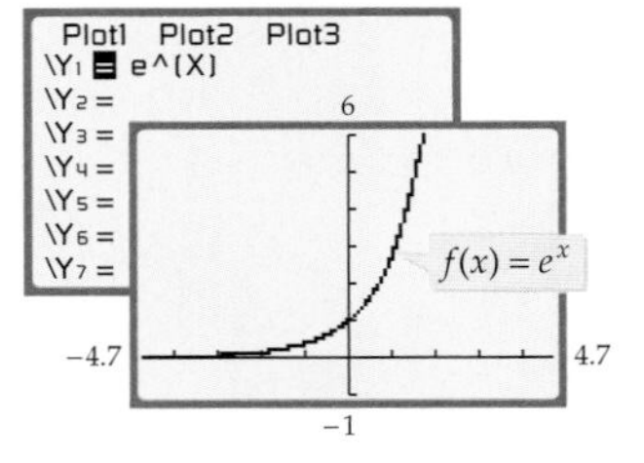

To graph $f(x) = e^x$, use a calculator to find the range values for a few domain values. The range values in **Table 7.6** have been rounded to the nearest tenth.

Table 7.6

x	−2	−1	0	1	2
f(x) = e^x	0.1	0.4	1.0	2.7	7.4

Plot the points given in **Table 7.6,** and then connect the points with a smooth curve. Because $e > 1$, we know that the graph is an increasing function. To the far left, the graph will approach the x-axis. The y-intercept is $(0, 1)$. See **Figure 7.11.** Note in **Figure 7.12** how the graph of $f(x) = e^x$ compares with the graphs of $g(x) = 2^x$ and $h(x) = 3^x$. You may have anticipated that the graph of $f(x) = e^x$ would lie between the two other graphs because e is between 2 and 3.

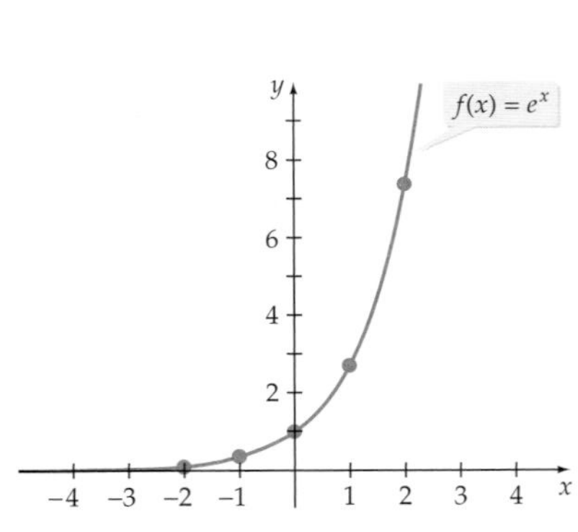

Figure 7.11

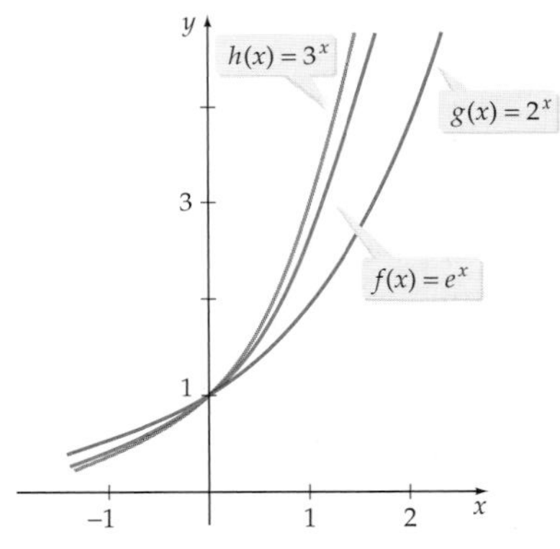

Figure 7.12

Many applications can be modeled effectively by functions that involve an exponential function. For instance, in Example 5 we make use of a function that involves an exponential function to model the temperature of a cup of coffee.

EXAMPLE 5 » Use a Mathematical Model

A cup of coffee is heated to 160°F and placed in a room that maintains a temperature of 70°F. The temperature T of the coffee, in degrees Fahrenheit, after t minutes is given by

$$T = 70 + 90e^{-0.0485t}$$

a. Find the temperature of the coffee, to the nearest degree, 20 minutes after it is placed in the room.

b. Use a graphing utility to determine when the temperature of the coffee will reach 90°F.

Continued ▶

take note

In Example 5b, we use a graphing utility to solve the equation $90 = 70 + 90e^{-0.0485t}$. Analytic methods of solving this type of equation without the use of a graphing utility will be developed in Section 7.4.

Solution

a. $T = 70 + 90e^{-0.0485t}$

$= 70 + 90e^{-0.0485 \cdot (20)}$ • Substitute 20 for t.

$\approx 70 + 34.1$

≈ 104.1

After 20 minutes the temperature of the coffee is about 104°F.

b. Graph $T = 70 + 90e^{-0.0485t}$ and $T = 90$. See the following figure.

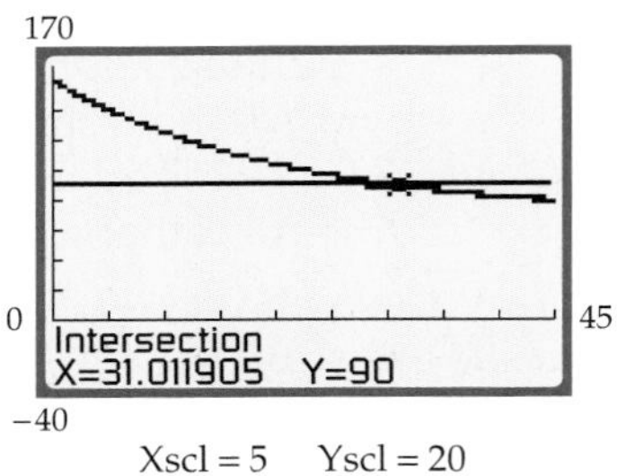

Xscl = 5 Yscl = 20

The graphs intersect at about (31.01, 90). It takes the coffee about 31 minutes to cool to 90°F.

Try Exercise 48, page 463

EXAMPLE 6 Use a Mathematical Model

The weekly revenue R, in dollars, from the sale of a product varies with time according to the function

$$R(x) = \frac{1760}{8 + 14e^{-0.03x}}$$

where x is the number of weeks that have passed since the product was put on the market. What will the weekly revenue approach as time goes by?

Solution

Method 1 Use a graphing utility to graph $R(x)$, and use the TRACE feature to see what happens to the revenue as the time increases. The graph on the right shows that as the weeks go by, the weekly revenue will increase and approach \$220.00 per week.

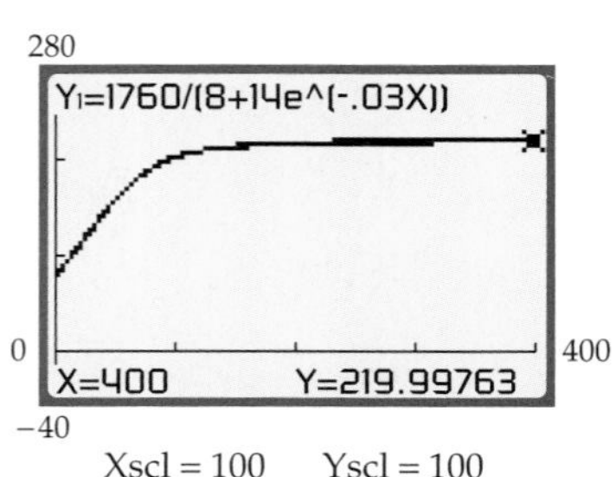

Xscl = 100 Yscl = 100

Method 2 Write the revenue function in the following form.

$$R(x) = \frac{1760}{8 + \dfrac{14}{e^{0.03x}}}$$ • $14e^{-0.03x} = \dfrac{14}{e^{0.03x}}$

As x increases without bound, $e^{0.03x}$ increases without bound, and the fraction $\dfrac{14}{e^{0.03x}}$ approaches 0. Therefore, as $x \to \infty, R(x) \to \dfrac{1760}{8 + 0} = 220$. Both methods indicate that as the number of weeks increases, the revenue approaches \$220 per week.

Try Exercise 54, page 464

Topics for Discussion

1. Explain how to use the graph of $f(x) = 2^x$ to produce the graph of $g(x) = 2^{(x-3)} + 4$.
2. At what point does the function $g(x) = e^{-x^2/2}$ take on its maximum value?
3. Without using a graphing utility, determine whether the revenue function $R(t) = 10 + e^{-0.05t}$ is an increasing function or a decreasing function.
4. Discuss the properties of the graph of $f(x) = b^x$ when $b > 1$.
5. What is the base of the natural exponential function? How is it calculated? What is its approximate value?

Exercise Set 7.1

In Exercises 1 to 8, evaluate the exponential function for the given *x*-values.

1. $f(x) = 3^x$; $x = 0$ and $x = 4$
2. $f(x) = 5^x$; $x = 3$ and $x = -2$
3. $g(x) = 10^x$; $x = -2$ and $x = 3$
4. $g(x) = 4^x$; $x = 0$ and $x = -1$
5. $h(x) = \left(\frac{3}{2}\right)^x$; $x = 2$ and $x = -3$
6. $h(x) = \left(\frac{2}{5}\right)^x$; $x = -1$ and $x = 3$
7. $j(x) = \left(\frac{1}{2}\right)^x$; $x = -2$ and $x = 4$
8. $j(x) = \left(\frac{1}{4}\right)^x$; $x = -1$ and $x = 5$

In Exercises 9 to 14, use a calculator to evaluate the exponential function for the given *x*-value. Round to the nearest hundredth.

9. $f(x) = 2^x$, $x = 3.2$
10. $f(x) = 3^x$, $x = -1.5$
11. $g(x) = e^x$, $x = 2.2$
12. $g(x) = e^x$, $x = -1.3$
13. $h(x) = 5^x$, $x = \sqrt{2}$
14. $h(x) = 0.5^x$, $x = \pi$

15. Examine the following four functions and the graphs labeled **a, b, c,** and **d.** For each graph, determine which function has been graphed.

$$f(x) = 5^x \qquad g(x) = 1 + 5^{-x}$$
$$h(x) = 5^{x+3} \qquad k(x) = 5^x + 3$$

a. 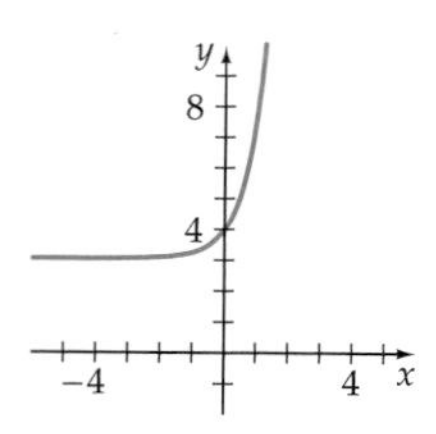
b. 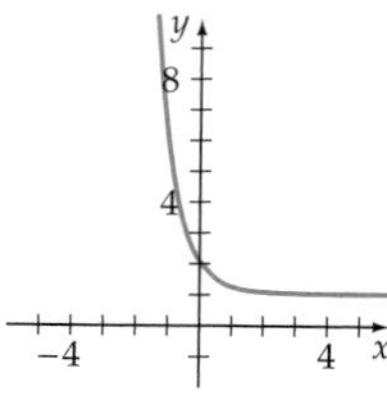
c. 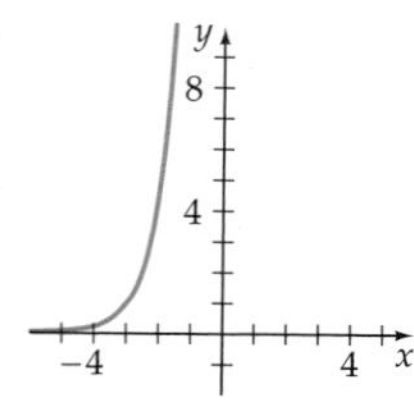
d. 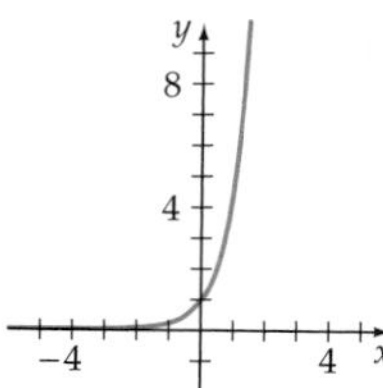

16. Examine the following four functions and the graphs labeled **a, b, c,** and **d.** For each graph, determine which function has been graphed.

$$f(x) = \left(\frac{1}{4}\right)^x \qquad g(x) = \left(\frac{1}{4}\right)^{-x}$$
$$h(x) = \left(\frac{1}{4}\right)^{x-2} \qquad k(x) = 3\left(\frac{1}{4}\right)^x$$

a. 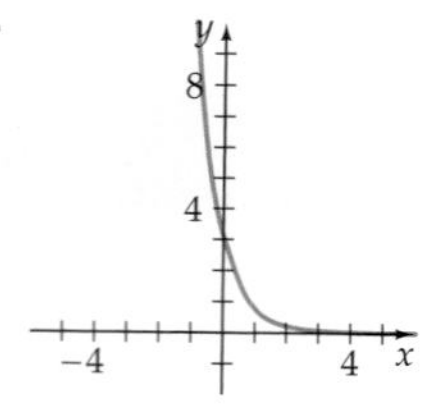
b. 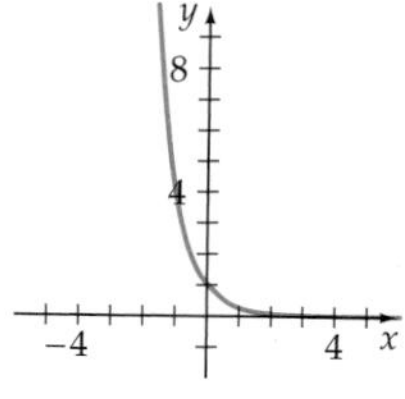
c. 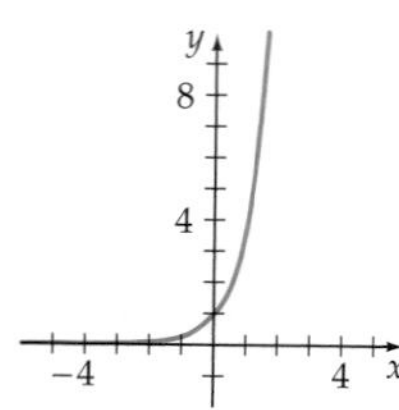
d. 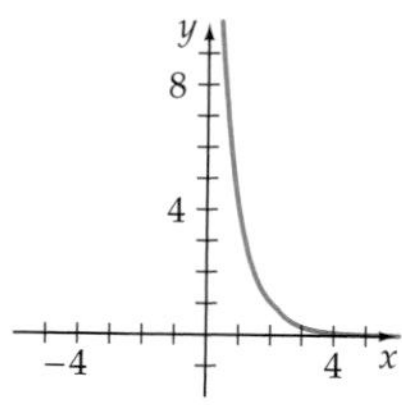

In Exercises 17 to 24, sketch the graph of each function.

17. $f(x) = 3^x$ **18.** $f(x) = 4^x$

19. $f(x) = 10^x$ **20.** $f(x) = 6^x$

21. $f(x) = \left(\frac{3}{2}\right)^x$ **22.** $f(x) = \left(\frac{5}{2}\right)^x$

23. $f(x) = \left(\frac{1}{3}\right)^x$ **24.** $f(x) = \left(\frac{2}{3}\right)^x$

In Exercises 25 to 38, explain how to use the graph of the first function *f* to produce the graph of the second function *F*.

25. $f(x) = 3^x, F(x) = 3^x + 2$

26. $f(x) = 4^x, F(x) = 4^x - 3$

27. $f(x) = 10^x, F(x) = 10^{x-2}$

28. $f(x) = 6^x, F(x) = 6^{x+5}$

29. $f(x) = \left(\frac{3}{2}\right)^x, F(x) = \left(\frac{3}{2}\right)^{-x}$

30. $f(x) = \left(\frac{5}{2}\right)^x, F(x) = -\left[\left(\frac{5}{2}\right)^x\right]$

31. $f(x) = \left(\frac{1}{3}\right)^x, F(x) = 2\left[\left(\frac{1}{3}\right)^x\right]$

32. $f(x) = \left(\frac{2}{3}\right)^x, F(x) = \frac{1}{2}\left[\left(\frac{2}{3}\right)^x\right]$

33. $f(x) = e^x, F(x) = e^{-x} + 2$

34. $f(x) = e^x, F(x) = e^{x-3} + 1$

35. $f(x) = 2^x, F(x) = -(2^{x-4})$

36. $f(x) = 2^x, F(x) = -(2^{-x})$

37. $f(x) = 0.5^x, F(x) = 3 + 0.5^{-x}$

38. $f(x) = 0.5^x, F(x) = 3(0.5^{x+2}) - 1$

In Exercises 39 to 46, use a graphing utility to graph each function. If the function has a horizontal asymptote, state the equation of the horizontal asymptote.

39. $f(x) = \dfrac{3^x + 3^{-x}}{2}$ **40.** $f(x) = 4 \cdot 3^{-x^2}$

41. $f(x) = \dfrac{e^x - e^{-x}}{2}$ **42.** $f(x) = \dfrac{e^x + e^{-x}}{2}$

43. $f(x) = -e^{(x-4)}$

44. $f(x) = 0.5e^{-x}$

45. $f(x) = \dfrac{10}{1 + 0.4e^{-0.5x}}$, $x \geq 0$

46. $f(x) = \dfrac{10}{1 + 1.5e^{-0.5x}}$, $x \geq 0$

47. **INTERNET CONNECTIONS** Data from Forrester Research suggest that the number of broadband [cable and digital subscriber line (DSL)] connections to the Internet can be modeled by $f(x) = 1.353(1.9025)^x$, where x is the number of years after January 1, 1998, and $f(x)$ is the number of connections in millions.

a. How many broadband Internet connections, to the nearest million, does this model predict will exist on January 1, 2007?

b. According to the model, in what year will the number of broadband connections first reach 1 billion? [*Hint:* Use the intersect feature of a graphing utility to determine the x-coordinate of the point of intersection of the graphs of $f(x)$ and $y = 1000$.]

48. **MEDICATION IN BLOODSTREAM** The function $A(t) = 200e^{-0.014t}$ gives the amount of medication, in milligrams, in a patient's bloodstream t minutes after the medication has been injected into the patient's bloodstream.

a. Find the amount of medication, to the nearest milligram, in the patient's bloodstream after 45 minutes.

b. Use a graphing utility to determine how long it will take, to the nearest minute, for the amount of medication in the patient's bloodstream to reach 50 milligrams.

49. **DEMAND FOR A PRODUCT** The demand d for a specific product, in items per month, is given by

$$d(p) = 25 + 880e^{-0.18p}$$

where p is the price, in dollars, of the product.

a. What will be the monthly demand, to the nearest unit, when the price of the product is \$8 and when the price is \$18?

b. What will happen to the demand as the price increases without bound?

50. **SALES** The monthly income I, in dollars, from a new product is given by

$$I(t) = 24{,}000 - 22{,}000e^{-0.005t}$$

where t is the time, in months, since the product was first put on the market.

a. What was the monthly income after the 10th month and after the 100th month?

b. What will the monthly income from the product approach as the time increases without bound?

51. ***E. COLI* INFECTION** *Escherichia coli (E. coli)* is a bacterium that can reproduce at an exponential rate. The *E. coli* reproduce by dividing. A small number of *E. coli* bacteria in the large intestine of a human can trigger a serious infection within a few hours. Consider a particular *E. coli* infection that starts with 100 *E. coli* bacteria. Each bacterium splits into two parts every half hour. Assuming none of the bacteria die, the size of the *E. coli* population after t hours is given by $P(t) = 100 \cdot 2^{2t}$, where $0 \leq t \leq 16$.

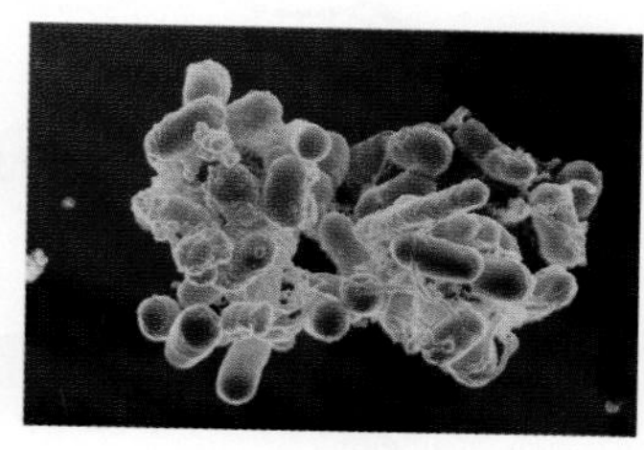

a. Find $P(3)$ and $P(6)$.

b. Use a graphing utility to find the time, to the nearest tenth of an hour, it takes for the *E. coli* population to number 1 billion.

52. **RADIATION** Lead shielding is used to contain radiation. The percentage of a certain radiation that can penetrate x millimeters of lead shielding is given by $I(x) = 100e^{-1.5x}$.

a. What percentage of radiation, to the nearest tenth of a percent, will penetrate a lead shield that is 1 millimeter thick?

b. How many millimeters of lead shielding are required so that less than 0.05% of the radiation penetrates the shielding? Round to the nearest millimeter.

53. **PHOTOCHROMATIC EYEGLASS LENSES** Photochromatic eyeglass lenses contain molecules of silver chloride or silver halide. These molecules are transparent in the absence of UV rays. UV rays are normally absent in artificial lighting. However, when the lenses are exposed to UV rays, as in direct sunlight, the molecules take on a new molecular structure, which causes the lenses to darken. The number of molecules affected varies with the intensity of the UV rays. The intensity of UV rays is measured using a scale called the UV index. On this scale, a value near 0 indicates a low UV intensity and a value near 10 indicates a high UV intensity.

For the photochromatic lenses shown below, the function $P(x) = (0.9)^x$ models the transparency P of the lenses as a function of the UV index x.

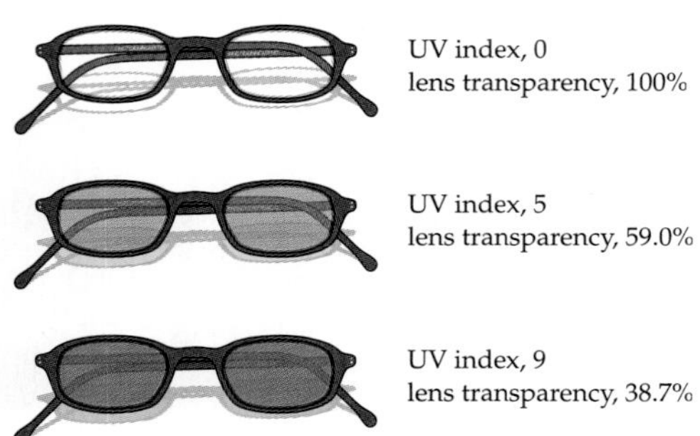

a. Find the transparency of these lenses, to the nearest tenth of a percent, when they are exposed to light rays with a UV index of 3.5.

b. What is the UV index of light rays that cause these photochromatic lenses to have a transparency of 45%? Round to the nearest tenth.

54. **Fish Population** The number of bass in a lake is given by

$$P(t) = \frac{3600}{1 + 7e^{-0.05t}}$$

where t is the number of months that have passed since the lake was stocked with bass.

a. How many bass were in the lake immediately after it was stocked?

b. How many bass were in the lake 1 year after the lake was stocked? Round to the nearest bass.

c. What will happen to the bass population as t increases without bound?

55. **The Pay It Forward Model** In the movie *Pay It Forward,* Trevor McKinney, played by Haley Joel Osment, is given a school assignment to "think of an idea to change the world—and then put it into action." In response to this assignment, Trevor develops a *pay it forward* project. In this project, anyone who benefits from another person's good deed must do a good deed for three additional people. Each of these three people is then obligated to do a good deed for another three people, and so on.

The following diagram shows the number of people who have been a beneficiary of a good deed after 1 round and after 2 rounds of this project.

Three beneficiaries after one round.

A total of 12 beneficiaries after two rounds $(3 + 9 = 12)$.

A mathematical model for the number of pay-it-forward beneficiaries after n rounds is given by $B(n) = \dfrac{3^{n+1} - 3}{2}$. Use this model to determine

a. the number of beneficiaries after 5 rounds and after 10 rounds. Assume that no person is a beneficiary of more than one good deed.

b. how many rounds are required to produce at least 2 million beneficiaries.

56. **Intensity of Light** The percent $I(x)$ of the original intensity of light striking the surface of a lake that is available x feet below the surface of the lake is given by $I(x) = 100e^{-0.95x}$.

a. What percentage of the light, to the nearest tenth of a percent, is available 2 feet below the surface of the lake?

b. At what depth, to the nearest hundredth of a foot, is the intensity of the light one-half the intensity at the surface?

57. **A TEMPERATURE MODEL** A cup of coffee is heated to 180°F and placed in a room that maintains a temperature of 65°F. The temperature of the coffee after t minutes is given by $T(t) = 65 + 115e^{-0.042t}$.

a. Find the temperature, to the nearest degree, of the coffee 10 minutes after it is placed in the room.

b. Use a graphing utility to determine when, to the nearest tenth of a minute, the temperature of the coffee will reach 100°F.

58. **A TEMPERATURE MODEL** Soup that is at a temperature of 170°F is poured into a bowl in a room that maintains a constant temperature. The temperature of the soup decreases according to the model given by $T(t) = 75 + 95e^{-0.12t}$, where t is time in minutes after the soup is poured.

a. What is the temperature, to the nearest tenth of a degree, of the soup after 2 minutes?

b. A certain customer prefers soup at a temperature of 110°F. How many minutes, to the nearest 0.1 minute, after the soup is poured does the soup reach that temperature?

c. What is the temperature of the room?

59. **MUSICAL SCALES** Starting on the left side of a standard 88-key piano, the frequency, in vibrations per second, of the nth note is given by $f(n) = (27.5)2^{(n-1)/12}$.

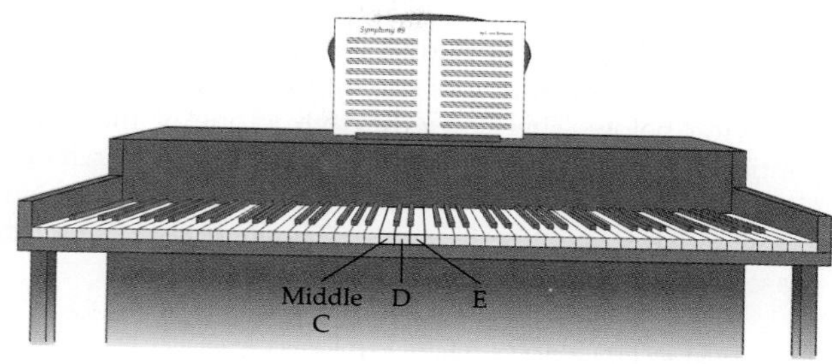

a. Using this formula, determine the frequency, to the nearest hundredth of a vibration per second, of middle C, key number 40 on an 88-key piano.

b. Is the difference in frequency between middle C (key number 40) and D (key number 42) the same as the difference in frequency between D (key number 42) and E (key number 44)? Explain.

Connecting Concepts

60. Verify that the hyperbolic cosine function $\cosh(x) = \dfrac{e^x + e^{-x}}{2}$ is an even function.

61. Verify that the hyperbolic sine function $\sinh(x) = \dfrac{e^x - e^{-x}}{2}$ is an odd function.

62. Graph $g(x) = 10^x$, and then sketch the graph of g reflected across the line given by $y = x$.

63. Graph $f(x) = e^x$, and then sketch the graph of f reflected across the line given by $y = x$.

In Exercises 64 to 67, determine the domain of the given function. Write the domain using interval notation.

64. $f(x) = \dfrac{e^x - e^{-x}}{e^x + e^{-x}}$

65. $f(x) = \dfrac{e^{|x|}}{1 + e^x}$

66. $f(x) = \sqrt{1 - e^x}$

67. $f(x) = \sqrt{e^x - e^{-x}}$

68. Let $h(x) = 2^{(x^2+4)}$. Express h as the composition of an exponential function f and a polynomial function g.

69. Let $h(x) = e^{(2x-5)}$. Express h as the composition of an exponential function f and a polynomial function g.

70. Explain why the graph of

$$f(x) = \frac{e^x + e^{-x}}{2}$$

can be produced by plotting the average height of $g(x) = e^x$ and $h(x) = e^{-x}$ for each value of x.

Projects

1. THE SAINT LOUIS GATEWAY ARCH The Gateway Arch in Saint Louis was designed in the shape of an inverted **catenary,** as shown by the red curve in the drawing below. The Gateway Arch is one of the largest optical illusions ever created. As you look at the arch (and its basic shape defined by the catenary curve), it appears to be much taller than it is wide. However, this is not the case. The height of the catenary is given by

$$h(x) = 693.8597 - 68.7672\left(\frac{e^{0.0100333x} + e^{-0.0100333x}}{2}\right)$$

where x and $h(x)$ are measured in feet and $x = 0$ represents the position at ground level that is directly below the highest point of the catenary.

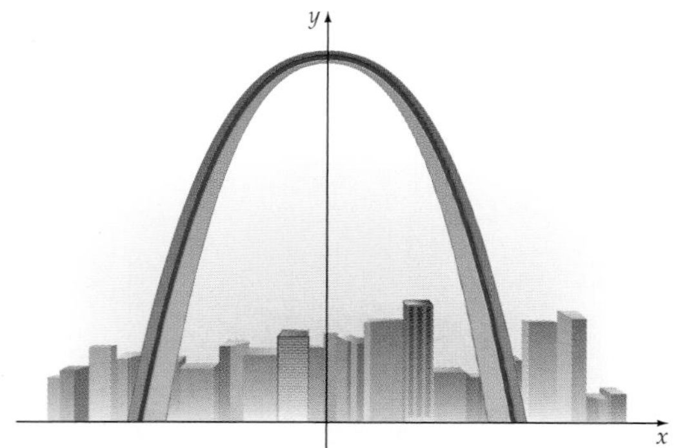

a. Use a graphing utility to graph $h(x)$.

b. Use your graph to find the height of the catenary for $x = 0$, 100, 200, and 299 feet. Round each result to the nearest tenth of a foot.

c. What is the width of the catenary at ground level and what is the maximum height of the catenary? Round each result to the nearest tenth of a foot.

d. By how much does the maximum height of the catenary exceed its width at ground level? Round to the nearest tenth of a foot.

2. AN EXPONENTIAL REWARD According to legend, when Sissa Ben Dahir of India invented the game of chess, King Shirham was so impressed with the game that he summoned the game's inventor and offered him the reward of his choosing. The inventor pointed to the chessboard and requested, for his reward, one grain of wheat on the first square, two grains of wheat on the second square, four grains on the third square, eight grains on the fourth square, and so on for all 64 squares on the chessboard. The king considered this a very modest reward and said he would grant the inventor's wish. The following table shows how many grains of wheat are on each of the first six squares and the total number of grains of wheat needed to cover squares 1 to n for $n \le 6$.

Square number, n	Number of grains of wheat on square n	Total number of grains of wheat on squares 1 through n
1	1	1
2	2	3
3	4	7
4	8	15
5	16	31
6	32	63

a. If all 64 squares of the chessboard are piled with wheat as requested by Sissa Ben Dahir, how many grains of wheat are on the board?

b. A grain of wheat weighs approximately 0.000008 kilogram. Find the total weight of the wheat requested by Sissa Ben Dahir.

c. In a recent year, a total of 6.5×10^8 metric tons of wheat were produced in the world. At this level, how many years, to the nearest year, of wheat production would be required to fill the request of Sissa Ben Dahir? One metric ton equals 1000 kilograms.

Section 7.2

- Logarithmic Functions
- Graphs of Logarithmic Functions
- Domains of Logarithmic Functions
- Common and Natural Logarithms

Logarithmic Functions and Their Applications

PREPARE FOR THIS SECTION

Prepare for this section by completing the following exercises. The answers can be found on page A37.

PS1. If $2^x = 16$, determine the value of x. [7.1]

PS2. If $3^{-x} = \frac{1}{27}$, determine the value of x. [7.1]

PS3. If $x^4 = 625$, determine the value of x. [7.1]

PS4. Find the inverse of $f(x) = \frac{2x}{x+3}$. [1.6]

PS5. State the domain of $g(x) = \sqrt{x-2}$. [1.3]

PS6. If the range of $h(x)$ is the set of all positive real numbers, then what is the domain of $h^{-1}(x)$? [1.6]

Logarithmic Functions

Every exponential function of the form $g(x) = b^x$ is a one-to-one function and therefore has an inverse function. Sometimes we can determine the inverse of a function represented by an equation by interchanging the variables of its equation and then solving for the dependent variable. If we attempt to use this procedure for $g(x) = b^x$, we obtain

$$g(x) = b^x$$
$$y = b^x$$
$$x = b^y \qquad \text{• Interchange the variables.}$$

None of our previous methods can be used to solve the equation $x = b^y$ for the exponent y. Thus we need to develop a new procedure. One method would be to merely write

$$y = \text{the power of } b \text{ that produces } x$$

Although this would work, it is not very concise. We need a compact notation to represent "y is the power of b that produces x." This more compact notation is given in the following definition.

Math Matters

Logarithms were developed by John Napier (1550–1617) as a means of simplifying the calculations of astronomers. One of his ideas was to devise a method by which the product of two numbers could be determined by performing an addition.

Definition of a Logarithm and a Logarithmic Function

If $x > 0$ and b is a positive constant ($b \neq 1$), then

$$y = \log_b x \quad \text{if and only if} \quad b^y = x$$

The notation $\log_b x$ is read "the **logarithm** (or log) base b of x." The function defined by $f(x) = \log_b x$ is a **logarithmic function** with base b. This function is the inverse of the exponential function $g(x) = b^x$.

It is essential to remember that $f(x) = \log_b x$ is the inverse function of $g(x) = b^x$. Because these functions are inverses and because functions that are inverses have the property that $f(g(x)) = x$ and $g(f(x)) = x$, we have the following important relationships.

Composition of Logarithmic and Exponential Functions

Let $g(x) = b^x$ and $f(x) = \log_b x$ $(x > 0, b > 0, b \neq 1)$. Then

$$g(f(x)) = b^{\log_b x} = x \qquad \text{and} \qquad f(g(x)) = \log_b b^x = x$$

take note

The notation $\log_b x$ replaces the phrase "the power of b that produces x." For instance, "3 is the power of 2 that produces 8" is abbreviated $3 = \log_2 8$. In your work with logarithms, remember that a logarithm is an *exponent*.

As an example of these relationships, let $g(x) = 2^x$ and $f(x) = \log_2 x$. Then

$$2^{\log_2 x} = x \qquad \text{and} \qquad \log_2 2^x = x$$

The equations

$$y = \log_b x \qquad \text{and} \qquad b^y = x$$

are different ways of expressing the same concept.

Definition of Exponential Form and Logarithmic Form

The **exponential form** of $y = \log_b x$ is $b^y = x$.

The **logarithmic form** of $b^y = x$ is $y = \log_b x$.

These concepts are illustrated in the next two examples.

EXAMPLE 1 ›› Change from Logarithmic to Exponential Form

Write each equation in its exponential form.

a. $3 = \log_2 8$ b. $2 = \log_{10}(x + 5)$ c. $\log_e x = 4$ d. $\log_b b^3 = 3$

Solution

Use the definition $y = \log_b x$ if and only if $b^y = x$.

Logarithms are exponents.

a. $3 = \log_2 8$ if and only if $2^3 = 8$

Base

b. $2 = \log_{10}(x + 5)$ if and only if $10^2 = x + 5$.

c. $\log_e x = 4$ if and only if $e^4 = x$.

d. $\log_b b^3 = 3$ if and only if $b^3 = b^3$.

›› *Try Exercise 4, page 476*

EXAMPLE 2 Change from Exponential to Logarithmic Form

Write each equation in its logarithmic form.

a. $3^2 = 9$ b. $5^3 = x$ c. $a^b = c$ d. $b^{\log_b 5} = 5$

Solution

The logarithmic form of $b^y = x$ is $y = \log_b x$.

a. $3^2 = 9$ if and only if $2 = \log_3 9$ (Exponent: 2; Base: 3)

b. $5^3 = x$ if and only if $3 = \log_5 x$.

c. $a^b = c$ if and only if $b = \log_a c$.

d. $b^{\log_b 5} = 5$ if and only if $\log_b 5 = \log_b 5$.

Try Exercise 14, page 476

The definition of a logarithm and the definition of an inverse function can be used to establish many properties of logarithms. For instance:

- $\log_b b = 1$ because $b = b^1$.
- $\log_b 1 = 0$ because $1 = b^0$.
- $\log_b(b^x) = x$ because $b^x = b^x$.
- $b^{\log_b x} = x$ because $f(x) = \log_b x$ and $g(x) = b^x$ are inverse functions. Thus $g[f(x)] = x$.

We will refer to the preceding properties as the *basic logarithmic properties.*

Basic Logarithmic Properties

1. $\log_b b = 1$ 2. $\log_b 1 = 0$ 3. $\log_b(b^x) = x$ 4. $b^{\log_b x} = x$

EXAMPLE 3 Apply the Basic Logarithmic Properties

Evaluate each of the following logarithms.

a. $\log_8 1$ b. $\log_5 5$ c. $\log_2(2^4)$ d. $3^{\log_3 7}$

Solution

a. By Property 2, $\log_8 1 = 0$.

b. By Property 1, $\log_5 5 = 1$.

c. By Property 3, $\log_2(2^4) = 4$.

d. By Property 4, $3^{\log_3 7} = 7$.

Try Exercise 32, page 476

Some logarithms can be evaluated just by remembering that a logarithm is an exponent. For instance, $\log_5 25$ equals 2 because the base 5 raised to the second power equals 25.

- $\log_{10} 100 = 2$ because $10^2 = 100$.
- $\log_4 64 = 3$ because $4^3 = 64$.
- $\log_7 \frac{1}{49} = -2$ because $7^{-2} = \frac{1}{7^2} = \frac{1}{49}$.

QUESTION What is the value of $\log_5 625$?

Graphs of Logarithmic Functions

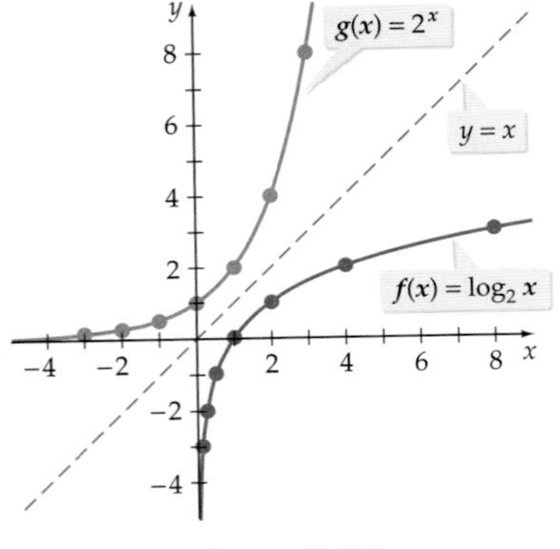

Figure 7.13

Because $f(x) = \log_b x$ is the inverse function of $g(x) = b^x$, the graph of f is a reflection of the graph of g across the line given by $y = x$. The graph of $g(x) = 2^x$ is shown in **Figure 7.13. Table 7.7** below shows some of the ordered pairs of the graph of g.

Table 7.7

x	-3	-2	-1	0	1	2	3
$g(x) = 2^x$	$\frac{1}{8}$	$\frac{1}{4}$	$\frac{1}{2}$	1	2	4	8

The graph of the inverse of g, which is $f(x) = \log_2 x$, is also shown in **Figure 7.13**. Some of the ordered pairs of f are shown in **Table 7.8.** Note that if (x, y) is a point on the graph of g, then (y, x) is a point on the graph of f. Also notice that the graph of f is a reflection of the graph of g across the line given by $y = x$.

Table 7.8

x	$\frac{1}{8}$	$\frac{1}{4}$	$\frac{1}{2}$	1	2	4	8
$f(x) = \log_2 x$	-3	-2	-1	0	1	2	3

The graph of a logarithmic function can be drawn by first rewriting the function in its exponential form. This procedure is illustrated in Example 4.

ANSWER $\log_5 625 = 4$ because $5^4 = 625$.

EXAMPLE 4 》 Graph a Logarithmic Function

Graph $f(x) = \log_3 x$.

Solution

To graph $f(x) = \log_3 x$, consider the equivalent exponential equation $x = 3^y$. Because this equation is solved for x, choose values of y and calculate the corresponding values of x, as shown in **Table 7.9.**

Table 7.9

$x = 3^y$	$\frac{1}{9}$	$\frac{1}{3}$	1	3	9
y	−2	−1	0	1	2

Now plot the ordered pairs and connect the points with a smooth curve, as shown in **Figure 7.14.**

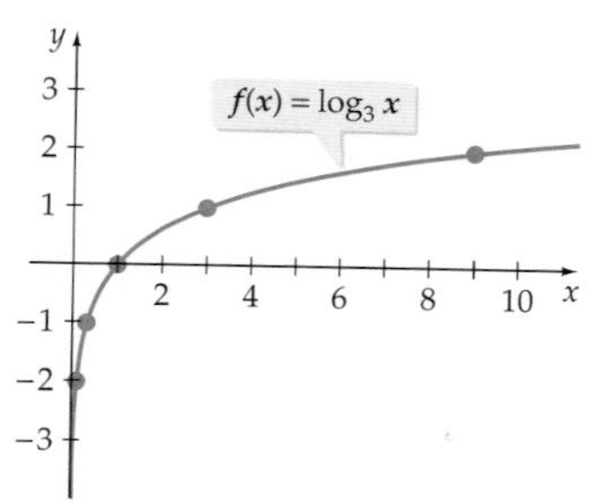

Figure 7.14

》 *Try Exercise 44, page 476*

We can use a similar procedure to draw the graph of a logarithmic function with a fractional base. For instance, consider $y = \log_{2/3} x$. Rewriting this in exponential form gives us $\left(\frac{2}{3}\right)^y = x$. Choose values of y and calculate the corresponding x values. See **Table 7.10**. Plot the points corresponding to the ordered pairs (x, y), and then draw a smooth curve through the points, as shown in **Figure 7.15.**

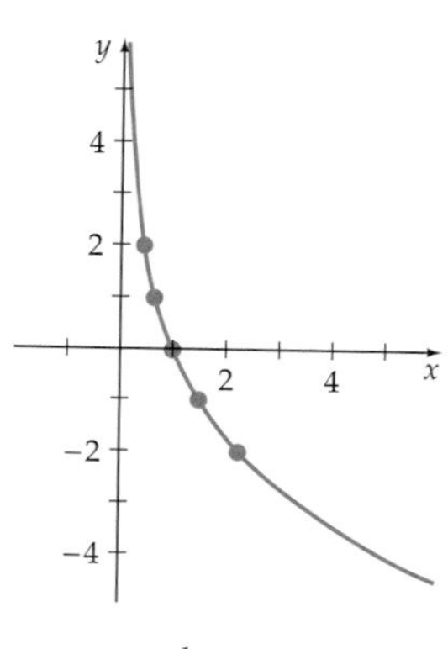

$y = \log_{2/3} x$

Figure 7.15

Table 7.10

$x = \left(\frac{2}{3}\right)^y$	$\left(\frac{2}{3}\right)^{-2} = \frac{9}{4}$	$\left(\frac{2}{3}\right)^{-1} = \frac{3}{2}$	$\left(\frac{2}{3}\right)^{0} = 1$	$\left(\frac{2}{3}\right)^{1} = \frac{2}{3}$	$\left(\frac{2}{3}\right)^{2} = \frac{4}{9}$
y	−2	−1	0	1	2

Properties of $f(x) = \log_b x$

For all positive real numbers b, $b \neq 1$, the function $f(x) = \log_b x$ has the following properties:

1. The domain of f consists of the set of positive real numbers, and its range consists of the set of all real numbers.
2. The graph of f has an x-intercept of $(1, 0)$ and passes through $(b, 1)$.
3. If $b > 1$, f is an increasing function and its graph is asymptotic to the negative y-axis. [As $x \to \infty$, $f(x) \to \infty$, and as $x \to 0$ from the right, $f(x) \to -\infty$.] See **Figure 7.16a.**
4. If $0 < b < 1$, f is a decreasing function and its graph is asymptotic to the positive y-axis. [As $x \to \infty$, $f(x) \to -\infty$, and as $x \to 0$ from the right, $f(x) \to \infty$.] See **Figure 7.16b.**

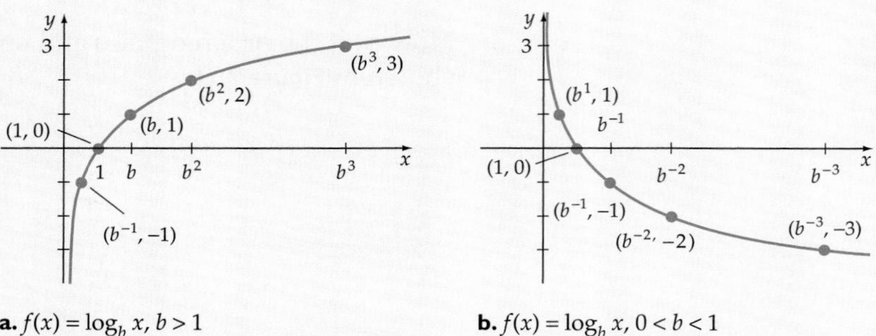

Figure 7.16

Domains of Logarithmic Functions

The function $f(x) = \log_b x$ has as its domain the set of positive real numbers. The function $f(x) = \log_b(g(x))$ has as its domain the set of all x for which $g(x) > 0$. To determine the domain of a function such as $f(x) = \log_b(g(x))$, we must determine the values of x that make $g(x)$ positive. This process is illustrated in Example 5.

EXAMPLE 5 » Find the Domain of a Logarithmic Function

Find the domain of each of the following logarithmic functions.

a. $f(x) = \log_6(x - 3)$ b. $F(x) = \log_2|x + 2|$ c. $R(x) = \log_5\left(\frac{x}{8 - x}\right)$

Solution

a. Solving $(x - 3) > 0$ for x gives us $x > 3$. The domain of f consists of all real numbers greater than 3. In interval notation, the domain is $(3, \infty)$.

b. The solution set of $|x + 2| > 0$ consists of all real numbers x except $x = -2$. The domain of F consists of all real numbers $x \neq -2$. In interval notation, the domain is $(-\infty, -2) \cup (-2, \infty)$.

c. Solving $\left(\frac{x}{8-x}\right) > 0$ yields the set of all real numbers x between 0 and 8. The domain of R is all real numbers x such that $0 < x < 8$. In interval notation, the domain is $(0, 8)$.

›› Try Exercise 52, page 476

Some logarithmic functions can be graphed by using horizontal and/or vertical translations of a previously drawn graph.

EXAMPLE 6 ›› Use Translations to Graph Logarithmic Functions

Graph. a. $f(x) = \log_4(x + 3)$ b. $f(x) = \log_4 x + 3$

Solution

a. The graph of $f(x) = \log_4(x + 3)$ can be obtained by shifting the graph of $g(x) = \log_4 x$ to the left 3 units. See **Figure 7.17.** Note that the domain of f consists of all real numbers x greater than -3 because $x + 3 > 0$ for $x > -3$. The graph of f is asymptotic to the vertical line $x = -3$.

b. The graph of $f(x) = \log_4 x + 3$ can be obtained by shifting the graph of $g(x) = \log_4 x$ upward 3 units. See **Figure 7.18.**

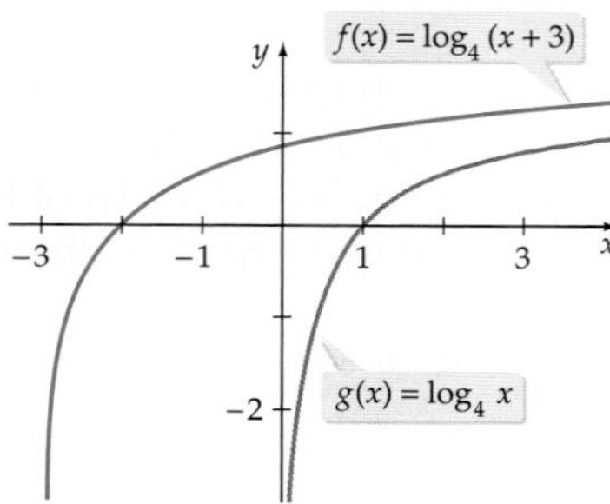

Figure 7.17

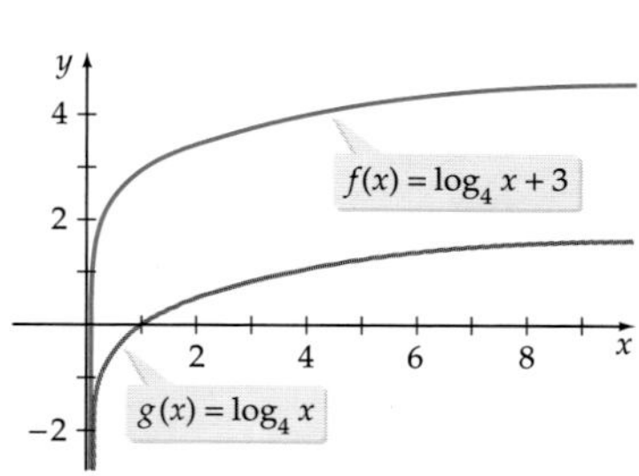

Figure 7.18

›› Try Exercise 66, page 476

Common and Natural Logarithms

Two of the most frequently used logarithmic functions are *common logarithms*, which have base 10, and *natural logarithms*, which have base e (the base of the natural exponential function).

Definition of Common and Natural Logarithms

The function defined by $f(x) = \log_{10} x$ is called the **common logarithmic function.** It is customarily written without stating the base as $f(x) = \log x$.

The function defined by $f(x) = \log_e x$ is called the **natural logarithmic function.** It is customarily written as $f(x) = \ln x$.

Most scientific or graphing calculators have a LOG key for evaluating common logarithms and an LN key to evaluate natural logarithms. For instance, using a graphing calculator,

$$\log 24 \approx 1.3802112 \qquad \text{and} \qquad \ln 81 \approx 4.3944492$$

The graphs of $f(x) = \log x$ and $f(x) = \ln x$ can be drawn using the same techniques we used to draw the graphs in the preceding examples. However, these graphs also can be produced with a graphing calculator by entering $\log x$ and $\ln x$ into the Y= menu. See **Figure 7.19** and **Figure 7.20.**

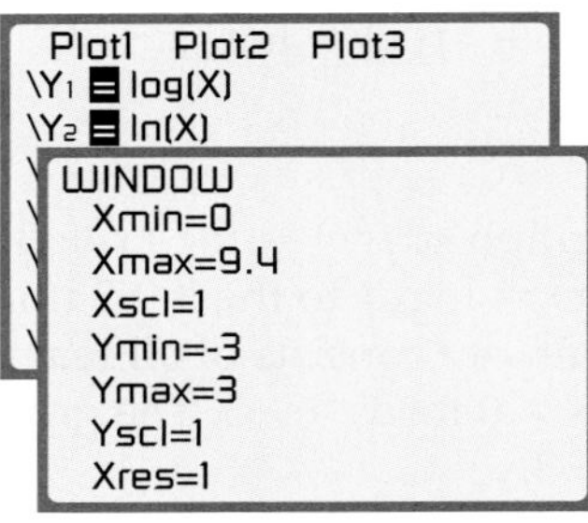

Figure 7.19

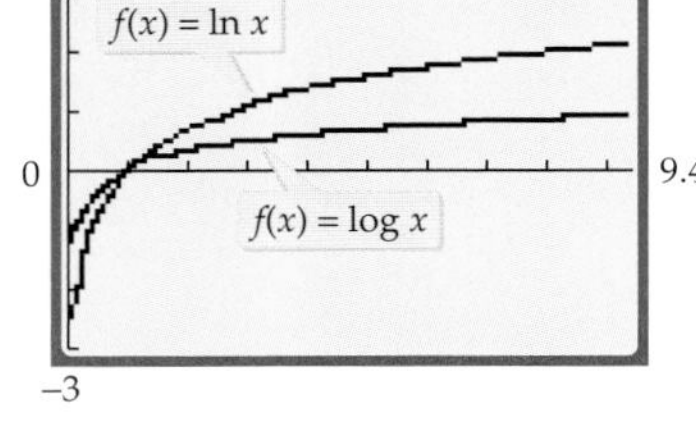

Figure 7.20

Observe that each graph passes through $(1, 0)$. Also note that as $x \to 0$ from the right, the functional values $f(x) \to -\infty$. Thus the y-axis is a vertical asymptote for each of the graphs. The domain of both $f(x) = \log x$ and $f(x) = \ln x$ is the set of positive real numbers. Each of these functions has a range consisting of the set of real numbers.

Many applications can be modeled by logarithmic functions.

EXAMPLE 7 ›› Applied Physiology

In the study *The Pace of Life*, M. H. Bornstein and H. G. Bornstein (*Nature*, Vol. 259, pp. 557–558, 1976) reported that as the population of a city increases, the average walking speed of a pedestrian also increases. An approximate relation between the average pedestrian walking speed s, in miles per hour, and the population x, in thousands, of a city is given by the function

$$s(x) = 0.37 \ln x + 0.05$$

a. Determine the average walking speed, to the nearest tenth of a mile per hour, in San Francisco, which has a population of 780,000, and in Round Rock, Texas, which has a population of 62,000.

b. Estimate the population of a city for which the average pedestrian walking speed is 3.1 miles per hour. Round to the nearest hundred-thousand.

Math Matters

Although logarithms were originally developed to assist with computations, logarithmic functions have a much broader use today. They are often used in such disciplines as geology, acoustics, chemistry, physics, and economics, to name a few.

Solution

a. The population of San Francisco, in thousands, is 780.

$$s(x) = 0.37 \ln x + 0.05$$
$$s(780) = 0.37 \ln 780 + 0.05 \quad \text{• Substitute 780 for } x.$$
$$\approx 2.5 \quad \text{• Use a calculator to evaluate.}$$

The average walking speed in San Francisco is about 2.5 miles per hour.

The population of Round Rock, in thousands, is 62.

$$s(x) = 0.37 \ln x + 0.05$$
$$s(62) = 0.37 \ln 62 + 0.05 \quad \text{• Substitute 62 for } x.$$
$$\approx 1.6 \quad \text{• Use a calculator to evaluate.}$$

The average walking speed in Round Rock is about 1.6 miles per hour.

b. Graph $s(x) = 0.37 \ln x + 0.05$ and $s = 3.1$ in the same viewing window.

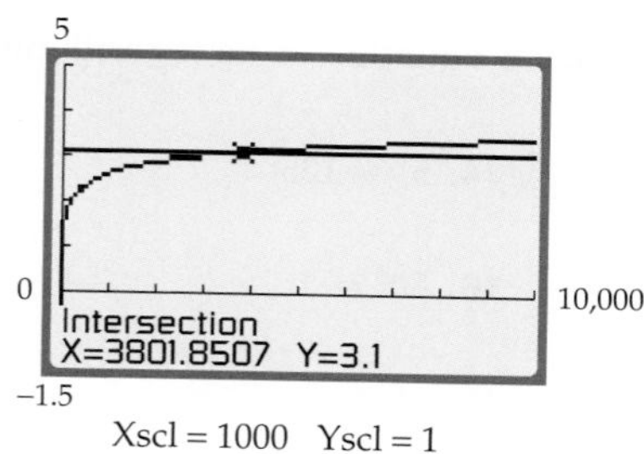

Xscl = 1000 Yscl = 1

The x value of the intersection point represents the population in thousands. The function indicates that a city with an average pedestrian walking speed of 3.1 miles per hour should have a population of about 3,800,000.

Try Exercise 86, page 477

Topics for Discussion

1. If $m > n$, must $\log_b m > \log_b n$?
2. For what values of x is $\ln x > \log x$?
3. What is the domain of $f(x) = \log(x^2 + 1)$? Explain why the graph of f does not have a vertical asymptote.
4.  The subtraction $3 - 5$ does not have an answer if we require that the answer be positive. Keep this idea in mind as you work the rest of this discussion exercise.

   Press the MODE key of a TI-83/TI-83 Plus/TI-84 Plus graphing calculator, and choose "Real" from the menu. Now use the calculator to evaluate $\log(-2)$. What output is given by the calculator? Press the MODE key, and choose "a + bi" from the menu. Now use the calculator to evaluate $\log(-2)$. What output is given by the calculator? Explain why the output is different for these two evaluations.

Exercise Set 7.2

In Exercises 1 to 12, write each equation in its exponential form.

1. $1 = \log 10$
2. $4 = \log 10{,}000$
3. $2 = \log_8 64$
4. $3 = \log_4 64$
5. $0 = \log_7 x$
6. $-4 = \log_3 \dfrac{1}{81}$
7. $\ln x = 4$
8. $\ln e^2 = 2$
9. $\ln 1 = 0$
10. $\ln x = -3$
11. $2 = \log(3x + 1)$
12. $\dfrac{1}{3} = \ln\left(\dfrac{x+1}{x^2}\right)$

In Exercises 13 to 24, write each equation in its logarithmic form. Assume $y > 0$ and $b > 0$.

13. $3^2 = 9$
14. $5^3 = 125$
15. $4^{-2} = \dfrac{1}{16}$
16. $10^0 = 1$
17. $b^x = y$
18. $2^x = y$
19. $y = e^x$
20. $5^1 = 5$
21. $100 = 10^2$
22. $2^{-4} = \dfrac{1}{16}$
23. $e^2 = x + 5$
24. $3^x = 47$

In Exercises 25 to 42, evaluate each logarithm. Do not use a calculator.

25. $\log_4 16$
26. $\log_{3/2} \dfrac{8}{27}$
27. $\log_3 \dfrac{1}{243}$
28. $\log_b 1$
29. $\ln e^3$
30. $\log_b b$
31. $\log \dfrac{1}{100}$
32. $\log 1{,}000{,}000$
33. $\log_{0.5} 16$
34. $\log_{0.3} \dfrac{100}{9}$
35. $4 \log 1000$
36. $\log_5 125^2$
37. $2 \log_7 2401$
38. $3 \log_{11} 161{,}051$
39. $\log_3 \sqrt[5]{9}$
40. $\log_6 \sqrt[3]{36}$
41. $5 \log_{13} \sqrt[3]{169}$
42. $2 \log_7 \sqrt[7]{343}$

In Exercises 43 to 50, graph each function by using its exponential form.

43. $f(x) = \log_4 x$
44. $f(x) = \log_6 x$
45. $f(x) = \log_{12} x$
46. $f(x) = \log_8 x$
47. $f(x) = \log_{1/2} x$
48. $f(x) = \log_{1/4} x$
49. $f(x) = \log_{5/2} x$
50. $f(x) = \log_{7/3} x$

In Exercises 51 to 64, find the domain of the function. Write the domain using interval notation.

51. $f(x) = \log_5(x - 3)$
52. $k(x) = \log_4(5 - x)$
53. $k(x) = \log_{2/3}(11 - x)$
54. $H(x) = \log_{1/4}(x^2 + 1)$
55. $P(x) = \ln(x^2 - 4)$
56. $J(x) = \ln\left(\dfrac{x-3}{x}\right)$
57. $h(x) = \ln\left(\dfrac{x^2}{x-4}\right)$
58. $R(x) = \ln(x^4 - x^2)$
59. $N(x) = \log_2(x^3 - x)$
60. $s(x) = \log_7(x^2 + 7x + 10)$
61. $g(x) = \log \sqrt{2x - 11}$
62. $m(x) = \log |4x - 8|$
63. $t(x) = 2 \ln(3x - 7)$
64. $v(x) = \ln(x - 4)^2$

In Exercises 65 to 72, use translations of the graphs in Exercises 43 to 50 to produce the graph of the given function.

65. $f(x) = \log_4(x - 3)$
66. $f(x) = \log_6(x + 3)$
67. $f(x) = \log_{12} x + 2$
68. $f(x) = \log_8 x - 4$
69. $f(x) = 3 + \log_{1/2} x$
70. $f(x) = 2 + \log_{1/4} x$
71. $f(x) = 1 + \log_{5/2}(x - 4)$
72. $f(x) = \log_{7/3}(x - 3) - 1$

73. Examine the following four functions and the graphs labeled **a**, **b**, **c**, and **d**. For each graph, determine which function has been graphed.

$f(x) = \log_5(x - 2)$ $\quad g(x) = 2 + \log_5 x$

$h(x) = \log_5(-x)$ $\quad k(x) = -\log_5(x + 3)$

a.

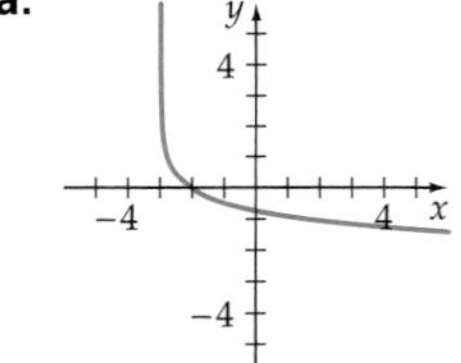

b.

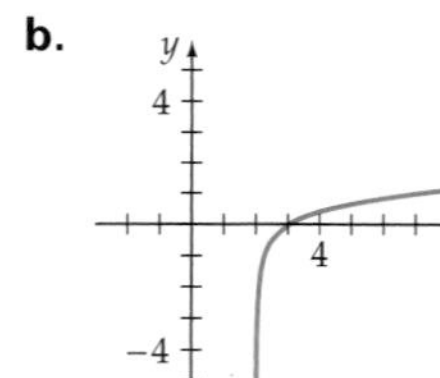

c.

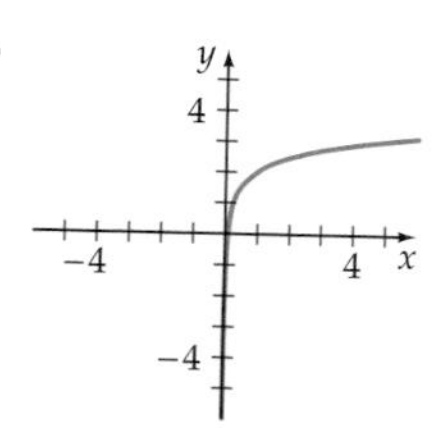

d.

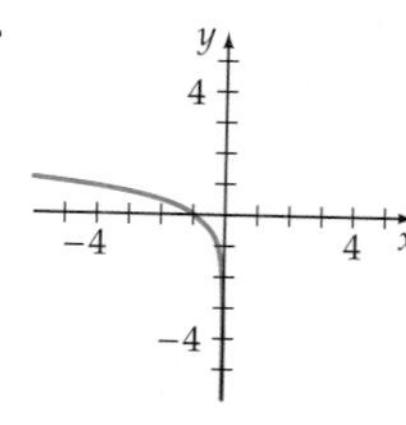

74. Examine the following four functions and the graphs labeled **a**, **b**, **c**, and **d**. For each graph, determine which function has been graphed.

$$f(x) = \ln x + 3 \qquad g(x) = \ln(x - 3)$$
$$h(x) = \ln(3 - x) \qquad k(x) = -\ln(-x)$$

a.

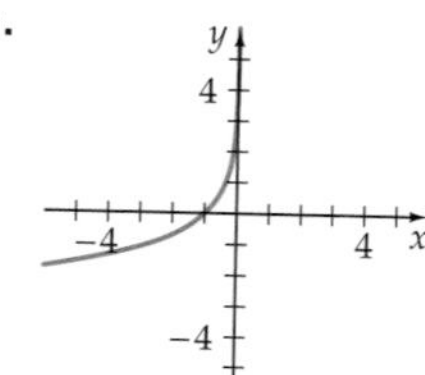

b.

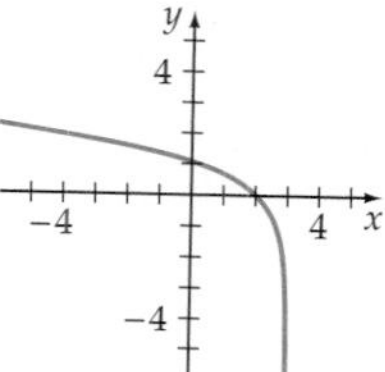

c.

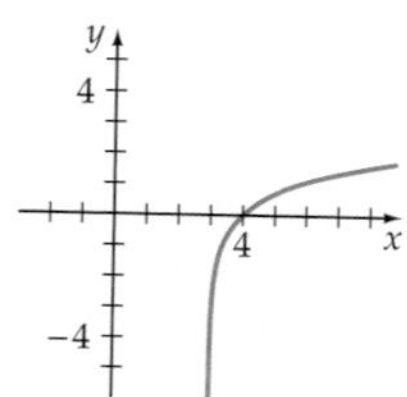

d.

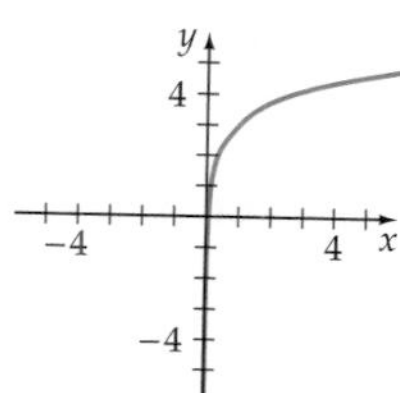

In Exercises 75 to 84, use a graphing utility to graph the function.

75. $f(x) = -2 \ln x$

76. $f(x) = -\log x$

77. $f(x) = |\ln x|$

78. $f(x) = \ln |x|$

79. $f(x) = \log \sqrt[3]{x}$

80. $f(x) = \ln \sqrt{x}$

81. $f(x) = \log(x + 10)$

82. $f(x) = \ln(x + 3)$

83. $f(x) = 3 \log |2x + 10|$

84. $f(x) = \dfrac{1}{2} \ln |x - 4|$

85. MONEY MARKET RATES The function

$$r(t) = 0.69607 + 0.60781 \ln t$$

gives the annual interest rate r, as a percent, a bank will pay on its money market accounts, where t is the term (the time the money is invested) in months.

a. What interest rate, to the nearest tenth of a percent, will the bank pay on a money market account with a term of 9 months?

b. What is the minimum number of complete months during which a person must invest to receive an interest rate of at least 3%?

86. AVERAGE TYPING SPEED The following function models the average typing speed S, in words per minute, of a student who has been typing for t months.

$$S(t) = 5 + 29 \ln(t + 1), \quad 0 \le t \le 16$$

a. What was the student's average typing speed, to the nearest word per minute, when the student first started to type? What was the student's average typing speed, to the nearest word per minute, after 3 months?

b. Use a graph of S to determine how long, to the nearest tenth of a month, it will take the student to achieve an average typing speed of 65 words per minute.

87. ADVERTISING COSTS AND SALES The function

$$N(x) = 2750 + 180 \ln\left(\frac{x}{1000} + 1\right)$$

models the relationship between the dollar amount x spent on advertising a product and the number of units N that a company can sell.

a. Find the number of units that will be sold with advertising expenditures of \$20,000, \$40,000, and \$60,000.

b. How many units will be sold if the company does not pay to advertise the product?

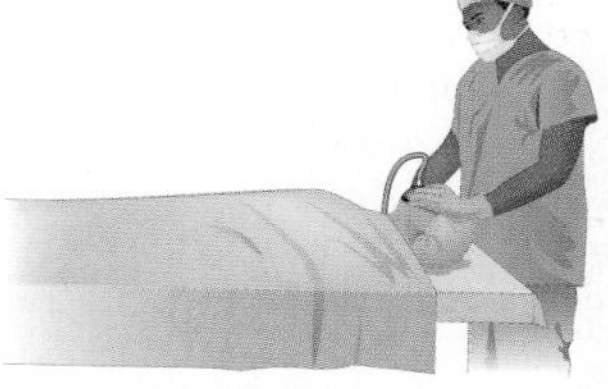

MEDICINE In anesthesiology it is necessary to accurately estimate the body surface area of a patient. One formula for estimating body surface area (BSA) was developed by Edith Boyd (University of Minnesota Press, 1935). Her formula for the BSA (in square meters) of a patient of height H (in centimeters) and weight W (in grams) is

$$BSA = 0.0003207 \cdot H^{0.3} \cdot W^{(0.7285 - 0.0188 \log W)}$$

In Exercises 88 and 89, use Boyd's formula to estimate the body surface area of a patient with the given weight and height. Round to the nearest hundredth of a square meter.

88. $W = 110$ pounds (49,895.2 grams); $H = 5$ feet 4 inches (162.56 centimeters)

89. $W = 180$ pounds (81,646.6 grams); $H = 6$ feet 1 inch (185.42 centimeters)

90. ASTRONOMY Astronomers measure the apparent brightness of a star by a unit called the **apparent magnitude.** This unit was created in the second century B.C. when the Greek astronomer Hipparchus classified the relative brightness of several stars. In his list he assigned the number 1 to the stars that appeared to be the brightest (Sirius, Vega, and Deneb). They are

first-magnitude stars. Hipparchus assigned the number 2 to all the stars in the Big Dipper. They are second-magnitude stars. The following table shows the relationship between a star's brightness relative to a first-magnitude star and the star's apparent magnitude. Notice from the table that a first-magnitude star appears to be about 2.51 times as bright as a second-magnitude star.

Brightness, x relative to a first-magnitude star	Apparent magnitude $M(x)$
1	1
$\frac{1}{2.51}$	2
$\frac{1}{6.31} \approx \frac{1}{2.51^2}$	3
$\frac{1}{15.85} \approx \frac{1}{2.51^3}$	4
$\frac{1}{39.82} \approx \frac{1}{2.51^4}$	5
$\frac{1}{100} \approx \frac{1}{2.51^5}$	6

The following logarithmic function gives the apparent magnitude $M(x)$ of a star as a function of its brightness x.

$$M(x) = -2.51 \log x + 1, \quad 0 < x \le 1$$

a. Use $M(x)$ to find the apparent magnitude of a star that is $\frac{1}{10}$ as bright as a first-magnitude star. Round to the nearest hundredth.

b. Find the approximate apparent magnitude of a star that is $\frac{1}{400}$ as bright as a first-magnitude star. Round to the nearest hundredth.

c. Which star appears brighter: a star with an apparent magnitude of 12 or a star with an apparent magnitude of 15?

d. Is $M(x)$ an increasing function or a decreasing function?

91. **NUMBER OF DIGITS IN b^x** An engineer has determined that the number of digits N in the expansion of b^x, where both b and x are positive integers, is $N = \text{int}(x \log b) + 1$, where $\text{int}(x \log b)$ denotes the greatest integer of $x \log b$. (*Note:* See pages 41–44 for information on the greatest integer function.)

a. Because $2^{10} = 1024$, we know that 2^{10} has four digits. Use the equation $N = \text{int}(x \log b) + 1$ to verify this result.

b. Find the number of digits in 3^{200}.

c. Find the number of digits in 7^{4005}.

d. The largest known prime number as of November 17, 2003, was $2^{20996011} - 1$. Find the number of digits in this prime number. (*Hint:* Because $2^{20996011}$ is not a power of 10, both $2^{20996011}$ and $2^{20996011} - 1$ have the same number of digits.)

92. **NUMBER OF DIGITS IN $9^{(9^9)}$** A science teacher has offered 10 points extra credit to any student who will write out all the digits in the expansion of $9^{(9^9)}$.

a. Use the formula from Exercise 91 to determine the number of digits in this number.

b. Assume that you can write 1000 digits per page and that 500 pages of paper are in a ream of paper. How many reams of paper, to the nearest tenth of a ream, are required to write out the expansion of $9^{(9^9)}$? Assume that you write on only one side of each page.

Connecting Concepts

93. Use a graphing utility to graph $f(x) = \frac{e^x - e^{-x}}{2}$ and $g(x) = \ln\left(x + \sqrt{x^2 + 1}\right)$ on the same screen. Use a square viewing window. What appears to be the relationship between f and g?

94. Use a graphing utility to graph $f(x) = \frac{e^x + e^{-x}}{2}$ for $x \ge 0$ and $g(x) = \ln\left(x + \sqrt{x^2 - 1}\right)$ for $x \ge 1$ on the same screen. Use a square viewing window. What appears to be the relationship between f and g?

95. The functions $f(x) = \dfrac{e^x - e^{-x}}{e^x + e^{-x}}$ and $g(x) = \dfrac{1}{2}\ln\dfrac{1+x}{1-x}$ are inverse functions. The domain of f is the set of all real numbers. The domain of g is $\{x \mid -1 < x < 1\}$. Use this information to determine the range of f and the range of g.

96. Use a graph of $f(x) = \dfrac{2}{e^x + e^{-x}}$ to determine the domain and range of f.

Projects

1. **Benford's Law** The authors of this text know some interesting details about your finances. For instance, of the last 150 checks you have written, about 30% are for amounts that start with the number 1. Also, you have written about 3 times as many checks for amounts that start with the number 2 as you have for amounts that start with the number 7.

We are sure of these results because of a mathematical formula known as **Benford's Law.** This law was first discovered by the mathematician Simon Newcomb in 1881 and then rediscovered by the physicist Frank Benford in 1938. Benford's Law states that the probability P that the first digit of a number selected from a wide range of numbers is d is given by

$$P(d) = \log\left(1 + \frac{1}{d}\right)$$

a. Use Benford's Law to complete the table below and the bar graph at the top of the next column.

d	$P(d) = \log\left(1 + \dfrac{1}{d}\right)$
1	0.301
2	0.176
3	0.125
4	
5	
6	
7	
8	
9	

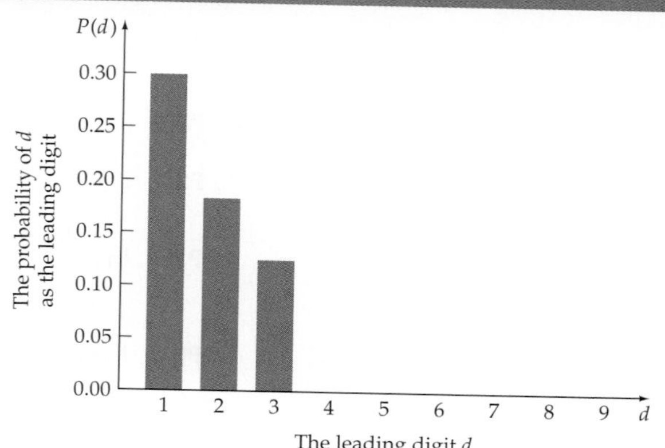

Benford's Law applies to many sets of data with a wide range. For instance, it applies to the populations of the cities in the U.S., the numbers of dollars in the savings accounts at your local bank, and the number of miles driven during a month by each person in a state.

b. Use the table in **a.** to find the probability that in a U.S. city selected at random, the number of telephones in that city will be a number starting with 6.

c. Use the table in **a.** to estimate how many times as many purchases you have made for dollar amounts that start with a 1 than for dollar amounts that start with a 9.

d. Explain why Benford's Law would not apply to the set of all the ages, in years, of students at a local high school.

An Application of Benford's Law Benford's Law has been used to identify fraudulent accountants. In most cases these accountants are unaware of Benford's Law and have replaced valid numbers with numbers selected at random. Their numbers do not conform to Benford's Law. Hence an audit is warranted.

Section 7.3

- Properties of Logarithms
- Change-of-Base Formula
- Logarithmic Scales

Properties of Logarithms and Logarithmic Scales

PREPARE FOR THIS SECTION

Prepare for this section by completing the following exercises. The answers can be found on page A38.

In Exercises PS1 to PS6, use a calculator to compare the values of the given expressions.

PS1. $\log 3 + \log 2; \log 6$ [7.2]

PS2. $\ln 8 - \ln 3; \ln\left(\frac{8}{3}\right)$ [7.2]

PS3. $3 \log 4; \log(4^3)$ [7.2]

PS4. $2 \ln 5; \ln(5^2)$ [7.2]

PS5. $\ln 5; \frac{\log 5}{\log e}$ [7.2]

PS6. $\log 8; \frac{\ln 8}{\ln 10}$ [7.2]

Properties of Logarithms

In Section 7.2 we introduced the following basic properties of logarithms.

$$\log_b b = 1 \qquad \text{and} \qquad \log_b 1 = 0$$

Also, because exponential functions and logarithmic functions are inverses of each other, we observed the relationships

$$\log_b(b^x) = x \qquad \text{and} \qquad b^{\log_b x} = x$$

We can use the properties of exponents to establish the following additional logarithmic properties.

take note

Pay close attention to these properties. Note that

$$\log_b(MN) \neq \log_b M \cdot \log_b N$$

and

$$\log_b \frac{M}{N} \neq \frac{\log_b M}{\log_b N}$$

Also,

$$\log_b(M + N) \neq \log_b M + \log_b N$$

In fact, the expression $\log_b(M + N)$ cannot be expanded at all.

Properties of Logarithms

In the following properties, b, M, and N are positive real numbers ($b \neq 1$).

Product property	$\log_b(MN) = \log_b M + \log_b N$
Quotient property	$\log_b \frac{M}{N} = \log_b M - \log_b N$
Power property	$\log_b(M^p) = p \log_b M$
Logarithm-of-each-side property	$M = N$ implies $\log_b M = \log_b N$
One-to-one property	$\log_b M = \log_b N$ implies $M = N$

Here is a proof of the product property.

Proof: Let $r = \log_b M$ and $s = \log_b N$. These equations can be written in exponential form as

$$M = b^r \quad \text{and} \quad N = b^s$$

Now consider the product MN.

$$MN = b^r b^s$$ • **Substitute for M and N.**

$$MN = b^{r+s}$$ • **Product property of exponents**

$$\log_b MN = r + s$$ • **Write in logarithmic form.**

$$\log_b MN = \log_b M + \log_b N$$ • **Substitute for r and s.**

The last equation is our desired result. ◆

The quotient property and the power property can be proved in a similar manner. See Exercises 87 and 88 on page 492.

The properties of logarithms are often used to rewrite logarithmic expressions in an equivalent form. The process of using the product or quotient rules to rewrite a single logarithm as the sum or difference of two or more logarithms, or using the power property to rewrite $\log_b(M^p)$ in its equivalent form $p \log_b M$, is called **expanding the logarithmic expression.** We illustrate this process in Example 1.

EXAMPLE 1 ⟫ Expand Logarithmic Expressions

Use the properties of logarithms to expand the following logarithmic expressions. Assume all variable expressions represent positive real numbers. When possible, evaluate logarithmic expressions.

a. $\log_5(xy^2)$ **b.** $\ln\left(\dfrac{e\sqrt{y}}{z^3}\right)$

Solution

a. $\log_5(xy^2) = \log_5 x + \log_5 y^2$ • **Product property**

$= \log_5 x + 2\log_5 y$ • **Power property**

b. $\ln\left(\dfrac{e\sqrt{y}}{z^3}\right) = \ln(e\sqrt{y}) - \ln z^3$ • **Quotient property**

$= \ln e + \ln\sqrt{y} - \ln z^3$ • **Product property**

$= \ln e + \ln y^{1/2} - \ln z^3$ • **Write $\sqrt{y}$ as $y^{1/2}$.**

$= \ln e + \dfrac{1}{2}\ln y - 3\ln z$ • **Power property**

$= 1 + \dfrac{1}{2}\ln y - 3\ln z$ • **Evaluate ln e.**

⟫ *Try Exercise 2, page 489*

The properties of logarithms are also used to *condense* expressions that involve the sum or difference of logarithms into a single logarithm. For instance, we can use the product property to rewrite $\log_b M + \log_b N$ as $\log_b(MN)$, and the quotient property to rewrite $\log_b M - \log_b N$ as $\log_b \frac{M}{N}$. Before applying the product or quotient properties, use the power property to write all expressions of the form $p \log_b M$ in their equivalent $\log_b M^p$ form. See Example 2.

QUESTION Does $\log 2 + \log 5 = 1$?

EXAMPLE 2 Condense Logarithmic Expressions

Use the properties of logarithms to rewrite each expression as a single logarithm with a coefficient of 1. Assume all variable expressions represent positive real numbers.

a. $2 \ln x + \frac{1}{2} \ln(x + 4)$ b. $\log_5(x^2 - 4) + 3 \log_5 y - \log_5(x - 2)^2$

Solution

a. $2 \ln x + \frac{1}{2} \ln(x + 4) = \ln x^2 + \ln(x + 4)^{1/2}$ • Power property

$= \ln[x^2(x + 4)^{1/2}]$ • Product property

$= \ln[x^2\sqrt{(x + 4)}]$ • Rewriting $(x + 4)^{1/2}$ as $\sqrt{x + 4}$ is an optional step.

b. $\log_5(x^2 - 4) + 3 \log_5 y - \log_5(x - 2)^2$

$= \log_5(x^2 - 4) + \log_5 y^3 - \log_5(x - 2)^2$ • Power property

$= [\log_5(x^2 - 4) + \log_5 y^3] - \log_5(x - 2)^2$ • Order of Operations Agreement

$= \log_5[(x^2 - 4)y^3] - \log_5(x - 2)^2$ • Product property

$= \log_5\left[\frac{(x^2 - 4)y^3}{(x - 2)^2}\right]$ • Quotient property

$= \log_5\left[\frac{(x + 2)(x - 2)y^3}{(x - 2)^2}\right]$ • Factor.

$= \log_5\left[\frac{(x + 2)y^3}{x - 2}\right]$ • Simplify.

Try Exercise 18, page 490

Change-of-Base Formula

Recall that to determine the value of y in $\log_3 81 = y$, we ask the question, "What power of 3 is equal to 81?" Because $3^4 = 81$, we have $\log_3 81 = 4$. Now suppose

ANSWER Yes. By the product property, $\log 2 + \log 5 = \log(2 \cdot 5) = \log 10 = 1$.

that we need to determine the value of $\log_3 50$. In this case we need to find the power of 3 that produces 50. Because $3^3 = 27$ and $3^4 = 81$, the value we are seeking is somewhere between 3 and 4. The following procedure can be used to produce an estimate of $\log_3 50$.

The exponential form of $\log_3 50 = y$ is $3^y = 50$. Applying logarithmic properties gives us

$$3^y = 50$$

$$\ln 3^y = \ln 50$$ • **Logarithm-of-each-side property**

$$y \ln 3 = \ln 50$$ • **Power property**

$$y = \frac{\ln 50}{\ln 3} \approx 3.56088$$ • **Solve for *y*.**

Thus $\log_3 50 \approx 3.56088$. In the above procedure we could just as well have used logarithms of any base and arrived at the same value. Thus any logarithm can be expressed in terms of logarithms of any base we wish. This general result is summarized in the following formula.

Change-of-Base Formula

If x, a, and b are positive real numbers with $a \neq 1$ and $b \neq 1$, then

$$\log_b x = \frac{\log_a x}{\log_a b}$$

Because most calculators use only common logarithms ($a = 10$) or natural logarithms ($a = e$), the change-of-base formula is used most often in the following form.

If x and b are positive real numbers and $b \neq 1$, then

$$\log_b x = \frac{\log x}{\log b} = \frac{\ln x}{\ln b}$$

EXAMPLE 3 ⟫ Use the Change-of-Base Formula

Evaluate each logarithm. Round to the nearest ten thousandth.

a. $\log_3 18$ b. $\log_{12} 400$

Solution

To approximate these logarithms, we may use the change-of-base formula with $a = 10$ or $a = e$. For this example we choose to use the change-of-base formula with $a = e$. That is, we will evaluate these logarithms by using the LN key on a scientific or graphing calculator.

a. $\log_3 18 = \dfrac{\ln 18}{\ln 3} \approx 2.6309$ b. $\log_{12} 400 = \dfrac{\ln 400}{\ln 12} \approx 2.4111$

⟫ *Try Exercise 34, page 490*

take note

If common logarithms had been used for the calculations in Example 3, the final results would have been the same.

$$\log_3 18 = \frac{\log 18}{\log 3} \approx 2.6309$$

$$\log_{12} 400 = \frac{\log 400}{\log 12} \approx 2.4111$$

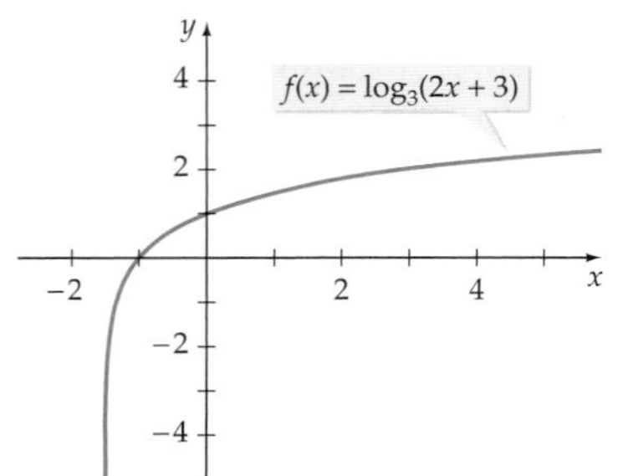

Figure 7.21

The change-of-base formula and a graphing calculator can be used to graph logarithmic functions that have a base other than 10 or e. For instance, to graph $f(x) = \log_3(2x + 3)$, we rewrite the function in terms of base 10 or base e. Using base 10 logarithms, we have $f(x) = \log_3(2x + 3) = \dfrac{\log(2x + 3)}{\log 3}$. The graph is shown in **Figure 7.21.**

EXAMPLE 4 » Use the Change-of-Base Formula to Graph a Logarithmic Function

Graph $f(x) = \log_2|x - 3|$.

Solution

Rewrite f using the change-of-base formula. We will use the natural logarithm function; however, the common logarithm function could be used instead.

$$f(x) = \log_2|x - 3| = \frac{\ln|x - 3|}{\ln 2}$$

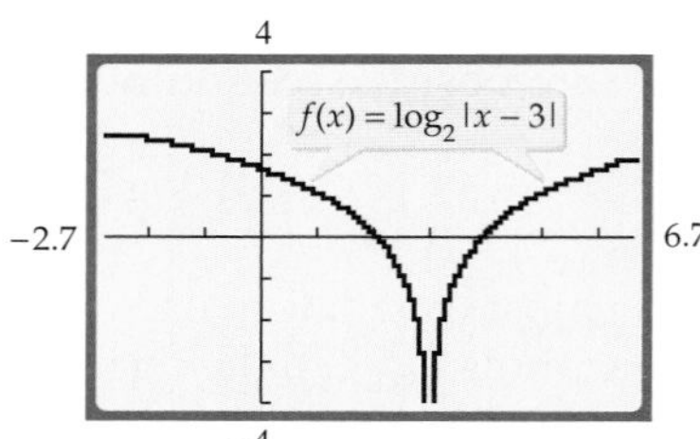

• Enter $\dfrac{\ln|x - 3|}{\ln 2}$ into Y1. Note that the domain of $f(x) = \log_2|x - 3|$ is all real numbers except 3 because $|x - 3| = 0$ when $x = 3$ and $|x - 3|$ is positive for all other values of x.

» *Try Exercise 46, page 490*

Logarithmic Scales

Logarithmic functions are often used to scale very large (or very small) numbers into numbers that are easier to comprehend. For instance, the *Richter scale* magnitude of an earthquake uses a logarithmic function to convert the intensity of the earthquake's shock waves I into a number M, which for most earthquakes is in the range of 0 to 10. The intensity I of an earthquake is often given in terms of the constant I_0, where I_0 is the intensity of the smallest earthquake (called a **zero-level earthquake**) that can be measured on a seismograph near the earthquake's epicenter. The following formula is used to compute the Richter scale magnitude of an earthquake.

Math Matters

The Richter scale was created by the seismologist Charles F. Richter in 1935. Notice that a tenfold increase in the intensity level of an earthquake increases the Richter scale magnitude of the earthquake by only 1.

The Richter Scale Magnitude of an Earthquake

An earthquake with an intensity of I has a **Richter scale magnitude** of

$$M = \log\left(\frac{I}{I_0}\right)$$

where I_0 is the measure of the intensity of a zero-level earthquake.

EXAMPLE 5 ⟫ Determine the Magnitude of an Earthquake

Find the Richter scale magnitude (to the nearest 0.1) of the 1999 Joshua Tree, California, earthquake that had an intensity of $I = 12{,}589{,}254I_0$.

Solution

$$M = \log\left(\frac{I}{I_0}\right) = \log\left(\frac{12{,}589{,}254I_0}{I_0}\right) = \log(12{,}589{,}254) \approx 7.1$$

The 1999 Joshua Tree earthquake had a Richter scale magnitude of 7.1.

⟫ *Try Exercise 78, page 491*

> **take note**
>
> Notice in Example 5 that we didn't need to know the value of I_0 to determine the Richter scale magnitude of the quake.

If you know the Richter scale magnitude of an earthquake, you can determine the intensity of the earthquake.

EXAMPLE 6 ⟫ Determine the Intensity of an Earthquake

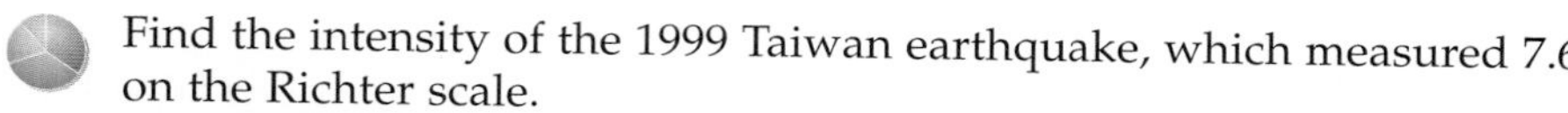

Find the intensity of the 1999 Taiwan earthquake, which measured 7.6 on the Richter scale.

Solution

$$\log\left(\frac{I}{I_0}\right) = 7.6$$

$$\frac{I}{I_0} = 10^{7.6} \qquad \text{• Write in exponential form.}$$

$$I = 10^{7.6}I_0 \qquad \text{• Solve for } I.$$

$$I \approx 39{,}810{,}717I_0$$

The 1999 Taiwan earthquake had an intensity that was approximately 39,811,000 times the intensity of a zero-level earthquake.

⟫ *Try Exercise 80, page 491*

In Example 7 we make use of the Richter scale magnitudes of two earthquakes to compare the intensities of the earthquakes.

EXAMPLE 7 ⟫ Compare Earthquakes

The 1960 Chile earthquake had a Richter scale magnitude of 9.5. The 1989 San Francisco earthquake had a Richter scale magnitude of 7.1. Compare the intensities of the earthquakes.

Continued ▶

take note

The results of Example 7 show that if an earthquake has a Richter scale magnitude of M_1 and a smaller earthquake has a Richter scale magnitude of M_2, then the larger earthquake is $10^{M_1-M_2}$ times as intense as the smaller earthquake.

Solution

Let I_1 be the intensity of the Chilean earthquake and I_2 the intensity of the San Francisco earthquake. Then

$$\log\left(\frac{I_1}{I_0}\right) = 9.5 \quad \text{and} \quad \log\left(\frac{I_2}{I_0}\right) = 7.1$$

$$\frac{I_1}{I_0} = 10^{9.5} \qquad \frac{I_2}{I_0} = 10^{7.1}$$

$$I_1 = 10^{9.5}I_0 \qquad I_2 = 10^{7.1}I_0$$

To compare the intensities of the earthquakes, we compute the ratio I_1/I_2.

$$\frac{I_1}{I_2} = \frac{10^{9.5}I_0}{10^{7.1}I_0} = \frac{10^{9.5}}{10^{7.1}} = 10^{9.5-7.1} = 10^{2.4} \approx 251$$

The earthquake in Chile was approximately 251 times as intense as the San Francisco earthquake.

Try Exercise 82, page 491

Seismologists generally determine the Richter scale magnitude of an earthquake by examining a **seismogram**. See **Figure 7.22**.

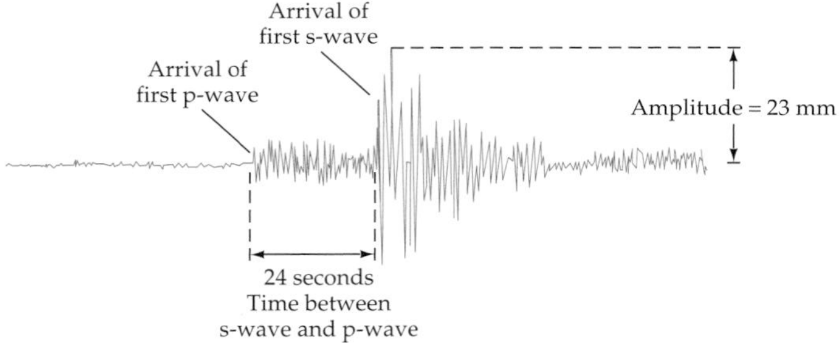

Figure 7.22

The magnitude of an earthquake cannot be determined just by examining the amplitude of a seismogram because this amplitude decreases as the distance between the epicenter of the earthquake and the observation station increases. To account for the distance between the epicenter and the observation station, a seismologist examines a seismogram for both small waves called **p-waves** and larger waves called **s-waves**. The Richter scale magnitude M of the earthquake is a function of both the amplitude A of the s-waves and the difference in time t between the occurrence of the s-waves and the occurrence of the p-waves. In the 1950s, Charles Richter developed the following formula to determine the magnitude of an earthquake from the data in a seismogram.

Amplitude-Time-Difference Formula

The Richter scale magnitude M of an earthquake is given by

$$M = \log A + 3 \log 8t - 2.92$$

where A is the amplitude, in millimeters, of the s-waves on a seismogram and t is the difference in time, in seconds, between the s-waves and the p-waves.

EXAMPLE 8 Determine the Magnitude of an Earthquake from Its Seismogram

Find the Richter scale magnitude of the earthquake that produced the seismogram in **Figure 7.22**.

take note

The Richter scale magnitude is usually rounded to the nearest tenth.

Solution

$$\begin{aligned} M &= \log A + 3 \log 8t - 2.92 \\ &= \log 23 + 3 \log[8 \cdot 24] - 2.92 && \text{• Substitute 23 for } A \text{ and 24 for } t. \\ &\approx 1.36173 + 6.84990 - 2.92 \\ &\approx 5.3 \end{aligned}$$

The earthquake had a magnitude of about 5.3 on the Richter scale.

Try Exercise 86, page 492

Logarithmic scales are also used in chemistry. One example concerns the pH of a liquid, which is a measure of the liquid's **acidity** or **alkalinity**. (You may have tested the pH of the water in a swimming pool or an aquarium.) Pure water, which is considered neutral, has a pH of 7.0. The pH scale ranges from 0 to 14, with 0 corresponding to the most acidic solutions and 14 to the most alkaline. Lemon juice has a pH of about 2, whereas household ammonia measures about 11.

Specifically, the pH of a solution is a function of the hydronium-ion concentration of the solution. Because the hydronium-ion concentration of a solution can be very small (with values such as 0.00000001), pH uses a logarithmic scale.

take note

One mole is equivalent to 6.022×10^{23} ions.

Definition of the pH of a Solution

The **pH of a solution** with a hydronium-ion concentration of H^+ moles per liter is given by

$$\text{pH} = -\log[H^+]$$

EXAMPLE 9 Find the pH of a Solution

Find the pH of each liquid. Round to the nearest tenth.

a. Orange juice with $H^+ = 2.8 \times 10^{-4}$ mole per liter

b. Milk with $H^+ = 3.97 \times 10^{-7}$ mole per liter

c. Rainwater with $H^+ = 6.31 \times 10^{-5}$ mole per liter

d. A baking soda solution with $H^+ = 3.98 \times 10^{-9}$ mole per liter

Math Matters

The pH scale was created by the Danish biochemist Søren Sørensen in 1909 to measure the acidity of water used in the brewing of beer. pH is an abbreviation for pondus hydrogenii, which translates as "potential hydrogen."

Solution

a. $\text{pH} = -\log[H^+] = -\log(2.8 \times 10^{-4}) \approx 3.6$
The orange juice has a pH of 3.6.

b. $\text{pH} = -\log[H^+] = -\log(3.97 \times 10^{-7}) \approx 6.4$
The milk has a pH of 6.4.

c. $\text{pH} = -\log[H^+] = -\log(6.31 \times 10^{-5}) \approx 4.2$
The rainwater has a pH of 4.2.

d. $\text{pH} = -\log[H^+] = -\log(3.98 \times 10^{-9}) \approx 8.4$
The baking soda solution has a pH of 8.4.

Try Exercise 70, page 491

Figure 7.23 illustrates the pH scale, along with the corresponding hydronium-ion concentrations. A solution on the left half of the scale, with a pH of less than 7, is an **acid,** and a solution on the right half of the scale is an **alkaline solution** or a **base.** Because the scale is logarithmic, a solution with a pH of 5 is 10 times more acidic than a solution with a pH of 6. From Example 9 we see that the orange juice, rainwater, and milk are acids, whereas the baking soda solution is a base.

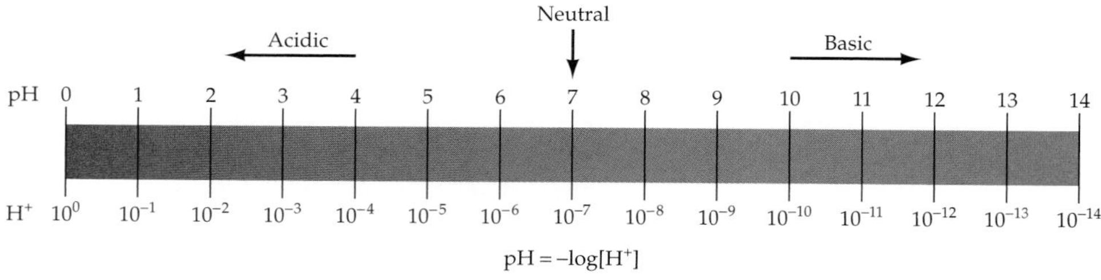

Figure 7.23

EXAMPLE 10 Find the Hydronium-Ion Concentration

A sample of blood has a pH of 7.3. Find the hydronium-ion concentration of the blood.

Solution

$$\begin{aligned} pH &= -\log[H^+] \\ 7.3 &= -\log[H^+] && \text{• Substitute 7.3 for pH.} \\ -7.3 &= \log[H^+] && \text{• Multiply both sides by } -1. \\ 10^{-7.3} &= H^+ && \text{• Change to exponential form.} \\ 5.0 \times 10^{-8} &\approx H^+ \end{aligned}$$

The hydronium-ion concentration of the blood is about 5.0×10^{-8} mole per liter.

Try Exercise 72, page 491

Topics for Discussion

1. The function $f(x) = \log_b x$ is defined only for $x > 0$. Explain why this condition is imposed.
2. If p and q are positive numbers, explain why $\ln(p + q)$ isn't normally equal to $\ln p + \ln q$.
3. If $f(x) = \log_b x$ and $f(c) = f(d)$, can we conclude that $c = d$?
4. Give examples of situations in which it is advantageous to use logarithmic scales.

Exercise Set 7.3

In Exercises 1 to 16, expand the given logarithmic expression. Assume all variable expressions represent positive real numbers. When possible, evaluate logarithmic expressions. Do not use a calculator.

1. $\log_b(xyz)$

2. $\ln \dfrac{z^3}{\sqrt{xy}}$

3. $\ln \dfrac{x}{z^4}$

4. $\log_5 \dfrac{xy^2}{z^4}$

5. $\log_2 \dfrac{\sqrt{x}}{y^3}$

6. $\log_b(x\sqrt[3]{y})$

7. $\log_7 \dfrac{\sqrt{xz}}{y^2}$

8. $\ln \sqrt[3]{x^2\sqrt{y}}$

9. $\ln(e^2z)$

10. $\ln(x^{1/2}y^{2/3})$

11. $\log_4\left(\dfrac{\sqrt[3]{z}}{16y^3}\right)$

12. $\log_5\left(\dfrac{\sqrt{xz^4}}{125}\right)$

13. $\log \sqrt{x\sqrt{z}}$

14. $\ln\left(\dfrac{\sqrt[3]{x^2}}{z^2}\right)$

15. $\ln(\sqrt[3]{z\sqrt{e}})$

16. $\ln\left[\dfrac{x^2\sqrt{z}}{y^{-3}}\right]$

In Exercises 17 to 32, write each expression as a single logarithm with a coefficient of 1. Assume all variable expressions represent positive real numbers.

17. $\log(x + 5) + 2\log x$

18. $3\log_2 t - \dfrac{1}{3}\log_2 u + 4\log_2 v$

19. $\ln(x^2 - y^2) - \ln(x - y)$

20. $\dfrac{1}{2}\log_8(x + 5) - 3\log_8 y$

21. $3\log x + \dfrac{1}{3}\log y + \log(x + 1)$

22. $\ln(xz) - \ln(x\sqrt{y}) + 2\ln\dfrac{y}{z}$

23. $\log(xy^2) - \log z$

24. $\ln(y^{1/2}z) - \ln z^{1/2}$

25. $2(\log_6 x + \log_6 y^2) - \log_6(x + 2)$

26. $\dfrac{1}{2}\log_3 x - \log_3 y + 2\log_3(x + 2)$

27. $2\ln(x + 4) - \ln x - \ln(x^2 - 3)$

28. $\log(3x) - (2\log x - \log y)$

29. $\ln(2x + 5) - \ln y - 2\ln z + \dfrac{1}{2}\ln w$

30. $\log_b x + \log_b(y + 3) + \log_b(y + 2) - \log_b(y^2 + 5y + 6)$

31. $\ln(x^2 - 9) - 2\ln(x - 3) + 3\ln y$

32. $\log_b(x^2 + 7x + 12) - 2\log_b(x + 4)$

In Exercises 33 to 44, use the change-of-base formula to approximate the logarithm accurate to the nearest ten thousandth.

33. $\log_7 20$

34. $\log_5 37$

35. $\log_{11} 8$

36. $\log_{50} 22$

37. $\log_6 \dfrac{1}{3}$

38. $\log_3 \dfrac{7}{8}$

39. $\log_9 \sqrt{17}$

40. $\log_4 \sqrt{7}$

41. $\log_{\sqrt{2}} 17$

42. $\log_{\sqrt{3}} 5.5$

43. $\log_\pi e$

44. $\log_\pi \sqrt{15}$

In Exercises 45 to 52, use a graphing utility and the change-of-base formula to graph the logarithmic function.

45. $f(x) = \log_4 x$

46. $g(x) = \log_8(5 - x)$

47. $g(x) = \log_8(x - 3)$

48. $t(x) = \log_9(5 - x)$

49. $h(x) = \log_3(x - 3)^2$

50. $J(x) = \log_{12}(-x)$

51. $F(x) = -\log_5|x - 2|$

52. $n(x) = \log_2\sqrt{x - 8}$

In Exercises 53 to 62, determine if the statement is true or false for all $x > 0$, $y > 0$. If it is false, write an example that disproves the statement.

53. $\log_b(x + y) = \log_b x + \log_b y$

54. $\log_b(xy) = \log_b x \cdot \log_b y$

55. $\log_b(xy) = \log_b x + \log_b y$

56. $\log_b x \cdot \log_b y = \log_b x + \log_b y$

57. $\log_b x - \log_b y = \log_b(x - y), \quad x > y$

58. $\log_b \dfrac{x}{y} = \dfrac{\log_b x}{\log_b y}$

59. $\dfrac{\log_b x}{\log_b y} = \log_b x - \log_b y$

60. $\log_b(x^n) = n\log_b x$

61. $(\log_b x)^n = n\log_b x$

62. $\log_b\sqrt{x} = \dfrac{1}{2}\log_b x$

63. Evaluate the following *without* using a calculator.

$$\log_3 5 \cdot \log_5 7 \cdot \log_7 9$$

64. Evaluate the following *without* using a calculator.

$$\log_5 20 \cdot \log_{20} 60 \cdot \log_{60} 100 \cdot \log_{100} 125$$

65. Which is larger, 500^{501} or 506^{500}? These numbers are too large for most calculators to handle. (They each have 1353 digits!) (*Hint:* Compare the logarithms of each number.)

66. Which number is smaller, $\frac{1}{50^{300}}$ or $\frac{1}{151^{233}}$? See hint in Exercise 65.

67. ANIMATED MAPS A software company that creates interactive maps for websites has designed an animated zooming feature such that when a user selects the zoom-in option, the map appears to expand on a location. This is accomplished by displaying several intermediate maps to give the illusion of motion. The company has determined that zooming in on a location is more informative and pleasing to observe when the scale of each step of the animation is determined using the equation

$$S_n = S_0 \cdot 10^{\frac{n}{N}(\log S_f - \log S_0)}$$

where S_n represents the scale of the current step n ($n = 0$ corresponds to the initial scale), S_0 is the starting scale of the map, S_f is the final scale, and N is the number of steps in the animation following the initial scale. (If the initial scale of the map is 1:200, then $S_0 = 200$.) Determine the scales to be used at each intermediate step if a map is to start with a scale of 1:1,000,000 and proceed through five intermediate steps to end with a scale of 1:500,000.

68. ANIMATED MAPS Use the equation in Exercise 67 to determine the scales for each stage of an animated map zoom that goes from a scale of 1:250,000 to a scale of 1:100,000 in four steps (following the initial scale).

69. pH Milk of magnesia has a hydronium-ion concentration of about 3.97×10^{-11} mole per liter. Determine the pH of milk of magnesia and state whether milk of magnesia is an acid or a base.

70. pH Vinegar has a hydronium-ion concentration of 1.26×10^{-3} mole per liter. Determine the pH of vinegar and state whether vinegar is an acid or a base.

71. HYDRONIUM-ION CONCENTRATION A morphine solution has a pH of 9.5. Determine the hydronium-ion concentration of the morphine solution.

72. HYDRONIUM-ION CONCENTRATION A rainstorm in New York City produced rainwater with a pH of 5.6. Determine the hydronium-ion concentration of the rainwater.

DECIBEL LEVEL **The range of sound intensities that the human ear can detect is so large that a special decibel scale (named after Alexander Graham Bell) is used to measure and compare sound intensities. The decibel level *dB* of a sound is given by**

$$dB(I) = 10 \log\left(\frac{I}{I_0}\right)$$

where I_0 is the intensity of sound that is barely audible to the human ear. Use the decibel level formula to work exercises 73-76.

73. Find the decibel level for the following sounds. Round to the nearest tenth of a decibel.

	Sound	*Intensity*
a.	Automobile traffic	$I = 1.58 \times 10^8 \cdot I_0$
b.	Quiet conversation	$I = 10{,}800 \cdot I_0$
c.	Fender guitar	$I = 3.16 \times 10^{11} \cdot I_0$
d.	Jet engine	$I = 1.58 \times 10^{15} \cdot I_0$

74. COMPARISON OF SOUND INTENSITIES A team in Arizona installed a 48,000-watt sound system in a Ford Bronco that it claims can output 175-decibel sound. The human pain threshold for sound is 125 decibels. How many times as great is the intensity of the sound from the Bronco than the human pain threshold?

75. COMPARISON OF SOUND INTENSITIES How many times as great is the intensity of a sound that measures 120 decibels than a sound that measures 110 decibels?

76. DECIBEL LEVEL If the intensity of a sound is doubled, what is the increase in the decibel level? (*Hint:* Find $dB(2I) - dB(I)$.)

77. EARTHQUAKE MAGNITUDE What is the Richter scale magnitude of an earthquake with an intensity of $I = 100{,}000I_0$?

78. EARTHQUAKE MAGNITUDE The Colombia earthquake of 1906 had an intensity of $I = 398{,}107{,}000I_0$. What did it measure on the Richter scale?

79. EARTHQUAKE INTENSITY The Coalinga, California, earthquake of 1983 had a Richter scale magnitude of 6.5. Find the intensity of this earthquake.

80. EARTHQUAKE INTENSITY The earthquake that occurred just south of Concepción, Chile, in 1960 had a Richter scale magnitude of 9.5. Find the intensity of this earthquake.

81. COMPARISON OF EARTHQUAKES Compare the intensity of an earthquake that measures 5.0 on the Richter scale to the intensity of an earthquake that measures 3.0 on the Richter scale by finding the ratio of the larger intensity to the smaller intensity.

82. COMPARISON OF EARTHQUAKES How many times as great was the intensity of the 1960 earthquake in Chile, which measured 9.5 on the Richter scale, than the San Francisco earthquake of 1906, which measured 8.3 on the Richter scale?

83. **COMPARISON OF EARTHQUAKES** On March 2, 1933, an earthquake of magnitude 8.9 on the Richter scale struck Japan. In October 1989, an earthquake of magnitude 7.1 on the Richter scale struck San Francisco, California. Compare the intensity of the larger earthquake to the intensity of the smaller earthquake by finding the ratio of the larger intensity to the smaller intensity.

84. **COMPARISON OF EARTHQUAKES** An earthquake that occurred in China in 1978 measured 8.2 on the Richter scale. In 1988, an earthquake in California measured 6.9 on the Richter scale. Compare the intensity of the larger earthquake to the intensity of the smaller earthquake by finding the ratio of the larger intensity to the smaller intensity.

85. **EARTHQUAKE MAGNITUDE** Find the Richter scale magnitude of the earthquake that produced the seismogram in the following figure.

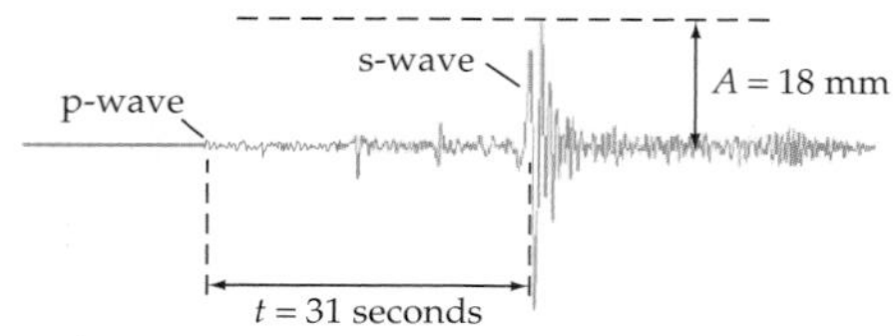

86. **EARTHQUAKE MAGNITUDE** Find the Richter scale magnitude of the earthquake that produced the seismogram in the following figure.

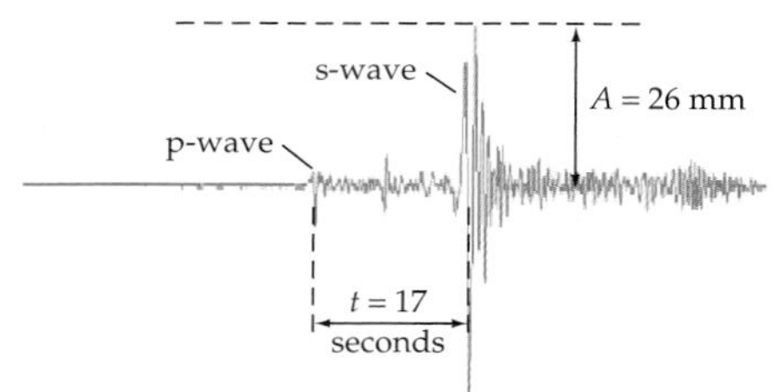

87. Prove the quotient property of logarithms:

$$\log_b \frac{M}{N} = \log_b M - \log_b N$$

(*Hint:* See the proof of the product property of logarithms on page 481.)

88. Prove the power property of logarithms:

$$\log_b(M^p) = p \log_b M$$

See the hint given in exercise 87.

Connecting Concepts

89. **NOMOGRAMS AND LOGARITHMIC SCALES** A **nomogram** is a diagram used to determine a numerical result by drawing a line across numerical scales. The following nomogram, used by Richter, determines the magnitude of an earthquake from its seismogram. To use the nomogram, mark the amplitude of a seismogram on the amplitude scale and mark the time between the s-wave and the p-wave on the S-P scale. Draw a line between these marks. The Richter scale magnitude of the earthquake that produced the seismogram is shown by the intersection of the line and the center scale. The nomogram on the right shows that an earthquake with a seismogram amplitude of 23 millimeters and an S-P time of 24 seconds has a Richter scale magnitude of about 5.

The amplitude and the S-P time are shown on logarithmic scales. On the amplitude scale, the distance from 1 to 10 is the same as the distance from 10 to 100, because $\log 100 - \log 10 = \log 10 - \log 1$.

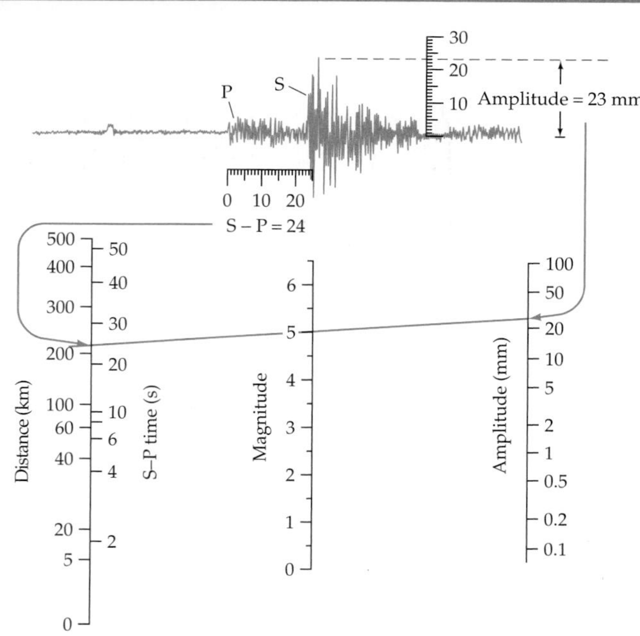

Richter's earthquake nomogram

Use the nomogram on page 492 to determine the Richter scale magnitude of an earthquake with a seismogram that has

a. amplitude of 50 millimeters and S-P time of 40 seconds

b. amplitude of 1 millimeter and S-P time of 30 seconds

c. How do the results in **a.** and **b.** compare with the Richter scale magnitude produced by using the amplitude-time-difference formula on page 487?

Projects

1. **LOGARITHMIC SCALES** Sometimes **logarithmic scales** are used to better view a collection of data that span a wide range of values. For instance, consider the table below, which lists the approximate masses of various marine creatures in grams. Next we have attempted to plot the masses on a number line.

Animal	Mass (g)
Rotifer	0.000000006
Dwarf goby	0.30
Lobster	15,900
Leatherback turtle	851,000
Giant squid	1,820,000
Whale shark	4,700,000
Blue whale	120,000,000

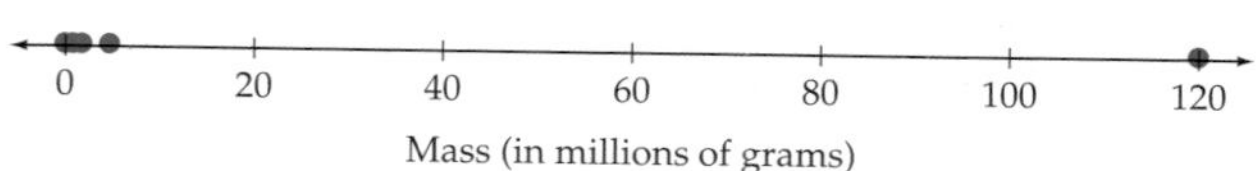

As you can see, we had to use such a large span of numbers that the data for most of the animals are bunched up at the left. Visually, this number line isn't very helpful for comparisons.

a. Make a new number line, this time plotting the logarithm (base 10) of each of the masses.

b. Which number line is more helpful to compare the masses of the different animals?

c. If the data points for two animals on the logarithmic number line are 1 unit apart, how do the animals' masses compare? What if the points are 2 units apart?

2. **LOGARITHMIC SCALES** The distances of the planets in our solar system from the sun are given in the table in the next column.

Planet	Distance (million km)
Mercury	58
Venus	108
Earth	150
Mars	228
Jupiter	778
Saturn	1427
Uranus	2871
Neptune	4497

a. Draw a number line with an appropriate scale to plot the distances.

b. Draw a second number line, this time plotting the logarithm (base 10) of each distance.

c. Which number line do you find more helpful to compare the different distances?

d. If two distances are 3 units apart on the logarithmic number line, how do the distances of the corresponding planets compare?

3. **BIOLOGIC DIVERSITY** To discuss the variety of species that live in a certain environment, a biologist needs a precise definition of *diversity*. Let $p_1, p_2, \ldots, p_n$ be the proportions of n species that live in an environment. The biologic diversity D of this system is

$$D = -(p_1 \log_2 p_1 + p_2 \log_2 p_2 + \cdots + p_n \log_2 p_n)$$

Suppose that an ecosystem has exactly five different varieties of grass: rye (R), bermuda (B), blue (L), fescue (F), and St. Augustine (A).

a. Calculate the diversity of this ecosystem if the proportions of these grasses are as shown in Table 1. Round to the nearest hundredth.

Table 1

R	B	L	F	A
$\frac{1}{5}$	$\frac{1}{5}$	$\frac{1}{5}$	$\frac{1}{5}$	$\frac{1}{5}$

b. Because bermuda and St. Augustine are virulent grasses, after a time the proportions will be as shown in Table 2. Calculate the diversity of this system. Does this system have more or less diversity than the system given in Table 1?

Table 2

R	B	L	F	A
$\frac{1}{8}$	$\frac{3}{8}$	$\frac{1}{16}$	$\frac{1}{8}$	$\frac{5}{16}$

c. After an even longer time period, the bermuda and St. Augustine grasses completely overrun the environment and the proportions are as shown in Table 3. Calculate the diversity of this system. (*Note:* Although the equation is not technically correct, for purposes of the diversity definition, we may say that $0 \log_2 0 = 0$. By using very small values of p_i, we can demonstrate that this definition makes sense.) Does this system have more or less diversity than the system given in Table 2?

Table 3

R	B	L	F	A
0	$\frac{1}{4}$	0	0	$\frac{3}{4}$

d. Finally, the St. Augustine grasses overrun the bermuda grasses and the proportions are as shown in Table 4. Calculate the diversity of this system. Write a sentence that explains the meaning of the value you obtained.

Table 4

R	B	L	F	A
0	0	0	0	1

Section 7.4

- Solve Exponential Equations
- Solve Logarithmic Equations

Exponential and Logarithmic Equations

PREPARE FOR THIS SECTION

Prepare for this section by completing the following exercises. The answers can be found on page A39.

PS1. Use the definition of a logarithm to write the exponential equation $3^6 = 729$ in logarithmic form. [7.2]

PS2. Use the definition of a logarithm to write the logarithmic equation $\log_5 625 = 4$ in exponential form. [7.2]

PS3. Use the definition of a logarithm to write the exponential equation $a^{x+2} = b$ in logarithmic form. [7.2]

PS4. Solve for x: $4a = 7bx + 2cx$. [1.1]

PS5. Solve for x: $165 = \dfrac{300}{1 + 12x}$. [1.1]

PS6. Solve for x: $A = \dfrac{100 + x}{100 - x}$. [1.1]

Solve Exponential Equations

If a variable appears in the exponent of a term of an equation, such as in $2^{x+1} = 32$, then the equation is called an **exponential equation**. Example 1 uses the following Equality of Exponents Theorem to solve $2^{x+1} = 32$.

Equality of Exponents Theorem

If $b^x = b^y$, then $x = y$, provided $b > 0$ and $b \neq 1$.

EXAMPLE 1 Solve an Exponential Equation

Use the Equality of Exponents Theorem to solve $2^{x+1} = 32$.

Solution

$$
\begin{aligned}
2^{x+1} &= 32 \\
2^{x+1} &= 2^5 && \text{• Write each side as a power of 2.} \\
x + 1 &= 5 && \text{• Equate the exponents.} \\
x &= 4 && \text{• Solve for } x.
\end{aligned}
$$

Check: Let $x = 4$. Then $2^{x+1} = 2^{4+1}$

$$= 2^5$$

$$= 32$$

Try Exercise 2, page 501

A graphing utility can also be used to find the solutions of an equation of the form $f(x) = g(x)$. Either of the following two methods can be employed.

Using a Graphing Utility to Find the Solutions of $f(x) = g(x)$

Intersection Method Graph $y_1 = f(x)$ and $y_2 = g(x)$ on the same screen. The solutions of $f(x) = g(x)$ are the x-coordinates of the points of intersection of the graphs.

Intercept Method The solutions of $f(x) = g(x)$ are the x-coordinates of the x-intercepts of the graph of $y = f(x) - g(x)$.

Figures 7.24 and **7.25** illustrate the graphical methods for solving $2^{x+1} = 32$.

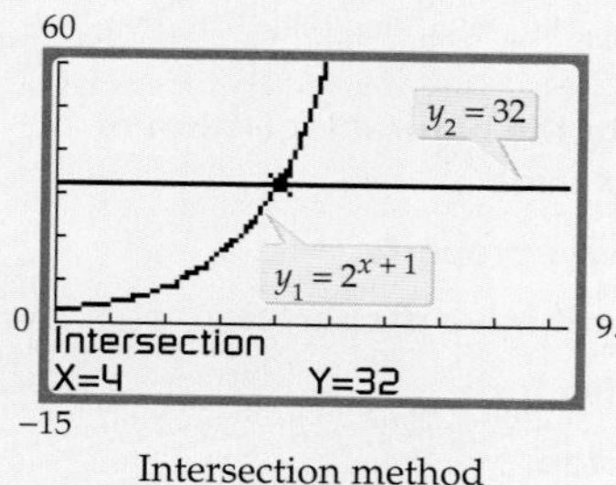

Intersection method

Figure 7.24

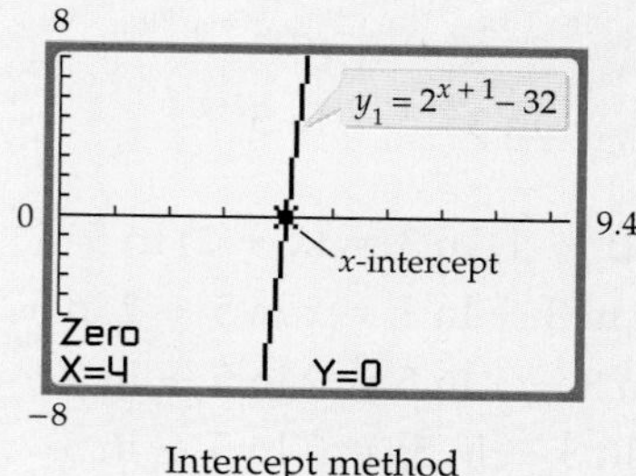

Intercept method

Figure 7.25

In Example 1 we were able to write both sides of the equation as a power of the same base. If you find it difficult to write both sides of an exponential equation in terms of the same base, then try the procedure of taking the logarithm of each side of the equation. This procedure is used in Example 2.

EXAMPLE 2 Solve an Exponential Equation

Solve: $5^x = 40$

ALGEBRAIC SOLUTION

$$5^x = 40$$

$$\log(5^x) = \log 40 \quad \text{• Take the logarithm of each side.}$$

$$x \log 5 = \log 40 \quad \text{• Power property}$$

$$x = \frac{\log 40}{\log 5} \quad \text{• Exact solution}$$

$$x \approx 2.3 \quad \text{• Decimal approximation}$$

To the nearest tenth, the solution is 2.3.

VISUALIZE THE SOLUTION

Intersection Method The solution of $5^x = 40$ is the x-coordinate of the point of intersection of $y = 5^x$ and $y = 40$.

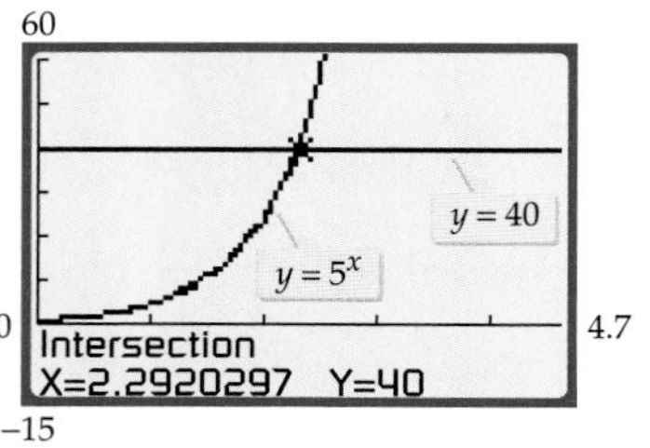

Try Exercise 10, page 501

An alternative approach to solving the equation in Example 2 is to rewrite the exponential equation in logarithmic form: $5^x = 40$ is equivalent to the logarithmic equation $\log_5 40 = x$. Using the change-of-base formula, we find that $x = \log_5 40 = \dfrac{\log 40}{\log 5}$. In the following example, however, we must take logarithms of both sides to reach a solution.

EXAMPLE 3 Solve an Exponential Equation

Solve: $3^{2x-1} = 5^{x+2}$

ALGEBRAIC SOLUTION

$$3^{2x-1} = 5^{x+2}$$

$$\ln 3^{2x-1} = \ln 5^{x+2} \quad \text{• Take the natural logarithm of each side.}$$

$$(2x - 1)\ln 3 = (x + 2)\ln 5 \quad \text{• Power property}$$

$$2x \ln 3 - \ln 3 = x \ln 5 + 2 \ln 5 \quad \text{• Distributive property}$$

$$2x \ln 3 - x \ln 5 = 2 \ln 5 + \ln 3 \quad \text{• Solve for } x.$$

$$x(2 \ln 3 - \ln 5) = 2 \ln 5 + \ln 3 \quad \text{• Factor.}$$

$$x = \frac{2 \ln 5 + \ln 3}{2 \ln 3 - \ln 5} \quad \text{• Exact solution}$$

$$x \approx 7.3 \quad \text{• Decimal approximation}$$

To the nearest tenth, the solution is 7.3.

VISUALIZE THE SOLUTION

Intercept Method The solution of $3^{2x-1} = 5^{x+2}$ is the x-coordinate of the x-intercept of $y = 3^{2x-1} - 5^{x+2}$.

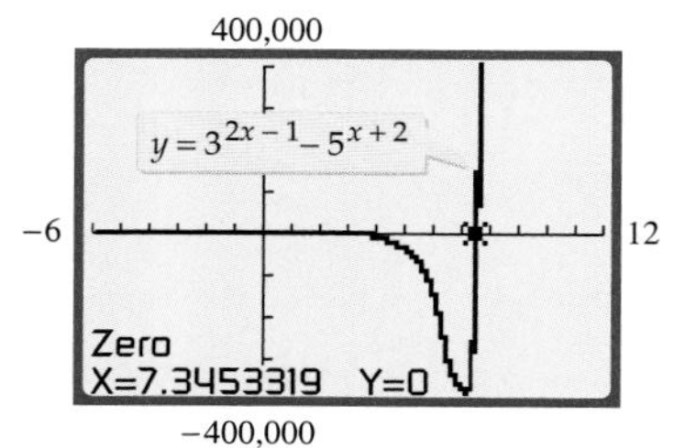

Try Exercise 18, page 501

In Example 4 we solve an exponential equation that has two solutions.

EXAMPLE 4 ›› Solve an Exponential Equation Involving $b^x + b^{-x}$

Solve: $\dfrac{2^x + 2^{-x}}{2} = 3$

ALGEBRAIC SOLUTION

Multiplying each side by 2 produces

$$2^x + 2^{-x} = 6$$

$$2^{2x} + 2^0 = 6(2^x)$$ • Multiply each side by 2^x to clear negative exponents.

$$(2^x)^2 - 6(2^x) + 1 = 0$$ • Write in quadratic form.

$$(u)^2 - 6(u) + 1 = 0$$ • Substitute u for 2^x.

By the quadratic formula,

$$u = \frac{6 \pm \sqrt{36 - 4}}{2} = \frac{6 \pm 4\sqrt{2}}{2} = 3 \pm 2\sqrt{2}$$

$$2^x = 3 \pm 2\sqrt{2}$$ • Replace u with 2^x.

$$\log 2^x = \log(3 \pm 2\sqrt{2})$$ • Take the common logarithm of each side.

$$x \log 2 = \log(3 \pm 2\sqrt{2})$$ • Power property

$$x = \frac{\log(3 \pm 2\sqrt{2})}{\log 2} \approx \pm 2.54$$ • Solve for x.

The approximate solutions are -2.54 and 2.54.

VISUALIZE THE SOLUTION

Intersection Method The solutions of $\dfrac{2^x + 2^{-x}}{2} = 3$ are the x-coordinates of the points of intersection of $y = \dfrac{2^x + 2^{-x}}{2}$ and $y = 3$.

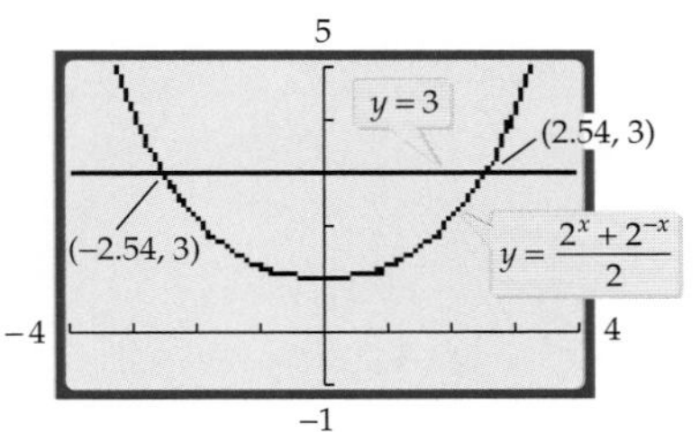

›› Try Exercise 42, page 501

Solve Logarithmic Equations

Equations that involve logarithms are called **logarithmic equations**. The properties of logarithms, along with the definition of a logarithm, are often used to find the solutions of a logarithmic equation.

EXAMPLE 5 ›› Solve a Logarithmic Equation

Solve: $\log(3x - 5) = 2$

Solution

$$\log(3x - 5) = 2$$

$$3x - 5 = 10^2$$ • Definition of a logarithm

$$3x = 105$$ • Solve for x.

$$x = 35$$

Check: $\log[3(35) - 5] = \log 100 = 2$

›› Try Exercise 22, page 501

EXAMPLE 6 Solve a Logarithmic Equation

Solve: $\log 2x - \log(x - 3) = 1$

Solution

$$\log 2x - \log(x - 3) = 1$$

$$\log \frac{2x}{x - 3} = 1 \quad \text{• Quotient property}$$

$$\frac{2x}{x - 3} = 10^1 \quad \text{• Definition of a logarithm}$$

$$2x = 10x - 30 \quad \text{• Multiply each side by } x - 3.$$

$$-8x = -30 \quad \text{• Solve for } x.$$

$$x = \frac{15}{4}$$

Check the solution by substituting $\frac{15}{4}$ into the original equation.

Try Exercise 26, page 501

In Example 7 we make use of the one-to-one property of logarithms to find the solution of a logarithmic equation.

EXAMPLE 7 Solve a Logarithmic Equation

Solve: $\ln(3x + 8) = \ln(2x + 2) + \ln(x - 2)$

ALGEBRAIC SOLUTION

$$\ln(3x + 8) = \ln(2x + 2) + \ln(x - 2)$$

$$\ln(3x + 8) = \ln[(2x + 2)(x - 2)] \quad \text{• Product property}$$

$$\ln(3x + 8) = \ln(2x^2 - 2x - 4)$$

$$3x + 8 = 2x^2 - 2x - 4 \quad \text{• One-to-one property of logarithms}$$

$$0 = 2x^2 - 5x - 12 \quad \text{• Subtract } 3x - 8 \text{ from each side.}$$

$$0 = (2x + 3)(x - 4) \quad \text{• Factor.}$$

$$x = -\frac{3}{2} \quad \text{or} \quad x = 4 \quad \text{• Solve for } x.$$

A check will show that 4 is a solution, but $-\frac{3}{2}$ is not a solution.

VISUALIZE THE SOLUTION

The graph of

$y = \ln(3x + 8) - \ln(2x + 2) - \ln(x - 2)$

has only one x-intercept. Thus there is only one real solution.

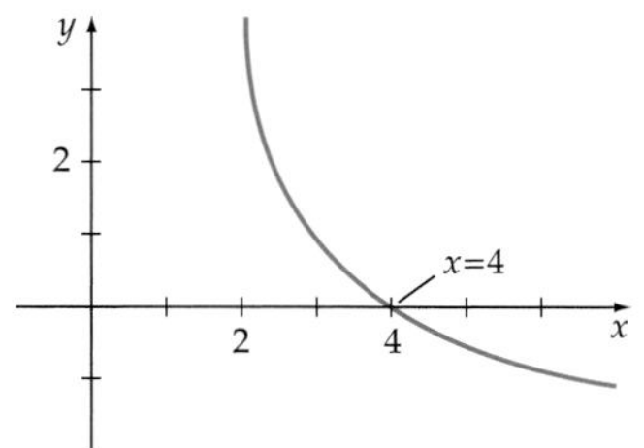

Try Exercise 36, page 501

QUESTION Why does $x = -\frac{3}{2}$ not check in Example 7?

EXAMPLE 8 » Velocity of a Sky Diver Experiencing Air Resistance

During the free-fall portion of a jump, the time t in seconds required for a sky diver to reach a velocity v in feet per second is given by

$$t = -\frac{175}{32}\ln\left(1 - \frac{v}{175}\right), 0 \le v < 175$$

a. Determine the velocity of the diver after 5 seconds.

b. The graph of t has a vertical asymptote at $v = 175$. Explain the meaning of the vertical asymptote in the context of this example.

take note

If air resistance is not considered, then the time in seconds required for a sky diver to reach a given velocity (in feet per second) is $t = \frac{v}{32}$. The function in Example 8 is a more realistic model of the time required to reach a given velocity during the free-fall of a sky diver who is experiencing air resistance.

Solution

a. Substitute 5 for t and solve for v.

$$t = -\frac{175}{32}\ln\left(1 - \frac{v}{175}\right)$$

$$5 = -\frac{175}{32}\ln\left(1 - \frac{v}{175}\right)$$ • Replace t with 5.

$$\left(-\frac{32}{175}\right)5 = \ln\left(1 - \frac{v}{175}\right)$$ • Multiply each side by $-\frac{32}{175}$.

$$-\frac{32}{35} = \ln\left(1 - \frac{v}{175}\right)$$ • Simplify.

$$e^{-32/35} = 1 - \frac{v}{175}$$ • Write in exponential form.

$$e^{-32/35} - 1 = -\frac{v}{175}$$ • Subtract 1 from each side.

$$v = 175(1 - e^{-32/35})$$ • Multiply each side by -175.

$$v \approx 104.86$$

Continued ▶

ANSWER If $x = -\frac{3}{2}$, the original equation becomes $\ln\left(\frac{7}{2}\right) = \ln(-1) + \ln\left(-\frac{7}{2}\right)$. This cannot be true, because the function $f(x) = \ln x$ is not defined for negative values of x.

After 5 seconds the velocity of the sky diver will be about 104.9 feet per second. See **Figure 7.26.**

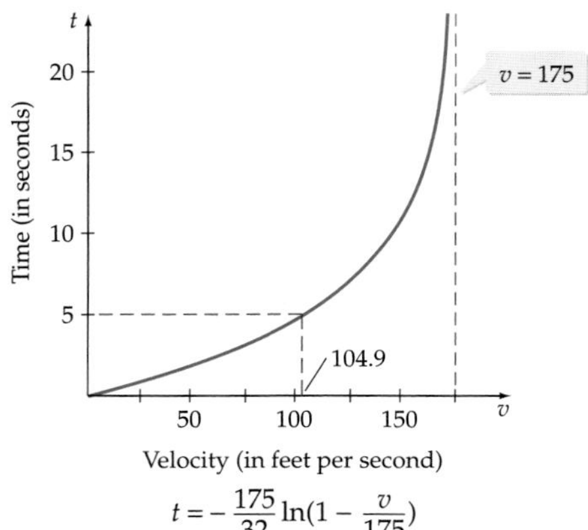

$t = -\frac{175}{32}\ln(1 - \frac{v}{175})$

Figure 7.26

b. The vertical asymptote $v = 175$ indicates that the velocity of the sky diver approaches, but never reaches or exceeds, 175 feet per second. In **Figure 7.26,** note that as $v \to 175$ from the left, $t \to \infty$.

Try Exercise 74, page 504

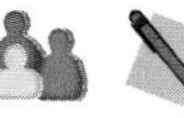

Topics for Discussion

1. Discuss how to solve the equation $a = \log_b x$ for x.
2. What is the domain of $y = \log_4(2x - 5)$? Explain why this means that the equation $\log_4(x - 3) = \log_4(2x - 5)$ has no real number solution.
3. -8 is not a solution of the equation $\log_2 x + \log_2(x + 6) = 4$. Discuss at which step in the following solution the extraneous solution -8 was introduced.

$$\begin{aligned} \log_2 x + \log_2(x + 6) &= 4 \\ \log_2 x(x + 6) &= 4 \\ x(x + 6) &= 2^4 \\ x^2 + 6x &= 16 \\ x^2 + 6x - 16 &= 0 \\ (x + 8)(x - 2) &= 0 \\ x = -8 \quad \text{or} \quad x &= 2 \end{aligned}$$

Exercise Set 7.4

In Exercises 1 to 48, use algebraic procedures to find the exact solution(s) of the equation.

1. $2^x = 64$
2. $3^x = 243$
3. $49^x = \frac{1}{343}$
4. $9^x = \frac{1}{243}$
5. $2^{5x+3} = \frac{1}{8}$
6. $3^{4x-7} = \frac{1}{9}$
7. $\left(\frac{2}{5}\right)^x = \frac{8}{125}$
8. $\left(\frac{2}{5}\right)^x = \frac{25}{4}$
9. $5^x = 70$
10. $6^x = 50$
11. $3^{-x} = 120$
12. $7^{-x} = 63$
13. $10^{2x+3} = 315$
14. $10^{6-x} = 550$
15. $e^x = 10$
16. $e^{x+1} = 20$
17. $2^{1-x} = 3^{x+1}$
18. $3^{x-2} = 4^{2x+1}$
19. $2^{2x-3} = 5^{-x-1}$
20. $5^{3x} = 3^{x+4}$
21. $\log(4x - 18) = 1$
22. $\log(x^2 + 19) = 2$
23. $\ln(x^2 - 12) = \ln x$
24. $\log(2x^2 + 3x) = \log(10x + 30)$
25. $\log_2 x + \log_2(x - 4) = 2$
26. $\log_3 x + \log_3(x + 6) = 3$
27. $\log(5x - 1) = 2 + \log(x - 2)$
28. $1 + \log(3x - 1) = \log(2x + 1)$
29. $\ln(1 - x) + \ln(3 - x) = \ln 8$
30. $\log(4 - x) = \log(x + 8) + \log(2x + 13)$
31. $\log \sqrt{x^3 - 17} = \frac{1}{2}$
32. $\log(x^3) = (\log x)^2$
33. $\log(\log x) = 1$
34. $\ln(\ln x) = 2$
35. $\ln(e^{3x}) = 6$
36. $\ln x = \frac{1}{2}\ln\left(2x + \frac{5}{2}\right) + \frac{1}{2}\ln 2$
37. $\log_7(5x) - \log_7 3 = \log_7(2x + 1)$
38. $\log_4 x + \log_4(x - 2) = \log_4 15$
39. $e^{\ln(x-1)} = 4$
40. $10^{\log(2x+7)} = 8$
41. $\frac{10^x - 10^{-x}}{2} = 20$
42. $\frac{10^x + 10^{-x}}{2} = 8$
43. $\frac{10^x + 10^{-x}}{10^x - 10^{-x}} = 5$
44. $\frac{10^x - 10^{-x}}{10^x + 10^{-x}} = \frac{1}{2}$
45. $\frac{e^x + e^{-x}}{2} = 15$
46. $\frac{e^x - e^{-x}}{2} = 15$
47. $\frac{1}{e^x - e^{-x}} = 4$
48. $\frac{e^x + e^{-x}}{e^x - e^{-x}} = 3$

In Exercises 49 to 58, use a graphing utility to approximate the solution(s) of the equation to the nearest hundredth.

49. $2^{-x+3} = x + 1$
50. $3^{x-2} = -2x - 1$
51. $e^{3-2x} - 2x = 1$
52. $2e^{x+2} + 3x = 2$
53. $3\log_2(x - 1) = -x + 3$
54. $2\log_3(2 - 3x) = 2x - 1$
55. $\ln(2x + 4) + \frac{1}{2}x = -3$
56. $2\ln(3 - x) + 3x = 4$
57. $2^{x+1} = x^2 - 1$
58. $\ln x = -x^2 + 4$
59. **Population Growth** The population P of a city grows exponentially according to the function

$$P(t) = 8500(1.1)^t, \quad 0 \le t \le 8$$

where t is measured in years.

a. Find the population at time $t = 0$ and also at time $t = 2$.

b. When, to the nearest year, will the population reach 15,000?

60. PHYSICAL FITNESS After a race, a runner's pulse rate R in beats per minute decreases according to the function

$$R(t) = 145e^{-0.092t}, \quad 0 \le t \le 15$$

where t is measured in minutes.

a. Find the runner's pulse rate at the end of the race and also 1 minute after the end of the race.

b. How long, to the nearest minute, after the end of the race will the runner's pulse rate be 80 beats per minute?

61. RATE OF COOLING A can of soda at 79°F is placed in a refrigerator that maintains a constant temperature of 36°F. The temperature T of the soda t minutes after it is placed in the refrigerator is given by

$$T(t) = 36 + 43e^{-0.058t}$$

a. Find the temperature, to the nearest degree, of the soda 10 minutes after it is placed in the refrigerator.

b. When, to the nearest minute, will the temperature of the soda be 45°F?

62. MEDICINE During surgery, a patient's circulatory system requires at least 50 milligrams of an anesthetic. The amount of anesthetic present t hours after 80 milligrams of anesthetic is administered is given by

$$T(t) = 80(0.727)^t$$

a. How much, to the nearest milligram, of the anesthetic is present in the patient's circulatory system 30 minutes after the anesthetic is administered?

b. How long, to the nearest minute, can the operation last if the patient does not receive additional anesthetic?

BERTALANFFY'S EQUATION In 1938, the biologist Ludwig von Bertalanffy developed the equation

$$L = m - (m - L_0)e^{-rx}$$

which models the length *L*, in centimeters, of a fish as it grows under optimal conditions for a period of *x* years. In Bertalanffy's equation, *m* represents the maximum length, in centimeters, the fish is expected to attain, L_0 is the length, in centimeters, of the fish at birth, and *r* is a constant related to the growth rate of the fish species. Use Bertalanffy's equation to predict the age of the fish described in Exercises 63 and 64.

63. A barracuda has a length of 114 centimeters. Use Bertalanffy's equation to predict, to the nearest tenth of a year, the age of the barracuda. Assume $m = 198$ centimeters, $L_0 = 0.9$ centimeter, and $r = 0.23$.

64. A haddock has a length of 21 centimeters. Use Bertalanffy's equation to predict, to the nearest tenth of a year, the age of the haddock. Assume $m = 94$ centimeters, $L_0 = 0.6$ centimeters, and $r = 0.21$.

65. TYPING SPEED The following function models the average typing speed S, in words per minute, for a student who has been typing for t months.

$$S(t) = 5 + 29\ln(t + 1), \quad 0 \le t \le 9$$

Use S to determine how long it takes the student to achieve an average typing speed of 65 words per minute. Round to the nearest tenth of a month.

66. WALKING SPEED An approximate relation between the average pedestrian walking speed s, in miles per hour, and the population x, in thousands, of a city is given by the formula

$$s(x) = 0.37\ln x + 0.05$$

Use s to estimate the population of a city for which the average pedestrian walking speed is 2.9 miles per hour. Round to the nearest hundred thousand.

67. DRAG RACING The quadratic function

$$s_1(x) = -2.25x^2 + 56.26x - 0.28, \quad 0 \le x \le 10$$

models the speed of a dragster from the start of a race until the dragster crosses the finish line 10 seconds later. This is the acceleration phase of the race.

The exponential function

$$s_2(x) = 8320(0.73)^x, \quad 10 < x \le 20$$

models the speed of the dragster during the 10-second period immediately following the time when the dragster crosses the finish line. This is the deceleration period.

How long after the start of the race did the dragster attain a speed of 275 miles per hour? Round to the nearest hundredth of a second.

68. EIFFEL TOWER The functions

$$h_1(x) = 363.4 - 88.4\ln x, \quad 16.47 < x \le 61.0$$

and

$$h_2(x) = 568.2 - 161.5\ln x, \quad 6.1 \le x \le 16.47$$

approximate the height, in meters, of the Eiffel Tower x meters to the right of the center line, shown by the y-axis in the following figure (on page 503).

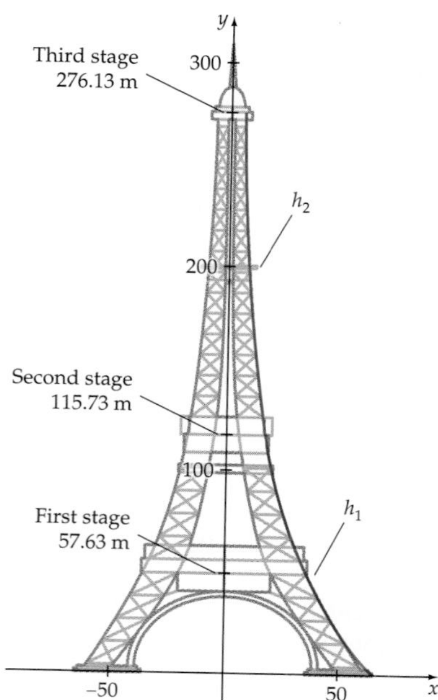

The graph of h_1 models the shape of the tower from ground level up to the second stage in the figure, and the graph of h_2 models the shape of the tower from the second stage up to the third stage.

Determine the horizontal distance across the Eiffel Tower, rounded to the nearest tenth of a meter, at a height of

a. 50 meters

b. 125 meters

69. **PSYCHOLOGY** Industrial psychologists study employee training programs to assess the effectiveness of the instruction. In one study, the percent score P on a test for a person who had completed t hours of training was given by

$$P = \frac{100}{1 + 30e^{-0.088t}}$$

a. Use a graphing utility to graph the equation for $t \geq 0$.

b. Use the graph to estimate (to the nearest hour) the number of hours of training necessary to achieve a 70% score on the test.

c. From the graph, determine the horizontal asymptote.

d. Write a sentence that explains the meaning of the horizontal asymptote.

70. **PSYCHOLOGY** An industrial psychologist has determined that the average percent score for an employee on a test of the employee's knowledge of the company's product is given by

$$P = \frac{100}{1 + 40e^{-0.1t}}$$

where t is the number of weeks on the job and P is the percent score.

a. Use a graphing utility to graph the equation for $t \geq 0$.

b. Use the graph to estimate (to the nearest week) the expected number of weeks of employment that are necessary for an employee to earn a 70% score on the test.

c. Determine the horizontal asymptote of the graph.

d. Write a sentence that explains the meaning of the horizontal asymptote.

71. **ECOLOGY** A herd of bison was placed in a wildlife preserve that can support a maximum of 1000 bison. A population model for the bison is given by

$$B = \frac{1000}{1 + 30e^{-0.127t}}$$

where B is the number of bison in the preserve and t is time in years, with the year 1999 represented by $t = 0$.

a. Use a graphing utility to graph the equation for $t \geq 0$.

b. Use the graph to estimate (to the nearest year) the number of years before the bison population reaches 500.

c. Determine the horizontal asymptote of the graph.

d. Write a sentence that explains the meaning of the horizontal asymptote.

72. **POPULATION GROWTH** A yeast culture grows according to the equation

$$Y = \frac{50{,}000}{1 + 250e^{-0.305t}}$$

where Y is the number of yeast and t is time in hours.

a. Use a graphing utility to graph the equation for $t \geq 0$.

b. Use the graph to estimate (to the nearest hour) the number of hours before the yeast population reaches 35,000.

c. From the graph, estimate the horizontal asymptote.

d. Write a sentence that explains the meaning of the horizontal asymptote.

73. **Consumption of Natural Resources** A model for how long our coal resources will last is given by

$$T = \frac{\ln(300r + 1)}{\ln(r + 1)}$$

where r is the percent increase in consumption from current levels of use and T is the time (in years) before the resources are depleted.

a. Use a graphing utility to graph this equation.

b. If our consumption of coal increases by 3% per year, in how many years will we deplete our coal resources?

c. What percent increase in consumption of coal will deplete the resources in 100 years? Round to the nearest tenth of a percent.

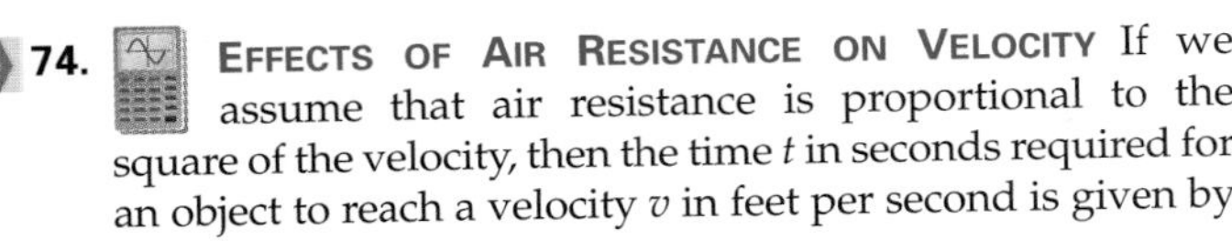

74. **Effects of Air Resistance on Velocity** If we assume that air resistance is proportional to the square of the velocity, then the time t in seconds required for an object to reach a velocity v in feet per second is given by

$$t = \frac{9}{24}\ln\frac{24 + v}{24 - v}, \quad 0 \le v < 24$$

a. Determine the velocity, to the nearest hundredth of a foot per second, of the object after 1.5 seconds.

b. Determine the vertical asymptote for the graph of this function.

c. Write a sentence that explains the meaning of the vertical asymptote in the context of this application.

75. **Terminal Velocity with Air Resistance** The velocity v of an object t seconds after it has been dropped from a height above the surface of the earth is given by the equation $v = 32t$ feet per second, assuming no air resistance. If we assume that air resistance is proportional to the square of the velocity, then the velocity after t seconds is given by

$$v = 100\left(\frac{e^{0.64t} - 1}{e^{0.64t} + 1}\right)$$

a. In how many seconds will the velocity be 50 feet per second?

b. Determine the horizontal asymptote for the graph of this function.

c. Write a sentence that explains the meaning of the horizontal asymptote in the context of this application.

76. **Effects of Air Resistance on Distance** The distance s, in feet, that the object in Exercise 75 will fall in t seconds is given by

$$s = \frac{100^2}{32}\ln\left(\frac{e^{0.32t} + e^{-0.32t}}{2}\right)$$

a. Use a graphing utility to graph this equation for $t \ge 0$.

b. How long does it take for the object to fall 100 feet? Round to the nearest tenth of a second.

77. **Retirement Planning** The retirement account for a graphic designer contains \$250,000 on January 1, 2006, and earns interest at a rate of 0.5% per month. On February 1, 2006, the designer withdraws \$2000 and plans to continue these withdrawals as retirement income each month. The value V of the account after x months is

$$V = 400{,}000 - 150{,}000(1.005)^x$$

If the designer wishes to leave \$100,000 to a scholarship foundation, what is the maximum number of withdrawals the designer can make from this account and still have \$100,000 to donate?

78. **Hanging Cable** The height h, in feet, of any point P on the cable shown is given by

$$h(x) = 10(e^{x/20} + e^{-x/20}), \quad -15 \le x \le 15$$

where $|x|$ is the horizontal distance in feet between P and the y-axis.

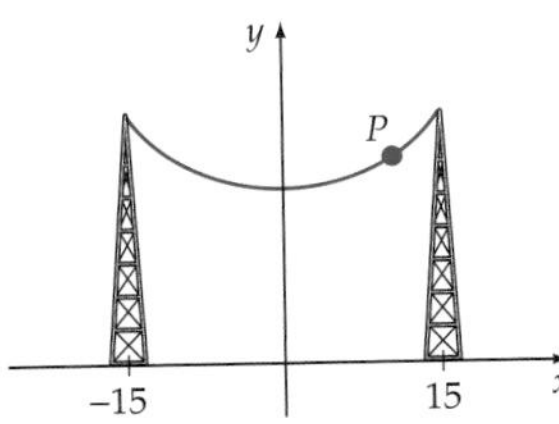

a. What is the lowest height of the cable?

b. What is the height of the cable 10 feet to the right of the y-axis? Round to the nearest tenth of a foot.

c. How far to the right of the y-axis is the cable 24 feet in height? Round to the nearest tenth of a foot.

Connecting Concepts

79. The following argument seems to indicate that $0.125 > 0.25$. Find the first incorrect statement in the argument.

$$3 > 2$$
$$3(\log 0.5) > 2(\log 0.5)$$
$$\log 0.5^3 > \log 0.5^2$$
$$0.5^3 > 0.5^2$$
$$0.125 > 0.25$$

80. The following argument seems to indicate that $4 = 6$. Find the first incorrect statement in the argument.

$$4 = \log_2 16$$
$$4 = \log_2(8 + 8)$$
$$4 = \log_2 8 + \log_2 8$$
$$4 = 3 + 3$$
$$4 = 6$$

81. A common mistake that students make is to write $\log(x + y)$ as $\log x + \log y$. For what values of x and y does $\log(x + y) = \log x + \log y$? (*Hint:* Solve for x in terms of y.)

82. Let $f(x) = 2 \ln x$ and $g(x) = \ln x^2$. Does $f(x) = g(x)$ for all real numbers x?

83. Explain why the functions $F(x) = 1.4^x$ and $G(x) = e^{0.336x}$ represent essentially the same function.

84. Find the constant k that will make $f(t) = 2.2^t$ and $g(t) = e^{-kt}$ represent essentially the same function.

Projects

1. **NAVIGATING** The pilot of a boat is trying to cross a river to a point O two miles due west of the boat's starting position by always pointing the nose of the boat toward O. Suppose the speed of the current is w miles per hour and the speed of the boat is v miles per hour. If point O is the origin and the boat's starting position is $(2, 0)$ (see the diagram at the right), then the equation of the boat's path is given by

$$y = \left(\frac{x}{2}\right)^{1-(w/v)} - \left(\frac{x}{2}\right)^{1+(w/v)}$$

a. If the speed of the current and the speed of the boat are the same, can the pilot reach point O by always having the nose of the boat pointed toward O? If not, at what point will the pilot arrive? Explain your answer.

b. If the speed of the current is greater than the speed of the boat, can the pilot reach point O by always pointing the nose of the boat toward point O? If not, where will the pilot arrive? Explain.

c. If the speed of the current is less than the speed of the boat, can the pilot reach point O by always pointing the nose of the boat toward point O? If not, where will the pilot arrive? Explain.

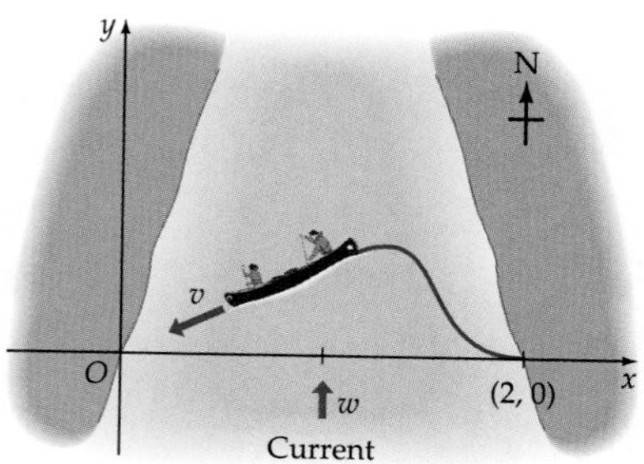

Section 7.5

- Exponential Growth and Decay
- Carbon Dating
- Compound Interest Formulas
- Restricted Growth Models

Exponential Growth and Decay

PREPARE FOR THIS SECTION

Prepare for this section by completing the following exercises. The answers can be found on page A40.

PS1. Evaluate $A = 1000\left(1 + \dfrac{0.1}{12}\right)^{12t}$ for $t = 2$. Round to the nearest hundredth. [7.1]

PS2. Evaluate $A = 600\left(1 + \dfrac{0.04}{4}\right)^{4t}$ for $t = 8$. Round to the nearest hundredth. [7.1]

PS3. Solve $0.5 = e^{14k}$ for k. Round to the nearest ten-thousandth. [7.4]

PS4. Solve $0.85 = 0.5^{t/5730}$ for t. Round to the nearest ten. [7.4]

PS5. Solve $6 = \dfrac{70}{5 + 9e^{-k \cdot 12}}$ for k. Round to the nearest thousandth. [7.4]

PS6. Solve $2{,}000{,}000 = \dfrac{3^{n+1} - 3}{2}$ for n. Round to the nearest tenth. [7.4]

Exponential Growth and Decay

In many applications, a quantity changes at a rate proportional to the amount present. In these applications, the amount present at time t is given by a special function called an *exponential growth function* or an *exponential decay function.*

Definition of Exponential Growth and Decay Functions

If a quantity N increases or decreases at a rate proportional to the amount present at time t, then the quantity can be modeled by

$$N(t) = N_0 e^{kt}$$

where N_0 is the value of N at time $t = 0$, and k is a constant called the **growth rate constant.**

- If k is positive, N increases as t increases and $N(t) = N_0 e^{kt}$ is called an **exponential growth function.** See **Figure 7.27.**
- If k is negative, N decreases as t increases and $N(t) = N_0 e^{kt}$ is called an **exponential decay function.** See **Figure 7.28.**

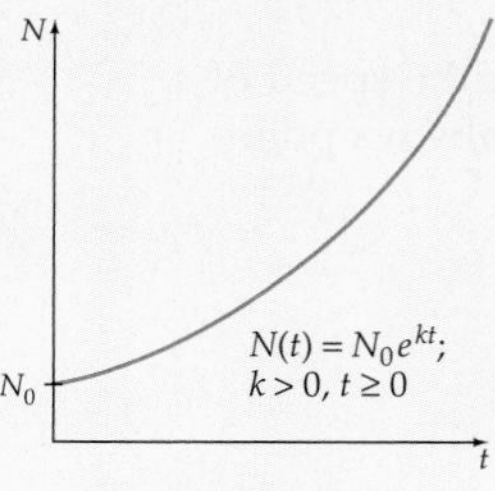

Figure 7.27
Exponential growth function

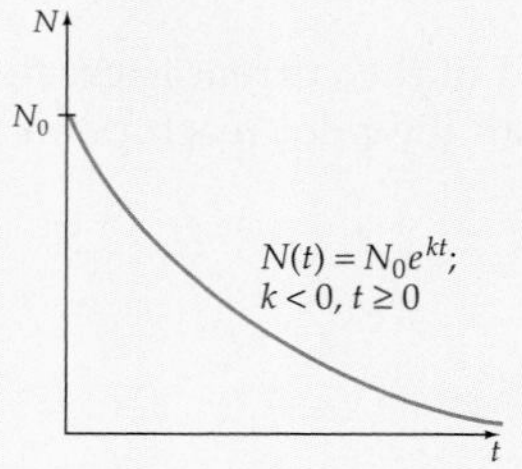

Figure 7.28
Exponential decay function

? QUESTION Is $N(t) = 1450e^{0.05t}$ an exponential growth function or an exponential decay function?

In Example 1 we find an exponential growth function that models the population growth of a city.

EXAMPLE 1 Find the Exponential Growth Function That Models Population Growth

a. The population of a city is growing exponentially. The population of the city was 16,400 in 1995 and 20,200 in 2005. Find the exponential growth function that models the population growth of the city.

b. Use the function from **a.** to predict, to the nearest 100, the population of the city in 2010.

Solution

a. We need to determine N_0 and k in $N(t) = N_0e^{kt}$. If we represent the year 1995 by $t = 0$, then our given data are $N(0) = 16{,}400$ and $N(10) = 20{,}200$. Because N_0 is defined to be $N(0)$, we know that $N_0 = 16{,}400$. To determine k, substitute $t = 10$ and $N_0 = 16{,}400$ into $N(t) = N_0e^{kt}$ to produce

$$N(10) = 16{,}400e^{k\cdot 10}$$

$$20{,}200 = 16{,}400e^{10k}$$ • Substitute 20,200 for $N(10)$.

$$\frac{20{,}200}{16{,}400} = e^{10k}$$ • Solve for e^{10k}.

$$\ln\frac{20{,}200}{16{,}400} = 10k$$ • Write in logarithmic form.

$$\frac{1}{10}\ln\frac{20{,}200}{16{,}400} = k$$ • Solve for k.

$$0.0208 \approx k$$

The exponential growth function is $N(t) \approx 16{,}400e^{0.0208t}$.

b. The year 1995 was represented by $t = 0$, so we will use $t = 15$ to represent the year 2010.

$$N(t) \approx 16{,}400e^{0.0208t}$$

$$N(15) \approx 16{,}400e^{0.0208\cdot 15}$$

$$\approx 22{,}400 \quad \text{(nearest 100)}$$

The exponential growth function yields 22,400 as the approximate population of the city in 2010.

Try Exercise 6, page 516

? ANSWER Because the growth rate constant $k = 0.05$ is positive, the function is an exponential growth function.

Many radioactive materials *decrease* in mass exponentially over time. This decrease, called radioactive decay, is measured in terms of **half-life,** which is defined as the time required for the disintegration of half the atoms in a sample of a radioactive substance. **Table 7.11** shows the half-lives of selected radioactive isotopes.

Table 7.11

Isotope	Half-Life
Carbon (^{14}C)	5730 years
Radium (^{226}Ra)	1660 years
Polonium (^{210}Po)	138 days
Phosphorus (^{32}P)	14 days
Polonium (^{214}Po)	1/10,000th of a second

EXAMPLE 2 ›› Find an Exponential Decay Function

Find the exponential decay function for the amount of phosphorus (^{32}P) that remains in a sample after t days.

take note

Because $e^{-0.0495} \approx (0.5)^{1/14}$, the decay function $N(t) = N_0e^{-0.0495t}$ can also be written as $N(t) = N_0(0.5)^{t/14}$. In this form it is easy to see that if t is increased by 14, then N will decrease by a factor of 0.5.

Solution

When $t = 0$, $N(0) = N_0e^{k(0)} = N_0$. Thus $N(0) = N_0$. Also, because the phosphorus has a half-life of 14 days (from **Table 7.11**), $N(14) = 0.5N_0$. To find k, substitute $t = 14$ into $N(t) = N_0e^{kt}$ and solve for k.

$$N(14) = N_0 \cdot e^{k \cdot 14}$$

$$0.5N_0 = N_0e^{14k}$$ • **Substitute $0.5N_0$ for $N(14)$.**

$$0.5 = e^{14k}$$ • **Divide each side by N_0.**

$$\ln 0.5 = 14k$$ • **Write in logarithmic form.**

$$\frac{1}{14}\ln 0.5 = k$$ • **Solve for k.**

$$-0.0495 \approx k$$

The exponential decay function is $N(t) \approx N_0e^{-0.0495t}$.

›› *Try Exercise 8, page 516*

Carbon Dating

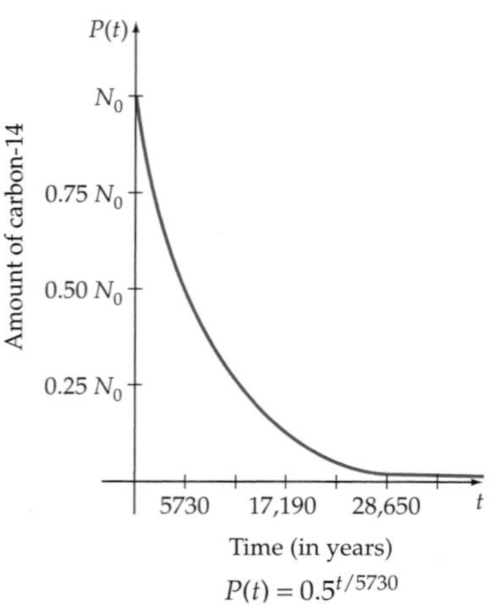

Figure 7.29

The bone tissue in all living animals contains both carbon-12, which is nonradioactive, and carbon-14, which is radioactive with a half-life of approximately 5730 years. See **Figure 7.29**. As long as the animal is alive, the ratio of carbon-14 to carbon-12 remains constant. When the animal dies ($t = 0$), the carbon-14 begins to decay. Thus a bone that has a smaller ratio of carbon-14 to carbon-12 is older than a bone that has a larger ratio. The percent of carbon-14 present at time t is

$$P(t) = 0.5^{t/5730}$$

Math Matters

The chemist Willard Frank Libby developed the carbon dating process in 1947. In 1960 he was awarded the Nobel Prize in chemistry for this achievement.

The process of using the percent of carbon-14 present at a given time to estimate the age of a bone is called **carbon dating.**

EXAMPLE 3 A Carbon Dating Application

Estimate the age of a bone if it now has 85% of the carbon-14 it had at time $t = 0$.

Solution

Let t be the time at which $P(t) = 0.85$.

$$0.85 = 0.5^{t/5730}$$

$$\ln 0.85 = \ln 0.5^{t/5730}$$ • Take the natural logarithm of each side.

$$\ln 0.85 = \frac{t}{5730} \ln 0.5$$ • Power property.

$$5730\left(\frac{\ln 0.85}{\ln 0.5}\right) = t$$ • Solve for t.

$$1340 \approx t$$

The bone is about 1340 years old.

Try Exercise 12, page 516

Compound Interest Formulas

Interest is money paid for the use of money. The interest I is called **simple interest** if it is a fixed percent r, per time period t, of the amount of money invested. The amount of money invested is called the **principal** P. Simple interest is computed using the formula $I = Prt$. For example, if \$1000 is invested at 12% for 3 years, the simple interest is

$$I = Prt = \$1000(0.12)(3) = \$360$$

The balance after t years is $A = P + I = P + Prt$. In the preceding example, the \$1000 invested for 3 years produced \$360 interest. Thus the balance after 3 years is \$1000 + \$360 = \$1360.

In many financial transactions, interest is added to the principal at regular intervals so that interest is paid on interest as well as on the principal. Interest earned in this manner is called **compound interest**. For example, if \$1000 is invested at 12% annual interest compounded annually for 3 years, then the total interest after 3 years is

First-year interest	$1000(0.12) = $120.00	
Second-year interest	$1120(0.12) = $134.40	
Third-year interest	$1254.40(0.12) ≈ $150.53	
	$404.93	• Total interest

This method of computing the balance can be tedious and time-consuming. A *compound interest formula* that can be used to determine the balance due after t years of compounding can be developed as follows.

Table 7.12

Number of Years	Balance
3	$A_3 = P(1 + r)^3$
4	$A_4 = P(1 + r)^4$
⋮	⋮
t	$A_t = P(1 + r)^t$

Note that if P dollars is invested at an interest rate of r per year, then the balance after 1 year is $A_1 = P + Pr = P(1 + r)$, where Pr represents the interest earned for the year. Observe that A_1 is the product of the original principal P and $(1 + r)$. If the amount A_1 is reinvested for another year, then the balance after the second year is

$$A_2 = (A_1)(1 + r) = P(1 + r)(1 + r) = P(1 + r)^2$$

Successive reinvestments lead to the results shown in **Table 7.12**. The equation $A_t = P(1 + r)^t$ is valid if r is the annual interest rate paid during each of the t years.

If r is an annual interest rate and n is the number of compounding periods per year, then the interest rate each period is r/n, and the number of compounding periods after t years is nt. Thus the compound interest formula is as follows:

The Compound Interest Formula

A principal P invested at an annual interest rate r, expressed as a decimal and compounded n times per year for t years, produces the balance

$$A = P\left(1 + \frac{r}{n}\right)^{nt}$$

EXAMPLE 4 ⟫ Solve a Compound Interest Application

Find the balance if \$1000 is invested at an annual interest rate of 10% for 2 years compounded

a. monthly **b.** daily

Solution

a. Because there are 12 months in a year, use $n = 12$.

$$A = \$1000\left(1 + \frac{0.1}{12}\right)^{12\cdot 2} \approx \$1000(1.008333333)^{24} \approx \$1220.39$$

b. Because there are 365 days in a year, use $n = 365$.

$$A = \$1000\left(1 + \frac{0.1}{365}\right)^{365\cdot 2} \approx \$1000(1.000273973)^{730} \approx \$1221.37$$

⟫ ***Try Exercise 16, page 517***

To **compound continuously** means to increase the number of compounding periods per year, n, without bound.

To derive a continuous compounding interest formula, substitute $\frac{1}{m}$ for $\frac{r}{n}$ in the compound interest formula

$$A = P\left(1 + \frac{r}{n}\right)^{nt}$$

to produce

$$A = P\left(1 + \frac{1}{m}\right)^{nt} \tag{1}$$

This substitution is motivated by the desire to express $\left(1 + \frac{r}{n}\right)^n$ as $\left[\left(1 + \frac{1}{m}\right)^m\right]^r$, which approaches e^r as m gets larger without bound.

Solving the equation $\frac{1}{m} = \frac{r}{n}$ for n yields $n = mr$, so the exponent nt can be written as mrt. Therefore Equation (1) can be expressed as

$$A = P\left(1 + \frac{1}{m}\right)^{mrt} = P\left[\left(1 + \frac{1}{m}\right)^m\right]^{rt} \tag{2}$$

By the definition of e, we know that as m increases without bound,

$$\left(1 + \frac{1}{m}\right)^m \quad \text{approaches} \quad e$$

Thus, using continuous compounding, Equation (2) simplifies to $A = Pe^{rt}$.

Continuous Compounding Interest Formula

If an account with principal P and annual interest rate r is compounded continuously for t years, then the balance is $A = Pe^{rt}$.

EXAMPLE 5 » Solve a Continuous Compound Interest Application

Find the balance after 4 years on \$800 invested at an annual rate of 6% compounded continuously.

ALGEBRAIC SOLUTION

Use the continuous compounding formula with $P = 800$, $r = 0.06$, and $t = 4$.

$$\begin{aligned} A &= Pe^{rt} \\ &= 800e^{0.06(4)} && \text{• Substitute given values.} \\ &= 800e^{0.24} && \text{• Simplify.} \\ &\approx 800(1.27124915) \\ &\approx 1017.00 && \text{• Round to the nearest cent.} \end{aligned}$$

The balance after 4 years will be \$1017.00.

VISUALIZE THE SOLUTION

The following graph of $A = 800e^{0.06t}$ shows that the balance is about \$1017.00 when $t = 4$.

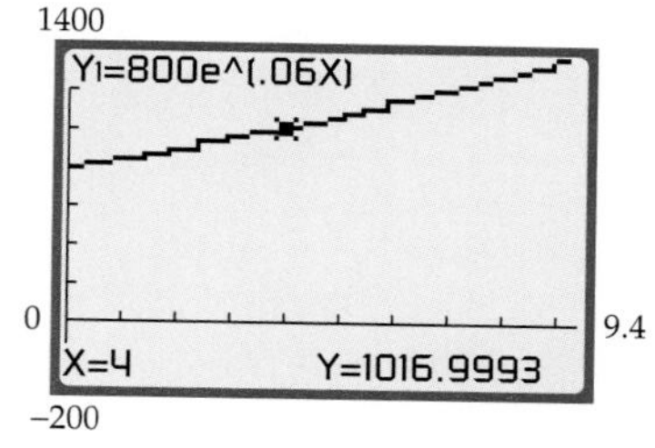

» Try Exercise 18, page 517

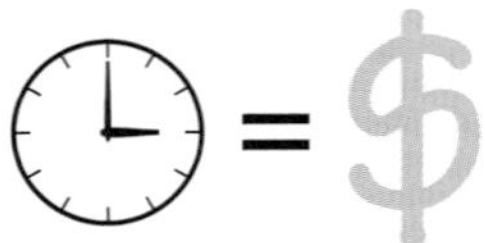

You have probably heard it said that time is money. In fact, many investors ask the question "How long will it take to double my money?" The following example answers this question for two different investments.

EXAMPLE 6 Double Your Money

Find the time required for money invested at an annual rate of 6% to double in value if the investment is compounded.

a. semiannually **b.** continuously

Solution

a. Use $A = P\left(1 + \frac{r}{n}\right)^{nt}$ with $r = 0.06$, $n = 2$, and the balance A equal to twice the principal ($A = 2P$).

$$2P = P\left(1 + \frac{0.06}{2}\right)^{2t}$$

$$2 = \left(1 + \frac{0.06}{2}\right)^{2t}$$ • Divide each side by P.

$$\ln 2 = \ln\left(1 + \frac{0.06}{2}\right)^{2t}$$ • Take the natural logarithm of each side.

$$\ln 2 = 2t \ln\left(1 + \frac{0.06}{2}\right)$$ • Apply the power property.

$$2t = \frac{\ln 2}{\ln\left(1 + \frac{0.06}{2}\right)}$$ • Solve for t.

$$t = \frac{1}{2} \cdot \frac{\ln 2}{\ln\left(1 + \frac{0.06}{2}\right)}$$

$$t \approx 11.72$$

If the investment is compounded semiannually, it will double in value in about 11.72 years.

b. Use $A = Pe^{rt}$ with $r = 0.06$ and $A = 2P$.

$$2P = Pe^{0.06t}$$

$$2 = e^{0.06t}$$ • Divide each side by P.

$$\ln 2 = 0.06t$$ • Write in logarithmic form.

$$t = \frac{\ln 2}{0.06}$$ • Solve for t.

$$t \approx 11.55$$

If the investment is compounded continuously, it will double in value in about 11.55 years.

Try Exercise 22, page 517

Restricted Growth Models

The exponential growth function $N(t) = N_0e^{kt}$ is an *unrestricted growth model* that does not consider any limited resources that eventually will curb population growth.

The **logistic model** is a *restricted growth model* that takes into consideration the effects of limited resources. The logistic model was developed by Pierre Verhulst in 1836.

Definition of the Logistic Model (A Restricted Growth Model)

The magnitude of a population at time $t \geq 0$ is given by

$$P(t) = \frac{c}{1 + ae^{-bt}}$$

where c is the **carrying capacity** (the maximum population that can be supported by available resources as $t \to \infty$) and b is a positive constant called the **growth rate constant.**

The **initial population** is $P_0 = P(0)$. The constant a is related to the initial population P_0 and the carrying capacity c by the formula

$$a = \frac{c - P_0}{P_0}$$

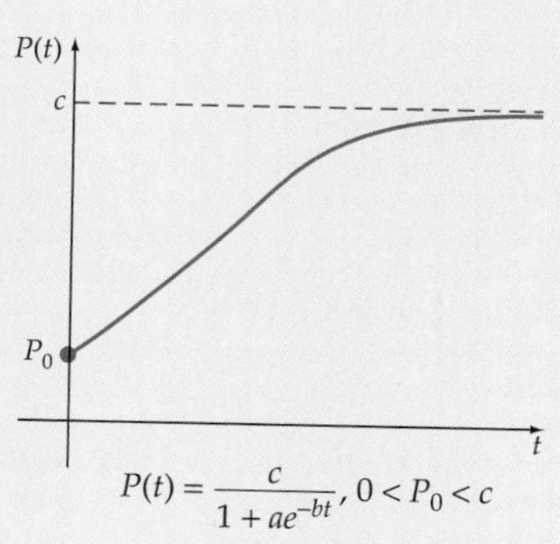

$P(t) = \dfrac{c}{1 + ae^{-bt}}, \; 0 < P_0 < c$

In the following example we determine a logistic growth model for a coyote population.

EXAMPLE 7 ⟩⟩ Find and Use a Logistic Model

At the beginning of 2005, the coyote population in a wilderness area was estimated at 200. By the beginning of 2007, the coyote population had increased to 250. A park ranger estimates that the carrying capacity of the wilderness area is 500 coyotes.

a. Use the given data to determine the growth rate constant for the logistic model of this coyote population.

b. Use the logistic model determined in **a.** to predict the year in which the coyote population will first reach 400.

Solution

a. If we represent the beginning of the year 2005 by $t = 0$, then the beginning of the year 2007 will be represented by $t = 2$. In the logistic model, make the following substitutions: $P(2) = 250$, $c = 500$, and

$$a = \frac{c - P_0}{P_0} = \frac{500 - 200}{200} = 1.5.$$

Continued ▶

$$P(t) = \frac{c}{1 + ae^{-bt}}$$

$$P(2) = \frac{500}{1 + 1.5e^{-b \cdot 2}}$$ • **Substitute the given values.**

$$250 = \frac{500}{1 + 1.5e^{-b \cdot 2}}$$ • **$P(2) = 250$**

$$250(1 + 1.5e^{-b \cdot 2}) = 500$$ • **Solve for the growth rate constant b.**

$$1 + 1.5e^{-b \cdot 2} = \frac{500}{250}$$

$$1.5e^{-b \cdot 2} = 2 - 1$$

$$e^{-b \cdot 2} = \frac{1}{1.5}$$

$$-2b = \ln\left(\frac{1}{1.5}\right)$$ • **Write in logarithmic form.**

$$b = -\frac{1}{2}\ln\left(\frac{1}{1.5}\right)$$

$$b \approx 0.20273255$$

Using $a = 1.5$, $b = 0.20273255$, and $c = 500$ gives us the following logistic model.

$$P(t) = \frac{500}{1 + 1.5e^{-0.20273255t}}$$

b. To determine during what year the logistic model predicts the coyote population will first reach 400, replace $P(t)$ with 400 and solve for t.

$$400 = \frac{500}{1 + 1.5e^{-0.20273255t}}$$

$$400(1 + 1.5e^{-0.20273255t}) = 500$$

$$1 + 1.5e^{-0.20273255t} = \frac{500}{400}$$

$$1.5e^{-0.20273255t} = 1.25 - 1$$

$$e^{-0.20273255t} = \frac{0.25}{1.5}$$

$$-0.20273255t = \ln\left(\frac{0.25}{1.5}\right)$$ • **Write in logarithmic form.**

$$t = \frac{1}{-0.20273255}\ln\left(\frac{0.25}{1.5}\right)$$ • **Solve for t.**

$$\approx 8.8$$

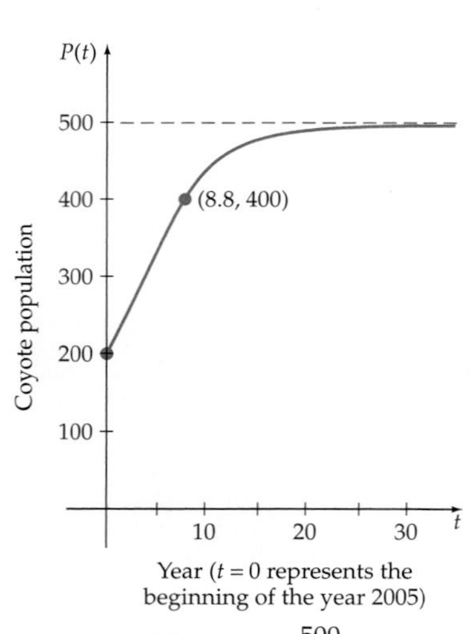

$P(t) = \dfrac{500}{1 + 1.5e^{-0.20273255t}}$

Figure 7.30

According to the logistic model, the coyote population will reach 400 about 8.8 years after the beginning of 2005, which is during the year 2013. The graph of the logistic model is shown in **Figure 7.30**. Note that $P(8.8) \approx 400$ and that as $t \to \infty$, $P(t) \to 500$.

Try Exercise 38, page 518

In Example 8, we use a function of the form $v = a(1 - e^{-kt})$ to model the velocity of an object that has been dropped from a high elevation.

EXAMPLE 8 » Application to Air Resistance

Assuming that air resistance is proportional to the velocity of a falling object, the velocity (in feet per second) of the object t seconds after it has been dropped is given by $v = 82(1 - e^{-0.39t})$.

a. Determine when the velocity will be 70 feet per second.

b. The graph of v has $v = 82$ as a horizontal asymptote. Explain the meaning of this asymptote in the context of this example.

ALGEBRAIC SOLUTION

a.

$$v = 82(1 - e^{-0.39t})$$

$$70 = 82(1 - e^{-0.39t}) \qquad \text{• Substitute 70 for } v.$$

$$\frac{70}{82} = 1 - e^{-0.39t} \qquad \text{• Divide each side by 82.}$$

$$e^{-0.39t} = 1 - \frac{70}{82} \qquad \text{• Solve for } e^{-0.39t}.$$

$$-0.39t = \ln\frac{6}{41} \qquad \text{• Write in logarithmic form.}$$

$$t = \frac{\ln(6/41)}{-0.39} \approx 4.9277246 \qquad \text{• Solve for } t.$$

The velocity will be 70 feet per second after approximately 4.9 seconds.

b. The horizontal asymptote $v = 82$ means that as time increases, the velocity of the object will approach, but never reach or exceed, 82 feet per second.

VISUALIZE THE SOLUTION

a. A graph of $y = 82(1 - e^{-0.39x})$ and $y = 70$ shows that the x-coordinate of the point of intersection is about 4.9.

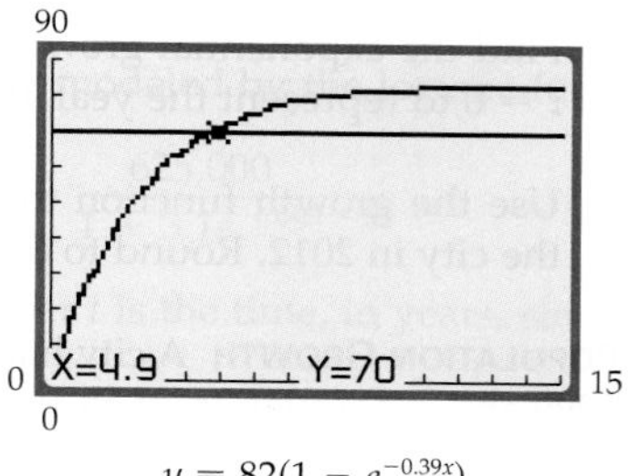

$y = 82(1 - e^{-0.39x})$

Note: The x value shown is rounded to the nearest tenth.

» Try Exercise 48, page 519

Topics for Discussion

1. Explain the difference between compound interest and simple interest.
2. What is an exponential growth function? Give an example of an application for which an exponential growth function might be an appropriate model.
3. What is an exponential decay function? Give an example of an application for which an exponential decay function might be an appropriate model.
4. Consider the exponential function $P(t) = P_0e^{kt}$ and the logistic function $P(t) = \dfrac{c}{1 + ae^{-bt}}$. Explain the similarities and differences between the graphs of the two functions.

Model Data

The methods used to model data using exponential or logarithmic functions are similar to the methods used in Section 1.7 to model data using linear or quadratic functions. Here is a summary of the modeling process.

> **Integrating Technology**
>
> Most graphing utilities have built-in routines that can be used to determine the exponential or logarithmic regression function that models a set of data. On a TI-83/TI-83 Plus/TI-84 Plus, the ExpReg instruction is used to find the exponential regression function and the LnReg instruction is used to find the logarithmic regression function. The TI-83/TI-83 Plus/TI-84 Plus does not show the value of the regression coefficient r or the coefficient of determination unless the DiagnosticOn command has been entered. The DiagnosticOn command is in the CATALOG menu.

The Modeling Process

Use a graphing utility to:

1. **Construct a *scatter plot* of the data** to determine which type of function will effectively model the data.
2. **Find the *equation*** of the modeling function and the correlation coefficient or the coefficient of determination for the equation.
3. **Examine the *correlation coefficient* or the *coefficient of determination*** and ***view a graph*** that displays both the modeling function and the scatter plot to determine how well your function fits the data.

In the following example we use the modeling process to find an exponential function that closely models the value of a diamond as a function of its weight.

EXAMPLE 2 》 Model an Application with an Exponential Function

A diamond merchant has determined the values of several white diamonds that have different weights (measured in carats), but are *similar in quality*. See **Table 7.13**.

Table 7.13

0.50 ct	0.75 ct	1.00 ct	1.25 ct	1.50 ct	1.75 ct	2.00 ct	3.00 ct	4.00 ct
$4,600	$5,000	$5,800	$6,200	$6,700	$7,300	$7,900	$10,700	$14,500

Find a function that models the values of the diamonds as a function of their weights, and use the function to predict the value of a 3.5-carat diamond of similar quality.

> **take note**
>
> The value of a diamond is generally determined by its color, cut, clarity, and carat weight. These characteristics of a diamond are known as the four c's. In Example 2 we have assumed that the color, cut, and clarity of all the diamonds are similar. This assumption enables us to model the value of each diamond as a function of just its carat weight.

Solution

1. **Construct a scatter plot of the data.**

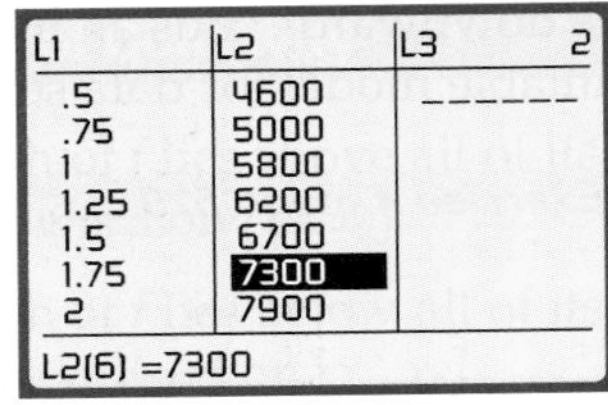

L1	L2	L3 2
.5	4600	------
.75	5000	
1	5800	
1.25	6200	
1.5	6700	
1.75	7300	
2	7900	

L2(6) =7300

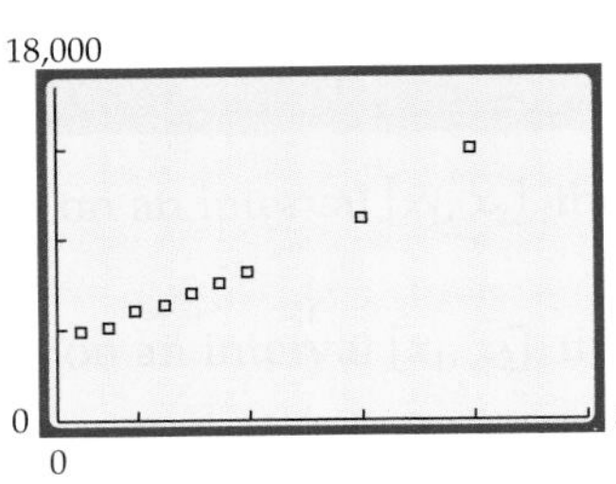

Figure 7.34

Math Matters

The Hope Diamond, shown below, is the world's largest deep blue diamond. It has a weight of 45.52 carats. We should not expect the function $y \approx 4067.6 \times 1.3816^x$ in Example 2 to yield an accurate value of the Hope Diamond because the Hope Diamond is not the same type of diamond as the diamonds in **Table 7.13,** and its weight is much larger than the weights of the diamonds in **Table 7.13.**

The Hope Diamond is on display at the Smithsonian Museum of Natural History in Washington, D.C.

From the scatter plot in **Figure 7.34,** it appears that the data can be closely modeled by an exponential function of the form $y = ab^x$, $a > 0$ and $b > 1$.

2. **Find the equation of the model.** The calculator display in **Figure 7.35** shows that the exponential regression equation is $y \approx 4067.6(1.3816)^x$, where x is the carat weight of the diamond and y is the value of the diamond.

```
ExpReg
 y=a*b^x
 a=4067.641145
 b=1.381644186
 r²=.994881215
 r=.9974373238
```

Figure 7.35
ExpReg display (DiagnosticOn)

3. **Examine the correlation coefficient or the coefficient of determination.** The correlation coefficient $r \approx 0.9974$ is close to 1. This indicates that the exponential regression function $y \approx 4067.6(1.3816)^x$ provides a good fit for the data. The graph in **Figure 7.36** also shows that the exponential regression function provides a good model for the data.

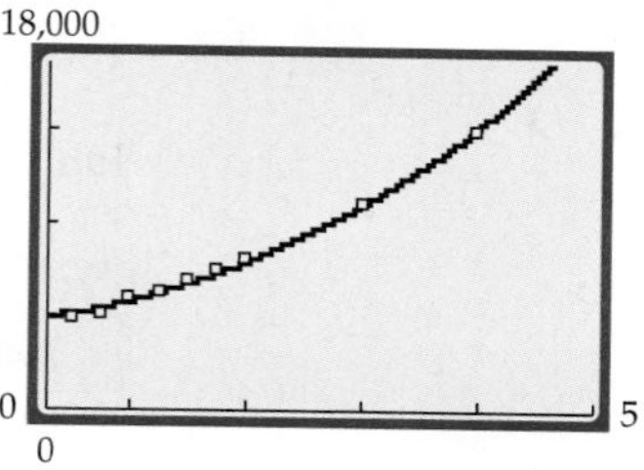

Figure 7.36

To estimate the value of a 3.5-carat diamond, substitute 3.5 for x in the exponential regression function or use the VALUE command in the CALCULATE menu to evaluate the exponential regression function at $x = 3.5$. See **Figure 7.37.**

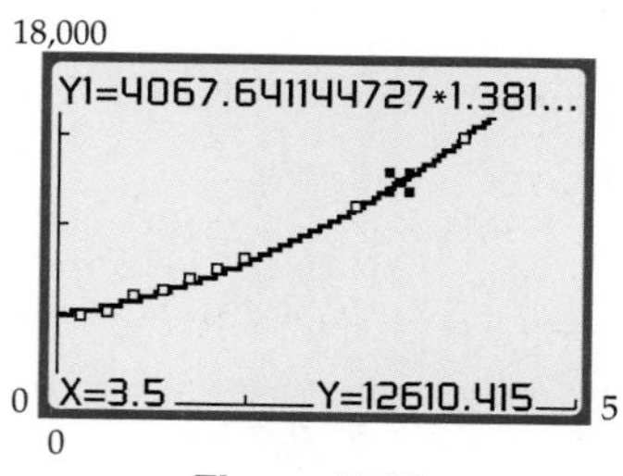

Figure 7.37

According to the exponential regression function, the value of a 3.5-carat diamond of similar quality is about $12,610.

Try Exercise 22, page 531

When you are selecting a function to model a given set of data, try to find a function that provides a good fit to the data *and* is likely to produce realistic predictions. The following guidelines may facilitate the selection process.

Guidelines for Selecting a Modeling Function

1. Use a graphing utility to construct a scatter plot of the data.
2. Compare the graphical features of the scatter plot with the graphical features of the basic modeling functions available on the graphing utility: linear, quadratic, cubic, exponential, logarithmic, or logistic. Pay particular attention to the concave nature of each function. Eliminate those functions that do not display the desired concavity.
3. Use the graphing utility to find the equation of each type of function you identified in Step 2 as a possible model.
4. Determine how well each function fits the given data, and compare the graphs of the functions to determine which function is most likely to produce realistic predictions.

EXAMPLE 3 Select a Modeling Function and Make a Prediction

Table 7.14 shows the winning times in the women's Olympic 100-meter freestyle event for the years 1960 to 2004.

Table 7.14 Women's Olympic 100-Meter Freestyle, 1960 to 2004

Year	Time (in seconds)	Year	Time (in seconds)
1960	61.2	1984	55.92
1964	59.5	1988	54.93
1968	60.0	1992	54.64
1972	58.59	1996	54.50
1976	55.65	2000	53.83
1980	54.79	2004	53.84

Source: Time Almanac 2006.

Find a function to model the data, and use the function to predict the winning time in the women's Olympic 100-meter freestyle event for the year 2012.

take note

When you use a graphing utility to find a logarithmic model, remember that the domain of $y = a + b \ln x$ is the set of positive numbers. Thus zero must not be used as an x value of a data point. This is the reason we have used $x = 1$ to represent the year 1960 in Example 3.

Solution

Construct a scatter plot of the data. See **Figure 7.38**. (*Note:* This scatter plot was produced using $x = 1$ to represent the year 1960, $x = 2$ to represent the year 1964, ..., and $x = 12$ to represent the year 2004.) The general shape of the scatter plot suggests that we consider functions whose graphs are decreasing and concave upward. Thus we

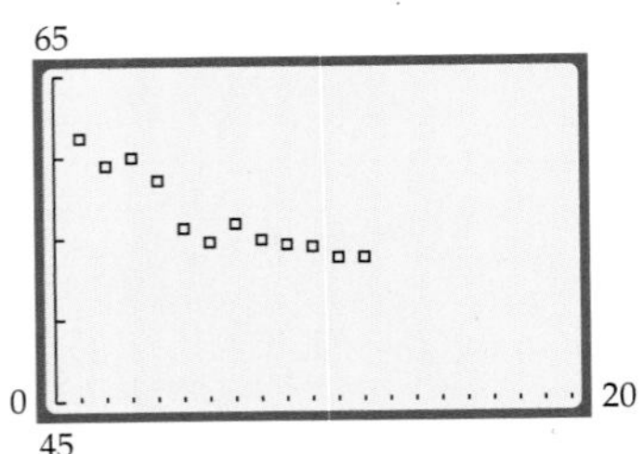

Figure 7.38

consider a decreasing exponential function and a decreasing logarithmic function as possible models. Use a graphing utility to find the exponential regression function and the logarithmic regression function for the data. See **Figures 7.39 and 7.40.**

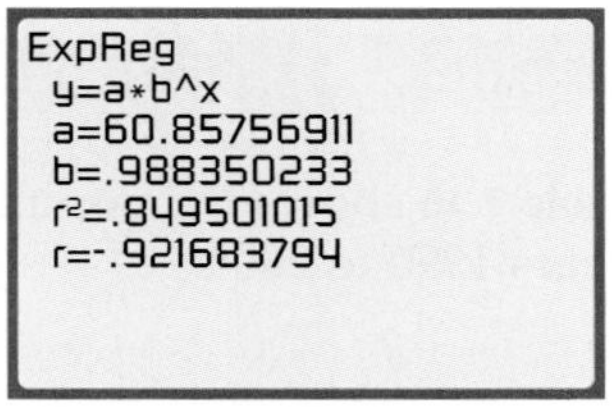

Figure 7.39

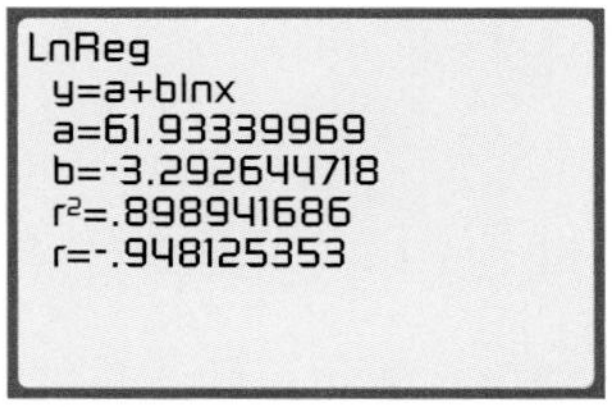

LnReg
y=a+blnx
a=61.93339969
b=-3.292644718
r²=.898941686
r=-.948125353

Figure 7.40

The exponential function is $y \approx 60.85757(0.98835)^x$ and the logarithmic function is $y \approx 61.93340 - 3.29264 \ln x$. The coefficient of determination r^2 for the logarithmic regression is larger than the coefficient of determination for the exponential regression. See **Figures 7.39 and 7.40.** Thus the logarithmic regression function provides a better fit to the data than the exponential regression function. The correlation coefficients r can also be used to determine which function provides the better fit. For decreasing functions, the function with correlation coefficient closest to -1 provides the better fit.

Notice that the graph of the logarithmic function has the desired behavior to the right of the scatter plot. That is, it is a *gradually decreasing* curve, and this is the *general* behavior we would expect for future winning times in the 100-meter freestyle event. The graph of the exponential function is almost linear and is decreasing at a rapid pace, which is not what we would expect for results in an established Olympic event. See **Figure 7.41.** Thus we select the logarithmic function as our modeling function.

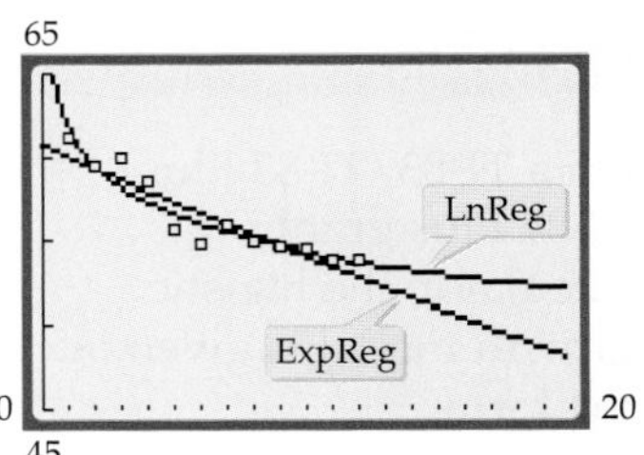

Figure 7.41

To predict the winning time for this event in the year 2012 (represented by $x = 14$), substitute 14 for x in the equation of the logarithmic function or use the VALUE command in the CALCULATE menu to produce the approximate time of 53.24 seconds, as shown in **Figure 7.42.**

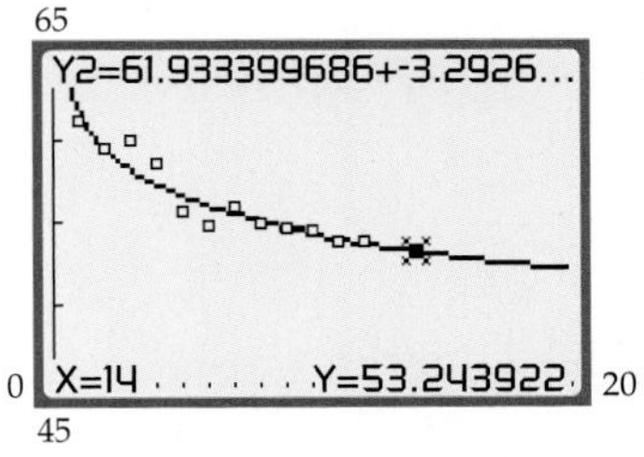

Figure 7.42

Try Exercise 24, page 531

Find a Logistic Growth Model

If a scatter plot of a set of data suggests that the data can be effectively modeled by a logistic growth model, then you can use the Logistic feature of a graphing utility to find the logistic growth model. This process is illustrated in Example 4.

EXAMPLE 4 》 Find a Logistic Growth Model

Table 7.15 shows the population of deer in an animal preserve for the years 1990 to 2004.

Table 7.15 Deer Population at the Wild West Animal Preserve

Year	Population	Year	Population	Year	Population
1990	320	1995	1150	2000	2620
1991	410	1996	1410	2001	2940
1992	560	1997	1760	2002	3100
1993	730	1998	2040	2003	3300
1994	940	1999	2310	2004	3460

Find a logistic model that approximates the deer population as a function of the year. Use the model to predict the deer population in the year 2010.

Integrating Technology

On a TI-83/TI-83 Plus/TI-84 Plus graphing calculator, the logistic growth model is given in the form

$$y = \frac{c}{1 + ae^{-bx}}$$

Think of the variable x as the time t and the variable y as $P(t)$.

Solution

1. **Construct a scatter plot of the data.** Enter the data into a graphing utility, and then use the utility to display a scatter plot of the data. In this example we represent the year 1990 by $x = 0$, the year 2004 by $x = 14$, and the deer population by y.
   Figure 7.43 shows that the data can be closely approximated by a logistic growth model.

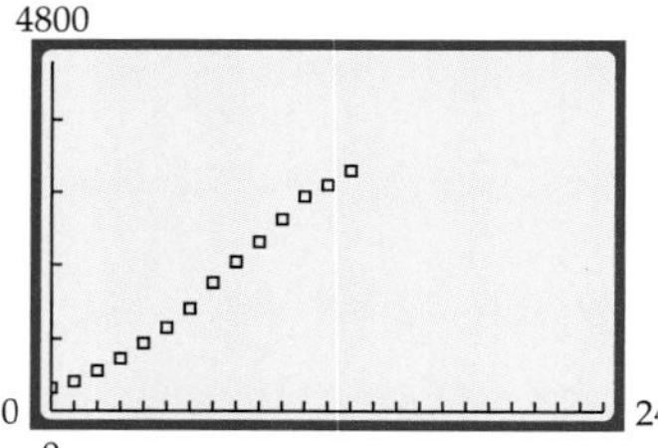

Figure 7.43

2. **Find the equation of the model.** On a TI-83/TI-83 Plus/TI-84 Plus graphing calculator, select B: Logistic, which is in the STAT CALC menu. The logistic function for the data is
   $$y \approx \frac{3965.3}{1 + 11.445e^{-0.31152x}}.$$ See **Figure 7.44**.

```
Logistic
 y=c/(1+ae^(-bx))
 a=11.44466821
 b=.3115234553
 c=3965.337214
```

Figure 7.44

3. **Examine the fit.** A TI-83/TI-83 Plus/TI-84 Plus calculator does not compute the coefficient of determination or the correlation coefficient for a logistic model. However, **Figure 7.45** shows that the logistic model provides a good fit to the data. The VALUE command in the CALCULATE menu shows that the logistic model predicts a deer population of about 3878 in the year 2010 ($x = 20$). See **Figure 7.46**.

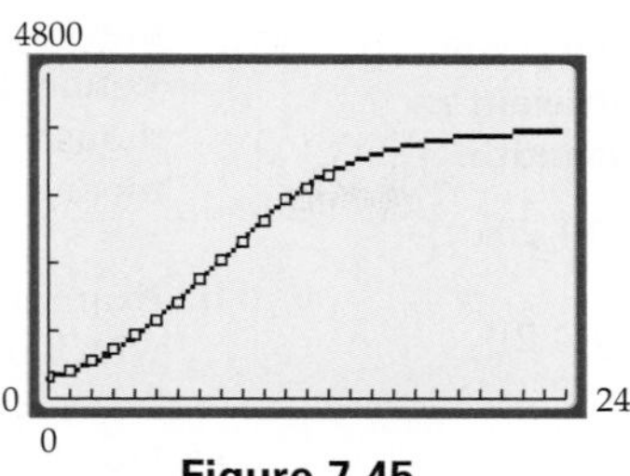

Figure 7.45

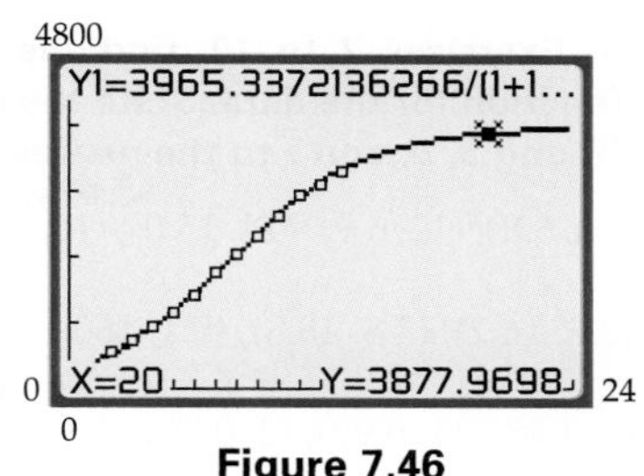

Figure 7.46

Try Exercise 26, page 532

Topics for Discussion

1. A student tries to determine the exponential regression equation for the following data.

x	1	2	3	4	5
y	8	2	0	−1.5	−2

The student's calculator displays an ERROR message. Explain why the calculator was unable to determine the exponential regression equation for the data.

2. Consider the logarithmic model $h(x) = 6 - 2 \ln x$.
   a. Is h an increasing or a decreasing function?
   b. Is h concave up or concave down on the interval $(0, \infty)$?
   c. Find, if possible, $h(0)$ and $h(e)$.
   d. Does h have a horizontal asymptote? Explain.

Exercises Set 7.6

In Exercises 1 to 6, use a scatter plot of the given data to determine which of the following types of functions might provide a suitable model of the data.

- **An increasing exponential function**
  $y = ab^x, \quad a > 0, \quad b > 1$ (See **Figure 7.31a.**)
- **An increasing logarithmic function**
  $y = a + b \ln x, \quad b > 0$ (See **Figure 7.31c.**)
- **A decreasing exponential function**
  $y = ab^x, \quad a > 0, \quad 0 < b < 1$ (See **Figure 7.31b.**)
- **A decreasing logarithmic function**
  $y = a + b \ln x, \quad b < 0$ (See **Figure 7.31d.**)

(*Note:* Some data sets can be closely modeled by more than one type of function.)

1. $\{(1, 3), (1.5, 4), (2, 6), (3, 13), (3.5, 19), (4, 27)\}$

2. $\{(1.0, 1.12), (2.1, 0.87), (3.2, 0.68), (3.5, 0.63), (4.4, 0.52)\}$

3. $\{(1, 2.4), (2, 1.1), (3, 0.5), (4, 0.2), (5, 0.1)\}$

4. $\{(5, 2.3), (7, 3.9), (9, 4.5), (12, 5.0), (16, 5.4), (21, 5.8), (26, 6.1)\}$

5. $\{(1, 2.5), (1.5, 1.7), (2, 0.7), (3, -0.5), (3.5, -1.3), (4, -1.5)\}$

6. $\{(1, 3), (1.5, 3.8), (2, 4.4), (3, 5.2), (4, 5.8), (6, 6.6)\}$

In Exercises 7 to 10, find the exponential regression function for the data. State the correlation coefficient *r*. Round *a*, *b*, and *r* to the nearest hundred thousandth.

7. {(10, 6.8), (12, 6.9), (14, 15.0), (16, 16.1), (18, 50.0), (19, 20.0)}

8. {(2.6, 16.2), (3.8, 48.8), (5.1, 160.1), (6.5, 590.2), (7, 911.2)}

9. {(0, 1.83), (1, 0.92), (2, 0.51), (3, 0.25), (4, 0.13), (5, 0.07)}

10. {(4.5, 1.92), (6.0, 1.48), (7.5, 1.14), (10.2, 0.71), (12.3, 0.49)}

In Exercises 11 to 14, find the logarithmic regression function for the data. State the correlation coefficient *r*. Round *a*, *b*, and *r* to the nearest hundred thousandth.

11. {(5, 2.7), (6, 2.5), (7.2, 2.2), (9.3, 1.9), (11.4, 1.6), (14.2, 1.3)}

12. {(11, 15.75), (14, 15.52), (17, 15.34), (20, 15.18), (23, 15.05)}

13. {(3, 16.0), (4, 16.5), (5, 16.9), (7, 17.5), (8, 17.7), (9.8, 18.1)}

14. {(8, 67.1), (10, 67.8), (12, 68.4), (14, 69.0), (16, 69.4)}

In Exercises 15 to 18, find the logistic regression function for the data. Round the constants *a*, *b*, and *c* to the nearest hundred thousandth.

15. {(0, 81), (2, 87), (6, 98), (10, 110), (15, 125)}

16. {(0, 175), (5, 195), (10, 217), (20, 264), (35, 341)}

17. {(0, 955), (10, 1266), (20, 1543), (30, 1752)}

18. {(0, 1588), (5, 2598), (10, 3638), (25, 5172)}

19. **Movie Ticket Prices** The following table shows the average U.S. movie theater ticket prices for selected years from 1994 to 2004.

Year	Price, *P*
1994	\$4.08
1996	\$4.42
1998	\$4.69
2000	\$5.39
2002	\$5.80
2004	\$6.21

Source: National Association of Theatre Owners

a. Determine an exponential regression model and a linear regression model for the data. Use $t = 0$ to represent the year 1994 and $t = 2$ to represent the year 1996. Round the constants a and b to the nearest hundred thousandth. State the correlation coefficient r for each model.

b. Examine the correlation coefficients to determine which model provides a better fit for the data.

c. Use the model you selected in **b.** to predict the average U.S. movie theater ticket price for the year 2010.

20. **Generation of Garbage** According to the U.S. Environmental Protection Agency, the amount of garbage generated per person has been increasing over the last few decades. The following table shows the per capita garbage, in pounds per day, generated in the United States.

Year	1960	1970	1980	1990	2003
Pounds per day, *p*	2.7	3.3	3.7	4.5	4.5

a. Find a linear model and a logarithmic model for the data. Use t as the independent variable (domain) and p as the dependent variable (range). Represent the year 1960 by $t = 60$ and the year 1970 by $t = 70$.

b. Examine the correlation coefficients to determine which model provides a better fit for the data.

c. Use the model you selected in **b.** to predict the amount of garbage that will be generated per capita per day in 2009. Round to the nearest tenth of a pound.

21. **Hypothermia** The following table shows the time T, in hours, before a scuba diver wearing a 3-millimeter-thick wet suit reaches hypothermia (95°F) for various water temperatures F, in degrees Fahrenheit.

Water temperature, °F	Time *T*, hours
41	1.1
46	1.4
50	1.8
59	3.7

a. Find an exponential regression model for the data. Round the constants a and b to the nearest hundred thousandth.

b. Use the model from **a.** to estimate the time it takes for the diver to reach hypothermia in water that has a temperature of 65°F. Round to the nearest tenth of an hour.

22. **ATMOSPHERIC PRESSURE** The following table shows the earth's atmospheric pressure y (in newtons per square centimeter) at an altitude of x kilometers. Find a suitable function that models the atmospheric pressure as a function of the altitude. Use the function to estimate the atmospheric pressure at an altitude of 24 kilometers. Round to the nearest tenth of a newton per square centimeter.

Altitude x, kilometers	Pressure y, newtons/cm^2
0	10.3
2	8.0
4	6.4
6	5.1
8	4.0
10	3.2
12	2.5
14	2.0
16	1.6
18	1.3

23. **HYPOTHERMIA** The following table shows the time T, in hours, before a scuba diver wearing a 4-millimeter-thick wet suit reaches hypothermia (95° E) for various water temperatures F, in degress Fahrenheit.

Water temperature, °F	Time T, hours
41	1.5
46	1.9
50	2.4
59	5.2

a. Find an exponential regression model for the data. Round the constants a and b to the nearest hundred thousandth.

b. Use the model from **a.** to estimate the time it takes for the diver to reach hypothermia in water that has a temperature of 65°F. Round to the nearest tenth of an hour. How much greater is this result compared with the answer to Exercise 21**b.**?

24. **400-METER RACE** The following table lists the progression of world record times in the men's 400-meter race for the years from 1948 to 2005. (*Note*: No new world record times were set during the time period from 2000 to 2005.)

World Record Times in the Men's 400-Meter Race, 1948 to 2005

Year	Time, in seconds	Year	Time, in seconds
1948	45.9	1964	44.9
1950	45.8	1967	44.5
1955	45.4	1968	44.1
1956	45.2	1968	43.86
1960	44.9	1988	43.29
1963	44.9	1999	43.18

Source: Track and Field Statistics, http://trackfield.brinkster.net/Main.asp.

a. Determine whether the data can best be modeled by a decreasing exponential function or a decreasing logarithmic function. Let $x = 48$ represent the year 1948 and $x = 50$ represent the year 1950.

b. Assume a new world record time will be established in 2008. Use the function you chose in **a.** to predict the world record time in the men's 400-meter race for the year 2008. Round to the nearest hundredth of a second.

25. **TELECOMMUTING** The graph below shows the projected growth in the number of telecommuters.

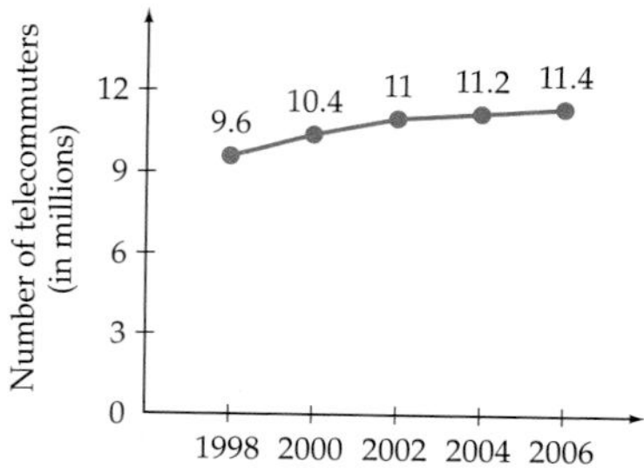

Which type of model, an increasing exponential model or an increasing logarithmic model, is more likely to provide a better fit for the data? Explain.

26.   **POPULATION OF HAWAII** The following table shows the population of the state of Hawaii for selected years from 1950 to 2004.

Population of the State of Hawaii

Year	Population, P
1950	499,000
1955	529,000
1960	642,000
1965	704,000
1970	762,920
1975	875,052
1980	967,710
1985	1,039,698
1990	1,113,491
1995	1,196,854
2000	1,212,670
2001	1,227,024
2002	1,244,898
2004	1,262,840

Source: economagic.com, http://www.economagic.com/em-cgi/data.exe/beapi/a15300.

a. Find a logistic growth model that approximates the population of the state of Hawaii as a function of the year. Use $t = 0$ to represent the year 1950.

b. Use the model from **a.** to predict the population of the state of Hawaii for the year 2010. Round to the nearest ten thousand.

c. What is the carrying capacity of the model? Round to the nearest thousand.

27. **OPTOMETRY** The *near point* p of a person is the closest distance at which the person can see an object distinctly. As one grows older, one's near point increases. The table below shows data for the average near point of various people with normal eyesight.

Age y, years	Near point p (cm)
15	11
20	13
25	15
30	17
35	20
40	23
50	26

a. Find an exponential regression model for these data. Round each constant to the nearest thousandth.

b. What near point does this equation predict for a person 60 years old? Round to the nearest centimeter.

28. **CHEMISTRY** The amount of oxygen x, in milliliters per liter, that can be absorbed by water at a certain temperature T, in degrees Fahrenheit, is given in the following table.

Temperature, °F	Oxygen absorbed, ml/L
32	10.5
38	8.4
46	7.6
52	7.1
58	6.8
64	6.5

a. Find a logarithmic regression equation for these data. Round each constant to the nearest thousandth.

b. Using your model, how much oxygen, to the nearest tenth of a milliliter per liter, can be absorbed in water that is 50°F?

29. **THE HENDERSON-HASSELBACH FUNCTION** The scientists Henderson and Hasselbach determined that the pH of blood is a function of the ratio q of the amounts of bicarbonate and carbonic acid in the blood.

a. Determine a linear model and a logarithmic model for the data. Use q as the independent variable (domain) and pH as the dependent variable (range). State the correlation coefficient for each model. Round a and b to five decimal places and r to six decimal places. Which model provides the better fit for the data?

q	7.9	12.6	31.6	50.1	79.4
pH	7.0	7.2	7.6	7.8	8.0

b. Use the model you chose in **a.** to find the q-value associated with a pH of 8.2. Round to the nearest tenth.

30. **WORLD POPULATION** The following table lists the years in which the world's population first reached 3, 4, 5, and 6 billion.

World Population Milestones

Year	Population
1960	3 billion
1974	4 billion
1987	5 billion
1999	6 billion

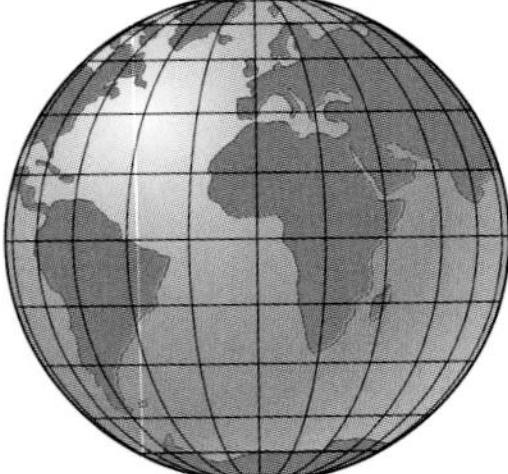

Source: The World Almanac 2006, p. 851.

a. Find an exponential model for the data in the table. Let $x = 0$ represent the year 1960.

b. Use the model to predict the year in which the world's population will first reach 8 billion.

31. **PANDA POPULATION** One estimate gives the world panda population as 3200 in 1980 and 590 in 2000.

a. Find an exponential model for the data and use the model to predict the year in which the panda population p will be reduced to 200. (Let $t = 0$ represent the year 1980.)

b. Because the exponential model in **a.** fits the data perfectly, does this mean that the model will accurately predict future panda populations? Explain.

32. **OLYMPIC HIGH JUMP** The following table shows the Olympic gold medal heights for the women's high jump from 1968 to 2004.

Women's Olympic High Jump, 1968 to 2004

Year	Height	Year	Height
1968	5 ft 11.75 in.	1988	6 ft 8 in.
1972	6 ft 3.65 in.	1992	6 ft 7.5 in.
1976	6 ft 4 in.	1996	6 ft 8.75 in.
1980	6 ft 5.5 in	2000	6 ft 7 in.
1984	6 ft 7.5 in.	2004	6 ft 9.1 in.

Source: Time Almanac 2006.

a. Determine a linear model and a logarithmic model for the data, with the height measured in inches. State the correlation coefficient r for each model. Represent the year 1968 by $x = 1$, the year 1972 by $x = 2$, and the year 2004 by $x = 10$.

b. Examine the correlation coefficients to determine which model provides a better fit for the data.

c. Use the model you selected in **b.** to predict the women's Olympic gold medal high jump height in 2012. Round to the nearest tenth of an inch.

33. **NUMBER OF CINEMA SITES** The following table shows the number of U.S. indoor cinema sites for the years 1999 to 2004.

Year	Number of Indoor Cinema Sites, S
1999	7031
2000	6550
2001	5813
2002	5712
2003	5700
2004	5629

Source: National Association of Theatre Owners

a. Determine an exponential regression model and a logarithmic regression model for the data. Use $t = 1$ to represent the year 1999. Round the constants a and b to the nearest hundred thousandth. State the correlation coefficient r for each model.

b. Examine the correlation coefficients to determine which model provides a better fit for the data.

c. Use the model you selected in **b.** to predict the number of U.S. indoor cinema sites for the year 2009.

34. **TEMPERATURE OF COFFEE** A cup of coffee is placed in a room that maintains a constant temperature of 70°F. The following table shows both the coffee temperature T after t minutes and the difference between the coffee temperature and the room temperature after t minutes.

Time t (minutes)	0	5	10	15	20	25
Coffee temp. T (°F)	165°	140°	121°	107°	97°	89°
$T - 70°$	95°	70°	51°	37°	27°	19°

a. Find an exponential model for the difference $T - 70°$ as a function of t.

b. Use the model in **a.** to predict how long it will take (to the nearest minute) for the coffee to cool to 80°F.

35.  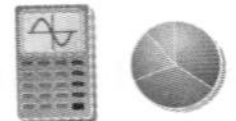 **World Population** The following table lists the years in which the world's population first reached 3, 4, 5, and 6 billion.

World Population Milestones

Year	Population
1960	3 billion
1974	4 billion
1987	5 billion
1999	6 billion

Source: The World Almanac 2006, p. 851.

a. Find a logistic growth model $P(t)$ for the data in the table. Let t represent the number of years after 1960 ($t = 0$ represents the year 1960).

b. According to the logistic growth model, what will the world's population approach as $t \to \infty$? Round to the nearest billion.

36. **A Correlation Coefficient of 1** A scientist uses a graphing calculator to model the data set $\{(2, 5), (4, 6)\}$ with a logarithmic function. The following display shows the results.

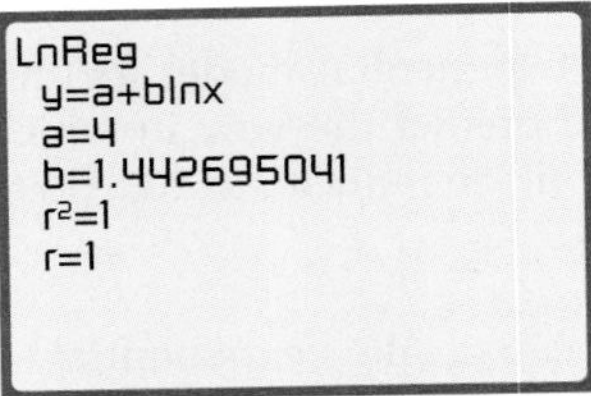

What is the significance of the fact that the correlation coefficient for the regression equation is $r = 1$?

Connecting Concepts

37. **Duplicate Data Points** An engineer needs to model the data in set A with an exponential function.

$$A = \{(2, 5), (3, 10), (4, 17), (4, 17), (5, 28)\}$$

Because the ordered pair (4, 17) is listed twice, the engineer decides to eliminate one of these ordered pairs and model the data in set B.

$$B = \{(2, 5), (3, 10), (4, 17), (5, 28)\}$$

Determine whether A and B both have the same exponential regression function.

38.  **Domain Error** A scientist needs to model the data in set A.

$$A = \{(0, 1.2), (1, 2.3), (2, 2.8), (3, 3.1), (4, 3.3), (5, 3.4)\}$$

The scientist views a scatter plot of the data and decides to model the data with a logarithmic function of the form $y = a + b \ln x$.

a. When the scientist attempts to use a graphing calculator to determine the logarithmic regression equation, the calculator displays the message

"ERR:DOMAIN"

Explain why the calculator was unable to determine the logarithmic regression equation for the data.

b. Explain what the scientist could do so that the data in set A could be modeled by a logarithmic function of the form $y = a + b \ln x$.

39. **Power Functions** A function that can be written in the form $y = ax^b$ is said to be a **power function.** Some data sets can best be modeled by a power function. On a TI-83/TI-83 Plus/TI-84 Plus, the PwrReg instruction is used to produce a power regression function for a set of data.

a. Find an exponential regression function and a power regression function for the following data. State the correlation coefficient r for each model.

x	1	2	3	4	5	6
y	2.1	5.5	9.8	14.6	20.1	25.8

b. Which of the two regression functions provides the better fit for the data?

40. **Period of a Pendulum** The following table shows the time t (in seconds) of the period of a pendulum of length l (in feet). (*Note*: The period of a pendulum is the time it takes the pendulum to complete a swing from the right to the left and back.)

Length l	1	2	3	4	6	8
Time t	1.11	1.57	1.92	2.25	2.72	3.14

a. Determine the equation of the best model for the data. Your model must be a power function or an exponential function.

b. According to the model you chose in **a.** what is the length of a pendulum, to the nearest tenth of a foot, that has a period of 12 seconds?

Projects

1.  **A MODELING PROJECT** The purpose of this Project is for you to find data that can be modeled by an exponential or a logarithmic function. Choose data from a *real-life* situation that you find interesting. Search for the data in a magazine, in a newspaper, in an almanac, or on the Internet. If you wish, you can collect your data by performing an experiment.

   a. List the source of your data. Include the date, page number, and any other specifics about the source. If your data were collected by performing an experiment, then provide all the details of the experiment.

   b. Explain what you have chosen as your variables. Which variable is the dependent variable and which variable is the independent variable?

   c. Use the three-step modeling process to find a regression equation that models the data.

   d. Graph the regression equation on the scatter plot of the data. What is the correlation coefficient for the model? Do you think your regression equation accurately models your data? Explain.

   e. Use the regression equation to predict the value of
   - the dependent variable for a specific value of the independent variable
   - the independent variable for a specific value of the dependent variable

   f. Write a few comments about what you have learned from this Project.

Exploring Concepts with Technology

Table 7.16

T	V
90	700
100	500
110	350
120	250
130	190
140	150
150	120

Using a Semilog Graph to Model Exponential Decay

Consider the data in **Table 7.16**, which shows the viscosity V of SAE 40 motor oil at various temperatures T in degrees Fahrenheit. The graph of these data is shown below, along with a curve that passes through the points. The graph in **Figure 7.47** appears to have the shape of an exponential decay model.

One way to determine whether the graph in **Figure 7.47** is the graph of an exponential function is to plot the data on *semilog* graph paper. On this graph paper, the horizontal axis remains the same, but the vertical axis uses a logarithmic scale.

The data in **Table 7.16** are graphed again in **Figure 7.48**, but this time the vertical axis is a natural logarithm axis. This graph is approximately a straight line.

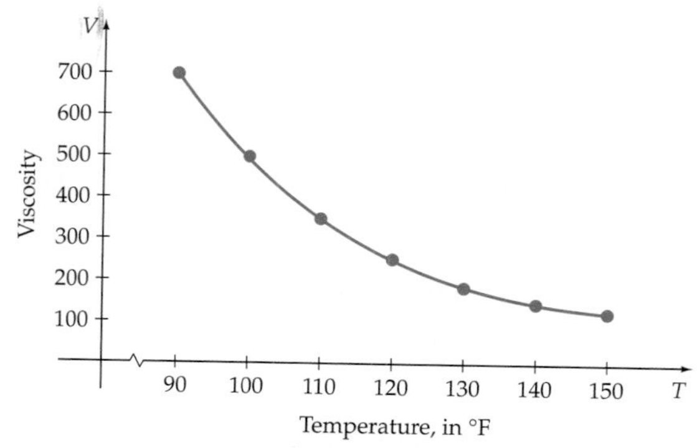

Figure 7.47

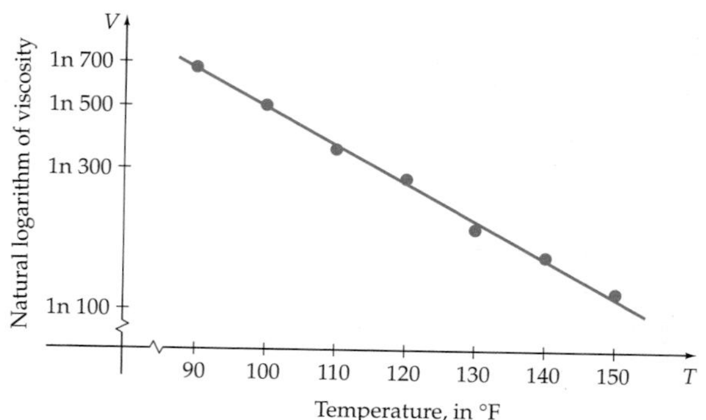

Figure 7.48

The slope of the line in **Figure 7.48**, to the nearest ten-thousandth, is

$$m = \frac{\ln 500 - \ln 120}{100 - 150} \approx -0.0285$$

Using this slope and the point-slope formula with V replaced by $\ln V$, we have

$$\ln V - \ln 120 = -0.0285(T - 150)$$
$$\ln V \approx -0.0285T + 9.062 \qquad (1)$$

Equation (1) is the equation of the line on a semilog coordinate grid.

Now solve Equation (1) for V.

$$e^{\ln V} = e^{-0.0285T+9.062}$$
$$V = e^{-0.0285T}e^{9.062}$$
$$V \approx 8621e^{-0.0285T} \qquad (2)$$

Equation (2) is a model of the data in the rectangular coordinate system shown in **Figure 7.47.**

1. A chemist wishes to determine the decay characteristics of iodine-131. A 100-milligram sample of iodine-131 is observed over a 30-day period. **Table 7.17** shows the amount A (in milligrams) of iodine-131 remaining after t days.

Table 7.17

t	A
1	91.77
4	70.92
8	50.30
15	27.57
20	17.95
30	7.60

   a. Graph the ordered pairs (t, A) on semilog paper. (*Note:* Semilog paper comes in different varieties. Our calculations are based on semilog paper that has a natural logarithm scale on the vertical axis.)

   b. Use the points (4, 4.3) and (15, 3.3) to approximate the slope of the line that passes through the points.

   c. Using the slope calculated in **b.** and the point (4, 4.3), determine the equation of the line.

   d. Solve the equation you derived in **c.** for A.

   e. Graph the equation you derived in **d.** in a rectangular coordinate system.

   f. What is the half-life of iodine-131?

2. The birth rates B per thousand people in the United States are given in **Table 7.18** for the years 1986 through 1990 ($t = 0$ corresponds to 1986).

Table 7.18

t	B
0	15.5
1	15.7
2	15.9
3	16.2
4	16.7

   a. Graph the ordered pairs $(t, \ln B)$. (You will need to adjust the scale so that you can discriminate between plotted points. A suggestion is given in **Figure 7.49.**)

   b. Use the points (1, 2.754) and (3, 2.785) to approximate the slope of the line that passes through the points.

   c. Using the slope calculated in **b.** and the point (1, 2.754), determine the equation of the line.

   d. Solve the equation you derived in **c.** for B.

   e. Graph the equation you derived in **d.** in a rectangular coordinate system.

   f. If the birth rate continues as predicted by your model, in what year will the birth rate be 17.5 per thousand?

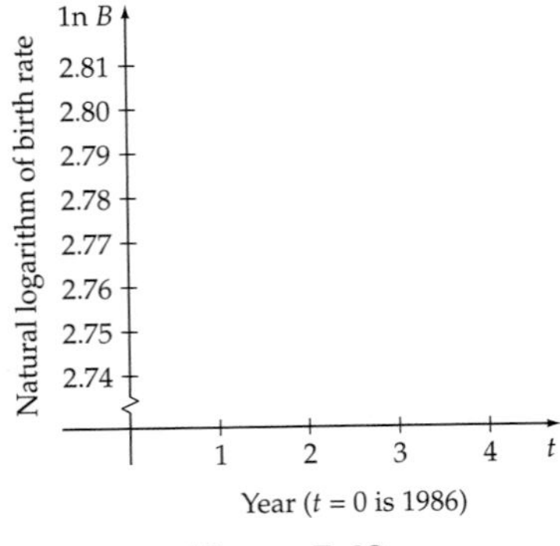

Figure 7.49

The difference in graphing strategies between Exercise 1 and Exercise 2 is that in Exercise 1, semilog paper was used. When a point is graphed on this coordinate paper, the y-coordinate is $\ln y$. In Exercise 2, graphing a point $(x, \ln y)$ in a rectangular coordinate system has the same effect as graphing (x, y) in a semilog coordinate system.

Chapter 7 Summary

7.1 Exponential Functions and Their Applications

- For all positive real numbers $b, b \neq 1$, the exponential function defined by $f(x) = b^x$ has the following properties:
  1. f has the set of real numbers as its domain.
  2. f has the set of positive real numbers as its range.
  3. f has a graph with a y-intercept of $(0, 1)$.
  4. f has a graph that is asymptotic to the x-axis.
  5. f is a one-to-one function.
  6. f is an increasing function if $b > 1$.
  7. f is a decreasing function if $0 < b < 1$.
- As n increases without bound, $(1 + 1/n)^n$ approaches an irrational number denoted by e. The value of e accurate to eight decimal places is 2.71828183.
- The function defined by $f(x) = e^x$ is called the natural exponential function.

7.2 Logarithmic Functions and Their Applications

- *Definition of a Logarithm* If $x > 0$ and b is a positive constant $(b \neq 1)$, then

$$y = \log_b x \quad \text{if and only if} \quad b^y = x$$

- For all positive real numbers $b, b \neq 1$, the function defined by $f(x) = \log_b x$ has the following properties:
  1. f has the set of positive real numbers as its domain.
  2. f has the set of real numbers as its range.
  3. f has a graph with an x-intercept of $(1, 0)$.
  4. f has a graph that is asymptotic to the y-axis.
  5. f is a one-to-one function.
  6. f is an increasing function if $b > 1$.
  7. f is a decreasing function if $0 < b < 1$.
- The exponential form of $y = \log_b x$ is $b^y = x$.
- The logarithmic form of $b^y = x$ is $y = \log_b x$.
- *Basic Logarithmic Properties*
  1. $\log_b b = 1$
  2. $\log_b 1 = 0$
  3. $\log_b (b^x) = x$
  4. $b^{\log_b x} = x$
- The function $f(x) = \log_{10} x$ is the common logarithmic function. It is customarily written as $f(x) = \log x$.
- The function $f(x) = \log_e x$ is the natural logarithmic function. It is customarily written as $f(x) = \ln x$.

7.3 Properties of Logarithms and Logarithmic Scales

- If b, M, and N are positive real numbers $(b \neq 1)$, and p is any real number, then

$$\log_b(MN) = \log_b M + \log_b N$$
$$\log_b \frac{M}{N} = \log_b M - \log_b N$$
$$\log_b(M^p) = p \log_b M$$
$$\log_b M = \log_b N \quad \text{implies} \quad M = N$$
$$M = N \quad \text{implies} \quad \log_b M = \log_b N$$

- *Change-of-Base Formula* If x, a, and b are positive real numbers with $a \neq 1$ and $b \neq 1$, then

$$\log_b x = \frac{\log_a x}{\log_a b}$$

- An earthquake with an intensity of I has a Richter scale magnitude of $M = \log\left(\frac{I}{I_0}\right)$, where I_0 is the measure of the intensity of a zero-level earthquake.
- The pH of a solution with a hydronium-ion concentration of H^+ mole per liter is given by $\text{pH} = -\log[H^+]$.

7.4 Exponential and Logarithmic Equations

- *Equality of Exponents Theorem* If b is a positive real number $(b \neq 1)$ such that $b^x = b^y$, then $x = y$.
- Exponential equations of the form $b^x = b^y$ can be solved by using the Equality of Exponents Theorem.
- Exponential equations of the form $b^x = c$ can be solved by taking either the common logarithm or the natural logarithm of each side of the equation.
- Logarithmic equations can often be solved by using the properties of logarithms and the definition of a logarithm.

7.5 Exponential Growth and Decay

- The function defined by $N(t) = N_0 e^{kt}$ is called an exponential growth function if k is a positive constant, and it is called an exponential decay function if k is a negative constant.
- *The Compound Interest Formula* A principal P invested at an annual interest rate r, expressed as a decimal and compounded n times per year for t years, produces the balance

$$A = P\left(1 + \frac{r}{n}\right)^{nt}$$

- *Continuous Compounding Interest Formula* If an account with principal P and annual interest rate r is compounded continuously for t years, then the balance is $A = Pe^{rt}$.
- *The Logistic Model* The magnitude of a population at time t is given by

$$P(t) = \frac{c}{1 + ae^{-bt}}$$

where $P_0 = P(0)$ is the population at time $t = 0$, c is the carrying capacity of the population, and b is a constant called the growth rate constant.

7.6 Modeling Data with Exponential and Logarithmic Functions

- If the graph of f lies above all of its tangents on $[x_1, x_2]$, then f is concave upward on $[x_1, x_2]$.
- If the graph of f lies below all of its tangents on $[x_1, x_2]$, then f is concave downward on $[x_1, x_2]$.
- *The Modeling Process* Use a graphing utility to
  1. construct a scatter plot of the data to determine which type of function will best model the data.
  2. find the equation of the modeling function and the correlation coefficient or the coefficient of determination for the equation.
  3. examine the correlation coefficient or the coefficient of determination and view a graph that displays both the function and the scatter plot to determine how well the function fits the data.

Chapter 7 Assessing Concepts

In Exercises 1 to 5, determine whether the statement is true or false. If the statement is false, give an example or state a reason to demonstrate that the statement is false.

1. Every function has an inverse function.
2. If $7^x = 40$, then $\log_7 40 = x$.
3. If $\log_4 x = 3.1$, then $4^{3.1} = x$.
4. The exponential function $h(x) = b^x$ is an increasing function.
5. The logarithmic function $j(x) = \log_b x$ is an increasing function.

In Exercises 6 to 12, match an expression to each description. A letter may be used more than once. Some letters may not be needed.

6. A function which is symmetric with respect to the y-axis. ______
7. An increasing function which is symmetric with respect to the origin. ______
8. A logistic function. ______
9. An increasing function with a vertical asymptote. ______
10. A decreasing function with a horizonal asymptote. ______
11. A decreasing function with a vertical asymptote. ______
12. An increasing function which is concave upward. ______

a. $f(x) = \ln(x - 4)$

b. $f(x) = \dfrac{e^x - e^{-x}}{2}$

c. $f(x) = \dfrac{e^x + e^{-x}}{2}$

d. $f(x) = \log_{1/2} x$

e. $f(x) = \dfrac{1}{2}e^{-x}$

f. $f(x) = \dfrac{485}{1 + 9.5e^{-0.019x}}$, $x \geq 0$

g. $f(x) = e^x$

Chapter 7 Review Exercises

In Exercises 1 to 12, solve each equation. Do not use a calculator.

1. $\log_5 25 = x$ **2.** $\log_3 81 = x$ **3.** $\ln e^3 = x$

4. $\ln e^{\pi} = x$ **5.** $3^{2x+7} = 27$ **6.** $5^{x-4} = 625$

7. $2^x = \frac{1}{8}$ **8.** $27(3^x) = 3^{-1}$ **9.** $\log x^2 = 6$

10. $\frac{1}{2}\log |x| = 5$ **11.** $10^{\log 2x} = 14$ **12.** $e^{\ln x^2} = 64$

In Exercises 13 to 22, sketch the graph of each function.

13. $f(x) = (2.5)^x$ **14.** $f(x) = \left(\frac{1}{4}\right)^x$

15. $f(x) = 3^{|x|}$ **16.** $f(x) = 4^{-|x|}$

17. $f(x) = 2^x - 3$ **18.** $f(x) = 2^{(x-3)}$

19. $f(x) = \frac{1}{3}\log x$ **20.** $f(x) = 3\log x^{1/3}$

21. $f(x) = -\frac{1}{2}\ln x$ **22.** $f(x) = -\ln |x|$

In Exercises 23 and 24, use a graphing utility to graph each function.

23. $f(x) = \dfrac{4^x + 4^{-x}}{2}$ **24.** $f(x) = \dfrac{3^x - 3^{-x}}{2}$

In Exercises 25 to 28, change each logarithmic equation to its exponential form.

25. $\log_4 64 = 3$ **26.** $\log_{1/2} 8 = -3$

27. $\log_{\sqrt{2}} 4 = 4$ **28.** $\ln 1 = 0$

In Exercises 29 to 32, change each exponential equation to its logarithmic form.

29. $5^3 = 125$ **30.** $2^{10} = 1024$

31. $10^0 = 1$ **32.** $8^{1/2} = 2\sqrt{2}$

In Exercises 33 to 36, expand the given logarithmic expression.

33. $\log_b \dfrac{x^2y^3}{z}$ **34.** $\log_b \dfrac{\sqrt{x}}{y^2z}$

35. $\ln xy^3$ **36.** $\ln \dfrac{\sqrt{xy}}{z^4}$

In Exercises 37 to 40, write each logarithmic expression as a single logarithm with a coefficient of 1.

37. $2\log x + \frac{1}{3}\log(x+1)$ **38.** $5\log x - 2\log(x+5)$

39. $\frac{1}{2}\ln 2xy - 3\ln z$ **40.** $\ln x - (\ln y - \ln z)$

In Exercises 41 to 44, use the change-of-base formula and a calculator to approximate each logarithm accurate to six significant digits.

41. $\log_5 101$ **42.** $\log_3 40$

43. $\log_4 0.85$ **44.** $\log_8 0.3$

In Exercises 45 to 60, solve each equation for *x*. Give exact answers. Do not use a calculator.

45. $4^x = 30$ **46.** $5^{x+1} = 41$

47. $\ln 3x - \ln(x-1) = \ln 4$ **48.** $\ln 3x + \ln 2 = 1$

49. $e^{\ln(x+2)} = 6$ **50.** $10^{\log(2x+1)} = 31$

51. $\dfrac{4^x + 4^{-x}}{4^x - 4^{-x}} = 2$ **52.** $\dfrac{5^x + 5^{-x}}{2} = 8$

53. $\log(\log x) = 3$ **54.** $\ln(\ln x) = 2$

55. $\log \sqrt{x-5} = 3$ **56.** $\log x + \log(x-15) = 1$

57. $\log_4(\log_3 x) = 1$ **58.** $\log_7(\log_5 x^2) = 0$

59. $\log_5 x^3 = \log_5 16x$ **60.** $25 = 16^{\log_4 x}$

61. EARTHQUAKE MAGNITUDE Determine, to the nearest 0.1, the Richter scale magnitude of an earthquake with an intensity of $I = 51{,}782{,}000I_0$.

62. **Earthquake Magnitude** A seismogram has an amplitude of 18 millimeters and the difference in time between the s-wave and the p-wave is 21 seconds. Find, to the nearest tenth, the Richter scale magnitude of the earthquake that produced the seismogram.

63. **Comparison of Earthquakes** An earthquake had a Richter scale magnitude of 7.2. Its aftershock had a Richter scale magnitude of 3.7. Compare the intensity of the earthquake to the intensity of the aftershock by finding, to the nearest unit, the ratio of the larger intensity to the smaller intensity.

64. **Comparison of Earthquakes** An earthquake has an intensity 600 times the intensity of a second earthquake. Find, to the nearest tenth, the difference between the Richter scale magnitudes of the earthquakes.

65. **Chemistry** Find the pH of tomatoes that have a hydronium-ion concentration of 6.28×10^{-5}. Round to the nearest tenth.

66. **Chemistry** Find the hydronium-ion concentration of rainwater that has a pH of 5.4.

67. **Compound Interest** Find the balance when $16,000 is invested at an annual rate of 8% for 3 years if the interest is compounded

a. monthly

b. continuously

68. **Compound Interest** Find the balance when $19,000 is invested at an annual rate of 6% for 5 years if the interest is compounded

a. daily

b. continuously

69. **Depreciation** The scrap value S of a product with an expected life span of n years is given by $S(n) = P(1 - r)^n$, where P is the original purchase price of the product and r is the annual rate of depreciation. A taxicab is purchased for $12,400 and is expected to last 3 years. What is its scrap value if it depreciates at a rate of 29% per year?

70. **Medicine** A skin wound heals according to the function given by $N(t) = N_0 e^{-0.12t}$, where N is the number of square centimeters of unhealed skin t days after the injury, and N_0 is the number of square centimeters covered by the original wound.

a. What percentage of the wound will be healed after 10 days?

b. How many days, to the nearest day, will it take for 50% of the wound to heal?

c. How long, to the nearest day, will it take for 90% of the wound to heal?

In Exercises 71 to 74, find the exponential growth or decay function $N(t) = N_0 e^{kt}$ that satisfies the given conditions.

71. $N(0) = 1, N(2) = 5$

72. $N(0) = 2, N(3) = 11$

73. $N(1) = 4, N(5) = 5$

74. $N(-1) = 2, N(0) = 1$

75. **Population Growth**

a. Find the exponential growth function for a city whose population was 25,200 in 2005 and 26,800 in 2006. Use $t = 0$ to represent the year 2005.

b. Use the growth function to predict, to the nearest hundred, the population of the city in 2012.

76. **Carbon Dating** Determine, to the nearest ten years, the age of a bone if it now contains 96% of its original amount of carbon-14. The half-life of carbon-14 is 5730 years.

77. **Cellular Telephone Subscribership** The following table shows the number of U.S. cellular telephone subscriptions, in thousands, for selected years from 1990 to 2004.

Year	Number of Cellular Telephone Subscriptions (in thousands)
1990	5283
1992	11,033
1994	24,134
1996	44,043
1998	69,209
2000	109,478
2002	140,767
2004	182,140

Source: The World Almanac 2006

Find the equation of the mathematical model that you believe will most accurately predict the number of U.S. cellular telephone subscriptions for the year 2008. Explain the reasoning you used to select your model.

78. **Mortality Rate** The following table shows the infant mortality rate in the United States for selected years from 1960 to 2003. (*Source: The World Almanac 2006.*)

U.S. Infant Mortality Rate, 1960–2003 (per 1000 live births)

Year	Rate, R
1960	26.0
1970	20.0
1980	12.6
1990	9.2
1995	7.6
2000	6.9
2001	6.8
2002	7.0
2003	6.9

a. Find an exponential model and a logarithmic model for the infant mortality rate, R, as a function of the year. Represent the year 1960 by $t = 60$.

b. Examine the correlation coefficients of the regression models to determine which model provides the better fit.

c. Use the model you selected in **b.** to predict, to the nearest 0.1, the infant mortality rate in 2008.

79. **Logistic Growth** The population of coyotes in a national park satisfies the logistic model with $P_0 = 210$ in 1997, $c = 1400$, and $P(3) = 360$ (the population in 2000).

a. Determine the logistic model.

b. Use the model to predict, to the nearest 10, the coyote population in 2010.

80. Consider the logistic function

$$P(t) = \frac{128}{1 + 5e^{-0.27t}}$$

a. Find P_0.

b. What does $P(t)$ approach as $t \to \infty$?

»»» Quantitative Reasoning: *Sales* »»»

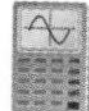

Digital Camera Sales The following table shows the worldwide sales, in millions, of digital cameras for the years 1999 to 2003. (*Source: Digital Photography Review*)

Year	1999	2000	2001	2002	2003
Worldwide Sales of Digital Cameras (in millions)	5.5	11.0	18.5	30.5	50.0

QR1. Find an exponential model and a logistic model for the data. Let $t = 0$ represent the year 1999. Round constants to the nearest hundred thousandth.

QR2. Use each of the models to predict the number of sales of digital cameras for the year 2009. Round to the nearest tenth of a million.

A business analyst thinks that the digital camera market has started to reach its saturation point. The analyst predicts that during the 2004 to 2007 period, the sales of digital cameras will be as shown in the following table.

Year	2004	2005	2006	2007
Projected Worldwide Sales of Digital Cameras (in millions)	59.3	69.2	77.2	82.5

QR3. Find a logarithmic model for the data in the above table. Let $t = 1$ represent the year 2004. Round constants to the nearest hundred thousandth.

QR4. Use the model from Exercise QR3 to predict the number of sales of digital cameras for the year 2009. Round to the nearest tenth of a million.

QR5. Find a logistic model for all the data by combining the tables. Let $t = 0$ represent the year 1999 and $t = 5$ represent the year 2004. Use this model to predict the number of sales for the year 2009.

QR6. The answers in Exercises QR2, QR4, and QR5 provide us with four predictions for the number of digital camera sales in the year 2009. Which of these predicted values do you think is the most realistic?

Chapter 7 Test

1. a. Write $\log_b(5x - 3) = c$ in exponential form.

   b. Write $3^{x/2} = y$ in logarithmic form.

2. Expand $\log_b \dfrac{z^2}{y^3\sqrt{x}}$.

3. Write $\log(2x + 3) - 3\log(x - 2)$ as a single logarithm with a coefficient of 1.

4. Use the change-of-base formula and a calculator to approximate $\log_4 12$. Round your result to the nearest ten thousandth.

5. Graph: $f(x) = 3^{-x/2}$

6. Graph: $f(x) = -\ln(x + 1)$

7. Solve: $5^x = 22$. Round your solution to the nearest ten thousandth.

8. Find the *exact* solution of $4^{5-x} = 7^x$.

9. Solve: $\log(x + 99) - \log(3x - 2) = 2$

10. Solve: $\ln(2 - x) + \ln(5 - x) = \ln(37 - x)$

11. Find the balance on \$20,000 invested at an annual interest rate of 7.8% for 5 years, compounded

    a. monthly

    b. continuously

12. COMPOUND INTEREST Find the time required for money invested at an annual rate of 4% to double in value if the investment is compounded monthly. Round to the nearest hundredth of a year.

13. EARTHQUAKE MAGNITUDE

    a. What, to the nearest tenth, will an earthquake measure on the Richter scale if it has an intensity of $I = 42{,}304{,}000I_0$?

    b. Compare the intensity of an earthquake that measures 6.3 on the Richter scale with the intensity of an earthquake that measures 4.5 on the Richter scale by finding the ratio of the larger intensity to the smaller intensity. Round to the nearest whole number.

14. a. Find the exponential growth function for a city whose population was 34,600 in 2001 and 39,800 in 2004. Use $t = 0$ to represent the year 2001.

b. Use the growth function in **a.** to predict the population of the city in 2011. Round to the nearest thousand.

15. Determine, to the nearest ten years, the age of a bone if it now contains 92% of its original amount of carbon-14. The half-life of carbon-14 is 5730 years.

16. a. Find the exponential regression function for the following data.

$\{(2.5, 16), (3.7, 48), (5.0, 155), (6.5, 571), (6.9, 896)\}$

b. Use the function to predict, to the nearest whole number, the y value associated with $x = 7.8$.

17. **WOMEN'S JAVELIN THROW** The following table shows the progression of the world record distances for the women's javelin throw from 1999 to 2005. (*Source:* http://www.athletix.org/Statistics/wrjt-women.html)

World Record Progression in the Women's Javelin Throw

Year	Distance in meters, d
1999	67.09
2000	68.22
2000	69.48
2001	71.54
2005	71.70

a. Find a logarithmic model and a logistic model for the data. Use $t = 1$ to represent the year 1999 and $t = 7$ to represent the year 2005.

b. Use each of the models from **a.** to predict the women's world record javelin throw distance for the year 2010. Round to the nearest hundredth of a meter.

18. **POPULATION GROWTH** The population of raccoons in a state park satisfies a logistic growth model with $P_0 = 160$ in 2003 and $P(1) = 190$. A park ranger has estimated the carrying capacity of the park to be 1100 raccoons.

a. Determine the logistic growth model for the raccoon population where t is the number of years after 2003.

b. Use the logistic model from **a.** to predict the raccoon population in 2010.

Cumulative Review Exercises

1. Given $f(x) = \cos x$ and $g(x) = x^2 + 1$, find $(f \circ g)(x)$.

2. Given $f(x) = 2x + 8$, find $f^{-1}(x)$.

3. For the right triangle shown at the right, find $\sin \theta$, $\cos \theta$, and $\tan \theta$.

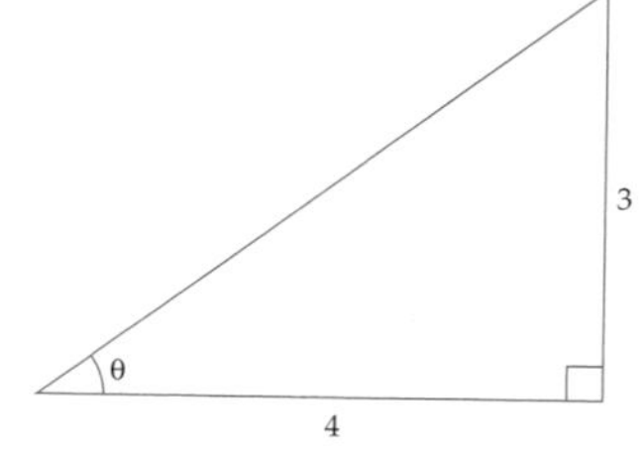

4. For the right triangle shown at the right, find a. Round to the nearest centimeter.

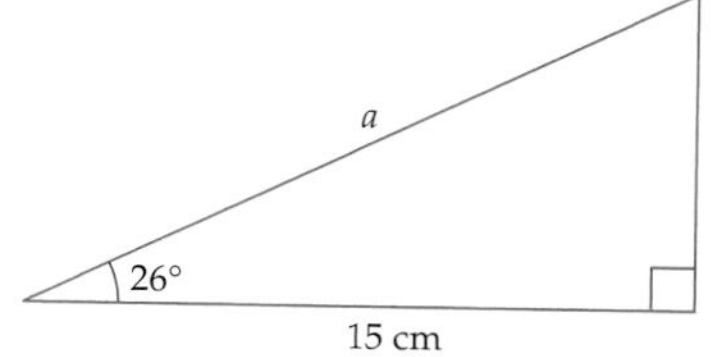

5. What are the amplitude, period, and phase shift of the graph of $y = 4\cos\left(2x - \dfrac{\pi}{2}\right)$?

6. What is the amplitude and period of the graph of $y = \sin x - \cos x$?

7. Is the sine function an even function, an odd function, or neither an even nor an odd function?

8. Verify the identity $\dfrac{1}{\sin x} - \sin x = \cos x \cot x$.

9. Evaluate $\tan\left(\sin^{-1}\left(\dfrac{12}{13}\right)\right)$.

10. Solve $2\cos^2 x + \sin x - 1 = 0$ for $0 \le x < 2\pi$.

11. Find the magnitude and direction angle for the vector $\langle -3, 4\rangle$. Round the angle to the nearest tenth of a degree.

12. Find the angle between the vectors $\mathbf{v} = \langle 2, -3\rangle$ and $\mathbf{w} = \langle -3, 4\rangle$. Round to the nearest tenth of a degree.

13. **Ground Speed and Course of a Plane** An airplane is traveling with an airspeed of 400 mph at a heading of 48°. A wind of 55 mph is blowing at a heading of 115°. Find the ground speed and the course of the plane.

14. For triangle ABC, $B = 32°$, $a = 42$ feet, and $b = 50$ feet. Find the measure of angle A. Round to the nearest degree.

15. Use De Moivre's Theorem to find $(1 - i)^8$.

16. Find the two square roots of the imaginary number i.

17. Transform the point $(1, 1)$ in a rectangular coordinate system to a point in a polar coordinate system.

18. Sketch the graph of the polar equation $r = 2 - 2\cos\theta$.

19. Solve $5^x = 10$. Round to the nearest hundredth.

20. **Radioactive Decay** The half-life of an isotope of polonium is 138 days. If an ore sample originally contained 3 milligrams of polonium, how many milligrams of polonium remain after 100 days? Round to the nearest tenth of a milligram.

Solutions to the Try Exercises

Exercise Set 1.1, page 12

6. $6(5s - 11) - 12(2s + 5) = 0$

$$30s - 66 - 24s - 60 = 0$$
$$6s - 126 = 0$$
$$s = \frac{126}{6} = 21$$

20. $3x - 5y = 15$

$-5y = -3x + 15$ • **Subtract 3x from each side.**

$y = \frac{3}{5}x - 3$ • **Divide each side by −5.**

30. $x^2 + x - 2 = 0$

$a = 1 \quad b = 1 \quad c = -2$

$$x = \frac{-1 \pm \sqrt{1^2 - 4(1)(-2)}}{2(1)}$$
$$= \frac{-1 \pm \sqrt{1 + 8}}{2} = \frac{-1 \pm 3}{2}$$
$$x = \frac{-1 + 3}{2} = 1 \quad \text{or} \quad x = \frac{-1 - 3}{2} = -2$$

The solutions are 1 and −2.

44. $12w^2 - 41w + 24 = 0$

$(4w - 3)(3w - 8) = 0$

$4w - 3 = 0$ or $3w - 8 = 0$

$w = \frac{3}{4} \qquad w = \frac{8}{3}$

The solutions are $\frac{3}{4}$ and $\frac{8}{3}$.

58. $3(x + 7) \le 5(2x - 8)$

$3x + 21 \le 10x - 40$

$-7x \le -61$

$x \ge \frac{61}{7}$

The solution set is $\left[\frac{61}{7}, \infty\right)$.

66. $12x^2 + 8x \ge 15$

$12x^2 + 8x - 15 \ge 0$

$(6x - 5)(2x + 3) \ge 0$

$x = \frac{5}{6}$ and $x = -\frac{3}{2}$ • **Critical values**

Use a test value from each of the intervals $\left(-\infty, -\frac{3}{2}\right)$, $\left(-\frac{3}{2}, \frac{5}{6}\right)$, and $\left(\frac{5}{6}, \infty\right)$ to determine where $12x^2 + 8x - 15$ is positive.

```
 ++ | - - - - - - - - | + + +
<---+---+---+---+---+---+---+--->
  -3/2        0       5/6
```

The solution set is $\left(-\infty, -\frac{3}{2}\right] \cup \left[\frac{5}{6}, \infty\right)$.

78. $|2x - 5| \ge 1$

$2x - 5 \le -1$ or $2x - 5 \ge 1$

$2x \le 4 \qquad 2x \ge 6$

$x \le 2 \qquad x \ge 3$

The solution set is $(-\infty, 2] \cup [3, \infty)$.

80. $|3 - 2x| \le 5$

$-5 \le 3 - 2x \le 5$

$-8 \le -2x \le 2$

$4 \ge x \ge -1$

The solution set is $[-1, 4]$.

Exercise Set 1.2, page 27

6. $d = \sqrt{(x_2 - x_1)^2 + (y_2 - y_1)^2}$

$$d = \sqrt{[-10 - (-5)]^2 + (14 - 8)^2}$$
$$= \sqrt{(-5)^2 + 6^2} = \sqrt{25 + 36}$$
$$= \sqrt{61}$$

26.

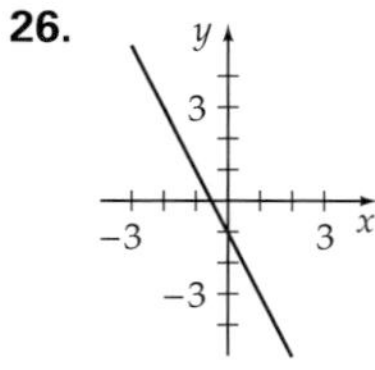

30.

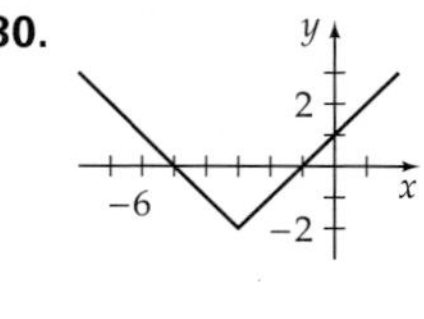

32.

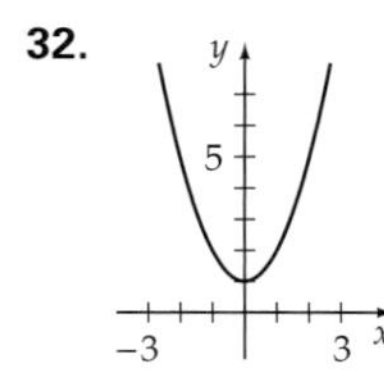

40. y-intercept: $\left(0, -\frac{15}{4}\right)$

x-intercept: $(5, 0)$

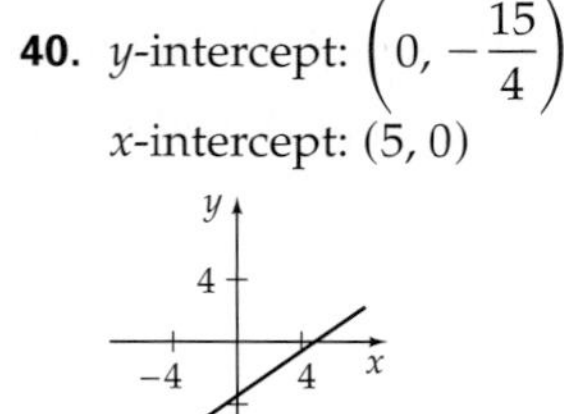

64. $r = \sqrt{(1-(-2))^2 + (7-5)^2}$
$= \sqrt{9+4} = \sqrt{13}$

Using the standard form

$(x-h)^2 + (y-k)^2 = r^2$

with $h = -2$, $k = 5$, and $r = \sqrt{13}$ yields

$(x+2)^2 + (y-5)^2 = (\sqrt{13})^2$

66.
$$x^2 + y^2 - 6x - 4y + 12 = 0$$
$$x^2 - 6x + y^2 - 4y = -12$$
$$x^2 - 6x + 9 + y^2 - 4y + 4 = -12 + 9 + 4$$
$$(x-3)^2 + (y-2)^2 = 1^2$$

center (3, 2), radius 1

Exercise Set 1.3, page 46

2. Given $g(x) = 2x^2 + 3$

a. $g(3) = 2(3)^2 + 3 = 18 + 3 = 21$

b. $g(-1) = 2(-1)^2 + 3 = 2 + 3 = 5$

c. $g(0) = 2(0)^2 + 3 = 0 + 3 = 3$

d. $g\left(\frac{1}{2}\right) = 2\left(\frac{1}{2}\right)^2 + 3 = \frac{1}{2} + 3 = \frac{7}{2}$

e. $g(c) = 2(c)^2 + 3 = 2c^2 + 3$

f. $g(c+5) = 2(c+5)^2 + 3 = 2c^2 + 20c + 50 + 3$
$= 2c^2 + 20c + 53$

10. **a.** Because $0 \le 0 \le 5$, $Q(0) = 4$.

b. Because $6 < a < 7$, $Q(e) = -a + 9$.

c. Because $1 < n < 2$, $Q(n) = 4$.

d. Because $1 < m \le 2$, $8 < m^2 + 7 \le 11$. Thus

$Q(m^2 + 7) = \sqrt{(m^2+7) - 7} = \sqrt{m^2} = m$

14. $x^2 - 2y = 2$ • **Solve for *y*.**
$-2y = -x^2 + 2$
$y = \frac{1}{2}x^2 - 1$

y is a function of x because each x value will yield one and only one y value.

28. Domain is the set of all real numbers.

40. Domain is the set of all real numbers.

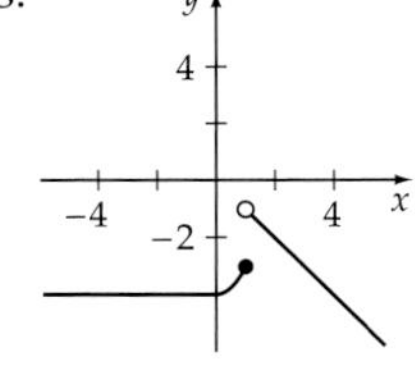

54. **a.** $[0, \infty)$

b. Since \$31,250 is between \$30,650 and \$74,200 use $T(x) = 0.25(x - 30{,}650) + 4220$. Then, $T(31{,}250) = 0.25(31{,}250 - 30{,}650) + 4220 = \$4370$.

c. Since \$78,900 is between \$74,200 and \$154,800, use $T(x) = 0.28(x - 74{,}200) + 15{,}107.50$ Then,
$T(78{,}900) = 0.28(78{,}900 - 74{,}200) + 15{,}107.50 =$ \$16,423.50.

56. **a.** This is the graph of a function. Every vertical line intersects the graph in at most one point.

b. This is not the graph of a function. Some vertical lines intersect the graph at two points.

c. This is not the graph of a function. The vertical line at $x = -2$ intersects the graph at more than one point.

d. This is the graph of a function. Every vertical line intersects the graph at exactly one point.

72. $v(t) = 44{,}000 - 4200t$, $0 \le t \le 8$

74. **a.** $V(x) = (30 - 2x)^2 x$
$= (900 - 120x + 4x^2)x$
$= 900x - 120x^2 + 4x^3$

b. Domain: $\{x \mid 0 < x < 15\}$

78. $AB = \sqrt{1 + x^2}$. The time required to swim from A to B at 2 mph is $\frac{\sqrt{1+x^2}}{2}$ hours.

$BC = 3 - x$. The time required to run from B to C at 8 mph is $\frac{3-x}{8}$ hours.

Thus the total time to reach point C is

$t = \frac{\sqrt{1+x^2}}{2} + \frac{3-x}{8}$ hours

Exercise Set 1.4, page 64

14. **a.** The graph is symmetric with respect to the x-axis because replacing y with $-y$ leaves the equation unaltered.

b. The graph is not symmetric with respect to the y-axis because replacing x with $-x$ alters the equation.

24. The graph is symmetric with respect to the origin because $(-y) = (-x)^3 - (-x)$ simplifies to $-y = -x^3 + x$, which is equivalent to the original equation $y = x^3 - x$.

44. Even, because $h(-x) = (-x)^2 + 1 = x^2 + 1 = h(x)$.

58.

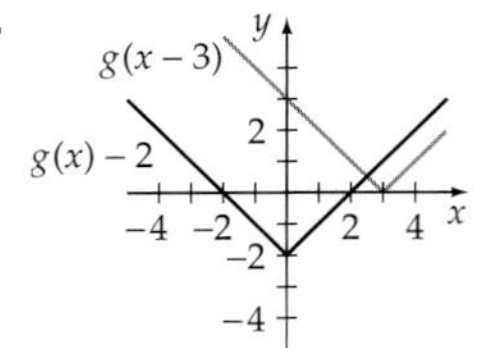

68.

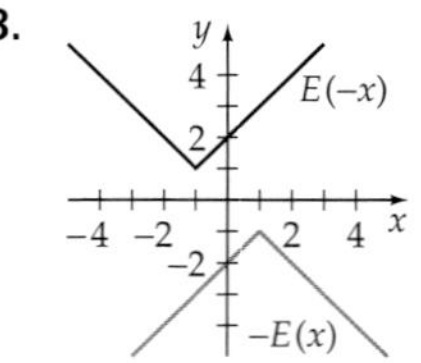

70.

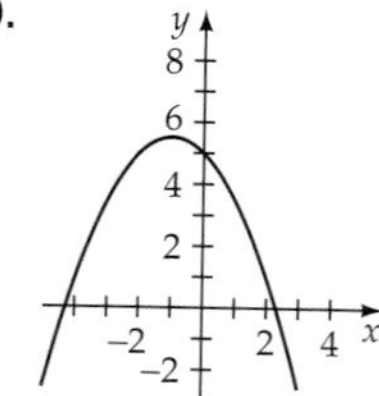

72. a.

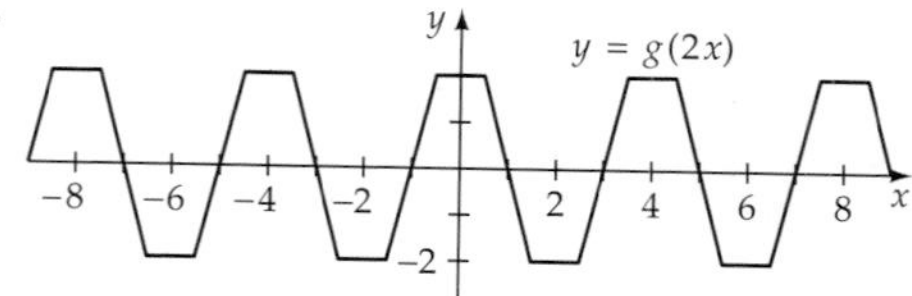

b.

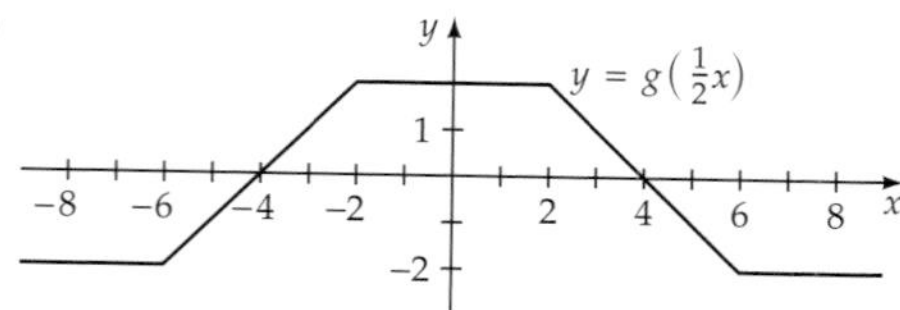

Exercise Set 1.5, page 77

10. $f(x) + g(x) = \sqrt{x-4} - x$ Domain: $\{x \mid x \geq 4\}$

$f(x) - g(x) = \sqrt{x-4} + x$ Domain: $\{x \mid x \geq 4\}$

$f(x) \cdot g(x) = -x\sqrt{x-4}$ Domain: $\{x \mid x \geq 4\}$

$\dfrac{f(x)}{g(x)} = -\dfrac{\sqrt{x-4}}{x}$ Domain: $\{x \mid x \geq 4\}$

14. $(f+g)(x) = (x^2 - 3x + 2) + (2x - 4) = x^2 - x - 2$

$(f+g)(-7) = (-7)^2 - (-7) - 2 = 49 + 7 - 2 = 54$

30. $$\frac{f(x+h) - f(x)}{h} = \frac{[4(x+h) - 5] - (4x - 5)}{h} = \frac{4x + 4(h) - 5 - 4x + 5}{h} = \frac{4(h)}{h} = 4$$

38. $(g \circ f)(x) = g[f(x)] = g[2x - 7]$

$= 3[2x - 7] + 2 = 6x - 19$

$(f \circ g)(x) = f[g(x)] = f[3x + 2]$

$= 2[3x + 2] - 7 = 6x - 3$

50. $(f \circ g)(4) = f[g(4)] = f[4^2 - 5(4)]$

$= f[-4] = 2(-4) + 3 = -5$

66. a. $l = 3 - 0.5t$ for $0 \leq t \leq 6$, and $l = -3 + 0.5t$ for $t > 6$. In either case, $l = |3 - 0.5t|$. The width is $w = |2 - 0.2t|$ as in Example 7.

b. $A(t) = |3 - 0.5t||2 - 0.2t|$

c. A is decreasing on $[0, 6]$ and on $[8, 10]$. A is increasing on $[6, 8]$ and on $[10, 14]$.

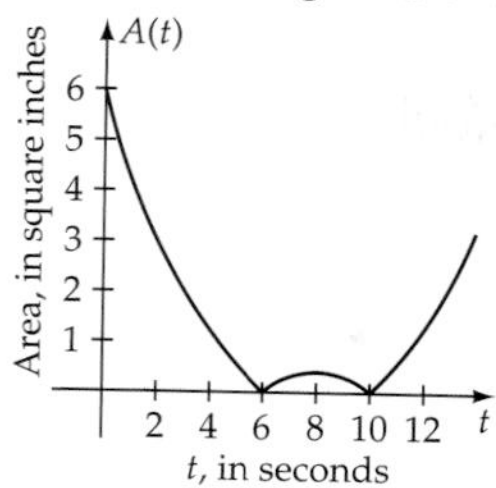

d. The highest point on the graph of A for $0 \leq t \leq 14$ occurs when $t = 0$ seconds.

72. a. On $[2, 3]$,

$a = 2$

$\Delta t = 3 - 2 = 1$

$s(a + \Delta t) = s(3) = 6 \cdot 3^2 = 54$

$s(a) = s(2) = 6 \cdot 2^2 = 24$

$$\text{Average velocity} = \frac{s(a + \Delta t) - s(a)}{\Delta t} = \frac{s(3) - s(2)}{1} = 54 - 24 = 30 \text{ feet per second}$$

This is identical to the slope of the line through $(2, s(2))$ and $(3, s(3))$ because

$$m = \frac{s(3) - s(2)}{3 - 2} = s(3) - s(2) = 54 - 24 = 30$$

b. On $[2, 2.5]$,

$a = 2$

$\Delta t = 2.5 - 2 = 0.5$

$s(a + \Delta t) = s(2.5) = 6(2.5)^2 = 37.5$

$$\text{Average velocity} = \frac{s(2.5) - s(2)}{0.5} = \frac{37.5 - 24}{0.5} = \frac{13.5}{0.5} = 27 \text{ feet per second}$$

c. On $[2, 2.1]$,

$a = 2$

$\Delta t = 2.1 - 2 = 0.1$

$s(a + \Delta t) = s(2.1) = 6(2.1)^2 = 26.46$

$$\text{Average velocity} = \frac{s(2.1) - s(2)}{0.1} = \frac{26.46 - 24}{0.1} = \frac{2.46}{0.1} = 24.6 \text{ feet per second}$$

Continued ▶

d. On $[2, 2.01]$,

$a = 2$

$\Delta t = 2.01 - 2 = 0.01$

$s(a + \Delta t) = s(2.01) = 6(2.01)^2 = 24.2406$

$$\text{Average velocity} = \frac{s(2.01) - s(2)}{0.01} = \frac{24.2406 - 24}{0.01} = \frac{0.2406}{0.01} = 24.06 \text{ feet per second}$$

e. On $[2, 2.001]$,

$a = 2$

$\Delta t = 2.001 - 2 = 0.001$

$s(a + \Delta t) = s(2.001) = 6(2.001)^2 = 24.024006$

$$\text{Average velocity} = \frac{s(2.001) - s(2)}{0.001} = \frac{24.024006 - 24}{0.001} = \frac{0.024006}{0.001} = 24.006 \text{ feet per second}$$

f. On $[2, 2 + \Delta t]$,

$$\frac{s(2 + \Delta t) - s(2)}{\Delta t} = \frac{6(2 + \Delta t)^2 - 24}{\Delta t} = \frac{6(4 + 4(\Delta t) + (\Delta t)^2) - 24}{\Delta t} = \frac{24 + 24(\Delta t) + 6(\Delta t)^2 - 24}{\Delta t} = \frac{24\Delta t + 6(\Delta t)^2}{\Delta t} = 24 + 6(\Delta t)$$

As Δt approaches zero, the average velocity approaches 24 feet per second.

Exercise Set 1.6, page 90

10. Because the graph of the given function is a line that passes through $(0, 6)$, $(2, 3)$, and $(6, -3)$, the graph of the inverse will be a line that passes through $(6, 0)$, $(3, 2)$, and $(-3, 6)$. See the following figure. Notice that the line shown below is a reflection of the line given in Exercise 10 across the line given by $y = x$. Yes, the inverse relation is a function.

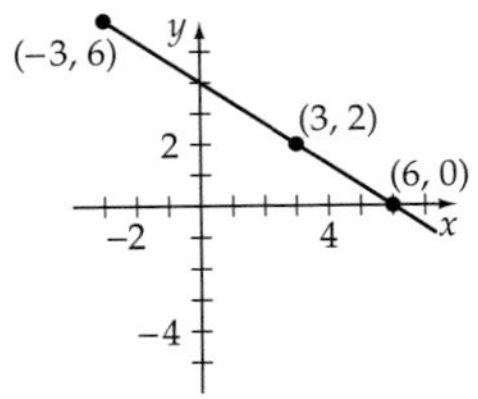

20. Check to see if $f[g(x)] = x$ for all x in the domain of g and $g[f(x)] = x$ for all x in the domain of f. The following shows that $f[g(x)] = x$ for all real numbers x.

$$f[g(x)] = f[2x + 3] = \frac{1}{2}(2x + 3) - \frac{3}{2} = x + \frac{3}{2} - \frac{3}{2} = x$$

The following shows that $g[f(x)] = x$ for all real numbers x.

$$g[f(x)] = g\left[\frac{1}{2}x - \frac{3}{2}\right] = 2\left(\frac{1}{2}x - \frac{3}{2}\right) + 3 = x - 3 + 3 = x$$

Thus f and g are inverses.

32.
$f(x) = 4x - 8$
$y = 4x - 8$ • Replace $f(x)$ by y.
$x = 4y - 8$ • Interchange x and y.
$x + 8 = 4y$ • Solve for y.
$\frac{1}{4}(x + 8) = y$
$y = \frac{1}{4}x + 2$
$f^{-1}(x) = \frac{1}{4}x + 2$ • Replace y by $f^{-1}(x)$.

38.
$f(x) = \frac{x}{x - 2}, x \neq 2$
$y = \frac{x}{x - 2}$ • Replace $f(x)$ by y.
$x = \frac{y}{y - 2}$ • Interchange x and y.
$x(y - 2) = y$
$xy - 2x = y$ • Solve for y.
$xy - y = 2x$
$y(x - 1) = 2x$
$y = \frac{2x}{x - 1}$
$f^{-1}(x) = \frac{2x}{x - 1}, x \neq 1$ • Replace y by $f^{-1}(x)$ and indicate any restrictions.

44.
$f(x) = \sqrt{4 - x}, x \leq 4$
$y = \sqrt{4 - x}$ • Replace $f(x)$ by y.
$x = \sqrt{4 - y}$ • Interchange x and y.
$x^2 = 4 - y$ • Solve for y.

$$x^2 - 4 = -y$$
$$-x^2 + 4 = y$$
$$f^{-1}(x) = -x^2 + 4, x \geq 0$$ • Replace y by $f^{-1}(x)$ and indicate any restrictions.

The range of f is $\{y \mid y \geq 0\}$. Therefore, the domain of f^{-1} is $\{x \mid x \geq 0\}$, as indicated above.

54. $K(x) = 1.3x - 4.7$

$y = 1.3x - 4.7$ • Replace $K(x)$ by y.

$x = 1.3y - 4.7$ • Interchange x and y.

$x + 4.7 = 1.3y$ • Solve for y.

$$\frac{x + 4.7}{1.3} = y$$

$K^{-1}(x) = \dfrac{x + 4.7}{1.3}$ • Replace y by $K^{-1}(x)$.

The function $K^{-1}(x) = \dfrac{x + 4.7}{1.3}$ can be used to convert a United Kingdom men's shoe size to its equivalent U.S. shoe size.

Exercise Set 1.7, page 103

18. a. Enter the data in the table. Then use your calculator to find the linear regression equation:

$$y = 3.410344828x + 65.09359606$$

b. Evaluate the linear regression equation when $x = 58$.

$$y = 3.410344828(58) + 65.09359606$$
$$\approx 263$$

The ball will travel approximately 263 feet.

32. a. Enter the data in the table. Then use your calculator to find the quadratic regression model:

$$y = 0.05208x^2 - 3.56026x + 82.32999$$

b. Evaluate the quadratic regression model when $x = 40$.

$$y = 0.05208(40^2) - 3.5602(40) + 82.3299$$
$$\approx 23$$

The bird will consume approximately 23 milliliters of oxygen per minute.

Exercise Set 2.1, page 130

2. The measure of the complement of an angle of 87° is

$(90° - 87°) = 3°$

The measure of the supplement of an angle of 87° is

$(180° - 87°) = 93°$

14. Because $765° = 2 \cdot 360° + 45°$, $\angle\alpha$ is coterminal with an angle that has a measure of 45°. $\angle\alpha$ is a Quadrant I angle.

32. $-45° = -45°\left(\dfrac{\pi \text{ radians}}{180°}\right) = -\dfrac{\pi}{4}$ radian

44. $\dfrac{\pi}{4}$ radian $= \dfrac{\pi}{4}$ radian $\left(\dfrac{180°}{\pi \text{ radians}}\right) = 45°$

68. $s = r\theta = 5\left(144° \cdot \dfrac{\pi}{180°}\right) = 4\pi \approx 12.57$ meters

72. Let θ_2 be the angle through which the pulley with a diameter of 0.8 meter turns. Let θ_1 be the angle through which the pulley with a diameter of 1.2 meters turns. Let $r_2 = 0.4$ meter be the radius of the smaller pulley, and let $r_1 = 0.6$ meter be the radius of the larger pulley.

$$\theta_1 = 240° = \frac{4}{3}\pi \text{ radians}$$

Thus $r_2\theta_2 = r_1\theta_1$

$$0.4\theta_2 = 0.6\left(\frac{4}{3}\pi\right)$$
$$\theta_2 = \frac{0.6}{0.4}\left(\frac{4}{3}\pi\right) = 2\pi \text{ radians or } 360°$$

74. The earth makes one revolution ($\theta = 2\pi$) in 1 day.

$$t = 24 \cdot 3600 = 86{,}400 \text{ seconds}$$
$$\omega = \frac{\theta}{t} = \frac{2\pi}{86{,}400} \approx 7.27 \times 10^{-5} \text{ radian/second}$$

80. $C = 2\pi r = 2\pi(18 \text{ inches}) = 36\pi$ inches

Thus one conversion factor is (36π inches/1 revolution).

$$\frac{500 \text{ revolutions}}{1 \text{ minute}} = \frac{500 \text{ revolutions}}{1 \text{ minute}}\left(\frac{36\pi \text{ inches}}{1 \text{ revolution}}\right)$$
$$= \frac{18{,}000\pi \text{ inches}}{1 \text{ minute}}$$

Now convert inches to miles and minutes to hours.

$$\frac{18{,}000\pi \text{ inches}}{1 \text{ minute}}$$
$$= \frac{18{,}000\pi \text{ inches}}{1 \text{ minute}}\left(\frac{1 \text{ foot}}{12 \text{ inches}}\right)\left(\frac{1 \text{ mile}}{5280 \text{ feet}}\right)\left(\frac{60 \text{ minutes}}{1 \text{ hour}}\right)$$
$$\approx 54 \text{ miles per hour}$$

Exercise Set 2.2, page 142

6. $\text{adj} = \sqrt{8^2 - 5^2}$

$\text{adj} = \sqrt{64 - 25} = \sqrt{39}$

$$\sin\theta = \frac{\text{opp}}{\text{hyp}} = \frac{5}{8} \qquad \csc\theta = \frac{\text{hyp}}{\text{opp}} = \frac{8}{5}$$
$$\cos\theta = \frac{\text{adj}}{\text{hyp}} = \frac{\sqrt{39}}{8} \qquad \sec\theta = \frac{\text{hyp}}{\text{adj}} = \frac{8}{\sqrt{39}} = \frac{8\sqrt{39}}{39}$$
$$\tan\theta = \frac{\text{opp}}{\text{adj}} = \frac{5}{\sqrt{39}} = \frac{5\sqrt{39}}{39} \qquad \cot\theta = \frac{\text{adj}}{\text{opp}} = \frac{\sqrt{39}}{5}$$

18. Because $\tan\theta = \dfrac{\text{opp}}{\text{adj}} = \dfrac{4}{3}$, let opp = 4 and adj = 3.

$\text{hyp} = \sqrt{3^2 + 4^2} = 5$

$\sec\theta = \dfrac{\text{hyp}}{\text{adj}} = \dfrac{5}{3}$

34. $\sin\dfrac{\pi}{3}\cos\dfrac{\pi}{4} - \tan\dfrac{\pi}{4} = \dfrac{\sqrt{3}}{2}\cdot\dfrac{\sqrt{2}}{2} - 1$

$= \dfrac{\sqrt{6}}{4} - 1 = \dfrac{\sqrt{6} - 4}{4}$

56. $\tan 68.9° = \dfrac{h}{116}$

$h = 116 \tan 68.9°$

$h \approx 301$ meters (three significant digits)

58.

$\sin 5° = \dfrac{80}{d}$

$d = \dfrac{80}{\sin 5°}$

$d \approx 917.9$ feet

Change 9 miles per hour to feet per minute.

$r = 9\,\dfrac{\text{miles}}{\text{hour}} = \dfrac{9 \text{ miles}}{1 \text{ hour}}\cdot\dfrac{5280 \text{ feet}}{1 \text{ mile}}\cdot\dfrac{1 \text{ hour}}{60 \text{ minutes}}$

$= \dfrac{9(5280)}{60}\,\dfrac{\text{feet}}{\text{minute}} = 792\,\dfrac{\text{feet}}{\text{minute}}$

$t = \dfrac{d}{r}$

$t \approx \dfrac{917.9 \text{ feet}}{792 \text{ feet per minute}} \approx 1.2$ minutes (to the nearest tenth of a minute)

66.

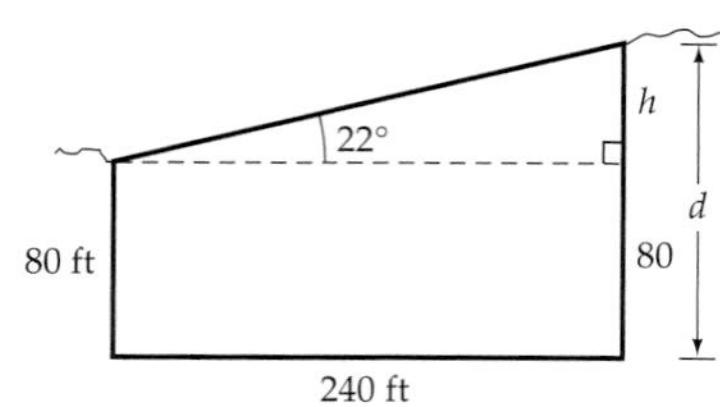

$\tan 22° = \dfrac{h}{240}$

$h = 240 \tan 22°$

$d = 80 + h$

$d = 80 + 240 \tan 22°$

$d \approx 180$ feet (two significant digits)

Exercise Set 2.3, page 153

6. $x = -6, y = -9, r = \sqrt{(-6)^2 + (-9)^2} = \sqrt{117} = 3\sqrt{13}$

$\sin\theta = \dfrac{y}{r} = \dfrac{-9}{3\sqrt{13}} = -\dfrac{3}{\sqrt{13}} = -\dfrac{3\sqrt{13}}{13}$ $\qquad \csc\theta = -\dfrac{\sqrt{13}}{3}$

$\cos\theta = \dfrac{x}{r} = \dfrac{-6}{3\sqrt{13}} = -\dfrac{2}{\sqrt{13}} = -\dfrac{2\sqrt{13}}{13}$ $\qquad \sec\theta = -\dfrac{\sqrt{13}}{2}$

$\tan\theta = \dfrac{y}{x} = \dfrac{-9}{-6} = \dfrac{3}{2}$ $\qquad \cot\theta = \dfrac{2}{3}$

30. $\sec\theta = \dfrac{2\sqrt{3}}{3} = \dfrac{r}{x}$ • **Let $r = 2\sqrt{3}$ and $x = 3$.**

$y = \pm\sqrt{(2\sqrt{3})^2 - 3^2} = \pm\sqrt{3}$

$y = -\sqrt{3}$ because $y < 0$ in Quadrant IV.

$\sin\theta = \dfrac{-\sqrt{3}}{2\sqrt{3}} = -\dfrac{1}{2}$

38. $\theta' = 255° - 180° = 75°$

50. $\cos 300° > 0$, $\theta' = 360° - 300° = 60°$

Thus $\cos 300° = \cos 60° = \dfrac{1}{2}$.

Exercise Set 2.4, page 166

10. $t = -\dfrac{7\pi}{4}$; $W(t) = P(x, y)$ where

$x = \cos t$ $\qquad y = \sin t$

$= \cos\left(-\dfrac{7\pi}{4}\right)$ $\qquad = \sin\left(-\dfrac{7\pi}{4}\right)$

$= \cos\dfrac{\pi}{4}$ $\qquad = \sin\dfrac{\pi}{4}$

$= \dfrac{\sqrt{2}}{2}$ $\qquad = \dfrac{\sqrt{2}}{2}$

$W\left(-\dfrac{7\pi}{4}\right) = \left(\dfrac{\sqrt{2}}{2}, \dfrac{\sqrt{2}}{2}\right)$

16. The reference angle for $-\dfrac{5\pi}{6}$ is $\dfrac{\pi}{6}$.

$\sec\left(-\dfrac{5\pi}{6}\right) = -\sec\dfrac{\pi}{6}$ • **sec $t < 0$ for t in Quadrant III**

$= -\dfrac{2\sqrt{3}}{3}$

44. $F(-x) = \tan(-x) + \sin(-x)$

$= -\tan x - \sin x$ • **tan x and sin x are odd functions.**

$= -(\tan x + \sin x)$

$= -F(x)$

Because $F(-x) = -F(x)$, the function defined by $F(x) = \tan x + \sin x$ is an odd function.

56.

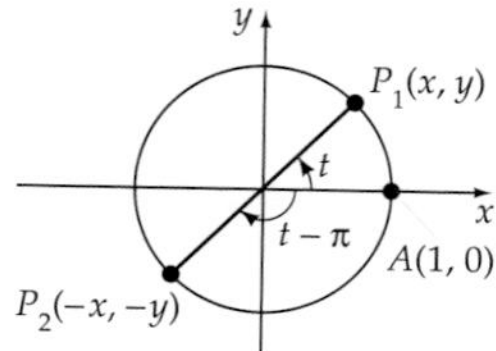

$\tan t = \dfrac{y}{x}$

$\tan(t - \pi) = \dfrac{-y}{-x} = \dfrac{y}{x}$ • **From the unit circle**

Therefore, $\tan t = \tan(t - \pi)$.

72. $\dfrac{1}{1 - \sin t} + \dfrac{1}{1 + \sin t} = \dfrac{1 + \sin t + 1 - \sin t}{(1 - \sin t)(1 + \sin t)}$

$= \dfrac{2}{1 - \sin^2 t}$

$= \dfrac{2}{\cos^2 t}$

$= 2\sec^2 t$

78. $1 + \tan^2 t = \sec^2 t$

$\tan^2 t = \sec^2 t - 1$

$\tan t = \pm\sqrt{\sec^2 t - 1}$

Because $\dfrac{3\pi}{2} < t < 2\pi$, $\tan t$ is negative. Thus $\tan t = -\sqrt{\sec^2 t - 1}$.

82. March 5 is represented by $t = 2$.

$T(2) = -41\cos\left(\dfrac{\pi}{6} \cdot 2\right) + 36$

$= -41\cos\left(\dfrac{\pi}{3}\right) + 36$

$= -41(0.5) + 36$

$= 15.5°\text{F}$

July 20 is represented by $t = 6.5$.

$T(6.5) = -41\cos\left(\dfrac{\pi}{6} \cdot 6.5\right) + 36$

$\approx -41(-0.9659258263) + 36$

$\approx 75.6°\text{F}$

Exercise Set 2.5, page 177

22. $y = -\dfrac{3}{2}\sin x$

$a = \left|-\dfrac{3}{2}\right| = \dfrac{3}{2}$

period $= 2\pi$

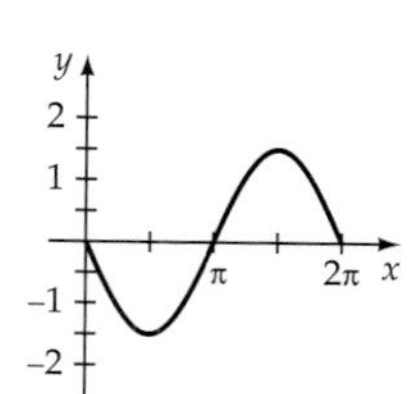

32. $y = \sin\dfrac{3\pi}{4}x$

$a = 1$

period $= \dfrac{2\pi}{b} = \dfrac{2\pi}{3\pi/4} = \dfrac{8}{3}$

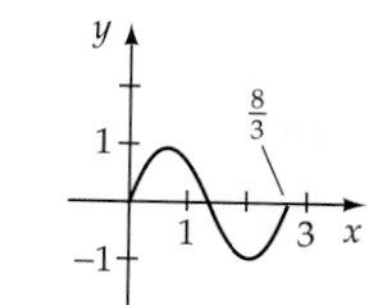

34. $y = \cos 3\pi x$

$a = 1$

period $= \dfrac{2\pi}{b} = \dfrac{2\pi}{3\pi} = \dfrac{2}{3}$

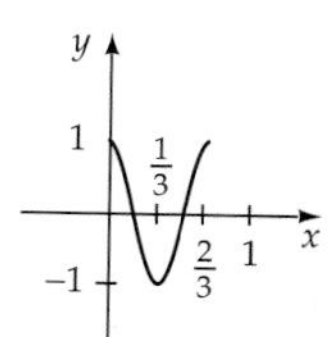

40. $y = \dfrac{1}{2}\sin\dfrac{\pi x}{3}$

$a = \dfrac{1}{2}$

period $= \dfrac{2\pi}{b} = \dfrac{2\pi}{\pi/3} = 6$

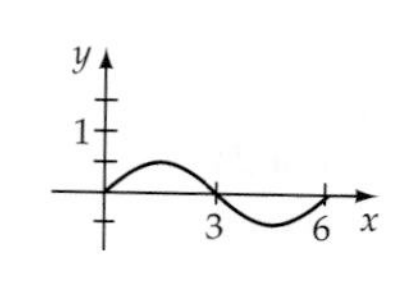

48. $y = -\dfrac{3}{4}\cos 5x$

$a = \left|-\dfrac{3}{4}\right| = \dfrac{3}{4}$

period $= \dfrac{2\pi}{b} = \dfrac{2\pi}{5}$

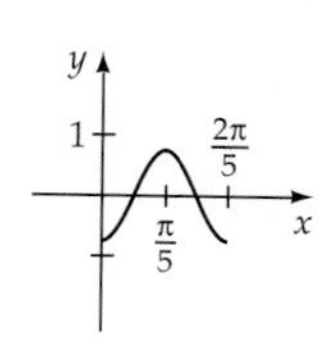

54. $y = -\left|3\sin\dfrac{2}{3}x\right|$

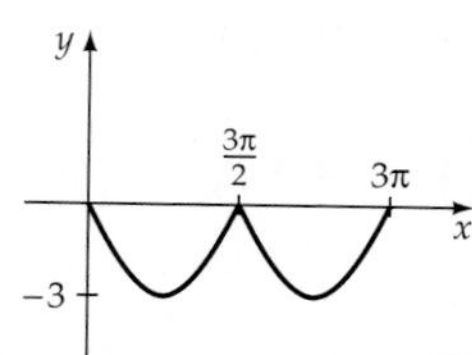

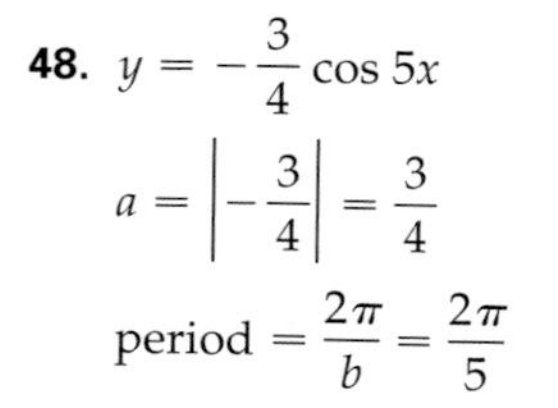

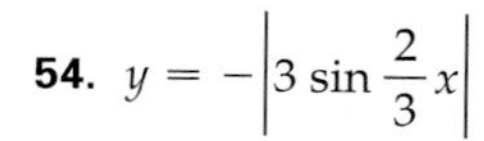

60. Because the graph passes through the origin, we start with an equation of the form $y = a\sin bx$. The graph completes one cycle in $\dfrac{4\pi}{3}$ units. Thus the period is $\dfrac{4\pi}{3}$.

Use the equation $\dfrac{2\pi}{b} = \dfrac{4\pi}{3}$ to solve for b.

$\dfrac{2\pi}{b} = \dfrac{4\pi}{3}$

$6\pi = 4b\pi$ • **Multiply each side by $3b$.**

$\dfrac{3}{2} = b$ • **Divide each side by 4π.**

The graph has a maximum height of $\dfrac{3}{2}$ and a minimum height of $-\dfrac{3}{2}$. Thus its amplitude is $a = \dfrac{3}{2}$.

Substitute $\dfrac{3}{2}$ for a and $\dfrac{3}{2}$ for b in $y = a\sin bx$ to produce

$y = \dfrac{3}{2}\sin\dfrac{3}{2}x$.

Exercise Set 2.6, page 189

24. $y = \frac{1}{3} \tan x$

period $= \frac{\pi}{b} = \pi$

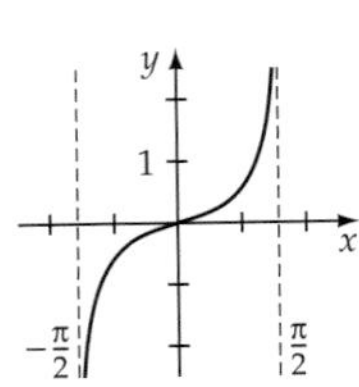

32. $y = -3 \tan 3x$

period $= \frac{\pi}{b} = \frac{\pi}{3}$

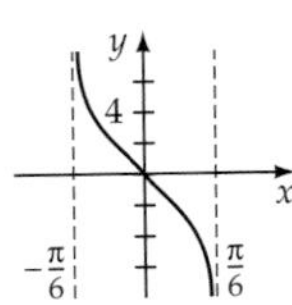

34. $y = \frac{1}{2} \cot 2x$

period $= \frac{\pi}{b} = \frac{\pi}{2}$

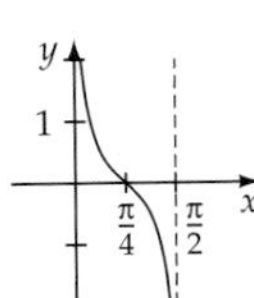

40. $y = 3 \csc \frac{\pi x}{2}$

period $= \frac{2\pi}{b} = \frac{2\pi}{\pi/2} = 4$

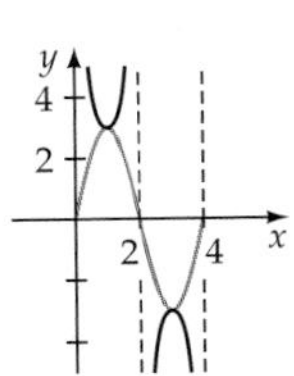

44. $y = \sec \frac{x}{2}$

period $= \frac{2\pi}{b} = \frac{2\pi}{1/2} = 4\pi$

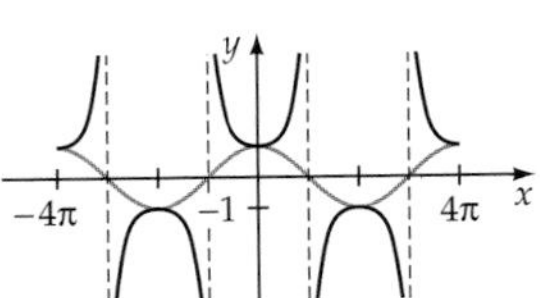

Exercise Set 2.7, page 199

20. $y = \cos\left(2x - \frac{\pi}{3}\right)$

$a = 1$

period $= \pi$

phase shift $= -\frac{c}{b}$

$= -\frac{-\pi/3}{2} = \frac{\pi}{6}$

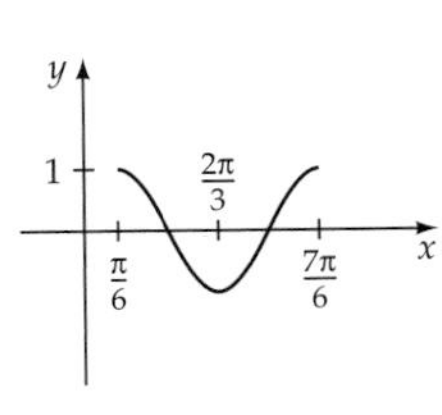

22. $y = \tan(x - \pi)$

period $= \pi$

phase shift $= -\frac{c}{b}$

$= -\frac{-\pi}{1} = \pi$

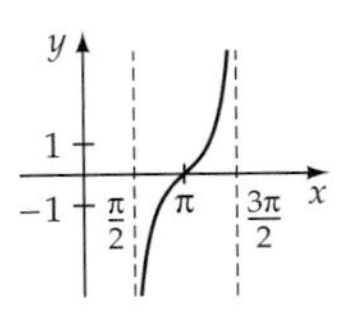

40. $y = 2 \sin\left(\frac{\pi x}{2} + 1\right) - 2$

$a = 2$

period $= 4$

phase shift $= -\frac{c}{b}$

$= -\frac{1}{2/\pi} = -\frac{2}{\pi}$

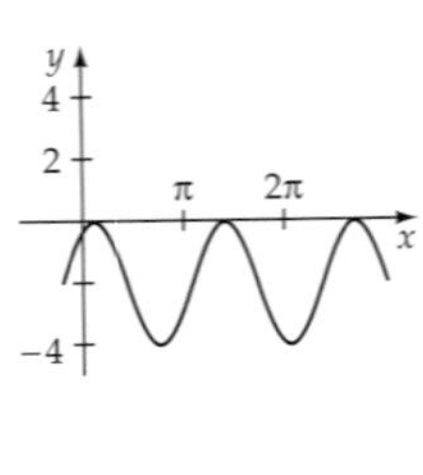

42. $y = -3 \cos(2\pi x - 3) + 1$

$a = 3$

period $= 1$

phase shift $= -\frac{c}{b} = \frac{3}{2\pi}$

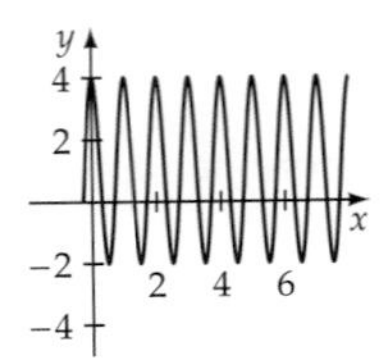

48. $y = \csc \frac{x}{3} + 4$

period $= 6\pi$

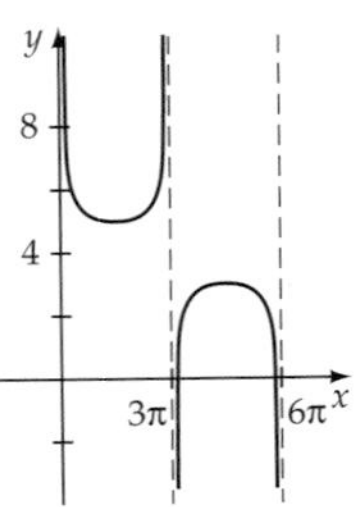

52. $y = \frac{x}{2} + \cos x$

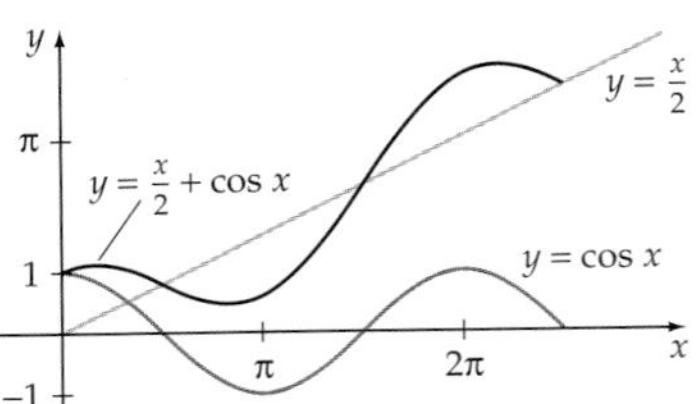

56. $y = \cos x - \sin x$

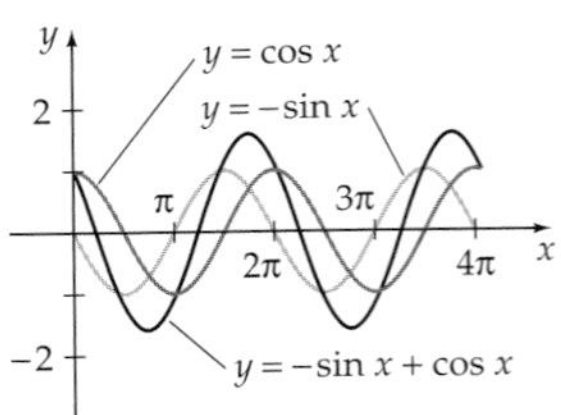

64. a. Phase shift $= -\dfrac{c}{b} = -\dfrac{\left(-\dfrac{7}{12}\pi\right)}{\left(\dfrac{\pi}{6}\right)} = 3.5$ months,

period $= \dfrac{2\pi}{b} = \dfrac{2\pi}{\pi/6} = 12$ months

b. First graph $y_1 = 2.7\cos\left(\dfrac{\pi}{6}t\right)$. Because the phase shift is 3.5 months, shift the graph of y_1 3.5 units to the right to produce the graph of y_2. Now shift the graph of y_2 upward 4 units to produce the graph of S.

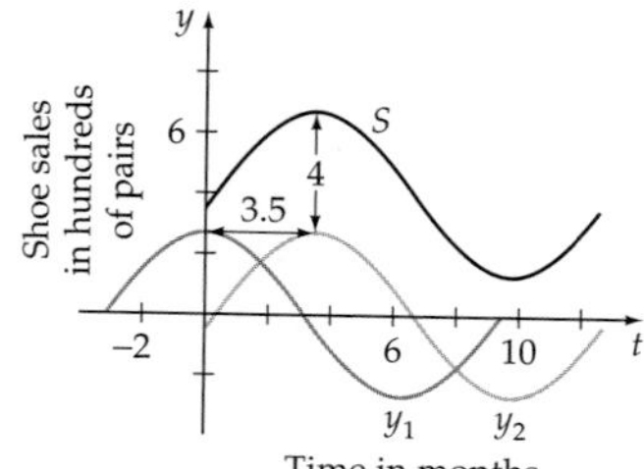

c. 3.5 months after January 1 is the middle of April.

78. $y = x\cos x$

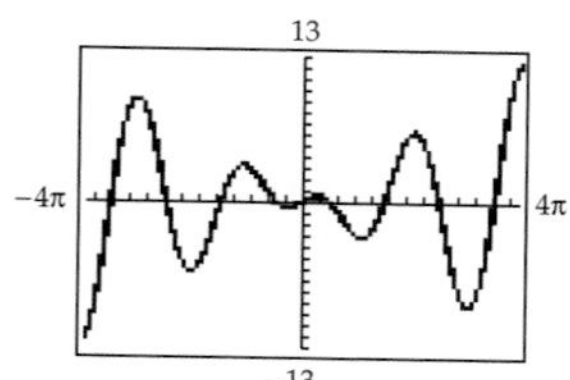

Exercise Set 2.8, page 208

20. Amplitude = 3, frequency $= \dfrac{1}{\pi}$, period $= \pi$

Because $\dfrac{2\pi}{b} = \pi$, we have $b = 2$. Thus $y = 3\cos 2t$.

28. Amplitude $= |-1.5| = 1.5$

$$f = \frac{1}{2\pi}\sqrt{\frac{k}{m}} = \frac{1}{2\pi}\sqrt{\frac{3}{27}} = \frac{1}{2\pi}\cdot\frac{1}{3} = \frac{1}{6\pi}, \text{ period} = 6\pi$$

$$y = a\cos 2\pi f t = -1.5\cos\left[2\pi\left(\frac{1}{6\pi}\right)t\right] = -1.5\cos\frac{1}{3}t$$

34.

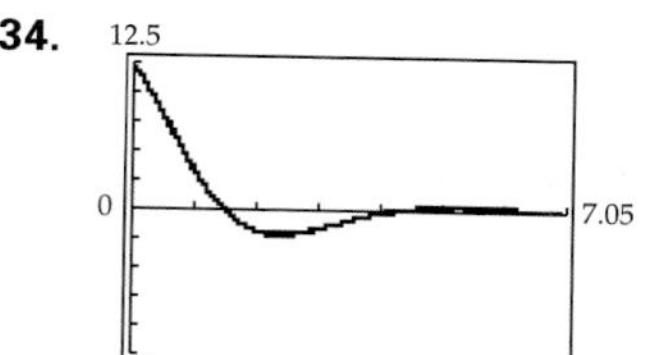

a. f has pseudoperiod $\dfrac{2\pi}{1} = 2\pi$.

$10 \div (2\pi) \approx 1.59$

Thus f completes only one full oscillation on $0 \le t \le 10$.

b. The following graph of f shows that $|f(t)| < 0.01$ for $t > 10.5$.

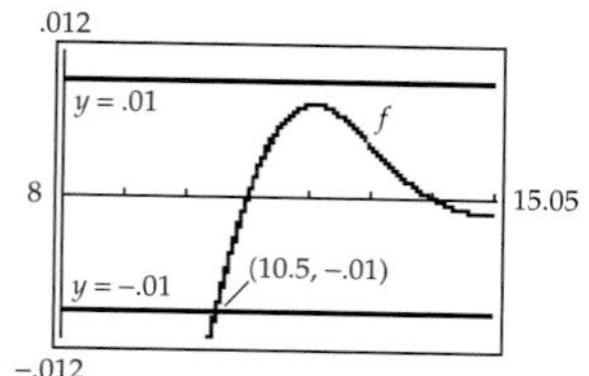

Exercise Set 3.1, page 222

2. We will try to verify the identity by rewriting the left side so that it involves only sines and cosines.

$$\begin{aligned}\tan x\sec x\sin x &= \frac{\sin x}{\cos x}\cdot\frac{1}{\cos x}\cdot\sin x\\ &= \frac{\sin^2 x}{\cos^2 x}\\ &= \left(\frac{\sin x}{\cos x}\right)^2\\ &= (\tan x)^2\\ &= \tan^2 x\end{aligned}$$

12. $\sin^4 x - \cos^4 x = (\sin^2 x + \cos^2 x)(\sin^2 x - \cos^2 x)$
$= 1(\sin^2 x - \cos^2 x) = \sin^2 x - \cos^2 x$

24.
$$\begin{aligned}&\frac{2\sin x\cot x + \sin x - 4\cot x - 2}{2\cot x + 1}\\ &= \frac{(\sin x)(2\cot x + 1) - 2(2\cot x + 1)}{2\cot x + 1}\\ &= \frac{(2\cot x + 1)(\sin x - 2)}{2\cot x + 1} = \sin x - 2\end{aligned}$$

34. $$\frac{\frac{1}{\sin x}+\frac{1}{\cos x}}{\frac{1}{\sin x}-\frac{1}{\cos x}} = \frac{\frac{1}{\sin x}+\frac{1}{\cos x}}{\frac{1}{\sin x}-\frac{1}{\cos x}} \cdot \frac{\sin x \cos x}{\sin x \cos x}$$
$$= \frac{\cos x + \sin x}{\cos x - \sin x}$$
$$= \frac{\cos x + \sin x}{\cos x - \sin x} \cdot \frac{\cos x - \sin x}{\cos x - \sin x}$$
$$= \frac{\cos^2 x - \sin^2 x}{\cos^2 x - 2\sin x \cos x + \sin^2 x}$$
$$= \frac{\cos^2 x - \sin^2 x}{1 - 2\sin x \cos x}$$

44. Rewrite the left side so that it involves only sines and cosines.

$$\frac{2\cot x}{\cot x + \tan x} = \frac{2 \cdot \frac{\cos x}{\sin x}}{\frac{\cos x}{\sin x} + \frac{\sin x}{\cos x}}$$
$$= \frac{2 \cdot \frac{\cos x}{\sin x}}{\frac{\cos x \cos x}{\sin x \cos x} + \frac{\sin x \sin x}{\cos x \sin x}}$$
$$= \frac{2 \cdot \frac{\cos x}{\sin x}}{\frac{\cos^2 x + \sin^2 x}{\sin x \cos x}}$$
$$= \frac{2 \cdot \frac{\cos x}{\sin x}}{\frac{1}{\sin x \cos x}}$$
$$= 2 \cdot \frac{\cos x}{\sin x} \cdot \frac{\sin x \cos x}{1}$$
$$= 2\cos^2 x$$

Exercise Set 3.2, page 233

4. Use the identity $\cos(\alpha - \beta) = \cos\alpha\cos\beta + \sin\alpha\sin\beta$ with $\alpha = 120°$ and $\beta = 45°$.

$$\cos(120° - 45°) = \cos 120° \cos 45° + \sin 120° \sin 45°$$
$$= \left(-\frac{1}{2}\right)\left(\frac{\sqrt{2}}{2}\right) + \left(\frac{\sqrt{3}}{2}\right)\left(\frac{\sqrt{2}}{2}\right)$$
$$= -\frac{\sqrt{2}}{4} + \frac{\sqrt{6}}{4}$$
$$= \frac{\sqrt{6} - \sqrt{2}}{4}$$

20. The value of a given trigonometric function of θ, measured in degrees, is equal to its cofunction of $90° - \theta$. Thus

$$\cos 80° = \sin(90° - 80°)$$
$$= \sin 10°$$

26. $\sin x \cos 3x + \cos x \sin 3x = \sin(x + 3x) = \sin 4x$

38. $\tan\alpha = \frac{24}{7}$, with $0° < \alpha < 90°$; $\sin\alpha = \frac{24}{25}$, $\cos\alpha = \frac{7}{25}$

$\sin\beta = -\frac{8}{17}$, with $180° < \beta < 270°$

$\cos\beta = -\frac{15}{17}$, $\tan\beta = \frac{8}{15}$

a. $$\sin(\alpha + \beta) = \sin\alpha\cos\beta + \cos\alpha\sin\beta$$
$$= \left(\frac{24}{25}\right)\left(-\frac{15}{17}\right) + \left(\frac{7}{25}\right)\left(-\frac{8}{17}\right)$$
$$= -\frac{360}{425} - \frac{56}{425} = -\frac{416}{425}$$

b. $$\cos(\alpha + \beta) = \cos\alpha\cos\beta - \sin\alpha\sin\beta$$
$$= \left(\frac{7}{25}\right)\left(-\frac{15}{17}\right) - \left(\frac{24}{25}\right)\left(-\frac{8}{17}\right)$$
$$= -\frac{105}{425} + \frac{192}{425} = \frac{87}{425}$$

c. $$\tan(\alpha - \beta) = \frac{\tan\alpha - \tan\beta}{1 + \tan\alpha\tan\beta}$$
$$= \frac{\frac{24}{7} - \frac{8}{15}}{1 + \left(\frac{24}{7}\right)\left(\frac{8}{15}\right)} = \frac{\frac{24}{7} - \frac{8}{15}}{1 + \frac{192}{105}} \cdot \frac{105}{105}$$
$$= \frac{360 - 56}{105 + 192} = \frac{304}{297}$$

50. $$\cos(\theta + \pi) = \cos\theta\cos\pi - \sin\theta\sin\pi$$
$$= (\cos\theta)(-1) - (\sin\theta)(0) = -\cos\theta$$

62. $$\cos 5x \cos 3x + \sin 5x \sin 3x = \cos(5x - 3x) = \cos 2x$$
$$= \cos(x + x) = \cos x \cos x - \sin x \sin x$$
$$= \cos^2 x - \sin^2 x$$

76. $$\sin(\theta + 2\pi) = \sin\theta\cos 2\pi + \cos\theta\sin 2\pi$$
$$= (\sin\theta)(1) + (\cos\theta)(0) = \sin\theta$$

Exercise Set 3.3, page 243

2. $2\sin 3\theta \cos 3\theta = \sin[2(3\theta)] = \sin 6\theta$

10. $\cos\alpha = \dfrac{24}{25}$ with $270° < \alpha < 360°$

$$\sin\alpha = -\sqrt{1-\left(\frac{24}{25}\right)^2} = -\frac{7}{25}$$

$$\tan\alpha = \frac{-7/25}{24/25} = -\frac{7}{24}$$

$$\sin 2\alpha = 2\sin\alpha\cos\alpha = 2\left(-\frac{7}{25}\right)\left(\frac{24}{25}\right) = -\frac{336}{625}$$

$$\cos 2\alpha = \cos^2\alpha - \sin^2\alpha = \left(\frac{24}{25}\right)^2 - \left(-\frac{7}{25}\right)^2 = \frac{527}{625}$$

$$\tan 2\alpha = \frac{2\tan\alpha}{1-\tan^2\alpha} = \frac{2\left(-\frac{7}{24}\right)}{1-\left(-\frac{7}{24}\right)^2} = \frac{-\frac{7}{12}}{1-\frac{49}{576}}\cdot\frac{576}{576} = -\frac{336}{527}$$

22. $\sin^2 x\cos^4 x$

$$= \sin^2 x\,(\cos^2 x)^2$$

$$= \left(\frac{1-\cos 2x}{2}\right)\left(\frac{1+\cos 2x}{2}\right)^2$$

$$= \left(\frac{1-\cos 2x}{2}\right)\left(\frac{1+2\cos 2x+\cos^2 2x}{4}\right)$$

$$= \frac{1}{8}(1-\cos 2x)\left(1+2\cos 2x+\frac{1+\cos 4x}{2}\right)$$

$$= \frac{1}{8}(1-\cos 2x)\left(\frac{2+4\cos 2x+1+\cos 4x}{2}\right)$$

$$= \frac{1}{16}(1-\cos 2x)(3+4\cos 2x+\cos 4x)$$

$$= \frac{1}{16}(3+4\cos 2x+\cos 4x-3\cos 2x-4\cos^2 2x - \cos 2x\cos 4x)$$

$$= \frac{1}{16}(3+\cos 2x+\cos 4x-4\cos^2 2x-\cos 2x\cos 4x)$$

$$= \frac{1}{16}\left(3+\cos 2x+\cos 4x-4\left(\frac{1+\cos 4x}{2}\right)-\cos 2x\cos 4x\right)$$

$$= \frac{1}{16}(3+\cos 2x+\cos 4x-2(1+\cos 4x)-\cos 2x\cos 4x)$$

$$= \frac{1}{16}(3+\cos 2x+\cos 4x-2-2\cos 4x-\cos 2x\cos 4x)$$

$$= \frac{1}{16}(1+\cos 2x-\cos 4x-\cos 2x\cos 4x)$$

28. Because $165° = \dfrac{1}{2}(330°)$, we can find $\cos 165°$ by using the half-angle identity for $\cos\dfrac{\alpha}{2}$ with $\alpha = 330°$. The angle $\dfrac{\alpha}{2} = 165°$ lies in Quadrant II and the cosine function is negative in Quadrant II. Thus $\cos 165° < 0$, and we must select the minus sign that precedes the radical in $\cos\dfrac{\alpha}{2} = \pm\sqrt{\dfrac{1+\cos\alpha}{2}}$ to produce the correct result.

$$\cos 165° = -\sqrt{\frac{1+\cos 330°}{2}} = -\sqrt{\frac{1+\frac{\sqrt{3}}{2}}{2}} = -\sqrt{\frac{\frac{2}{2}+\frac{\sqrt{3}}{2}}{2}} = -\sqrt{\left(\frac{2+\sqrt{3}}{2}\right)\cdot\frac{1}{2}} = -\sqrt{\frac{2+\sqrt{3}}{4}} = -\frac{\sqrt{2+\sqrt{3}}}{2}$$

38. Because α is in Quadrant III, $\cos\alpha < 0$. We now solve for $\cos\alpha$.

$$\cos\alpha = -\sqrt{1-\sin^2\alpha} = -\sqrt{1-\left(-\frac{7}{25}\right)^2} = -\sqrt{1-\frac{49}{625}} = -\frac{24}{25}$$

Because $180° < \alpha < 270°$, we know that $90° < \dfrac{\alpha}{2} < 135°$. Thus $\dfrac{\alpha}{2}$ is in Quadrant II, $\sin\dfrac{\alpha}{2} > 0$, $\cos\dfrac{\alpha}{2} < 0$, and $\tan\dfrac{\alpha}{2} < 0$.

Use the half-angle formulas.

$$\sin\frac{\alpha}{2} = \sqrt{\frac{1-\cos\alpha}{2}} = \sqrt{\frac{1-\left(-\frac{24}{25}\right)}{2}} = \sqrt{\frac{25+24}{50}} = \sqrt{\frac{49}{50}} = \frac{7\sqrt{2}}{10}$$

Continued ▶

$$\cos\frac{\alpha}{2} = -\sqrt{\frac{1+\cos\alpha}{2}} = -\sqrt{\frac{1+\left(-\frac{24}{25}\right)}{2}}$$
$$= -\sqrt{\frac{25-24}{50}} = -\sqrt{\frac{1}{50}} = -\frac{\sqrt{2}}{10}$$
$$\tan\frac{\alpha}{2} = \frac{\sin\alpha}{1+\cos\alpha} = \frac{-\frac{7}{25}}{1+\left(-\frac{24}{25}\right)}$$
$$= \frac{-\frac{7}{25}}{\frac{1}{25}} = -7$$

50. $$\frac{1}{1-\cos 2x} = \frac{1}{1-1+2\sin^2 x}$$
$$= \frac{1}{2\sin^2 x} = \frac{1}{2}\csc^2 x$$

68. $$\cos^2\frac{x}{2} = \left[\pm\sqrt{\frac{1+\cos x}{2}}\right]^2$$
$$= \frac{1+\cos x}{2}$$
$$= \frac{1+\cos x}{2}\cdot\frac{\sec x}{\sec x}$$
$$= \frac{\sec x + 1}{2\sec x}$$

Exercise Set 3.4, page 251

22. $$\cos 3\theta + \cos 5\theta = 2\cos\frac{3\theta+5\theta}{2}\cos\frac{3\theta-5\theta}{2}$$
$$= 2\cos 4\theta\cos(-\theta) = 2\cos 4\theta\cos\theta$$

36. $$\sin 5x\cos 3x = \frac{1}{2}[\sin(5x+3x)+\sin(5x-3x)]$$
$$= \frac{1}{2}(\sin 8x + \sin 2x)$$
$$= \frac{1}{2}(2\sin 4x\cos 4x + 2\sin x\cos x)$$
$$= \sin 4x\cos 4x + \sin x\cos x$$

44. $$\frac{\cos 5x - \cos 3x}{\sin 5x + \sin 3x} = \frac{-2\sin\frac{5x+3x}{2}\sin\frac{5x-3x}{2}}{2\sin\frac{5x+3x}{2}\cos\frac{5x-3x}{2}}$$
$$= -\frac{\sin 4x\sin x}{\sin 4x\cos x} = -\tan x$$

62. $a = 1, b = \sqrt{3}, k = \sqrt{(\sqrt{3})^2 + (1)^2} = 2$. Thus α is a first-quadrant angle.

$$\sin\alpha = \frac{\sqrt{3}}{2} \quad\text{and}\quad \cos\alpha = \frac{1}{2}$$

Thus $\alpha = \frac{\pi}{3}$.

$$y = k\sin(x+\alpha)$$
$$y = 2\sin\left(x+\frac{\pi}{3}\right)$$

70. From Exercise 62, we know that

$$y = \sin x + \sqrt{3}\cos x = 2\sin\left(x+\frac{\pi}{3}\right)$$

The graph of $y = \sin x + \sqrt{3}\cos x$ has an amplitude of 2, and a phase shift of $-\frac{\pi}{3}$. It is the graph of $y = 2\sin x$ shifted $\frac{\pi}{3}$ units to the left.

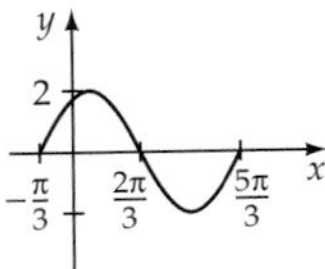

Exercise Set 3.5, page 265

2. $y = \sin^{-1}\frac{\sqrt{2}}{2}$ implies

$$\sin y = \frac{\sqrt{2}}{2} \quad\text{for } -\frac{\pi}{2} \le y \le \frac{\pi}{2}$$

Thus $y = \frac{\pi}{4}$.

28. Because $\tan(\tan^{-1} x) = x$ for all real numbers x, we have $\tan\left[\tan^{-1}\left(\frac{1}{2}\right)\right] = \frac{1}{2}$.

50. Let $x = \cos^{-1}\frac{3}{5}$. Thus

$$\cos x = \frac{3}{5} \quad\text{and}\quad \sin x = \sqrt{1-\left(\frac{3}{5}\right)^2} = \frac{4}{5}$$
$$y = \tan\left(\cos^{-1}\frac{3}{5}\right) = \tan x = \frac{\sin x}{\cos x} = \frac{4/5}{3/5} = \frac{4}{3}$$

56. $$y = \cos\left(\sin^{-1}\frac{3}{4} + \cos^{-1}\frac{5}{13}\right)$$

Let $\alpha = \sin^{-1}\frac{3}{4}$, $\sin\alpha = \frac{3}{4}$, $\cos\alpha = \sqrt{1-\left(\frac{3}{4}\right)^2} = \frac{\sqrt{7}}{4}$.

$$\beta = \cos^{-1}\frac{5}{13}, \cos\beta = \frac{5}{13}, \sin\beta = \sqrt{1 - \left(\frac{5}{13}\right)^2} = \frac{12}{13}.$$

$$\begin{aligned} y &= \cos(\alpha + \beta) \\ &= \cos\alpha\cos\beta - \sin\alpha\sin\beta \\ &= \frac{\sqrt{7}}{4}\cdot\frac{5}{13} - \frac{3}{4}\cdot\frac{12}{13} = \frac{5\sqrt{7}}{52} - \frac{36}{52} = \frac{5\sqrt{7} - 36}{52} \end{aligned}$$

66. $\sin^{-1}x + \cos^{-1}\frac{4}{5} = \frac{\pi}{6}$

$$\begin{aligned} \sin^{-1}x &= \frac{\pi}{6} - \cos^{-1}\frac{4}{5} \\ \sin(\sin^{-1}x) &= \sin\left(\frac{\pi}{6} - \cos^{-1}\frac{4}{5}\right) \\ x &= \sin\frac{\pi}{6}\cos\left(\cos^{-1}\frac{4}{5}\right) - \cos\frac{\pi}{6}\sin\left(\cos^{-1}\frac{4}{5}\right) \\ &= \frac{1}{2}\cdot\frac{4}{5} - \frac{\sqrt{3}}{2}\cdot\frac{3}{5} = \frac{4 - 3\sqrt{3}}{10} \end{aligned}$$

72. Let $\alpha = \cos^{-1}x$ and $\beta = \cos^{-1}(-x)$. Thus $\cos\alpha = x$ and $\cos\beta = -x$. We know that $\sin\alpha = \sqrt{1 - x^2}$ and $\sin\beta = \sqrt{1 - x^2}$ because α is in Quadrant I and β is in Quadrant II.

$$\begin{aligned} &\cos^{-1}x + \cos^{-1}(-x) \\ &= \alpha + \beta \\ &= \cos^{-1}[\cos(\alpha + \beta)] \\ &= \cos^{-1}(\cos\alpha\cos\beta - \sin\alpha\sin\beta) \\ &= \cos^{-1}\left[x(-x) - \sqrt{1 - x^2}\cdot\sqrt{1 - x^2}\right] \\ &= \cos^{-1}(-x^2 - 1 + x^2) \\ &= \cos^{-1}(-1) = \pi \end{aligned}$$

76. The graph of $y = f(x - a)$ is a horizontal shift of the graph of $y = f(x)$. Therefore, the graph of $y = \cos^{-1}(x - 1)$ is the graph of $y = \cos^{-1}x$ shifted 1 unit to the right.

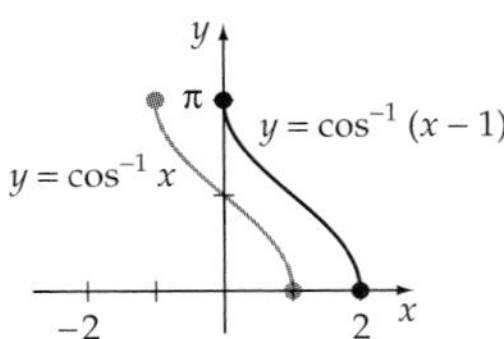

84. a.

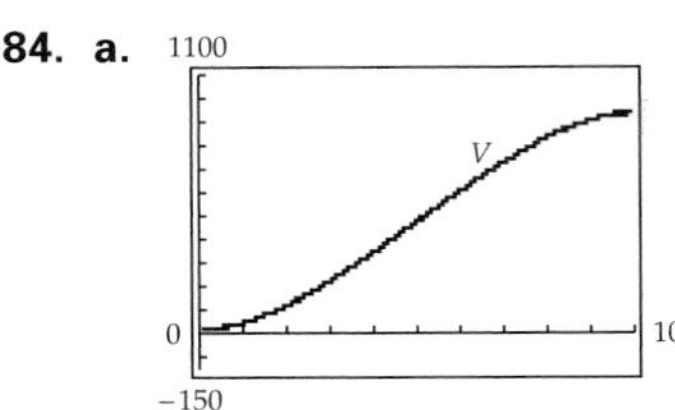

b. Although the water rises 0.1 foot in each case, there is more surface area (and thus more volume of water) at the 4.9- to 5.0-foot level near the diameter of the cylinder than at the 0.1- to 0.2-foot level near the bottom.

c.

$$\begin{aligned} V(4) &= 12\left[25\cos^{-1}\left(\frac{5 - (4)}{5}\right) - [5 - (4)]\sqrt{10(4) - (4)^2}\right] \\ &= 12\left[25\cos^{-1}\left(\frac{1}{5}\right) - \sqrt{24}\right] \\ &\approx 352.04 \text{ cubic feet} \end{aligned}$$

d. 1100

V

V = 288

0 Intersection X=3.4476991 Y=288 10

−150

When $V = 288$ cubic feet, $x \approx 3.45$ feet.

Exercise Set 3.6, page 278

14.

$$\begin{aligned} 2\cos^2 x + 1 &= -3\cos x \\ 2\cos^2 x + 3\cos x + 1 &= 0 \\ (2\cos x + 1)(\cos x + 1) &= 0 \end{aligned}$$

$$2\cos x + 1 = 0 \quad \text{or} \quad \cos x + 1 = 0$$

$$\cos x = -\frac{1}{2} \qquad \cos x = -1$$

$$x = \frac{2\pi}{3}, \frac{4\pi}{3} \qquad x = \pi$$

The solutions in the interval $0 \le x < 2\pi$ are $\frac{2\pi}{3}$, π, and $\frac{4\pi}{3}$.

52. $\sin x + 2\cos x = 1$

$$\begin{aligned} \sin x &= 1 - 2\cos x \\ (\sin x)^2 &= (1 - 2\cos x)^2 \\ \sin^2 x &= 1 - 4\cos x + 4\cos^2 x \\ 1 - \cos^2 x &= 1 - 4\cos x + 4\cos^2 x \\ 0 &= \cos x(5\cos x - 4) \end{aligned}$$

$$\cos x = 0 \quad \text{or} \quad 5\cos x - 4 = 0$$

$$x = 90^\circ, 270^\circ \qquad \cos x = \frac{4}{5}$$

$$x \approx 36.9^\circ, 323.1^\circ$$

The solutions in the interval $0 \le x < 360^\circ$ are 90° and 323.1°. (*Note:* $x = 270^\circ$ and $x = 36.9^\circ$ are extraneous solutions. Neither of these values satisfies the original equation.)

69. $(Y \circ F)(x)$ converts x inches to yards. **71. a.** 99.8; this is identical to the slope of the line through $(0, C(0))$ and $(1, C(1))$. **b.** 156.2 **c.** −49.7 **d.** −30.8 **e.** −16.4 **f.** 0

Prepare for This Section (1.6), page 81

PS1. $y = -\frac{2}{5}x + 3$ **PS2.** $y = \frac{1}{x-1}$ **PS3.** −1 **PS4.** (3, 7) **PS5.** all real numbers **PS6.** $\{x \mid x \geq -2\}$

Exercise Set 1.6, page 90

1. 3 **3.** −3 **5.** 3 **7.** range **9.** Yes

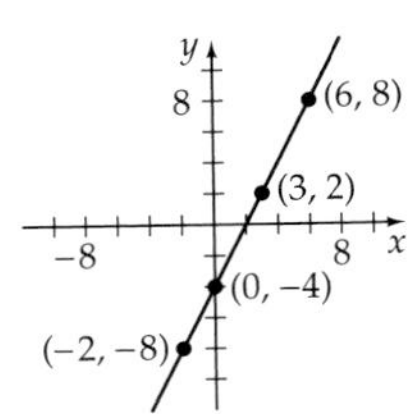

11. Yes

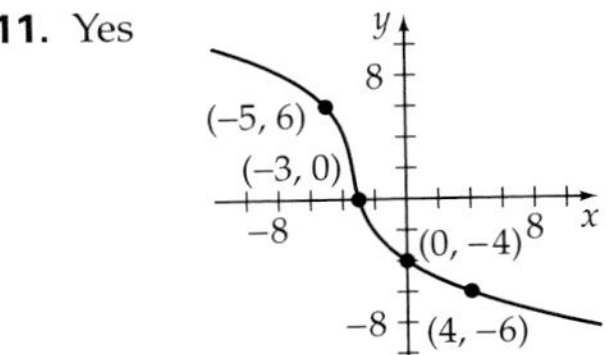

13. Yes

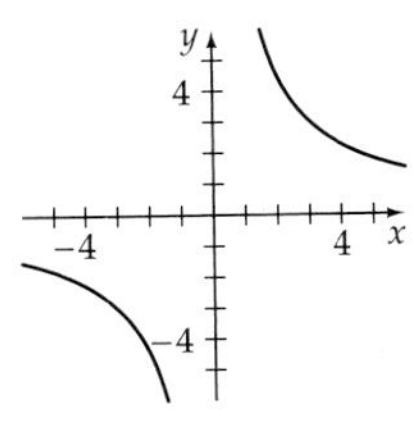

15. No

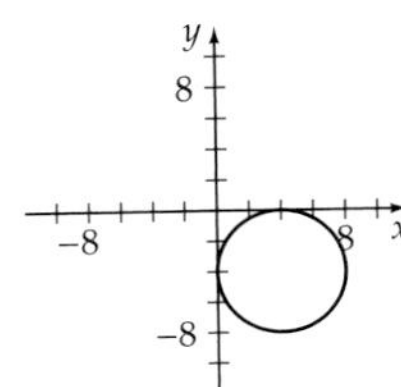

17. Yes **19.** Yes **21.** No **23.** Yes **25.** Yes **27.** $\{(1, -3), (2, -2), (5, 1), (-7, 4)\}$

29. $\{(1, 0), (2, 1), (4, 2), (8, 3), (16, 4)\}$ **31.** $f^{-1}(x) = \frac{1}{2}x - 2$ **33.** $f^{-1}(x) = \frac{1}{3}x + \frac{7}{3}$ **35.** $f^{-1}(x) = -\frac{1}{2}x + \frac{5}{2}$ **37.** $f^{-1}(x) = \frac{x}{x-2}, x \neq 2$ **39.** $f^{-1}(x) = \frac{x+1}{1-x}, x \neq 1$ **41.** $f^{-1}(x) = \sqrt{x-1}, x \geq 1$ **43.** $f^{-1}(x) = x^2 + 2, x \geq 0$ **45.** $f^{-1}(x) = \sqrt{x+4} - 2, x \geq -4$ **47.** $f^{-1}(x) = -\sqrt{x+5} - 2, x \geq -5$ **49.** $V^{-1}(x) = \sqrt[3]{x}$. V^{-1} finds the length of a side of a cube given the volume. **51.** $f^{-1}(x) = \frac{9}{5}x + 32$; $f^{-1}(x)$ is used to convert x degrees Celsius to its equivalent Fahrenheit temperature. **53.** $s^{-1}(x) = \frac{1}{2}x - 12$ **55.** $E^{-1}(s) = 20s - 50{,}000$. The executive can determine the value of the software that must be sold in order to achieve a given monthly income. **57. a.** $p(10) \approx 0.12 = 12\%$; $p(30) \approx 0.71 = 71\%$ **b.** The graph of p, for $1 \leq n \leq 60$, is an increasing function. Thus p (with $1 \leq n \leq 60$) has an inverse that is a function. **c.** Answers will vary. **59. a.** 25 47 71 67 47 59 53 71 33 47 43 27 63 47 53 39 **b.** PHONE HOME **c.** Answers will vary. **61.** Because the function is increasing and 4 is between 2 and 5, c must be between 7 and 12. **63.** between 2 and 5 **65.** between 3 and 7 **67.** slope: $\frac{1}{m}$; y-intercept: $\left(0, -\frac{b}{m}\right)$ **69.** The reflection of f across the line given by $y = x$ yields f. Thus f is its own inverse. **71.** Yes **73.** No

Prepare for This Section (1.7), page 95

PS1. slope: $-\frac{3}{4}$; y-intercept: 4 **PS2.** slope: $\frac{3}{4}$; y-intercept: −3 **PS3.** $y = -0.45x + 2.3$ **PS4.** 19 **PS5.** 69 **PS6.** 3

Exercise Set 1.7, page 103

1. no linear relationship **3.** linear **5.** Figure A **7.** $y = 2.00862069x + 0.5603448276$ **9.** $y = -0.7231182796x + 9.233870968$ **11.** $y = 2.222641509x - 7.364150943$ **13.** $y = 1.095779221x^2 - 2.69642857x + 1.136363636$ **15.** $y = -0.2987274717x^2 - 3.20998141x + 3.416463667$ **17. a.** $y = 23.55706665x - 24.4271215$ **b.** 1248 centimeters **19. a.** $y = 0.1094224924x + 0.7978723404$ **b.** 4.3 meters per second **21. a.** $y = 0.1628623408x - 0.6875682232$ **b.** 25 **23.** No, because the linear correlation coefficient is close to 0. **25. a.** Yes, there is a strong linear correlation. **b.** $y = -0.9033088235x + 78.62573529$ **c.** 56 years **27. a.** positively **b.** 1098 calories **29.** $y = -0.6328671329x^2 + 33.61608392x - 379.4405594$
31. a. $y = -0.0165034965x^2 + 1.366713287x + 5.685314685$ **b.** 32.8 miles per gallon
33. a. 5-pound: $s = 0.6130952381t^2 - 0.0714285714t + 0.1071428571$
10-pound: $s = 0.6091269841t^2 - 0.0011904762t - 0.3$
15-pound: $s = 0.5922619048t^2 + 0.3571428571t - 1.520833333$
b. All the regression equations are approximately the same. Therefore, the equations of motion of the three masses are the same.

Chapter 1 Assessing Concepts, page 111

1. a, c, d, and e **2.** No **3.** 3 **4.** $|x + 2| < 3$ **5.** It is the slope of the line between $(a, f(a))$ and $(b, f(b))$. **6.** $(3, -2)$ **7.** $(7, 3)$ **8.** $(-3, 6)$ **9.** $(3, 4)$ **10.** Yes. The slope of the regression line in negative.

Chapter 1 Review Exercises, page 112

1. $-\frac{9}{4}$ [1.1] **2.** -4 [1.1] **3.** -2 [1.1] **4.** $-\frac{2}{3}$ [1.1] **5.** $-3, 6$ [1.1] **6.** $\frac{1}{2}, 4$ [1.1] **7.** $\frac{-1 \pm \sqrt{13}}{6}$ [1.1] **8.** $\frac{-3 \pm \sqrt{41}}{4}$ [1.1] **9.** $c \geq -6$ [1.1] **10.** $a > 1$ [1.1] **11.** $(-\infty, -3] \cup [4, \infty)$ [1.1] **12.** $-\frac{1}{2} < x < 1$ [1.1] **13.** $(-\infty, 1) \cup (4, \infty)$ [1.1] **14.** $-1 \leq x \leq \frac{5}{3}$ [1.1] **15.** 10 [1.2] **16.** $3\sqrt{31}$ [1.2] **17.** 5 [1.2] **18.** $7\sqrt{5}$ [1.2] **19.** $\sqrt{181}$ [1.2] **20.** $4\sqrt{5}$ [1.2] **21.** $\left(-\frac{1}{2}, 10\right)$ [1.2] **22.** $(2, -2)$ [1.2]

23. [1.4] **24. a.** **b.** **c.** [1.4]

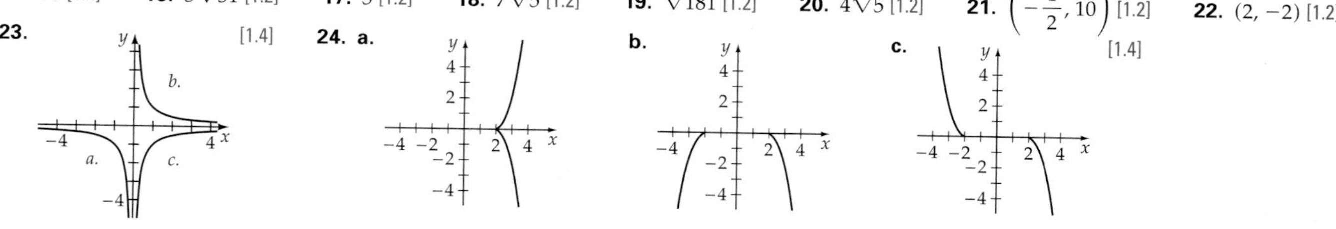

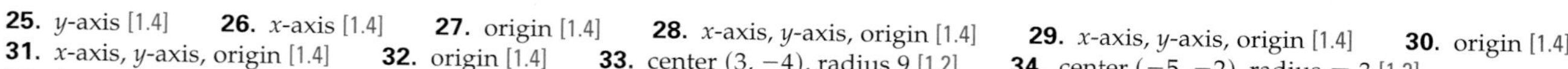

25. y-axis [1.4] **26.** x-axis [1.4] **27.** origin [1.4] **28.** x-axis, y-axis, origin [1.4] **29.** x-axis, y-axis, origin [1.4] **30.** origin [1.4] **31.** x-axis, y-axis, origin [1.4] **32.** origin [1.4] **33.** center $(3, -4)$, radius 9 [1.2] **34.** center $(-5, -2)$, radius $= 3$ [1.2] **35.** $(x - 2)^2 + (y + 3)^2 = 5^2$ [1.2] **36.** $(x + 5)^2 + (y - 1)^2 = 64$ [1.2] **37. a.** 2 **b.** 10 **c.** $3t^2 + 4t - 5$ **d.** $3x^2 + 6xh + 3h^2 + 4x + 4h - 5$ **e.** $9t^2 + 12t - 15$ **f.** $27t^2 + 12t - 5$ [1.3] **38. a.** $\sqrt{55}$ **b.** $\sqrt{39}$ **c.** 0 **d.** $\sqrt{64 - x^2}$ **e.** $2\sqrt{64 - t^2}$ **f.** $2\sqrt{16 - t^2}$ [1.3] **39. a.** 5 **b.** -11 **c.** $x^2 - 12x + 32$ **d.** $x^2 + 4x - 8$ [1.5] **40. a.** 79 **b.** 56 **c.** $2x^2 - 4x + 9$ **d.** $2x^2 + 6$ [1.5] **41.** $8x + 4h - 3$ [1.5] **42.** $3x^2 + 3xh + h^2 - 1$ [1.5] **43.** $(-\infty, \infty)$ [1.3] **44.** $x \leq 6$ [1.3] **45.** $[-5, 5]$ [1.3] **46.** all real numbers except -3 and 5 [1.3]

47. [1.3] **48.** [1.3] **49.** [1.3] **50.** [1.3]

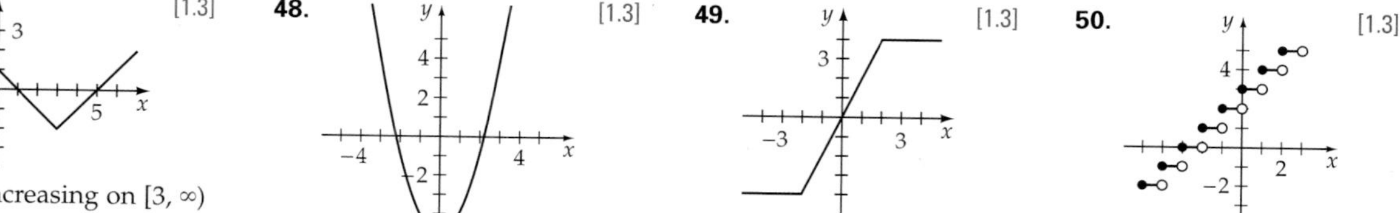

47. increasing on $[3, \infty)$; decreasing on $(-\infty, 3]$

48. decreasing on $(-\infty, 0]$; increasing on $[0, \infty)$

49. increasing on $[-2, 2]$; constant on $(-\infty, -2] \cup [2, \infty)$

50. constant on $[n, n + 1)$, where n is an integer

51. [1.3] **52.** [1.3] **53.** [1.3/1.4] **54.** [1.3/1.4]

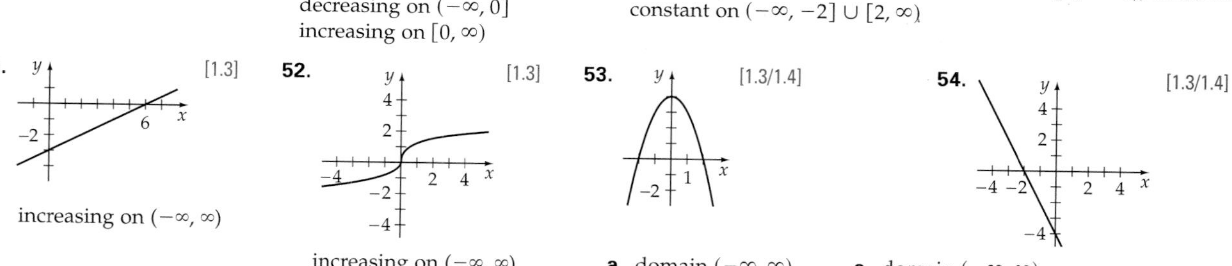

51. increasing on $(-\infty, \infty)$

52. increasing on $(-\infty, \infty)$

53. a. domain $(-\infty, \infty)$; range $\{y \mid y \leq 4\}$ **b.** even

54. a. domain $(-\infty, \infty)$; range $(-\infty, \infty)$ **b.** neither

55. [1.3/1.4] **56.** [1.3/1.4] **57.** [1.3/1.4] **58.** [1.3/1.4]

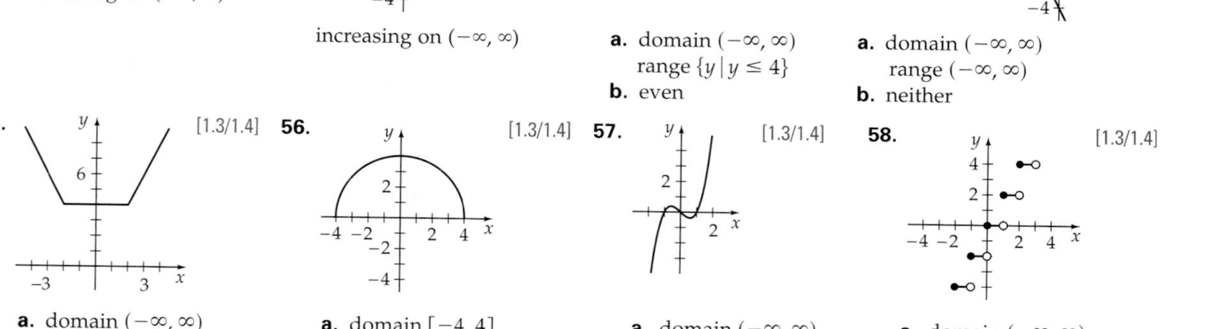

55. a. domain $(-\infty, \infty)$; range $\{y \mid y \geq 4\}$ **b.** even

56. a. domain $[-4, 4]$; range $[0, 4]$ **b.** even

57. a. domain $(-\infty, \infty)$; range $(-\infty, \infty)$ **b.** odd

58. a. domain $(-\infty, \infty)$; range: even integers **b.** neither

59. $f(x) + g(x) = x^2 + x - 6$, domain: $(-\infty, \infty)$
$f(x) - g(x) = x^2 - x - 12$, domain: $(-\infty, \infty)$
$f(x) \cdot g(x) = x^3 + 3x^2 - 9x - 27$, domain: $(-\infty, \infty)$
$\dfrac{f(x)}{g(x)} = x - 3$, domain $\{x \mid x \neq -3\}$ [1.5]

60. $f(x) + g(x) = x^3 + x^2 - 2x + 12$, domain: $(-\infty, \infty)$
$f(x) - g(x) = x^3 - x^2 + 2x + 4$, domain: $(-\infty, \infty)$
$f(x) \cdot g(x) = x^5 - 2x^4 + 4x^3 + 8x^2 - 16x + 32$, domain: $(-\infty, \infty)$
$\dfrac{f(x)}{g(x)} = x + 2$, domain: $(-\infty, \infty)$ [1.5]

61. Yes [1.6] **62.** Yes [1.6]

63. Yes [1.6] **64.** No [1.6] **65.** $f^{-1}(x) = \dfrac{x+4}{3}$ [1.6] **66.** $g^{-1}(x) = -\dfrac{1}{2}x + \dfrac{3}{2}$ [1.6] **67.** $h^{-1}(x) = -2x - 4$ [1.6] **68.** $k^{-1}(x) = \dfrac{1}{x}$ [1.6]

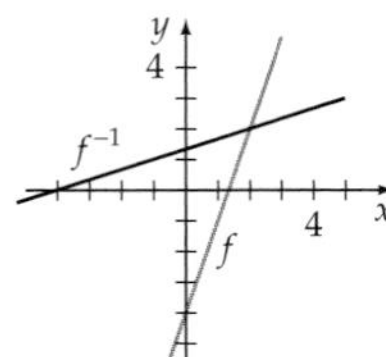

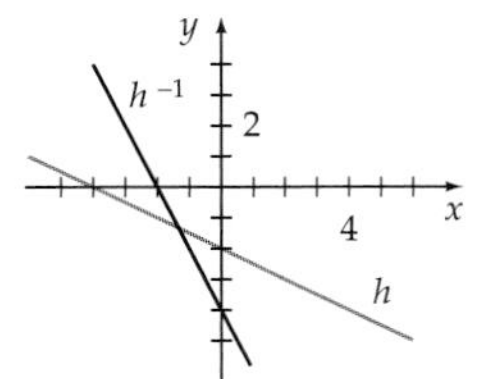

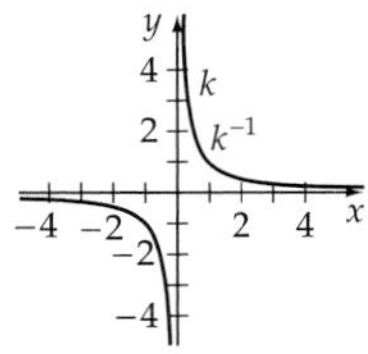

69. $t \approx 3.7$ seconds [1.3] **70. a.** 150 feet **b.** 525 feet [1.3] **71. a.** $y = 0.018024687x + 0.00050045744$ **b.** Yes. $r \approx 0.999$, which is very close to 1. **c.** 1.8 seconds [1.7] **72. a.** $y = 0.0047952048x^2 - 1.756843157x + 180.4065934$ **b.** The graph of the regression equation never crosses the x-axis. Therefore, the model predicts that the can will never be empty. **c.** A regression model only approximates a situation. [1.7]

Chapter 1 Quantitative Reasoning Exercises, page 114

QR1. a. 3 **b.** 2 **c.** 52 **QR2.** Factor the modulus, m. **QR3.** Answer will vary.

Chapter 1 Test, page 116

1. $\dfrac{1}{2}$ [1.1] **2.** $x \leq 1$ [1.1] **3.** $-\dfrac{1}{2}, 2$ [1.1] **4.** $-\dfrac{2}{3}, 1$ [1.1] **5.** $\left(-\infty, -\dfrac{2}{5}\right) \cup (2, \infty)$ [1.1] **6.** $\sqrt{85}$ [1.2] **7.** midpoint $(1, 1)$; length $2\sqrt{13}$ [1.2]

8. $(0, \sqrt{2}), (0, -\sqrt{2}), (-4, 0)$ [1.2] **9.** [1.2] **10.** center $(2, -1)$; radius 3 [1.2]
11. -4 [1.3]
12. domain $\{x \mid x \geq 4 \text{ or } x \leq -4\}$ [1.3]

13. [1.3/1.4]
increasing on $(-\infty, 2]$
decreasing on $[2, \infty)$

14. [1.4] **15.** b [1.4] **16.** $x^2 + x - 3; \dfrac{x^2 - 1}{x - 2}, x \neq 2$ [1.5] **17.** $2x + h$ [1.5]

18. $x - 2\sqrt{x-2} - 1$ [1.5] **19.** $f^{-1}(x) = \dfrac{x}{1-x}$ [1.6] **20. a.** $y = -7.98245614x + 767.122807$ **b.** 56.7 calories [1.7]

Exercise Set 2.1, page 130

1. 75°, 165° **3.** 19°45′, 109°45′ **5.** 33°26′45″, 123°26′45″ **7.** $\dfrac{\pi}{2} - 1, \pi - 1$ **9.** $\dfrac{\pi}{4}, \dfrac{3\pi}{4}$ **11.** $\dfrac{\pi}{10}, \dfrac{3\pi}{5}$ **13.** 250°, Quadrant III
15. 105°, Quadrant II **17.** 296°, Quadrant IV **19.** 24°33′36″ **21.** 64°9′28.8″ **23.** 3°24′7.2″ **25.** 25.42° **27.** 183.56° **29.** 211.78°
31. $\dfrac{\pi}{6}$ **33.** $\dfrac{\pi}{2}$ **35.** $\dfrac{11\pi}{12}$ **37.** $\dfrac{7\pi}{3}$ **39.** $\dfrac{13\pi}{4}$ **41.** $-\dfrac{\pi}{20}$ **43.** 420° **45.** 36° **47.** 30° **49.** 67.5° **51.** 660° **53.** −75° **55.** 85.94°
57. 2.32 **59.** 472.69° **61.** 4, 229.18° **63.** 2.38, 136.63° **65.** 6.28 inches **67.** 18.33 centimeters **69.** 3π **71.** $\dfrac{5\pi}{12}$ radians or 75°

73. $\frac{\pi}{30}$ radian per second $\approx$ 0.105 radian per second **75.** $\frac{5\pi}{3}$ radians per second $\approx$ 5.24 radians per second **77.** $\frac{10\pi}{9}$ radians per second $\approx$ 3.49 radians per second **79.** 40 mph **81.** 1885 feet **83.** 6.9 mph **85.** 840,000 miles **87. a.** 3.9 radians per hour **b.** 27,300 kilometers per hour **89. a.** B **b.** Both points have the same linear velocity. **91. a.** 1.15 statute miles **b.** 10% **93.** 13 square inches **95.** 4680 square centimeters **97.** 1780 miles

Prepare for This Section (2.2), page 134

PS1. $\frac{\sqrt{3}}{3}$ **PS2.** $\sqrt{2}$ **PS3.** 2 **PS4.** $\frac{\sqrt{3}}{3}$ **PS5.** 3.54 **PS6.** 10.39

Exercise Set 2.2, page 142

1. $\sin\theta = \frac{12}{13}$ $\csc\theta = \frac{13}{12}$ $\cos\theta = \frac{5}{13}$ $\sec\theta = \frac{13}{5}$ $\tan\theta = \frac{12}{5}$ $\cot\theta = \frac{5}{12}$

3. $\sin\theta = \frac{4}{7}$ $\csc\theta = \frac{7}{4}$ $\cos\theta = \frac{\sqrt{33}}{7}$ $\sec\theta = \frac{7\sqrt{33}}{33}$ $\tan\theta = \frac{4\sqrt{33}}{33}$ $\cot\theta = \frac{\sqrt{33}}{4}$

5. $\sin\theta = \frac{5\sqrt{29}}{29}$ $\csc\theta = \frac{\sqrt{29}}{5}$ $\cos\theta = \frac{2\sqrt{29}}{29}$ $\sec\theta = \frac{\sqrt{29}}{2}$ $\tan\theta = \frac{5}{2}$ $\cot\theta = \frac{2}{5}$

7. $\sin\theta = \frac{\sqrt{21}}{7}$ $\csc\theta = \frac{\sqrt{21}}{3}$ $\cos\theta = \frac{2\sqrt{7}}{7}$ $\sec\theta = \frac{\sqrt{7}}{2}$ $\tan\theta = \frac{\sqrt{3}}{2}$ $\cot\theta = \frac{2\sqrt{3}}{3}$

9. $\sin\theta = \frac{\sqrt{3}}{2}$ $\csc\theta = \frac{2\sqrt{3}}{3}$ $\cos\theta = \frac{1}{2}$ $\sec\theta = 2$ $\tan\theta = \sqrt{3}$ $\cot\theta = \frac{\sqrt{3}}{3}$

11. $\sin\theta = \frac{6\sqrt{61}}{61}$ $\csc\theta = \frac{\sqrt{61}}{6}$ $\cos\theta = \frac{5\sqrt{61}}{61}$ $\sec\theta = \frac{\sqrt{61}}{5}$ $\tan\theta = \frac{6}{5}$ $\cot\theta = \frac{5}{6}$

13. $\frac{3}{4}$ **15.** $\frac{4}{5}$ **17.** $\frac{3}{4}$ **19.** $\frac{12}{13}$ **21.** $\frac{13}{5}$ **23.** $\frac{3}{2}$ **25.** $\sqrt{2}$ **27.** $-\frac{3}{4}$ **29.** $\frac{5}{4}$ **31.** $\sqrt{3}$ **33.** $\frac{3\sqrt{2}+2\sqrt{3}}{6}$ **35.** $\frac{3-\sqrt{3}}{3}$ **37.** $2\sqrt{2}-\sqrt{3}$ **39.** 0.6249 **41.** 0.4488 **43.** 0.8221 **45.** 1.0053 **47.** 0.4816 **49.** 1.0729 **51.** 9.5 feet **53.** 92.9 inches **55.** 5.1 feet **57.** 1.7 miles **59.** 74.6 feet **61.** 686,000,000 kilometers **65.** 612 feet **67.** 560 feet **69. a.** 559 feet **b.** 193 feet **71.** $\sqrt{27}$ meters$\approx$5.2 meters **73.** $\approx$8.5 feet

Prepare for This Section (2.3), page 147

PS1. $-\frac{4}{3}$ **PS2.** $\frac{\sqrt{5}}{2}$ **PS3.** 60 **PS4.** $\frac{\pi}{5}$ **PS5.** π **PS6.** $\sqrt{34}$

Exercise Set 2.3, page 153

1. $\sin\theta = \frac{3\sqrt{13}}{13}$ $\csc\theta = \frac{\sqrt{13}}{3}$ $\cos\theta = \frac{2\sqrt{13}}{13}$ $\sec\theta = \frac{\sqrt{13}}{2}$ $\tan\theta = \frac{3}{2}$ $\cot\theta = \frac{2}{3}$

3. $\sin\theta = \frac{3\sqrt{13}}{13}$ $\csc\theta = \frac{\sqrt{13}}{3}$ $\cos\theta = -\frac{2\sqrt{13}}{13}$ $\sec\theta = -\frac{\sqrt{13}}{2}$ $\tan\theta = -\frac{3}{2}$ $\cot\theta = -\frac{2}{3}$

5. $\sin\theta = -\frac{5\sqrt{89}}{89}$ $\csc\theta = -\frac{\sqrt{89}}{5}$ $\cos\theta = -\frac{8\sqrt{89}}{89}$ $\sec\theta = -\frac{\sqrt{89}}{8}$ $\tan\theta = \frac{5}{8}$ $\cot\theta = \frac{8}{5}$

7. $\sin\theta = 0$ $\csc\theta$ is undefined. $\cos\theta = -1$ $\sec\theta = -1$ $\tan\theta = 0$ $\cot\theta$ is undefined.

9. 0 **11.** 0 **13.** 1 **15.** 0 **17.** undefined **19.** 1 **21.** Quadrant I **23.** Quadrant IV **25.** Quadrant III **27.** $\frac{\sqrt{3}}{3}$ **29.** -1 **31.** $-\frac{\sqrt{3}}{3}$ **33.** $\frac{2\sqrt{3}}{3}$ **35.** $-\frac{\sqrt{3}}{3}$ **37.** 20° **39.** 9° **41.** $\frac{\pi}{5}$ **43.** $\pi - \frac{8}{3}$ **45.** 34° **47.** 65° **49.** $-\frac{\sqrt{2}}{2}$ **51.** 1 **53.** $-\frac{2\sqrt{3}}{3}$ **55.** $\frac{\sqrt{2}}{2}$ **57.** $\sqrt{2}$ **59.** cot 540° is undefined. **61.** 0.798636 **63.** -0.438371 **65.** -1.26902 **67.** -0.587785 **69.** -1.70130 **71.** -3.85522 **73.** 0 **75.** 1 **77.** $-\frac{3}{2}$ **79.** 1 **81.** 30°, 150° **83.** 150°, 210° **85.** 225°, 315° **87.** $\frac{3\pi}{4}, \frac{7\pi}{4}$ **89.** $\frac{5\pi}{6}, \frac{11\pi}{6}$ **91.** $\frac{\pi}{3}, \frac{2\pi}{3}$

Prepare for This Section (2.4), page 155

PS1. Yes **PS2.** Yes **PS3.** No **PS4.** 2π **PS5.** even function **PS6.** neither

Exercise Set 2.4, page 166

1. $\left(\frac{\sqrt{3}}{2}, \frac{1}{2}\right)$ **3.** $\left(-\frac{\sqrt{3}}{2}, -\frac{1}{2}\right)$ **5.** $\left(\frac{1}{2}, -\frac{\sqrt{3}}{2}\right)$ **7.** $\left(\frac{\sqrt{3}}{2}, -\frac{1}{2}\right)$ **9.** $(-1, 0)$ **11.** $\left(-\frac{1}{2}, -\frac{\sqrt{3}}{2}\right)$ **13.** $-\frac{\sqrt{3}}{3}$ **15.** $-\frac{1}{2}$ **17.** $-\frac{2\sqrt{3}}{3}$ **19.** -1 **21.** $-\frac{2\sqrt{3}}{3}$ **23.** 0.9391 **25.** -1.1528 **27.** -0.2679 **29.** 0.8090 **31.** 48.0889 **33. a.** 0.9 **b.** -0.4 **35. a.** -0.8 **b.** 0.6 **37.** 0.4, 2.7 **39.** 3.4, 6.0 **41.** odd **43.** neither **45.** even **47.** odd **49.** 2π **51.** π **53.** 2π **61.** $\sin t$ **63.** $\sec t$ **65.** $-\tan^2 t$ **67.** $-\cot t$ **69.** $\cos^2 t$ **71.** $2\csc^2 t$ **73.** $\csc^2 t$ **75.** 1 **77.** $\sqrt{1-\cos^2 t}$ **79.** $\sqrt{1+\cot^2 t}$ **81.** 750 miles **83.** $-\frac{\sin^2 t}{\cos t}$ **85.** $\csc t \sec t$ **87.** $1 - 2\sin t + \sin^2 t$ **89.** $1 - 2\sin t \cos t$ **91.** $\cos^2 t$ **93.** $2\csc t$ **95.** $(\cos t - \sin t)(\cos t + \sin t)$ **97.** $(\tan t + 2)(\tan t - 3)$ **99.** $(2\sin t + 1)(\sin t - 1)$ **101.** $\frac{\sqrt{2}}{2}$ **103.** $-\frac{\sqrt{3}}{3}$

Prepare for This Section (2.5), page 169

PS1. 0.7 **PS2.** -0.7 **PS3.** Reflect the graph of $y = f(x)$ across the x-axis. **PS4.** Compress each point on the graph of $y = f(x)$ toward the y-axis by a factor of $\frac{1}{2}$. **PS5.** 6π **PS6.** 5π

Exercise Set 2.5, page 177

1. $2, 2\pi$ **3.** $1, \pi$ **5.** $\frac{1}{2}, 1$ **7.** $2, 4\pi$ **9.** $\frac{1}{2}, 2\pi$ **11.** $1, 8\pi$ **13.** 2, 6 **15.** $3, 3\pi$ **17.** 4.7, 2.5

19.

21.

23.

25.

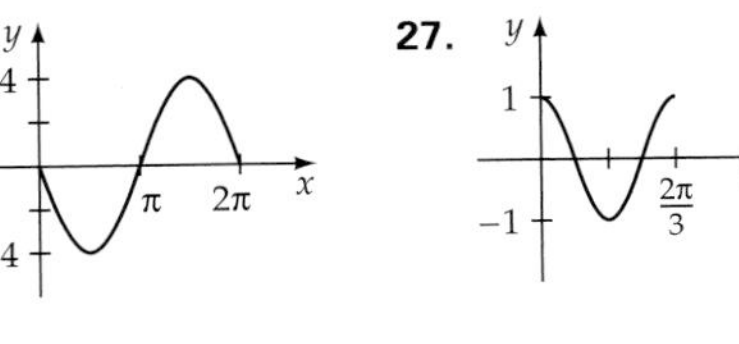

27.

29.

31.

33.

35.

37.

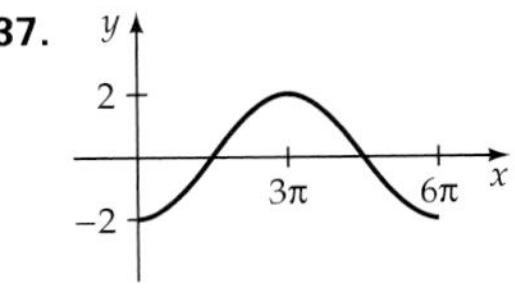

39.

41.

43.

45.

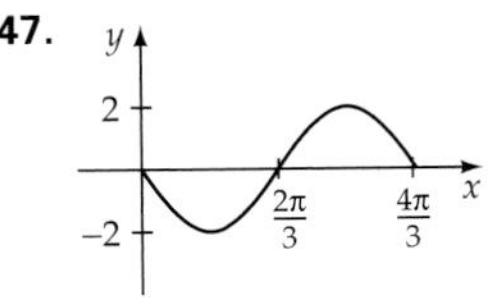

47.

49.

51.

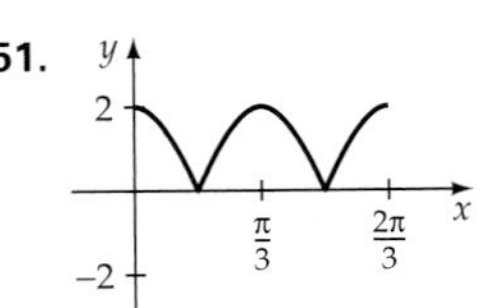

53.

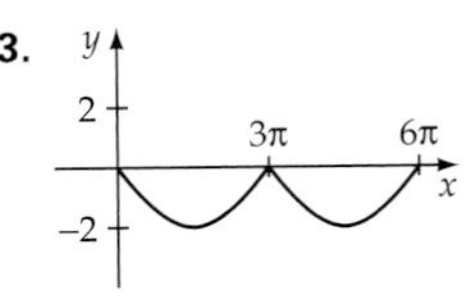

55.

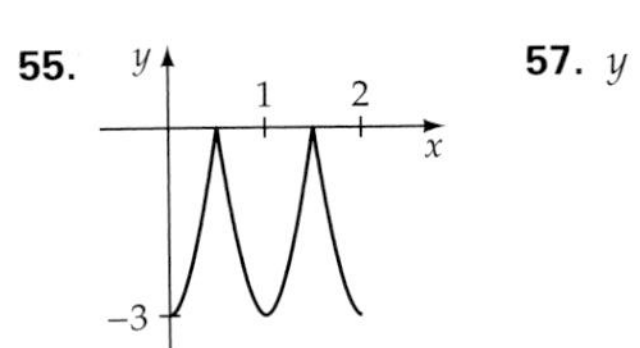

57. $y = \cos 2x$

59. $y = 2 \sin \frac{2}{3}x$ **61.** $y = -2 \cos \pi x$ **63. a.** $V = 4 \sin \pi t,\ 0 \le t \le 8$ milliseconds **b.** $\frac{1}{2}$ cycle per millisecond

65.

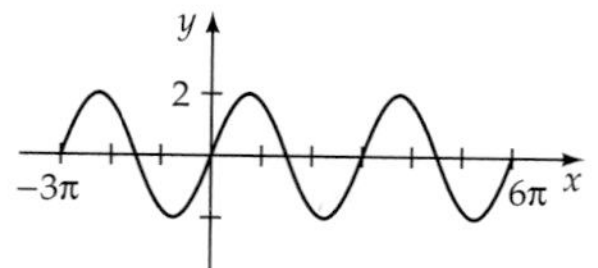

67.

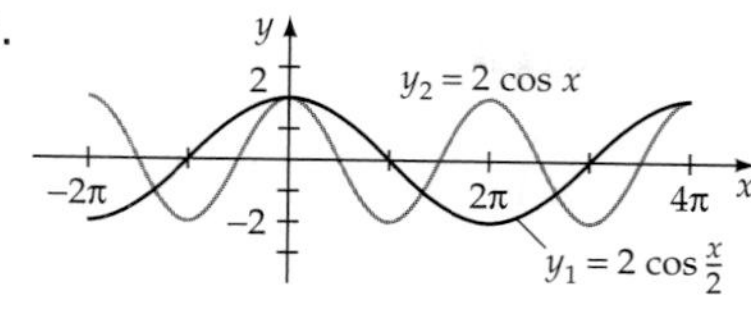

69.

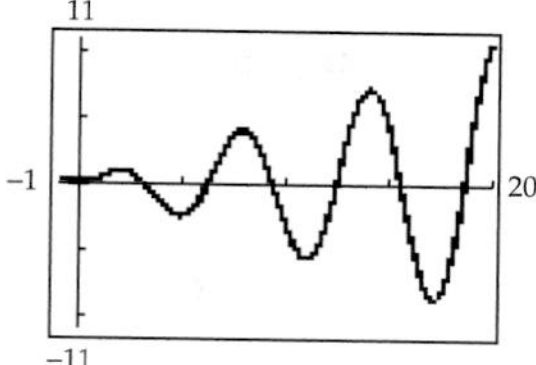

71.

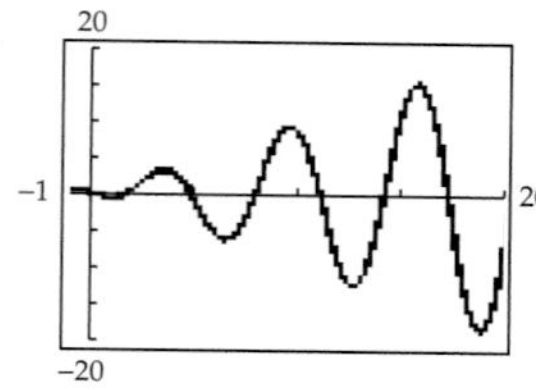

73.

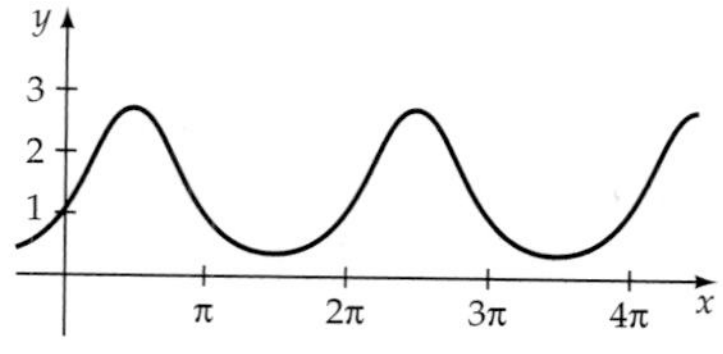

maximum $= e$, minimum $= \frac{1}{e} \approx 0.3679$, period $= 2\pi$

75. $y = 2 \sin \frac{2}{3}x$ **77.** $y = 2.5 \sin \frac{5\pi}{8}x$

79. $y = 3 \cos \frac{4\pi}{5}x$

Prepare for This Section (2.6), page 180

PS1. 1.7 **PS2.** 0.6 **PS3.** Stretch each point on the graph of $y = f(x)$ away from the x-axis by a factor of 2.
PS4. Shift the graph of $y = f(x)$ 2 units to the right and 3 units up. **PS5.** 2π **PS6.** $\frac{4}{3}\pi$

Exercise Set 2.6, page 189

1. $\frac{\pi}{2} + k\pi$, k an integer **3.** $\frac{\pi}{2} + k\pi$, k an integer **5.** 2π **7.** π **9.** 2π **11.** $\frac{2\pi}{3}$ **13.** $\frac{\pi}{3}$ **15.** 8π **17.** 1 **19.** 5 **21.** 8.5

23.

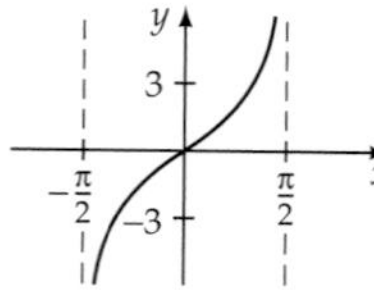

25.

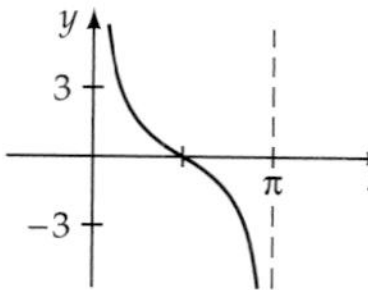

27.

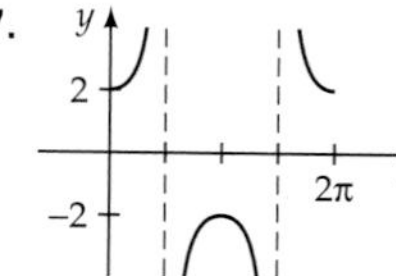

29.

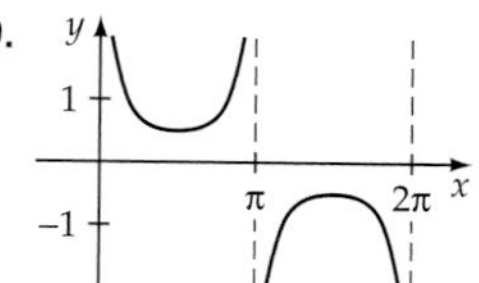

31.

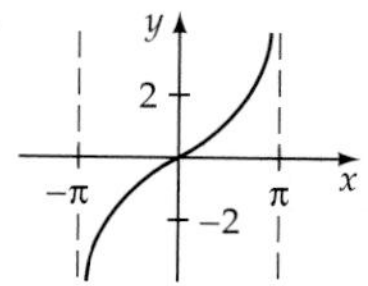

33.

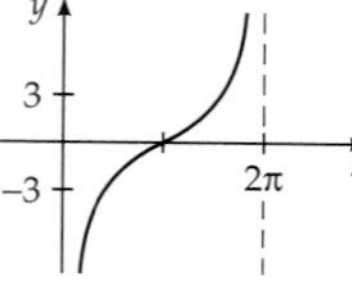

35.

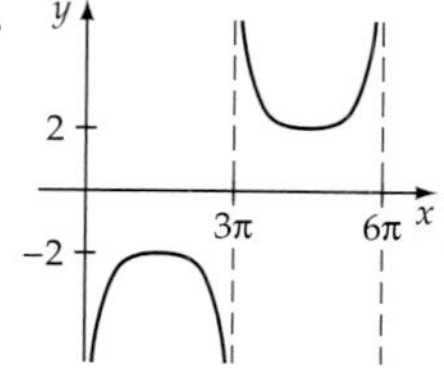

37.

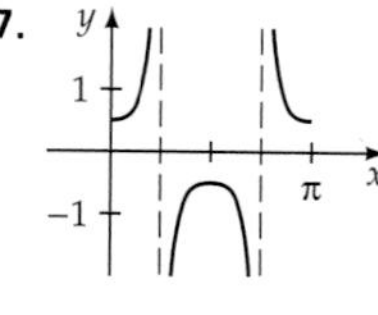

39.

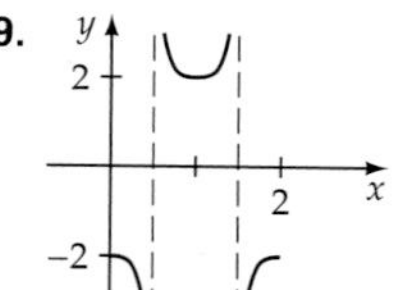

41.

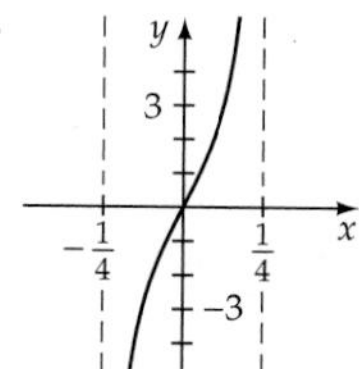

43.

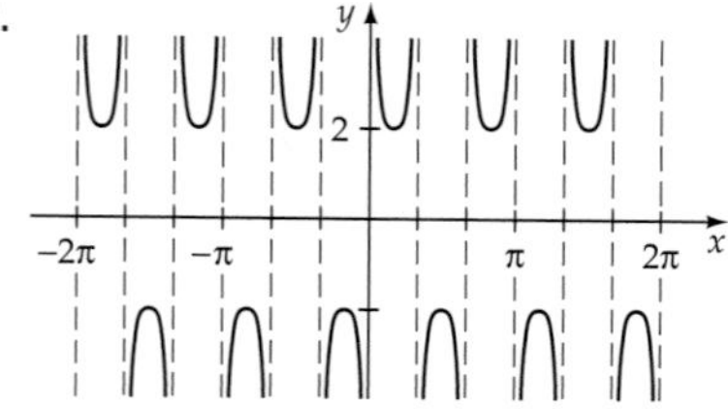

45.

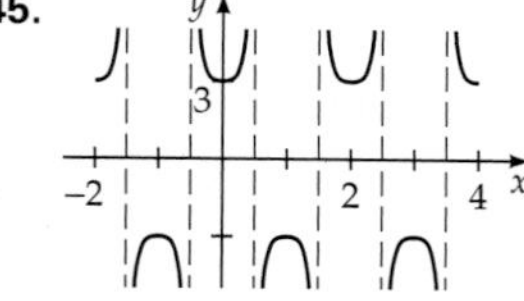

47.

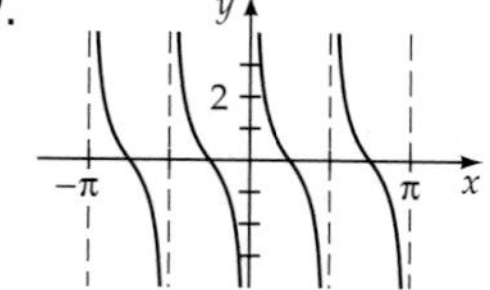

49.

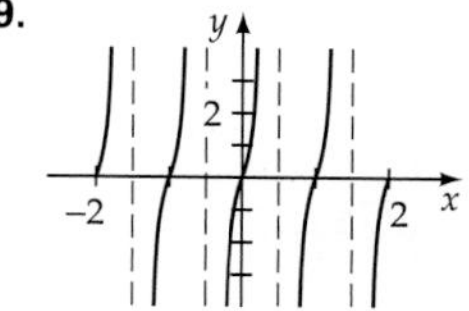

51. $y = \cot \frac{3}{2}x$ **53.** $y = \csc \frac{2}{3}x$ **55.** $y = \sec \frac{3}{4}x$ **57.**

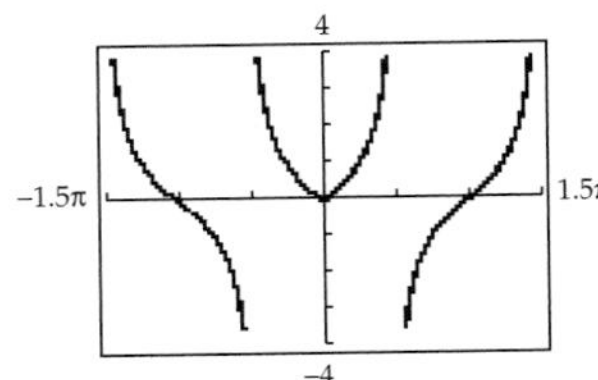

59.

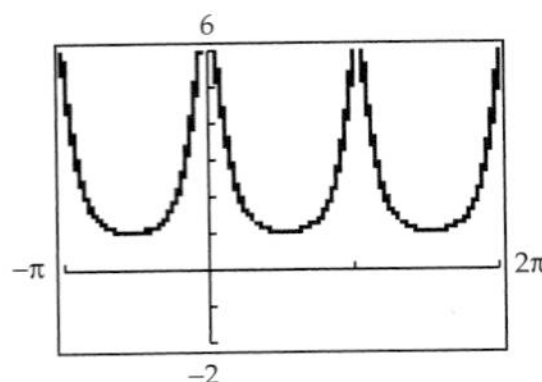

61. a. $h = 1.4 \tan x$ **b.** $d = 1.4 \sec x$ **c.**

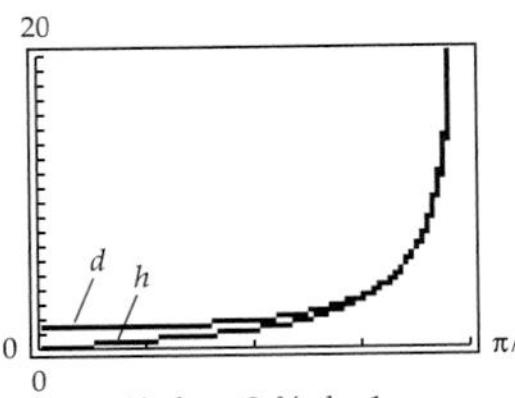

d. The graph of d is above the graph of h, but the distance between the graphs approaches 0 as x approaches $\frac{\pi}{2}$.

63. $y = \tan 3x$ **65.** $y = \sec \frac{8}{3}x$ **67.** $y = \cot \frac{\pi}{2}x$ **69.** $y = \csc \frac{4\pi}{3}x$

Prepare for This Section (2.7), page 192

PS1. amplitude 2, period π **PS2.** amplitude $\frac{2}{3}$, period 6π **PS3.** amplitude 4, period 1 **PS4.** 2 **PS5.** -3 **PS6.** y-axis

Exercise Set 2.7, page 199

1. $2, \frac{\pi}{2}, 2\pi$ **3.** $1, \frac{\pi}{8}, \pi$ **5.** $4, -\frac{\pi}{4}, 3\pi$ **7.** $\frac{5}{4}, \frac{2\pi}{3}, \frac{2\pi}{3}$ **9.** $\frac{\pi}{8}, \frac{\pi}{2}$ **11.** $-3\pi, 6\pi$ **13.** $\frac{\pi}{16}, \pi$ **15.** $-12\pi, 4\pi$

17.

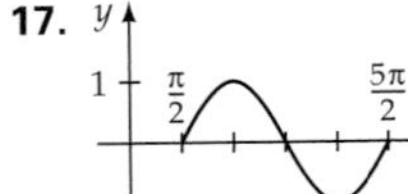
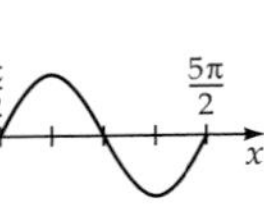

19.

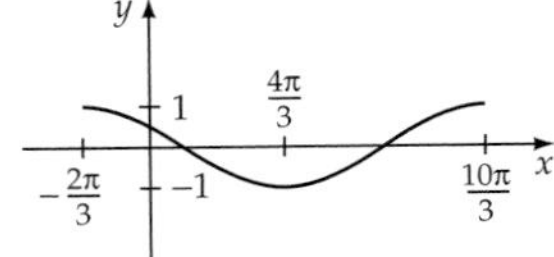

21.

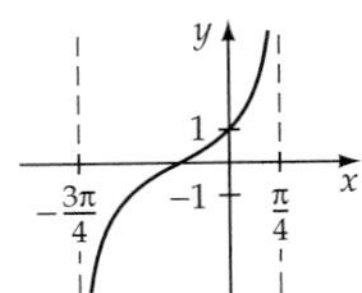

23.

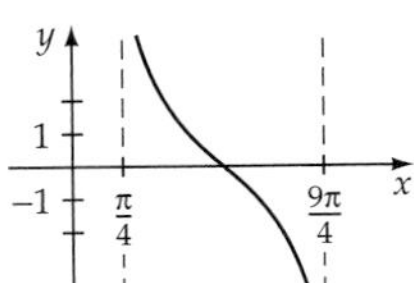

25.

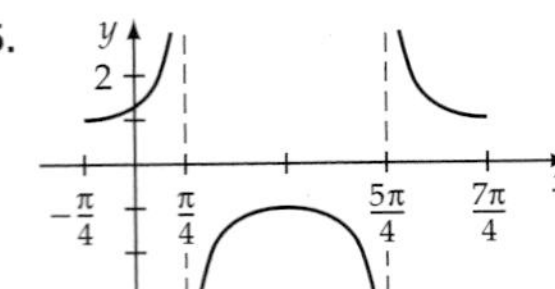

27.

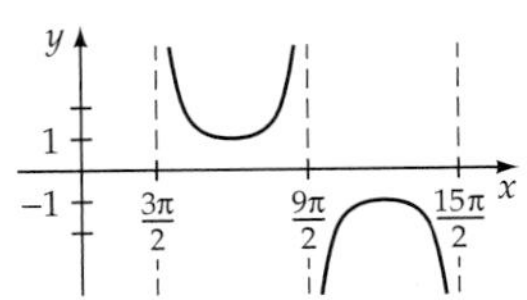

29.

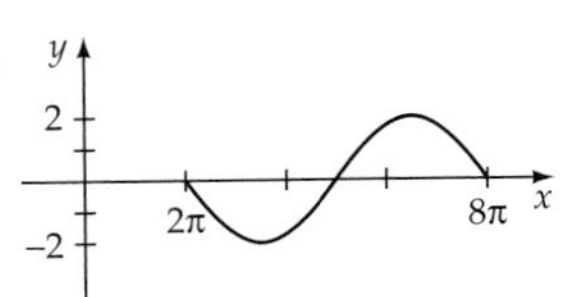

31.

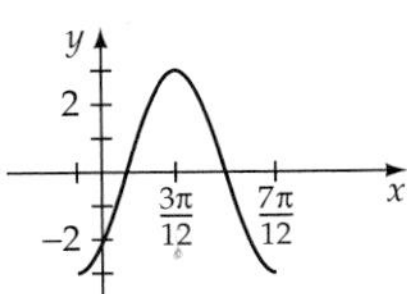

33.

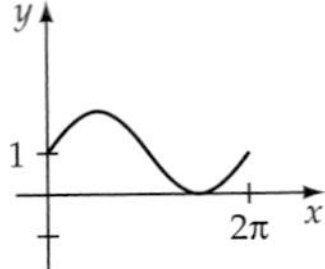

35.

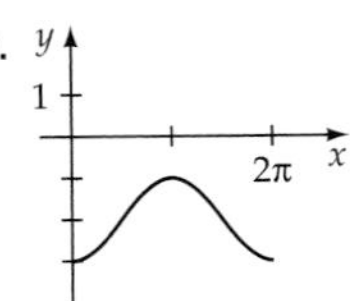

37.

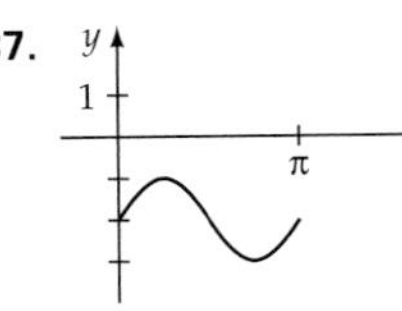

39.

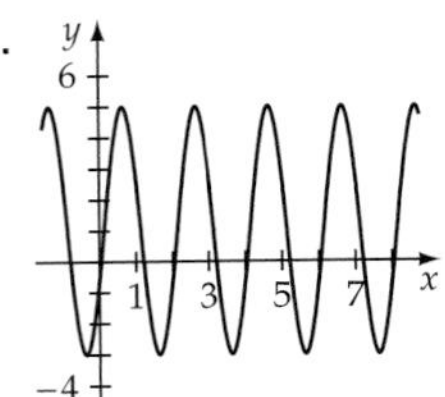

41.

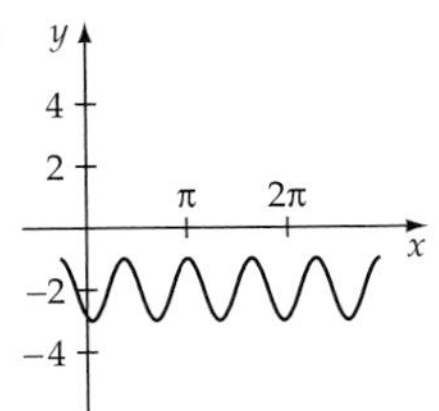

43.

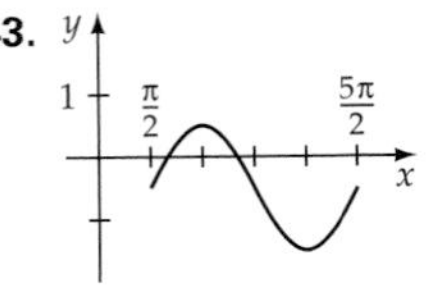

45.

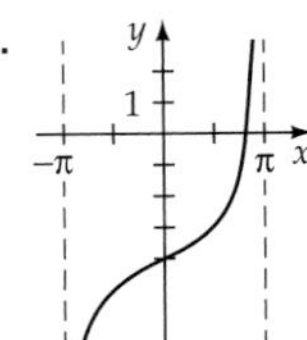

47.

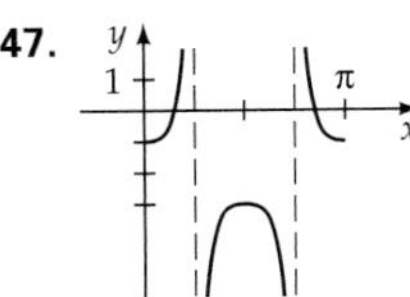

49.

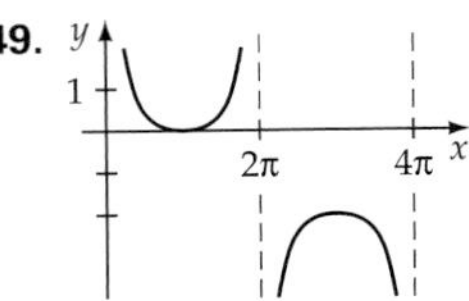

51.

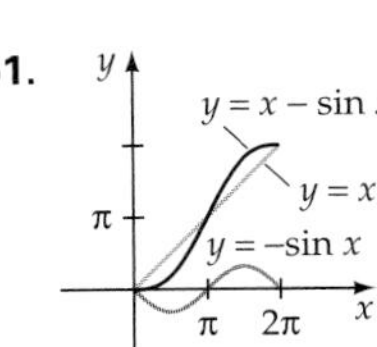

53.

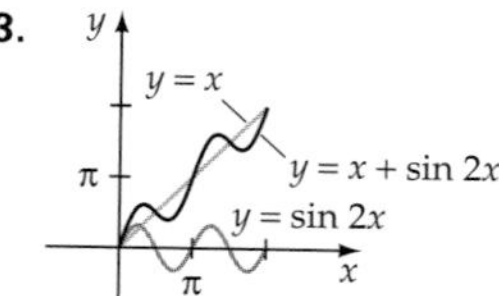

55.

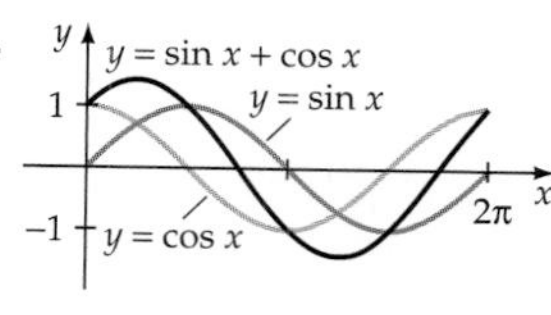

57. $y = \sin\left(2x - \frac{\pi}{3}\right)$ **59.** $y = \csc\left(\frac{x}{2} - \pi\right)$ **61.** $y = \sec\left(x - \frac{\pi}{2}\right)$

63. a. 7.5 months, 12 months **b.**

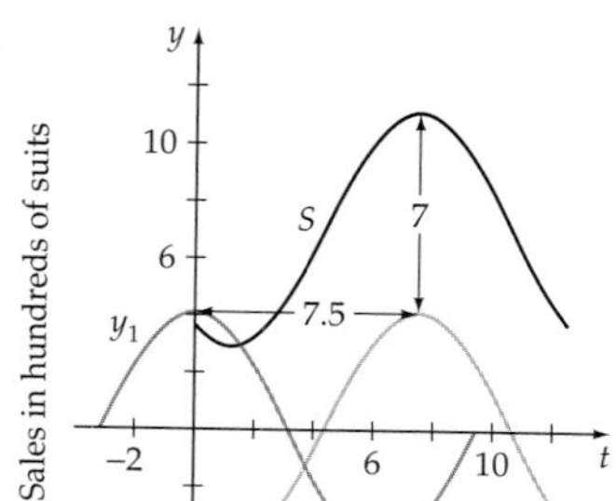

c. August **65.** ≈ 20 parts per million

67. $s = 7 \cos 10\pi t + 5$ **69.** $s = 400 \tan \frac{\pi}{5} t$, t in seconds **71.** $y = 3 \cos \frac{\pi}{6} t + 9$, 12 feet at 6:00 P.M. **73.**

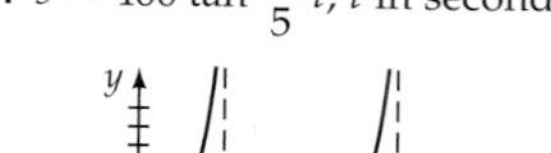

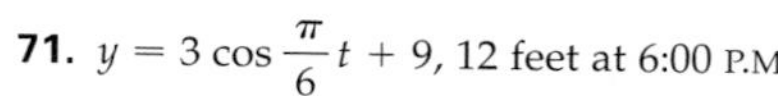

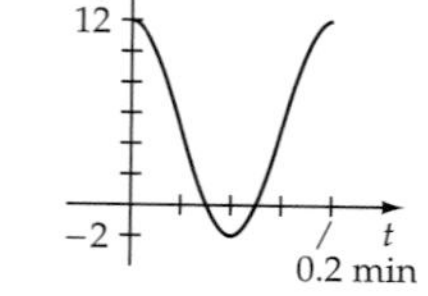

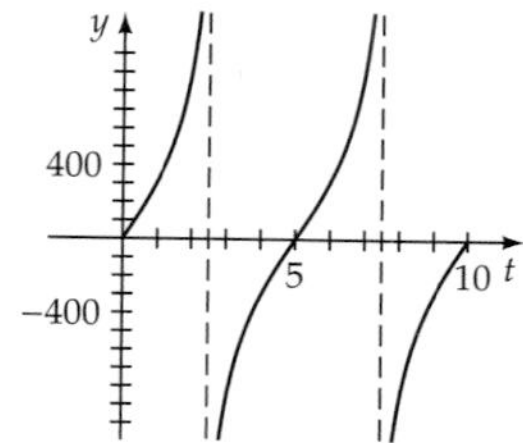

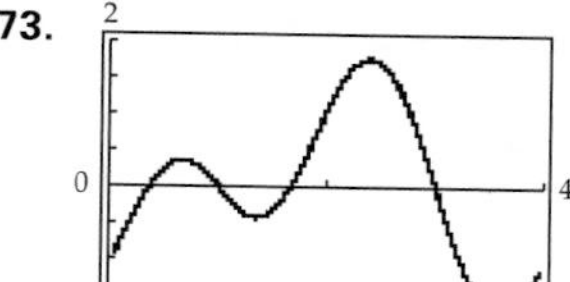

75. (window: 0 to 4π; −3 to 3)

77. (window: 0 to 4π; −6 to 6)

79.

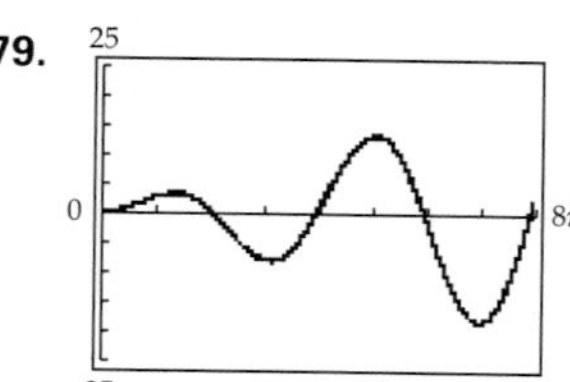

81.

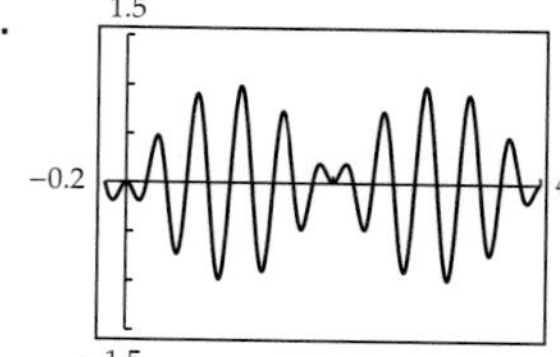

83. $y = 2\sin\left(2x - \frac{2\pi}{3}\right)$ **85.** $y = \tan\left(\frac{x}{2} - \frac{\pi}{4}\right)$ **87.** $\cos^2 x + 2$ **89.**

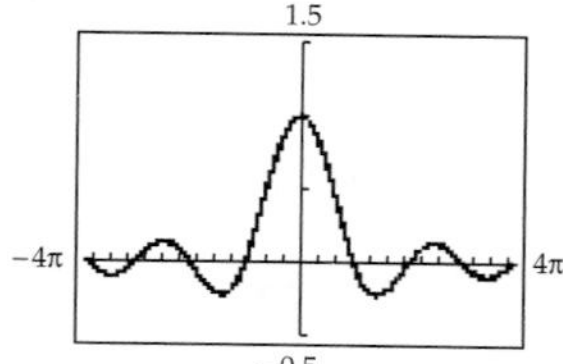

The graph above does not show that the function is undefined at $x = 0$.

91.

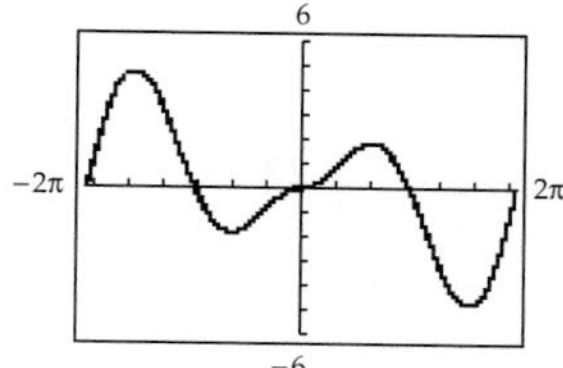

Prepare for This Section (2.8), page 203

PS1. $\frac{3}{2\pi}$ **PS2.** $\frac{5}{2}$ **PS3.** 4 **PS4.** 3 **PS5.** 4 **PS6.** $y = 4 \cos \pi x$

Exercise Set 2.8, page 208

1. $2, \pi, \frac{1}{\pi}$ **3.** $3, 3\pi, \frac{1}{3\pi}$ **5.** $4, 2, \frac{1}{2}$ **7.** $\frac{3}{4}, 4, \frac{1}{4}$

9. $y = 4\cos 3\pi t$ **11.** $y = \frac{3}{2}\cos\frac{4\pi}{3}t$ **13.** $y = 2\sin 2t$ **15.** $y = \sin \pi t$ **17.** $y = 2\sin 2\pi t$

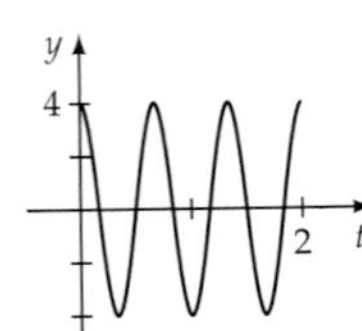

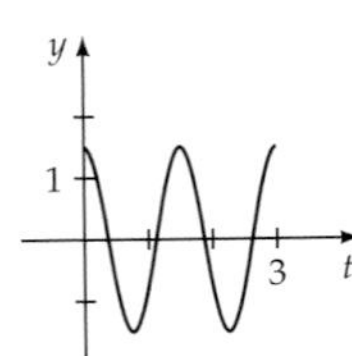

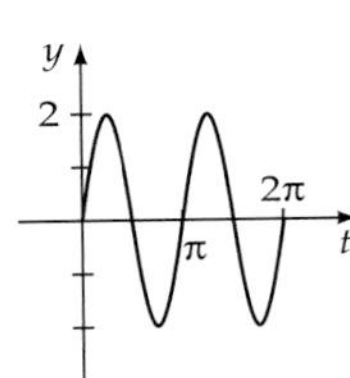

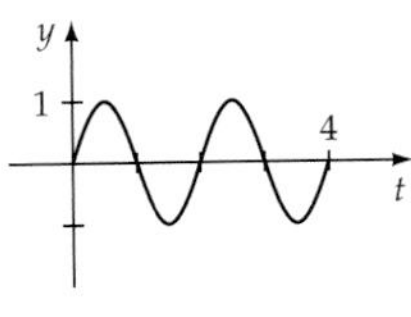

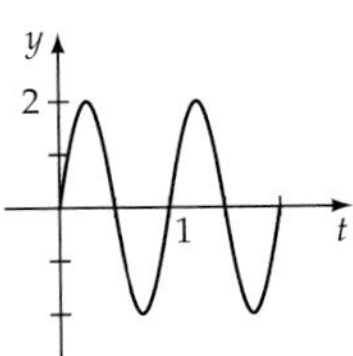

19. $y = \frac{1}{2}\cos 4t$ **21.** $y = 2.5\cos \pi t$ **23.** $y = \frac{1}{2}\cos\frac{2\pi}{3}t$ **25.** $y = 4\cos 4t$ **27.** $4\pi, \frac{1}{4\pi}$, 2 feet; $y = -2\cos\frac{t}{2}$

29. a. 196 cycles per second; $\frac{1}{196}$ second **b.** The amplitude needs to increase. **31.** $h = -37\cos\left(\frac{\pi}{22.5}t\right) + 41$ **33. a.** 3 **b.** 59.8 seconds

35. a. 10 **b.** 71.0 seconds **37. a.** 10 **b.** 9.1 seconds **39. a.** 10 **b.** 6.1 seconds **41.** The new period is 3 times the original period.

43. Yes **45.** Yes

Chapter 2 Assessing Concepts, page 212

1. True **2.** False **3.** False **4.** True **5.** $\frac{\pi}{4}$ **6.** $(0, 1)$ **7.** $\frac{8}{3}$ **8.** Shift the graph of y_1 to the left $\frac{\pi}{2}$ units.

9. all real numbers except multiples of π **10.** $x = \frac{\pi}{2}$ and $x = \frac{3\pi}{2}$

Chapter 2 Review Exercises, page 212

1. complement measures 25°; supplement measures 115°. [2.1] **2.** 80° [2.3] **3.** 114.59° [2.1] **4.** $\frac{7\pi}{4}$ [2.1] **5.** 3.93 meters [2.1] **6.** 0.3 [2.1]

7. 55 radians per second [2.1] **8.** $\frac{\sqrt{5}}{3}$ [2.2] **9.** $\frac{\sqrt{5}}{2}$ [2.2] **10.** $\frac{2}{3}$ [2.2] **11.** $\frac{3\sqrt{5}}{5}$ [2.2] **12.** $\sin\theta = -\frac{3\sqrt{10}}{10}$ $\quad \csc\theta = -\frac{\sqrt{10}}{3}$ [2.3]

$\cos\theta = \frac{\sqrt{10}}{10}$ $\quad \sec\theta = \sqrt{10}$

$\tan\theta = -3$ $\quad \cot\theta = -\frac{1}{3}$

13. a. $-\frac{2\sqrt{3}}{3}$ **b.** 1 **c.** -1 **d.** $-\frac{1}{2}$ [2.3] **14. a.** -0.5446 **b.** 0.5365 **c.** -3.2361 **d.** 3.0777 [2.3] **15. a.** $-\frac{1}{2}$ **b.** $\frac{\sqrt{3}}{3}$ [2.3]

16. a. $-\frac{2\sqrt{3}}{3}$ **b.** 2 [2.3] **17. a.** $\frac{\sqrt{2}}{2}$ **b.** -1 [2.3] **18. a.** $(-1, 0)$ **b.** $\left(\frac{1}{2}, -\frac{\sqrt{3}}{2}\right)$ **c.** $\left(-\frac{\sqrt{2}}{2}, -\frac{\sqrt{2}}{2}\right)$ **d.** $(1, 0)$ [2.4]

19. even [2.4] **22.** $\sec^2\phi$ [2.4] **23.** $\tan\phi$ [2.4] **24.** $\sin\phi$ [2.4] **25.** $\tan^2\phi$ [2.4] **26.** $\csc^2\phi$ [2.4] **27.** 0 [2.4] **28.** $3, \pi, \frac{\pi}{2}$ [2.5]

29. no amplitude, $\frac{\pi}{3}$, 0 [2.6] **30.** $2, \frac{2\pi}{3}, -\frac{\pi}{9}$ [2.5] **31.** $1, \pi, \frac{\pi}{3}$ [2.5] **32.** no amplitude, $\frac{\pi}{2}, \frac{3\pi}{8}$ [2.6] **33.** no amplitude, $2\pi, \frac{\pi}{4}$ [2.6]

34. [2.5] **35.** [2.5] **36.** [2.5] **37.** [2.7]

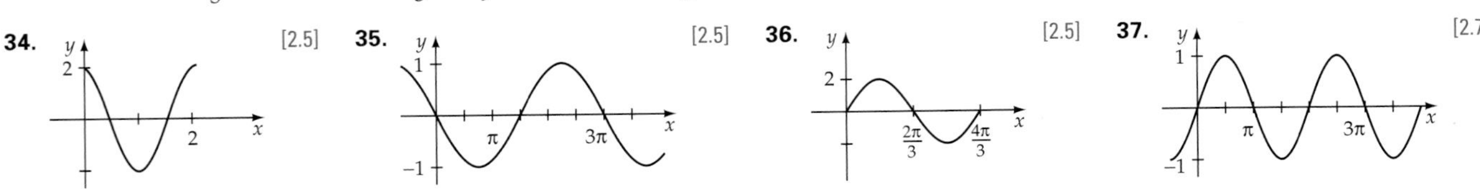

38. 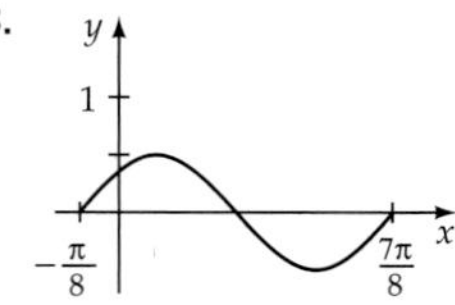

[2.7]

39. 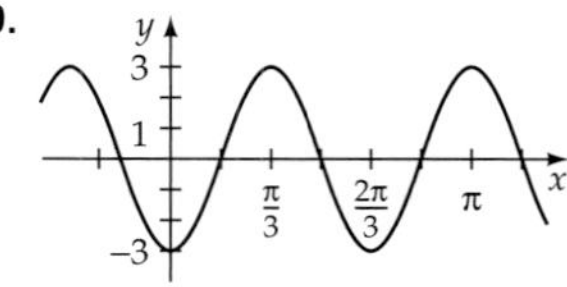

[2.7]

40. 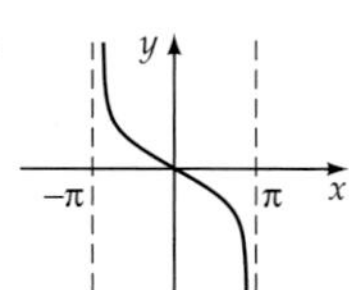

[2.6]

41. 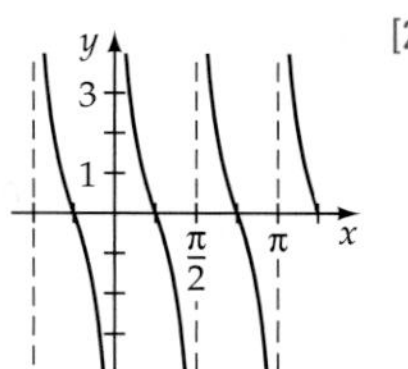

[2.6]

42. 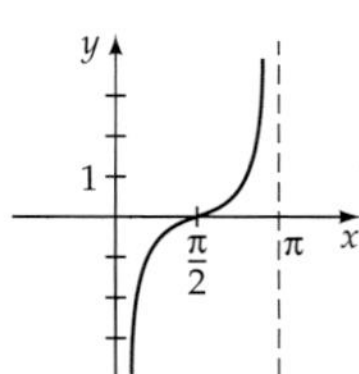

[2.7]

43. 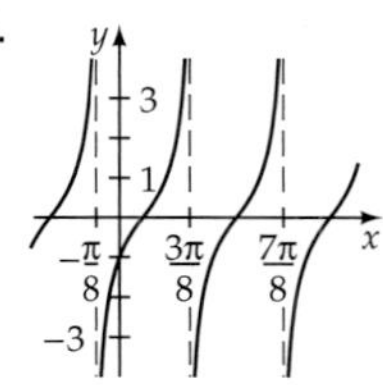

[2.7]

44. 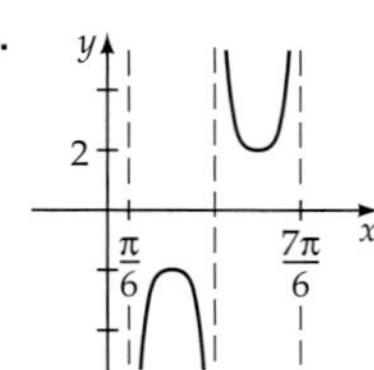

[2.7]

45. 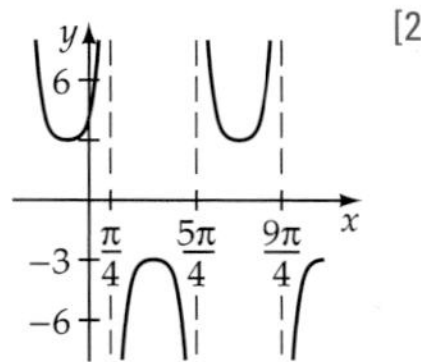

[2.7]

46. 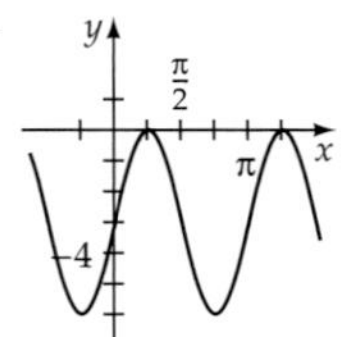

[2.7]

47. 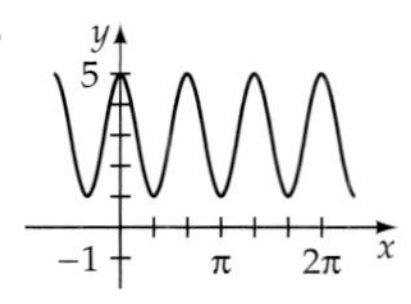

[2.7]

48. 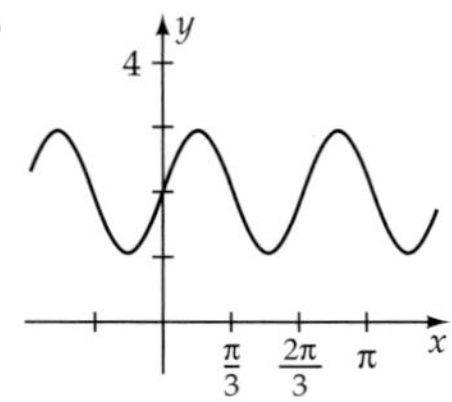

[2.7]

49. 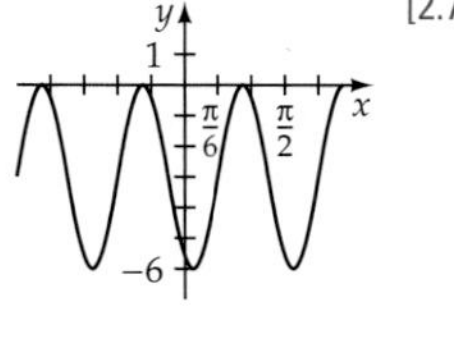

[2.7]

50. 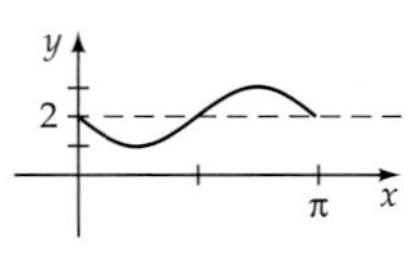

[2.7]

51. 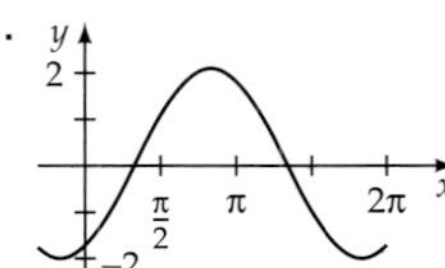

[2.7]

52. 0.089 mile [2.2] **53.** 12.3 feet [2.2] **54.** 1.7 feet per second [2.1]

55. 46 feet [2.2] **56.** 2.5, $\frac{\pi}{25}$, $\frac{25}{\pi}$ [2.8] **57.** amplitude = 0.5, $f = \frac{1}{\pi}$, $p = \pi$, $y = -0.5 \cos 2t$ [2.8] **58.** 7.2 seconds [2.8]

Chapter 2 Quantitative Reasoning Exercises, page 214

QR1. a. 6π **b.** 4 **c.** 2π  **d.** 24π **e.** 7.5 **f.** 4π **QR2.** 15 seconds **QR3.** 11.25 seconds **QR4.** 54 seconds

Chapter 2 Test, page 214

1. $\frac{5\pi}{6}$ [2.1] **2.** $\frac{\pi}{12}$ [2.1] **3.** 13.1 centimeters [2.1] **4.** 12π radians/second [2.1] **5.** 80 centimeters/second [2.1] **6.** $\frac{\sqrt{58}}{7}$ [2.2]

7. 1.0864 [2.2] **8.** $\frac{\sqrt{3} - 6}{6}$ [2.3] **9.** $\left(\frac{\sqrt{3}}{2}, -\frac{1}{2}\right)$ [2.4] **10.** $\sin^2 t$ [2.4] **11.** $\frac{\pi}{3}$ [2.6] **12.** amplitude 3, period π, phase shift $-\frac{\pi}{4}$ [2.7]

13. period 3, phase shift $-\frac{1}{2}$ [2.7]

14. 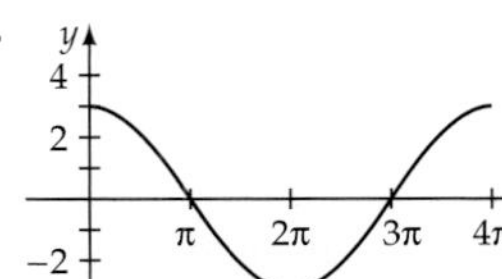

[2.5]

15. 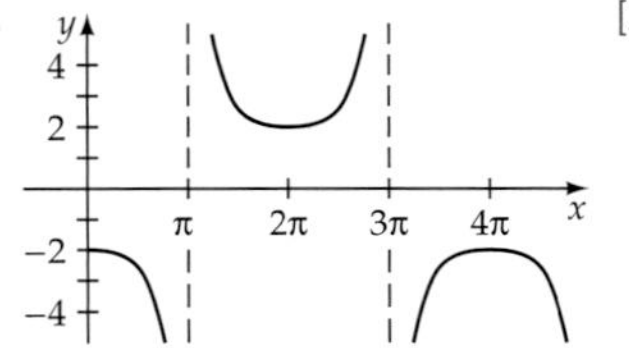

[2.6]

16. Shift the graph of $y = 2\sin 2x$, $\frac{\pi}{4}$ units to the right and 1 unit down. [2.7] **17.**

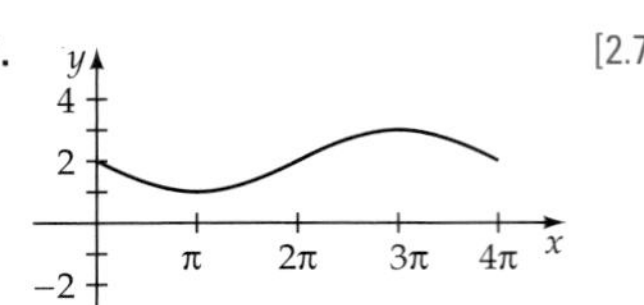

[2.7]

18.

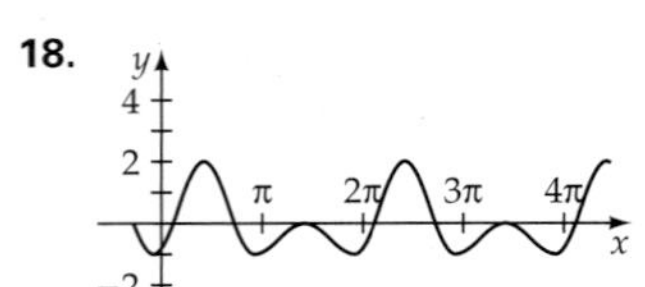

[2.7] **19.** 25.5 meters [2.2] **20.** $y = 13\sin\frac{2\pi}{5}t$ [2.8]

Cumulative Review Exercises, page 215

1. $5\sqrt{2}$ [1.2] **2.** $\frac{\sqrt{3}}{2}$ [1.2] **3.** y-intercept $(0, -9)$, x-intercepts $(-3, 0)$ and $(3, 0)$ [1.2] **4.** odd function [1.4] **5.** $f^{-1}(x) = \frac{3x}{2x - 1}$ [1.6] **6.** $(-\infty, 4) \cup (4, \infty)$ [1.3] **7.** 2, −3 [1.1] **8.** Shift the graph of $y = f(x)$ horizontally 3 units to the right. [1.4] **9.** Reflect the graph of $y = f(x)$ across the y-axis. [1.4] **10.** $\frac{5\pi}{3}$ [2.1] **11.** 225° [2.1] **12.** 1 [2.3] **13.** $\frac{\sqrt{3}+1}{2}$ [2.2] **14.** $\frac{5}{4}$ [2.2] **15.** negative [2.3] **16.** 30° [2.3] **17.** $\frac{\pi}{3}$ [2.3] **18.** $(-\infty, \infty)$ [2.5] **19.** $[-1, 1]$ [2.5] **20.** $\frac{3}{5}$ [2.2]

Exercise Set 3.1, page 222

57. identity **59.** identity **61.** identity **63.** not an identity **65.** If $x = \frac{\pi}{4}$, the left side is 2 and the right side is 1. **67.** If $x = 0°$, the left side is $\frac{\sqrt{3}}{2}$ and the right side is $\frac{2+\sqrt{3}}{2}$. **69.** If $x = 0$, the left side is −1 and the right side is 1.

Prepare for This Section (3.2), page 225

PS1. Both function values equal $\frac{1}{2}$. **PS2.** Both function values equal $\frac{1}{2}$. **PS3.** For each of the given values of θ, the function values are equal. **PS4.** For each of the given values of θ, the function values are equal. **PS5.** Both function values equal $\frac{\sqrt{3}}{3}$. **PS6.** 0

Exercise Set 3.2, page 233

1. $\frac{\sqrt{6}+\sqrt{2}}{4}$ **3.** $\frac{\sqrt{6}+\sqrt{2}}{4}$ **5.** $2 - \sqrt{3}$ **7.** $\frac{-\sqrt{6}+\sqrt{2}}{4}$ **9.** $-\frac{\sqrt{6}+\sqrt{2}}{4}$ **11.** $2 + \sqrt{3}$ **13.** 0 **15.** $\frac{1}{2}$ **17.** $\sqrt{3}$ **19.** cos 48° **21.** cot 75° **23.** csc 65° **25.** $\sin 5x$ **27.** $\cos x$ **29.** $\sin 4x$ **31.** $\cos 2x$ **33.** $\sin x$ **35.** $\tan 7x$ **37. a.** $-\frac{77}{85}$ **b.** $\frac{84}{85}$ **c.** $\frac{77}{36}$ **39. a.** $-\frac{63}{65}$ **b.** $-\frac{56}{65}$ **c.** $-\frac{63}{16}$ **41. a.** $\frac{63}{65}$ **b.** $\frac{56}{65}$ **c.** $\frac{33}{56}$ **43. a.** $-\frac{77}{85}$ **b.** $-\frac{84}{85}$ **c.** $-\frac{13}{84}$ **45. a.** $-\frac{33}{65}$ **b.** $-\frac{16}{65}$ **c.** $\frac{63}{16}$ **47. a.** $-\frac{56}{65}$ **b.** $-\frac{63}{65}$ **c.** $\frac{16}{63}$ **75.** $-\cos\theta$ **77.** $\tan\theta$ **79.** $\sin\theta$ **81.** identity **83.** identity

Prepare for This Section (3.3), page 236

PS1. $2\sin\alpha\cos\alpha$ **PS2.** $\cos^2\alpha - \sin^2\alpha$ **PS3.** $\frac{2\tan\alpha}{1 - \tan^2\alpha}$ **PS4.** For each of the given values of α, the function values are equal. **PS5.** Let $\alpha = 45°$; then the left side of the equation is 1, and the right side of the equation is $\sqrt{2}$. **PS6.** Let $\alpha = 60°$; then the left side of the equation is $\frac{\sqrt{3}}{2}$, and the right side of the equation is $\frac{1}{4}$.

Exercise Set 3.3, page 243

1. $\sin 4\alpha$ **3.** $\cos 10\beta$ **5.** $\cos 6\alpha$ **7.** $\tan 6\alpha$ **9.** $\sin 2\alpha = -\frac{24}{25}$, $\cos 2\alpha = \frac{7}{25}$, $\tan 2\alpha = -\frac{24}{7}$ **11.** $\sin 2\alpha = -\frac{240}{289}$, $\cos 2\alpha = \frac{161}{289}$, $\tan 2\alpha = -\frac{240}{161}$

13. $\sin 2\alpha = -\frac{336}{625}$, $\cos 2\alpha = -\frac{527}{625}$, $\tan 2\alpha = \frac{336}{527}$ **15.** $\sin 2\alpha = \frac{240}{289}$, $\cos 2\alpha = -\frac{161}{289}$, $\tan 2\alpha = -\frac{240}{161}$

17. $\sin 2\alpha = -\frac{720}{1681}$, $\cos 2\alpha = \frac{1519}{1681}$, $\tan 2\alpha = -\frac{720}{1519}$ **19.** $3(1 + \cos 2x)$ **21.** $\frac{1}{8}(3 + 4\cos 2x + \cos 4x)$

23. $\frac{1}{16}(1 - \cos 2x - \cos 4x + \cos 2x \cos 4x)$ **25.** $\frac{\sqrt{2+\sqrt{3}}}{2}$ **27.** $\sqrt{2} + 1$ **29.** $-\frac{\sqrt{2+\sqrt{2}}}{2}$ **31.** $\frac{\sqrt{2-\sqrt{2}}}{2}$ **33.** $\frac{\sqrt{2-\sqrt{2}}}{2}$

35. $\frac{\sqrt{2-\sqrt{3}}}{2}$ **37.** $\sin\frac{\alpha}{2} = \frac{5\sqrt{26}}{26}$, $\cos\frac{\alpha}{2} = \frac{\sqrt{26}}{26}$, $\tan\frac{\alpha}{2} = 5$ **39.** $\sin\frac{\alpha}{2} = \frac{5\sqrt{34}}{34}$, $\cos\frac{\alpha}{2} = -\frac{3\sqrt{34}}{34}$, $\tan\frac{\alpha}{2} = -\frac{5}{3}$

41. $\sin\frac{\alpha}{2} = \frac{\sqrt{5}}{5}$, $\cos\frac{\alpha}{2} = \frac{2\sqrt{5}}{5}$, $\tan\frac{\alpha}{2} = \frac{1}{2}$ **43.** $\sin\frac{\alpha}{2} = \frac{\sqrt{2}}{10}$, $\cos\frac{\alpha}{2} = -\frac{7\sqrt{2}}{10}$, $\tan\frac{\alpha}{2} = -\frac{1}{7}$

91. a. $\frac{2}{\sqrt{2-\sqrt{2}}} \approx 2.61$ **b.** $\alpha = 2\sin^{-1}\left(\frac{1}{M}\right)$ **c.** α decreases. **93.** identity **95.** identity

Prepare for This Section (3.4), page 246

PS1. $\sin\alpha\cos\beta$ **PS2.** $\cos\alpha\cos\beta$ **PS3.** Both function values equal $-\frac{1}{2}$. **PS4.** $\sin x + \cos x$ **PS5.** Answers will vary. **PS6.** 2

Exercise Set 3.4, page 251

1. $\sin 3x - \sin x$ **3.** $\frac{1}{2}(\sin 8x - \sin 4x)$ **5.** $\sin 8x + \sin 2x$ **7.** $\frac{1}{2}(\cos 4x - \cos 6x)$ **9.** $\frac{1}{4}$ **11.** $-\frac{\sqrt{2}}{4}$ **13.** $-\frac{1}{4}$ **15.** $\frac{\sqrt{3}-2}{4}$

17. $2\sin 3\theta\cos\theta$ **19.** $2\cos 2\theta\cos\theta$ **21.** $-2\sin 4\theta\sin 2\theta$ **23.** $2\cos 4\theta\cos 3\theta$ **25.** $2\sin 7\theta\cos 2\theta$ **27.** $-2\sin\frac{3}{2}\theta\sin\frac{1}{2}\theta$

29. $2\sin\frac{3}{4}\theta\sin\frac{\theta}{4}$ **31.** $2\cos\frac{5}{12}\theta\sin\frac{1}{12}\theta$ **49.** $y = \sqrt{2}\sin(x - 135°)$ **51.** $y = \sin(x - 60°)$ **53.** $y = \frac{\sqrt{2}}{2}\sin(x - 45°)$

55. $y = 3\sqrt{2}\sin(x + 135°)$ **57.** $y = \pi\sqrt{2}\sin(x - 45°)$ **59.** $y = \sqrt{2}\sin\left(x + \frac{3\pi}{4}\right)$ **61.** $y = \sin\left(x + \frac{\pi}{6}\right)$ **63.** $y = 20\sin\left(x + \frac{2\pi}{3}\right)$

65. $y = 5\sqrt{2}\sin\left(x + \frac{3\pi}{4}\right)$

67.
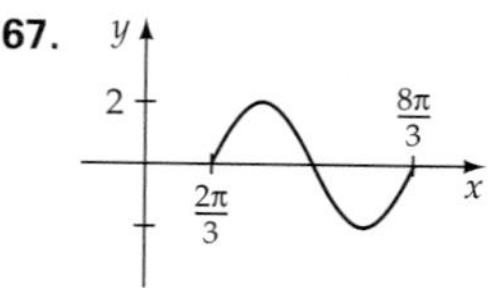

69.
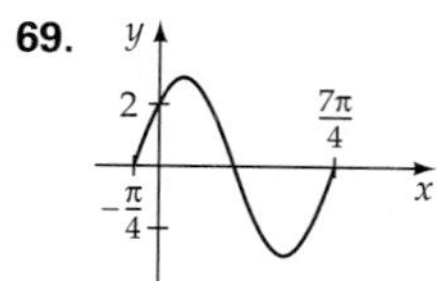

71.
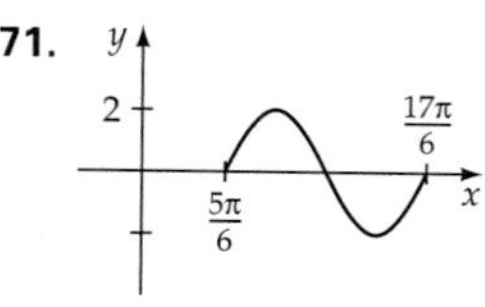

73.
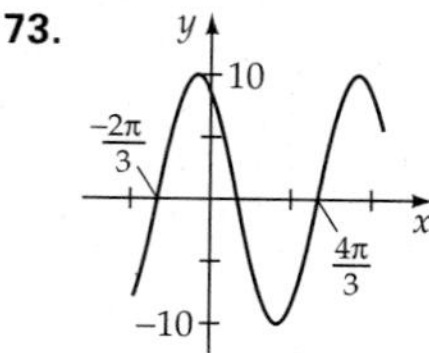

75.
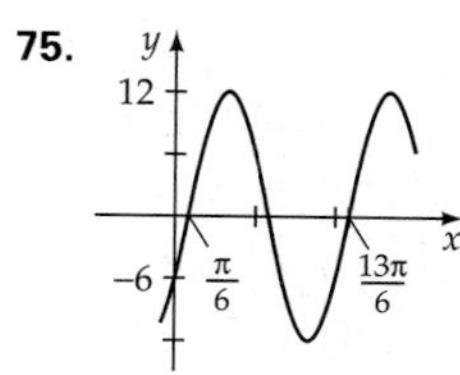

77. a. $p(t) = \sin(2\pi \cdot 1336t) + \sin(2\pi \cdot 770t)$ **b.** $p(t) = 2\sin(2106\pi t)\sin(566\pi t)$ **c.** 1053 cycles per second **79.** identity **81.** identity **83.** identity

Prepare for This Section (3.5), page 255

PS1. A one-to-one function is a function for which each range value (y value) is paired with one and only one domain value (x value).
PS2. If every horizontal line intersects the graph of a function at most once, then the function is a one-to-one function.
PS3. $f[g(x)] = x$ **PS4.** $f[f^{-1}(x)] = x$ **PS5.** The graph of f^{-1} is the reflection of the graph of f across the line given by $y = x$. **PS6.** No

Exercise Set 3.5, page 265

1. $\frac{\pi}{2}$ **3.** $\frac{5\pi}{6}$ **5.** $-\frac{\pi}{4}$ **7.** $\frac{\pi}{3}$ **9.** $\frac{\pi}{3}$ **11.** $-\frac{\pi}{4}$ **13.** $-\frac{\pi}{3}$ **15.** $\frac{2\pi}{3}$ **17.** $\frac{\pi}{6}$ **19. a.** 1.0014 **b.** 0.2341 **21. a.** 1.1102 **b.** 0.2818

23. $\theta = \cos^{-1}\left(\frac{x}{7}\right)$ **25.** $\frac{1}{2}$ **27.** 2 **29.** $\frac{3}{5}$ **31.** 1 **33.** $\frac{1}{2}$ **35.** $\frac{\pi}{6}$ **37.** $\frac{\pi}{4}$ **39.** not defined **41.** 0.4636 **43.** $-\frac{\pi}{6}$ **45.** $\frac{\sqrt{3}}{3}$

47. $\frac{4\sqrt{15}}{15}$ **49.** $\frac{24}{25}$ **51.** 0 **53.** $\frac{24}{25}$ **55.** $\frac{2+\sqrt{15}}{6}$ **57.** $\frac{1}{5}(3\sqrt{7} - 4\sqrt{3})$ **59.** $\frac{12}{13}$ **61.** 2 **63.** $\frac{2-\sqrt{2}}{2}$ **65.** $\frac{7\sqrt{2}}{10}$

67. $\cos\frac{5\pi}{12} \approx 0.2588$ **69.** $\frac{\sqrt{1-x^2}}{x}$

75. 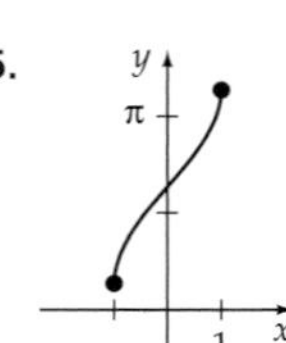

77. 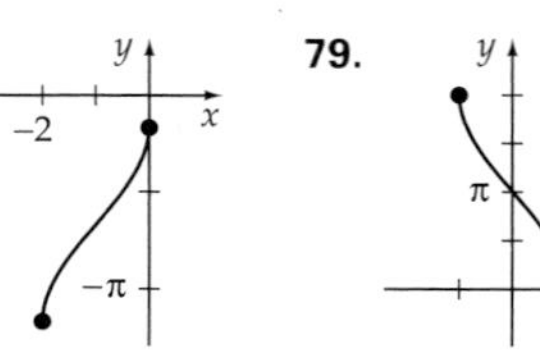

79. 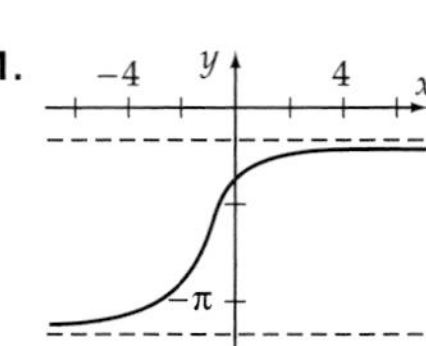

81. −4 y 4 x −π

83. a. $s = 3960\cos^{-1}\left(\dfrac{3960}{a+3960}\right)$ **b.** 17,930 miles

85. 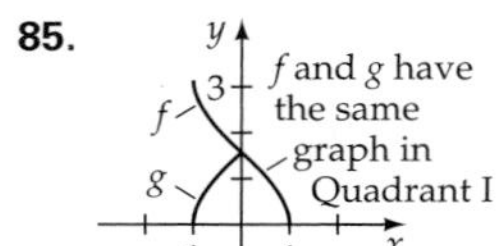

$f \neq g$

87. 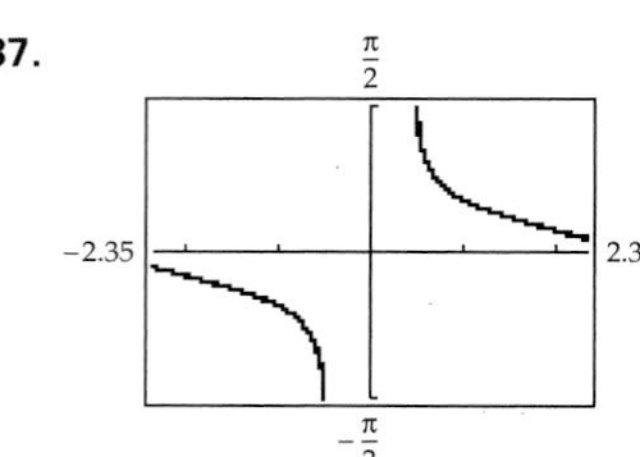

89. 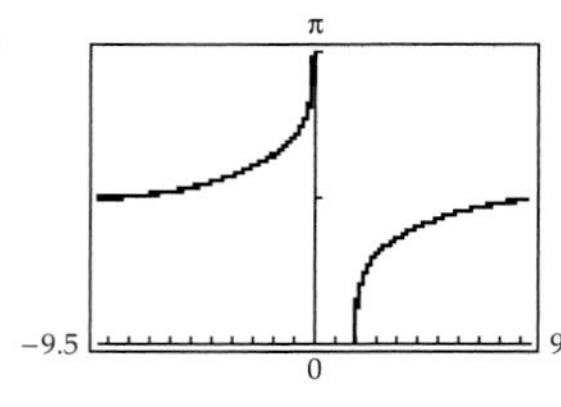

91. 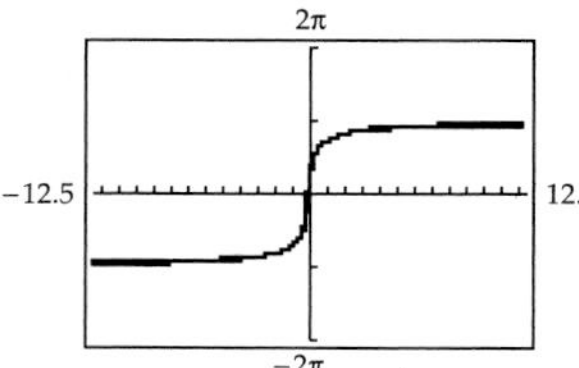

97. $y = \dfrac{1}{3}\tan 5x$ **99.** $y = 3 + \cos\left(x - \dfrac{\pi}{3}\right)$

Prepare for This Section (3.6), page 268

PS1. $x = \dfrac{5 \pm \sqrt{73}}{6}$ **PS2.** $1 - \cos^2 x$ **PS3.** $\dfrac{5}{2}\pi, \dfrac{9}{2}\pi$, and $\dfrac{13}{2}\pi$ **PS4.** $(x+1)\left(x - \dfrac{\sqrt{3}}{2}\right)$ **PS5.** 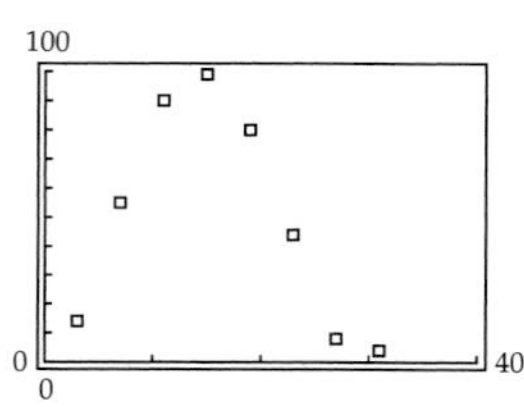

PS6. 0, 1

Exercise Set 3.6, page 278

1. $\frac{\pi}{4}, \frac{7\pi}{4}$ **3.** $\frac{\pi}{3}, \frac{4\pi}{3}$ **5.** $\frac{\pi}{4}, \frac{\pi}{2}, \frac{3\pi}{4}, \frac{3\pi}{2}$ **7.** $\frac{\pi}{2}, \frac{3\pi}{2}$ **9.** $\frac{\pi}{6}, \frac{\pi}{4}, \frac{3\pi}{4}, \frac{11\pi}{6}$ **11.** $\frac{\pi}{4}, \frac{3\pi}{4}$ **13.** $\frac{\pi}{6}, \frac{\pi}{2}, \frac{5\pi}{6}$ **15.** $\frac{\pi}{6}, \frac{5\pi}{6}, \frac{7\pi}{6}, \frac{11\pi}{6}$ **17.** $0, \frac{\pi}{4}, \frac{3\pi}{4}, \pi, \frac{5\pi}{4}, \frac{7\pi}{4}$ **19.** $\frac{\pi}{6}, \frac{5\pi}{6}, \frac{4\pi}{3}, \frac{5\pi}{3}$ **21.** $0, \frac{\pi}{2}, \pi, \frac{3\pi}{2}$ **23.** 41.4°, 318.6° **25.** no solution **27.** 68.0°, 292.0° **29.** no solution **31.** 12.8°, 167.2° **33.** 15.5°, 164.5° **35.** 0°, 33.7°, 180°, 213.7° **37.** no solution **39.** no solution **41.** 0°, 120°, 240° **43.** 70.5°, 289.5° **45.** 68.2°, 116.6°, 248.2°, 296.6° **47.** 19.5°, 90°, 160.5°, 270° **49.** 60°, 90°, 300° **51.** 53.1°, 180° **53.** 72.4°, 220.2° **55.** 50.1°, 129.9°, 205.7°, 334.3° **57.** no solution **59.** 22.5°, 157.5° **61.** $\frac{\pi}{8} + \frac{k\pi}{2}$, where k is an integer **63.** $\frac{\pi}{10} + \frac{2k\pi}{5}$, where k is an integer **65.** $0 + 2k\pi$, $\frac{\pi}{3} + 2k\pi, \pi + 2k\pi, \frac{5\pi}{3} + 2k\pi$, where k is an integer **67.** $\frac{\pi}{2} + k\pi, \frac{5\pi}{6} + k\pi$, where k is an integer **69.** $0 + 2k\pi$, where k is an integer **71.** $0, \pi$ **73.** $0, \frac{\pi}{6}, \frac{\pi}{2}, \frac{5\pi}{6}, \pi, \frac{7\pi}{6}, \frac{3\pi}{2}, \frac{11\pi}{6}$ **75.** $0, \frac{\pi}{2}, \frac{3\pi}{2}$ **77.** $0, \frac{\pi}{3}, \frac{2\pi}{3}, \pi, \frac{4\pi}{3}, \frac{5\pi}{3}$ **79.** $\frac{4\pi}{3}, \frac{5\pi}{3}$ **81.** $0, \frac{\pi}{4}, \frac{3\pi}{4}, \pi, \frac{5\pi}{4}, \frac{7\pi}{4}$ **83.** $\frac{\pi}{6}, \frac{5\pi}{6}, \pi$ **85.** 0.7391 **87.** −3.2957, 3.2957 **89.** 1.16 **91.** 14.99° and 75.01°

The sine regression functions in Exercises 93, 95, and 97 were obtained on a TI-83/TI-83 Plus/TI-84 Plus calculator by using an iteration factor of 16. The use of a different iteration factor may produce a sine regression function that varies from the regression functions listed below.

93. a. $y \approx 1.1213\sin(0.01595x + 1.8362) + 6.6257$ **b.** 6:49 **95. a.** $y \approx 49.2125\sin(0.2130x - 1.4576) + 48.0550$ **b.** 3% **97. a.** $y \approx 35.185\sin(0.30395x - 2.1630) + 2.1515$ **b.** 24.8° **99. b.** 42° and 79° **c.** 60° **101.** 0.93 foot, 1.39 feet **103.** $\frac{\pi}{6}, \frac{\pi}{2}$ **105.** $\frac{5\pi}{3}, 0$ **107.** $0, \frac{\pi}{4}, \frac{\pi}{2}, \frac{3\pi}{4}, \pi, \frac{5\pi}{4}, \frac{3\pi}{2}, \frac{7\pi}{4}$

Chapter 3 Assessing Concepts, page 286

1. True **2.** False **3.** False **4.** True **5.** 4 **6.** $-\frac{1}{2} \le x \le \frac{1}{2}$ **7.** $0 \le y \le \pi$ **8.** $\frac{\pi}{3}$ **9.** $\sqrt{2}$ **10.** $2 + \sqrt{3}$

Chapter 3 Review Exercises, page 286

1. $\dfrac{\sqrt{6}-\sqrt{2}}{4}$ [3.2] **2.** $\sqrt{3}-2$ [3.2] **3.** $\dfrac{\sqrt{6}-\sqrt{2}}{4}$ [3.2] **4.** $\sqrt{2}-\sqrt{6}$ [3.2] **5.** $-\dfrac{\sqrt{6}+\sqrt{2}}{4}$ [3.2] **6.** $\dfrac{\sqrt{2}+\sqrt{6}}{4}$ [3.2] **7.** $\dfrac{\sqrt{2-\sqrt{2}}}{2}$ [3.3] **8.** $-\dfrac{\sqrt{2-\sqrt{3}}}{2}$ [3.3] **9.** $\sqrt{2}+1$ [3.3] **10.** $\dfrac{\sqrt{2+\sqrt{2}}}{2}$ [3.3] **11. a.** 0 **b.** $\sqrt{3}$ **c.** $\dfrac{1}{2}$ [3.2/3.3] **12. a.** 0 **b.** -2 **c.** $\dfrac{1}{2}$ [3.2/3.3] **13. a.** $\dfrac{\sqrt{3}}{2}$ **b.** $-\sqrt{3}$ **c.** $-\dfrac{\sqrt{2-\sqrt{3}}}{2}$ [3.2/3.3] **14. a.** $\dfrac{\sqrt{6}-\sqrt{2}}{4}$ **b.** $-\sqrt{3}$ **c.** 1 [3.2/3.3] **15.** $\sin 6x$ [3.3] **16.** $\tan 3x$ [3.2] **17.** $\sin 3x$ [3.2] **18.** $\cos 4\theta$ [3.3] **19.** $\tan 2\theta$ [3.1] **20.** $\tan\theta$ [3.3] **21.** $2\sin 3\theta\,\sin\theta$ [3.4] **22.** $-2\cos 4\theta\,\sin\theta$ [3.4] **23.** $2\sin 4\theta\cos 2\theta$ [3.4] **24.** $2\cos 3\theta\sin 2\theta$ [3.4] **43.** $\dfrac{13}{5}$ [3.5] **44.** $\dfrac{4}{5}$ [3.5] **45.** $\dfrac{56}{65}$ [3.5] **46.** $\dfrac{7}{25}$ [3.5] **47.** $\dfrac{3}{2}$ [3.6] **48.** $\dfrac{4}{5}$ [3.6] **49.** 30°, 150°, 240°, 300° [3.6] **50.** 0°, 45°, 135° [3.6] **51.** $\dfrac{\pi}{2}+2k\pi$, $3.8713+2k\pi$, $5.553+2k\pi$, where k is an integer. [3.6] **52.** $-\dfrac{\pi}{4}+k\pi$, $1.2490+k\pi$, where k is an integer. [3.6] **53.** $\dfrac{\pi}{12}, \dfrac{5\pi}{12}, \dfrac{13\pi}{12}, \dfrac{17\pi}{12}$ [3.6] **54.** $\dfrac{7\pi}{12}, \dfrac{19\pi}{12}, \dfrac{3\pi}{4}, \dfrac{7\pi}{4}$ [3.6]

55. $y=2\sin\left(x+\dfrac{\pi}{6}\right)$ [3.4] **56.** $y=2\sqrt{2}\sin\left(x+\dfrac{5\pi}{4}\right)$ [3.4] **57.** $y=2\sin\left(x+\dfrac{4\pi}{3}\right)$ [3.4] **58.** $y=\sin\left(x-\dfrac{\pi}{6}\right)$ [3.4]

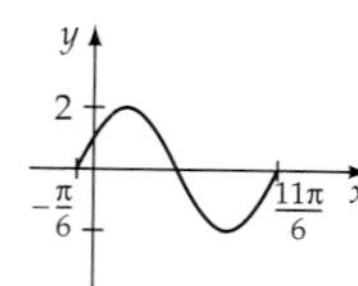

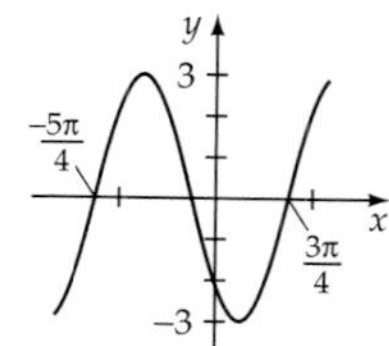

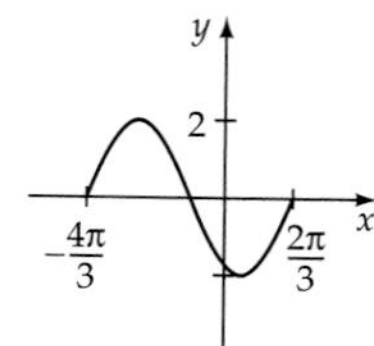

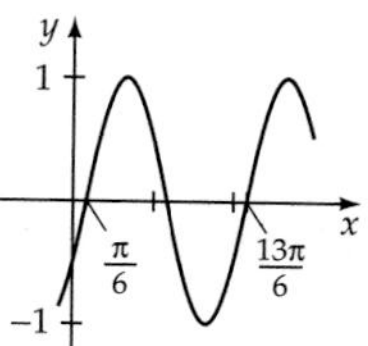

59. [3.5] **60.** [3.5] **61.** [3.5] **62.** [3.5]

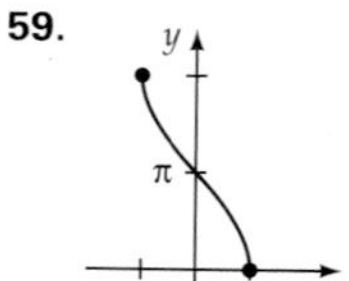

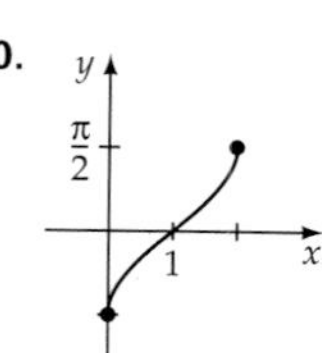

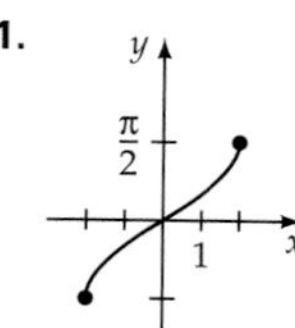

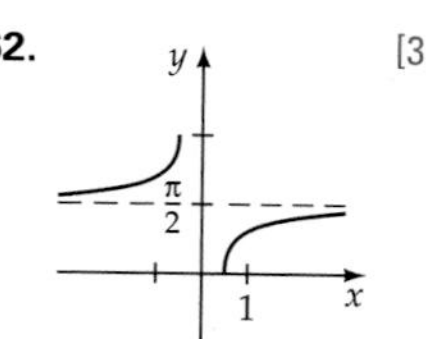

63. a. $y \approx 1.1835\sin(0.01600x+1.8497)+6.4394$ **b.** 6:01 [3.6]

Chapter 3 Quantitative Reasoning Exercises, page 288

QR1. a.

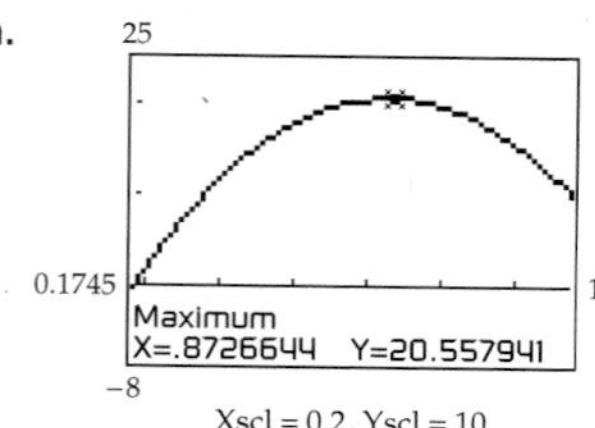

Xscl = 0.2, Yscl = 10

$\alpha \approx 0.87266$

QR2. a.

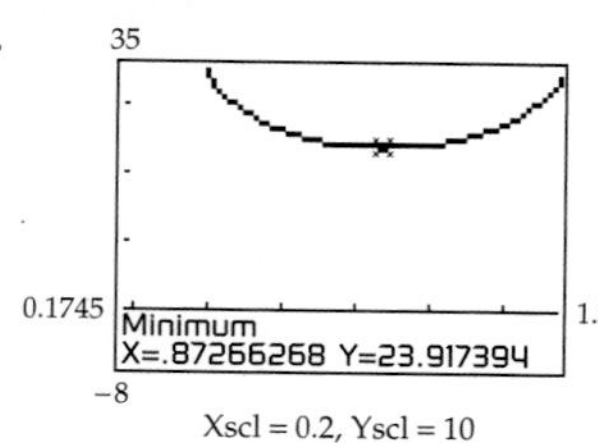

Xscl = 0.2, Yscl = 10

$\alpha \approx 0.87266$

Chapter 3 Test, page 289

5. $\dfrac{-\sqrt{6}+\sqrt{2}}{4}$ [3.2] **6.** $-\dfrac{\sqrt{2}}{10}$ [3.2] **8.** $\sin 9x$ [3.2] **9.** $-\dfrac{7}{25}$ [3.3] **12.** $\dfrac{2-\sqrt{3}}{4}$ [3.4] **13.** $y=\sin\left(x+\dfrac{5\pi}{6}\right)$ [3.4] **14.** 0.701 [3.5] **15.** $\dfrac{5}{13}$ [3.5] **16.** [3.5]

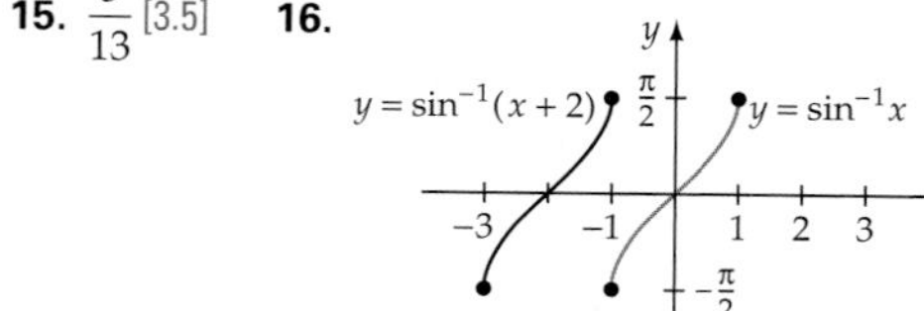

17. 41.8°, 138.2° [3.6] **18.** $0, \dfrac{\pi}{6}, \pi, \dfrac{11\pi}{6}$ [3.6] **19.** $\dfrac{\pi}{2}, \dfrac{2\pi}{3}, \dfrac{4\pi}{3}$ [3.6]

20. a. $y \approx 1.7569\sin(0.01675x+1.3056)+12.1100$ **b.** 12 hours 1 minute [3.6]

Cumulative Review Exercises, page 290

1. $x > -3$ [1.1] **2.** Shift the graph of $y = f(x)$ horizontally 1 unit to the left and up 2 units. [1.4] **3.** Reflect the graph of $y = f(x)$ across the x-axis. [1.4] **4.** odd function [1.4/2.5] **5.** $f^{-1}(x) = \frac{x}{x-5}$ [1.6] **6.** $\frac{4\pi}{3}$ [2.1] **7.** 300° [2.1] **8.** $\frac{\sqrt{3}}{2}$ [2.2] **9.** $\frac{2\sqrt{3}}{3}$ [2.2] **10.** $\frac{2\sqrt{5}}{5}$ [2.2] **11.** positive [2.3] **12.** 50° [2.3] **13.** $\frac{\pi}{3}$ [2.3] **14.** $x = \frac{1}{2}, y = \frac{\sqrt{3}}{2}$ [2.4] **15.** 0.43, π, $\frac{\pi}{12}$ [2.7] **16.** $\frac{\pi}{6}$ [3.5] **17.** 2.498 [3.5] **18.** $[-1, 1]$ [3.5] **19.** $\left(-\frac{\pi}{2}, \frac{\pi}{2}\right)$ [3.5] **20.** $\frac{\pi}{2}, \frac{7\pi}{6}, \frac{11\pi}{6}$ [3.6]

Exercise Set 4.1, page 298

1. $C = 77°, b \approx 16, c \approx 17$ **3.** $B = 38°, a \approx 18, c \approx 10$ **5.** $C \approx 15°, B \approx 33°, c \approx 7.8$ **7.** $C = 45.1°, b \approx 39.4, c \approx 30.2$ **9.** $C = 32.6°, c \approx 21.6, a \approx 39.8$ **11.** $B = 47.7°, a \approx 57.4, b \approx 76.3$ **13.** $A \approx 58.5°, B \approx 7.3°, a \approx 81.5$ **15.** $C = 59°, B = 84°, b \approx 46$ or $C = 121°, B = 22°, b \approx 17$ **17.** No triangle is formed. **19.** No triangle is formed. **21.** $C = 19.8°, B = 145.4°, b \approx 10.7$ or $C = 160.2°, B = 5.0°, b \approx 1.64$ **23.** No triangle is formed. **25.** $C = 51.21°, A = 11.47°, c \approx 59.00$ **27.** $B \approx 130.9°, C \approx 28.6°, b \approx 22.2$ or $B \approx 8.1°, C \approx 151.4°, b \approx 4.17$ **29.** $\approx$ 68.8 miles **31.** 231 yards **33.** $\approx$ 110 feet **35.** 4840 feet **37.** $\approx$ 96 feet **39.** $\approx$ 33 feet **41.** $\approx$ 8.1 miles **43.** $\approx$ 1200 miles **45.** $\approx$ 260 meters **49.**

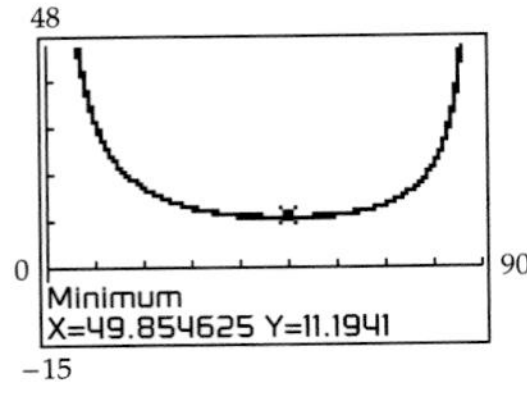

minimum value of $L \approx 11.19$ meters

Prepare for This Section (4.2), page 302

PS1. 20.7 **PS2.** 25.5 square inches **PS3.** $C = \cos^{-1}\left(\frac{a^2 + b^2 - c^2}{2ab}\right)$ **PS4.** 12.5 meters **PS5.** 6 **PS6.** $c^2 = a^2 + b^2$

Exercise Set 4.2, page 308

1. $\approx$ 13 **3.** $\approx$ 150 **5.** $\approx$ 29 **7.** $\approx$ 9.5 **9.** $\approx$ 10 **11.** $\approx$ 40.1 **13.** $\approx$ 90.7 **15.** $\approx$ 39° **17.** $\approx$ 90° **19.** $\approx$ 47.9° **21.** $\approx$ 116.67° **23.** $\approx$ 80.3° **25.** $a \approx 11.1, B \approx 62.0°, C \approx 78.6°$ **27.** $A \approx 34.2°, B \approx 104.6°, C \approx 41.3°$ **29.** $\approx$ 140 square units **31.** $\approx$ 53 square units **33.** $\approx$ 81 square units **35.** $\approx$ 299 square units **37.** $\approx$ 36 square units **39.** $\approx$ 7.3 square units **41.** $\approx$ 710 miles **43.** $\approx$ 74 feet **45.** $\approx$ 60.9° **47.** $\approx$ 350 miles **49.** 40 centimeters **51.** $\approx$ 2800 feet **53.** 402 miles, S62.6°E **55.** $\approx$ 47,500 square meters **57.** 162 square inches **59.** $\approx$ \$41,000 **61.** $\approx$ 6.23 acres **63.** Triangle *DEF* has an incorrect dimension. **65.** $\approx$ 12.5° **69.** $\approx$ 140 cubic inches

Prepare for This Section (4.3), page 312

PS1. 1 **PS2.** −6.691 **PS3.** 30° **PS4.** 157.6° **PS5.** $\frac{\sqrt{5}}{5}$ **PS6.** $\frac{14\sqrt{17}}{17}$

Exercise Set 4.3, page 325

1. $a = 4, b = 2; \langle 4, 2\rangle$ **3.** $a = -5, b = 4; \langle -5, 4\rangle$ **5.** $a = 7, b = -1; \langle 7, -1\rangle$ **7.** $a = -7, b = -5; \langle -7, -5\rangle$ **9.** $a = 0, b = 8; \langle 0, 8\rangle$ **11.** 5, $\approx 126.9°, \left\langle -\frac{3}{5}, \frac{4}{5}\right\rangle$ **13.** $\approx 44.7, \approx 296.6°, \left\langle \frac{\sqrt{5}}{5}, \frac{-2\sqrt{5}}{5}\right\rangle$ **15.** $\approx 4.5, \approx 296.6°, \left\langle \frac{\sqrt{5}}{5}, \frac{-2\sqrt{5}}{5}\right\rangle$ **17.** $\approx 45.7, \approx 336.8°, \left\langle \frac{7\sqrt{58}}{58}, \frac{-3\sqrt{58}}{58}\right\rangle$ **19.** $\langle -6, 12\rangle$ **21.** $\langle -1, 10\rangle$ **23.** $\left\langle -\frac{11}{6}, \frac{7}{3}\right\rangle$ **25.** $2\sqrt{5}$ **27.** $2\sqrt{109}$ **29.** $-8\mathbf{i} + 12\mathbf{j}$ **31.** $14\mathbf{i} - 6\mathbf{j}$ **33.** $\frac{11}{12}\mathbf{i} + \frac{1}{2}\mathbf{j}$ **35.** $\sqrt{113}$ **37.** $a_1 \approx 4.5, a_2 \approx 2.3, 4.5\mathbf{i} + 2.3\mathbf{j}$ **39.** $a_1 \approx 2.8, a_2 \approx 2.8, 2.8\mathbf{i} + 2.8\mathbf{j}$ **41.** $\approx$ 380 miles per hour **43.** $\approx$ 250 miles per hour at a heading of 86° **45.** 293 pounds **47.** **a.** 131 pounds **b.** 319 pounds **49.** The forces are in equilibrium. **51.** The forces are not in equilibrium. $\mathbf{F}_4 = 0\mathbf{i} + 10\mathbf{j}$ **53.** The forces are in equilibrium. **55.** −3 **57.** 0 **59.** 1 **61.** 0 **63.** $\approx$ 79.7° **65.** 45° **67.** 90°, orthogonal **69.** 180° **71.** $\frac{46}{5}$ **73.** $\frac{14\sqrt{29}}{29} \approx 2.6$ **75.** $\sqrt{5} \approx 2.2$ **77.** $-\frac{11\sqrt{5}}{5} \approx -4.9$ **79.** $\approx$ 954 foot-pounds **81.** $\approx$ 779 foot-pounds

83. 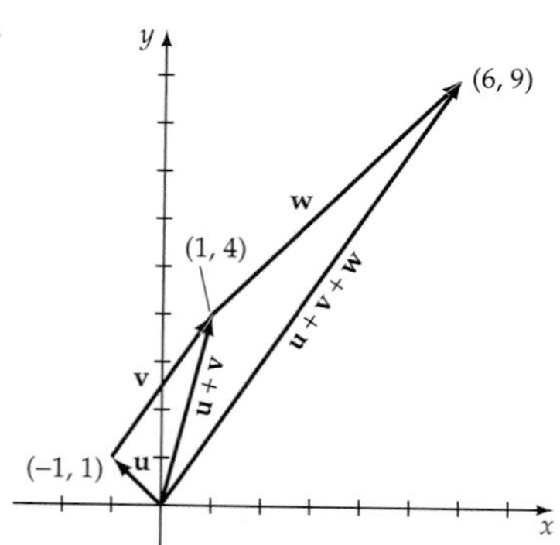

$\langle 6, 9\rangle$ **85.** the vector from $P_1(3, -1)$ to $P_2(5, -4)$

87. Because $\mathbf{v} \cdot \mathbf{w} = 0$, the vectors are perpendicular. **89.** $\langle 7, 2\rangle$ is one example. **91.** No **95.** The same amount of work is done.

Chapter 4 Assessing Concepts, page 329

1. a triangle that does not contain a right angle **2.** the Law of Cosines **3.** SSA **4.** the semiperimeter of a triangle **5.** a scalar **6.** a scalar **7.** True **8.** False **9.** True **10.** True

Chapter 4 Review Exercises, page 329

1. $B = 51°, a \approx 11, c \approx 18$ [4.1] **2.** $A = 8.6°, a \approx 1.77, b \approx 11.5$ [4.1] **3.** $B \approx 48°, C \approx 95°, A \approx 37°$ [4.2] **4.** $A \approx 47°, B \approx 76°, C \approx 58°$ [4.2] **5.** $c \approx 13, A \approx 55°, B \approx 90°$ [4.2] **6.** $a \approx 169, B \approx 37°, C \approx 61°$ [4.2] **7.** No triangle is formed. [4.1] **8.** No triangle is formed. [4.1] **9.** $C = 45°, a \approx 29, b \approx 35$ [4.1] **10.** $A = 115°, a \approx 56, b \approx 26$ [4.1] **11.** ≈ 360 square units [4.2] **12.** ≈ 31 square units [4.2] **13.** ≈ 920 square units [4.2] **14.** ≈ 46 square units [4.2] **15.** ≈ 790 square units [4.2] **16.** ≈ 210 square units [4.2] **17.** ≈ 170 square units [4.2] **18.** ≈ 140 square units [4.2] **19.** $a_1 = 5, a_2 = 3, \langle 5, 3\rangle$ [4.3] **20.** $a_1 = 1, a_2 = 6, \langle 1, 6\rangle$ [4.3]

21. $\approx 4.5, 153.4°$ [4.3] **22.** $\approx 6.7, 333.4°$ [4.3] **23.** $\approx 3.6, 123.7°$ [4.3] **24.** $\approx 8.1, 240.3°$ [4.3] **25.** $\left\langle -\frac{8\sqrt{89}}{89}, \frac{5\sqrt{89}}{89}\right\rangle$ [4.3]

26. $\left\langle \frac{7\sqrt{193}}{193}, -\frac{12\sqrt{193}}{193}\right\rangle$ [4.3] **27.** $\frac{5\sqrt{26}}{26}\mathbf{i} + \frac{\sqrt{26}}{26}\mathbf{j}$ [4.3] **28.** $\frac{3\sqrt{34}}{34}\mathbf{i} - \frac{5\sqrt{34}}{34}\mathbf{j}$ [4.3] **29.** $\langle -7, -3\rangle$ [4.3] **30.** $\langle 18, 7\rangle$ [4.3]

31. $-6\mathbf{i} - \frac{17}{2}\mathbf{j}$ [4.3] **32.** $-\frac{13}{6}\mathbf{i} - \frac{47}{6}\mathbf{j}$ [4.3] **33.** 420 mph [4.3] **34.** $\approx 7°$ [4.3] **35.** 18 [4.3] **36.** -21 [4.3] **37.** -9 [4.3]

38. 20 [4.3] **39.** $\approx 86°$ [4.3] **40.** $\approx 138°$ [4.3] **41.** $\approx 125°$ [4.3] **42.** $\approx 157°$ [4.3] **43.** $\frac{10\sqrt{41}}{41}$ [4.3] **44.** $\frac{27\sqrt{29}}{29}$ [4.3]

45. ≈ 662 foot-pounds [4.3]

Chapter 4 Quantitative Reasoning Exercises, page 330

QR1. 2210 miles **QR2.** 289° **QR3.** 1620 miles **QR4.** 78°

Chapter 4 Test, page 332

1. $B = 94°, a \approx 48, b \approx 51$ [4.1] **2.** $\approx 11°$ [4.1] **3.** ≈ 14 [4.2] **4.** $\approx 48°$ [4.2] **5.** ≈ 39 square units [4.2] **6.** ≈ 93 square units [4.2] **7.** ≈ 260 square units [4.2] **8.** $\sqrt{13}$ [4.3] **9.** $-9.193\mathbf{i} - 7.713\mathbf{j}$ [4.3] **10.** $-19\mathbf{i} - 29\mathbf{j}$ [4.3] **11.** -1 [4.3] **12.** 103° [4.3] **13.** ≈ 27 miles [4.2] **14.** ≈ 21 miles [4.1] **15.** $\approx \$65{,}800$ [4.2]

Cumulative Review Exercises, page 332

1. $\sqrt{74}$ units [1.2] **2.** $\sin x + \cos x$ [1.5] **3.** $\sec(\cos x)$ [1.5] **4.** $f^{-1}(x) = 2x + 6$ [1.6] **5.** shifted 2 units to the right and 3 units up [1.4]

6. 33 centimeters [2.2] **7.** 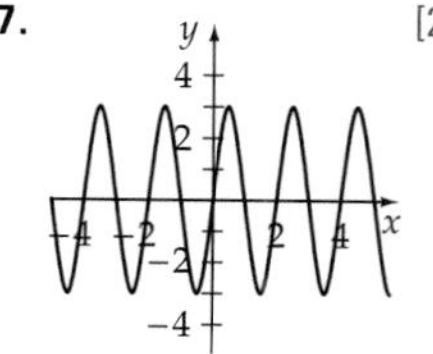

[2.5] **8.** 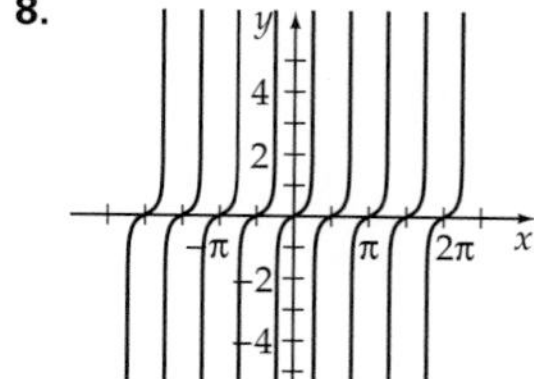

[2.6] **9.** 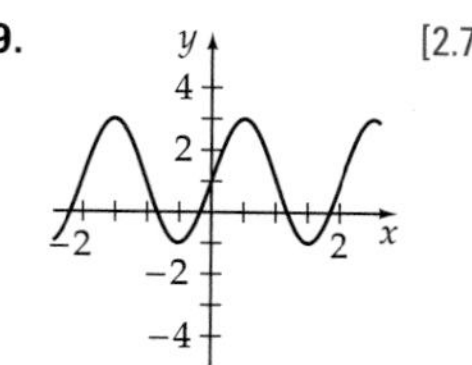

[2.7]

10. amplitude: 3; period: 6π, phase shift: $\frac{3\pi}{2}$ [2.7] **11.** amplitude: $\sqrt{2}$, period: 2π, phase shift: $\frac{\pi}{4}$ [2.7] **12.** 9.5 centimeters [4.2] **13.** See [3.1] **14.** $\frac{\pi}{3}$ [3.5] **15.** $\frac{5}{12}$ [3.5] **16.** $0, \frac{\pi}{6}, \frac{5\pi}{6}, \pi$ [3.6] **17.** magnitude: 5; angle: 323.1° [4.3] **18.** 60.3° [4.3] **19.** 289.5° [4.3] **20.** ground speed: 592 mph, heading: 67.4° [4.3]

Exercise Set 5.1, page 340

1. $9i$ **3.** $7i\sqrt{2}$ **5.** $4 + 9i$ **7.** $5 + 7i$ **9.** $8 - 3i\sqrt{2}$ **11.** $11 - 5i$ **13.** $-7 + 4i$ **15.** $8 - 5i$ **17.** -10 **19.** $-2 + 16i$ **21.** -40 **23.** -10 **25.** $19i$ **27.** $20 - 10i$ **29.** $22 - 29i$ **31.** 41 **33.** $12 - 5i$ **35.** $-114 + 42i\sqrt{2}$ **37.** $-6i$ **39.** $3 - 6i$ **41.** $\frac{7}{53} - \frac{2}{53}i$ **43.** $1 + i$ **45.** $\frac{15}{41} - \frac{29}{41}i$ **47.** $\frac{5}{13} + \frac{12}{13}i$ **49.** $2 + 5i$ **51.** $-16 - 30i$ **53.** $-11 - 2i$ **55.** $-i$ **57.** -1 **59.** $-i$ **61.** -1

63. $z_1 = 2 + 4i, z_2 = 6 + 7i, z_3 = 10 + 10i, z_4 = 14 + 13i, z_5 = 18 + 16i$ **65.** $z_1 = 1 + i, z_2 = -1 + i, z_3 = -1 - i, z_4 = 1 - i, z_5 = z_1, z_6 = z_2, z_7 = z_3, z_8 = z_4$ **67.** $\frac{1}{2} + \frac{\sqrt{3}}{2}i$ **69.** $-\frac{3}{2} + \frac{\sqrt{3}}{2}i$ **71.** $\frac{1}{2} + \frac{1}{2}i$ **73.** $(x + 4i)(x - 4i)$ **75.** $(z + 5i)(z - 5i)$ **77.** $(2x + 9i)(2x - 9i)$ **83.** 0

Prepare for This Section (5.2), page 343

PS1. $1 + 3i$ **PS2.** $\frac{1}{2} + \frac{1}{2}i$ **PS3.** $2 - 3i$ **PS4.** $3 + 5i$ **PS5.** $-\frac{1}{2} \pm \frac{\sqrt{3}}{2}i$ **PS6.** $-3i, 3i$

Exercise Set 5.2, page 349

1–7.

Im

Re

$-2 - 2i$ $\sqrt{3} - i$ $-2i$ $3 - 5i$

$|-2 - 2i| = 2\sqrt{2}$
$|\sqrt{3} - i| = 2$
$|-2i| = 2$
$|3 - 5i| = \sqrt{34}$

9. $\sqrt{2}$ cis 315° **11.** 2 cis 330° **13.** 3 cis 90° **15.** 5 cis 180° **17.** 16 cis 120° **19.** 4 cis 240° **21.** $\sqrt{2} + i\sqrt{2}$ **23.** $\frac{\sqrt{2}}{2} - \frac{\sqrt{2}}{2}i$ **25.** $-3\sqrt{2} + 3i\sqrt{2}$ **27.** 8 **29.** $-\sqrt{3} + i$ **31.** $-3i$ **33.** $-4\sqrt{2} + 4i\sqrt{2}$ **35.** $\frac{9\sqrt{3}}{2} - \frac{9}{2}i$ **37.** $\approx -0.832 + 1.819i$ **39.** 6 cis 255° **41.** 12 cis 335° **43.** 10 cis $\frac{16\pi}{15}$ **45.** 24 cis 6.5 **47.** $-4 - 4i\sqrt{3}$ **49.** $3i$ **51.** $-\frac{3\sqrt{3}}{2} + \frac{3i}{2}$ **53.** $\approx -2.081 + 4.546i$ **55.** $\approx 2.732 - 0.732i$ **57.** $6 + 0i = 6$ **59.** $-\frac{1}{2} + \frac{\sqrt{3}}{2}i$ **61.** $0 - \sqrt{2}i = -\sqrt{2}i$ **63.** $16 - 16i$ **65.** $-\frac{3}{8} + \frac{\sqrt{3}}{8}i$ **67.** $59.0 + 43.0i$ **69.** r^2 or $a^2 + b^2$

Prepare for This Section (5.3), page 350

PS1. i **PS2.** 2 **PS3.** 3 **PS4.** $2\sqrt{2}$ cis $\frac{\pi}{4}$ **PS5.** $-\sqrt{3} + i$ **PS6.** 1

Exercise Set 5.3, page 354

1. $-128 - 128i\sqrt{3}$ **3.** $-16 + 16i\sqrt{3}$ **5.** $16\sqrt{2} + 16i\sqrt{2}$ **7.** $64 + 0i = 64$ **9.** $0 - 32i = -32i$ **11.** $-4 + 0i = -4$ **13.** $1024 - 1024i$

15. $0 - 1i = -i$

17. $3 + 0i = 3$, $-3 + 0i = -3$

19. $2 + 0i = 2$, $1 + i\sqrt{3}$, $-1 + i\sqrt{3}$, $-2 + 0i = -2$, $-1 - i\sqrt{3}$, $1 - i\sqrt{3}$

21. $0.809 + 0.588i$, $-0.309 + 0.951i$, $-1 + 0i = -1$, $-0.309 - 0.951i$, $0.809 - 0.588i$

23. $1 + 0i = 1$, $-\frac{1}{2} + \frac{i\sqrt{3}}{2}$, $-\frac{1}{2} - \frac{i\sqrt{3}}{2}$

25. $1.070 + 0.213i$, $-0.213 + 1.070i$, $-1.070 - 0.213i$, $0.213 - 1.070i$

27. $-0.276 + 1.563i$, $-1.216 - 1.020i$, $1.492 - 0.543i$

29. $2\sqrt{2} + 2i\sqrt{6}$, $-2\sqrt{2} - 2i\sqrt{6}$

31. 2 cis 60°, 2 cis 180°, 2 cis 300°

33. cis 67.5°, cis 157.5°, cis 247.5°, cis 337.5°

35. 3 cis 0°, 3 cis 120°, 3 cis 240°

37. 3 cis 45°, 3 cis 135°, 3 cis 225°, 3 cis 315°

39. $\sqrt[4]{2}$ cis 75°, $\sqrt[4]{2}$ cis 165°, $\sqrt[4]{2}$ cis 255°, $\sqrt[4]{2}$ cis 345°

41. $\sqrt[3]{2}$ cis 80°, $\sqrt[3]{2}$ cis 200°, $\sqrt[3]{2}$ cis 320°

47. For $n \geq 2$, the sum of the nth roots of 1 is 0.

Chapter 5 Assessing Concepts, page 358

1. True **2.** True **3.** True **4.** False **5.** The four roots are equally spaced around a circle with center (0, 0) and radius 1. **6.** 1 **7.** 5 **8.** No **9.** $-3 + 5i$ **10.** 2

Chapter 5 Review Exercises, page 358

1. $3 - 8i, 3 + 8i$ [5.1] **2.** $6 + 2i, 6 - 2i$ [5.1] **3.** $-2 + i\sqrt{5}, -2 - i\sqrt{5}$ [5.1] **4.** $-5 - 3i\sqrt{3}, -5 + 3i\sqrt{3}$ [5.1] **5.** -4 [5.1] **6.** 9 [5.1] **7.** $5 + 2i$ [5.1] **8.** $-3 + 4i$ [5.1] **9.** $-3 + 3i$ [5.1] **10.** $-5 - 15i$ [5.1] **11.** $25 - 19i$ [5.1] **12.** $29 - 2i$ [5.1] **13.** $\frac{8}{25} - \frac{6}{25}i$ [5.1] **14.** $\frac{26}{53} + \frac{15}{53}i$ [5.1] **15.** $-2 - 2i$ [5.1] **16.** $2 - 11i$ [5.1] **17.** $6 + 6i$ [5.1] **18.** $-5 - 6i$ [5.1] **19.** 7 [5.1] **20.** $11 + i\sqrt{5}$ [5.1] **21.** $-i$ [5.1] **22.** i [5.1] **23.** 1 [5.1] **24.** -1 [5.1] **25.** 8 [5.1] **26.** $\sqrt{13}$ [5.1] **27.** $\sqrt{41}$ [5.1] **28.** $\sqrt{2}$ [5.1] **29.** $2\sqrt{2}$ cis 315° [5.2] **30.** 2 cis 150° [5.2] **31.** $\approx \sqrt{13}$ cis 146.3° [5.2] **32.** $\sqrt{17}$ cis 345.96° [5.2] **33.** $\frac{5\sqrt{2}}{2} - \frac{5\sqrt{2}}{2}i \approx 3.536 - 3.536i$ [5.2] **34.** $-3 - 3i\sqrt{3}$ [5.2] **35.** $\approx -0.832 + 1.819i$ [5.2] **36.** $-1.27 + 2.72i$ [5.2] **37.** $0 - 30i = -30i$ [5.2] **38.** $-5\sqrt{2} - 5i\sqrt{2}$ [5.2] **39.** $\approx -8.918 + 8.030i$ [5.2] **40.** $\approx -0.968 + 3.881i$ [5.2] **41.** $\approx -6.012 - 13.742i$ [5.2] **42.** $\approx 13.2 + 27.0i$ [5.2] **43.** 3 cis (−100°) [5.2] **44.** 3 cis 110° [5.2] **45.** 5 cis (−59°) [5.2] **46.** $\sqrt{2}$ cis (−50°) [5.2] **47.** $\frac{5}{3}$ cis 1.9 [5.2] **48.** $\frac{1}{2}$ cis (−4) [5.2] **49.** $-\frac{243\sqrt{2}}{2} - \frac{243\sqrt{2}}{2}i \approx -171.827 - 171.827i$ [5.3] **50.** -1 [5.3] **51.** $64 - 64\sqrt{3}i \approx 64 - 110.851i$ [5.3] **52.** $32{,}768i$ [5.3] **53.** $-16\sqrt{2} + 16i\sqrt{2} \approx -22.627 + 22.627i$ [5.3] **54.** $-237 + 3116i$ [5.3] **55.** 3 cis 30°, 3 cis 150°, 3 cis 270° [5.3] **56.** $\sqrt[4]{8}$ cis 22.5°, $\sqrt[4]{8}$ cis 112.5°, $\sqrt[4]{8}$ cis 202.5°, $\sqrt[4]{8}$ cis 292.5° [5.3] **57.** 4 cis 0°, 4 cis 90°, 4 cis 180°, 4 cis 270° [5.3] **58.** 2 cis 45°, 2 cis 117°, 2 cis 189°, 2 cis 261°, 2 cis 333° [5.3] **59.** 3 cis 0°, 3 cis 90°, 3 cis 180°, 3 cis 270° [5.3] **60.** 5 cis 60°, 5 cis 180°, 5 cis 300° [5.3]

Chapter 5 Quantitative Reasoning Exercises, page 359

QR1.

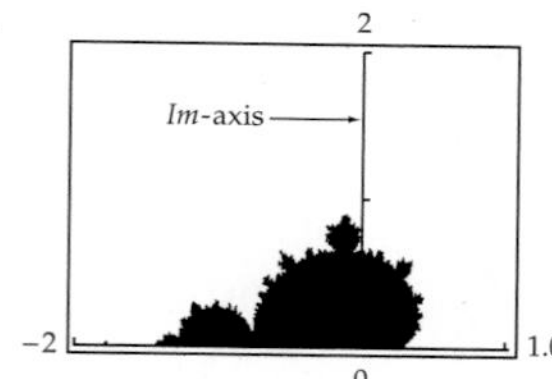

QR2. $-0.25 + 0.25i$, $-1 + 0.1i$, and $0.1 + 0.2i$
QR3. -2 is an element of the Mandelbrot set because all of its iterates equal 2.
QR4. The first iterate of $2i$ is $z_1 = -4 + 2i$. $2i$ is not an element of the Mandelbrot set because $|-4 + 2i| > 2$.
QR5. $(Z^2 + S) \rightarrow Z$
QR6. Answers will vary.

Chapter 5 Test, page 360

1. $6 + 3i$ [5.1] **2.** $3i\sqrt{2}$ [5.1] **3.** $10 - i$ [5.1] **4.** $-9 + 9i$ [5.1] **5.** -6 [5.1] **6.** $-i$ [5.1] **7.** $5 + 16i$ [5.1] **8.** $3 - 54i$ [5.1] **9.** $16 + 30i$ [5.1] **10.** $-5 - 4i$ [5.1] **11.** $-\frac{13}{25} - \frac{34}{25}i$ [5.1] **12.** $2 + 4i$ [5.1] **13.** $\sqrt{34}$ [5.1] **14.** $3\sqrt{2}$ cis 315° [5.2] **15.** 6 cis 270° [5.2] **16.** $-2 + 2i\sqrt{3}$ [5.2] **17.** $-\frac{5\sqrt{2}}{2} - \frac{5\sqrt{2}i}{2}$ [5.2] **18.** $6\sqrt{2} + 6i\sqrt{2}$ [5.2] **19.** $\approx -11.472 + 16.383i$ [5.2] **20.** -4 [5.2] **21.** $-6i$ [5.2] **22.** 16,777,216 [5.3] **23.** 2 cis 0°, 2 cis 60°, 2 cis 120°, 2 cis 180°, 2 cis 240°, 2 cis 300° [5.3] **24.** $\sqrt[3]{2}$ cis 40°, $\sqrt[3]{2}$ cis 160°, $\sqrt[3]{2}$ cis 280° [5.3] **25.** 2 cis 36°, 2 cis 108°, 2 cis 180°, 2 cis 252°, 2 cis 324° [5.3]

Cumulative Review Exercises, page 361

1. $[-2, 3]$ [1.1] **2.** All real numbers except -2 and 2 [1.3] **3.** -2 [1.3] **4.** $(f \circ g)(x) = \sin(x^2 - 1)$ [1.5] **5.** $\frac{3}{2}$ [1.6] **6.** 270° [2.1] **7.** ≈25 centimeters [2.2] **8.** $a = -1, b = 1$ [2.4] **9.**

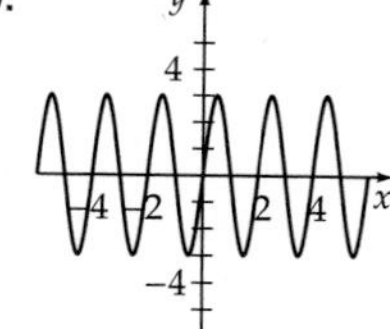

[2.5] **10.** [2.6]

11. See [3.1] **12.** $-\sin x$ [3.2] **13.** $\frac{56}{65}$ [3.2] **14.** $\frac{33}{65}$ [3.5] **15.** $0, \frac{\pi}{6}, \pi, \frac{11\pi}{6}$ [3.6] **16.** ≈170 centimeters [4.2] **17.** 64.7° [4.3] **18.** 1449 foot-pounds [4.3] **19.** $2\sqrt{2}$ cis 45° [5.2] **20.** $-3, \frac{3}{2} + \frac{3\sqrt{3}}{2}i, \frac{3}{2} - \frac{3\sqrt{3}}{2}i$ [5.3]

Exercise Set 6.1, page 370

1. a. iii **b.** i **c.** iv **d.** ii

3. vertex: $(0, 0)$
focus: $(0, -1)$
directrix: $y = 1$

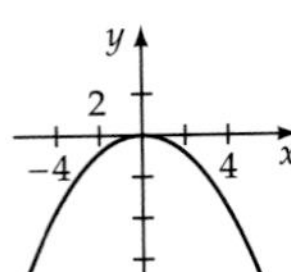

5. vertex: $(0, 0)$
focus: $\left(\frac{1}{12}, 0\right)$
directrix: $x = -\frac{1}{12}$

7. vertex: $(2, -3)$
focus: $(2, -1)$
directrix: $y = -5$

9. vertex: $(2, -4)$
focus: $(1, -4)$
directrix: $x = 3$

11. vertex: $(-4, 1)$
focus: $\left(-\frac{7}{2}, 1\right)$
directrix: $x = -\frac{9}{2}$

13. vertex: $(2, 2)$
focus: $\left(2, \frac{5}{2}\right)$
directrix: $y = \frac{3}{2}$

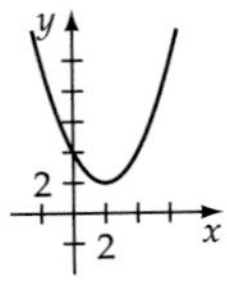

15. vertex: $(-4, -10)$
focus: $\left(-4, -\frac{39}{4}\right)$
directrix: $y = -\frac{41}{4}$

17. vertex: $\left(-\frac{7}{4}, \frac{3}{2}\right)$
focus: $\left(-2, \frac{3}{2}\right)$
directrix: $x = -\frac{3}{2}$

19. vertex: $(-5, -3)$
focus: $\left(-\frac{9}{2}, -3\right)$
directrix: $x = -\frac{11}{2}$

21. vertex: $\left(-\frac{3}{2}, \frac{13}{12}\right)$
focus: $\left(-\frac{3}{2}, \frac{1}{3}\right)$
directrix: $y = \frac{11}{6}$

23. vertex: $\left(2, -\frac{5}{4}\right)$
focus: $\left(2, -\frac{3}{4}\right)$
directrix: $y = -\frac{7}{4}$

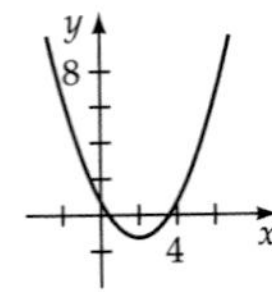

25. vertex: $\left(\frac{9}{2}, -1\right)$
focus: $\left(\frac{35}{8}, -1\right)$
directrix: $x = \frac{37}{8}$

27. vertex: $\left(1, \frac{1}{9}\right)$
focus: $\left(1, \frac{31}{36}\right)$
directrix: $y = -\frac{23}{36}$

29. $x^2 = -16y$ **31.** $(x + 1)^2 = 4(y - 2)$

33. $(x - 3)^2 = 4(y + 4)$ **35.** $(x + 4)^2 = 4(y - 1)$ **37.** vertex: $(250, 20)$, focus: $\left(\frac{3240}{13}, 20\right)$ **39.** on axis of symmetry 4 feet above vertex **41.** 6.0 inches **43. a.** 5900 square feet **b.** 56,800 square feet **45.** $a = 1.5$ inches

47. a. $4\left(\frac{800}{84}\right)^2(y - 32) = x^2$ **b.** 117 millimeters **49.** 4 **51.** $4|p|$ **53.**

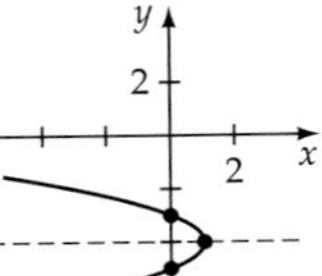

55.

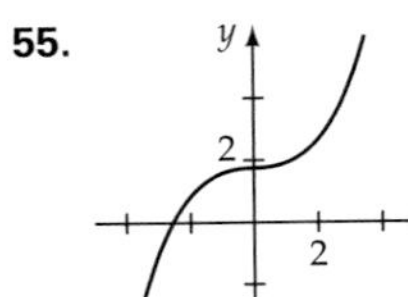

57. $x^2 + y^2 - 8x - 8y - 2xy = 0$

Prepare for This Section (6.2), page 374

PS1. midpoint: $(2, 3)$; length: $2\sqrt{13}$ **PS2.** $-8, 2$ **PS3.** $1 \pm \sqrt{3}$ **PS4.** $x^2 - 8x + 16 = (x - 4)^2$ **PS5.** $y = \pm\sqrt{4 - (x - 2)^2}$

PS6.

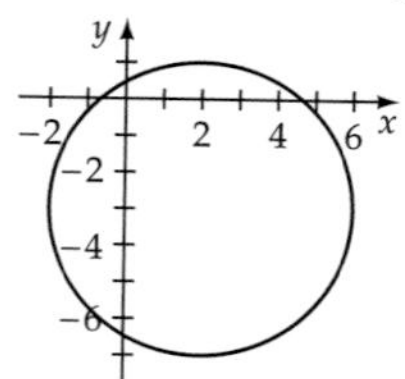

Exercise Set 6.2, page 385

1. a. iv **b.** i **c.** ii **d.** iii

3. center: $(0, 0)$
vertices: $(0, 5), (0, -5)$
foci: $(0, 3), (0, -3)$

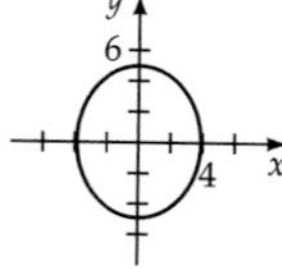

5. center: $(0, 0)$
vertices: $(3, 0), (-3, 0)$
foci: $(\sqrt{5}, 0), (-\sqrt{5}, 0)$

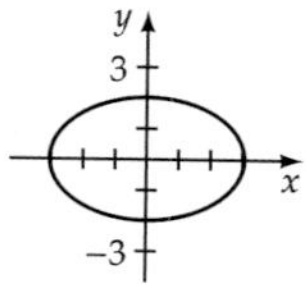

7. center: $(0, 0)$
vertices: $(3, 0), (-3, 0)$
foci: $(\sqrt{2}, 0), (-\sqrt{2}, 0)$

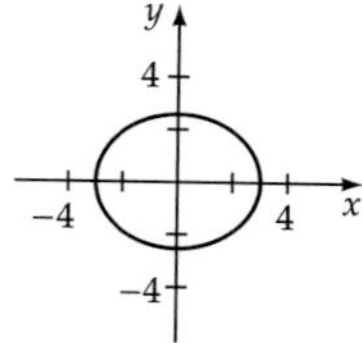

9. center: $(0, 0)$
vertices: $(0, 4), (0, -4)$
foci: $\left(0, \frac{\sqrt{55}}{2}\right), \left(0, -\frac{\sqrt{55}}{2}\right)$

11. center: $(3, -2)$
vertices: $(8, -2), (-2, -2)$
foci: $(6, -2), (0, -2)$

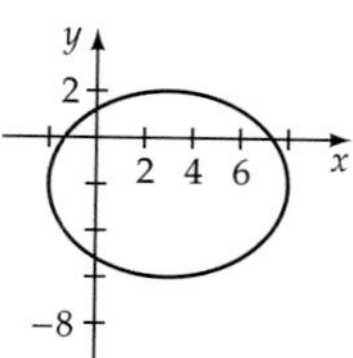

13. center: $(-2, 0)$
vertices: $(-2, 5), (-2, -5)$
foci: $(-2, 4), (-2, -4)$

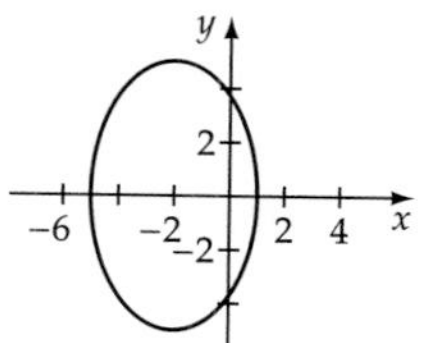

15. center: $(1, 3)$
vertices: $(1 + \sqrt{21}, 3), (1 - \sqrt{21}, 3)$
foci: $(1 + \sqrt{17}, 3), (1 - \sqrt{17}, 3)$

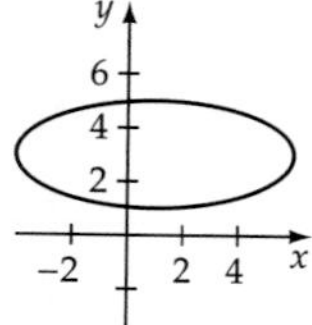

17. center: $(1, -1)$
vertices: $(1, 2), (1, -4)$
foci: $\left(1, -1 + \frac{\sqrt{65}}{3}\right), \left(1, -1 - \frac{\sqrt{65}}{3}\right)$

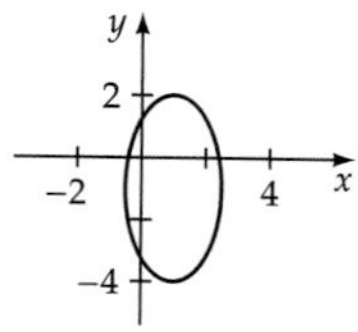

19. center: $(0, 0)$
vertices: $(2, 0), (-2, 0)$
foci: $(1, 0), (-1, 0)$

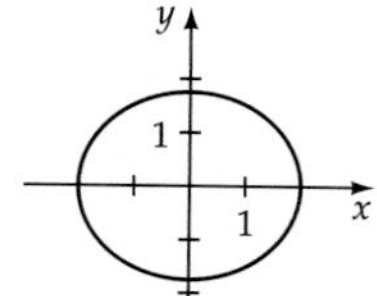

21. center: $(0, 0)$
vertices: $(0, 5), (0, -5)$
foci: $(0, 3), (0, -3)$

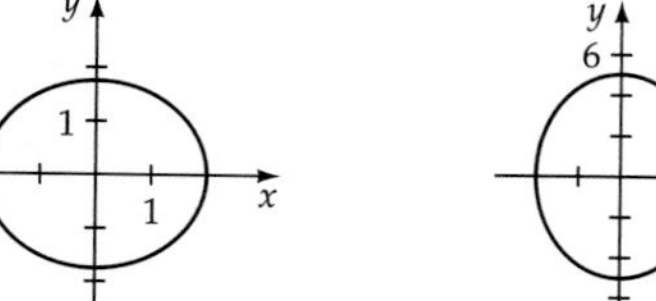

23. center: $(0, 0)$
vertices: $(0, 4), (0, -4)$
foci: $\left(0, \frac{\sqrt{39}}{2}\right), \left(0, -\frac{\sqrt{39}}{2}\right)$

25. center: $(3, 4)$
vertices: $(3, 6), (3, 2)$
foci: $(3, 4 + \sqrt{3}), (3, 4 - \sqrt{3})$

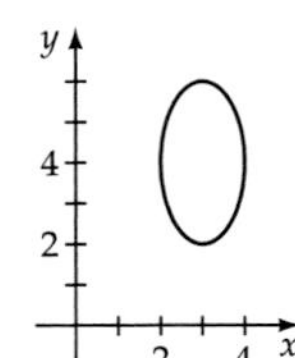

27. center: $(2, -3)$
vertices: $(-1, -3), (5, -3)$
foci: $(0, -3), (4, -3)$

29. center: $(2, 0)$
vertices: $(2, 4), (2, -4)$
foci: $(2, \sqrt{7}), (2, -\sqrt{7})$

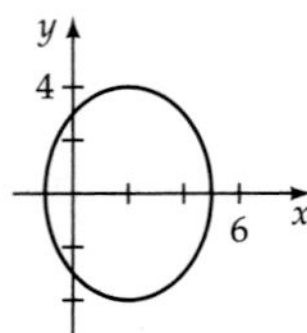

31. center: $(-1, 1)$
vertices: $(-1, 6), (-1, -4)$
foci: $(-1, 4), (-1, -2)$

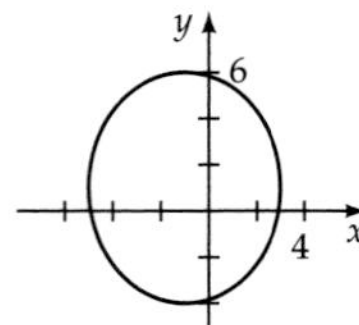

33. center: $(3, -1)$
vertices: $\left(\frac{11}{2}, -1\right), \left(\frac{1}{2}, -1\right)$
foci: $\left(3 + \frac{\sqrt{17}}{2}, -1\right), \left(3 - \frac{\sqrt{17}}{2}, -1\right)$

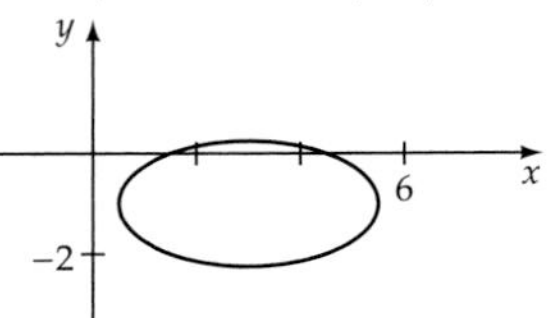

35. $\frac{x^2}{25} + \frac{y^2}{9} = 1$ **37.** $\frac{x^2}{36} + \frac{y^2}{16} = 1$ **39.** $\frac{x^2}{36} + \frac{y^2}{81/8} = 1$ **41.** $\frac{(x + 2)^2}{16} + \frac{(y - 4)^2}{7} = 1$ **43.** $\frac{(x - 2)^2}{25/24} + \frac{(y - 4)^2}{25} = 1$
45. $\frac{(x - 5)^2}{16} + \frac{(y - 1)^2}{25} = 1$ **47.** $\frac{x^2}{25} + \frac{y^2}{21} = 1$ **49.** $\frac{x^2}{20} + \frac{y^2}{36} = 1$ **51.** $\frac{(x - 1)^2}{25} + \frac{(y - 3)^2}{21} = 1$ **53.** $\frac{x^2}{80} + \frac{y^2}{144} = 1$
55. on the major axis of the ellipse, 41 centimeters from the emitter **57.** $\frac{x^2}{884.74^2} + \frac{y^2}{883.35^2} = 1$ **59.** 40 feet
61. $\frac{\left(x - \frac{9\sqrt{15}}{2}\right)^2}{324} + \frac{y^2}{81/4} = 1$ **63.** 24 feet **65. a.** $\sqrt{7}$ feet to the right and left of O. **b.** 8 feet

67. $y = \frac{-36 \pm \sqrt{1296 - 36(16x^2 - 108)}}{18}$

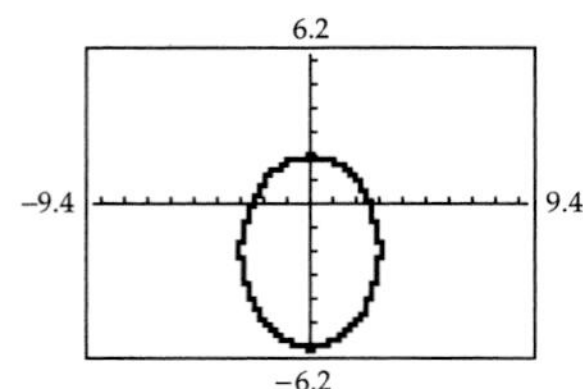

69. $y = \frac{-18 \pm \sqrt{324 - 36(4x^2 + 24x + 44)}}{18}$

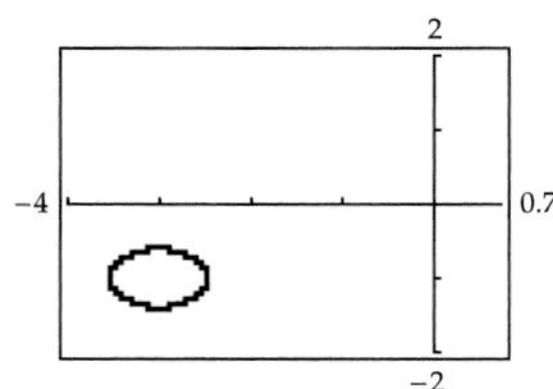

71. $\frac{x^2}{36} + \frac{y^2}{27} = 1$ **73.** $\frac{(x - 1)^2}{16} + \frac{(y - 2)^2}{12} = 1$
75. $\frac{9}{2}$

Prepare for This Section (6.3), page 390

PS1. midpoint: $(1, -1)$; length: $2\sqrt{13}$ **PS2.** $-4, 2$ **PS3.** $\sqrt{2}$ **PS4.** $4(x^2 + 6x + 9) = 4(x + 3)^2$ **PS5.** $y = \pm\frac{3}{2}\sqrt{x^2 - 4}$

PS6.

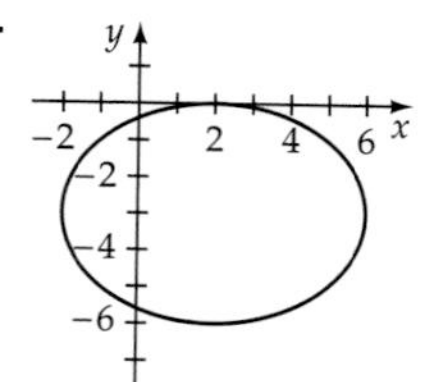

Exercise Set 6.3 page 399

1. a. iii **b.** ii **c.** i **d.** iv

3. center: $(0, 0)$

vertices: $(\pm 4, 0)$

foci: $\left(\pm\sqrt{41}, 0\right)$

asymptotes: $y = \pm\dfrac{5}{4}x$

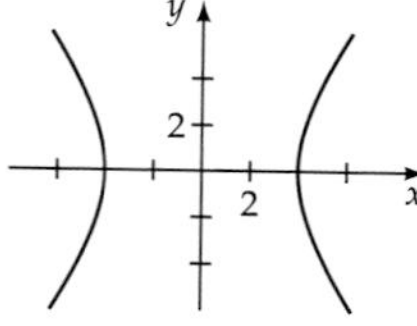

5. center: $(0, 0)$

vertices: $(0, \pm 2)$

foci: $\left(0, \pm\sqrt{29}\right)$

asymptotes: $y = \pm\dfrac{2}{5}x$

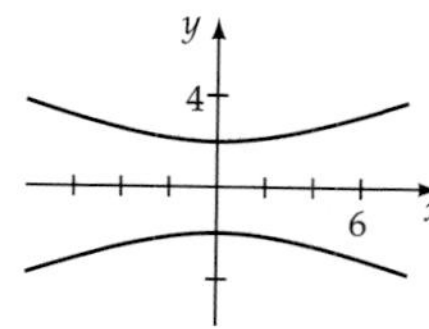

7. center: $(0, 0)$

vertices: $\left(\pm\sqrt{7}, 0\right)$

foci: $(\pm 4, 0)$

asymptotes: $y = \pm\dfrac{3\sqrt{7}}{7}x$

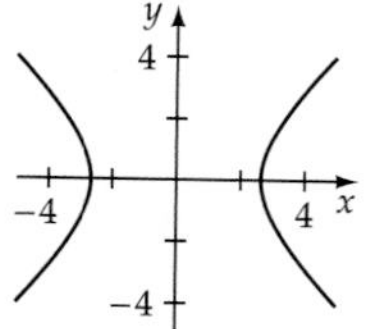

9. center: $(0, 0)$

vertices: $\left(\pm\dfrac{3}{2}, 0\right)$

foci: $\left(\pm\dfrac{\sqrt{73}}{2}, 0\right)$

asymptotes: $y = \pm\dfrac{8}{3}x$

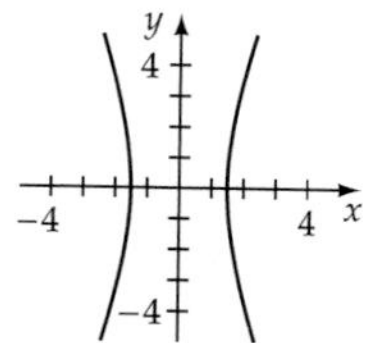

11. center: $(3, -4)$
vertices: $(7, -4), (-1, -4)$
foci: $(8, -4), (-2, -4)$
asymptotes: $y + 4 = \pm\dfrac{3}{4}(x - 3)$

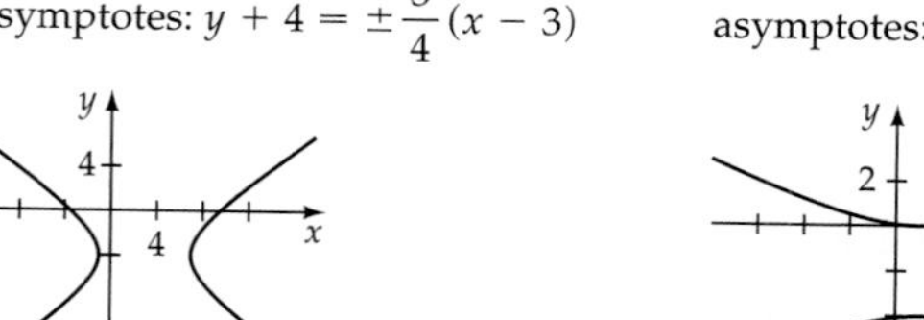

13. center: $(1, -2)$
vertices: $(1, 0), (1, -4)$
foci: $\left(1, -2 \pm 2\sqrt{5}\right)$
asymptotes: $y + 2 = \pm\dfrac{1}{2}(x - 1)$

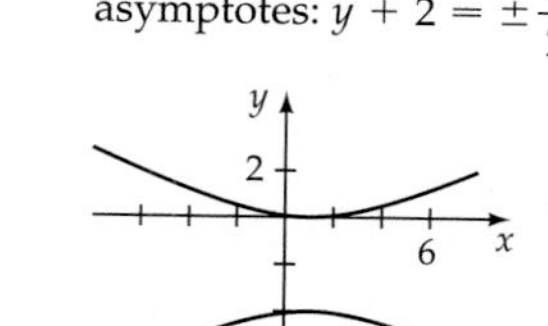

15. center: $(-2, 0)$
vertices: $(1, 0), (-5, 0)$
foci: $\left(-2 \pm \sqrt{34}, 0\right)$
asymptotes: $y = \pm\dfrac{5}{3}(x + 2)$

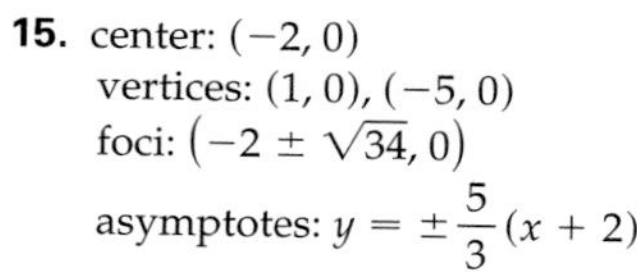

17. center: $(1, -1)$

vertices: $\left(\dfrac{7}{3}, -1\right), \left(-\dfrac{1}{3}, -1\right)$

foci: $\left(1 \pm \dfrac{\sqrt{97}}{3}, -1\right)$

asymptotes: $y + 1 = \pm\dfrac{9}{4}(x - 1)$

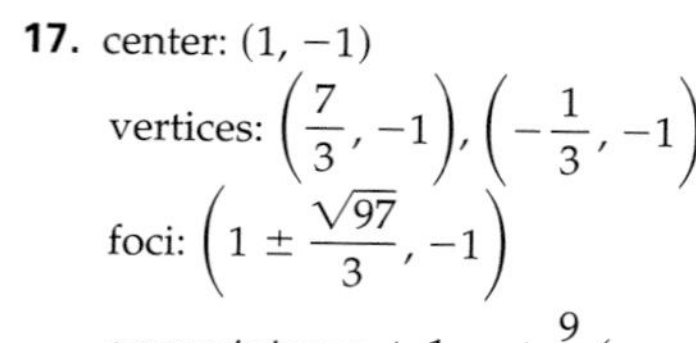

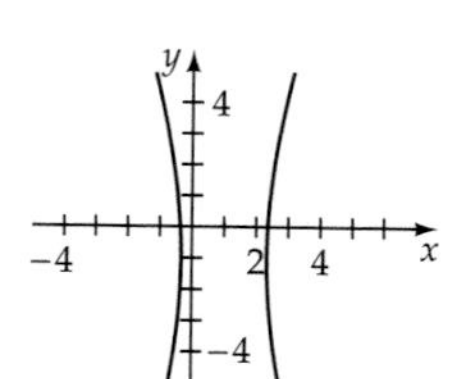

19. center: $(0, 0)$

vertices: $(\pm 3, 0)$

foci: $\left(\pm 3\sqrt{2}, 0\right)$

asymptotes: $y = \pm x$

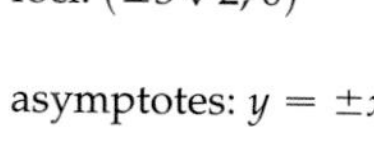

21. center: $(0, 0)$

vertices: $(0, \pm 3)$

foci: $(0, \pm 5)$

asymptotes: $y = \pm\dfrac{3}{4}x$

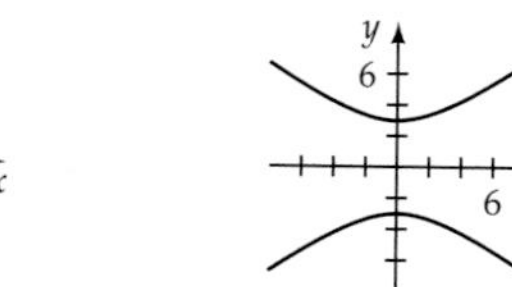

23. center: $(0, 0)$

vertices: $\left(0, \pm\dfrac{2}{3}\right)$

foci: $\left(0, \pm\dfrac{\sqrt{5}}{3}\right)$

asymptotes: $y = \pm 2x$

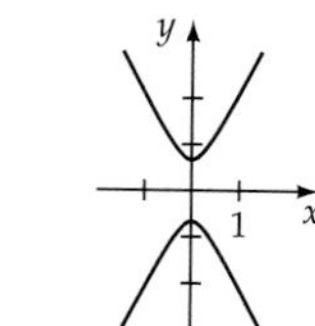

25. center: $(3, 4)$
vertices: $(3, 6), (3, 2)$
foci: $\left(3, 4 \pm 2\sqrt{2}\right)$
asymptotes: $y - 4 = \pm(x - 3)$

27. center: $(-2, -1)$
vertices: $(-2, 2), (-2, -4)$
foci: $\left(-2, -1 \pm \sqrt{13}\right)$
asymptotes: $y + 1 = \pm\dfrac{3}{2}(x + 2)$

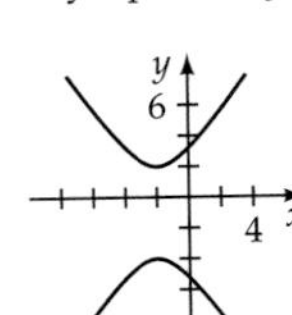

29. $y = \dfrac{-6 \pm \sqrt{36 + 4(4x^2 + 32x + 39)}}{-2}$

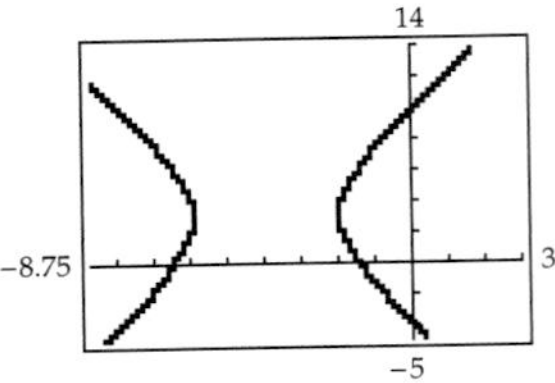

31. $y = \dfrac{64 \pm \sqrt{4096 + 64(9x^2 - 36x + 116)}}{-32}$

33. $y = \dfrac{18 \pm \sqrt{324 + 36(4x^2 + 8x - 6)}}{-18}$

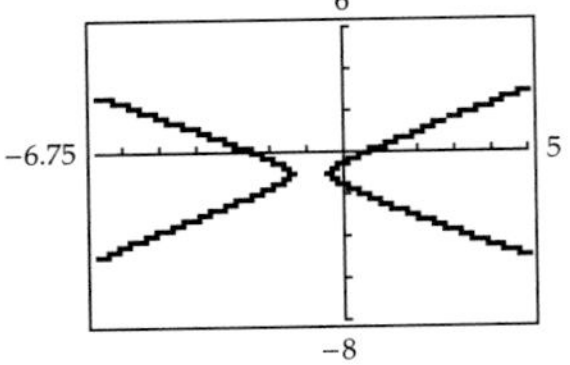

35. $\dfrac{x^2}{9} - \dfrac{y^2}{7} = 1$ **37.** $\dfrac{y^2}{20} - \dfrac{x^2}{5} = 1$

39. $\dfrac{y^2}{9} - \dfrac{x^2}{36/7} = 1$ **41.** $\dfrac{y^2}{16} - \dfrac{x^2}{64} = 1$ **43.** $\dfrac{(x - 4)^2}{4} - \dfrac{(y - 3)^2}{5} = 1$ **45.** $\dfrac{(x - 4)^2}{144/41} - \dfrac{(y + 2)^2}{225/41} = 1$ **47.** $\dfrac{(y - 2)^2}{3} - \dfrac{(x - 7)^2}{12} = 1$

49. $\dfrac{(y - 7)^2}{1} - \dfrac{(x - 1)^2}{3} = 1$ **51.** $\dfrac{x^2}{4} - \dfrac{y^2}{12} = 1$ **53.** $\dfrac{(x - 4)^2}{36/7} - \dfrac{(y - 1)^2}{4} = 1$ and $\dfrac{(y - 1)^2}{36/7} - \dfrac{(x - 4)^2}{4} = 1$ **55. a.** $\dfrac{x^2}{2162.25} - \dfrac{y^2}{13{,}462.75} = 1$

b. 221 miles **57.** $y^2 - x^2 = 10{,}000^2$, hyperbola **59. a.** $\dfrac{x^2}{2^2} - \dfrac{y^2}{0.5^2} = 1$ **b.** 6.25 inches

61. ellipse **63.** parabola **65.** parabola **67.** ellipse **69.** $\dfrac{x^2}{1} - \dfrac{y^2}{3} = 1$

71. $\dfrac{y^2}{9} - \dfrac{x^2}{7} = 1$ **73.**

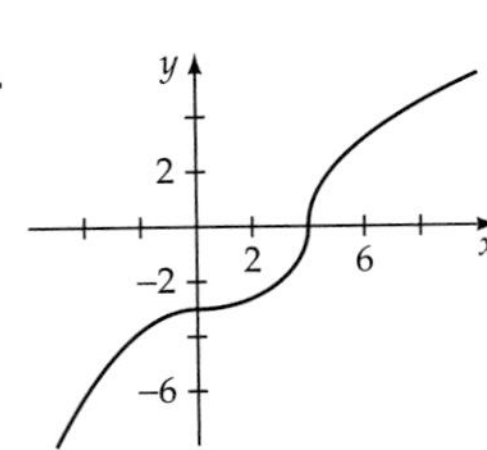

Prepare for This Section (6.4), page 403

PS1. $\cos\alpha\cos\beta - \sin\alpha\sin\beta$ **PS2.** $\sin\alpha\cos\beta + \cos\alpha\sin\beta$ **PS3.** $\dfrac{\pi}{6}$ **PS4.** $150°$ **PS5.** hyperbola

PS6.

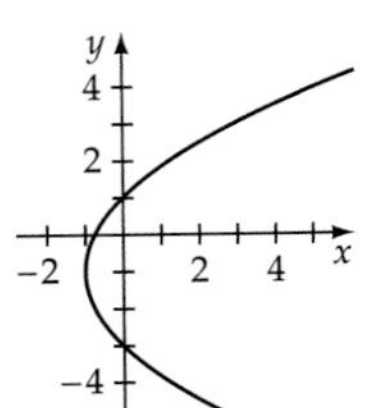

Exercise Set 6.4, page 410

1. 45° **3.** 36.9° **5.** 73.5° **7.** 22.5° **9.** 45°, $\frac{(x')^2}{8} - \frac{(y')^2}{8} = 1$ **11.** 18.4°, $\frac{(x')^2}{9} + \frac{(y')^2}{3} = 1$ **13.** 26.6°, $\frac{(x')^2}{1/2} - \frac{(y')^2}{1/3} = 1$

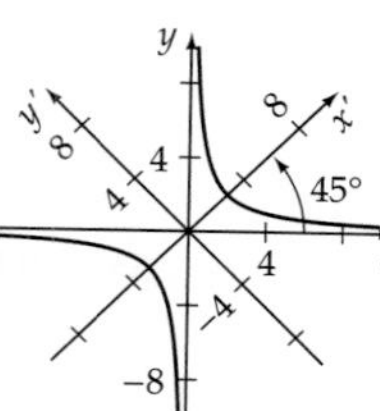

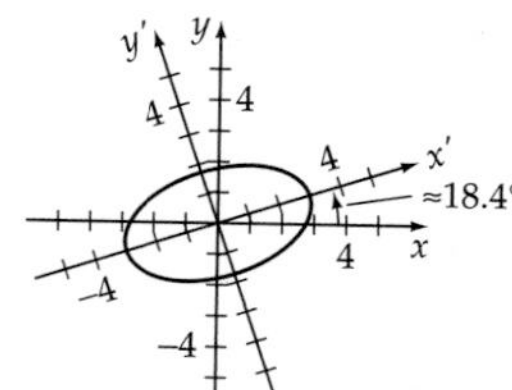

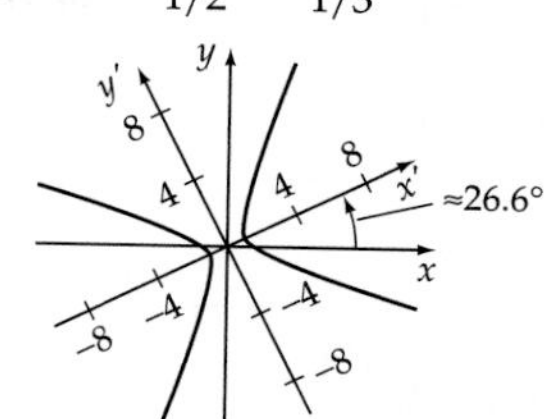

15. 30°, $y' = (x')^2 + 4$ **17.** 36.9°, $(y')^2 = 2(x' - 2)$ **19.** 36.9°, $15(x')^2 - 10(y')^2 + 6x' + 28y' + 11 = 0$

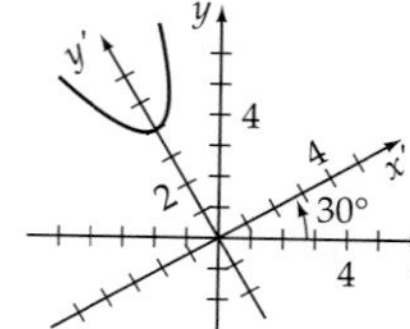

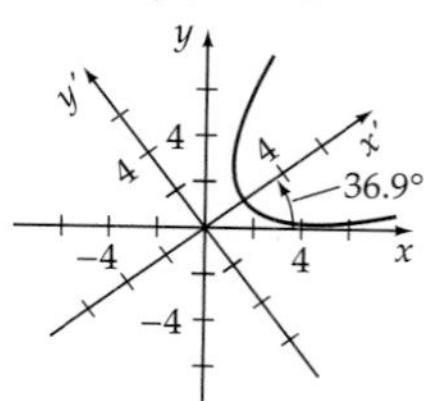

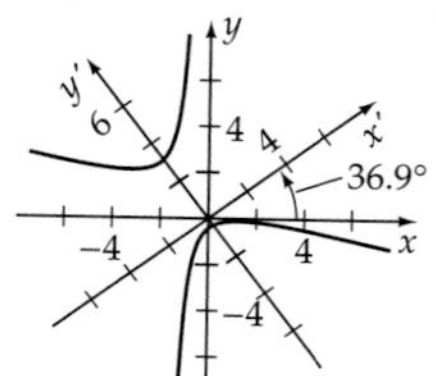

21.

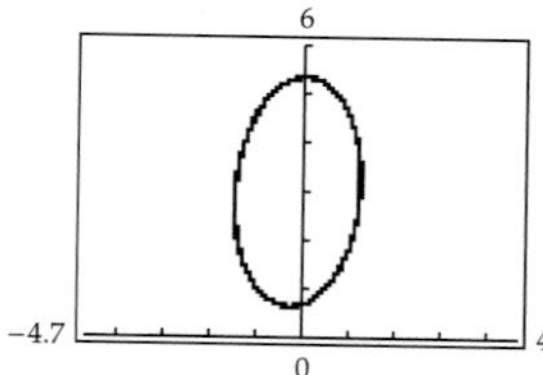

23.

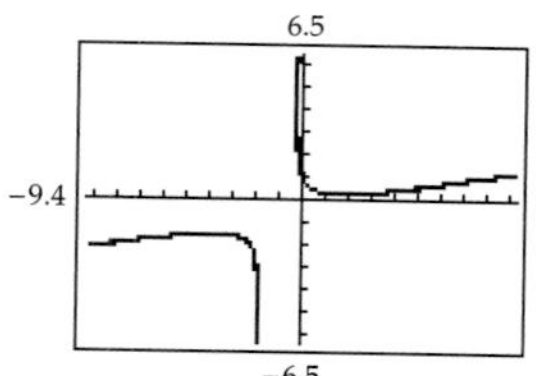

25.

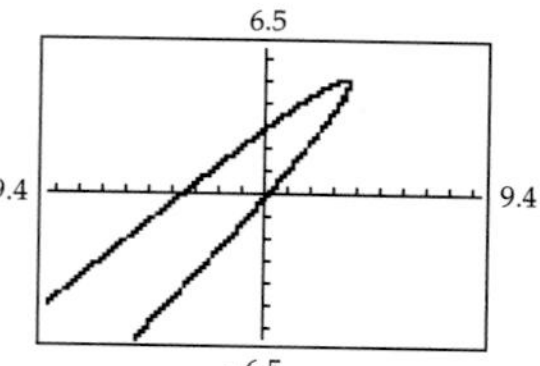

27. $y = \frac{\sqrt{3} + 2\sqrt{2}}{2\sqrt{3} - \sqrt{2}}x$ and $y = \frac{\sqrt{3} - 2\sqrt{2}}{2\sqrt{3} + \sqrt{2}}x$ **29.** $\left(\frac{3\sqrt{15}}{5}, \frac{\sqrt{15}}{5}\right)$ and $\left(-\frac{3\sqrt{15}}{5}, -\frac{\sqrt{15}}{5}\right)$ **31.** hyperbola **33.** parabola **35.** parabola **37.** hyperbola **39.** ellipse **43.** $9x^2 - 4xy + 6y^2 = 100$

Prepare for This Section (6.5), page 412

PS1. odd **PS2.** even **PS3.** $\frac{2\pi}{3}, \frac{5\pi}{3}$ **PS4.** 240° **PS5.** r^2 **PS6.** (4.2, 2.6)

Exercise Set 6.5, page 423

1–7.

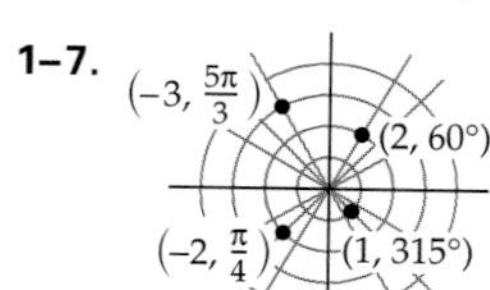

9.

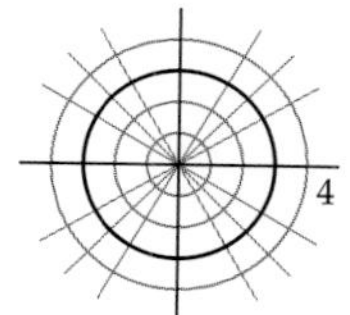

$0 \le \theta \le 2\pi$

11.

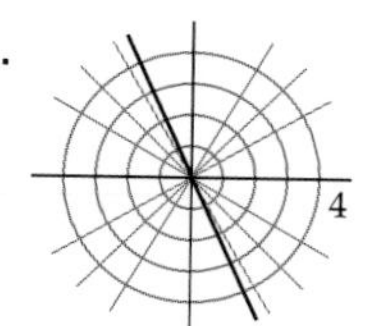

13.

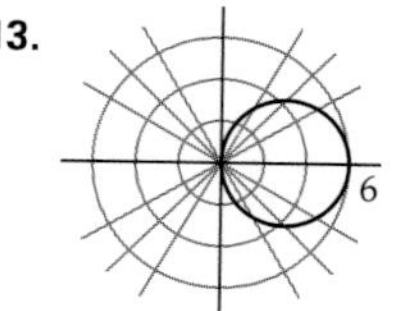

$0 \le \theta \le \pi$

15.

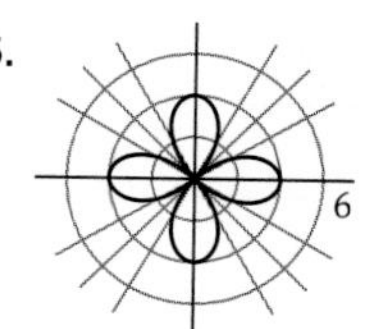

$0 \le \theta \le 2\pi$

17.

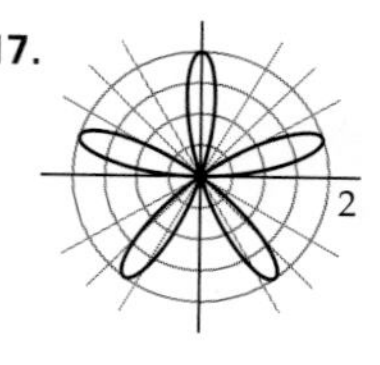

$0 \le \theta \le \pi$

19.

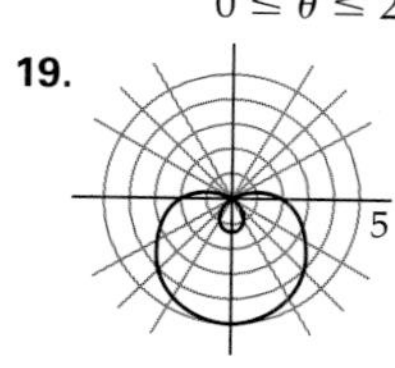

$0 \le \theta \le 2\pi$

21.

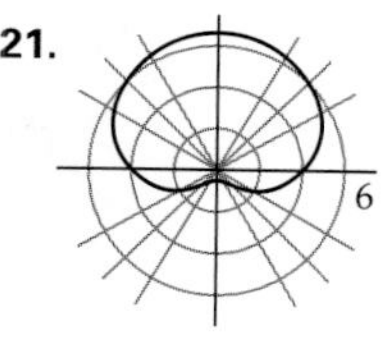

$0 \le \theta \le 2\pi$

23.

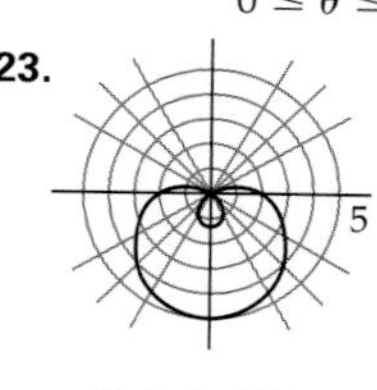

$0 \le \theta \le 2\pi$

25.

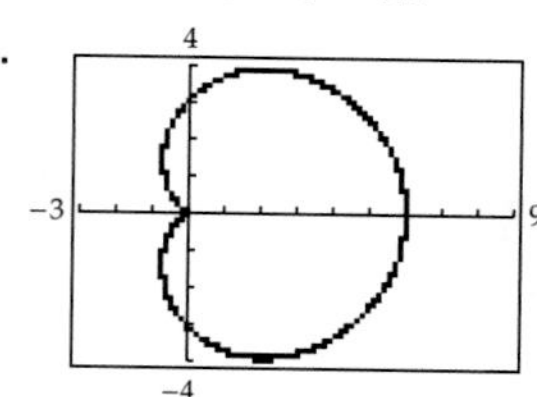

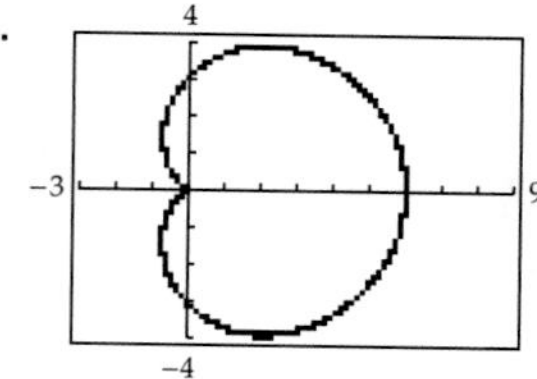

$0 \le \theta \le 2\pi$

27. 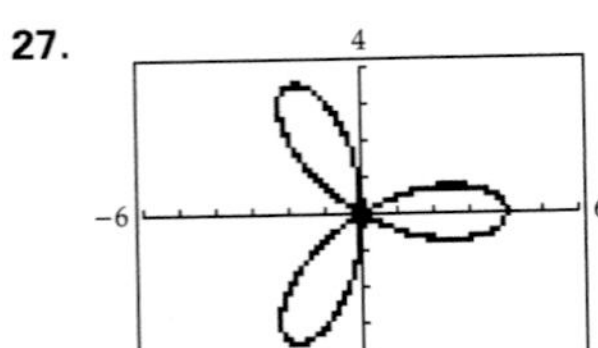

$0 \le \theta \le \pi$

29. 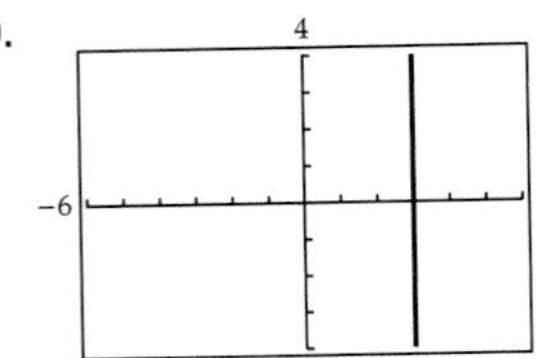

$0 \le \theta \le \pi$

31. 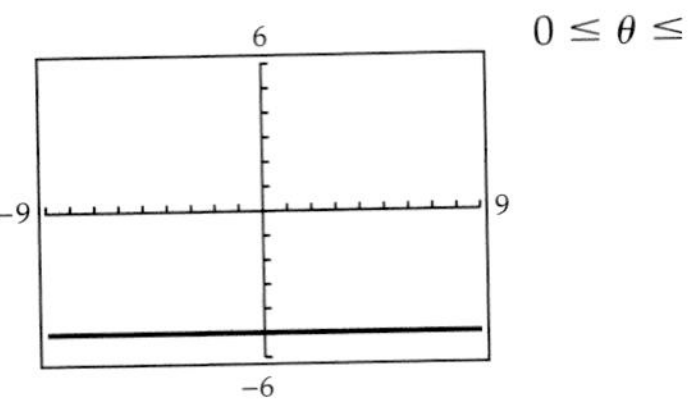

$0 \le \theta \le \pi$

33. 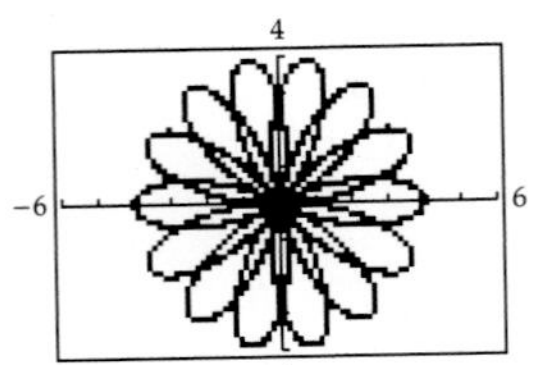

$0 \le \theta \le 4\pi$

35. 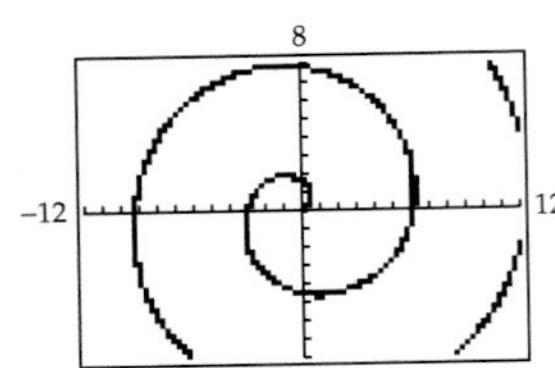

$0 \le \theta \le 6\pi$

37. 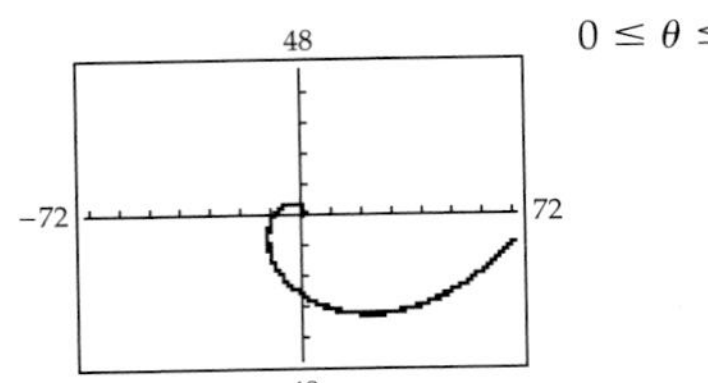

$0 \le \theta \le 2\pi$

39. 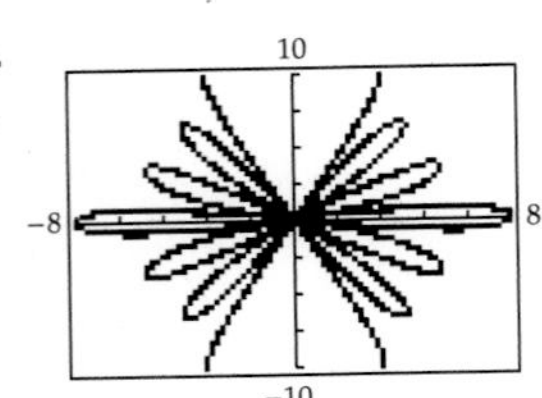

$0 \le \theta \le 2\pi$

41. $(2, -60°)$ **43.** $\left(\frac{3}{2}, -\frac{3\sqrt{3}}{2}\right)$ **45.** $(0, 0)$ **47.** $(5, 53.1°)$ **49.** $x^2 + y^2 - 3x = 0$

51. $x = 3$ **53.** $x^2 + y^2 = 16$ **55.** $y = \frac{\sqrt{3}}{3}x$ **57.** $x^4 - y^2 + x^2y^2 = 0$ **59.** $y^2 + 4x - 4 = 0$ **61.** $y = 2x + 6$ **63.** $r = 2 \csc \theta$ **65.** $\theta = \frac{\pi}{3}$

67. $r = 3 \sec \theta$ **69.** $r = 2$ **71.** $r \cos^2 \theta = 8 \sin \theta$ **73.** $r^2(\cos 2\theta) = 25$

75. 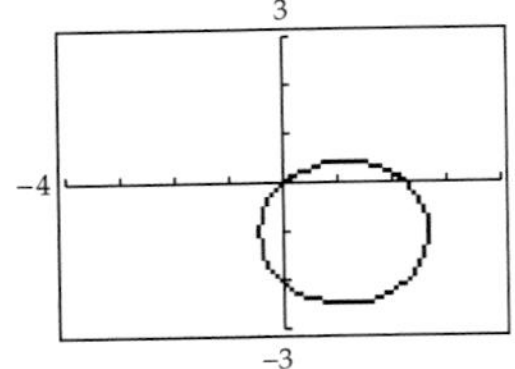

77. 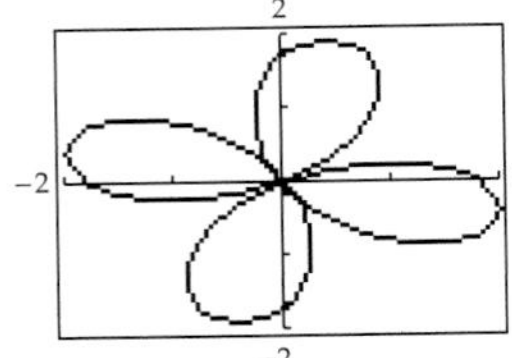

79. 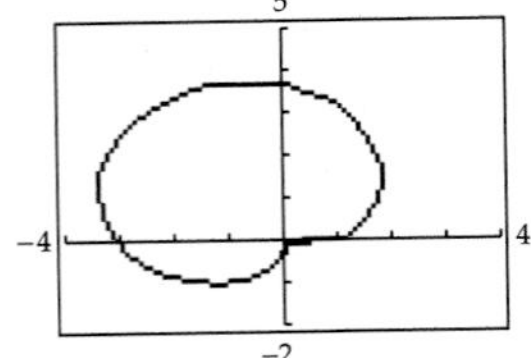

81. 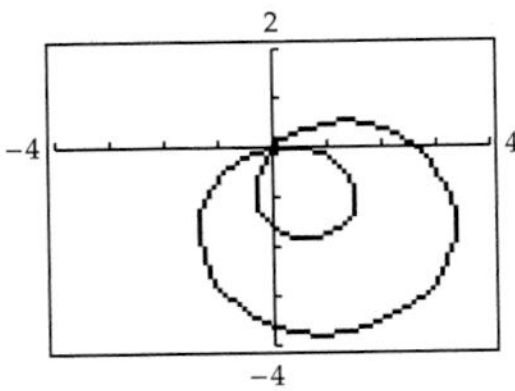

85. 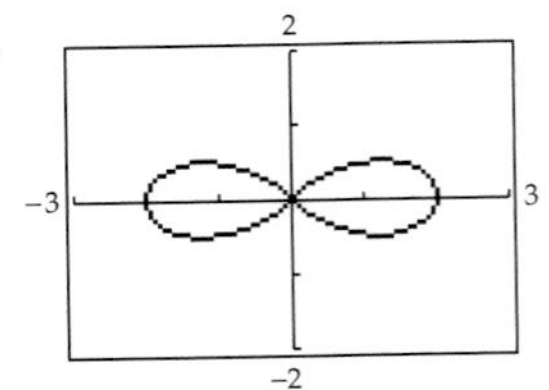

$0 \le \theta \le \pi$

87. 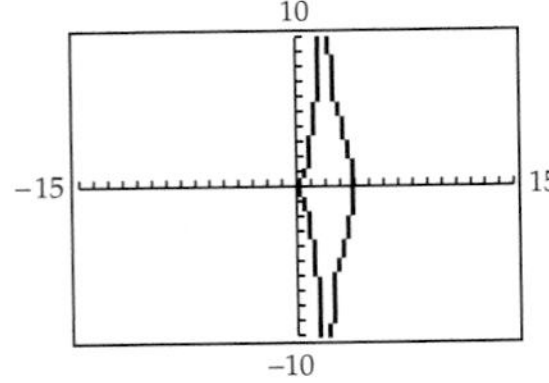

$0 \le \theta \le 2\pi$

89. 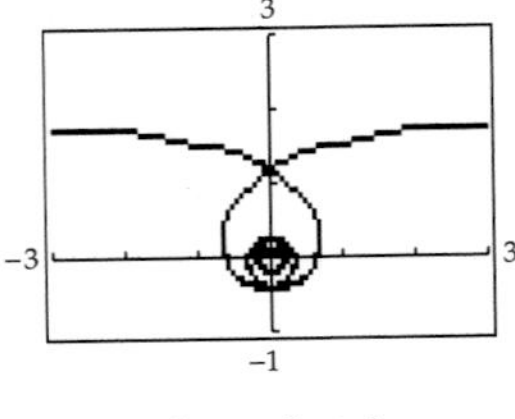

$-4\pi \le \theta \le 4\pi$

91. 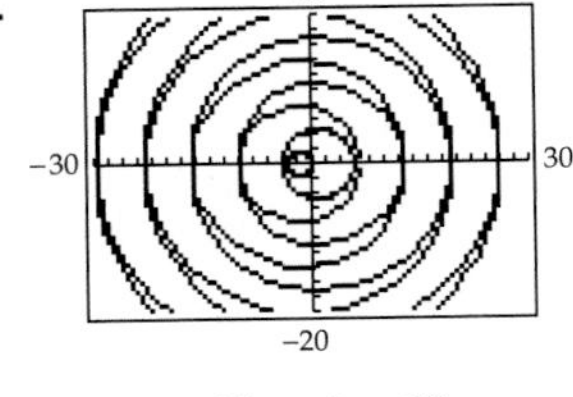

$-30 \le \theta \le 30$

93. a. 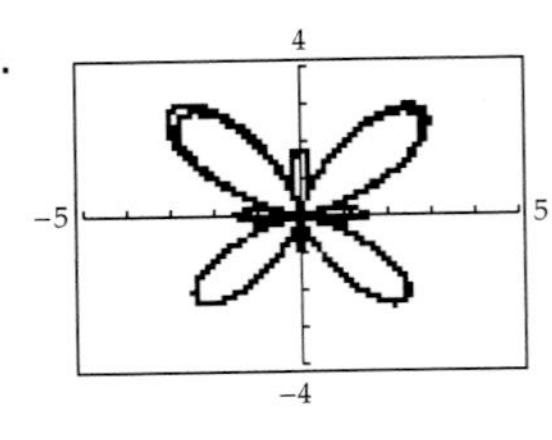

$0 \le \theta \le 5\pi$

b. 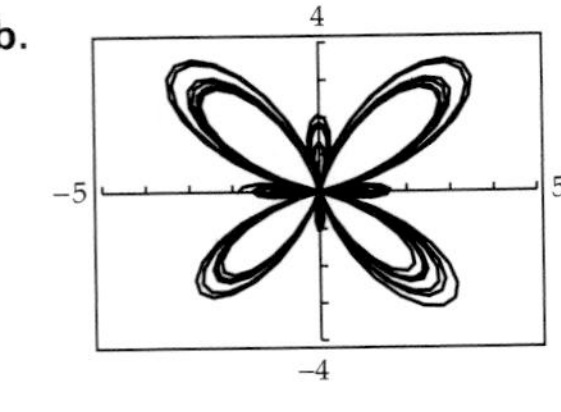

$0 \le \theta \le 20\pi$

Prepare for This Section (6.6), page 426

PS1. $\frac{3}{5}$ **PS2.** $x = -1$ **PS3.** $y = \frac{2}{1 - 2x}$ **PS4.** $\frac{3\pi}{2}$ **PS5.** $e > 1$ **PS6.** $\frac{4}{2 - \cos x}$

Exercise Set 6.6, page 431

1. hyperbola

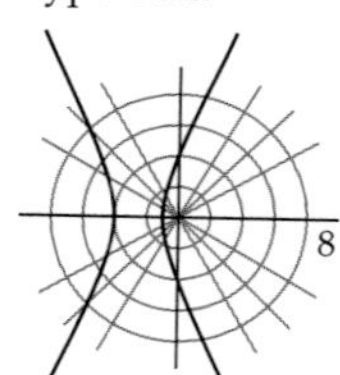

3. ellipse

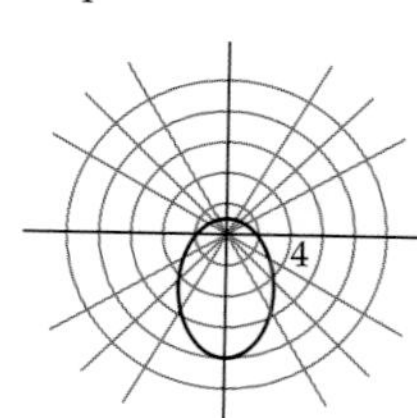

5. parabola

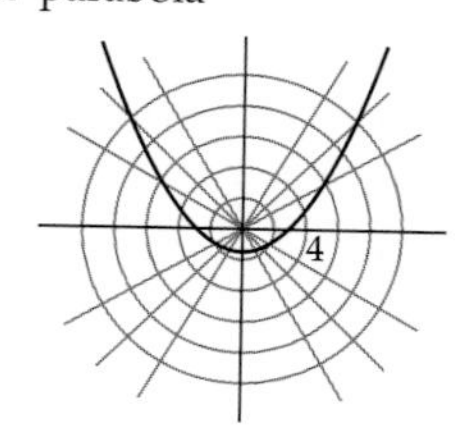

7. hyperbola

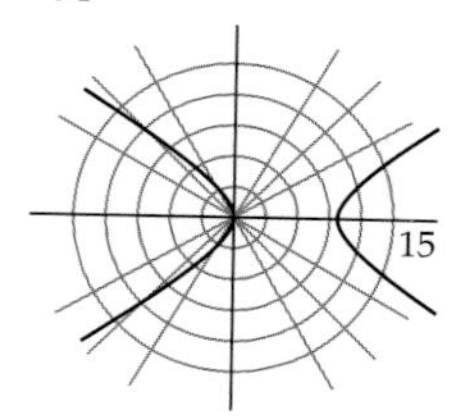

9. ellipse with holes at $\left(2, \frac{\pi}{2}\right)$ and $\left(2, \frac{3\pi}{2}\right)$

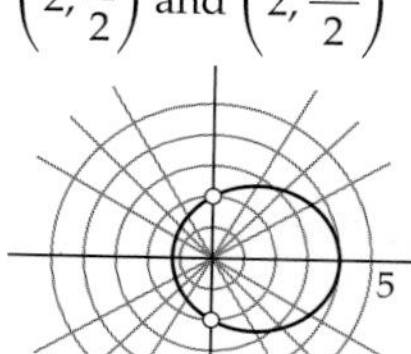

11. ellipse with holes at $(2, 0)$ and $(2, \pi)$

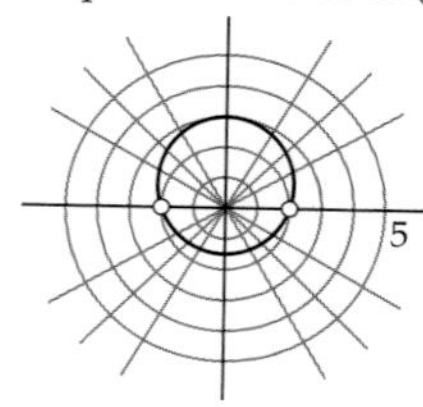

13. parabola

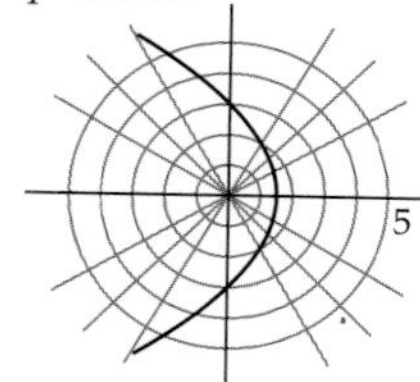

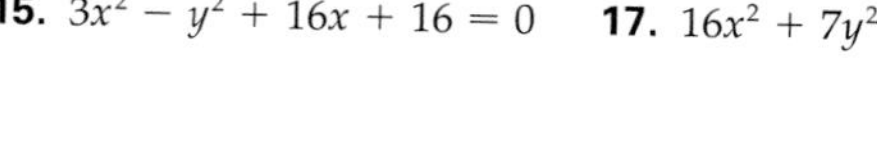

15. $3x^2 - y^2 + 16x + 16 = 0$ **17.** $16x^2 + 7y^2 + 48y - 64 = 0$

19. $x^2 - 6y - 9 = 0$ **21.** $r = \frac{2}{1 - 2\cos\theta}$ **23.** $r = \frac{2}{1 + \sin\theta}$ **25.** $r = \frac{8}{3 - 2\sin\theta}$ **27.** $r = \frac{6}{2 + 3\cos\theta}$ **29.** $r = \frac{4}{1 - \cos\theta}$

31. $r = \frac{3}{1 - 2\sin\theta}$

33.

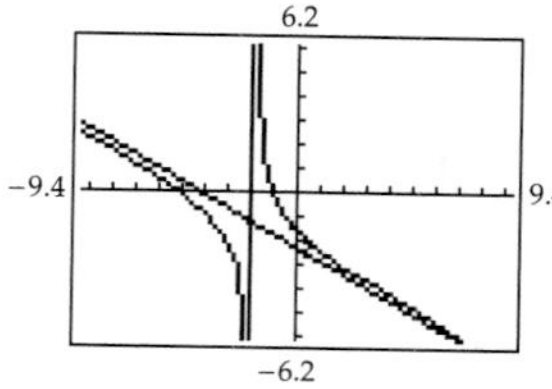

Rotate the graph in Exercise 1 counterclockwise $\frac{\pi}{6}$ radian.

35.

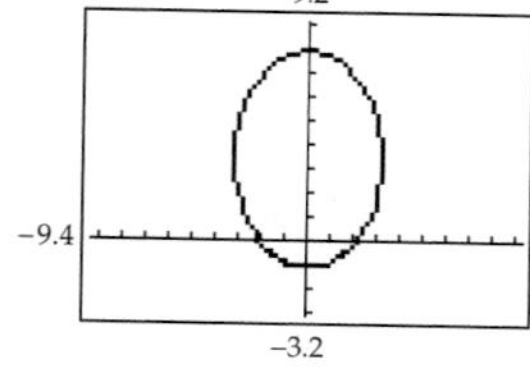

Rotate the graph in Exercise 3 counterclockwise π radians.

37.

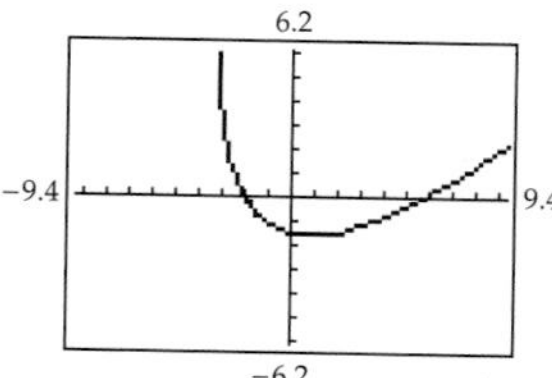

Rotate the graph in Exercise 5 clockwise $\frac{\pi}{6}$ radian.

39.

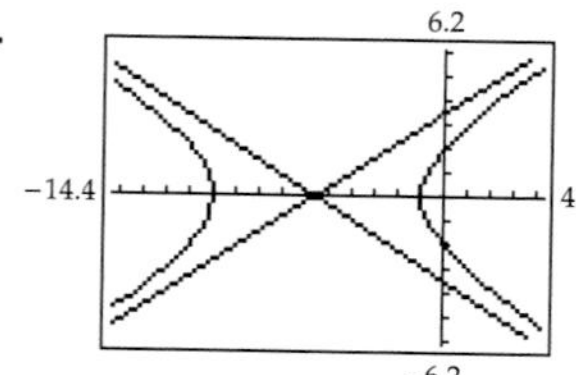

Rotate the graph in Exercise 7 clockwise π radians.

41.

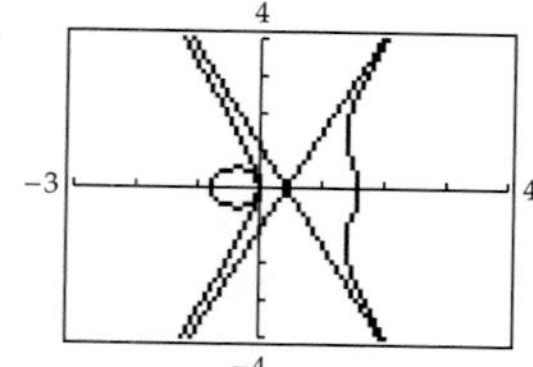

43.

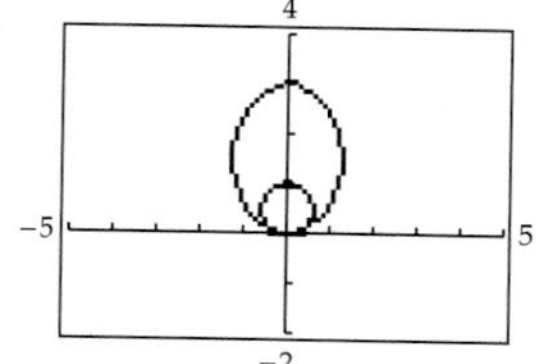

45.

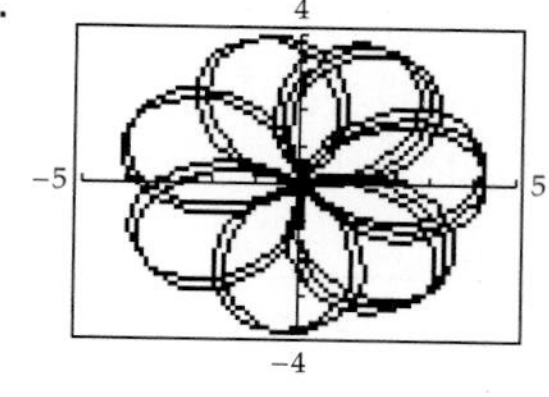

$0 \le \theta \le 12\pi$

Prepare for This Section (6.7), page 432

PS1. $y^2 + 3y + \frac{9}{4} = \left(y + \frac{3}{2}\right)^2$ **PS2.** $y = 4t^2 + 4t + 1$ **PS3.** ellipse **PS4.** 1 **PS5.** $(6, -2)$ **PS6.** domain: $(-\infty, \infty)$; range: $[-3, 3]$

Exercise Set 6.7, page 440

1.

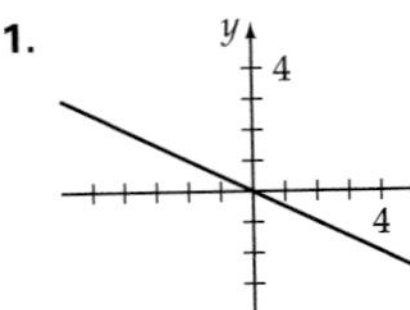

3.

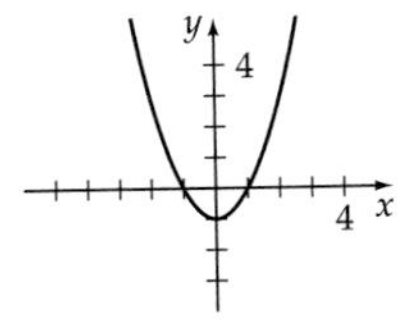

5.

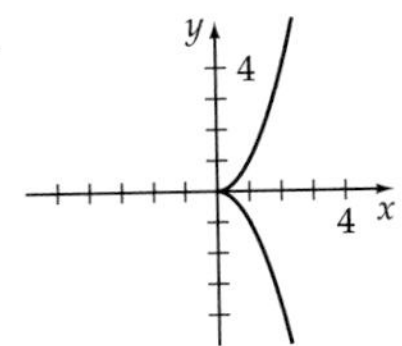

7.

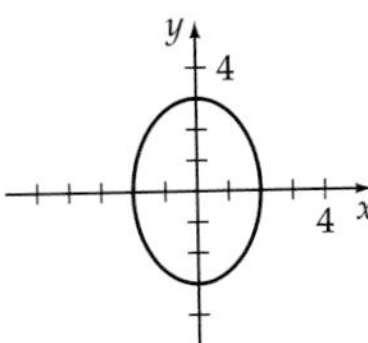

9.

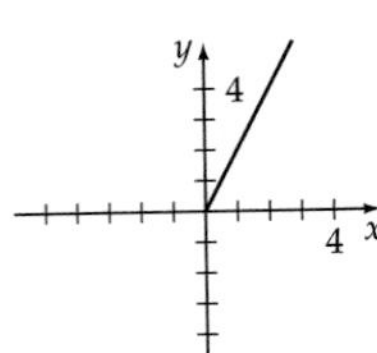

11. $x^2 - y^2 - 1 = 0$
$x \geq 1$
$y \in R$

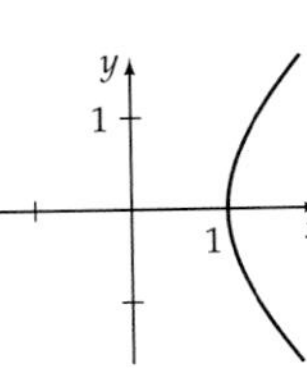

13. $y = -2x + 7$
$x \leq 2$
$y \geq 3$

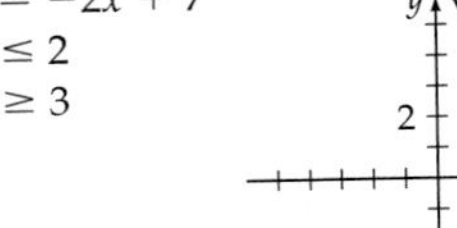

15. $x^{2/3} + y^{2/3} = 1$
$-1 \leq x \leq 1$
$-1 \leq y \leq 1$

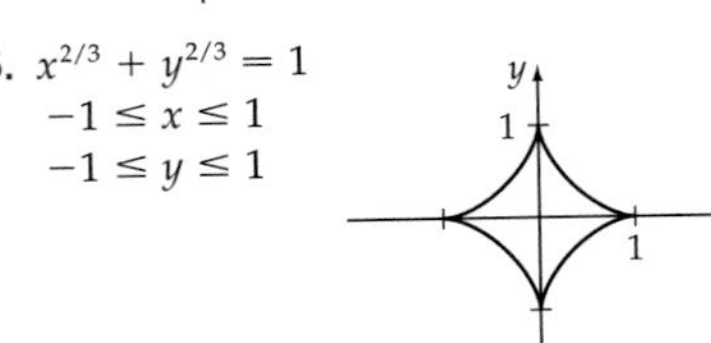

17. $y = x^2 - 1$
$x \geq 0$
$y \geq -1$

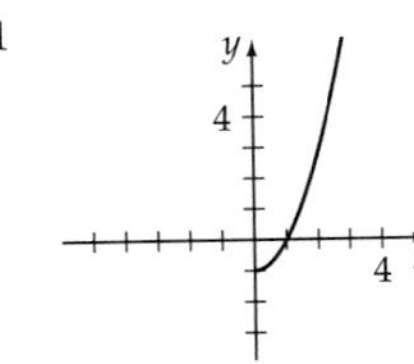

19. The point traces the top half of the ellipse $\frac{(x-2)^2}{3^2} + \frac{(y-3)^2}{2^2} = 1$, as shown in the figure. The point starts at (5, 3) and moves counterclockwise along the ellipse until it reaches the point $(-1, 3)$ at time $t = \pi$.

21. The point traces a line segment, as shown in the figure. The point starts at $(-1, 1)$ and moves along the line segment until it reaches the point (5, 4) at time $t = 3$.

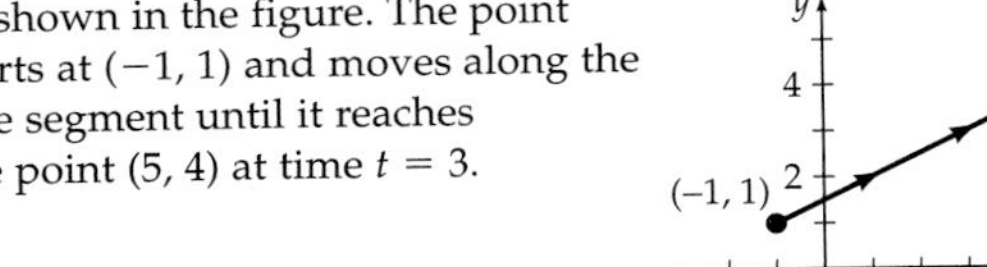

23. The point traces a portion of the top branch of the hyperbola $y^2 - x^2 = 1$, as shown in the figure. The point starts at $(1, \sqrt{2})$ and moves along the hyperbola until it reaches the point $(-1, \sqrt{2})$ at time $t = \frac{\pi}{2}$.

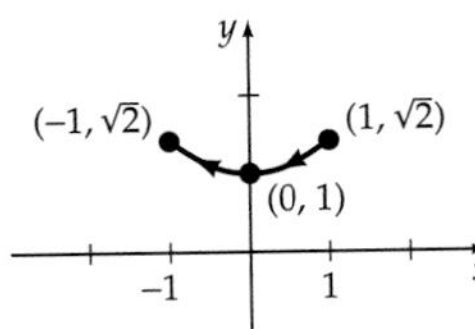

25. C_1: $y = -2x + 5, x \geq 2$; C_2: $y = -2x + 5, x \in R$. C_2 is a line. C_1 is a ray.

27.

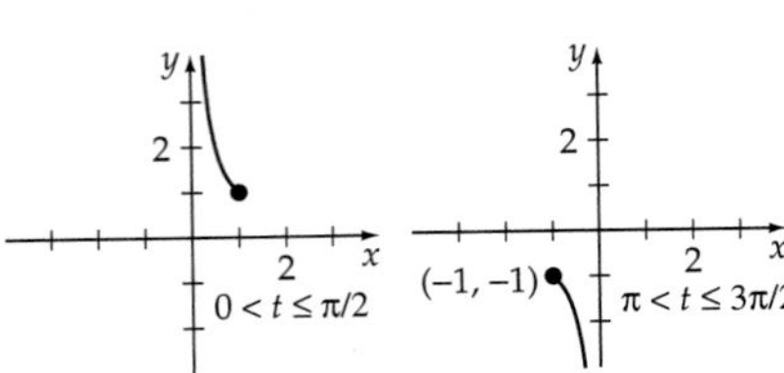

29.

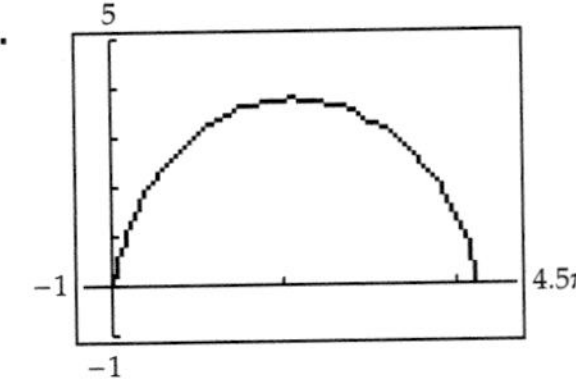

31. a. $x = 6, y = 60t$, for $t \geq 0$ **b.** the Hummer

33.

Max height (nearest foot) of 462 feet is attained when $t \approx 5.38$ seconds. Range (nearest foot) of 1295 feet is attained when $t \approx 10.75$ seconds.

35.

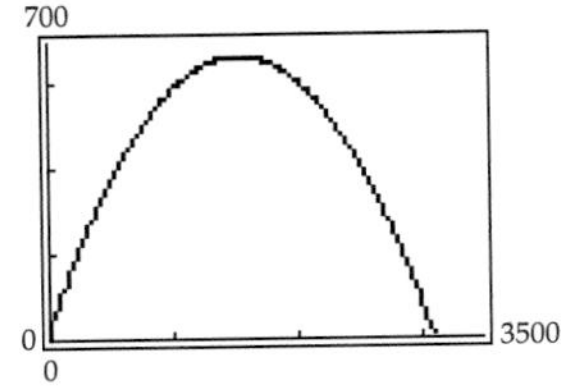

Max height (nearest foot) of 694 feet is attained when $t \approx 6.59$ seconds. Range (nearest foot) of 3084 feet is attained when $t \approx 13.17$ seconds.

37.

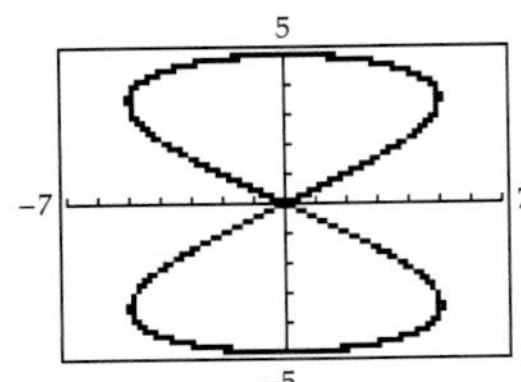

39.

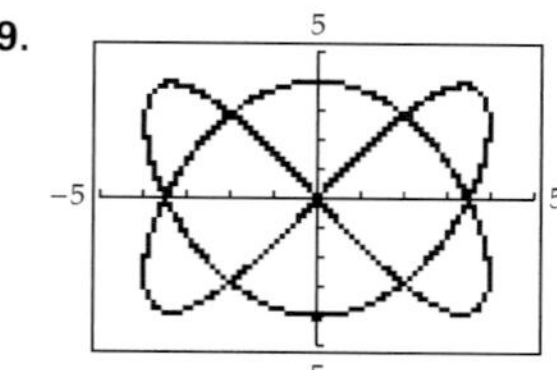

43. $x = a\cos\theta + a\theta\sin\theta$

$y = a\sin\theta - a\theta\cos\theta$

45. $x = (b - a)\cos\theta + a\cos\left(\frac{b-a}{a}\theta\right)$

$y = (b - a)\sin\theta - a\sin\left(\frac{b-a}{a}\theta\right)$

Chapter 6 Assessing Concepts, page 445

1. d **2.** b **3.** e **4.** c **5.** a **6.** f **7.** g **8.** i **9.** h **10.** j **11.** k **12.** k

Chapter 6 Review Exercises, page 446

1. center: $(0, 0)$
vertices: $(\pm 2, 0)$
foci: $(\pm 2\sqrt{2}, 0)$
asymptotes: $y = \pm x$

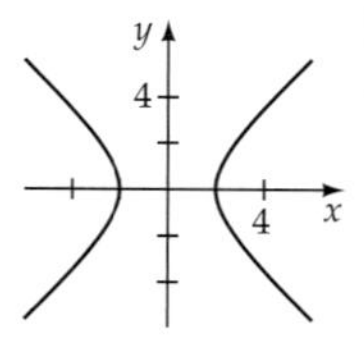

[6.3]

2. vertex: $(0, 0)$
focus: $(4, 0)$
directrix: $x = -4$

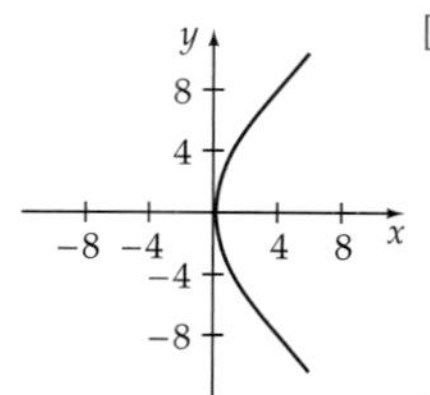

[6.1]

3. center: $(3, -1)$
vertices: $(-1, -1), (7, -1)$
foci: $(3 \pm 2\sqrt{3}, -1)$

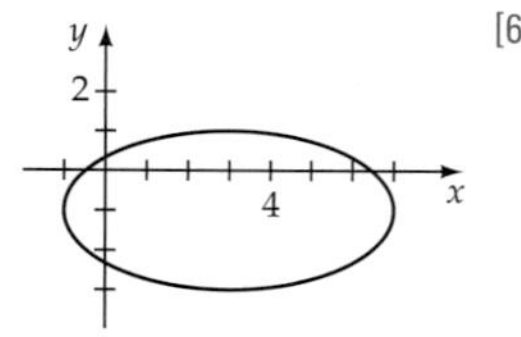

[6.2]

4. center: $(-2, -3)$
vertices: $(0, -3), (-4, -3)$
foci: $(-2 \pm \sqrt{7}, -3)$
asymptotes: $y + 3 = \pm\frac{\sqrt{3}}{2}(x + 2)$

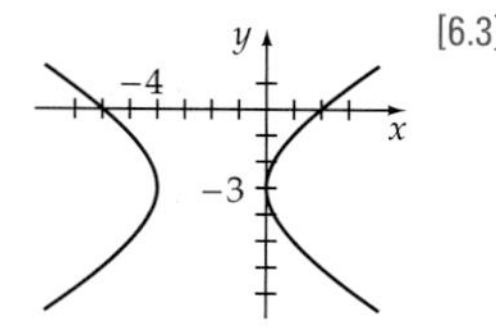

[6.3]

5. vertex: $(-2, 1)$
focus: $\left(-\frac{29}{16}, 1\right)$
directrix: $x = -\frac{35}{16}$

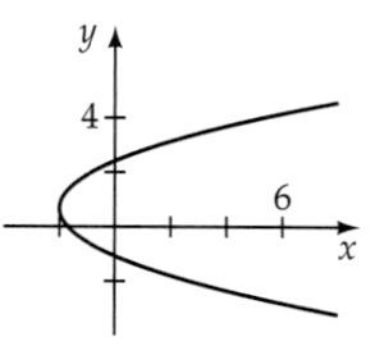

[6.1]

6. vertex: $(3, 1)$
focus: $\left(\frac{21}{8}, 1\right)$
directrix: $x = \frac{27}{8}$

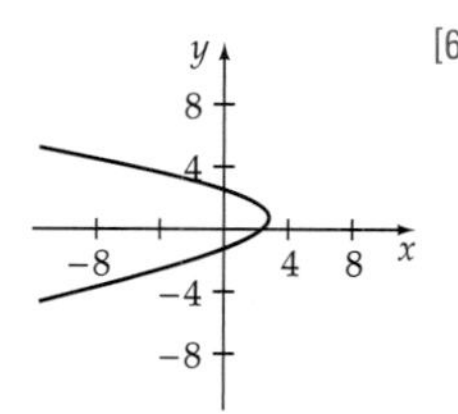

[6.1]

7. center: $(-2, 1)$
vertices: $(-2, -2), (-2, 4)$
foci: $(-2, 1 \pm \sqrt{5})$

[6.2]

8. center: $(2, -1)$
vertices: $(7, -1), (-3, -1)$
foci: $(8, -1), (-4, -1)$
asymptotes: $y + 1 = \pm\frac{\sqrt{11}}{5}(x - 2)$

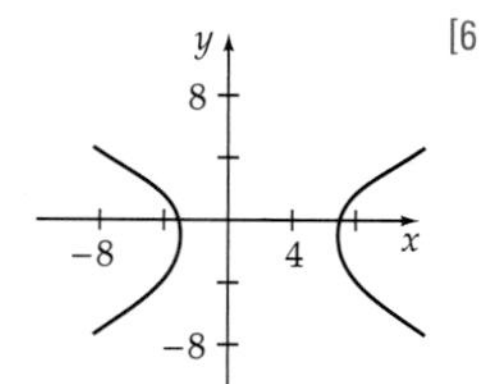

[6.3]

9. center: $\left(1, \frac{2}{3}\right)$
vertices: $\left(-5, \frac{2}{3}\right), \left(7, \frac{2}{3}\right)$
foci: $\left(1 \pm 2\sqrt{13}, \frac{2}{3}\right)$
asymptotes: $y - \frac{2}{3} = \pm\frac{2}{3}(x - 1)$

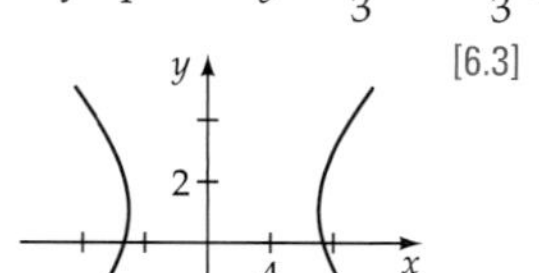

[6.3]

10. center: $\left(-2, \frac{1}{2}\right)$
vertices: $\left(2, \frac{1}{2}\right), \left(-6, \frac{1}{2}\right)$
foci: $\left(-2 \pm \sqrt{7}, \frac{1}{2}\right)$

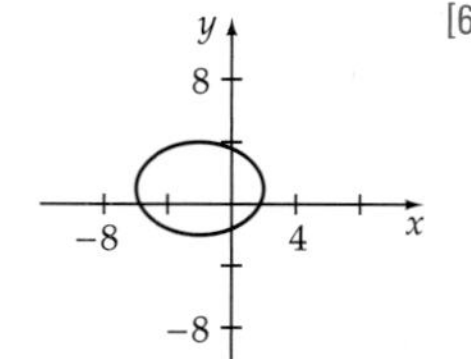

[6.2]

11. vertex: $\left(-\frac{7}{2}, -1\right)$
focus: $\left(-\frac{7}{2}, -3\right)$
directrix: $y = 1$

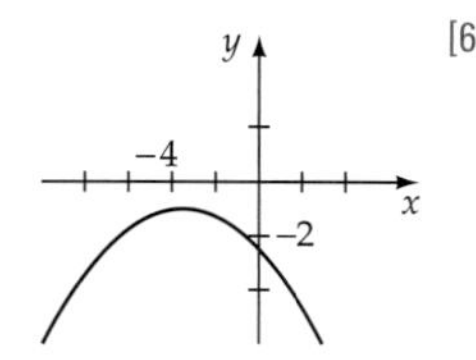

[6.1]

12. vertex: $(3, 2)$
focus: $\left(3, \frac{17}{4}\right)$
directrix: $y = -\frac{1}{4}$

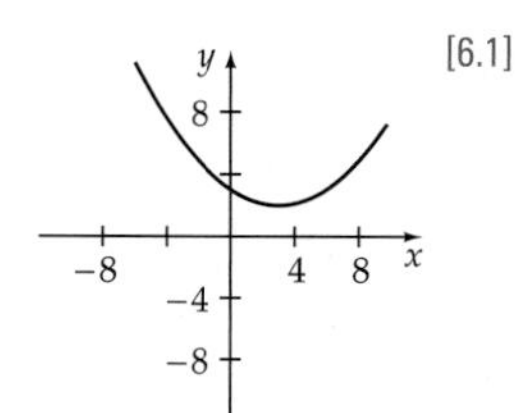

[6.1]

13. $\dfrac{(x-2)^2}{25}+\dfrac{(y-3)^2}{16}=1$ [6.2] **14.** $\dfrac{(x-1)^2}{9}-\dfrac{(y-1)^2}{7}=1$ [6.3] **15.** $\dfrac{(x+2)^2}{4}-\dfrac{(y-2)^2}{5}=1$ [6.3] **16.** $(y+3)^2=-8(x-4)$ [6.1]

17. $x^2=\dfrac{3(y+2)}{2}$ or $(y+2)^2=12x$ [6.1] **18.** $\dfrac{(x+2)^2}{9}+\dfrac{(y+1)^2}{5}=1$ [6.2] **19.** $\dfrac{x^2}{36}-\dfrac{y^2}{4/9}=1$ [6.3] **20.** $y=(x-1)^2$ [6.1]

21. $(x')^2+2(y')^2-4=0$, ellipse [6.4] **22.** $6(x')^2+\dfrac{\sqrt{2}}{2}x'+\dfrac{9\sqrt{2}}{2}y'-12=0$, parabola [6.4] **23.** $(x')^2-4y'+8=0$, parabola [6.4]

24. $\dfrac{1}{2}(x')^2-\dfrac{1}{2}(y')^2-\sqrt{2}x'-1=0$, hyperbola [6.4]

25. [6.5]

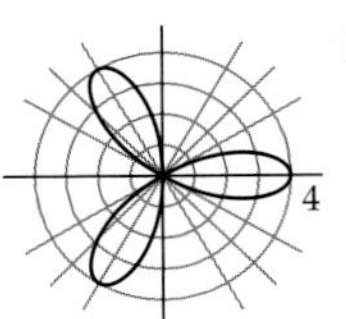

26. [6.5]

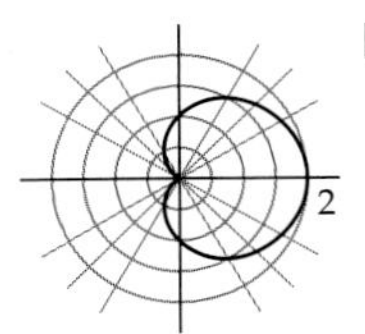

27. [6.5]

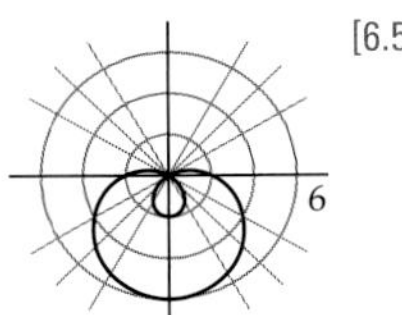

28. [6.5]

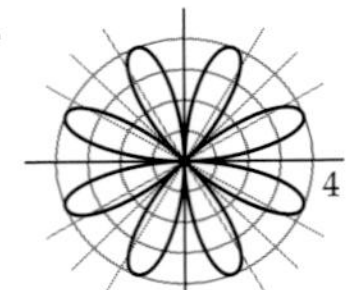

29. [6.5]

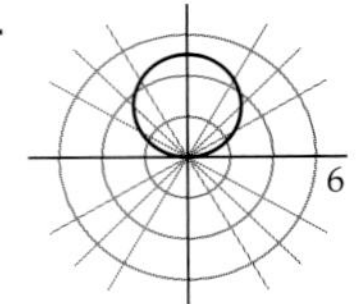

30. [6.5]

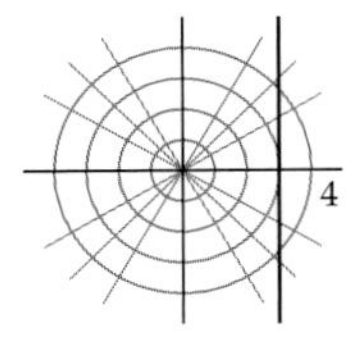

31. [6.5]

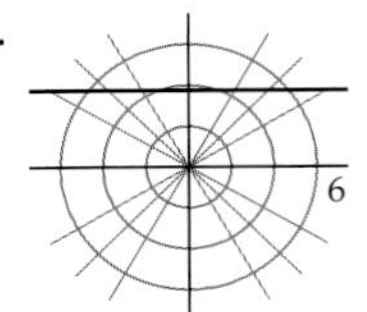

32. [6.5]

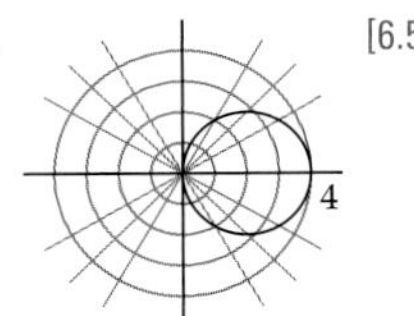

33. [6.5]

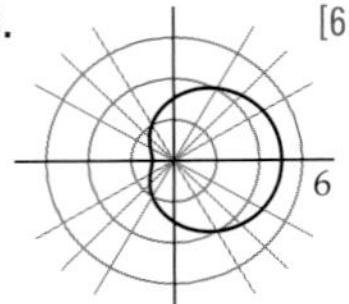

34. [6.5]

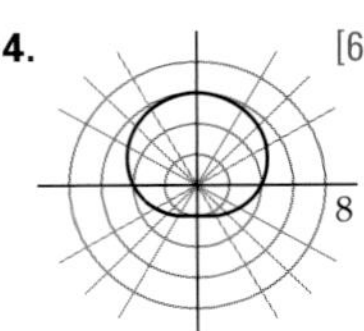

35. $r\sin^2\theta=16\cos\theta$ [6.5] **36.** $r+4\cos\theta+3\sin\theta=0$ [6.5] **37.** $3r\cos\theta-2r\sin\theta=6$ [6.5]

38. $r^2\sin 2\theta=8$ [6.5] **39.** $y^2=8x+16$ [6.5] **40.** $x^2-3x+y^2+4y=0$ [6.5] **41.** $x^4+y^4+2x^2y^2-x^2+y^2=0$ [6.5]

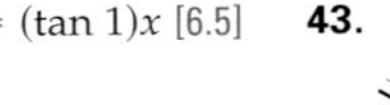

42. $y=(\tan 1)x$ [6.5]

43. [6.6]

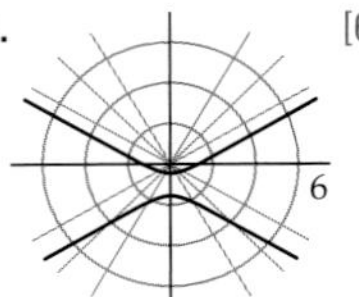

44. [6.6]

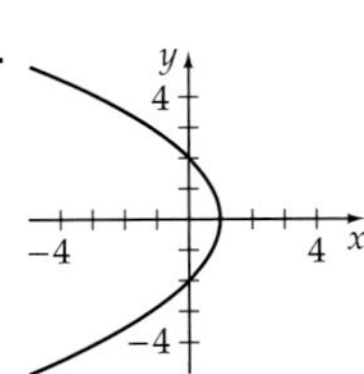

45. [6.6]

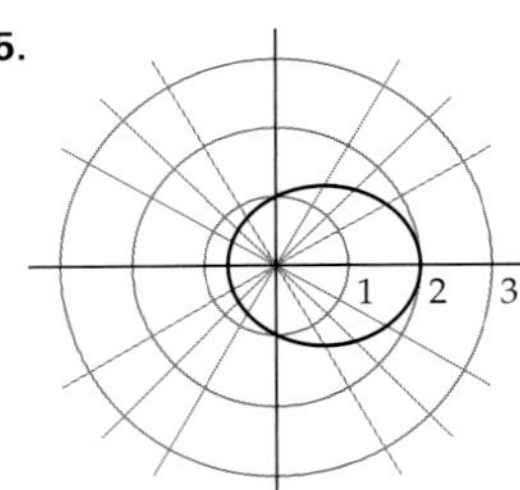

46. [6.6]

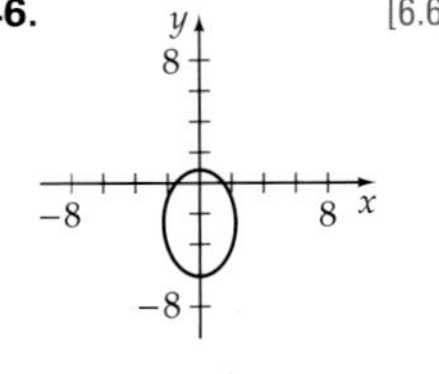

47. $y=\dfrac{3}{4}x+\dfrac{5}{2}$ [6.7]

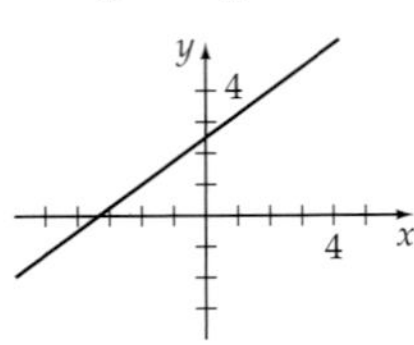

48. $y=2x+1, x\le 1$ [6.7]

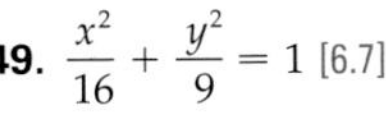

49. $\dfrac{x^2}{16}+\dfrac{y^2}{9}=1$ [6.7]

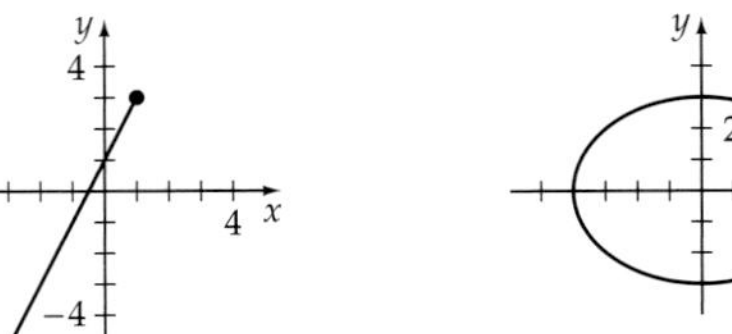

50. $\dfrac{x^2}{1}-\dfrac{y^2}{16}=1$ [6.7]

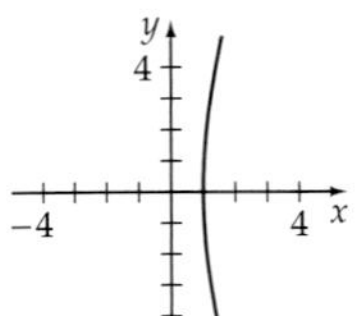

51. $y=-2x, x>0$ [6.7]

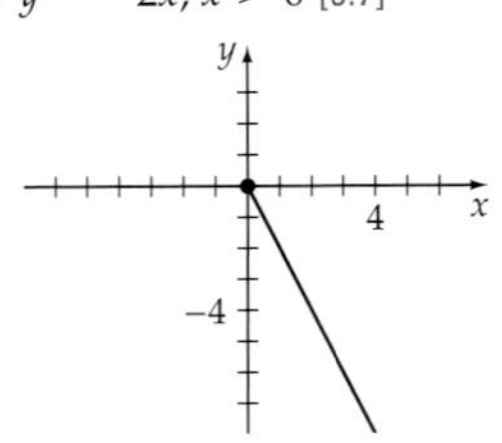

52. $(x-1)^2+(y-2)^2=1$ [6.7] **53.** $y=2^{-x^2}, x\ge 0$ [6.7]

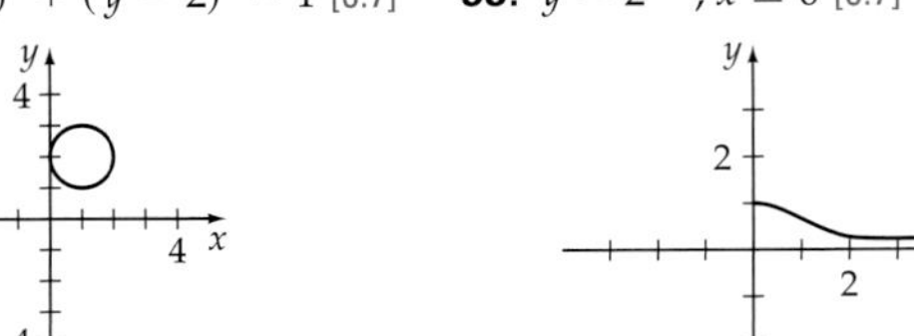

54. [6.7]

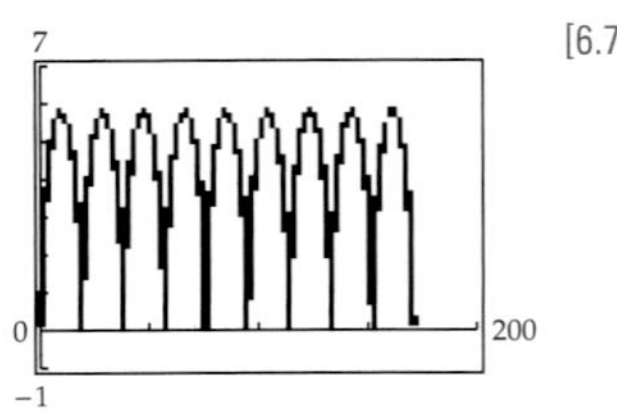

55. 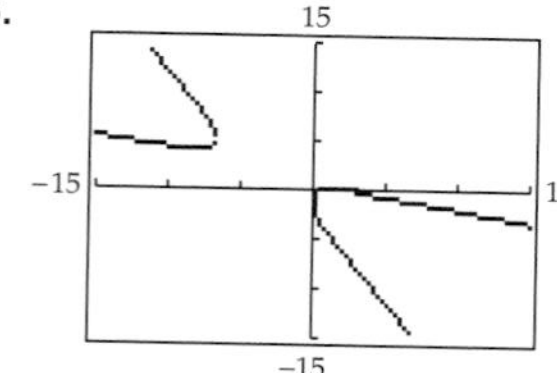

[6.4]

56. 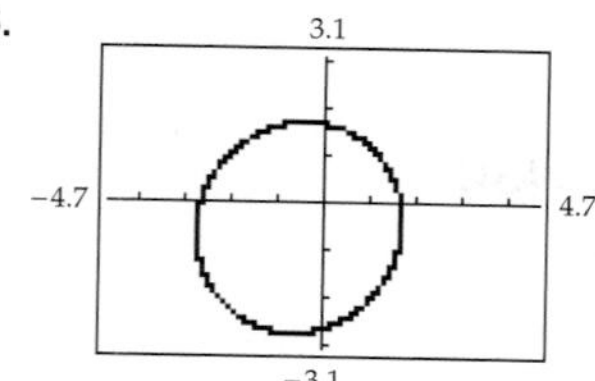

[6.5]

57. 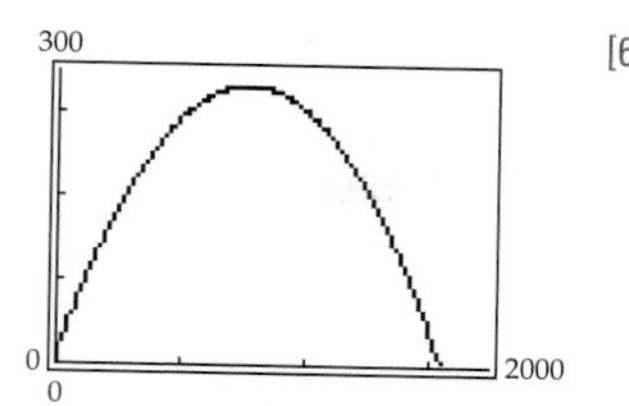

[6.7]

Max. height (nearest foot) of 278 feet is attained when $t \approx 4.17$ seconds.

Chapter 6 Quantitative Reasoning Exercises, page 447

QR1. They appear to be the same. **QR2.** They appear to be the same. **QR3.** 5.2 units

Chapter 6 Test, page 449

1. vertex: (0, 0)
focus: (0, 2)
directrix: $y = -2$ [6.1]

2. 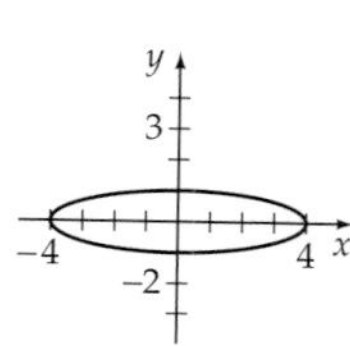

[6.2]

3. vertices: (3, 4), (3, −6)
foci: (3, 3), (3, −5) [6.2]

4. $\dfrac{x^2}{45} + \dfrac{(y+3)^2}{9} = 1$ [6.2]

5. 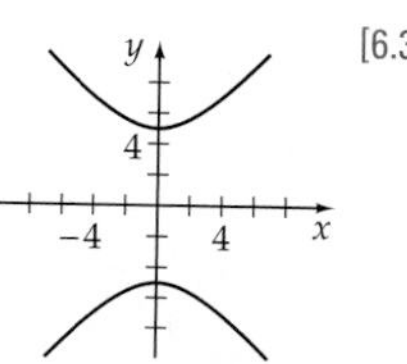

[6.3]

6. vertices: (6, 0), (−6, 0)
foci: (−10, 0), (10, 0)
asymptotes: $y = \pm\dfrac{4x}{3}$ [6.3]

7. 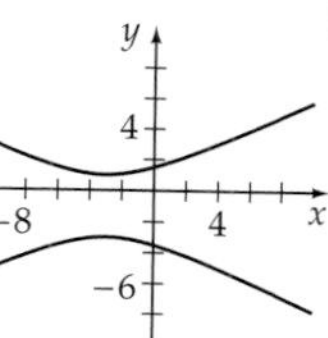 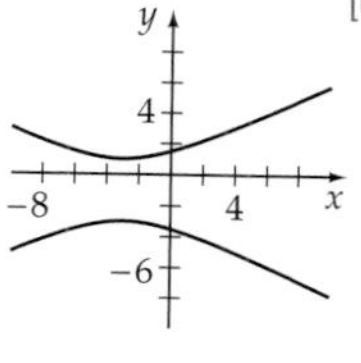

[6.3]

8. 73.15° [6.4] **9.** ellipse [6.4] **10.** $P(2, 300°)$ [6.5]

11. 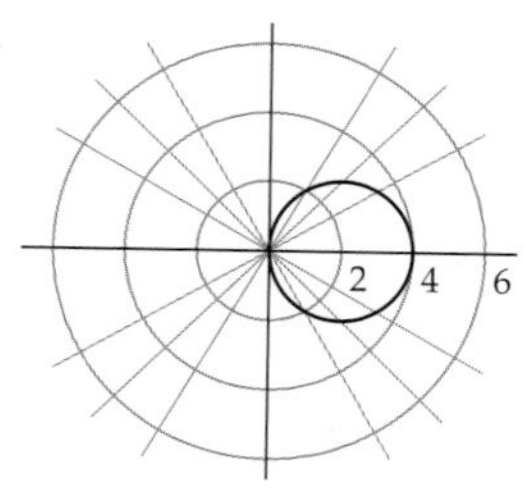

[6.5]

12. 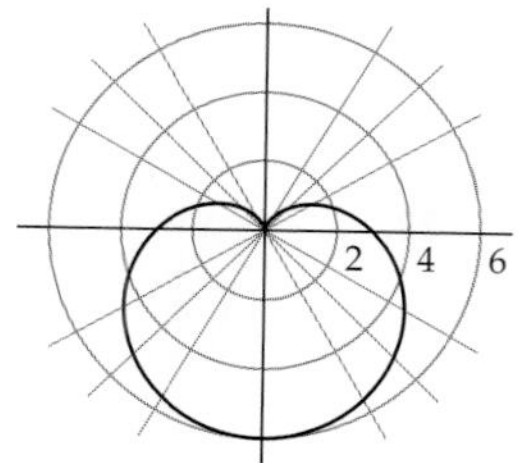

[6.5]

13. 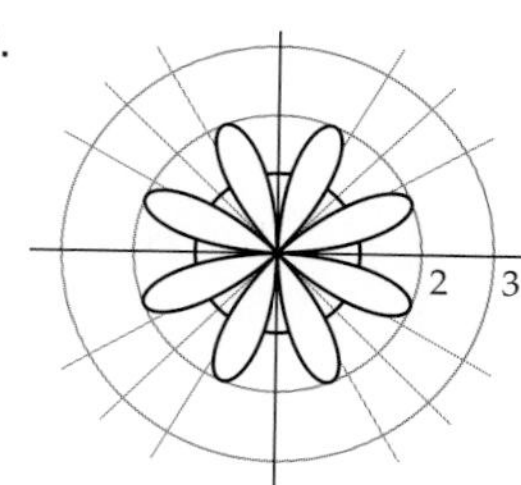

[6.5]

14. $\left(\dfrac{5}{2}, \dfrac{5\sqrt{3}}{2}\right)$ [6.5]

15. $y^2 - 8x - 16 = 0$ [6.5] **16.** $x^2 + 8y - 16 = 0$ [6.6] **17.** $(x+3)^2 = \dfrac{1}{2}y$ [6.7]

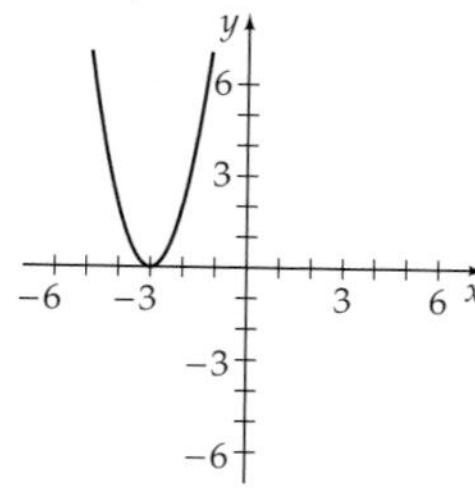

18. 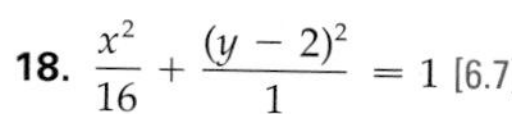 $\dfrac{x^2}{16} + \dfrac{(y-2)^2}{1} = 1$ [6.7]

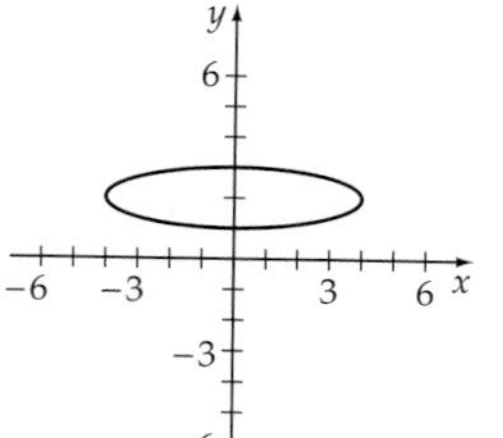

19. 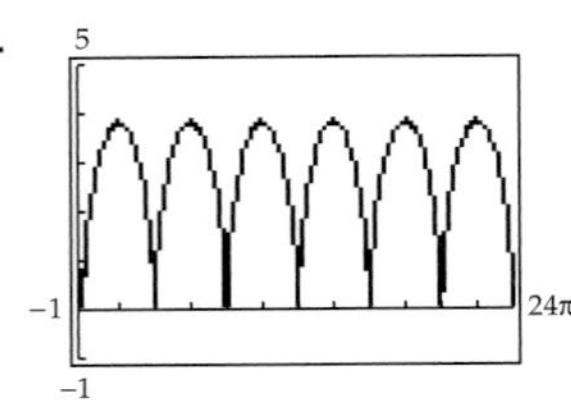

Xscl = 2π

[6.7] **20.** 443 feet [6.7]

Cumulative Review Exercises, page 450

1. $-2 \pm \sqrt{10}$ [1.1] **2.** origin [1.4] **3.** $(g \circ f)(x) = 3 \sin x - 2$ [1.5] **4.** $\frac{4\pi}{3}$ [2.1] **5.** 11 mph [2.1] **6.** $\tan(t) = -\sqrt{3}$ [2.4] **7.** 19 centimeters [2.2] **8.** amplitude: $\frac{1}{2}$, period: 6 [2.5] **9.** 3 [2.6] **10.** See [3.1]. **11.** $-\frac{63}{65}$ [3.2] **12.** $\frac{2\sqrt{6}}{5}$ [3.5] **13.** $\frac{5}{13}$ [3.6] **14.** 298 feet [4.1] **15.** 38° [4.2] **16.** $-24.6\mathbf{i} + 17.2\mathbf{j}$ [4.3] **17.** No. $\mathbf{v} \cdot \mathbf{w} = 1 \neq 0$ [4.3] **18.** $4 \operatorname{cis}\left(\frac{2\pi}{3}\right)$ [5.2]

19. 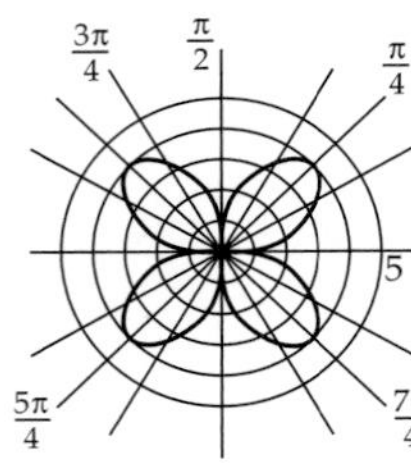

[6.5] **20.** $y = x^2 + 2x + 2$ [6.7]

Exercise Set 7.1, page 461

1. $f(0) = 1; f(4) = 81$ **3.** $g(-2) = \frac{1}{100}; g(3) = 1000$ **5.** $h(2) = \frac{9}{4}; h(-3) = \frac{8}{27}$ **7.** $j(-2) = 4; j(4) = \frac{1}{16}$ **9.** 9.19 **11.** 9.03 **13.** 9.74 **15. a.** $k(x)$ **b.** $g(x)$ **c.** $h(x)$ **d.** $f(x)$

17. 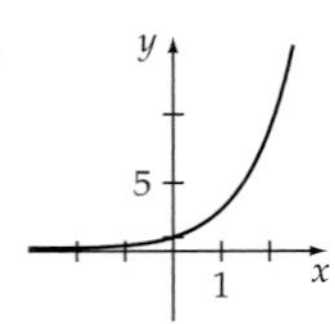

19. 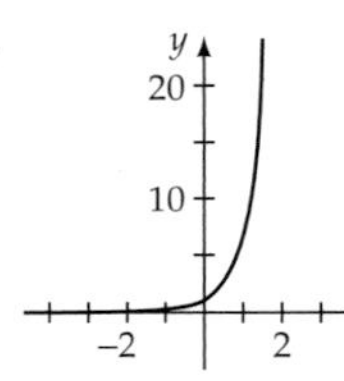

21. 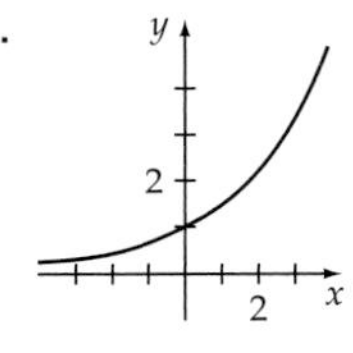

23. 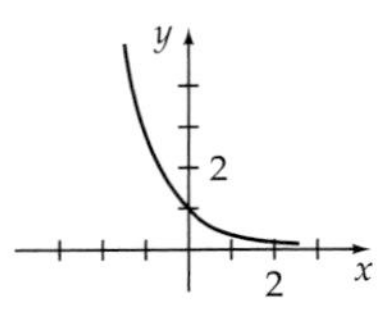

25. Shift the graph of f vertically upward 2 units. **27.** Shift the graph of f horizontally to the right 2 units. **29.** Reflect the graph of f across the y-axis. **31.** Stretch the graph of f vertically away from the x-axis by a factor of 2. **33.** Reflect the graph of f across the y-axis and then shift this graph vertically upward 2 units. **35.** Shift the graph of f horizontally to the right 4 units and then reflect this graph across the x-axis. **37.** Reflect the graph of f across the y-axis and then shift this graph vertically upward 3 units.

39. no horizontal asymptote

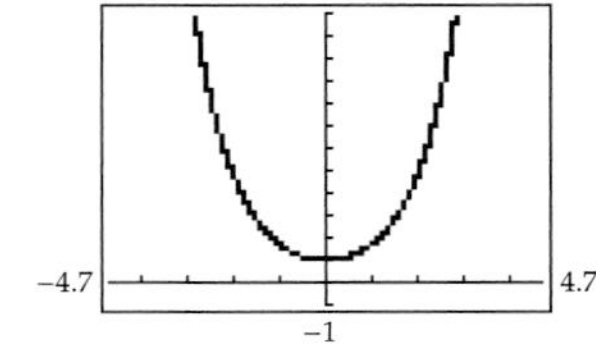

41. no horizontal asymptote

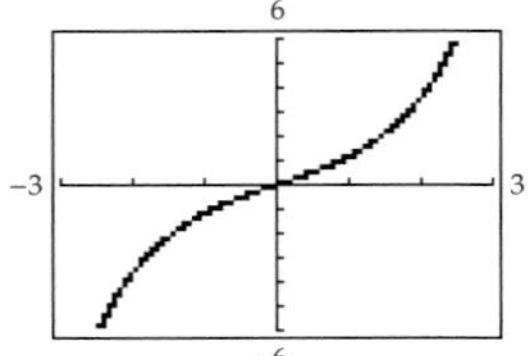

43. horizontal asymptote: $y = 0$

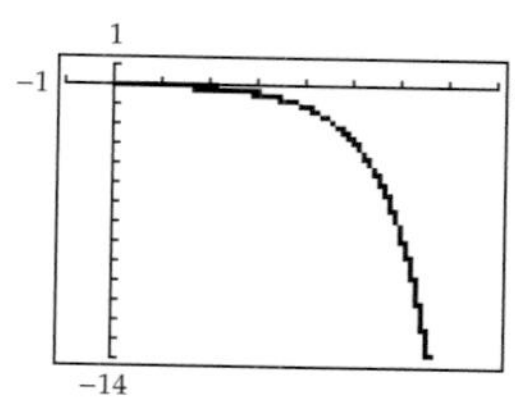

45. horizontal asymptote: $y = 10$

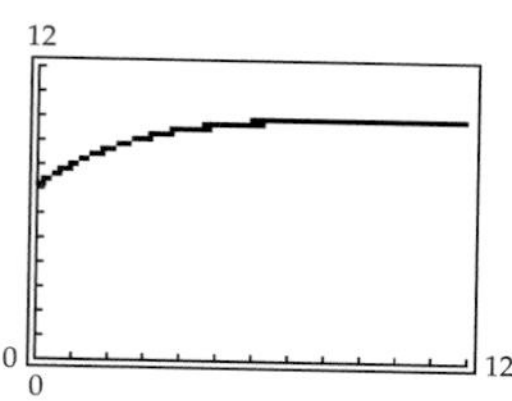

47. a. 442 million connections **b.** 2008 **49. a.** 233 items per month; 59 items per month **b.** The demand will approach 25 items per month. **51. a.** 6400 bacteria; 409,600 bacteria **b.** 11.6 hours **53. a.** 69.2% **b.** 7.6 **55. a.** 363 beneficiaries; 88,572 beneficiaries **b.** 13 rounds **57. a.** 141°F **b.** after 28.3 minutes **59. a.** 261.63 vibrations per second **b.** No. The function $f(n)$ is not a linear function. Therefore, the graph of $f(n)$ does not increase at a constant rate. **63.**

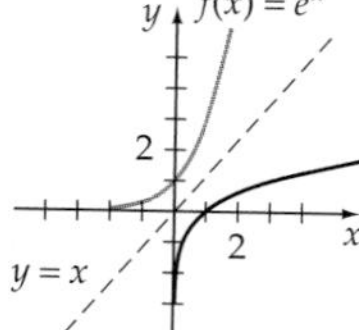

65. $(-\infty, \infty)$ **67.** $[0, \infty)$

69. Let $f(x) = e^x$ and $g(x) = 2x - 5$. Then $h(x) = e^{2x-5} = f[2x - 5] = f[g(x)] = (f \circ g)(x)$.

Prepare for This Section (7.2), page 467

PS1. 4 **PS2.** 3 **PS3.** 5 **PS4.** $f^{-1}(x) = \dfrac{3x}{2 - x}$ **PS5.** $\{x \mid x \geq 2\}$ **PS6.** the set of all positive real numbers

Exercise Set 7.2, page 476

1. $10^1 = 10$ **3.** $8^2 = 64$ **5.** $7^0 = x$ **7.** $e^4 = x$ **9.** $e^0 = 1$ **11.** $10^2 = 3x + 1$ **13.** $\log_3 9 = 2$ **15.** $\log_4 \dfrac{1}{16} = -2$ **17.** $\log_b y = x$

19. $\ln y = x$ **21.** $\log 100 = 2$ **23.** $2 = \ln(x + 5)$ **25.** 2 **27.** −5 **29.** 3 **31.** −2 **33.** −4 **35.** 12 **37.** 8 **39.** $\dfrac{2}{5}$ **41.** $\dfrac{10}{3}$

43. **45.** **47.**

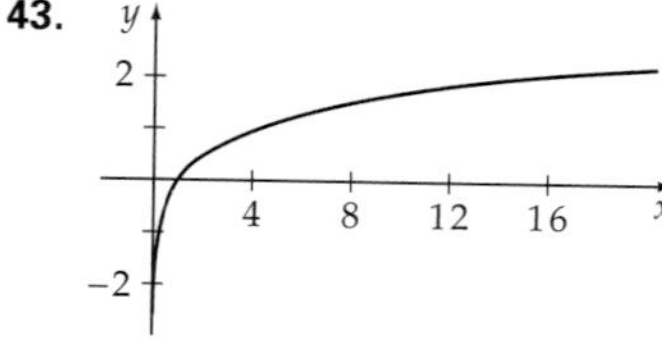

49.

51. $(3, \infty)$ **53.** $(-\infty, 11)$ **55.** $(-\infty, -2) \cup (2, \infty)$ **57.** $(4, \infty)$ **59.** $(-1, 0) \cup (1, \infty)$ **61.** $\left(\dfrac{11}{2}, \infty\right)$

63. $\left(\dfrac{7}{3}, \infty\right)$ **65.**

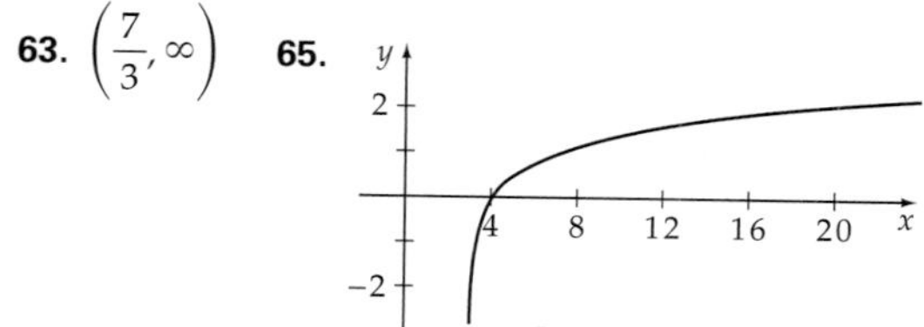

67.

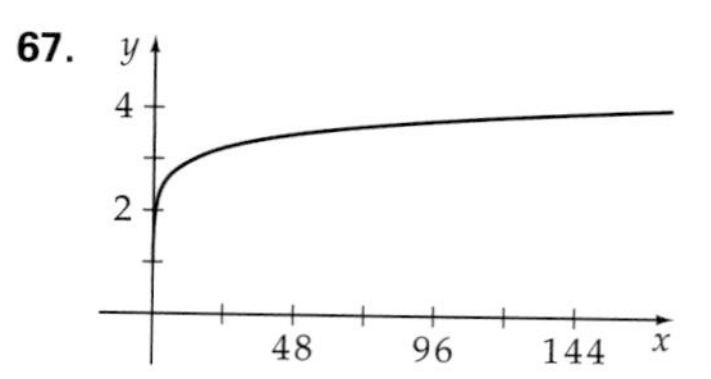

69.

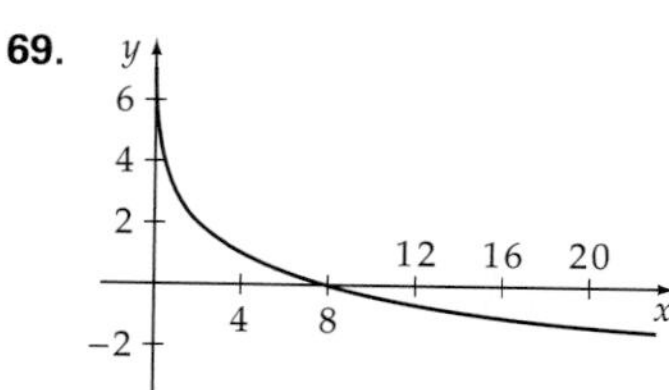

71.

73. a. $k(x)$ **b.** $f(x)$ **c.** $g(x)$ **d.** $h(x)$

75.

77.

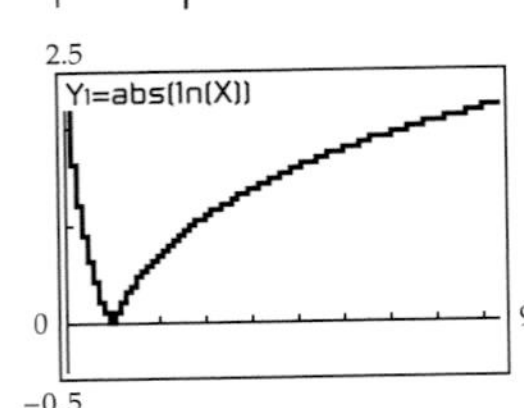

79.

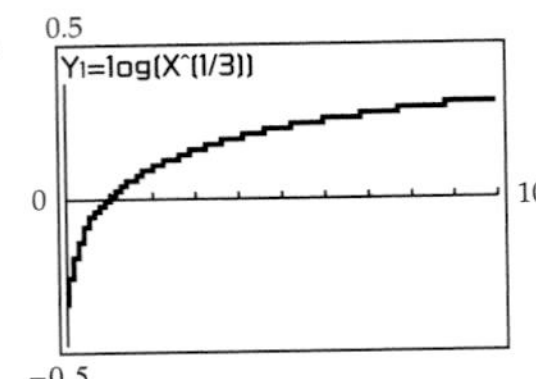

81.

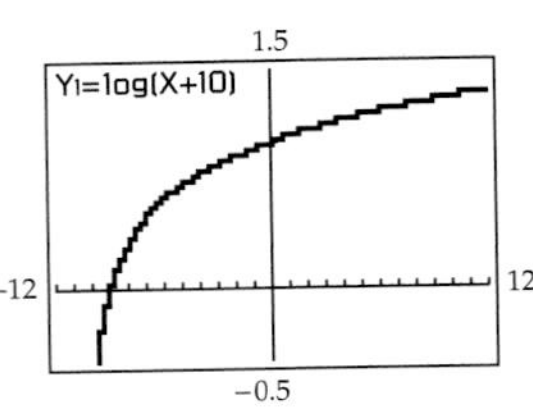

83.

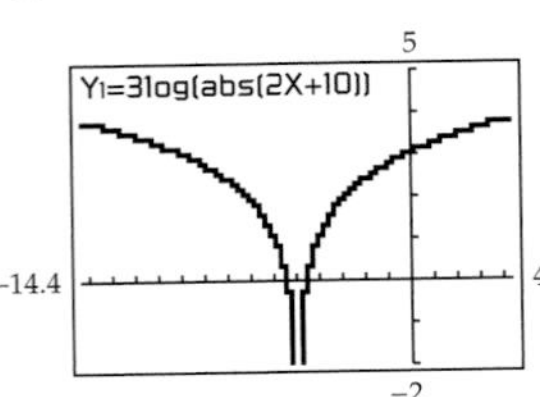

85. a. 2.0% **b.** 45 months **87. a.** 3298 units; 3418 units; 3490 units **b.** 2750 units

89. 2.05 square meters **91. a.** Answers will vary. **b.** 96 digits **c.** 3385 digits **d.** 6,320,430 digits **93.** f and g are inverse functions.

95. range of f: $\{y|-1 < y < 1\}$; range of g: all real numbers

Prepare for This Section (7.3), page 480

PS1. ≈ 0.77815 for each expression **PS2.** ≈ 0.98083 for each expression **PS3.** ≈ 1.80618 for each expression

PS4. ≈ 3.21888 for each expression **PS5.** ≈ 1.60944 for each expression **PS6.** ≈ 0.90309 for each expression

Exercise Set 7.3, page 489

1. $\log_b x + \log_b y + \log_b z$ **3.** $\ln x - 4 \ln z$ **5.** $\frac{1}{2}\log_2 x - 3\log_2 y$ **7.** $\frac{1}{2}\log_7 x + \frac{1}{2}\log_7 z - 2\log_7 y$ **9.** $2 + \ln z$

11. $\frac{1}{3}\log_4 z - 2 - 3\log_4 z$ **13.** $\frac{1}{2}\log x + \frac{1}{4}\log z$ **15.** $\frac{1}{3}\ln z + \frac{1}{6}$ **17.** $\log[x^2(x+5)]$ **19.** $\ln(x+y)$

21. $\log\left[x^3 \cdot \sqrt[3]{y}(x+1)\right]$ **23.** $\log\left(\frac{xy^2}{z}\right)$ **25.** $\log_6\left(\frac{x^2y^4}{x+2}\right)$ **27.** $\ln\left[\frac{(x+4)^2}{x(x^2-3)}\right]$ **29.** $\ln\left[\frac{(2x+5)\sqrt{w}}{yz^2}\right]$ **31.** $\ln\left[\frac{(x+3)y^3}{x-3}\right]$

33. 1.5395 **35.** 0.8672 **37.** −0.6131 **39.** 0.6447 **41.** 8.1749 **43.** 0.8735

45.

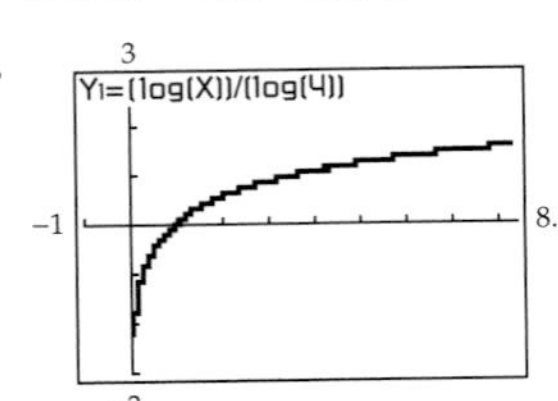

47.

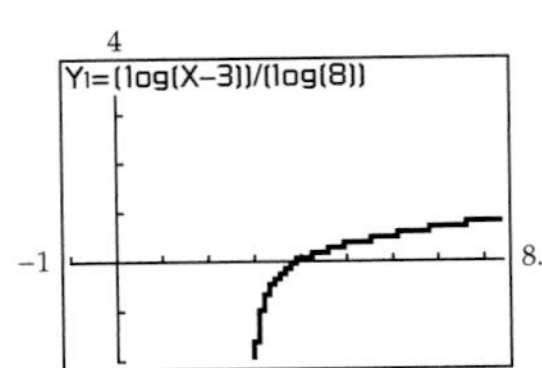

49.

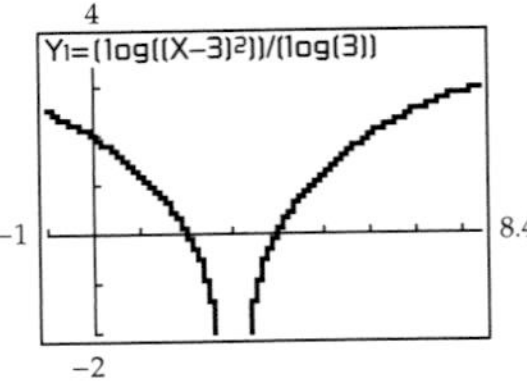

51.

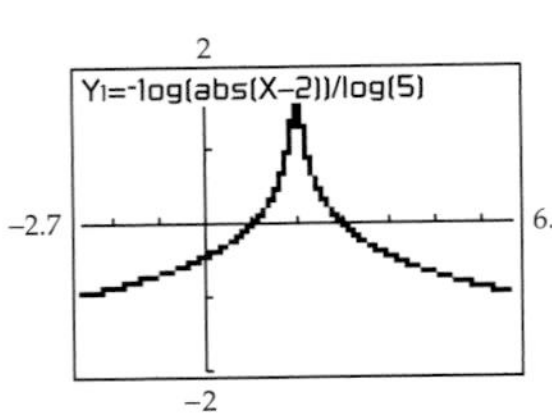

53. False; $\log 10 + \log 10 = 2$ but $\log(10 + 10) = \log 20 \neq 2$. **55.** True

57. False; $\log 100 - \log 10 = 1$ but $\log(100 - 10) = \log 90 \neq 1$. **59.** False; $\dfrac{\log 100}{\log 10} = \dfrac{2}{1} = 2$ but $\log 100 - \log 10 = 1$.

61. False; $(\log 10)^2 = 1$ but $2 \log 10 = 2$. **63.** 2 **65.** 500^{501} **67.** 1:870,551; 1:757,858; 1:659,754; 1:574,349; 1:500,000 **69.** 10.4; base **71.** 3.16×10^{-10} mole per liter **73. a.** 82.0 decibels **b.** 40.3 decibels **c.** 115.0 decibels **d.** 152.0 decibels **75.** 10 times as great **77.** 5 **79.** $10^{6.5}I_0$ or about $3{,}162{,}277.7I_0$ **81.** 100 to 1 **83.** $10^{1.8}$ to 1 or about 63 to 1 **85.** 5.5 **89. a.** $M \approx 6$ **b.** $M \approx 4$ **c.** The results are close to the magnitudes produced by using the amplitude-time-difference formula.

Prepare for This Section (7.4), page 494

PS1. $\log_3 729 = 6$ **PS2.** $5^4 = 625$ **PS3.** $\log_a b = x + 2$ **PS4.** $x = \dfrac{4a}{7b + 2c}$ **PS5.** $x = \dfrac{3}{44}$ **PS6.** $x = \dfrac{100(A - 1)}{A + 1}$

Exercise Set 7.4, page 501

1. 6 **3.** $-\dfrac{3}{2}$ **5.** $-\dfrac{6}{5}$ **7.** 3 **9.** $\dfrac{\log 70}{\log 5}$ **11.** $-\dfrac{\log 120}{\log 3}$ **13.** $\dfrac{\log 315 - 3}{2}$ **15.** $\ln 10$ **17.** $\dfrac{\ln 2 - \ln 3}{\ln 6}$ **19.** $\dfrac{3\log 2 - \log 5}{2\log 2 + \log 5}$

21. 7 **23.** 4 **25.** $2 + 2\sqrt{2}$ **27.** $\dfrac{199}{95}$ **29.** -1 **31.** 3 **33.** 10^{10} **35.** 2 **37.** no solution **39.** 5 **41.** $\log(20 + \sqrt{401})$

43. $\dfrac{1}{2}\log\left(\dfrac{3}{2}\right)$ **45.** $\ln(15 \pm 4\sqrt{14})$ **47.** $\ln(1 + \sqrt{65}) - \ln 8$ **49.** 1.61 **51.** 0.96 **53.** 2.20 **55.** -1.93 **57.** -1.34

59. a. 8500, 10,285 **b.** in 6 years **61. a.** 60°F **b.** 27 minutes **63.** 3.7 years **65.** 6.9 months

67. 6.67 seconds and 10.83 seconds

69. a.

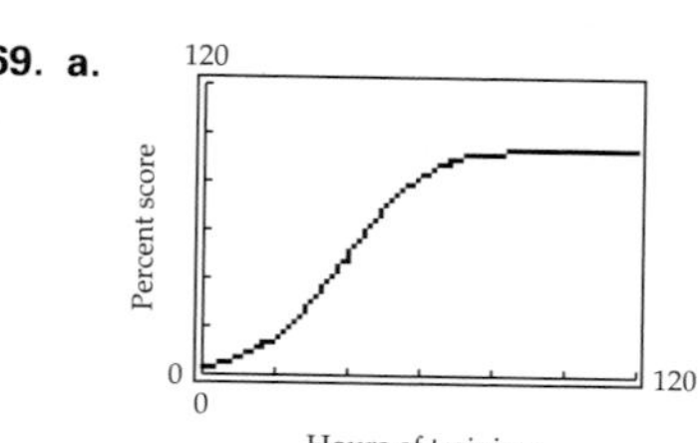

b. 48 hours
c. $P = 100$
d. As the number of hours of training increases, the test scores approach 100%.

71. a.

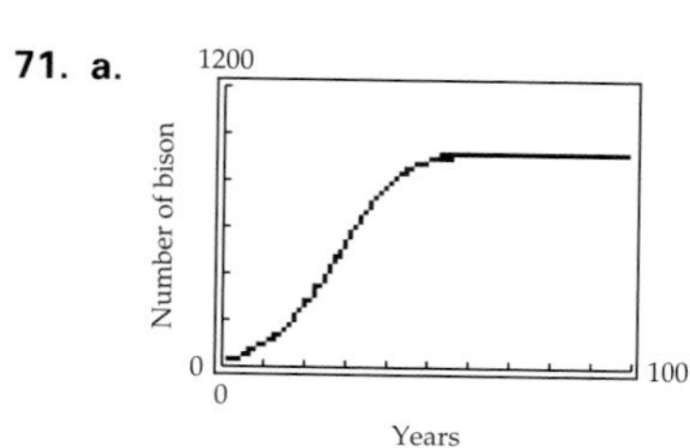

b. in 27 years, or the year 2026 **c.** $B = 1000$ **d.** As the number of years increases, the bison population approaches, but never reaches or exceeds, 1000.

73. a.

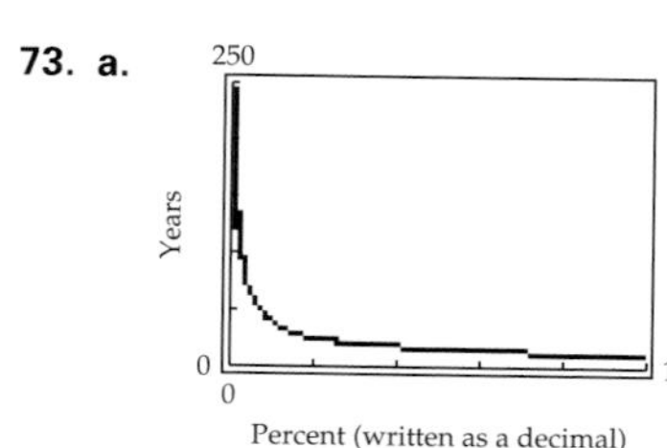

b. 78 years **c.** 1.9% **75. a.** 1.72 seconds **b.** $v = 100$ **c.** The object cannot fall faster than 100 feet per second.

77. 138 withdrawals **79.** The second step. Because $\log 0.5 < 0$, the inequality sign must be reversed. **81.** $x = \dfrac{y}{y - 1}$ **83.** $e^{0.336} \approx 1.4$

Prepare for This Section (7.5), page 506

PS1. 1220.39 **PS2.** 824.96 **PS3.** −0.0495 **PS4.** 1340 **PS5.** 0.025 **PS6.** 12.8

Exercise Set 7.5, page 516

1. a. 2200 bacteria **b.** 17,600 bacteria **3. a.** $N(t) \approx 22{,}600e^{0.01368t}$ **b.** 27,700 **5.** $N(t) \approx 362{,}300\, e^{0.011727t}$; 402,600

7. a. **b.** 3.18 micrograms **c.** ≈15.07 hours **d.** ≈30.14 hours **9.** ≈6601 years ago

11. ≈2378 years old **13. a.** \$9724.05 **b.** \$11,256.80 **15. a.** \$48,885.72 **b.** \$49,282.20 **c.** \$49,283.30 **17.** \$24,730.82 **19.** 8.8 years **21.** $t = \dfrac{\ln 3}{r}$ **23.** 14 years **25. a.** 1900 **b.** 0.16 **c.** 200 **27. a.** 157,500 **b.** 0.04 **c.** 45,000 **29. a.** 2400 **b.** 0.12 **c.** 300 **31.** $P(t) \approx \dfrac{5500}{1 + 12.75e^{-0.37263t}}$ **33.** $P(t) \approx \dfrac{100}{1 + 4.55556e^{-0.22302t}}$ **35. a.** \$158,000, \$163,000 **b.** \$625,000 **37. a.** $P(t) \approx \dfrac{1600}{1 + 4.12821e^{-0.06198t}}$ **b.** 497 wolves **39. a.** $P(t) \approx \dfrac{8500}{1 + 4.66667e^{-0.14761t}}$ **b.** 2016 **41. a.** 0.056 **b.** 42°F **c.** after 54 minutes **43. a.** 211 hours **b.** 1386 hours

45. 3.1 years **47. a.**

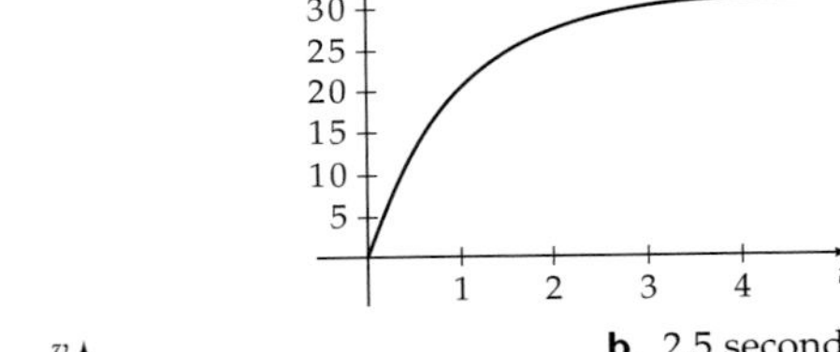

b. 0.98 second **c.** $v = 32$ **d.** As time increases, the velocity approaches, but never reaches or exceeds, 32 feet per second.

49. a. **b.** 2.5 seconds **c.** ≈24.56 feet per second **d.** The average speed of the object was approximately 24.56 feet per second during the period from $t = 1$ to $t = 2$ seconds.

51. 45 hours **53. a.** 0.71 gram **b.** 0.96 gram **c.** 0.52 gram **55.** 2.91%

Prepare for This Section (7.6), page 521

PS1. decreasing **PS2.** decreasing **PS3.** 36 **PS4.** 840 **PS5.** 15.8 **PS6.** $P = 55$

Exercise Set 7.6, page 529

1. increasing exponential function

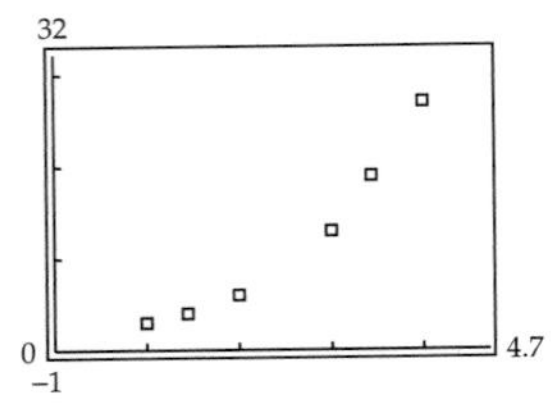

3. decreasing exponential function; decreasing logarithmic function

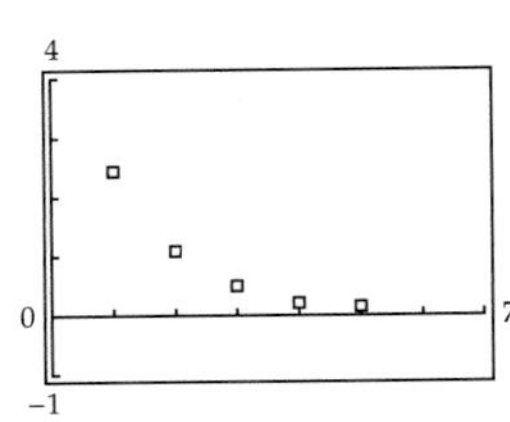

5. decreasing logarithmic function

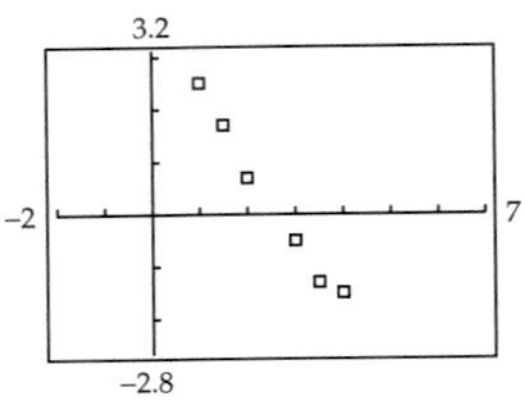

7. $y \approx 0.99628(1.20052)^x$; $r \approx 0.85705$ **9.** $y \approx 1.81505(0.51979)^x$; $r \approx -0.99978$ **11.** $y \approx 4.89060 - 1.35073 \ln x$; $r \approx -0.99921$
13. $y \approx 14.05858 + 1.76393 \ln x$; $r \approx 0.99983$ **15.** $y \approx \dfrac{235.58598}{1 + 1.90188e^{-0.05101x}}$ **17.** $y \approx \dfrac{2098.68307}{1 + 1.19794e^{-0.06004x}}$
19. a. LinReg: $P \approx 0.22129t + 3.99190$, $r \approx 0.99288$; ExpReg: $P \approx 4.05326(1.04460)^t$, $r = 0.99412$ **b.** the exponential model **c.** \$8.15
21. a. $T \approx 0.06273(1.07078)^F$ **b.** 5.3 hours **23. a.** $T \approx 0.07881(1.07259)^F$ **b.** 7.5 hours; 2.2 hours **25.** An increasing logarithmic model provides a better fit because of the concave-downward nature of the graph. **27. a.** $p \approx 7.862(1.026)^y$ **b.** 36 centimeters
29. a. LinReg: pH $\approx 0.01353q + 7.02852$, $r \approx 0.956627$; LnReg: pH $\approx 6.10251 + 0.43369 \ln q$, $r \approx 0.999998$. The logarithmic model provides a better fit. **b.** 126.0 **31. a.** $p \approx 3200(0.91894)^t$; 2012 **b.** No. The model fits the data perfectly because there are only two data points.
33. a. ExpReg: $S \approx 7062.46390(0.956776)^t$, $r \approx -0.89618$; LnReg: $S \approx 6995.50673 - 841.74326 \ln t$, $r \approx -0.96240$ **b.** the logarithmic model **c.** 4977 **35. a.** $P(t) \approx \dfrac{11.26828}{1 + 2.74965e^{-0.02924t}}$ **b.** 11 billion people **37.** A and B have different exponential regression functions.
39. a. ExpReg: $y \approx 1.81120(1.61740)^x$, $r \approx 0.96793$; PwrReg: $y \approx 2.09385(x)^{1.40246}$, $r \approx 0.99999$ **b.** The power regression function provides the better fit.

Chapter 7 Assessing Concepts, page 538

1. False; $f(x) = x^2$ does not have an inverse function. **2.** True **3.** True **4.** False; $h(x)$ is not an increasing function for $0 < b < 1$. **5.** False; $j(x)$ is not an increasing function for $0 < b < 1$. **6.** c **7.** b **8.** f **9.** a **10.** e **11.** d **12.** g

Chapter 7 Review Exercises, page 539

1. 2 [7.2] **2.** 4 [7.2] **3.** 3 [7.2] **4.** π [7.2] **5.** -2 [7.4] **6.** 8 [7.4] **7.** -3 [7.4] **8.** -4 [7.4] **9.** ± 1000 [7.4] **10.** $\pm 10^{10}$ [7.4] **11.** 7 [7.4]
12. ± 8 [7.4] **13.** [7.1] **14.** [7.1]

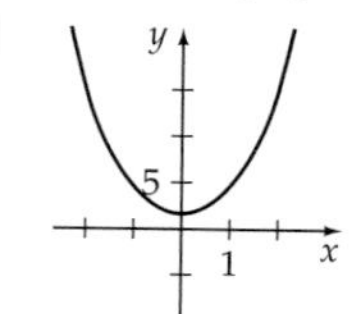

15. [7.1] **16.** [7.1]

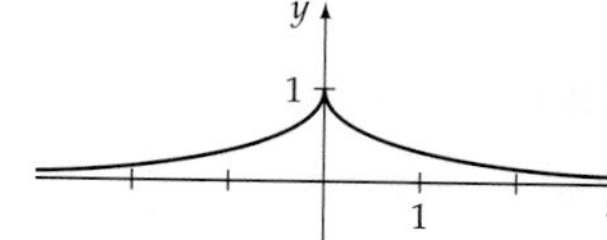

17. [7.1]

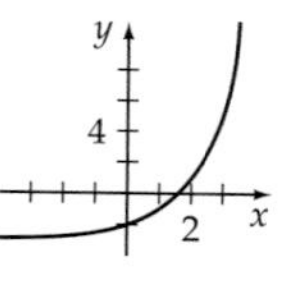

18. [7.1]

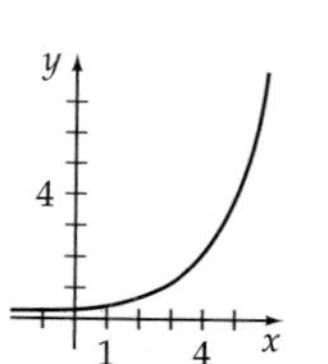

19. [7.2]

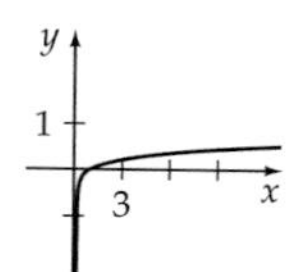

20. [7.2]

21. [7.2]

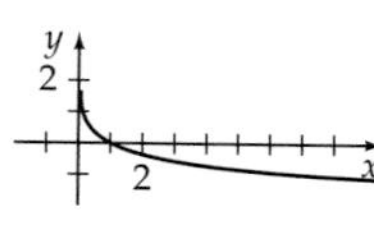

22. [7.2]

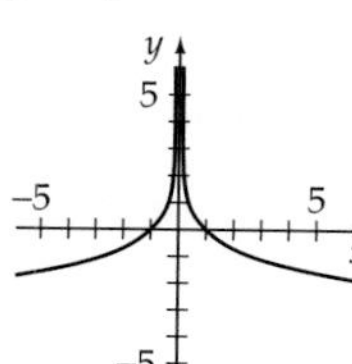

23. [7.1]

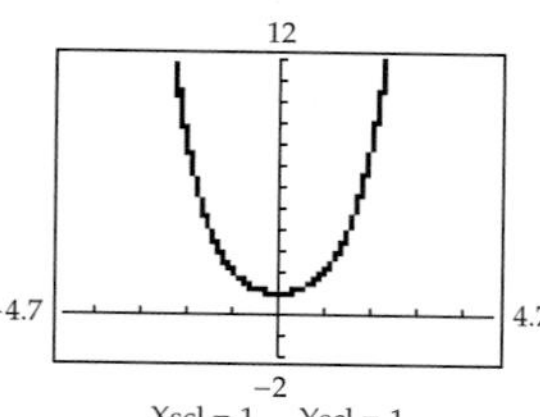

24. [7.1]

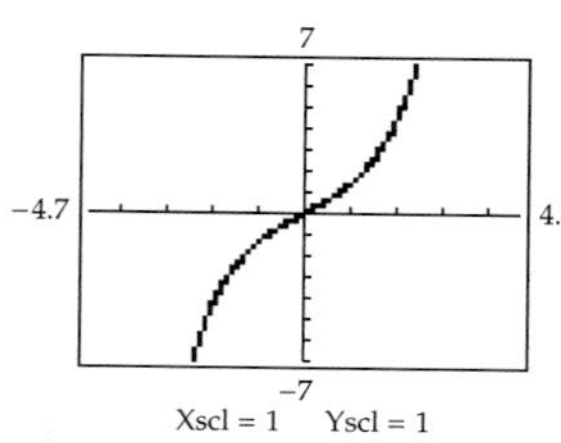

25. $4^3 = 64$ [7.2] **26.** 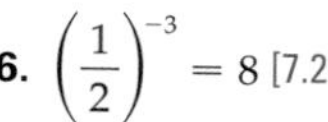 $\left(\frac{1}{2}\right)^{-3} = 8$ [7.2] **27.** $(\sqrt{2})^4 = 4$ [7.2] **28.** $e^0 = 1$ [7.2]

29. $\log_5 125 = 3$ [7.2] **30.** $\log_2 1024 = 10$ [7.2] **31.** $\log_{10} 1 = 0$ [7.2] **32.** $\log_8 2\sqrt{2} = \frac{1}{2}$ [7.2] **33.** $2 \log_b x + 3 \log_b y - \log_b z$ [7.3]
34. $\frac{1}{2} \log_b x - 2 \log_b y - \log_b z$ [7.3] **35.** $\ln x + 3 \ln y$ [7.3] **36.** $\frac{1}{2} \ln x + \frac{1}{2} \ln y - 4 \ln z$ [7.3] **37.** $\log(x^2 \sqrt[3]{x+1})$ [7.3]
38. $\log \dfrac{x^5}{(x+5)^2}$ [7.3] **39.** $\ln \dfrac{\sqrt{2xy}}{z^3}$ [7.3] **40.** $\ln \dfrac{xz}{y}$ [7.3] **41.** 2.86754 [7.3] **42.** 3.35776 [7.3] **43.** -0.117233 [7.3] **44.** -0.578989 [7.3]

45. $\dfrac{\ln 30}{\ln 4}$ [7.4] **46.** $\dfrac{\log 41}{\log 5} - 1$ [7.4] **47.** 4 [7.4] **48.** $\dfrac{1}{6}e$ [7.4] **49.** 4 [7.4] **50.** 15 [7.4] **51.** $\dfrac{\ln 3}{2\ln 4}$ [7.4] **52.** $\dfrac{\ln(8 \pm 3\sqrt{7})}{\ln 5}$ [7.4]
53. 10^{1000} [7.4] **54.** $e^{(e^e)}$ [7.4] **55.** 1,000,005 [7.4] **56.** $\dfrac{15 + \sqrt{265}}{2}$ [7.4] **57.** 81 [7.4] **58.** $\pm\sqrt{5}$ [7.4] **59.** 4 [7.4] **60.** 5 [7.4]
61. 7.7 [7.3] **62.** 5.0 [7.3] **63.** 3162 to 1 [7.3] **64.** 2.8 [7.3] **65.** 4.2 [7.3] **66.** $\approx 3.98 \times 10^{-6}$ [7.3] **67. a.** \$20,323.79 **b.** \$20,339.99 [7.5]
68. a. \$25,646.69 **b.** \$25,647.32 [7.5] **69.** \$4,438.10 [7.5] **70. a.** 69.9% **b.** 6 days **c.** 19 days [7.5] **71.** $N(t) \approx e^{0.8047t}$ [7.5]
72. $N(t) \approx 2e^{0.5682t}$ [7.5] **73.** $N(t) \approx 3.783e^{0.0558t}$ [7.5] **74.** $N(t) \approx e^{-0.6931t}$ [7.5] **75. a.** $P(t) \approx 25{,}200e^{0.06155789t}$ **b.** 38,800 [7.5]
76. 340 years [7.5] **77.** Answers will vary. [7.6] **78. a.** ExpReg: $R \approx 179.94943(0.968094)^t$, $r \approx -0.99277$;
LnReg: $R \approx 171.19665 - 35.71341 \ln t$, $r \approx -0.98574$ **b.** The exponential equation provides a better fit for the data. **c.** 5.4 per 1000 live births [7.6] **79. a.** $P(t) \approx \dfrac{1400}{1 + \frac{17}{3}e^{-0.22458t}}$ **b.** 1070 coyotes [7.5] **80. a.** $21\frac{1}{3}$ **b.** $P(t) \to 128$ [7.5]

Chapter 7 Quantitative Reasoning Exercises, page 541

QR1. ExpReg: $S \approx 5.94860(1.72192)^t$; logistic: $S \approx \dfrac{244.56468}{1 + 39.38651e^{-0.57829t}}$ [7.6] **QR2.** exponential: 1363.1 million; logistic: 218.1 million [7.6]
QR3. $S \approx 58.73554 + 16.75801 \ln t$ [7.6] **QR4.** 88.8 million [7.6] **QR5.** $y \approx \dfrac{85.24460}{1 + 14.0040e^{-0.70591t}}$, 84.2 million [7.6]
QR6. Answers will vary. [7.6]

Chapter 7 Test, page 542

1. a. $b^c = 5x - 3$ [7.2] **b.** $\log_3 y = \dfrac{x}{2}$ [7.2] **2.** $2\log_b z - 3\log_b y - \dfrac{1}{2}\log_b x$ [7.3] **3.** $\log \dfrac{2x+3}{(x-2)^3}$ [7.3] **4.** 1.7925 [7.3]
5. [7.1]

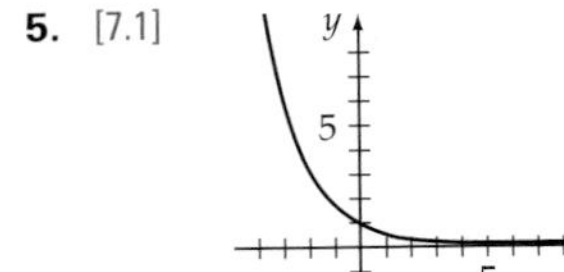

6. [7.2]

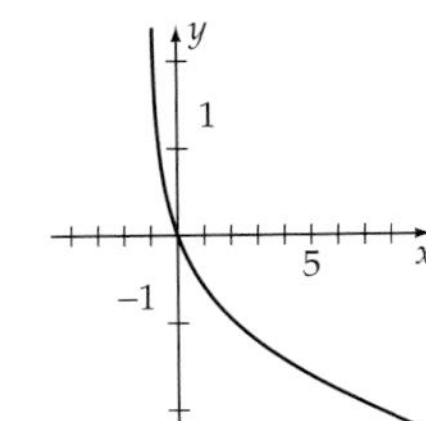

7. 1.9206 [7.4] **8.** $\dfrac{5\ln 4}{\ln 28}$ [7.4] **9.** 1 [7.4] **10.** −3 [7.4]
11. a. \$29,502.36 **b.** \$29,539.62 [7.5] **12.** 17.36 years [7.5] **13. a.** 7.6 **b.** 63 to 1 [7.3] **14. a.** $P(t) \approx 34{,}600e^{0.04667108t}$ **b.** 55,000 [7.5]
15. 690 years [7.5] **16. a.** $y \approx 1.67199(2.47188)^x$ **b.** 1945 [7.6] **17. a.** LnReg: $d \approx 67.35501 + 2.54015 \ln t$; logistic: $d \approx \dfrac{72.03783}{1 + 0.15279e^{-0.67752t}}$
b. logarithmic: 73.67 meters; logistic: 72.03 meters [7.6] **18. a.** $P(t) \approx \dfrac{1100}{1 + 5.875e^{-0.20429t}}$ **b.** about 457 raccoons [7.5]

Cumulative Review Exercises, page 543

1. $\cos(x^2 + 1)$ [1.5] **2.** $f^{-1}(x) = \dfrac{1}{2}x - 4$ [1.6] **3.** $\sin\theta = \dfrac{3}{5}$, $\cos\theta = \dfrac{4}{5}$, $\tan\theta = \dfrac{3}{4}$ [2.2] **4.** 17 centimeters [2.2]
5. amplitude: 4, period: π, phase shift: $\dfrac{\pi}{4}$ [2.5] **6.** amplitude: $\sqrt{2}$, period: 2π [3.4] **7.** odd [2.4] **9.** $\dfrac{12}{5}$ [3.5] **10.** $\dfrac{\pi}{2}, \dfrac{7\pi}{6}, \dfrac{11\pi}{6}$ [3.6]
11. magnitude: 5, direction: 126.9° [4.3] **12.** 176.8° [4.3] **13.** ground speed: ≈420 mph; course: ≈55° [4.1/4.2] **14.** 26° [4.1] **15.** 16 [5.3]
16. $\dfrac{\sqrt{2}}{2} + i\dfrac{\sqrt{2}}{2}, -\dfrac{\sqrt{2}}{2} - i\dfrac{\sqrt{2}}{2}$ [5.3] **17.** $(\sqrt{2}, 45°)$ [6.5] **18.** [6.5]

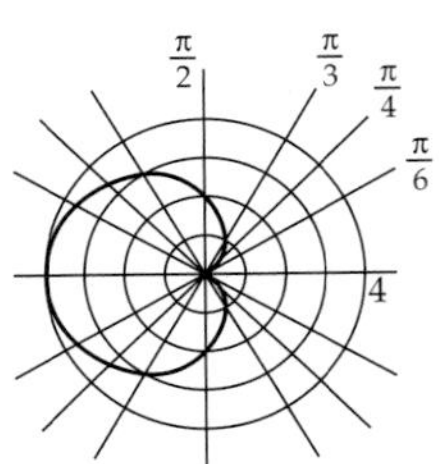

19. 1.43 [7.4] **20.** 1.8 milligrams [7.5]

Index

Important Formulas

Pythagorean Theorem
$c^2 = a^2 + b^2$

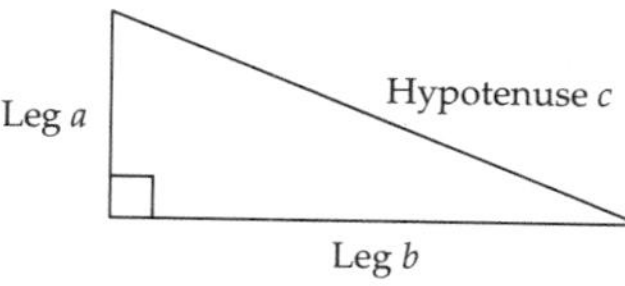

The *distance* between $P_1(x_1, y_1)$ and $P_2(x_2, y_2)$ is

$$d(P_1, P_2) = \sqrt{(x_2 - x_1)^2 + (y_2 - y_1)^2}$$

The *slope m* of a line through $P_1(x_1, y_1)$ and $P_2(x_2, y_2)$ is

$$m = \frac{y_2 - y_1}{x_2 - x_1}, \quad x_1 \neq x_2$$

Quadratic Formula
If $a \neq 0$, the solutions of $ax^2 + bx + c = 0$ are

$$x = \frac{-b \pm \sqrt{b^2 - 4ac}}{2a}$$

Properties of Functions

A *function* is a set of ordered pairs in which no two ordered pairs that have the same first coordinate have different second coordinates.

If a and b are elements of an interval I that is a subset of the domain of a function f, then

- f is an *increasing* function on I if $f(a) < f(b)$ whenever $a < b$.
- f is a *decreasing* function on I if $f(a) > f(b)$ whenever $a < b$.
- f is a *constant* function on I if $f(a) = f(b)$ for all a and b.

A *one-to-one* function satisfies the additional condition that given any y, there is one and only one x that can be paired with that given y.

Graphing Concepts

Odd Functions
A function f is an odd function if $f(-x) = -f(x)$ for all x in the domain of f. The graph of an odd function is symmetric with respect to the origin.

Even Functions
A function is an even function if $f(-x) = f(x)$ for all x in the domain of f. The graph of an even function is symmetric with respect to the y-axis.

Vertical and Horizontal Translations
If f is a function and c is a positive constant, then the graph of

- $y = f(x) + c$ is the graph of $y = f(x)$ shifted up *vertically* c units.
- $y = f(x) - c$ is the graph of $y = f(x)$ shifted down *vertically* c units.
- $y = f(x + c)$ is the graph of $y = f(x)$ shifted left *horizontally* c units.
- $y = f(x - c)$ is the graph of $y = f(x)$ shifted right *horizontally* c units.

Reflections
If f is a function then the graph of

- $y = -f(x)$ is the graph of $y = f(x)$ reflected across the x-axis.
- $y = f(-x)$ is the graph of $y = f(x)$ reflected across the y-axis.

Vertical Shrinking and Stretching

- If $c > 0$ and the graph of $y = f(x)$ contains the point (x, y), then the graph of $y = c \cdot f(x)$ contains the point (x, cy).
- If $c > 1$, the graph of $y = c \cdot f(x)$ is obtained by stretching the graph of $y = f(x)$ away from the x-axis by a factor of c.
- If $0 < c < 1$, the graph of $y = c \cdot f(x)$ is obtained by shrinking the graph of $y = f(x)$ toward the x-axis by a factor of c.

Horizontal Shrinking and Stretching

- If $a > 0$ and the graph of $y = f(x)$ contains the point (x, y), then the graph of $y = f(ax)$ contains the point $\left(\frac{1}{a}x, y\right)$.
- If $a > 1$, the graph of $y = f(ax)$ is a *horizontal shrinking* of the graph of $y = f(x)$.
- If $0 < a < 1$, the graph of $y = f(ax)$ is a *horizontal stretching* of the graph of $y = f(x)$.

Definitions of Trigonometric Functions

$$\sin\theta = \frac{b}{r} \qquad \csc\theta = \frac{r}{b}$$

$$\cos\theta = \frac{a}{r} \qquad \sec\theta = \frac{r}{a}$$

$$\tan\theta = \frac{b}{a} \qquad \cot\theta = \frac{a}{b}$$

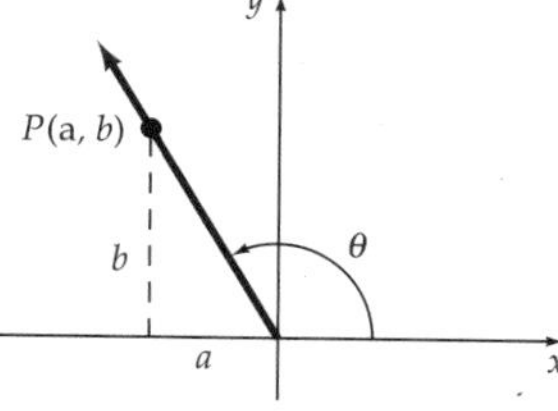

where $r = \sqrt{a^2 + b^2}$

Definitions of Circular Functions

$\sin t = y$ $\qquad \csc t = \dfrac{1}{y}$

$\cos t = x$ $\qquad \sec t = \dfrac{1}{x}$

$\tan t = \dfrac{y}{x}$ $\qquad \cot t = \dfrac{x}{y}$

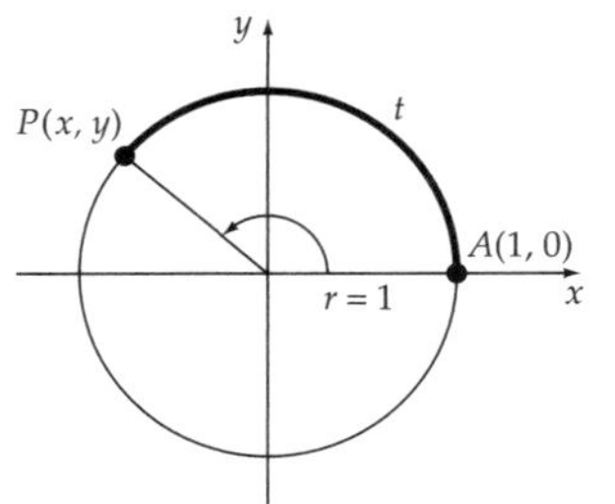

Formulas for Triangles

For any triangle ABC, the following formulas can be used.

Law of Sines

$$\frac{a}{\sin A} = \frac{b}{\sin B} = \frac{c}{\sin C}$$

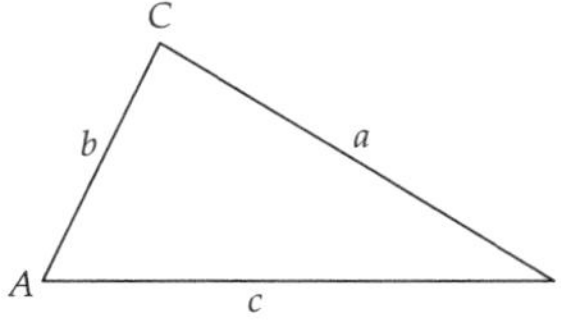

Law of Cosines

$$c^2 = a^2 + b^2 - 2ab \cos C$$

Area of a Triangle

$$K = \frac{1}{2} ab \sin C \qquad K = \frac{a^2 \sin B \sin C}{2 \sin A}$$

Heron's Formula

$$K = \sqrt{s(s-a)(s-b)(s-c)}, \text{ where } s = \frac{a+b+c}{2}$$

Fundamental Identities

$$\tan \theta = \frac{\sin \theta}{\cos \theta} \qquad \cot \theta = \frac{\cos \theta}{\sin \theta}$$

$$\sin^2 \theta + \cos^2 \theta = 1 \qquad 1 + \tan^2 \theta = \sec^2 \theta$$

$$1 + \cot^2 \theta = \csc^2 \theta$$

Formulas for Negatives

$$\sin(-\theta) = -\sin \theta \qquad \cos(-\theta) = \cos \theta$$

$$\tan(-\theta) = -\tan \theta$$

Reciprocal Identities

$$\csc \theta = \frac{1}{\sin \theta} \qquad \sec \theta = \frac{1}{\cos \theta} \qquad \cot \theta = \frac{1}{\tan \theta}$$

Sum of Two Angle Identities

$$\sin(\alpha + \beta) = \sin \alpha \cos \beta + \cos \alpha \sin \beta$$

$$\cos(\alpha + \beta) = \cos \alpha \cos \beta - \sin \alpha \sin \beta$$

$$\tan(\alpha + \beta) = \frac{\tan \alpha + \tan \beta}{1 - \tan \alpha \tan \beta}$$

Difference of Two Angle Identities

$$\sin(\alpha - \beta) = \sin \alpha \cos \beta - \cos \alpha \sin \beta$$

$$\cos(\alpha - \beta) = \cos \alpha \cos \beta + \sin \alpha \sin \beta$$

$$\tan(\alpha - \beta) = \frac{\tan \alpha - \tan \beta}{1 + \tan \alpha \tan \beta}$$

Double-Angle Identities

$$\sin 2\alpha = 2 \sin \alpha \cos \alpha$$

$$\cos 2\alpha = \cos^2 \alpha - \sin^2 \alpha = 1 - 2 \sin^2 \alpha = 2 \cos^2 \alpha - 1$$

$$\tan 2\alpha = \frac{2 \tan \alpha}{1 - \tan^2 \alpha}$$

Power-Reducing Identities

$$\sin^2 \alpha = \frac{1 - \cos 2\alpha}{2}$$

$$\cos^2 \alpha = \frac{1 + \cos 2\alpha}{2}$$

$$\tan^2 \alpha = \frac{1 - \cos 2\alpha}{1 + \cos 2\alpha}$$

Half-Angle Identities

$$\sin \frac{\alpha}{2} = \pm \sqrt{\frac{1 - \cos \alpha}{2}}$$

$$\cos \frac{\alpha}{2} = \pm \sqrt{\frac{1 + \cos \alpha}{2}}$$

$$\tan \frac{\alpha}{2} = \frac{\sin \alpha}{1 + \cos \alpha} = \frac{1 - \cos \alpha}{\sin \alpha}$$

SKYLANDERS

THE KAOS TRAP™

FORGETTING FLYNN

Story by:
RON MARZ and DAVID A. RODRIGUEZ
Art by:
MIKE BOWDEN
Colors by:
FERNANDO PENICHE
Letters by:
DERON BENNETT & TOM B. LONG
Edited by:
DAVID HEDGECOCK

ABDOPUBLISHING.COM

Reinforced library bound edition published in 2016 by Spotlight, a division of ABDO, PO Box 398166, Minneapolis, Minnesota 55439. Spotlight produces high-quality reinforced library bound editions for schools and libraries. Published by agreement with IDW.

Printed in the United States of America, North Mankato, Minnesota.
042015
092015

THIS BOOK CONTAINS RECYCLED MATERIALS

LIBRARY OF CONGRESS CATALOGING-IN-PUBLICATION DATA

Marz, Ron, author.
Forgetting Flynn / writer: Ron Marz and David A. Rodriguez ; artist: Mike Bowden ; colors: Fernando Peniche.
pages cm. -- (Skylanders: the Kaos trap)
Summary: "Flynn is flying all the young students off to Skylanders Academy, but gets his memory wiped when he runs into an enemy in disguise. What will become of Flynn? Will Kaos' troublesome plan work?"-- Provided by publisher.
ISBN 978-1-61479-385-4
1. Graphic novels. I. Rodriguez, David A., author. II. Bowden, Mike, illustrator. III. Skylanders (Game) IV. Title.
PZ7.7.M3754For 2016
741.5'973--dc23
2015001610

Spotlight

A Division of ABDO
abdopublishing.com

ARE THERE T? ARE THERE YET?!
GO FASTER, MR. FLYNN!
CAN I STEER? I WANNA STEER! WHY WON'T YOU LET ME STEER?
WHERE'S THE BATHROOM?
HOW DID I EVER GET ROPED INTO FERRYING THE FRESHMAN CLASS TO SKYLANDER ACADEMY?
YOU VOLUNTEERED, MR. FLYNN!

YOU KNOW, IT'S STILL GOING TO BE A LITTLE WHILE BEFORE WE REACH SKYLANDER ACADEMY...

YOU KNOW WHAT? THAT'S A GREAT IDEA.
YOU GUYS STEER, JUST KEEP US POINTED STRAIGHT FOR A FEW MINUTES.

I'LL NIP DOWN TO THE DREADYACHT'S GALLEY AND GET A LITTLE SNACKEROO.
I'M THINKING... ENCHILADA.
A LOT OF ENCHILADA.

WEERUPTOR? SORRY, LITTLE GUY, YOU'RE NOT SUPPOSED TO BE IN HERE. HOPEFULLY YOU DIDN'T EAT MY ENCHILADA.
WHAT'S TAKING THIS SPELL SO LONG?
OH, I KNOW EXACTLY WHAT YOU MEAN. IN SCHOOL, I WAS TERRIBLE AT SPELLING.

I SAID "SPELL," YOU IDIOT!
YAAA!

WHAT'S UP WITH YOUR FACE? YOU LOOK JUST LIKE...
...KAOS?!

I'VE GOTTA WARN SKYLANDER ACADEMY!
YOU'RE NOT WARNING ANYBODY!
MY PLAN TO RUIN SKYLANDER ACADEMY IS FOOLPROOF, SO IT WON'T BE STOPPED BY A FOOL LIKE YOU, FLYNN!
SKRAK
BUT ALL I WANTED WAS AN ENCHILADA!
SPLANG
ZORK!
MY AMNESIA SPELL WILL MAKE YOU FORGET YOU EVER SAW ME.
SNEAKING INTO SKYLANDER ACADEMY DISGUISED AS A NEW STUDENT WILL REVEAL ALL YOUR SECRETS TO ME!
OOBA...
...OOBA DOOBA.

WUBBA DUBBA.
I'LL *DESTROY* THE SKYLANDERS FROM WITHIN...

...AND NO ONE WILL *EVER* SUSPECT A THING!

ENCHILADA.
GOTTA GOTTA GET AN ENCHILADA.

MY TURN TO STEER!
ENCHILADA?

FLYNN?
KITCHEN MUST BE THIS WAY...
FLYNN!

PILOTS ARE SUPPOSED TO KNOW BETTER THAN TO *WALK OFF* THEIR OWN SHIPS!
WINDY IN THIS KITCHEN.

I'LL SAVE YOU, FLYNN...

...OR MY NAME'S NOT *CHOPPER!*
SOMEBODY MUST'VE LEFT THE *WINDOW* OPEN.

GOTCHA!

YOU'RE TOO *HEAVY!* WE'RE GOING *DOWN!*
UH, *FLYNN?* HAVE YOU PUT ON A LITTLE *WEIGHT?*
THE *BONEY ISLANDS* ARE OVER THERE, BUT I'M NOT SURE WE CAN REACH 'EM!
ENCHILADA ISLANDS?
STOP SAYING *ENCHILADA!* AT LEAST HELP ME STEER!
MY TURN TO STEER!
WHUMP

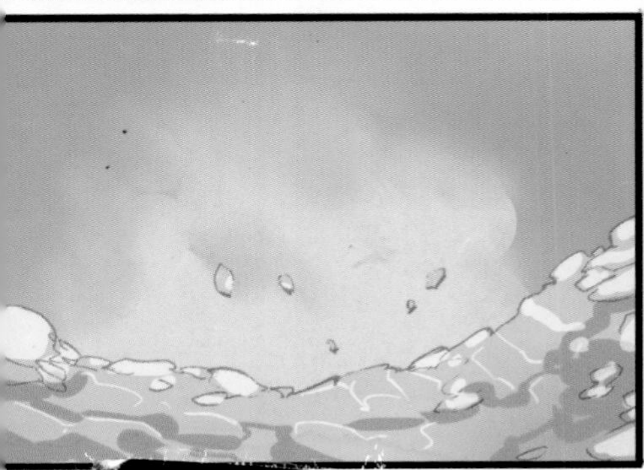

WHO'S FLYNN?
YOU'RE FLYNN! I'M CHOPPER! WE JUST CRASH-LANDED IN THE BONEY ISLANDS!
WHY ARE YOU SO OUT OF IT? I MEAN, EVEN MORE THAN USUAL.
YOU'RE OBVIOUSLY NOT GOING TO BE MUCH HELP.
WE HAVE TO FIND SOME WAY OFF THIS ISLAND, AND YOU'RE A LITTLE TOO BEEFY FOR ME TO FLY YOU.

HEY! WHERE DO YOU THINK YOU'RE GOING?!
TAKE ME TO YOUR ENCHILADAS.

THIS ISN'T THE TIME TO TAKE A TOUR!

COME ON, FLYNN, GET OUT OF THERE!
NEXT STOP, ERNESTO'S ENCHILADA EMPORIUM.

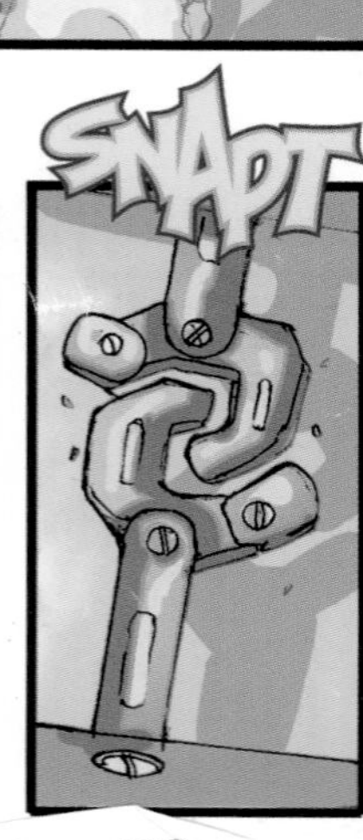
SNAPT

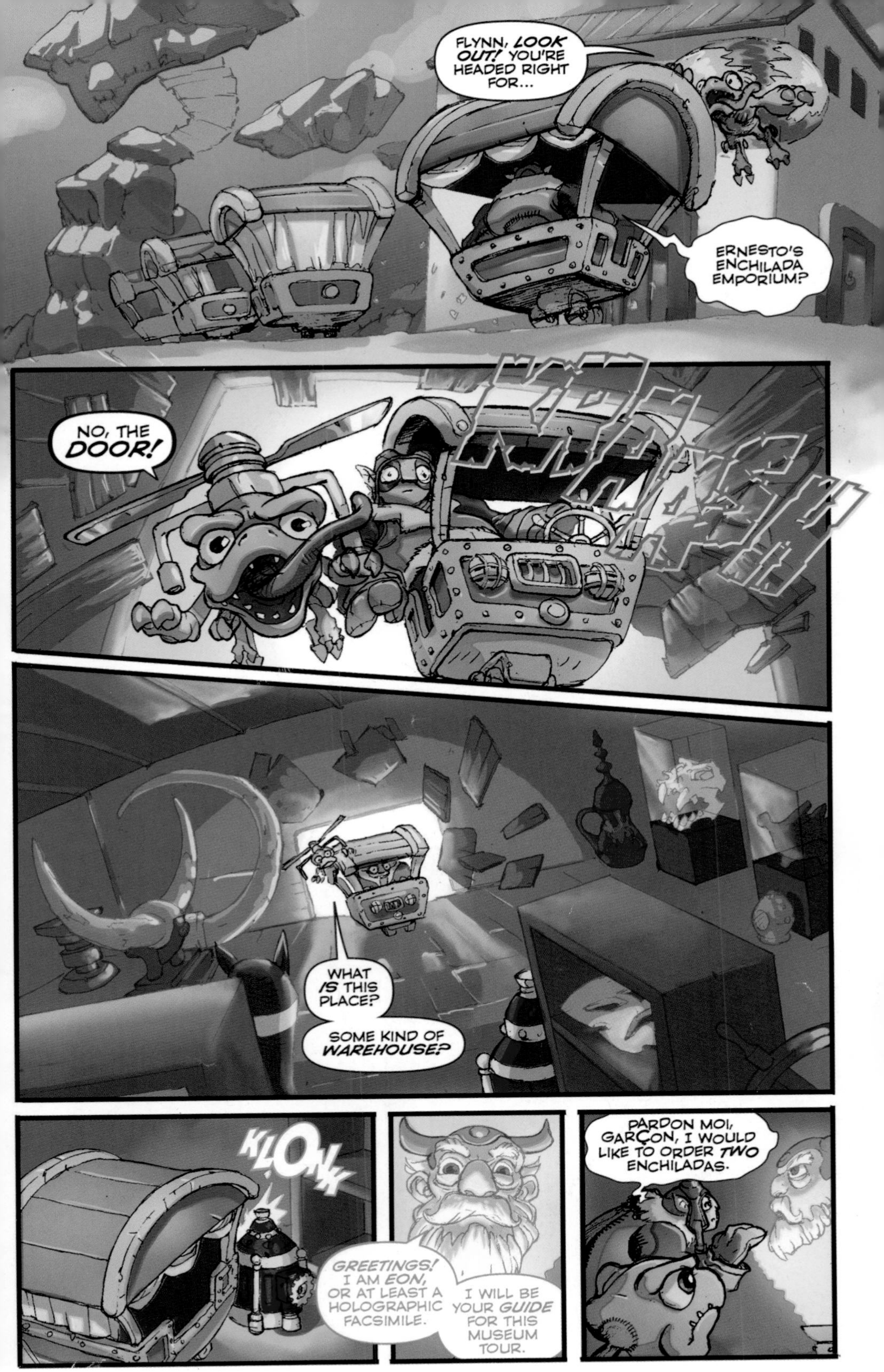

FLYNN, *LOOK OUT!* YOU'RE HEADED RIGHT FOR...

ERNESTO'S ENCHILADA EMPORIUM?

NO, THE *DOOR!*

WHAT *IS* THIS PLACE?

SOME KIND OF *WAREHOUSE?*

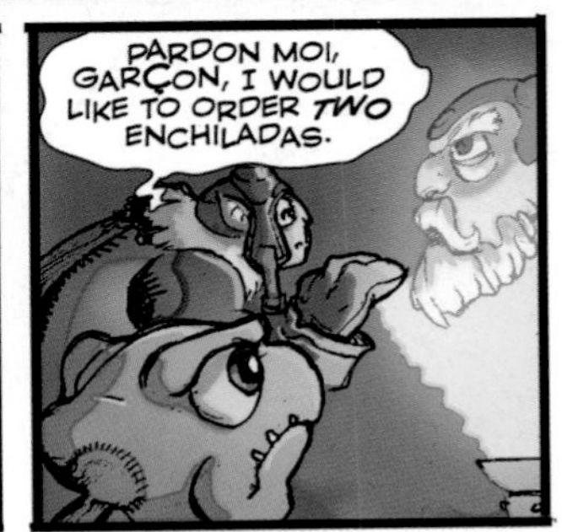

WELCOME TO THE SKYLANDERS MUSEUM! WE HOPE THIS INTERACTIVE TOUR WILL MAKE YOUR VISIT MEMORABLE!
THE SKYLANDS ARE A VERY SPECIAL PLACE...
...WATCHED OVER BY VERY SPECIAL HEROES. THE SKYLANDERS ARE MIGHTY WARRIORS OF ALL SHAPES AND SIZES, WHO FACE ALL MANNER OF FOES.
I RECRUITED MANY OF THE FIRST SKYLANDERS TO HELP ME IN MY ROLE AS A PORTAL MASTER, BATTLING EVIL...
...ESPECIALLY THE EVIL OF THAT VILE VILLAIN KAOS, WHO HAS TRIED AGAIN AND AGAIN TO CONQUER THE SKYLANDS.

KAOS AND HIS HORDES *ATTACKED* THE *CORE OF LIGHT,* CONSTRUCTED BY THE ANCIENTS TO KEEP THE DARKNESS AT BAY.
EVENTUALLY, KAOS UNLEASHED A GREAT BEAST CALLED THE *HYDRA* AND SUCCEEDED IN DESTROYING THE CORE.
THE SKYLANDERS WERE SCATTERED FAR AND WIDE TO *OTHER* REALMS. ALL SEEMED VERY LITERALLY *LOST*...
...BUT *NEW* PORTAL MASTERS HELPED BRING THEM *HOME,* WHERE THEY ONCE AGAIN PROTECT THE SKYLANDS AND ALL WHO LIVE HERE.

REMEMBER, SKYLANDERS ARE ALWAYS READY FOR ADVENTURE!
PLEASE ENJOY THE REST OF YOUR TOUR...

ZZZORF

HE DIDN'T MENTION ENCHILADAS EVEN ONCE.
GIVE THAT A REST, FLYNN.
WE HAVE TO FIND A WAY OFF THIS ISLAND, SO WE CAN GET BACK TO YOUR SHIP AND GO TO SKYLANDER ACADEMY.

ENCHILADA ACADEMY?

THEN THERE MUST BE ENCHILADAS AROUND HERE SOMEWHERE...
FLYNN! WE'RE IN A WAREHOUSE! THERE ARE NO ENCHILADAS!
OH, WHY DO I EVEN BOTHER?

JUST NEED SOME *ELEVATION* TO GET A BETTER LOOK AROUND.
GET *DOWN* FROM THERE BEFORE YOU HURT YOURSELF!
UH-OH...

BOY, THAT FEELS BETTER!
HI, CHOPPER. WHAT ARE YOU DOING HERE?

AND, UH, WHERE IS HERE? THIS ISN'T MY TRUSTY DREADYACHT.
THAT'S BECAUSE YOU WALKED OFF THE DREADYACHT, AND I CAUGHT YOU BEFORE YOU FELL TO CERTAIN DOOM.
WE'RE ON THE BONEY ISLANDS, IN SOME SORT OF STORAGE WAREHOUSE FILLED WITH OLD ITEMS FROM THE MUSEUM.
HONESTLY, YOUR MAIN CONCERN HAS BEEN GETTING YOUR HANDS ON AN ENCHILADA.

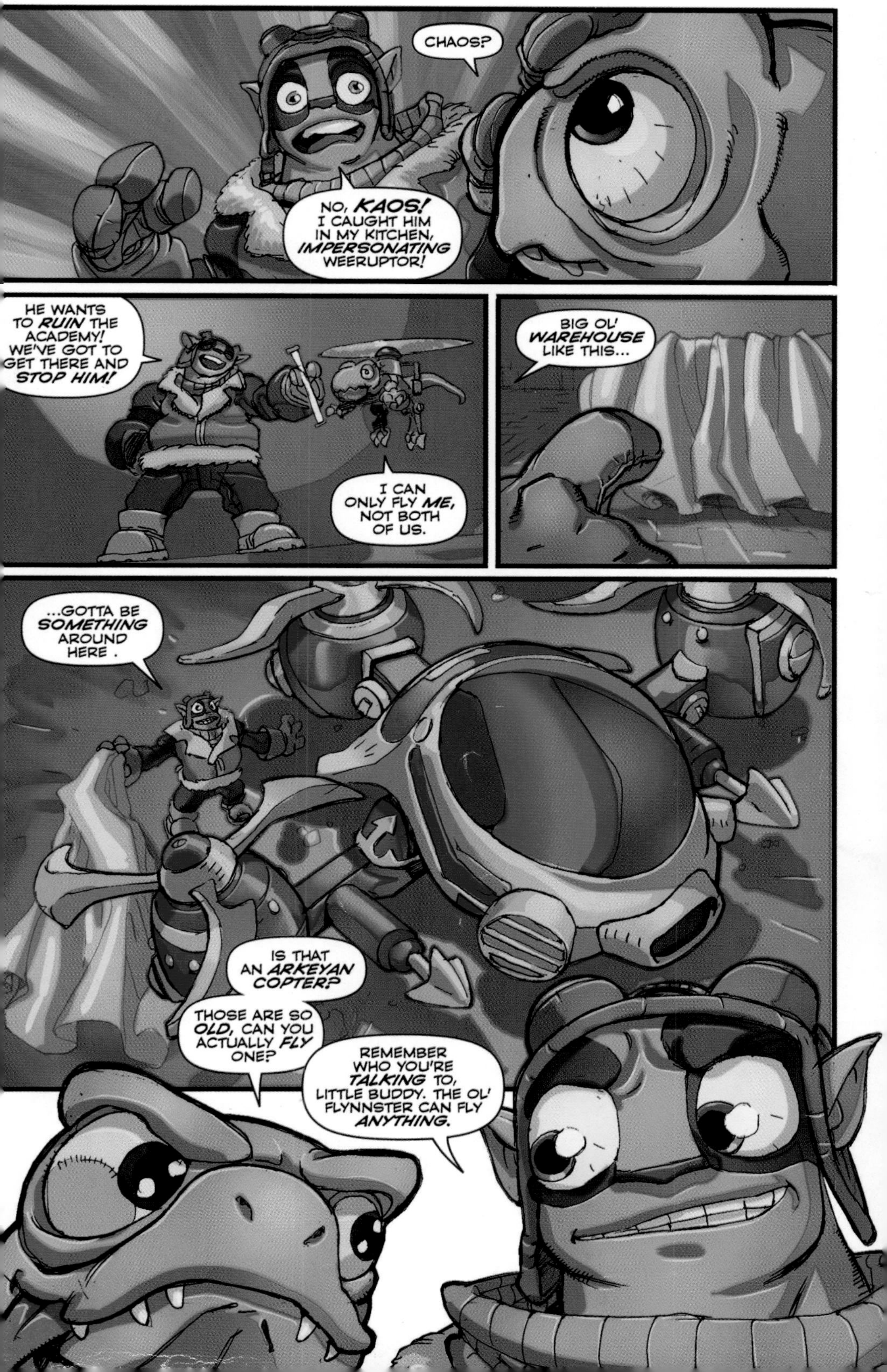

CHAOS?

NO, ***KAOS!*** I CAUGHT HIM IN MY KITCHEN, ***IMPERSONATING*** WEERUPTOR!

HE WANTS TO ***RUIN*** THE ACADEMY! WE'VE GOT TO GET THERE AND ***STOP HIM!***

I CAN ONLY FLY ***ME***, NOT BOTH OF US.

BIG OL' ***WAREHOUSE*** LIKE THIS...

...GOTTA BE ***SOMETHING*** AROUND HERE .

IS THAT AN ***ARKEYAN COPTER?***

THOSE ARE SO ***OLD***, CAN YOU ACTUALLY ***FLY*** ONE?

REMEMBER WHO YOU'RE ***TALKING*** TO, LITTLE BUDDY. THE OL' FLYNNSTER CAN FLY ***ANYTHING***.

SOMEBODY BROKE *INTO* THE WAREHOUSE?
WHO WOULD BE CRAZY ENOUGH TO DO *THAT?*
ARE YOU *SURE* THE OL' FLYNNSTER CAN FLY ANYTHING?
NO PROBLEMO...
RUN!
LOOK OUT!
...BECAUSE FLYING TO THE RESCUE AND STOPPING KAOS IS WHAT I DO *BEST!*
WELL, THAT AND EATING ENCHILADAS...

FOOD FIGHT

BIO

Food Fight does more than just play with his food, he battles with it! This tough little Veggie Warrior is the byproduct of a troll food experiment gone wrong. When the Troll Farmers Guild attempted to fertilize their soil with gunpowder, they got more than a super snack—they got an all-out Food Fight! Rising from the ground, he led the neighborhood Garden Patrol to victory. Later, he went on to defend his garden home against a rogue army of gnomes after they attempted to wrap the Asparagus people in bacon! His courage caught the eye of Master Eon, who decided that this was one veggie lover he needed on his side as a valued member of the Skylanders. When it comes to Food Fight, it's all you can eat for evil!

WILDFIRE

BIO

Wildfire was once a young lion of the Fire Claw Clan, about to enter into the Rite of Infernos—a test of survival in the treacherous fire plains. However, because he was made of gold, he was treated as an outcast and not allowed to participate. But this didn't stop him. That night, Wildfire secretly followed the path of the other lions, carrying only his father's enchanted shield. Soon he found them cornered by a giant flame scorpion. Using the shield, he protected the group from the beast's enormous stinging tail, giving them time to safely escape. And though Wildfire was injured in the fight, his father's shield magically changed him—magnifying the strength that was already in his heart—making him the mightiest of his clan. Now part of the Trap Team, Wildfire uses his enormous Traptanium-bonded shield to defend any and all who need it!

SNAP SHOT

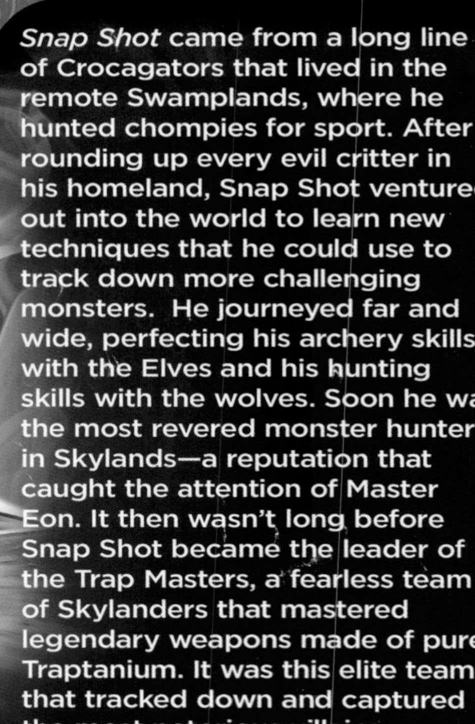

BIO

Snap Shot came from a long line of Crocagators that lived in the remote Swamplands, where he hunted chompies for sport. After rounding up every evil critter in his homeland, Snap Shot ventured out into the world to learn new techniques that he could use to track down more challenging monsters. He journeyed far and wide, perfecting his archery skills with the Elves and his hunting skills with the wolves. Soon he wa the most revered monster hunter in Skylands—a reputation that caught the attention of Master Eon. It then wasn't long before Snap Shot became the leader of the Trap Masters, a fearless team of Skylanders that mastered legendary weapons made of pure Traptanium. It was this elite team that tracked down and captured the most notorious villains Skylands had ever known!

WALLOP

BIO

For generations, *Wallop's* people used the volcanic lava pits of Mount Scorch to forge the most awesome weapons in all of Skylands. And Wallop was the finest apprentice any of the masters had ever seen. Using hammers in both of his mighty hands, he could tirelessly pound and shape the incredibly hot metal into the sharpest swords or the hardest axes. But on the day he was to demonstrate his skills to the masters of his craft, a fierce fire viper awoke from his deep sleep in the belly of the volcano. The huge snake erupted forth, attacking Wallop's village. But by bravely charging the beast with his two massive hammers, Wallop was able to bring down the creature and save his village. Now with his Traptanium-infused hammers, he fights with the Skylanders to protect the lands from any evil that rises to attack!

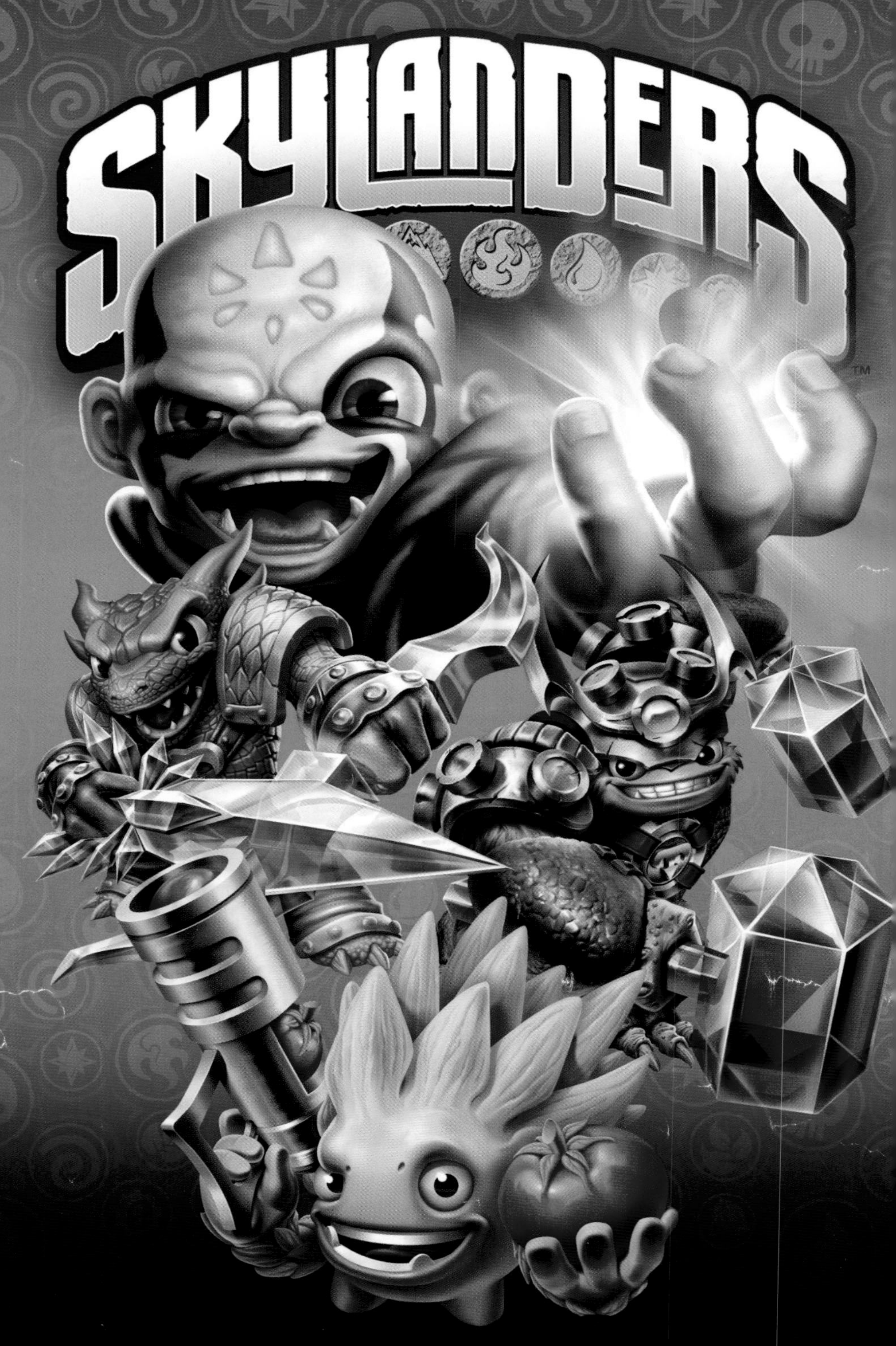

THE KAOS TRAP